P9-BZV-589

EARTH FROM SPACE

Views of the Earth from space can help us grasp how the atmosphere, oceans, land, and life itself are interconnected. In the atmosphere, swirling cloud patterns and storm systems reflect not only the distribution of moisture in the atmosphere, but the transfer of vast amounts of energy among regions of varying temperature and pressure. NASA's Earth Observatory created these images of Earth as part of its "Blue Marble" series. Both are composite images based on satellite data showing the land surface, clouds, and city lights, with a resolution of 1 km per pixel. The view on the left shows North America, Greenland, and the ice-covered Arctic Ocean, along with part of South America. The view on the right, centered on the Indian Ocean, shows much of the Eastern Hemisphere.

Understanding Weather and Climate

SIXTH EDITION

Edward Aguado

SAN DIEGO STATE UNIVERSITY

James E. Burt

UNIVERSITY OF WISCONSIN-MADISON

PEARSON

Boston Columbus Indianapolis New York San Francisco Upper Saddle River
Amsterdam Cape Town Dubai London Madrid Milan Munich Paris Montreal Toronto
Delhi Mexico City São Paulo Sydney Hong Kong Seoul Singapore Taipei Tokyo

Geography Editor: Christian Botting
Marketing Manager: Maureen McLaughlin
Project Editor: Crissy Dudonis
VP, Executive Director, Development: Carol Trueheart
Executive Development Editor: Jonathan Cheney
Senior Media Producer: Angela Bernhardt
Associate Media Producer: Tim Hainley
Assistant Editor: Sean Hale
Editorial Assistant: Bethany Sexton
Marketing Assistant: Nicola Houston
Managing Editor, Geosciences and Chemistry: Gina M. Cheselka
Production Project Manager: Maureen Pancza
Production Management: Thistle Hill Publishing Services, LLC
Compositor: Cenveo Publisher Services/Nesbitt Graphics, Inc.
Design Director: Mark Ong
Interior and Cover Designer: Gary Hespenheide
Senior Art Specialist: Connie Long
Art Studio: Precision Graphics
Photo Research Manager: Maya Melenchuk
Text Permissions Manager: Beth Wollar
Text Permissions Researcher: Jenny Bevington
Operations Specialist: Michael Penne
Cover Photo: *Photolibrary*

Credits and acknowledgments borrowed from other sources and reproduced, with permission, in this textbook appear on pages 540–541.

Copyright © 2013, 2010, 2007, 2004, 2001, 1998 Pearson Education, Inc. All rights reserved. Manufactured in the United States of America. This publication is protected by Copyright, and permission should be obtained from the publisher prior to any prohibited reproduction, storage in a retrieval system, or transmission in any form or by any means, electronic, mechanical, photocopying, recording, or likewise. To obtain permission(s) to use material from this work, please submit a written request to Pearson Education, Inc., Permissions Department, 1900 E. Lake Ave., Glenview, IL 60025. For information regarding permissions, call (847) 486-2635.

Many of the designations used by manufacturers and sellers to distinguish their products are claimed as trademarks. Where those designations appear in this book, and the publisher was aware of a trademark claim, the designations have been printed in initial caps or all caps.

CIP data on file with Library of Congress.

1 2 3 4 5 6 7 8 9 10—CRK—15 14 13 12

ISBN 10: **0-321-76963-5**; ISBN 13: **978-0-321-76963-3**   [Student Edition]
ISBN 10: **0-321-77323-3**; ISBN 13: **978-0-321-77323-4**   [Instructor's Review Copy]

www.pearsonhighered.com

Brief Contents

To Lauren, William, and Babsie June
—EA

To my parents, Martha F. Burt

and Robert L. Burt
—JEB

About Our Sustainability Initiatives

Pearson recognizes the environmental challenges facing this planet, as well as acknowledges our responsibility in making a difference. This book is carefully crafted to minimize environmental impact. The binding, cover, and paper come from facilities that minimize waste, energy consumption, and the use of harmful chemicals. Pearson closes the loop by recycling every out-of-date text returned to our warehouse.

Along with developing and exploring digital solutions to our market's needs, Pearson has a strong commitment to achieving carbon-neutrality. As of 2009, Pearson became the first carbon- and climate-neutral publishing company. Since then, Pearson remains strongly committed to measuring, reducing, and offsetting our carbon footprint.

The future holds great promise for reducing our impact on Earth's environment, and Pearson is proud to be leading the way. We strive to publish the best books with the most up-to-date and accurate content, and to do so in ways that minimize our impact on Earth. To learn more about our initiatives, please visit **www.pearson.com/responsibility**.

Contents

PART ONE
Energy and Mass xxvi

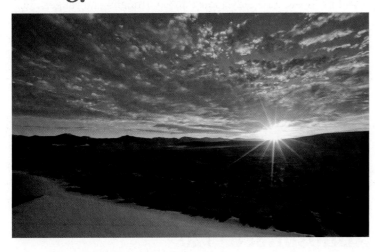

1 Composition and Structure of the Atmosphere 2

2 Solar Radiation and the Seasons 30

3 Energy Balance and Temperature 54

PART TWO
Water in the Atmosphere 120

PART THREE

Distribution and Movement of Air 212

PART FOUR
Disturbances 278

PART FIVE
Human Activities 378

13   Weather Forecasting and Analysis 380

PART SIX

Current, Past, and Future Climates 438

PART SEVEN
Special Topics and Appendices 502

MyMeteorologyLab™ Website Multimedia

Preface

The atmosphere is the most dynamic of all Earth's spheres. In no other realm do events routinely unfold so quickly, with so great a potential impact on humans. Some of the most striking atmospheric disturbances (such as tornadoes) can take place over time scales on the order of minutes—but nevertheless have permanent consequences. Events such the searing 2011 drought in Texas take longer, but can have much more widespread effects. In this case the state's entire hay crop was lost along with more than half of its cotton crop. Cotton and beef prices surged throughout the world, and by late summer the U.S. cattle herd was down to numbers not seen since 1963. Catastrophes such as this are momentous, but even the most mundane of atmospheric phenomena influence our lives on a daily basis (for instance, the beauty of blue skies or red sunsets, rain, or the daily cycle of temperature).

Atmospheric processes, despite their immediacy on a personal level and their importance in human affairs on a larger level, are not readily understood by most people. This is probably not surprising, given that the atmosphere consists primarily of invisible gases, along with suspended, frequently microscopic particles, water droplets, and ice crystals.

Understanding Weather and Climate is a college-level text intended for both science majors and nonmajors taking their first course in atmospheric science. We have attempted to write a text that is informative, timely, engaging to students, and easily used by professors. In this book, our overriding goal is to bridge the gap between abstract explanatory processes and the expression of those processes in everyday events. We have written the book so that students with little or no science background will be able to build a nonmathematical understanding of the atmosphere.

That said, we do not propose to abandon the foundations of physical science. We know from our own teaching experience that physical laws and principles can be mastered by students of widely varying backgrounds. In addition, we believe one of meteorology's great advantages is that reasoning from fundamental principles explains so much of the field. Compared to some other disciplines, this is one in which there is an enormous payoff for mastering a relatively small number of basic ideas.

Finally, our experience is that students are always excited to learn the "why" of things, and to do so gives real meaning to "what" and "where." For us, therefore, the idea of forsaking explanation in favor of a purely descriptive approach has no appeal whatsoever. Rather, we propose merely to replace mathematical proof (corroboration by formal argument) with qualitative reasoning and appeal to everyday occurrences. As the title implies, the goal remains understanding atmospheric behavior.

New to the Sixth Edition

- NEW **Big Picture part-opener spreads** open each unit, helping students preview and connect key concepts, visuals, learning goals, and media from the respective chapters.

- NEW **Learning Outcomes** are integrated into the chapter-opening spreads to help students organize and prioritize key concepts and content.

- NEW **Checkpoint in-line questions** (6–10 per chapter) offer conceptual "speed bumps" or reading questions after major sections to help students make sure that they understand or can apply the material they've just read.

- NEW and UPDATED *Focus On* feature essays better emphasize Severe and Hazardous Weather, meteorological impacts on Aviation, and Environment issues, accommodating departments and classes where these important and timely topics are of particular interest.

- Significant updates to the visual program, with many new or updated illustrations, maps, and photos, all housed in a wider and more colorful and engaging page layout and design.

- The latest data, case studies, applications, and examples from meteorology today are integrated to make the 6th edition the most current and relevant introduction to meteorology. For example, Chapter 3 includes the most up-to-date values of the global energy balance; Chapter 11 presents the most recent instructions on tornado safety as well as a review of the devastating 2011 tornado season; and Chapter 12 presents updated statistics on hurricane incidence and the revised criteria for the issuance of hurricane watches and warnings.

- NEW full color reference maps are now included inside the front and rear covers.

- **NEW MyMeteorologyLab with Pearson eText.** www.MyMeteorologyLab.com is a new resource both for student self-study and for instructors to manage their courses online and provide customizable assessments to students. MyMeteorologyLab's assignable content includes Interactive Tutorials, Weather in Motion videos, Geoscience Animations, MapMaster™ interactive maps, a variety of chapter quizzes, and more. Students can also access the Pearson eText for *Understanding Weather and Climate*, 6th edition, Carbone's *Exercises for Weather & Climate* interactive media, *In the News* RSS feeds, glossary flashcards, social networking features, and additional references and resources to extend learning beyond the text.

Distinguishing Features

Scientific Literacy and Currency We have emphasized scientific literacy throughout the book. This emphasis gives students an opportunity to develop a deeper understanding about the building blocks of atmospheric science and serves as tacit instruction regarding the workings of all the sciences. For instance, in Chapter 2 we cover the molecular changes that occur when radiation is absorbed or emitted, items that are often considered a "given" in introductory texts. In Chapter 3 these basic ideas are used to help build student understanding of why individual gases radiate and absorb particular wavelengths of radiation and illustrate how processes operating at a subatomic level can manifest themselves at global scales. Similarly, our discussion of anthropogenic warming in Chapter 16 includes cloud, water vapor, and lapse rate feedbacks in order to provide a more complete account of the uncertainties surrounding this critical environmental topic.

An emphasis on scientific literacy can be effectively implemented only if it is accompanied by careful attention to currency. We believe that two kinds of currency are required in a text: an integration of current *events* as they relate to the topic at hand, and an integration of current *scientific thinking*. For instance, the reader will find discussion of both recent hurricane activity and the most recent theories regarding the mechanisms that generate severe storms. Scientific literacy also calls for attention to language—after all, precision of language is an important distinguishing characteristic of science, one that sets it apart from other intellectual activities. With that in mind, we have tried to avoid some common statements of dubious accuracy, such as "Warm air is able to hold more water vapor than cold air."

Media A fundamental feature of this book is the integration of the classic print textbook model with instructional technology. These dynamic media resources are delivered through the new **www.MyMeteorologyLab.com** platform. The online media consist of several components. Perhaps most fundamental to our approach, the media package features 17 Interactive Tutorials covering basic principles of atmospheric science. These software modules have undergone considerable testing and have been used successfully by thousands of students. They rely heavily on three-dimensional diagrams and animations to present material not easily visualized using conventional media. The software modules follow a tutorial style, with explanations and new vocabulary introduced incrementally, building on what was presented earlier in the modules and what was presented in the text. The Tutorials are best used in conjunction with the assigned readings. In choosing topics for the modules, we have emphasized material that is both difficult to master and has the potential to benefit from digital technology. New to the sixth edition are modules on Vertical and Horizontal Pressure Variations (Chapter 1) and Earth's Climate (Chapter 15). Tutorial icons with brief descriptive text are placed in the book adjacent to the discussion of a topic. We advise that you first view a Tutorial in its entirety.

If additional review is needed, you can easily move within a Tutorial to the section under discussion.

In addition to the Tutorials, the Web site for the sixth edition contains movies called *Weather in Motion,* which depict events and phenomena discussed in the text. Examples include a satellite movie showing clouds and temperature across the globe, three-dimensional simulations of thunderstorm development, and animations depicting variations in Earth's orbit. Each animation is accompanied by critical-thinking questions calling for application of material presented in the text to new situations. Like the tutorials, each movie is listed at the end of every chapter.

The Web site at **www.MyMeteorologyLab.com** also includes MapMaster™ interactive maps, Geoscience Animations, *In the News* RSS feeds, review questions, quantitative exercises, social networking options, and other materials that allow users to query the Internet for timely atmospheric data. For instructors, MyMeteorologyLab also serves as a full-featured course management and homework system, with an easy-to-use gradebook and helpful diagnostic features.

Instructor Flexibility During the writing process, we have enjoyed interacting with many of our colleagues who teach courses in weather and climate on a regular basis. It was especially interesting to see how little consensus exists regarding topic order (truth be told, the authors of this book don't agree on the optimal sequence). With this in mind, we tried to minimize the degree to which individual chapters depend on material presented earlier. Thus, instructors who prefer a chapter order different from the one we ultimately chose will not be disadvantaged.

Emphasis on Climate Change In 2007 the Intergovernmental Panel on Climate Change (IPCC) released its landmark report on the current knowledge of climate change and human impacts. The sixth edition of *Understanding Weather and Climate* makes heavy reference to that work, but has updated climate statistics through 2010, and post-IPCC developments are included throughout. These sections present physically based explanations behind the changes that have occurred and are likely to occur in the future.

Emphasis on Forecasting In addition to a comprehensive chapter on the topic, this text contains numerous examples of how physical principles are employed in weather forecasting. We have included several discussions of the use of thermodynamic diagrams in weather forecasting and analysis. These charts are extremely valuable but not immediately comprehensible to most students. To alleviate this problem, we introduce thermodynamic diagrams in a sequential fashion. That is, their use for plotting vertical temperature profiles is presented in the chapter on temperature. We expand upon this in the chapter on atmospheric moisture to show how various measures of humidity can also be determined with the aid of the charts. Thus, instructors can teach their students how to use these diagrams without inundating them with excessive detail all at once.

Increased Use of Current Examples This edition presents a greater number of weather maps and images to illustrate how atmospheric phenomena occur in everyday settings. The new examples have been selected for currency and illustrative value. Special attention has been given to some of the most notable hurricanes and typhoons of recent years, along with 2011 tornado swarms and blizzards from the preceding winter.

Readability In contrast to the more formal scientific style used in many science textbooks, we have chosen to adopt more casual prose. Our goal is to present the material in language that is clear, readable, and friendly to the student reader. We employ frequent headings and subheadings to help students follow discussions and identify the most important ideas in each chapter. As a rule, we keep technical language to a minimum.

Focus on Learning The chapters offer a number of study aids:

- *Learning Outcomes.* These chapter-opening learning goals help prioritize key concepts and skills that students should master after reading the chapter.
- *Checkpoints.* These conceptual questions and tasks are integrated throughout the chapters after major sections, giving students a chance to stop, practice, and apply their understanding of key chapter content.
- *Did You Know.* This feature highlights interesting meteorological facts in every chapter.
- *Key Terms.* Key terms in each chapter are printed in boldface when first introduced. Most are also listed at the end of each chapter, along with the page number on which each first appears. All key terms are defined in the glossary at the end of the book and in the eText ebook on the MyMeteorologyLab Web site.
- *Focus on the Environment.* These features highlight environmental issues as they relate to the study of the atmosphere.
- *Focus on Severe Weather.* These features focus on dramatic and dangerous severe and hazardous weather phenomena, including coverage of many recent events like the deadly 2011 tornado season.
- *Focus on Aviation.* These features explore the impacts of various atmospheric phenomena on aviation. Examples include discussion of winter storms and air travel (Chapter 1); impacts of icing on aircraft (Chapter 5); recommended pilot responses to icing in different types of clouds (Chapter 6); and lightning and aircraft (Chapter 11).
- *Physical Principles.* More mathematical in nature than the rest of the text, these boxes accommodate students who have a more quantitative interest in the topic. An understanding of the material in these boxes is not essential to an understanding of the material presented in the body of the text.
- *Special Interest.* These features highlight various interesting topics related to the discussion at hand.
- *Forecasting.* These features describe how the principles discussed in the chapter can be used in forecasting and often include simple "rules of thumb" that help students make their own forecasts.
- *Chapter Summary and Key Terms.* Each chapter concludes with a chapter summary that highlights the main points in the chapter, followed by a listing of the chapter Key Terms with page references.
- *Review Questions.* At the end of each chapter, you will find a list of questions about the subject of that chapter. These review questions test reading comprehension and can be answered from information presented in the chapter.
- *Critical Thinking.* Each chapter concludes with Critical Thinking questions that require students to use material presented in the chapter to work out answers relevant to real-world questions.
- *Problems and Exercises.* These questions encourage students to work out solutions to numerical questions to gain a better understanding of chapter material.
- *Quantitative Problems.* The MyMeteorologyLab Web site features quantitative exercises to accompany each chapter.
- *Useful Web Sites.* These describe relevant and interesting Web sites with valuable information or current data at the end of each chapter. These links are active in the eText version of the book and in MyMeteorologyLab.

Supplements

The authors and publisher have been pleased to work with a number of talented people to produce an excellent supplements package for the text.

For The Student

- *MeteorologyLab* provides a robust suite of resources, including Interactive Tutorials, Weather in Motion videos, Geoscience Animations, MapMaster™ interactive maps, the Pearson eText for the 6th edition, Carbone's *Exercises for Weather & Climate* interactive media, *In the News* RSS feeds, glossary flashcards, quizzes, social networking features, and additional references and resources to extend learning beyond the text. **www.MyMeteorologyLab.com**
- ***Pearson eText*** Integrated with the Web site, **Pearson eText** is a full-featured electronic version of the book with linked media resources, which allows students to add personal notes and highlights to the pages and see any comments made by the instructor. The search function allows for full-text searching of book content.
- ***Exercises for Weather & Climate,*** **8th edition by Greg Carbone [0321769651]** This bestselling exercise manual's 17 exercises encourage students to review important ideas and concepts through problem solving, simulations, and guided thinking. The graphics program and computer-based simulations and tutorials help students grasp key concepts. This manual is designed to complement any introductory meteorology or weather and climate course.

- ***Encounter Meteorology: Interactive Explorations of Earth Using Google Earth*** [0321815912] This workbook and premium Web site provides rich, interactive explorations of meteorology concepts through Google Earth™ explorations. All chapter explorations are available in print format as well as via online quizzes and downloadable PDFs, accommodating different classroom needs. Each worksheet is accompanied by corresponding Google Earth™ KMZ media files containing the placemarks, overlays, and annotations referred to in the worksheets, available for download from **www.mygeoscienceplace.com**.

- ***Encounter Geosystems: Interactive Explorations of Earth Using Google Earth*** [0321636996] This workbook and premium Web site provides rich, interactive explorations of physical geography concepts through Google Earth™ explorations. All chapter explorations are available in print format as well as via online quizzes and downloadable PDFs, accommodating different classroom needs. Each worksheet is accompanied by corresponding Google Earth™ KMZ media files containing the placemarks, overlays, and annotations referred to in the worksheets, available for download from **www.mygeoscienceplace.com**.

- ***Goode's World Atlas,*** **22nd edition [0321652002]** *Goode's World Atlas* has been the world's premiere educational atlas since 1923—and for good reason. It features over 250 pages of maps, from definitive physical and political maps to important thematic maps that illustrate the spatial aspects of many important topics. The 22nd edition includes 160 pages of new, digitally produced reference maps, as well as new thematic maps on global climate change, sea-level rise, CO_2 emissions, polar ice fluctuations, deforestation, extreme weather events, infectious diseases, water resources, and energy production.

- ***Dire Predictions: Understanding Global Warming*** **[0136044352]** Periodic reports from the Intergovernmental Panel on Climate Change (IPCC) evaluate the risk of climate change brought on by humans. But the sheer volume of scientific data remains inscrutable to the general public, particularly to those who may still question the validity of climate change. In just over 200 pages, this practical text presents and expands upon the essential findings in a visually stunning and undeniably powerful way to the lay reader. Scientific findings that provide validity to the implications of climate change are presented in clear-cut graphic elements, striking images, and understandable analogies.

- ***Geoscience Animation Library on DVD,*** **5th Edition [0321716841] Geoscience Animations** illuminate the most difficult-to-visualize topics from across the physical geosciences, such as solar system formation, hydrologic cycle, plate tectonics, glacial advance and retreat, global warming, etc. Animations include audio narration and text transcript, with assignable multiple-choice quizzes to select animations in MyMeteorologyLab to help students master these core physical process concepts.

- ***Earth Report*** **Geography Videos on DVD [0321662989]** This three-DVD set is designed to help students visualize how human decisions and behavior have affected the environment and how individuals are taking steps toward recovery. With topics ranging from poor land management promoting the devastation of river systems in Central America to the struggles for electricity in China and Africa, these 13 videos from Television for the Environment's global *Earth Report* series recognize the efforts of individuals around the world to unite and protect the planet.

For the Professor

- ***Instructor Resource DVD*** **[0321773187]** The *Instructor Resource DVD* provides high-quality electronic versions of photos and illustrations from the book, as well as customizable PowerPoint™ lecture presentations, Classroom Response System questions in PowerPoint, and the *Instructor Resource Manual* and *Test Bank* in Microsoft Word and TestGen formats. The IRC also includes all the illustrations and photos from the text in presentation-ready JPEG files, as well as digital transparencies. For easy reference and identification, all resources are organized by chapter. All of the elements on the IRDVD are also available online to professors at **www.pearsonhighered.com/irc**.

- ***Instructor Resource Manual*** **by Lou McNally (Download only) [0321773373]** The *Instructor Resource Manual* is intended as a resource for both new and experienced instructors. It includes a variety of lecture outlines, additional source materials, teaching tips, advice about how to integrate visual supplements (including the Web-based resources), and various other ideas for the classroom. See **www.pearsonhighered.com/irc**.

- **TestGen® Computerized *Test Bank* by Tom Konvicka (Download only) [0321773179]** TestGen® is a computerized test generator that lets instructors view and edit *Test Bank* questions, transfer questions to tests, and print tests in a variety of customized formats. This *Test Bank* includes more than 2000 multiple-choice, fill-in-the-blank, and short-answer/essay questions. Questions are correlated to the revised U.S. National Geography Standards and Bloom's Taxonomy to help instructors better map the assessments against both broad and specific teaching and learning objectives. The *Test Bank* is also available in Microsoft Word and is importable into Blackboard. See **www.pearsonhighered.com/irc**.

- ***Earth Report*** **Geography Videos on DVD [0321662989]** This three-DVD set is designed to help visualize how human decisions and behavior have affected the environment and how individuals are taking steps toward recovery. With topics ranging from poor land management promoting the devastation of river systems in Central America to the struggles for electricity in China and Africa, these 13 videos from Television for the Environment's global *Earth Report* series recognize the efforts of individuals around the world to unite and protect the planet.

- *Geoscience Animation Library on DVD*, **5th Edition** **[0321716841]** **Geoscience Animations** illuminate the most difficult-to-visualize topics from across the physical geosciences, such as solar system formation, hydrologic cycle, plate tectonics, glacial advance and retreat, global warming, etc. Animations include audio narration and a text transcript, with assignable multiple-choice quizzes to select animations in MyMeteorologyLab to help students master these core physical process concepts.

Acknowledgments

The authors would like to extend their thanks to the editorial and production team at Pearson Education: our managing editor Christian Botting, executive development editor Jonathan Cheney, senior project editor Crissy Dudonis, media producer Tim Hainley, and senior media producer Angela Bernhardt. Among many other things, we greatly appreciate their management of a complicated schedule. The book is much improved for their efforts. A number of other people were extremely helpful on the production side of things, including Angela Urquhart at Thistle Hill Publishing Services.

We also benefited greatly from the advice of many professional educators and meteorologists. Mark Moede of the San Diego office of the National Weather Service provided valuable information on the day-to-day activities of weather forecasters.

We thank researchers and staff at government agencies and institutions throughout the world for creating many of the images and movies on www.MyMeteorologyLab.com and for their willingness to make them available to use in projects like this. The following deserve special mention in this regard: Space Science and Engineering Center, University of Wisconsin–Madison; Dr. Richard Orville, Texas A&M University; Dr. Gary Huffines, University of Northern Colorado; Diana DeRubertis, University of California, Berkeley; and Paul Morin, University of Minnesota.

We must also offer special thanks to the many colleagues who spent valuable time and energy preparing in-depth reviews of our early efforts, many of whom have continued in this role through multiple revisions. In that regard we are particularly grateful to Robert Rohli of Louisiana State University, who read the earliest editions of the manuscript with exceptional care and made many excellent suggestions. Additionally, we thank the following teachers who provided reviews of the current and previous editions:

Rafique Ahmed, *University of Wisconsin–LaCrosse*
Al Armendariz, *Southern Methodist University*
David Barclay, *SUNY Cortland*
Greg Bierly, *Indiana State University*
Mark Binkley, *Mississippi State University*
Gerald Brothen, *El Camino College*
David P. Brown, *University of Arizona*
Adam W. Burnett, *Colgate University*
Gregory Carbone, *University of South Carolina*
R. E. Carlson, *Iowa State University*
Donna J. Charlevoix, *University of Illinois, Urbana–Champaign*
Christopher R. Church, *Miami University of Ohio*
John H. E. Clark, *Penn State University*
Andrew Comrie, *University of Arizona*
William T. Corcoran, *Missouri State University*
Eugene Cordero, *San Jose State University*
Mario Daoust, *Southwest Missouri State University*
Arthur (Tim) Doggett, *Texas Tech University*
Dennis M. Driscoll, *Texas A&M University*
Ted Eckmann, *Bowling Green State University*
Neil I. Fox, *University of Missouri–Columbia*
Christopher M. Godfrey, *University of Oklahoma*
Thomas Guinn, *Embry Riddle Aeronautical University*
Rex J. Hess, *University of Utah*
Jay S. Hobgood, *Ohio State University*
Edward J. Hopkins, *University of Wisconsin–Madison*
Scott A. Isard, *University of Illinois*
Eric Johnson, *Illinois State University*
Jon Kahl, *University of Wisconsin–Milwaukee*
Scott Kirsch, *University of Memphis*
Daniel James Leathers, *University of Delaware*
Gong-Yuh Lin, *California State University–Northridge*
Anthony Lupo, *University of Missouri*
Jason Martinelli, *Creighton University*
Deborah Metzel, *University of Massachusetts–Boston*
Thomas L. Mote, *University of Georgia*
Gregory D. Nastrom, *St. Cloud State University*
Gerald R. North, *Texas A&M University*
Jim Norwine, *Texas A&M University–Kingsville*
John E. Oliver, *Indiana State University*
Stephen Podewell, *Western Michigan University*
David Privette, *Central Piedmont Community College*
Sarah Pryor, *Indiana University–Bloomington*
Azizur Rahman, *University of Minnesota, Crookston*
Robert V. Rohli, *Louisiana State University*
Steven A. Rutledge, *Colorado State University*
Arthur N. Samel, *Bowling Green State University*
Hans Peter Schmid, *Indiana University*
Justin Schoof, *Southern Illinois University–Carbondale*
Marshall Shepherd, *University of Georgia*
Brent Skeeter, *Salisbury University*
Stephen Stadler, *Oklahoma State University*
S. Elwynn Taylor, *Iowa State University*
Mingfang Ting, *University of Illinois*
Graham Tobin, *University of South Florida*
Paul E. Todunter, *University of North Dakota*
Liem Tran, *Florida Atlantic University*
Donna Tucker, *University of Kansas*
Barry Warmerdam, *Kings River Community College*
Thompson Webb III, *Brown University*
Jack Williams, *University of Wisconsin–Madison*
Thomas B. Williams, *Western Illinois University*
Morton Wurtele, *University of California–Los Angeles*
Douglas Yarger, *Iowa State University*
Charlie Zender, *University of California–Irvine*

About the Authors

Edward Aguado is Professor of Geography in the Department of Geography of San Diego State University. He received his Ph.D. from the University of Wisconsin–Madison, and his M.A. and B.A. from UCLA. His research interests are in the precipitation and hydrology of western U.S. mountains. He regularly teaches introductory and advanced meteorology, climatology, and physical geography, and often serves as a consultant and expert witness on climatology and weather.

Jim Burt is Professor of Geography in the Department of Geography of the University of Wisconsin–Madison. He received his M.A. and Ph.D. from UCLA. His research interests are in physical geography, climatology, and geographic information science. He regularly teaches introductory and advanced courses in all three.

Structured Learning

Powerful pedagogy equips students with the skills needed to master the science.

New! Big Picture Part Opener

Each part begins with a preview of the key concepts of the chapters, showcasing important visuals and dynamic tutorial media and posing critical thinking questions to encourage readers to engage with inquiry.

New! Learning Outcomes

Clear, testable learning outcomes at the beginning of each chapter focus students on key concepts and skills.

New! Checkpoints

These questions appear at the end of chapter sections, giving students a chance to stop, check, and practice their understanding of key chapter concepts before moving on with the reading.

LEARNING OUTCOMES

After reading this chapter, you should be able to:

► Explain the concept of air pressure.

► Describe how pressure changes vertically and horizontally in the atmosphere.

► Apply the equation of state to calculate air density.

► Identify devices used to measure air pressure.

► Describe the distribution of air pressure across the globe, at sea level, and in the upper atmosphere.

► Explain the factors that affect the wind speed and direction.

► Describe winds in relation to pressure gradients in the upper atmosphere and near the surface.

► Explain the forces that produce anticyclones, cyclones, troughs, and ridges.

► Describe how meteorologists measure wind.

Checkpoint

1. Define weather and climate in your own words.
2. Compare the concerns of the sciences of meteorology and climatology, giving some examples of different phenomena they might investigate.
3. List some places you have visited whose climate is affected by proximity to an ocean or position deep within a continent.

Forecasting

These features help students understand the concepts and processes involved in weather forecasting, and often include simple "rules of thumb" that help students make their own forecasts.

Physical Principles

These quantitative explorations extend learning with more mathematical treatments of concepts.

11–4 FORECASTING

Doppler Radar

Just as we are able to distinguish different colors of light by their wavelengths, so can we differentiate sounds by the length of their sound waves. If an object making a sound is moving away from a listener, the sound waves are stretched out and assume a lower pitch. Sound waves are compressed when an object moves toward the listener, making them higher pitched. Unconsciously, we use this principle, called the Doppler effect, to determine whether an ambulance siren is coming closer or moving away. If the pitch of the siren seems to become higher, we know the ambulance is getting nearer (of course, the siren would also sound louder). A similar process occurs when electromagnetic waves are reflected by a moving object. The light shifts to shorter wavelengths when reflected by an object moving toward the receiver and to longer wavelengths as it bounces off an object moving away from the receiver.

Doppler radar is a type of radar system that takes advantage of this principle. It allows the user to observe the movement of raindrops and ice particles (and thus determine wind speed and direction) from the shift in wavelength of the radar waves, as well as the intensity of precipitation. Like any other type of radar, Doppler radar has a transmitter that emits pulses of electromagnetic energy with wavelengths on the order of several centimeters. Depending on the wavelength used, water droplets and snow crystals above certain critical sizes reflect a portion of the radar's electromagnetic energy back to the transmitter/receiver. Doppler radar is special in its ability to observe the motion of the cloud constituents. If a cloud droplet is moving away from the radar unit, the wavelength of the beam is slightly elongated as it bounces off the reflector. Such reflections are normally indi-

3–2 PHYSICAL PRINCIPLES

Earth's Equilibrium Temperature

If Earth had no atmosphere, and therefore no greenhouse effect (see page 67), the mean temperature of the planet would be much colder. Using the principles discussed thus far, we can easily estimate the magnitude of the greenhouse effect. To do so, we will compute the equilibrium temperature for a planet having no atmosphere. By comparing the computed and observed temperatures, we will see how the atmosphere influences Earth's temperature.

First, assume the planet acts as a blackbody with regard to longwave radiation, that the planetary albedo is 30 percent, and that the solar constant is 1367 watts per square meter. If Earth were a flat disk perpendicular to incoming radiation, each square meter would receive 1367 joules/second. But Earth is not a flat disk; it is a sphere, the surface area of which is four times larger than that of a disk of the same radius. Thus the intensity of radiation averaged over the sphere is one-fourth as large as for the imaginary disk. Because of this, each square meter of Earth receives 1367/4, or 342 watts/m². (These 342 watts per square meter are the 100 units of solar radiation discussed in the section, "The Fate of Solar Radiation," earlier in this chapter.) Given the planetary albedo of 30 percent, it must be that 70 percent of this incoming radiation is absorbed. In other words, total absorbed radiation is

$$1367 \text{ watts/m}^2 \times 0.25 \times 0.7 = 239.2 \text{ watts/m}^2$$

The planet must lose exactly as much energy as it gains, and the intensity of radiation for a blackbody is determined by rearranging and applying the Stefan-Boltzmann law. Recall that the Stefan-Boltzmann law for a blackbody states that

$$I = \sigma T^4$$

We know, however, that the intensity of radiation for the planet without an atmosphere must be 239.2 watts per square meter. We then rearrange the equation to solve for T, rather than I, to get

$$T^4 = I/\sigma$$

which can be reduced to

$$T = (I/\sigma)^{0.25}$$

Using the values, $\sigma = 5.67 \times 10^{-8}$ watts/(m²K⁴) and $I = 239.2$ watts/m², the equilibrium temperature works out to 254.9 K (−18.3 °C, 0 °F) Thus, the mean temperature of Earth would be far colder without an atmosphere.

Note that our calculation is highly simplified and somewhat questionable. For example, we used 30 percent for the planetary albedo, but that value arises in part from the albedo of the atmosphere, which our imaginary planet lacks. Should we therefore have used present-day surface albedo in the computation? Perhaps, but surface albedo on the real Earth is in part the result of temperature, the very thing we are trying to compute! Ideally, we would treat albedo as a variable, allowing it to respond to changes in temperature.

We see then that even a beginning question about the global effect of greenhouse gases raises complications that are not easy to address with a simple model. Given this, it is not surprising that realistic computer models of atmospheric behavior are enormously complex, requiring huge computer resources. Nonetheless, computer models have become indispensable for daily weather forecasting (Chapter 13) and tell us much of what we know about potential climatic changes due to human activity (Chapter 16).

Observe and Apply

Tools to refine students' critical thinking skills and emphasize the relevance of meteorology today.

New! Focus on Aviation

These features explore the impacts of weather events and phenomena on aviation science.

1-1 FOCUS ON AVIATION

Winter Storms and Air Travel

Winter weather can play havoc with air travel in the United States. Naturally, some years are worse than others, and the winter of 2010–2011 was especially difficult. Between November and early February snow and ice conditions associated with four major storms caused U.S. airlines to cancel some 86,000 flights, with thousands of other flights encountering major delays. These events amounted to a significant portion of all scheduled flights for an industry recovering from some economic difficulties. In December 2010 3.7 percent of all U.S. flights were canceled because of winter storms, in contrast to the 2.89 percent canceled the previous December. Even airports not normally subject to crippling winter storms, such as Hartsfield-Jackson Atlanta International Airport, were subject to the cancellation of thousands of flights when a single January storm left behind 6 inches of snow in a city that usually receives half that much in an entire season (Figure 1).

One of the problems associated with winter storms is that aircraft loaded to capacity with passengers can be forced to wait on taxiways between terminal gates and runways for extended periods. An outcry of consumer criticism led to a policy taking effect in April 2010 allowing the U.S. Department of Transportation to levy fines

▶ FIGURE 1 Snowbound airplanes in Atlanta following the storm of January 2011.

more readily than in the past, causing greater inconvenience for passengers forced to wait hours or even days for another flight to get them to their destination. On the other hand, passengers are now much less likely to spend half a day on a crowded aircraft just waiting to get to the runway.

than it is to wait for scheduled personnel who are delayed by impassable highways to get to the airport. And of course, flights scheduled to depart San Diego on a warm, sunny day may be unable to do so because the aircraft needed for the flight is stranded at an East Coast airport.

Focus on the Environment

These features focus on important climate and weather issues affecting and being affected by our physical environment.

1-4 FOCUS ON THE ENVIRONMENT

Depletion of the Ozone Layer

In 1972 Sherwood Rowland and Mario Molina, atmospheric chemists then working at the University of California at Irvine, proposed that certain human-produced chemicals called chlorofluorocarbons (CFCs) could be carried naturally into the stratosphere and damage the ozone layer. CFCs are widely used in refrigeration and air conditioning, in the manufacture of plastic foams, and as solvents in the electronics industry.

Knowing that CFCs do not easily react with other molecules in the lower atmosphere, Rowland and Molina proposed that these molecules could reach the stratosphere intact, where they would break down and release free atoms of chlorine (Cl). Under certain circumstances, chlorine atoms can effectively destroy ozone molecules. In the first step of this process, a chlorine atom reacts with an ozone molecule to produce O_2 and chlorine monoxide (ClO). Next, an oxygen atom (O) reacts with ClO, creating another O_2 molecule while freeing the chlorine atom (Cl). Note that the chlorine atom that first reacted with the ozone molecule is still present and capable of reacting again with another ozone molecule. This fact makes chlorine atoms able to repeatedly break down ozone molecules. In fact, as many as 100,000 ozone molecules can be removed from the atmosphere for every chlo-

▲ FIGURE 1 A series of satellite images showing a reduction of the stratospheric ozone over Antarctica. Mapped values are Dobson Units (DU), a measure of total ozone amount in the atmospheric column, nearly all of which is found in the stratosphere. The global average is about 340 DU. The boundary of the "ozone hole" is considered to be 220 DU, shown in dark blue. Values that low were not observed prior to 1979.

these compounds to create free chlorine atoms (Cl). These Cl free radicals readily destroy ozone, as described previously. A very recent discovery shows that deple-

Government and private industry have taken action to help reduce the amount of CFCs entering the atmosphere. In accordance with the Montreal Protocol of 1987

New! Cloud Guide

A fold out cloud guide found at the back of the book provides students with a tool for real-world observation.

Cloud Guide

High Clouds: cloud bases above 6km (20,000 ft)   Middle Clouds: cloud bases 2–6km (6,500–20,000 ft)   Weather and Climate

Focus on Severe Weather

These features focus on the dramatic severe and hazardous weather phenomena that increasingly impact our world.

11-6 FOCUS ON SEVERE WEATHER

Deadly 2011 Tornado Season

By the end of June (as this is written) the 2011 season has already turned out to be remarkably deadly. The anomalous number of fatalities resulted from the combination of a high number of tornadoes, including very powerful ones, and the bad luck of several very large ones passing over populated areas.

April 2011

Though April is on average the third most active month for United States tornadoes (see Figure 11-34), this particular April was anything but normal. The year 2011 proved not only to have the most in the United States tornadoes ever recorded for an April (875), in fact it turned out to eclipse the record for the number of tornadoes recorded for any month (542 in May 2003). The number of fatalities was also brutally high, at 369. To put this in perspective, over the previous 11 years the average annual number of fatalities was 55.

There were six significant multiday tornado outbreaks over the United States

that month. One of them swept a path from Oklahoma to North Carolina on April 14–16, producing about 155 tornadoes—more than is typically encountered in an entire April—and killing 38 people. On April 22 an EF-4 tornado passed through parts of St. Louis, Missouri, and did substantial damage to Lambert St. Louis International Airport.

But the worst outbreak that month was that of April 25–28, which unleashed tornadoes—about 305 of them—all the way from Texas to New York State. Many of the tornadoes were particularly violent, with three of the classified as EF-5, 12 as EF-4, and 21 as EF-3. The deadliest tornado of the outbreak was an EF-4 that hit the cities

of Tuscaloosa and Birmingham, Alabama (Figure 1) on the 27th. With winds that peaked at 306 km/hr (190 mph) and a width of 2.4 km (1.5 miles), the tornado cut a path of 129 km (80 miles), killed at least 71 people and injured at least another 1000. Over the entire region on the 27th alone there were 317 fatalities, making it the worst single day for tornado fatalities since 1925.

MyMeteorologyLab™

Visualization and Practice

Assignable and assessable media bring concepts to life.

Interactive Tutorials

Dynamic learning modules with interactive media, student self-check questions, and assignable quizzes help students visualize difficult physical processes over space and time.

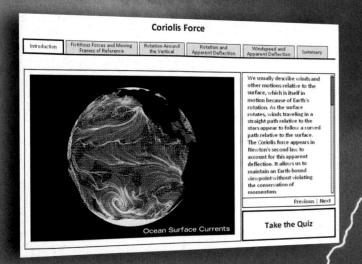

Weather in Motion Videos

These videos offer current, real-world visualizations and case studies of meteorology.

New! MapMaster™

These interactive maps act as a mini-GIS tool, allowing students to layer various thematic maps to analyze spatial patterns and data at regional and global scales.

Exercises for Weather & Climate

These interactive simulations and exercises give students practice with problem solving and guided critical thinking.

MyMeteorologyLab™

A Full Course Solution

Assignable media, an intuitive gradebook, and powerful diagnostics help you focus on teaching.

Course Management

MyMeteorologyLab™ is a full-featured online course management and homework system.

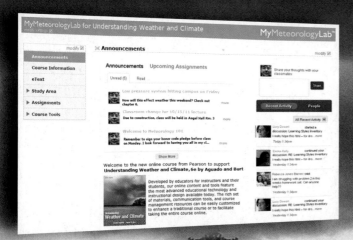

eText

MyMeteorologyLab™ includes the option of a Pearson eText of *Understanding Weather and Climate*, with full search, annotation, and highlighting capabilities.

Gradebook

An easy-to-use and flexible gradebook makes it easy to identify struggling students.

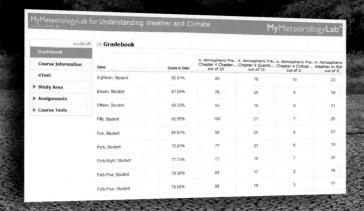

Study Area

(RSS feeds, flashcards, quizzes, social networking)

Students have access to a rich set of study tools to help extend learning, master concepts, and prepare for exams.

The atmosphere is remarkably variable. Its characteristics are quite different from place to place and from the surface to its upper reaches. It is also subject to subtle movements (such as a gentle breeze) or violent motions (such as tornadoes). These chapters look at the composition of the atmosphere and how it is distributed around the planet, how the Sun heats the air, and how pressure and wind patterns are created.

Sun rising over Namib Desert, Namibia

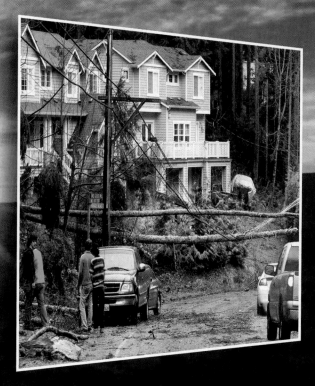

3 Energy Balance and Temperature

TUTORIAL Global Energy Balance

What energy fluxes are involved in the energy balance of the surface?

4 Atmospheric Pressure and Wind

TUTORIAL Atmospheric Forces and Wind

How do the various forces involved in wind combine to govern wind speed and direction?

1

Composition and Structure of the Atmosphere

After reading this chapter, you should be able to:

▶ Distinguish between weather and climate.

▶ Describe the composition of the atmosphere.

▶ Explain how air pressure arises and describe the vertical variation of pressure and density.

▶ Identify and describe the layers of Earth's atmosphere.

▶ Explain the evolution of the atmosphere during Earth's history.

▶ Identify the basic types of data found on weather maps.

▶ List major events in the history of meteorology.

"It sucked me out of my house and carried me across the road and dropped me.... I was Superman for a while.... You're just free-floating through the air. Trees are knocking you and smacking you down."

So it went for Chris Sizemore of Cincinnati on New Year's Eve 2010. Powered by unusually warm air, tornadoes struck the South and Midwest that day, destroying homes and killing four people. This storm foreshadowed what would be an extraordinary year for tornadoes in the United States. In just 6 months 1145 were counted, well above the normal total for a full year. The number of tornado deaths was also far higher than usual, with more than 548 fatalities in the same time period (compared to about 60 in a typical year). This high death count arose from the large number of storms, a high incidence of unusually strong storms, and the extraordinarily large number that struck urban areas. Undoubtedly the worst was the May 23 monster that devastated Joplin, Missouri. Perhaps 1.5 kilometer (km) (approximately 1 mile) wide, it killed 153, injured more than 900, and damaged 8000 buildings. More than 2000 buildings were completely destroyed (mostly homes), representing about 25 percent of the city structure. Although seldom rising to this level of drama, there is no denying the impact of weather on our day-to-day lives: We revel in the beauty of a sunrise, marvel at the power of distant hurricanes, rely on the rain to nourish our gardens, and complain about the heat and cold. And weather routinely disrupts our travel plans—in the United States weather is responsible for 40–50 percent of the time lost to delays in airports! (See Box 1–1, *Focus on Aviation: Winter Storms and Travel*, for more examples of weather impacts on travel.) The most extreme weather events come with enormous price tags, as Table 1–1 shows in its summary of three decades' worth of weather-related disasters in the United States. Though hurricanes and tornadoes can exact a terrible human cost, extreme heat and cold are the biggest killers in the United States, with floods and hurricanes producing the greatest financial costs. Table 1–1 actually understates the toll exerted by weather by not considering the approximately 6000 weather-related traffic fatalities occurring each year from mostly nonsevere events. See *Box 1–2, Focus on the Environment: Highlights of Billion-Dollar Weather Disasters,* on page 8 for a complete list of major weather-related disasters that occurred in the last few decades.

◄ Considering their small size and short lifetime, tornadoes are one of the most powerful and deadly natural hazards. This is one of an abnormally large number to strike the United States in the first half of 2011.

1–1 FOCUS ON AVIATION

Winter Storms and Air Travel

Winter weather can play havoc with air travel in the United States. Naturally, some years are worse than others, and the winter of 2010–2011 was especially difficult. Between November and early February snow and ice conditions associated with four major storms caused U.S. airlines to cancel some 86,000 flights, with thousands of other flights encountering major delays. These events amounted to a significant portion of all scheduled flights for an industry recovering from some economic difficulties. In December 2010 3.7 percent of all U.S. flights were canceled because of winter storms, in contrast to the 2.89 percent canceled the previous December. Even airports not normally subject to crippling winter storms, such as Hartsfield-Jackson Atlanta International Airport, were subject to the cancellation of thousands of flights when a single January storm left behind 6 inches of snow in a city that usually receives half that much in an entire season (Figure 1).

One of the problems associated with winter storms is that aircraft loaded to capacity with passengers can be forced to wait on taxiways between terminal gates and runways for extended periods. An outcry of consumer criticism led to a policy taking effect in April 2010 allowing the U.S. Department of Transportation to levy fines for air carriers of up to $27,500 per passenger when flights are forced to remain on the tarmac for more than 3 hours. Some believe this will motivate airlines to cancel flights

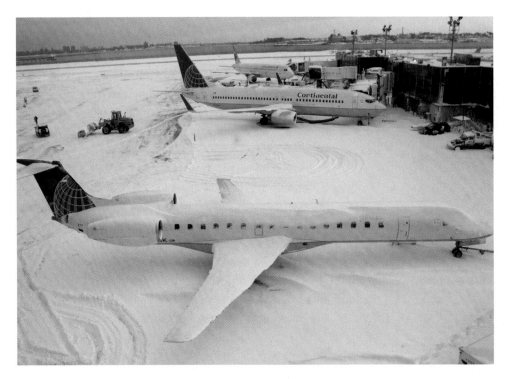

▲ **FIGURE 1** Snowbound airplanes in Atlanta following the storm of January 2011.

more readily than in the past, causing greater inconvenience for passengers forced to wait hours or even days for another flight to get them to their destination. On the other hand, passengers are now much less likely to spend half a day on a crowded aircraft just waiting to get to the runway.

Often the disruption of air travel is due to what happens outside the airport during winter weather. Sometimes it is easier for airlines to fly crews in from other areas

than it is to wait for scheduled personnel who are delayed by impassable highways to get to the airport. And of course, flights scheduled to depart San Diego on a warm, sunny day may be unable to do so because the aircraft needed for the flight is stranded at an East Coast airport.

Weather in other seasons poses different hazards to commercial aviation, as will be discussed in later chapters of this book.

Although we are continually surrounded and affected by the atmosphere, most of us know relatively little about how and why the atmosphere behaves as it does. In the chapters that follow, we hope to provide an account of both the how and the why, in ways that will lead you to understand the underlying physical processes. This chapter introduces the most basic elements of meteorology, laying the foundation for much of the rest of the book.

The Atmosphere, Weather, and Climate

The **atmosphere** is a mixture of gas molecules, small suspended particles of solid and liquid, and falling precipitation.

Meteorology is the study of the atmosphere and the processes (such as cloud formation, lightning, and wind movement) that cause what we refer to as the "weather." **Weather** is distinct from **climate** in that the former deals with short-term phenomena and the latter with characteristic long-term patterns. A rough analogy can help with the distinction. Most of us have an image of New York's Brooklyn Bridge during the morning rush hour. If our mental picture of slow-moving congestion is the bridge's "traffic climate," weather would be the particular combination of individual cars, buses, and trucks found there on a single day. Take a look outside your window and what you will see is weather. The current temperature, humidity, wind conditions, amount and type of cloud cover, and presence or absence of precipitation—these are all elements of weather.

TABLE 1–1

Three Decades of Billion-Dollar U.S. Weather Disasters
Dollar amounts are adjusted to 2007 values.

Year	Events Exceeding $1B					
2010	Northeast Flooding > $1.5B, 11 Deaths	East/South Flooding/ Severe Weather > $2.3B, 32 Deaths	Midwest Tornadoes & Severe Weather > $3.0B, 3 Deaths	South/Southeast Tornadoes & Severe Weather > $1.2B, 6 Deaths		
2009	Southeast/Ohio Valley Severe Weather > $1.4B, 10 Deaths	Midwest/Southeast Tornadoes > $1.0B, No Deaths	South/Southeast Tornadoes & Severe Weather > $1.2B, 6 Deaths	Midwest, South, East Severe Weather > $1.1B, No Deaths	Western Wild Fires > $1.0B, 10 Deaths	Southwest/G. Plains Drought > $5.0B, No Deaths
2008	Southeast / Midwest Tornadoes > $1.0, 57 Deaths	MW / Ohio Valley Svr Wx / Tornadoes > $2.4, 13 Deaths	MW / Mid-Atl. Svr Wx / Tornadoes > $1.1, 18 Deaths	Midwest Flooding e > $15.0, 24 Deaths	U.S. Wild Fires > $2.0, 16 Deaths	Hurricane Dolly > $1.2, 3 Deaths / Hurricane Gustav > $5.0, 53 Deaths
	Hurricane Ike > $27.0, > 112 Deaths	Widespread Drought > $2.0, No Deaths				
2007	Great Plains East Drought > $5.0, *Deaths	Western Wildfires > $1.0, 12 Deaths	Spring Freeze > $2.0, No Deaths	East/South Severe Weather > $1.5, 9 Deaths	California Freeze > $1.4, 1 Deaths	
2006	Numerous Wildfires > $1.0, 28 Deaths	Widespread Drought e > $6.2, *Deaths	Severe Storms Tornadoes e > $1.0, 10 Deaths	Northeast Flooding > $1.0, 20 Deaths	MW/SE Tornadoes > $1.5, 10 Deaths	MW/Ohio Valley Tornadoes ~ $1.1, 27 Deaths
2005	Hurricane Dennis e > $2.2, > 15 Deaths	Hurricane Katrina e ~ $133.8, > 1833 Deaths	Hurricane Rita e ~ $17.1, 119 Deaths	Midwest Drought e > $1.0, No Deaths	Hurricane Wilma e ~ $17.1, 35 Deaths	
2004	Hurricane Charley e ~ $16.5, 35 Deaths	Hurricane Frances e ~ $9.9, 48 Deaths	Hurricane Ivan e > $15.4, 57 Deaths	Hurricane Jeanne e > $7.7, 28 Deaths		
2003	Severe Wx/Hail > $1.8, 3 Deaths	Severe Wx/ Tornadoes > $3.8, 51 Deaths	Hurricane Isabel ~ $5.6, 55 Deaths	S California Wildfires > $2.8, 22 Deaths		
2002	30-State Drought e > $11.4, No Deaths	Western Fires > $2.3, ~21 Deaths				
2001	Tropical Strom Allison e ~ $5.6, 43 Deaths	Midwest/OH Valley Hail/Tornadoes > $2.2, > 3 Deaths				
2000	Drought/Heat Wave e > $4.8, ~ 140 Deaths	Western Fires > $2.4, No Deaths				
1999	AR - TN Tornadoes ~ $1.6, 17 Deaths	OK - KS Tornadoes > $2.0, 55 Deaths	E Drought/ Heat Wave > $1.2 e 502 Deaths	Hurricane Floyd e > $7.4, 77 Deaths		
1998	New England Ice Storm > $1.8, 16 Deaths	SE Severe Wx > $1.3, 132 Deaths	MN Severe Storms/ Hail > $1.9, 1 Death	S Drought/Heat Wave $9.5, > 200 Deaths	Hurricane Bonnie ~ $1.3, 3 Deaths	Hurricane Georges e $7.4, 16 Deaths / Texas Flooding ~ $1.3, 31 Deaths
1997	Midwest Flood/ Tornadoes e $1.3, 67 Deaths	N Plains Flooding ~ $4.8, 11 Deaths	W Coast Flooding ~ $3.9, 36 Deaths			
1996	Blizzard/Flooding ~ $4.0, 187 Deaths	Pacific NW Flooding ~ $1.3, 9 Deaths	S Plains Drought ~ $6.8, No Deaths	Hurricane Fran > $6.6, 37 Deaths		
1995	CA Flooding > $4.1, 27 Deaths	SE/SW Severe Wx $7.5, 32 Deaths	Hurricane Marilyn e $2.9, 13 Deaths	Hurricane Opal > $4.1, 27 Deaths		
1994	SE Ice Storm ~ $4.2, 9 Deaths	Tropical Storm Alberto ~ $1.4, 32 Deaths	Texas Flooding ~ $1.4, 19 Deaths	W Fire Season ~ $1.4, No Deaths		
1993	E Storm/Blizzard $7.9, ~ 270 Deaths	SE Drought/Heat Wave ~ $1.4, > 16 Deaths	Midwest Flooding ~ $30.2, 48 Deaths	CA Wildfires ~ $1.4, 4 Deaths		
1992	Hurricane Andrew ~ $40.0, 61 Deaths	Hurricane Iniki ~ $2.7, 7 Deaths	Nor'easter $2.3, 19 Deaths			
1991	Hurricane Bob $2.3, 18 Deaths	Oakland CA Firestorm ~ $3.9, 25 Deaths				

(Continued)

TABLE 1–1
(Continued)

Year	Events Exceeding $1B			
1990	S Plains Flooding > $1.6, 13 Deaths			
1989	Hurricane Hugo > $15.3, 86 Deaths	N Plains Drought > $1.7, No Deaths		
1988	Drought/Heat Wave e $71.2, ~7,500 Deaths			
1986	Drought/Heat Wave $2.4, ~100 Deaths			
1985	Florida Freeze ~ $2.3, No Deaths	Hurricane Elena $2.5, 4 Deaths	Hurricane Juan $2.9, 63 Deaths	
1983	Hurricane Alicia $6.3, 21 Deaths	Florida Freeze ~ $4.2, No Deaths	Gulf Storms/Flooding ~ $2.3, ~ 50 Deaths	W Storms/Flooding ~ $2.3, ~ 45 Deaths
1981	Hurricane Alicia $6.3B, 21 Deaths	Florida Freeze ~ $4.2, No Deaths	Gulf Storms/Flooding ~ $2.3, ~ 50 Deaths	W Storms/Flooding ~ $2.3, ~ 45 Deaths
1980	Drought/Heat Wave $55.4B, 10,000 Deaths			

Note: e = estimated; ~ = approximately/about; * = undetermined

Source: www.ncdc.noaa.gov/oa/reports/billionz.html

Climatology concerns itself with the same elements of the atmosphere that meteorology does, but on a different time scale. Rather than focusing on a single point in time, climatology relies on averages taken over a number of years in order to gauge typical atmospheric conditions for locations across Earth's surface. When people joke about summer conditions as "Sahara-like," they are implicitly making a climatological reference to average conditions in North Africa. Averages are very important, but climatologists also want to know the variability of the weather elements just as, in addition to the average speed, the bridge commuter wants to know about traffic variability. In the case of the atmosphere, it might be useful to know that Boulder, Colorado, has an average April temperature of 7 °C (45 °F), but this figure becomes more meaningful when one understands just how far the temperature might depart from the value on any given day. Frequencies of occurrence of weather events—such as extreme heat, hail, or lightning—are also aspects of climates. Finally, a particularly important part of climatology is concerned with changes in Earth's climate and the factors responsible for those changes.

This book's focus on weather and climate correctly suggests that the atmosphere is of primary interest, but there is no understanding our atmospheric environment without reference to land and ocean processes. For example, moist Pacific climates in western Oregon give way to desertlike conditions a short distance eastward because of mountain topography. Lush forests in the Amazon basin control a number of processes that are key to the region's climate. Similarly, bright, highly reflective snow and ice surfaces in polar locations contribute greatly to extreme cold. Hurricanes that batter east coast locations could not form without the fuel provided by a warm ocean surface.

Western Europe would be frigid if not for the heat imported by ocean currents, and on much longer time scales large shifts in ocean circulation have led to major climatic changes. Furthermore, the composition of the atmosphere and Earth climate cannot be explained without considering the exchange of material between the solid Earth and the atmosphere. Clearly, an integrated approach that considers all components of the Earth system is necessary. Figure 1–1 shows one approach to conceptualizing the components and their interactions. Note the presence of external natural processes and human activities as agents of change.

Checkpoint

1. Define weather and climate in your own words.
2. Compare the concerns of the sciences of meteorology and climatology, giving some examples of different phenomena they might investigate.
3. List some places you have visited whose climate is affected by proximity to an ocean or position deep within a continent.

Thickness of the Atmosphere

Every child has wondered, "How high is the sky?" There is no definitive answer to that question, however, because Earth's atmosphere becomes thinner at higher altitudes. A person in a rising hot-air balloon would be surrounded by an atmosphere that gradually becomes less dense. At some height, the air becomes so thin that the balloonist would pass out from a shortage of oxygen—but there would still be an atmosphere. At an altitude of 16 km, or 10 mi, the density of the air is only about 10 percent of that at sea level, and at 50 km (30 mi) it

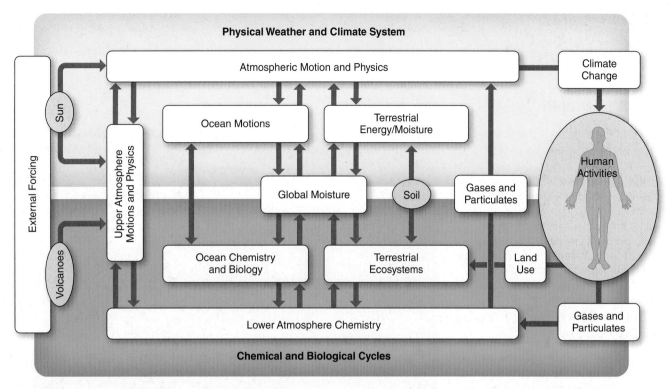

Physical Weather and Climate System

Chemical and Biological Cycles

▲ **FIGURE 1–1** A simplified view of the Earth system. The upper part of the diagram represents purely physical aspects of Earth, such as ocean currents, winds, cloud formations, and temperature distributions. The bottom half depicts the constant exchange of material throughout the system known as "cycling." These exchanges occur between and among the living and nonliving realms, and they both affect and are themselves affected by the physical components of the Earth system.

is only about 1 percent of what it is at sea level. But even at heights of several hundred kilometers above sea level, there is some air and, hence, an atmosphere. We have no way to establish its upper boundary, however, because no universally accepted definition exists of how much air in a given volume constitutes the presence of an atmosphere. You would probably not say only one molecule of air per cubic kilometer constitutes an atmosphere.

Viewed from Earth's surface, the atmosphere appears to be extremely deep. In reality, however, most of the atmosphere is contained within a relatively shallow envelope surrounding the oceans and continents. Let's assume for the sake of discussion that the upper limit of the atmosphere occurs at 100 km (60 mi) above sea level (in fact, 99.99997 percent of the atmosphere is below this height). By comparing this 100 km thickness with the 6400 km (4000 mi) radius of Earth, we see that the depth of the atmosphere adds less than 2 percent to Earth's cross-sectional size. This is evident in Figure 1–2, an image of Earth and its atmosphere taken from space. The top of the thunderstorm cloud probably has an altitude of about 12 km (7.5 mi), but when viewed from space, it appears to hug the ground. Though impressive when we look up at them,

◄ **FIGURE 1–2** Despite its appearance from the surface, the atmosphere is extremely thin relative to the rest of Earth.

1-2 FOCUS ON SEVERE WEATHER

Highlights of Billion-Dollar Weather Disasters

Between 1980 and 2010 the United States experienced 99 separate billion-dollar weather-related disasters, amounting to $728 billion in total losses adjusted to 2007 dollars (Table 1–1). Four such events—all hurricanes—occurred in 2004, causing $50 billion in damages and 168 fatalities. Hurricanes Charley, Frances, Ivan, and Jeanne were part of an unusually active Atlantic hurricane season made all the more unusual because all four hurricanes went through the state of Florida. Only once before had four hurricanes gone through a particular state in a given year, and three of the 2004 storms crossed over the same part of central Florida.

A year later, Floridians were still rebuilding when the 2005 hurricane season got off to a remarkable start: Hurricane Dennis set two significant milestones on July 5. It marked the earliest date ever that a fourth named storm had formed in a given season (hurricane season in the Atlantic does not usually become active until August or September), and it was the earliest date that a rare Category 4 Atlantic hurricane had ever occurred. In August 2005, Hurricane Katrina moved through southern Florida as a Category 1 hurricane, depositing rainfall amounts approaching 50 cm (20 in.) in places and bringing its share of wind damage. Within a few days, Katrina intensified to an extremely rare Category 5 hurricane in the Gulf of Mexico, turned northward toward the Gulf Coast, and then ravaged southern Louisiana (including New Orleans), Mississippi, and Alabama. Though reduced to a Category 4 hurricane

▲ **FIGURE 1** Wreckage from the deadly tornado that struck Joplin, Missouri on May 22, 2011. Three months later, estimates of the financial costs ranged from hundreds of millions to over one billion dollars.

just prior to landfall, Katrina was one of the biggest natural disasters ever to hit the United States. Chapter 12 contains analyses of the major hurricanes of 2004 and 2005.

Measured in dollar costs, hurricanes are the leading cause of U.S. weather-related events. But, in some years, they can be overshadowed by other types of weather disasters. This fact was highlighted in the years 2006 and 2007, in which there were a combined 11 billion-dollar weather-related events, none of which was a hurricane. Among these were major fires that spread across much of the western United States and some major tornado outbreaks.

Nonhurricane-related floods and heat waves and/or droughts tie for second among weather-related damages. By far the deadliest disasters to hit the United States in the last quarter-century were the heat/drought events of 1980 and 1988, which resulted in nearly $100 billion in economic losses and perhaps 20,000 fatalities from heat stress. Another heat wave in 1999 killed an additional 500 people. Clearly, more Americans and Canadians die from severe heat episodes than from hurricanes and tornadoes.

Did You Know?

The total mass of the atmosphere, 5.4×10^{15} kilograms (kg),[1] is equivalent to 5.65 billion million tons—the amount of water that would fill a lake the size of California to a depth of 13 km (7.7 mi).

---
[1] 5,140,000,000,000,000 kg

those clouds are no thicker than the skin of an apple, comparatively speaking.

Given the shallowness of the atmosphere, its motion over large areas must be primarily horizontal. Indeed, with some very notable exceptions, horizontal wind speeds are typically hundreds to thousands of times greater than vertical wind speeds. However, we cannot overlook the wind's vertical motions. As we shall see, even small vertical displacements of air have an impact on the state of the atmosphere. Paradoxically, the least impressive motions—vertical—despite being hardest to detect and forecast, turn out to determine much of atmospheric behavior.

Composition of the Atmosphere

The atmosphere is composed of a mixture of invisible gases and a large number of suspended microscopic solid particles and water droplets. Molecules of the gases can be exchanged between the atmosphere and Earth's surface by physical processes, such as volcanic eruptions, or by biological processes, such as plant and animal respiration. Molecules can also be produced and destroyed by purely internal processes, such as chemical reactions between the gases.

Consider a gas that is constantly exchanged between the atmosphere and Earth's surface. If we think of the atmosphere as a reservoir for this gas, the gas concentration in the reservoir will remain constant so long as the input rate (the rate at which the gas moves from surface to atmosphere) is equal to the output rate (the rate at which the gas moves from atmosphere to surface). Under such conditions, we say that the concentration of the gas exists in a *steady state* or *equilibrium* condition.

Although the atmospheric concentration of a gas remains constant under steady-state conditions, individual molecules stay in the atmosphere for only a finite period of time before they are removed by whatever output processes are active. An equilibrium exists, but it is a dynamic equilibrium with molecules cycling in and out of the atmosphere. The average length of time that individual molecules of a given substance remain in the atmosphere is called the *residence time*. The residence time is found by dividing the mass of the substance in the atmosphere (in kilograms) by the rate at which the substance enters and exits the atmosphere (in kilograms per year). Figure 1–3 illustrates both the concepts of steady-state and residence time. Parts (a) and (b) show the same mass constituting the "atmospheric reservoir," with the length of the input and output arrows indicating the rates at which gases are put into and removed from the reservoir. In part (a), the input and output rates are equal, indicating a steady state. Both rates are small, however, meaning that any particular molecule of the gas has a long residence time. In part (b), there is again a steady state, but the greater rate of input and output relative to reservoir size leads to a shorter residence time.

The lowest 80 km (50 mi) of the atmosphere is sufficiently stirred by mixing that differences in molecular weight have no role in the distribution of gases. In other words, in this region vertical motions are more important than gravitational settling, and thus processes other than settling under gravity must explain any variations present. This layer is known as the **homosphere**, as it reflects the homogenizing role of wind and other motions. Above that lies the **heterosphere**, where gases segregate according to molecular weight. Heavier gases are more abundant in the lower heterosphere, whereas lighter gases such as hydrogen and helium become relatively more abundant with increasing altitude. Considering that only a tiny fraction of atmosphere remains above 80 km, we will not pursue these details of the heterosphere but will instead deal almost exclusively with the homosphere.

(a)

(b)

▲ **FIGURE 1–3** The atmosphere can be thought of as a gas reservoir experiencing constant input and output by surface exchange and/or internal processes. If the inputs and outputs occur at the same rate, there is no net change in the content of the gas. In (a) the arrows depict a slow rate of exchange of a hypothetical gas. As a result, any molecule of that gas can be expected to remain for a long time before leaving the atmosphere. In (b) the reservoir size is the same, but the exchange rate is much more rapid, so the gas has a shorter residence time.

Within the homosphere atmospheric gases are often categorized as being permanent or variable, depending on whether or not their concentration is uniform. **Permanent gases** are found everywhere in nearly the same proportion, whereas **variable gases** are those whose distribution in the atmosphere is uneven in both time and space. Notice that describing the amount of a gas is not completely straightforward. It is natural to use a percentage, but how do we compute the percentage? Do we find the percent of the atmosphere's mass represented by a gas, or do we find the percent of the atmosphere's volume occupied by that gas? The answer is that both are used! Percent by volume is particularly attractive because—at least in the homosphere—percent by volume also represents percent by number of molecules. For example, if a gas is 20 percent by volume, 20 percent of the molecules in the atmosphere are that gas. The presence of variable gases, however, complicates the problem of calculating these percentages. If the quantity of some variable gas increases, the percentage of other gases must decrease even though no molecules are removed. In the face of this complication, atmospheric scientists calculate permanent gas percentages after excluding variable gases from the calculation. Fortunately, variable gases are so rare that global average values are hardly affected.

Did You Know?

The atmosphere teems with insects flying and riding on air currents at altitudes up to several thousand km (12,000 feet) in search of better locations for food, nesting places, and mating partners. The amount varies greatly by location and season, but a recent study in England in the summer found an average of 3 billion insects crossing 1 square km (0.39 square miles) each day.

The Permanent Gases

Whether considered by mass or volume, the atmosphere is more than 99 percent permanent gases. The most abundant by far are nitrogen and oxygen, followed by the inert gases argon and neon in small amounts and several other gases in even smaller amounts (Table 1–2). Atmospheric **nitrogen** occurs primarily as paired nitrogen atoms bonded together to form single molecules denoted N_2. Nitrogen gas has a molecular weight of 28.02, meaning that on average the mass of one N_2 molecule is slightly in excess of that of a combined total of 28 protons and neutrons. Almost all N atoms contain 7 protons and 7 neutrons, but some have 8 neutrons and are therefore slightly heavier. Variants of an element with different neutron counts are known as *isotopes* of that element. Each isotope has the same number of protons and thus has almost identical chemical properties, but as will be seen in later chapters, differences in molecular weight among isotopes can be important. Averaging the weights of the isotopes and accounting for their relative abundance gives the molecular weight of N_2 as slightly greater than 28. (Similarly, other gases appear in various isotopes, which explains why their molecular weights also have fractional parts.)

Nitrogen is a largely unreactive gas that accounts for 78 percent of the volume of all the permanent gases, or 75.5 percent of their mass. The processes that add and remove nitrogen from the atmosphere occur very slowly, so nitrogen has a very long residence time, measured in millions of years.

TABLE 1–2

Permanent Gases of the Atmosphere

Constituent	Formula	Percent by Volume	Molecular Weight
Nitrogen	N_2	78.08	28.01
Oxygen	O_2	20.95	32.00
Argon	Ar	0.93	39.95
Neon	Ne	0.002	20.18
Helium	He	0.0005	4.00
Krypton	Kr	0.0001	83.80
Xenon	Xe	0.00009	131.30
Hydrogen	H_2	0.00005	2.02

TABLE 1–3

Variable Gases of the Atmosphere

Constituent	Formula	Percent by Volume	Molecular Weight
Water Vapor	H_2O	0.25	18.01
Carbon Dioxide	CO_2	0.039	44.01
Ozone	O_3	0.01	48.00

The second most abundant gas, **oxygen** (O_2), constitutes 21 percent of the volume of the atmosphere and 23 percent of its mass. Oxygen is crucial to the existence of virtually all forms of life. Like nitrogen, the oxygen molecules of the atmosphere consist mostly of paired atoms, called *diatomic oxygen*. Their residence time is about 6000 years. Together, nitrogen and oxygen account for 99 percent of all the permanent gases, with **argon** making up most of the remainder. Removal processes are so slow for argon that its residence time is extremely long. Note that the abundance of a gas gives no clue about its importance in weather. Indeed, despite the fact that permanent gases in Table 1–2 make up nearly the entire atmosphere, none can explain storms, clouds, the deep cold that often follows a clear night, or other events we associate with weather.

Variable Gases

The variable gases account for only a small percentage of the total mass of the atmosphere (Table 1–3). Despite their relative scarcity, some of these gases profoundly affect the behavior of the atmosphere—and even your own physical comfort.

Water Vapor The most abundant of the variable gases, **water vapor,** accounts for about one-quarter of 1 percent of the total volume of the atmosphere on average. Because the source of water vapor in the atmosphere is evaporation from Earth's surface, its concentration normally decreases rapidly with altitude, and most atmospheric water vapor is found in the lowest 5 km (3 mi) of the atmosphere. As explained in Chapter 5, water vapor condenses to a liquid at relatively low levels in the homosphere. Thus although vertical mixing extends to great heights, water vapor is anything but uniformly distributed with altitude.

Water is constantly exchanged between the planet and the atmosphere in what is called the **hydrologic cycle**, which is described in Chapter 5. Water continuously evaporates from both open water and plant leaves into the atmosphere, where it eventually condenses to form liquid droplets and ice crystals. These liquid and solid particles are removed from the atmosphere by precipitation as rain, snow, sleet, or hail. Because of the rapidity of global evaporation, condensation, and precipitation, water vapor has a very short residence time of only 10 days.

We all know how damp and muggy the air feels when the water vapor content is high and how parched our skin can get when the air is dry (not to mention "bad hair days" when the moisture content is extremely high or low). Despite the wide range of physical comfort levels we experience in response to variations in water vapor content, the actual range of water vapor content is really quite limited. Near Earth's surface, the water vapor content ranges from just a fraction of 1 percent of the total atmosphere over deserts and polar regions to about 4 percent in the tropics. This means that at most we would find that 4 out of every 100 air molecules are water vapor. (Outside of the tropics, water vapor content does not usually exceed 2 percent.) At higher altitudes, water vapor is even rarer.

Despite being a relatively small portion of the atmosphere, water vapor is extremely important. Not only is it the source of the moisture needed to form clouds, it is a very effective absorber of energy emitted by Earth's surface. (We describe radiant energy in the next chapter.) Its ability to absorb Earth's thermal energy makes water vapor one of the "greenhouse gases" we discuss in Chapter 3.

Keep in mind that water vapor is not the same as small droplets of liquid water. Water vapor exists as individual gas molecules. Unlike the molecules of liquids and solids, water vapor molecules are not bound together. In that regard, water vapor is similar to N_2, O_2, and other atmospheric gases. Unlike other gases, however, it readily changes phase into liquid and solid forms both at Earth's surface and in the atmosphere. As will be seen in later chapters, such changes in phase involve large amounts of energy that are very important in weather and climate.

Although water vapor is an invisible gas, some satellite systems can indirectly measure the amount of water vapor in the air and display its varying content on images (Figure 1–4).

These images typically show that water vapor exhibits large changes over very short distances, more so than other variable gases. Displays of water vapor content can be quite valuable to forecasters, who use the images to determine broad-scale wind patterns in the middle and upper atmosphere. The images also help meteorologists identify the boundaries between adjoining bodies of air.

Carbon Dioxide Another variable gas is **carbon dioxide**, CO_2. We do not physically sense variations in the amount of carbon dioxide present in the atmosphere, as we do with water vapor, and yet, as you will see shortly, we cannot ignore these variations. Increases in the carbon dioxide content of the atmosphere may have some important climatic consequences that could greatly affect human societies.

Carbon dioxide currently accounts for about 0.039 percent of the atmosphere's volume. When a gas occupies such a small proportion of the atmosphere, we often express its content as parts per million (ppm) rather than percent (parts per hundred). Thus, the current atmospheric concentration of CO_2 is about 390 ppm. It is supplied to the atmosphere by plant and animal respiration, the decay of organic material, volcanic eruptions, and natural and *anthropogenic* (human-produced) combustion. Carbon dioxide is removed from the atmosphere by **photosynthesis**, the process by which green plants convert light energy to chemical energy. Photosynthesis uses energy from sunlight to convert water absorbed by the roots of plants and carbon dioxide taken in from the air into the chemical compounds known as carbohydrates, which both support plant processes and ultimately nourish the animal kingdom. In this process, oxygen molecules are released into the atmosphere as a by-product (see *Box 1–3, Focus on the Environment: Photosynthesis, Respiration, and Carbon Dioxide*).

(a)

(b)

▲ **FIGURE 1–4** Paired satellite images based on a system that detects the presence of water vapor (a) and one that gives a visible rendition of cloud tops (b). Notice that though the two patterns are somewhat similar, the water vapor image shows a broader distribution of moisture than does the image showing actual clouds.

Photosynthesis, Respiration, and Carbon Dioxide

Without the process of photosynthesis, Earth would have an entirely different atmosphere—and would probably be without life as we know it. Through photosynthesis, plants utilize light energy from the Sun and make food. Because it requires sunlight, photosynthesis occurs only during the day. It also requires chlorophyll, an organic substance found in green plants and some single-celled organisms. Photosynthesis converts solar energy, water, and carbon dioxide into simple carbohydrates. These can then be converted to complex carbohydrates, starches, and proteins, all of which supply plants with the material for their own growth. Plants in turn provide the basic nutrients for grazing and browsing animals.

In addition to photosynthesis, an exchange of gases takes place through the leaves of plants—respiration. **Respiration** provides the mechanism by which plants obtain the oxygen they need to perform their metabolic processes. (For animals, respiration is synonymous with breathing. Though plants do not have lungs, they take

in oxygen through their leaves.) Unlike photosynthesis, respiration occurs during both day and night. A plant must "fix" carbon into its cells in order to grow; to obtain this carbon, the photosynthesis rate in the plant must exceed the respiration rate. Thus, during periods of plant growth more carbon dioxide is being removed from the air than is being added through respiration. After a plant dies, however, it no longer takes in CO_2 from the air. Instead, its stored carbon is oxidized and released back to the atmosphere as CO_2 as the plant decomposes. If no major changes occur in the amount and distribution of vegetation, CO_2 intake (for photosynthesis) balances CO_2 output (from respiration and decay) over the course of a year, and the total store of atmospheric CO_2 is unaffected by plant growth.

There is, however, another factor governing the CO_2 balance. Under some circumstances, dead plant material is quickly buried beneath the surface, does not decompose, and therefore does not release its stored carbon back into the atmosphere. Instead, over millions of years the material is transformed into fossil fuels, such as petroleum or coal, which we extract from the ground and burn for heat and energy, yielding CO_2 as a combustion product. In

doing so, we release carbon that otherwise would have remained underground for eons. Although we speak of "adding" to the atmosphere, it is probably more accurate to think of "moving" carbon back to the atmosphere. Regardless, the result is certainly an increase in atmospheric CO_2, with potentially global consequences.

Recognizing the role of fossil fuel combustion on the carbon balance of the atmosphere, 160 of the world's nations met in 1997 at Kyoto, Japan, to formally agree on a multiyear plan to lower CO_2 emissions. Under this plan, averaged over the 5-year period from 2008 to 2012, industrialized countries would reduce CO_2 emissions by 6 to 8 percent of 1990 values. To be legally binding, the plan would not only have to be signed by the representatives of the participating countries, but also formally ratified by their governments. In March 2002 the European Union formally ratified the treaty, followed by Canada in December of that year. Ratification by Russia in November 2004 satisfied the remaining condition for approval and allowed the Kyoto Agreement to come into force on February 16, 2005. The United States, which has not signed this agreement, is the only major industrialized nation not to have done so.

For many decades, the rate of carbon dioxide input to the atmosphere has exceeded the rate of removal, leading to a global increase in concentration. Figure 1–5 plots the record of carbon dioxide content obtained from the Mauna Loa Observatory. (Taken at an elevation of 3400 m, or 11,150 ft, on one of the Hawaiian Islands, these observations are considered representative for the Northern Hemisphere.) Since the 1950s, the concentration of CO_2 has increased at a rate of about 2.15 ppm per year. The increase has occurred mainly because of an increase in anthropogenic combustion and, to a lesser extent, deforestation of large tracts of woodland (Figure 1–6). The increase has received considerable scientific and media attention because CO_2 (like water vapor) effectively absorbs radiation emitted by Earth's surface. The CO_2 increases projected to the year 2100 would double existing levels of carbon dioxide and could lead to a continued warming of the lower atmosphere that began more than a century ago.

Figure 1–5 shows not only the overall increase in CO_2 levels in the Northern Hemisphere but also shows seasonal oscillations. Specifically, the amount of carbon dioxide in the atmosphere is greatest in the early spring and lowest in late

summer. The springtime maximum occurs because during the winter plant growth is slow, so plants take less CO_2 from the air. In addition, leaf litter has been decomposing all winter, which means carbon in the litter has been oxidized to CO_2 and has entered the air. During the period of summer regrowth, carbon is removed from the atmosphere, and so carbon dioxide levels fall.

For every CO_2 molecule removed from the atmosphere by photosynthesis, plants produce one O_2 molecule. Atmospheric O_2 does not show any seasonal variation, however, because the photosynthetic contribution is so small relative to the huge atmospheric O_2 reservoir. Carbon dioxide has a residence time of about 150 years.

Ozone The form of oxygen in which three O atoms are joined to form a single molecule is called **ozone**, O_3. This substance is something of a paradox. The small amount of it that exists in the upper atmosphere is absolutely essential to life on Earth; near Earth's surface, however, it is a major component of air pollution, causing irritation to lungs and eyes and damage to vegetation. Fortunately, ozone occurs only in minute amounts

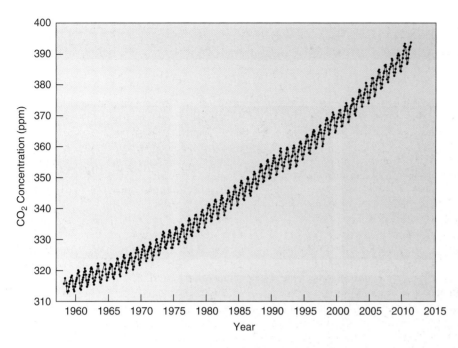

◄ **FIGURE 1-5** The carbon dioxide content of the atmosphere has been increasing for the last 150 years due to human activities. Only recent decades are shown in this figure. The data were obtained from the Mauna Loa Observatory on a high mountain in Hawaii and are representative of a broad band of Northern Hemisphere latitudes. The zig-zag line reveals the seasonal cycle in the growth and decay of plants. Comparable measurements taken in the Southern Hemisphere are similar, except the seasonal cycle is of course reversed.

in the lower atmosphere, so that even in highly polluted urban air its concentration may be only about 0.15 ppm (that is, 15 out of every 100 million molecules are ozone). Aloft, at altitudes of 25 km, its concentration might be 50 to 100 times higher. Yet with a concentration of only 15 ppm at that altitude, ozone is just a tiny part of the atmosphere. As we've said, the homosphere is essentially nothing but permanent gases.

Ozone in the part of the upper atmosphere called the *stratosphere* is vital to life on Earth because it absorbs lethal ultraviolet radiation from the Sun. Why is ozone mostly found in the stratosphere rather than at Earth's surface? The answer is complicated, as literally hundreds of chemical reactions govern its abundance. To simplify, it can be said that ozone

forms when atomic oxygen (O) collides with molecular oxygen (O_2). Atomic oxygen is produced in the very upper reaches of the atmosphere, but little ozone forms there because of low air density at those altitudes. Nearer Earth's surface (but still high up in the atmosphere), the chances of O atoms colliding with O_2 molecules are greater, so the highest ozone values are found there.

When it absorbs ultraviolet radiation, ozone splits into its constituent parts (O + O_2) which can then recombine to form another ozone molecule. Through these reactions, ozone is continually being broken down and reformed to yield a relatively constant concentration in the ozone layer (see *Box 1-4, Focus on the Environment: Depletion of the Ozone Layer*).

◄**FIGURE 1-6** Burning rain forest in Guatemala. Deforestation by humans has contributed to the increase of carbon dioxide in the atmosphere because it releases carbon previously held in plant tissues and reduces the amount of plant material that can carry out photosynthesis.

Depletion of the Ozone Layer

In 1972 Sherwood Rowland and Mario Molina, atmospheric chemists then working at the University of California at Irvine, proposed that certain human-produced chemicals called chlorofluorocarbons (CFCs) could be carried naturally into the stratosphere and damage the ozone layer. CFCs are widely used in refrigeration and air conditioning, in the manufacture of plastic foams, and as solvents in the electronics industry.

Knowing that CFCs do not easily react with other molecules in the lower atmosphere, Rowland and Molina proposed that these molecules could reach the stratosphere intact, where they would break down and release free atoms of chlorine (Cl). Under certain circumstances, chlorine atoms can effectively destroy ozone molecules. In the first step of this process, a chlorine atom reacts with an ozone molecule to produce O_2 and chlorine monoxide (ClO). Next, an oxygen atom (O) reacts with ClO, creating another O_2 molecule while freeing the chlorine atom (Cl). Note that the chlorine atom that first reacted with the ozone molecule is still present and capable of reacting again with another ozone molecule. This fact makes chlorine atoms able to repeatedly break down ozone molecules. In fact, as many as 100,000 ozone molecules can be removed from the atmosphere for every chlorine molecule present. Rowland and Molina's theory is now accepted as fact, and the two scientists were rewarded for their work with a Nobel prize in 1995.

The most severe ozone depletion occurs every October (spring in the Southern Hemisphere) and persists for several months. Why is the ozone hole found over the Antarctic during the spring? The answer is fairly complex, but a variety of factors can be cited here. First, air currents surrounding Antarctica isolate the region from the rest of the hemisphere, so there is less mixing with ozone-rich air from the north. Another factor is the unusual chemistry of clouds found in the Antarctic stratosphere. At the very low temperatures found there, clouds are largely made up of nitric acid and water, rather than ordinary water ice. Processes involving these clouds allow certain chlorine compounds to accumulate. When the dark Antarctic winter ends, the burst of ultraviolet radiation breaks apart

▲ **FIGURE 1** A series of satellite images showing a reduction of the stratospheric ozone over Antarctica. Mapped values are Dobson Units (DU), a measure of total ozone amount in the atmospheric column, nearly all of which is found in the stratosphere. The global average is about 340 DU. The boundary of the "ozone hole" is considered to be 220 DU, shown in dark blue. Values that low were not observed prior to 1979.

these compounds to create free chlorine atoms (Cl). These Cl free radicals readily destroy ozone, as described previously. A very recent discovery shows that depletion begins at the rim of Antarctica, where sunlight arrives first, and works its way poleward. Thus depletion begins in June at 65° S but not until late August at 75° S. We should emphasize that chlorine is not usually abundant in the Antarctic stratosphere. Rather, what chlorine exists is in a form able to destroy ozone, thanks to the unique conditions that take place in the spring (see Figure 1).

Significant stratospheric ozone depletion has also been detected over much of Europe and North America, including a less pronounced hole over the Arctic. Nobody knows how long the ozone will continue to decrease, or how depleted it may become, but in November 1999 satellite and land-based data revealed abnormally low ozone levels over northwestern Europe. Stratospheric zone values were 30 percent below normal for that time of year, resulting in levels nearly as low as those normally observed over the Antarctic.

Government and private industry have taken action to help reduce the amount of CFCs entering the atmosphere. In accordance with the Montreal Protocol of 1987 and subsequent conferences, the world's developed countries have ceased production of CFCs. Continued production in developing countries was permitted by the protocol and many have done so (by 2007 China was the world's largest producer). CFCs have lifetimes in the atmosphere of about 100 years, so an immediate reduction in the ozone hole will not occur. But progress has been made; since 1997 there has been a decline in the amount of chlorine in the stratosphere, and the size of the ozone hole appears to have stabilized, although significant year-to-year variations still occur due to weather conditions. A 2006 United Nations study predicted that Antarctic ozone levels may return to pre-1980 levels sometime between 2060 and 2075—a slow but significant improvement. Thus, curbing CFC emissions illustrates how solid science, combined with international cooperation, can have a major impact on the protection of the environment.

▲ **FIGURE 1-7** Methane is a gas that has varied greatly over long time scales. The most recent increase began about 5000 years ago from a value of 600 ppb. Methane amounts roughly doubled in the last 200 years, but have not increased much in the most recent decade.

Methane Another variable gas is **methane**, CH_4. For hundreds of thousands of years, the concentration of methane has varied cyclically, rising and falling on roughly a 23,000-year basis. The pattern was broken about 5000 years ago, when methane began increasing in a departure from the expected downward trend. With industrialization the rate of increase accelerated, so that values have more than doubled over the last 200 years, reaching about 1.8 ppm at present (Figure 1–7). Most of this increase is attributed to rice cultivation, biomass burning, and fossil fuel extraction (coal and petroleum mining). The roughly stable trend from 1998 to 2007 gave way to increasing values in 2008. The residence time for methane is about 10 years.

Despite its low concentration in the atmosphere, methane is an extremely effective absorber of thermal radiation emitted by earth's surface. Thus, increases in atmospheric methane levels could play a role in the warming of the atmosphere.

Aerosols

Small solid particles and liquid droplets in the air (excluding cloud droplets and precipitation) are collectively known as **aerosols**.[1] Although we associate them with polluted urban atmospheres, aerosols are formed by both human and natural processes. Thus, they are ever present, even in areas far

from human activity. They normally occur at concentrations of about 10,000 particles per cubic centimeter (cm^3) over land surfaces, which is equivalent to roughly 17,000 particles per cubic inch (in^3).

Did You Know?

On average, each breath a person takes brings into the lungs about cm^3 (1 liter, or 64 in^3) of air. Given the average size and concentration of aerosols, each of us draws about 1 trillion aerosols into our lungs several times each minute, or about two tablespoons of solids each day.

The smallest of these particles have radii on the order of 0.1 micrometer (μm; one-millionth of a meter) and are believed to form primarily from the chemical conversion of sulfate gases to solids or liquids. Larger particles are introduced into the air directly as wind-generated dust, volcanic ejections, sea spray, and combustion by-products. Because these particles are extremely small, most fall so slowly that even the smallest of vertical motions keep them suspended in the atmosphere. (The most effective mechanism for removing aerosols is their capture by falling precipitation.) Aerosols typically have life spans of a few days to several weeks.

Aerosols have some noticeable effects on the atmosphere. Urban smog, which includes aerosols, severely reduces visibility, and dust storms can reduce visibility to near zero when large volumes of soil are dislodged from the surface (Figure 1–8). Aerosols also play a major role in the formation of cloud droplets because virtually all cloud droplets that form

[1]The term **particulate** is often used interchangeably with *aerosol*. This terminology may be confusing for some people, who would take *particulate* to mean solid particles only. However, at these microscopically small sizes, it is difficult to make a distinction between the liquid and solid states; thus, there is little reason to draw a fine distinction.

▲ **FIGURE 1-8** A dust storm in Australia.

▲ **FIGURE 1-9** Because of compression, the atmosphere is more dense near its base and progressively "thins out" with altitude.

in nature do so on suspended aerosols called **condensation nuclei.** Condensation nuclei are discussed in more detail in Chapter 5.

Checkpoint

1. What are permanent gases? Variable gases?
2. Classify the gases in the atmosphere according to whether they are permanent or variable.
3. Only a few of the gases in Tables 1–2 and 1–3 are important in weather and climate. Which ones?

Vertical Structure of the Atmosphere

As we have seen, the atmosphere has no distinct upper boundary; the air simply becomes less and less dense with increasing altitude. We also know that its composition remains nearly constant up to a height of about 80 km (50 mi)—that is, within the homosphere. Yet despite this gradual change in density and nearly constant chemical composition with height, meteorologists still find it convenient to divide the atmosphere vertically into several distinct layers. Some layers are distinguished by electrical characteristics, some by chemical composition (this is the homosphere/heterosphere distinction discussed earlier), and some by temperature characteristics. Together with the change in density with height, this layering of the atmosphere gives it its **structure**. In this section, we look first at the changing density of the atmosphere, then pressure (a closely related variable), then temperature variation, and finally at the electrical properties of the layer called the *ionosphere*.

Density

The **density** of any substance is the amount of mass of the substance (expressed in kilograms)[2] contained in a unit of volume (1 cubic meter, m^3). In gases, individual molecules have no attachment to one another and move about randomly. One characteristic of gases is that no definite limit exists to the amount of mass that can fill a given volume—molecules can always be added to or removed from the volume, resulting in a density change. Alternatively, the volume containing a fixed mass of gas can decrease (as happens in a bicycle pump as the piston moves inward and compresses the air), resulting in a density increase as the constant mass is squeezed into a smaller and smaller volume.

Like any other assemblage of gases, therefore, the atmosphere is compressible. When we feel the weight of something, we are being subjected to the downward gravitational force exerted by the overlying mass. At lower altitudes, there is more overlying atmospheric mass than at higher altitudes. Because air is compressible and subjected to greater compression at lower elevations, the density of the air at lower levels is greater than that aloft. Figure 1–9 illustrates this principle. Despite its simplicity, this concept is a key to understanding concepts presented later.

Air density decreases gradually with increasing altitude. At sea level, the air density is normally about 1.2 kg/m^3. By comparison, at Denver, Colorado, the "Mile High City," the air density is only about 85 percent of that. As a result, punted footballs and batted baseballs experience a corresponding decrease in air resistance and travel farther than they would at sea level.

One way to think about density is in terms of the average distance a molecule travels before colliding with another, called the **mean free path.** Near Earth's surface, the mean free path

---

[2]Although a kilogram corresponds to about 2.2 pounds in the English system of measurement used in the United States, there is a difference between the two. A kilogram is a unit of mass. A pound, however, is a unit of weight equal to mass times the acceleration of gravity. On Earth 1 kilogram of a substance has a weight of 2.2 pounds. If the same kilogram goes to the Moon, its mass will not change but its weight will be only one-sixth of what it was on Earth. This is true because the acceleration of gravity on the Moon is one-sixth as strong as the acceleration of gravity on Earth.

is a mere 0.0001 millimeter (mm). In contrast, at 150 km (93 mi) above sea level, a molecule travels about 10 m (33 ft) before colliding with another; at 250 km (155 mi), a molecule is likely to travel a full kilometer (0.62 mi) before meeting another.

Pressure

We don't feel it, but the atmosphere's great mass is far from weightless as the crushed can in Figure 1–10 demonstrates. At sea level the weight is about 14.7 pounds per square inch, which in everyday language is known as sea level air pressure. The corresponding metric unit of pressure used in the United States is the **millibar** (mb) and in Canada the **kilopascal** (kPa). Specified in these terms sea level pressure is 1013 mb and 101.3 kPa. Chapter 4 will discuss this more thoroughly. The main points for now are that pressure arises because of the atmosphere's mass, and its value reflects the mass of atmosphere above a given point. We don't feel the atmosphere's weight because our body cavities are full of air matching the surrounding air pressure. The closest we get to a sensation of pressure is the "pop" in our ears as airplanes ascend and descend, reflecting a temporary pressure imbalance between the cabin air and the air inside the ear drum. It follows that as one goes upward through the atmosphere, the amount of overlying air is less and the pressure is necessarily lower. You can see this by inspecting Figure 1–9 and imagining a count of the molecules above any height. The count would decrease with height, and with that the corresponding pressure.

 TUTORIAL

VERTICAL AND HORIZONTAL PRESSURE VARIATIONS

Experiment with bricks, springs, and air in Sections 2 and 4 to understand the relations between height, pressure, and compressibility.

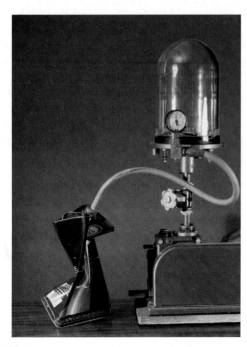

▲ **FIGURE 1–10** A can is crushed by atmospheric pressure after air is removed by a vacuum pump.

Pressure must decrease vertically, but the rate of decrease is not constant. Again looking at Figure 1–9, it seems clear that in the lower part of the column one passes through many molecules in a short vertical distance and pressure falls rapidly, but at higher altitudes lower density means one must go farther to find the same pressure decrease. Rather than a uniform decrease with height, the pressure falls rapidly near the surface and more slowly aloft as seen in Figure 1–11. Instead of a straight line corresponding to a constant absolute change with height, the curve suggests change by a constant percentage per unit distance. In fact, as a rough rule of thumb, pressure falls by about 50 percent for every 5 km of altitude. So the graph shows a pressure of about 500 mb at 5 km (1000 divided by 2), 250 mb at 10 km (500 divided by 2), and so forth. This nonlinear relationship between pressure and height arises because of the atmosphere's compressibility, just as the density variation with altitude arises because of compressibility. Moreover, vertical changes in density follow the same rough rule of thumb (ignoring small effects produced by temperature differences and assuming no vertical accelerations are present[3]). Pressure and density are clearly related, and for understanding the structure of the atmosphere there is no reason to prefer one over the other. That said, pressure does have an advantage in being much easier to measure, and it is much more commonly used in most aspects of meteorology.

[3]Chapter 4 will discuss these assumptions in more detail and will explain the remarkable reliability of this approximation. It will also make the connection between density and temperature explicit.

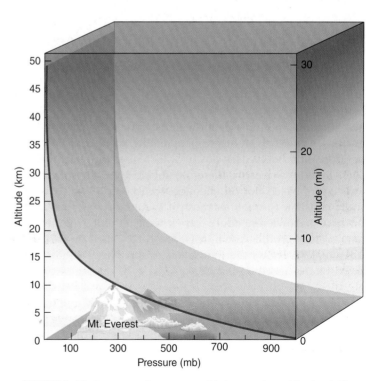

▲ **FIGURE 1–11** Pressure decreases with increasing altitude, falling by about 50 percent for every 5 km. Atmospheric density behaves similarly. The nonlinear trend in both is a consequence of the compressibility of air.

▲ **FIGURE 1–12** The temperature profile of the atmosphere results in four layers based on thermal characteristics.

Layering Based on Temperature Profiles

Compared to horizontal motions in the atmosphere (i.e., wind), vertical motions are usually slow and limited in range because of the shallowness of the atmosphere. Nonetheless, such motions greatly influence the likelihood of cloud development, precipitation, and thunderstorm activity. Air whose temperature decreases rapidly with height rises readily, while air whose temperature either decreases slowly or increases with height resists such motion. Scientists therefore divide the atmosphere into four layers based not on chemical composition (which is relatively constant throughout most of the atmosphere) but rather on how mean temperature varies with altitude. The average temperature profile shown in Figure 1–12, called the **standard atmosphere**, shows the four layers: troposphere, stratosphere, mesosphere, and thermosphere.

The Troposphere Unless you get a chance to fly in a military jet, you will spend your entire life in the **troposphere**, the lowest of the four temperature layers. The name is derived from the Greek word *tropos* ("turn") and implies an "overturning" of air resulting from the vertical mixing and turbulence characteristic of this layer. The troposphere is where the vast majority of weather events occur and is marked by a general pattern in which temperature decreases with height. Despite being the shallowest of the atmosphere's four layers,

the troposphere contains about 90 percent of its mass. This is possible, of course, because air is compressible.

The depth of the troposphere varies considerably, ranging from 8 to 20 km (3.6 to 12 mi), with a mean of about 15 km (9 mi). The altitude at which the troposphere ends depends largely on its average temperature, being highest where the air is warm and lowest in cold regions. The troposphere is therefore thicker over the tropics than over the polar regions and thicker during the summer than during the winter.

Temperatures vary greatly from bottom to top in the troposphere. The average global temperature is about 15 °C (59 °F) near the ground but only about −59 °C (−71 °F) at the top of the troposphere, an average decrease of about 6.5 °C/km (3.6 °F/1000 ft). Thus, you might feel perfectly comfortable inside an airplane at an altitude of 10 km (33,000 ft), but the temperature outside would likely be in the neighborhood of −50 °C (−58 °F). At the top of the troposphere, a transition zone called the **tropopause** marks the level at which temperature ceases to decrease with height.

An apparent paradox is associated with the trend toward decreasing temperature with height in the troposphere. We know that Earth is heated almost entirely by the Sun, which might make you think temperature should *increase* with height during the day (because as we move upward from the surface, we get closer to the Sun). The explanation for this paradox is that the atmosphere is relatively transparent to most types of radiant energy emitted by the Sun. In other words, a large portion of the sunlight passing through the atmosphere is not absorbed and therefore does not contribute greatly to its warming. As we shall see in Chapter 3, the most important direct source of energy for the atmosphere is not downward-moving solar radiation but rather energy emitted upward from Earth's surface.

Despite the strong tendency for temperature to decrease with altitude in the troposphere, it is not uncommon for the reverse situation to occur. Such situations, where temperature increases with height, are known as **inversions**. An inversion is significant because it inhibits upward motion and thereby allows high concentrations of pollutants to be confined to the lowest parts of the atmosphere.

The Stratosphere Above the tropopause is the **stratosphere**, a name derived from the Latin word for *layer*. Except for the penetration of some strong thunderstorms into the lower stratosphere, little weather occurs in this region (Figure 1–11). In the lowest part of the stratosphere, the temperature remains relatively constant at about −57 °C (−71 °F) up to a height of about 20 km (12 mi). From there to the top of the stratosphere (called the **stratopause**), about 50 km (30 mi) above sea level, the temperature increases with altitude until it reaches a mean value of −2 °C (28 °F).

In the upper stratosphere, heating is almost exclusively the result of ultraviolet radiation being absorbed by ozone. Therefore, as solar energy penetrates downward through the stratosphere, there is less and less ultraviolet radiation available and a resultant decrease in temperature. In the part of the stratosphere where temperature does not vary with height, heating is the result of both absorption of solar ultraviolet radiation and absorption of thermal radiation from below. Thus, as we move up or down in this region, the reduction in solar heat is offset by the increase in heat given off by Earth. The net result is the straight vertical line of no temperature change shown in Figure 1–12.

The stratosphere contains about 19.9 percent of the total mass of the atmosphere. Thus, the troposphere and stratosphere together account for 99.9 percent of the total mass of the atmosphere. The fact that gases are so compressible allows the major portion of the mass of the atmosphere to be contained in the two lowest layers. Note that despite being more than triple the depth of the troposphere, the stratosphere contains only one-quarter the mass: troposphere = 80 percent of total atmospheric mass; stratosphere = 20 percent.

Within the stratosphere is the **ozone layer**, a zone of increased ozone concentration at altitudes between 20 and 30 km (12 and 18 mi). Despite its name, the ozone layer is not composed primarily of ozone. In fact, at 25 km (15 mi) above sea level, where the percentage of ozone in this region is greatest, its concentration might be only about 10 ppm. Despite its scarcity, ozone is an extremely important constituent of the stratosphere. It is largely responsible for absorbing the solar energy that warms the stratosphere, and it also protects life on Earth from the lethal effects of ultraviolet radiation.

▲ **FIGURE 1–13** Most clouds exist in the troposphere, but the tops of strong thunderstorm clouds can extend kilometers into the stratosphere when violent updrafts occur. The flattened area at the top of this cumulonimbus cloud is in the stratosphere.

Being well removed from Earth's surface, where evaporation supplies water vapor to the atmosphere, the stratosphere has a very low moisture content. Moreover, as seen in Figure 1–12, the temperature characteristics of the stratosphere inhibit vertical motions that favor the formation of clouds (to be discussed in Chapter 6). These conditions also inhibit precipitation, and consequently particulates from volcanic eruptions can remain in the stratosphere for many months. Furthermore, the strong winds of the stratosphere can cause its aerosol content to be distributed across the globe, creating a veil of material that can affect the penetration of sunlight to the surface. For a couple of years after the 1991 eruption of Mount Pinatubo in the Philippines, for example, the Northern Hemisphere experienced redder-than-normal sunrises and sunsets as a result of aerosols in the stratosphere and accompanying that there was a measurable decrease in global temperature.

The Mesosphere and Thermosphere Of the 0.1 percent of the atmosphere not contained in the troposphere and stratosphere, 99.9 percent exists in the **mesosphere**, which extends to a height of about 80 km (50 mi) above sea level. As in the troposphere, temperature in the mesosphere decreases with altitude. The only significant source of heat is absorption of solar radiation near the base of the mesosphere. This is dispersed upward only weakly by vertical air motions, thus temperatures fall rapidly with increasing altitude.

Above the mesosphere is the **thermosphere**, where temperature increases vertically to values in excess of 1500 °C. These high temperatures can be misleading, however, if we overlook the distinction between high temperature and high heat content. The temperature of the air is an expression of its internal energy, which is related to the speed at which its molecules move. The amount of heat contained in a volume of air reflects not only its temperature but also its mass. Because there are so few gas molecules in this layer, the air cannot have

1–5 PHYSICAL PRINCIPLES

The Three Temperature Scales

Fahrenheit and Celsius

At one time, the temperature scale used all over the world was the Fahrenheit scale. Invented in the early 1700s by Gabriel Fahrenheit, it assigns values of 32° and 212° to the freezing and boiling points of water.* There are thus 180 Fahrenheit degrees between freezing and boiling. Fahrenheit developed his scale by assigning 0° and 100° to temperatures he could produce in his laboratory. For the lower temperature, he used a mixture of water, ice, and a salt, and (according to one account) his wife's armpit served to establish the other as the temperature of a human body. Although the Fahrenheit scale has been replaced in Canada and nearly every other country around the world, in the United States it is still the scale used by the general public.

The other familiar scale for measuring temperature is the Celsius scale, named for Anders Celsius, who formulated it in 1742. The Celsius scale assigns values of 0° and 100° to the freezing and boiling points of water, so there are only 100 Celsius degrees between the two points. This means that a Celsius degree is larger than a Fahrenheit degree. Thus, for example, a 2 °C change is larger than a 2 °F change. However, this is *not* to say a tem-

*To be precise, these values apply to pure water at sea level. In addition, water doesn't freeze spontaneously at 32 °F, so "melting point" is a better term than "freezing point."

perature expressed in °C is always higher than the same temperature expressed in °F; it can be higher or lower. To convert from Celsius to Fahrenheit, we use the following formula:

$$°F = 9/5 \, °C + 32$$

To convert from Fahrenheit to Celsius, we use

$$°C = 5/9 \, (°F - 32)$$

You can use these formulas to verify that −40 °F = −40 °C.

Kelvin

Though useful in everyday applications, both the Fahrenheit and Celsius scales have a significant shortcoming—they allow for negative values. This is not the case for other units of measurement. For example, buildings do not have negative heights and weights, cars do not travel at negative speeds, and children do not have negative ages. But negative temperature values give the impression that substances can have negative heat contents—a situation that is physically impossible. To overcome this problem, scientists use a different scale for the measurement of temperature, called the **Kelvin scale**. In this system, the temperature 0 K is the lowest possible temperature that can exist in the universe. (Notice that we omit the degree notation with this scale and just refer to the number of kelvins,[†] K). A temperature of 0 K implies no heat, and it is therefore impossible for subzero temperatures to exist with this scale.

373.16	100	212	Boiling point of water
273.16	0	32	Melting point of ice
0	−273.16	−459.6	Absolute zero
Kelvin (K)	Celsius (°C)	Fahrenheit (°F)	

▲ **FIGURE 1** The Kelvin, Celsius, and Fahrenheit scales.

The Kelvin scale is really a modified form of the Celsius scale insofar as the increments of the two are equal. Thus, if the temperature increases 1 degree Celsius, it also increases 1 kelvin. The only difference between the two is the starting point; 0 K corresponds to −273.16 °C Therefore, conversion from Celsius to Kelvin is simply

$$K = °C + 273.16$$

To convert from Kelvin to Celsius, we use

$$°C = K - 273.16$$

Figure 1 shows the Kelvin, Celsius, and Fahrenheit scales.

[†]The Kelvin scale is abbreviated with a capital K; the unit of measurement is spelled with a lowercase k.

a high heat content no matter what its temperature is. In fact, density is so low in the upper reaches of the thermosphere that a gas molecule will normally move as much as several kilometers before colliding with another. Thus, an ordinary thermometer in this part of the atmosphere would have little contact with the surrounding air. Under these circumstances, the concept of temperature loses meaning and cannot be associated with everyday terms such as *hot* and *cold*. (See *Box 1–5, Physical Principles: The Three Temperature Scales,* for important information on temperature scales.)

The thermosphere is also the main source of atomic oxygen (O) needed for the production of ozone. At these high altitudes intense ultraviolet solar radiation is absorbed by O_2, splitting the molecule into its constituent atoms (O + O). This process, known as **photodissociation**, occurs so rapidly in some parts of the thermosphere that the oxygen present is about equally divided between O and O_2. In addition to its role in the production of O, photodissociation is important for heating the thermosphere and—most significantly—shielding the surface from lethal doses of ultraviolet radiation.

Did You Know?

Although Venus is approximately the same size as Earth, its atmosphere is vastly different. The surface density of Venus's atmosphere is some 50 times greater than Earth's and it is 96.5 percent carbon dioxide. Our planetary neighbor has a vastly different temperature as well—averaging about 475 °C, or 890 °F!

A Layer Based on Electrical Properties: The Ionosphere

The four layers of the atmosphere described previously are delineated on the basis of temperature profiles. An additional layer, called the **ionosphere**, can be defined based on its electrical properties. This layer, which extends from the upper mesosphere into the thermosphere, contains large numbers of electrically charged particles called *ions*. **Ions** are formed when electrically neutral atoms or molecules lose one or more electrons and become positively charged ions or gain one or more electrons and become negatively charged ions. In the ionosphere, atoms and molecules lose electrons as they are bombarded by solar energy, thus creating positively charged ions and free electrons. The resulting ion-rich layers wax and wane on a daily basis with the rise and fall of the sun's ionizing radiation. (Like dissociation, ionization also warms the thermosphere, contributing further to the high temperatures found there.)

The ionosphere affects radio communication with satellites, including those of the Global Positioning System. Without accounting for ionospheric disturbances, GPS receivers would have much larger positional errors. The ionosphere is also responsible for the **aurora borealis,** or northern lights, and **aurora australis,** the southern lights (Figure 1–14). In the ionosphere, subatomic particles from the Sun are captured by Earth's magnetic field (the same field that makes compass needles point to the north). Collisions with atmospheric gases ionize some atoms and *excite* others (meaning the electrons of the atoms jump to greater orbital distances from their nuclei). Radiation is emitted when electrons fall back to lower orbits, or when ionized atoms regain a free electron. Thus, the aurora does not reflect radiant energy as do clouds but rather emits light that is much like a neon lamp.

Checkpoint

1. How does air pressure arise? How does it vary vertically?
2. Why do air pressure and density show much the same vertical trend?
3. List the four temperature layers of the atmosphere and describe their characteristics.

Origin and Evolution of the Atmosphere

It is generally believed that Earth was formed 4.6 billion years ago. If an atmosphere formed with Earth, it must have

▲ **FIGURE 1-14** An aurora borealis. Subatomic particles from the Sun are captured by Earth's magnetic field, causing an agitation of molecules and the emission of light with different colors.

consisted of the gases most abundant in the early solar system—including large amounts of hydrogen and helium, the two lightest elements. Today's atmosphere is much different, composed mostly of nitrogen and oxygen. So where did the original gases go and how were they replaced?

Earth's First Atmosphere

The gases of the earliest atmosphere were simply lost to space during the first half-billion years of the planet's existence. We are not sure exactly how, but two possible explanations have been offered. For many years, the conventional viewpoint was that Earth's gravitational field would not have been strong enough to keep hold of such an atmosphere, and the light elements therefore escaped into space. If a molecule moves with sufficient speed, known as the **escape velocity,** it can overcome gravity and leave the atmosphere. Gases having lower molecular weights are more likely to achieve escape velocity; thus, the hydrogen and helium of the primordial atmosphere were most readily lost.

Recently, an alternate explanation has gained acceptance. According to this view, gases drawn to Earth during its formation would have been removed by collisions between the growing Earth and other large bodies. Some of these other objects, so-called failed planets, would have been as large as Mars. Tremendous energy would have been released by collisions with failed planets, ejecting large amounts of planetary material along with any atmosphere that might be present. (In fact, the Moon is believed to have formed by condensation of matter dislodged by just such a collision.) If this hypothesis is correct, removal by slow leakage would have been insignificant.

Earth's Second Atmosphere

Over time a new, secondary atmosphere formed, made up of gases released from Earth's interior by volcanic eruptions—

a process called **outgassing**. The gases spewed out during volcanic events are predominantly water vapor and carbon dioxide, with lesser amounts of sulfur dioxide, nitrogen, and other gases. Volcanic outgassing was possibly augmented, maybe even dwarfed, by material brought to Earth in small comets on the order of 15 m (49 ft) in diameter. Recent satellite observations suggest that there is a steady rain of these water-bearing "cosmic snowballs," with from 5 to 30 striking Earth at any one time. If they have been falling at this rate since the formation of Earth, they might well be the source of most of the world's water vapor.

Further Evolution of the Atmosphere

However it arose, the secondary atmosphere has been greatly transformed, because water vapor and carbon dioxide now constitute only a small portion of the atmosphere. The removal of the water vapor is easily explained: Upon cooling in the prehistoric atmosphere, it readily condensed and precipitated to the surface to form the oceans. Precipitation also contributed to the removal of carbon dioxide from the atmosphere. As raindrops fell through the CO_2-rich atmosphere, some CO_2 was dissolved in every drop. (Soft drinks provide an everyday example of how CO_2 is dissolved in water, as do limestone gravestones dissolving in rainwater acidified by atmospheric CO_2). As they fell, the CO_2 drops transferred carbon from the atmosphere to the oceans, where it combined with material eroding from the continents and was buried in seafloor sediments.

The transformation to an atmosphere high in oxygen depended on the advent of primitive, anaerobic bacteria (those that survive in the absence of oxygen) about 3.5 billion years ago. These primitive life-forms were the first in a long line of organisms that removed carbon dioxide from the air and replaced it with oxygen (these gases are exchanged freely between the oceans and the atmosphere). Ultimately, plant and later animal material sank to the ocean floor (as it continues to do today), where the organic carbon was locked away in sediments. As a matter of fact, the vast majority of carbon released by volcanoes exists in neither the atmosphere nor the ocean; it is held in carbonate rock formations.

All of these processes gradually led to an increase in atmospheric oxygen at the expense of carbon dioxide. But one other transformation had to take place to create an atmosphere that could support life at the surface. Recall that without an ozone layer in the upper atmosphere sunlight reaching the surface would contain lethal levels of ultraviolet radiation. Thus, a protective ozone layer had to develop before life could exist outside the oceans. Fortunately, this occurs naturally when the ultraviolet radiation breaks down diatomic oxygen molecules into individual oxygen atoms. The oxygen atoms could then recombine with O_2 in the upper atmosphere to form the ozone layer. Once this happened, the amount of ultraviolet radiation reaching the surface was reduced sufficiently to allow plants to occur on land. This increase in plant cover in turn accelerated the rate at which photosynthesis replaced atmospheric carbon dioxide with oxygen.

None of this accounts for the high concentration of nitrogen in the atmosphere. Although it constitutes a small portion of the material released by outgassing, nitrogen is removed from the atmosphere very slowly. As a result, its concentration has gradually increased to the point that it is now the main constituent of the atmosphere. Finally, argon, the third most abundant gas, is explained by slow seepage from the solid Earth of argon-40 with its 18 protons and 22 neutrons. Argon-40 is produced by the radioactive decay of a rare form of potassium. Another isotope, argon-36, is abundant in the universe but nearly absent on Earth. This is additional evidence that any primordial atmosphere was lost. Argon is an inert gas, so if the current atmosphere had formed at the same time as Earth, argon-36 should be present in much higher amounts.

Some Weather Basics

All of us are familiar with daily weather forecasts, from which we routinely receive information on the present and predicted state of the atmosphere. In recent years, the amount of weather information available to the public has exploded, most notably via the Internet. Detailed maps and weather reports that were once available only to professional meteorologists can now be accessed by anybody with a computer, tablet, or smart phone connected to the Internet. Such access makes learning the fundamentals of weather and climate far more enjoyable, because we can now look at map, satellite, radar, and other resources and see how the principles of meteorology play out on a daily basis. We now present an overview of the fundamentals of weather, along with an introduction to weather maps.

Atmospheric Pressure and Wind

Atmospheric **pressure** is one of the most fundamental weather characteristics because of its role in producing wind. Three generalizations about surface pressure are particularly important. First, air tends to blow away from regions of high pressure toward areas of lower pressure. In other words, it is the horizontal variation in air pressure that generates **winds**. If the pressure were uniform from place to place, the air would be continually calm. Second, the greater the change in pressure over a given distance, the greater the resulting wind speed. The third generalization is that air tends to rise in areas of low surface pressure and sink in zones of high pressure. This is important because rising motions favor the formation of clouds, while sinking motions promote clear skies.

Mapping Air Pressure

For all these reasons maps of air pressure are an essential tool in understanding weather. However, if one were to map pressure as measured at the surface, the result would be almost useless! The decline with altitude is so strong that the resulting map would essentially be a map of elevation. Thus standard practice is to convert measured values to what would be expected at sea level. That is, for locations high above sea level, the measured value is increased substantially. Elevations near sea level get little correction. By using a common altitude (0 km) for all locations, mapped differences in pressure reflect horizontal gradients that can set the air in motion.

Sea level pressure is routinely plotted on maps with lines called **isobars**. Each isobar connects points having equal air pressure. Figure 1–15a illustrates how isobars depict the distribution of pressure. Notice that an isobar labeled 1032 encircles extreme northern Washington, northern Idaho, western Montana, and southern British Columbia. Any point on that line has a pressure that is equal to 1032 mb (which corresponds to 103.2 kPa). Another isobar, indicating a pressure of 1028 mb, surrounds the 1032 mb isobar. Not only do the two isobars tell us what the pressure is at any point along those lines, but they also allow us to infer what the approximate pressure is anywhere between them. The atmospheric pressure between any two isobars will always be a value between those pressures that the two isobars are labeled. Thus, over the region between the two isobars over the Pacific Northwest and southern British Columbia, the air pressure is between 1028 and 1032 mb. Because the pressure is greater within the area bounded by the 1032 mb isobar than it is surrounding it, we know that the air is flowing out from this area.

There are three major low pressure areas over North America in Figure 1–15a: one northwest and offshore of the high pressure centered over northern Washington, another northwest of the Great Lakes in Ontario, and one off the coast of the southeastern United States (Hurricane Noel). We know from the first generalization about high- and low-pressure systems given above that wind must be flowing into each of these areas from the surrounding regions. The spacing of the isobars also tells us where the wind is blowing most strongly. Notice that the isobars are closer together near the low-pressure system over central Canada and relatively far apart over the south central and southwestern United States. It therefore follows that at this point in time the wind speeds are generally greater over central Canada than farther to the south.

Station Models

More detailed information regarding wind speed and direction can be obtained on weather maps by looking at the **station models**, which contain symbols and numbers giving detailed weather information for particular locations. The station model for Cheyenne in southeastern Wyoming contains an open circle with a line pointing toward the northwest, which indicates the direction that the wind is blowing *from*. At the end of the line is a single tick mark. The number and type of tick marks at the end of the line give an approximate wind speed, using the conventions shown in Figure 1–16a. In this example, Cheyenne has winds coming from the northwest at 4–13 km per hour (9–14 mph).

TUTORIAL
VERTICAL AND HORIZONTAL PRESSURE VARIATIONS

Use the maps and 3D diagrams in Section 6 to visualize pressure gradients and the corresponding patterns in sea-level isobars.

Station models also give cloud cover information. Open circles, such as that shown for Cheyenne, indicate clear skies, while fully shaded circles indicate overcast conditions. Intermediate amounts of cloud cover are indicated by the patterns shown in Figure 1–16b.

In Figure 1–15a, the isobar pattern and station model information verify the generalizations about pressure distributions, wind movement, and cloud cover. The wind barbs show that the air does indeed move outward from the region of high pressure, while air is flowing into the low-pressure area north of the Great Lakes. Furthermore, the satellite images in Figures 1–15b and c show that much of the area around the high-pressure systems has clear skies, in contrast to the general overcast near the three low-pressure systems.

Temperature

Temperature is one of the most obvious and important weather components. Everyday experience indicates that temperature usually varies gradually from one place to another. In other words, as you take a 50-mile car trip over flat terrain you are not likely to experience temperatures that vary greatly. Instead, you will probably observe only gradual changes in temperature as you drive along. On the other hand, there are times when substantial temperature differences appear over short distances, or when major shifts in temperature occur over short time periods at a particular location. These major changes in temperature often occur due to the presence of **fronts**, fairly narrow boundary zones separating relatively warm and cold air. As you drive across the frontal zones, you will experience notable temperature shifts. Likewise, if a front passes over your location, the temperature will change substantially.

Four types of fronts exist, which are discussed later in this text. For now, let's concern ourselves with two particular types: *cold fronts* and *warm fronts*. A cold front (shown as a heavy blue line with triangles) is a boundary separating cold and warm air, with the cold air moving in the direction toward which the triangles are pointing. In a warm front (indicated by a red line with semicircles), warm air advances in the direction that the semicircles are facing and replaces the colder air ahead. In Figure 1–15a, a cold front extends southwestward from the low-pressure region over Ontario to northwest Kansas, and then westward to eastern Nevada. A warm front extends southeastward from the Ontario low to the northern Great Lakes.

Temperatures in degrees Fahrenheit are given as the uppermost of the two numbers just to the left of station models. Thus, the temperature for Cheyenne is 27 °F.

Humidity

You have undoubtedly heard the term **relative humidity**. Relative humidity is just one of several ways of expressing the amount of water vapor in the air. (Remember, water vapor is a gas!) It indicates the amount of water vapor present relative to the maximum possible; thus, it is usually

(a)

(b)

(c)

▲ **FIGURE 1–15** (a) A typical surface weather map. (b) Satellite image of the eastern United States and Canada. Areas in bright red depict heavy precipitation. (c) As in Figure b but for the western United States and Canada.

reported as a percentage. Though a commonly used indicator of water vapor content, relative humidity has some serious shortcomings. For this reason, another index called the **dew point temperature** (or simply the dew point) is often preferred. For now let's just say that the higher the dew point, the greater the amount of water vapor in the air. Dew points above about 15 °C (59 °F) or so indicate humid air, and dew points above about 20 °C (68 °F) are very uncomfortable. Dew points less than about 5 °C (41 °F) are relatively dry. Dew point values are given at the bottom left of the station models, just below the temperature readings. In Figure 1–15a, much of the south eastern and south central United States has humid air, while the interior western and north central United States are drier.

(a)

Wind Speed

	Miles per hour	Kilometers per hour
◎	Calm	Calm
	1–2	1–3
	3–8	4–13
	9–14	14–19
	15–20	20–32
	21–25	33–40
	26–31	41–50
	32–37	51–60
	38–43	61–69
	44–49	70–79
	50–54	80–87
	55–60	88–96
	61–66	97–106
	67–71	107–114
	72–77	115–124
	78–83	125–134
	84–89	135–143
	119–123	192–198

(b)

Cloud Cover

○	No clouds
	1/8
	Scattered
	3/8
	4/8
	5/8
	Broken
	7/8
●	Overcast
⊗	Sky obscured

▲ **FIGURE 1-16** Station model symbols describing (a) wind speed and (b) percentage of cloud cover.
Note: You might wonder why the number of tick marks don't correspond to wind speeds in increments of 5s and 10s, rather than the numbers you see here. Well, in fact they do, but the units are expressed as nautical miles per hour, or knots. Thus one small tick mark corresponds to 5 knot winds, a large tick mark 10 knots, and so on. However, this scale is not widely used outside of some aviation and maritime situations.

Checkpoint

> 1. How are wind and sea level pressure related?
> 2. Using the map in Figure 1–14a and referring to Figure 1–15, describe the weather along the coast of North Carolina.

A Brief History of Meteorology

The amount of information available to even casual weather observers is enormous compared to just a couple of decades ago. Computers and mobile devices almost instantly deliver better information than a professional weather forecaster had available in the late 1900s. But the information explosion did not happen overnight; for centuries the advances in meteorological knowledge came about in piecemeal increments.

For atmospheric science to become quantitative, basic instrumentation was required. Galileo led the way in 1593 with a prototype version of the basic thermometer, which ultimately led to the development of the Fahrenheit and Celsius scales in 1714 and 1736, respectively. In 1643, Evangelista Torricelli invented the mercury barometer, the instrument still used as a standard for measuring atmospheric pressure, but it was not until the late 1700s that an instrument became available for measuring water vapor.

Instruments alone cannot discern atmospheric patterns or make forecasts—a network of observers must also exist to operate the instruments and maintain permanent records of weather elements. The first network of this type in North America was authorized in 1847 when the Smithsonian Institution allocated the considerable sum of $1000 to purchase instruments for a network of volunteer weather observers across the United States. These observers submitted their subjective observations and instrument data to the Institution each month, providing useful climatological information. Still, the inability to collect, map, and rapidly disseminate the data made operational forecasting impossible. At about the same time, however, a new invention—the telegraph—enabled the rapid transmission of data from sites across the country to a central collecting station, thus making possible the forecasting of weather by simply noting the location and movement of current weather conditions.

The birth of U.S. weather forecasting occurred in 1870, when President Ulysses S. Grant authorized the establishment of the Army Signal Service. In 1891 the agency was renamed the Weather Bureau and transferred to the Department of Agriculture, where it remained until 1940. The agency was then transferred to the Department of Commerce and in 1970 was renamed the **National Weather Service** as part of the newly established National Oceanic and Atmospheric Administration.

The ever-growing network of surface stations was important to understanding and forecasting weather, but the greatest amount of the atmosphere lies well above the surface. The balloon ascents taken by J. L. Gay-Lussac and Jean Biot in 1804 were landmark events in the scientific understanding of the atmosphere and in human courage. The first ride, taken by both scientists, rose to a height of 3962 m (13,000 ft). A second ascent taken solo by Gay-Lussac went up to 7000 m (23,000 ft). At that height, the average temperature is some 30 degrees below zero Celsius ($-22\,°F$) and the air density is only 40 percent of its value at sea level. Gay-Lussac continued to make temperature and pressure observations until he passed out from oxygen deficiency. Unmanned weather balloons became a valuable source of data in the 1920s, and since the 1940s the tracking of balloon-carried instrument packages called *radiosondes* became a regular source of climatological and forecasting data.

The era following World War II brought with it enormous advances in meteorological knowledge and technology.

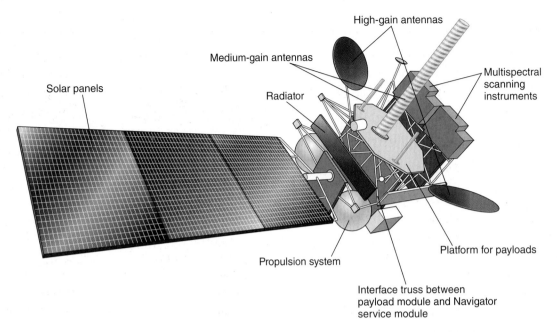

Solar panels

Medium-gain antennas

High-gain antennas

Radiator

Multispectral scanning instruments

Propulsion system

Platform for payloads

Interface truss between payload module and Navigator service module

▲ **FIGURE 1–17** The Russian EELEKTRO-L1 weather satellite launched January 20, 2011.

Weather radars, first developed during the war, have evolved into units that can peer into clouds and observe their internal characteristics and motions. Weather satellites have likewise become essential to meteorologists since the launching of Tiros I (Television and Infrared Observation Satellite) in 1960 (Figure 1–17). And technological advances in computer hardware and software that have revolutionized most other aspects of society have had an incredible impact on atmospheric understanding and forecasting since they were introduced to meteorology in the 1950s. Indeed, weather prediction was one of the first uses for digital computers and remains a driving force behind their continued development.

Summary

Meteorology and climatology study the elements of the atmosphere and the causes of atmospheric behavior. While these two sciences are closely intertwined, the former places a greater emphasis on the short-term events that we think of as weather, while the latter looks at longer-term characteristics. Earth's atmosphere is composed of a mixture of gases and contains an enormous number of suspended solids and liquids called *aerosols*. Three of the gases—nitrogen, oxygen, and argon—occur in nearly constant proportions and constitute the vast majority of the atmospheric mass. Other gases occur in slight amounts and some vary considerably in their concentration. These variable gases, especially water vapor and carbon dioxide, can be extremely important to life on Earth.

Atmospheric density and pressure decline with increasing distance from the surface. The decline is progressive, with no layering or interruption, but the decline with height is not uniform. The compressibility of air ensures a rapid decrease near the surface, and also contributes to the overall shallowness of the atmosphere. Four layers of the atmosphere—the troposphere, stratosphere, mesosphere, and thermosphere—can be distinguished on the basis of their temperature profiles, and one—the ionosphere—is designated for its electrical characteristics. Of the first four, the lowest two (the troposphere and stratosphere) contain almost all atmospheric mass. Moreover, nearly everything we consider weather occurs in just the lowest layer, the troposphere. The two higher layers, the mesosphere and thermosphere, account for less than 0.1 percent of the atmosphere's mass and are relatively unimportant to most of the processes described in this book. The ionosphere spans the upper mesosphere and thermosphere. It affects the transmission of AM radio waves and provides the locale for the auroras of both hemispheres.

The present atmosphere did not arrive with the formation of the planet some 4.6 billion years ago but rather evolved after the primordial atmosphere was lost to space. The process of outgassing released water vapor and carbon dioxide (along with other gases) from Earth's interior. Ultimately, photosynthesis reduced carbon dioxide levels and contributed oxygen to the atmosphere.

This chapter introduced some significant weather elements, demonstrated how they are depicted on surface weather maps, and provided an overview to the development of meteorology as we know it today.

Key Terms

atmosphere *page 4*
meteorology *page 4*
weather *page 4*
climate *page 4*
climatology *page 6*
homosphere *page 9*
heterosphere *page 9*
permanent gases
  page 9
variable gases *page 9*
nitrogen *page 10*
oxygen *page 10*
argon *page 10*
water vapor *page 10*
hydrologic cycle *page 10*

carbon dioxide *page 11*
photosynthesis *page 11*
respiration *page 12*
ozone *page 12*
methane *page 15*
aerosols *page 15*
particulate *page 15*
condensation
  nuclei *page 16*
structure *page 16*
density *page 16*
mean free path
  page 16
millibar *page 17*
kilopascal *page 17*

standard atmosphere
  page 18
troposphere *page 18*
tropopause *page 18*
inversion *page 19*
stratosphere *page 19*
stratopause *page 19*
ozone layer *page 19*
mesosphere *page 19*
thermosphere *page 19*
Kelvin scale *page 20*
photodissociation
  page 20
ionosphere *page 21*
ions *page 21*

aurora borealis *page 21*
aurora australis *page 21*
escape velocity *page 21*
outgassing *page 22*
pressure *page 22*
wind *page 22*
isobar *page 23*
station model *page 23*
front *page 23*
relative humidity
  page 23
dew point temperature
  page 24
National Weather
  Service *page 25*

Review Questions

1. Explain why television newscasts have weather segments but not climate segments.

2. Why is it difficult to define an absolute top of the atmosphere?

3. What are the homosphere and the heterosphere?

4. What is the difference between the permanent and variable gases of the atmosphere? Which gases contribute most to the total mass of the atmosphere?

5. Given that variable gases are so rare, why consider them at all?

6. Why has the concentration of carbon dioxide in the atmosphere been increasing over the last century?

7. What is ozone, and why is it both beneficial and harmful to life on Earth?

8. What are aerosols? Are they formed only by human activities or do they occur naturally?

9. Convert the following Fahrenheit temperatures to Celsius: $-22\,°F$, $50\,°F$, $113\,°F$.

10. Convert the following Celsius temperatures to Fahrenheit: $-20\,°C$, $10\,°C$, $40\,°C$.

11. How do photosynthesis, respiration, and decay affect the carbon dioxide balance of the atmosphere?

12. In what way does the density of the atmosphere vary with altitude?

13. What are the distinguishing characteristics of the troposphere, stratosphere, mesosphere, and thermosphere?

14. What is the tropopause?

15. In which thermal layer of the atmosphere is the ozone layer found? Why is the term "ozone layer" somewhat misleading?

16. What percentage of the total mass of the atmosphere is contained in the troposphere and the stratosphere?

17. The troposphere is less than half as thick as the stratosphere but contains 10 times as much mass. Please explain.

18. A station at elevation 900 m (3000 ft) has a surface pressure of 900 mb. An airplane is flying 5 km above the surface. Use the rule of thumb in the section on pressure to guess the air pressure surrounding the plane.

19. How is the ionosphere distinct from the layers of the atmosphere defined by their temperature profiles?

20. What is outgassing and why was it important?

21. Why were anaerobic bacteria important to the evolution of the atmosphere?

22. Briefly describe the effect that variations in pressure exert on other weather elements.

23. What are isobars?

24. What are station models and what useful information do they depict?

Critical Thinking

1. The Kyoto Treaty that limits carbon dioxide emissions has not been ratified by all countries that originally signed the treaty. What are the pros and cons of doing so?

2. Volcanic eruptions continue to occur and outgas water vapor, carbon dioxide, and other gases. Do you think that this will be a significant factor in increasing the concentration of these gases over the next century? Why or why not?

3. Temperatures usually decrease with height in the troposphere but increase with height in the stratosphere. Why do the two layers have such different profiles?

4. The thermosphere has extremely high temperatures, but a person exposed to the thermosphere would rapidly freeze. Explain the apparent contradiction in terms of what you know about heat and temperature.

Problems and Exercises

1. The National Climatic Data Center (NCDC) has a Web site at **lwf.ncdc.noaa.gov/oa/climate/severeweather/extremes.html** that provides a wealth of information on extreme weather and climate events. Examine the site to see how much information is available. Do any of the topics relate to weather events in your hometown?

2. Examine the map of National Weather Service Offices at **www.nws.noaa.gov/organization.php#maps**. Is there a Weather Service office near you? If so, consider visiting the office and seeing firsthand what goes into producing a forecast.

3. Keep a record of daily weather in your area and download current weather maps from one of the available Web sites. Do you notice any patterns on the maps that tend to be associated with particular weather conditions?

4. Look at today's weather map and observe the contrasting weather conditions across the United States and Canada. Do any areas exhibit significant changes in weather from adjacent regions? How well defined are the boundary zones? Repeat this exercise for several days and see if these transition regions show movement.

Quantitative Problems

Your comprehension of a great many concepts can be enhanced by working out numerical solutions to questions. Go to this book's Web site at **www.MyMeteorologyLab .com** and click on the Begin button. This will take you to the main page for Chapter 1, which offers a variety of valuable resources. Take a look at the options available. Along with the self-quizzes, links, and other resources on the page, the quantitative exercises should prove particularly valuable.

Useful Web Sites

www.weather.com

Home page for the Weather Channel. Contains current weather information along with reports related to travel, health, severe weather, and many other topics. A good first stop.

www.nws.noaa.gov/organization.php

Listing of all local National Weather Service offices, with links connecting the viewer directly to sites. Also lists Uniform Resource Locators (URLs)—Internet addresses—for regional and national support offices such as the National Hurricane Center. An excellent site for local information anywhere in the United States and for advisories on severe weather.

weatheroffice.ec.gc.ca/canada_e.html

Home page for the Meteorological Service of Canada. Provides forecasts and links to satellite, radar, and many other useful resources.

www.wunderground.com and **www.wunderground.com/ global/CN_ST_Index.html**

Multipurpose sites that provide detailed information for the United States and Canada, respectively. Offer standard and unusual weather maps and local weather data. Also good sources for weather information all over the world.

www.ozonelayer.noaa.gov

Up-to-date information on the status of the ozone layer.

MyMeteorologyLab™

The *Understanding Weather and Climate*, Sixth Edition, MyMeteorologyLab™ Web site contains numerous multimedia resources to aid in your study of **Composition and Structure of the Atmosphere.**
Visit **www.MyMeteorologyLab.com** to:

- **Review** key chapter concepts.
- **Read** the Pearson eText, *In the News* RSS feeds, and other chapter-specific Web resources.
- **Visualize** challenging topics with Interactive Tutorials, Geoscience Animations, MapMaster interactive maps, and Weather in Motion movies and visualizations, and more.
- **Test Yourself** with self-study quizzes.

 TUTORIAL

VERTICAL AND HORIZONTAL PRESSURE VARIATIONS

Use the interactive animations and quizzes in this tutorial to visualize and review key concepts related to air pressure.

WEATHER IN MOTION

View movies and visualizations of meteorology patterns and processes.

Improving Hurricane Predictions
Global Carbon Uptake by Plants
Global Changes in Carbon Dioxide Concentrations
The Influence of Volcanic Ash
Ozone Hole
What if We Hadn't Regulated Chlorofluorocarbons (CFCs)?
A Global Montage of Clouds and Sea Surface Temperature

2 Solar Radiation and the Seasons

After reading this chapter, you should be able to:

▸ Identify the types of energy and how they can be transmitted.

▸ Describe the characteristics of solar radiation.

▸ Explain the influences of solar angle on solar radiation receipt at the surface.

▸ Describe how seasonal changes in solar energy vary with latitude.

▸ Explain the laws governing the amount and type of radiation emitted by objects.

▸ Recognize the characteristics of Earth's orbit around the Sun and how they create the seasons.

The winter of 2010–2011 had already been a severe one for the eastern United States. Cities from Atlanta to Boston—and beyond—had already accumulated more snowfall by the end of January than normally falls in an entire winter. Up to that point, the central United States had had a different experience, with relatively little snowfall. Then things changed–drastically. On January 31, a major snowfall occurred over the southern Great Plains, with Tulsa setting a record one-day record snowfall accumulation of 36 cm (14 inches). Dallas, Texas, had travel disrupted for several days as it was preparing to host the Super Bowl. The storm migrated up the Mississippi River Valley and brought low temperatures, gusty winds, and heavy snow across a large swath of the nation's interior. By the afternoon of February 1, Chicago's roads and highways became impassable as heavy snow and strong winds caused very limited visibility and deep snow drifts. All this occurred while, thousands of miles away, a huge, powerful cyclone hit the northeastern coast of Australia.

We all know northern regions of North America are vulnerable to such events, and primarily during winter months. But there are reasons *why* these events are limited to certain regions and times of year. We know that Honolulu, Hawaii, is not subject to this type of weather—but exactly why may be a little unclear to many people. In fact, many people have misconceptions about what causes these differences in seasons and climates. For example, some people incorrectly believe that variations in the distance between Earth and the Sun cause the seasons, with summer occurring when Earth and the Sun are closest together. But Earth and the Sun are closest to each other on or about January 4—in the midst of the Northern Hemisphere winter! Likewise, when the Sun is farthest from Earth on July 4, it's summer in the Northern Hemisphere. The purpose of this chapter is to provide an explanation of how Earth's orientation toward the Sun creates the seasons and how the amount of solar energy Earth receives varies with latitude.

◂ After the Chicago blizzard of February 1, 2011, there was little traffic to direct on the city's snow-clogged streets.

Energy

Energy is traditionally defined as "the ability to do work." This definition isn't entirely accurate and raises its own questions (What is work?), but it is impossible to do better in just a few words. Rather than travel far afield in search of a precise definition, we'll assume that everyone has at least a vague idea of energy as an agent capable of setting an object in motion, warming a teapot, or otherwise manifesting itself in everyday events. The standard unit of energy in the International System (SI) used in scientific applications is the **joule** (abbreviated as J). Although students may be more familiar with the calorie as the unit for energy, the joule is preferred in this context (1 joule = 0.239 calories). A related term, **power**, is the rate at which energy is released, transferred, or received. The unit of power is the **watt** (W), which corresponds to 1 joule per second.

Even the simplest activity requires a transfer of energy. In fact, while you read these words, an energy transfer is occurring as the chemical energy from food you have eaten is converted into the kinetic energy (energy of motion) needed to move your eyes across this line of type. But your body, like any other machine, is not perfectly efficient; it loses some thermal energy (think of this as the heat contained within a body) as chemical energy is converted into kinetic energy. Thus, your eye muscles give off heat as they contract and relax.

The same concept applies to our atmosphere. About one two-billionth of the energy emitted by the Sun is transferred to Earth as **electromagnetic radiation**, some of which is directly absorbed by the atmosphere and surface. This radiation provides the energy for the movement of the atmosphere, the growth of plants, the evaporation of water, and an infinite variety of other activities.

Kinds of Energy

Energy can occur in a variety of forms. We often speak of radiant, electrical, nuclear, and chemical energy; but, strictly speaking, all forms of energy fall into the general categories of **kinetic energy** and **potential energy**. These are illustrated in Figure 2–1.

Kinetic energy can be viewed as energy in use and is often described as the energy of motion. Motion can occur on a large scale, as in the movement of an object from one place to another. Examples that occur in nature include falling raindrops (Figure 2–2a), water flowing through a river channel, and grains of dust transported by the wind. The motion of kinetic energy can also occur at a microscale, as in the case of molecular vibration or rotation (illustrated for water in Figure 2–2b).

A solid object may seem to be standing still, but its atoms or molecules undergo a certain amount of vibration. Gas and

▶ **FIGURE 2–1** Energy assumes several different forms, but each of these is a form of either kinetic energy (the energy of motion) or potential energy.

FORMS OF ENERGY

KINETIC ENERGY

Light and other forms of radiation

Heat

Motion

Electrical power

POTENTIAL ENERGY

Reservoir behind hydroelectric dam

High pressure

Five types of chemical potential energy

Battery

Gasoline

Firewood

Explosives

Food

▲ **FIGURE 2–2** Kinetic energy can occur as the motion associated with moving objects, such as the falling raindrop in (a), or as molecular vibration or rotation, as depicted for water molecules in (b). The greater the rate of vibration or rotation, the higher the temperature of the substance.

liquid molecules, in contrast, are not fixed in space but move about randomly (Figure 2–3). In solids, liquids, or gases, the rate of vibration or random movement determines the temperature of the object.

If kinetic energy is energy in use, potential energy is energy that hasn't yet been used. Potential energy can assume many forms. For example, a plant's carbohydrates have potential energy that can be consumed by animals (or by the plant itself) and then metabolized to yield the energy needed for all of its biological activity. When our own bodies metabolize food, we are using this potential energy, converting it to kinetic energy, and releasing heat as a by-product.

Another form of potential energy results from an object's position. Consider, for example, a cloud droplet that occupies some position above Earth's surface. Like all other objects, the droplet is subject to the effect of gravity. As it falls toward Earth's surface, the object's potential energy is converted to kinetic energy. Obviously, the higher the droplet's altitude, the greater the distance it is capable of falling and the greater its potential energy. It is important to recognize that the droplet did not attain its height by magic, because energy was used to elevate its mass in the first place.

Randomly moving air molecules

▲ **FIGURE 2–3** Air molecules move about in random motion.

Checkpoint

1. What is potential energy?
2. Explain what happens in terms of kinetic and potential energy as a raindrop falls from a cloud.

Energy Transfer Mechanisms

Energy can be transferred from one place to another by three processes: conduction, convection, and radiation.

Conduction **Conduction** is the movement of heat through a substance without appreciable movement of molecules. A simple example is a metal rod, one end of which is placed over a campfire. The part of the rod above the flame is warmed, and molecules there gain energy. Some of this energy is passed to neighboring molecules, which, in turn, heat adjacent molecules. (The exact mechanism of molecular "passing" depends on the substance—in metals, it is mainly accomplished by electrons.) This process occurs throughout the length of the rod so that after a few minutes the entire piece of metal becomes too hot to handle. The transfer of heat from the warmer to the colder part of the rod is conduction. Note that although heat travels through the rod, the molecules that make up the rod do not move. Conduction is most effective in solid materials, but as we will see in Chapter 3, it also is an important process in a very thin layer of air near Earth's surface.

Convection The transfer of heat by the mixing of a fluid is called **convection**. Unlike conduction, convection is accomplished by movement of the liquid or gas in which the process occurs. You can observe this process by watching a pot of water boil on a kitchen stove. The water at the bottom of the pot is closest to the source of energy and warms most rapidly. In warming, the water expands ever so slightly, becomes less dense, and rises to the surface. The rising water must, of course, be replaced from above, so water formerly at the surface sinks to the base of the pot. These rising and sinking motions cause a rapid movement not only of mass but also of the thermal energy within the circulating water.

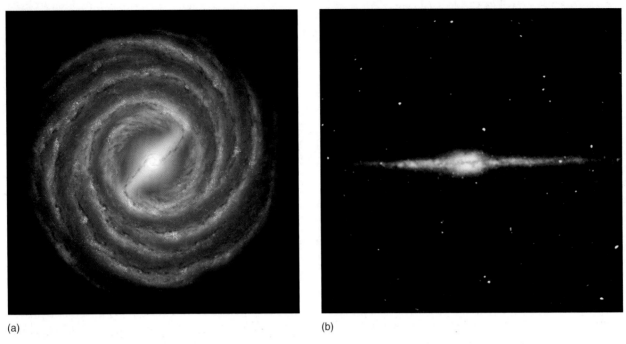

(a)

(b)

▲ **FIGURE 2–4** Our solar system is part of the Milky Way galaxy that contains more than 100 billion stars. (a) An artist's rendition. (b) A wide-angle infrared image of the plane and bulge of our Milky Way.

Convection in the atmosphere is not much different from that within a pot of boiling water. During the daytime, heating of Earth's surface warms a very thin layer of air (on the order of 1 mm thick) in contact with the surface. Above this thin *laminar layer*, air heated from below expands and rises upward because of the inherent **buoyancy** of warm air. Buoyancy is the tendency for a light fluid (a liquid or gas) to float upward when surrounded by a heavier fluid. Unlike water in a pot, the atmosphere can undergo convection even in the absence of buoyancy through a process called *forced convection*, the vertical mixing that happens as the wind blows. We discuss these processes in more detail in Chapter 3.

Radiation Of the three energy transfer mechanisms, **radiation** is the only one that can be propagated without a transfer medium. In other words, unlike conduction or convection, the transfer of energy by radiation can occur through empty space. Virtually all the energy available on Earth originates from the nearby (in astronomical terms) star we call the Sun, a member of the Milky Way galaxy (Figure 2–4). The atmosphere also has other sources of energy: Minute amounts of radiation are received from the billions of other stars in the universe, and some energy reaches the surface from Earth's interior. However, the contribution of these sources is minuscule compared to the energy from the Sun.

We will now examine the characteristics of radiation and the way Earth's orientation affects the radiation received. The spatial and seasonal variations in the receipt of solar energy are not mere abstractions; they are, in fact, the driving force for virtually all the processes discussed in the rest of this book. Figure 2–5 depicts radiation, along with the other types of energy transfer, conduction and convection.

Checkpoint

1. How is energy transferred in conduction, convection, and radiation?
2. Describe the types of energy transfer involved in heating the soup in an iron cooking pot suspended above the glowing coals of a campfire.

▲ **FIGURE 2–5** The fire below the pan transfers heat due to the radiation of energy, along with the convection of air circulating between the pot and the fire. The boiling water inside the pot circulates heat via convection. The person's hand holding the pot handle will feel it warm as conduction occurs.

Characteristics of Radiation

Radiation is emitted by all matter. Thus, *everything*—including the stars, Earth, ourselves, and this book—is constantly emitting electromagnetic energy. We are all familiar with electromagnetic energy in many of its forms. We see the environment around us because a type of radiation we call *visible light* impinges on our eyes, which then send signals to our brains to produce visual images. A different type of electromagnetic energy is used when we warm a meal in a microwave oven; the radiation agitates the molecules of the food and thereby increases its temperature. Other types of radiation may be less beneficial or even harmful, such as ultraviolet radiation, which can lead to sunburns, malignancies, or even death. Although different types of radiation have different effects, they are all very similar in that they are transmitted as a sequence of waves.

TUTORIAL
RADIATION

Use the tutorial to explore the characteristics of electromagnetic radiation and the effect of temperature on its intensity and wavelength distribution.

Think of a wave created by a rock tossed into a pond. The wave is revealed by an oscillation in the water surface with alternating crests (high points in the ripple) and troughs (low points). When you observe the regular rise and fall of the surface as the wave passes, you know energy is being transferred.

In the case of radiation, the waves are electrical and magnetic oscillations. That is, radiation consists of both an electrical and a magnetic wave. With the proper instruments, we would detect these electrical and magnetic variations—hence the term *electromagnetic radiation*. To put it differently, when an object emits radiation, both an electrical field and a magnetic field radiate outward. At a fixed point in space, the strength of both fields rises and falls rhythmically, thereby forming electric and magnetic waves, each with its own crest-to-trough pattern. The electric and magnetic waves are perpendicular to one another, as shown in Figure 2–6. More importantly, the electric and magnetic components are closely coupled—the two rise and fall in unison.

Radiation Quantity and Quality

To describe electromagnetic radiation completely, we need to provide information about the amount of energy transferred (quantity) and the type, or quality, of the energy. This is similar to describing someone's weight, where we might state quantity in pounds and indicate quality using words such as "mostly flab." In the case of radiation, quantity is associated with the height of the wave, or its *amplitude*. Everything else being equal, the amount of energy carried is directly proportional to wave amplitude.

(a)

(b)

▲ **FIGURE 2–6** Electromagnetic radiation consists of an electric wave (E) and a magnetic wave (M). As radiation travels, the waves migrate in the direction shown by the pink arrow. The waves in (a) and (b) have the same amplitude, so the radiation intensity is the same. However, (a) has a shorter wavelength, so it is qualitatively different from (b). Depending on the exact wavelengths involved, the radiation in (a) might pass through the atmosphere, whereas that in (b) might be absorbed.

The quality, or "type," of radiation is related to another property of the wave, the distance between wave crests. Figure 2–6 shows waves of electromagnetic radiation moving in the same direction. All have the same amplitude, but the distance between the individual wave crests is smaller for the waves depicted at the top. The upper waves therefore have a shorter **wavelength**, which is the distance between any two corresponding points along the wave (crest-to-crest, trough-to-trough, etc.). Because of their shorter wavelengths, the waves in Figure 2–6a are qualitatively different, and might produce different effects, from the waves in Figure 2–6b. For example, X-ray radiation has an extremely short wavelength and is able to penetrate soft tissues. On the other hand, ordinary light, having a somewhat longer wavelength, is absorbed by the skin. Compared to everyday objects, the radiation of interest here has very small wavelengths. It is therefore convenient to specify wavelengths using small units called **micrometers** (or **microns**). One micrometer—signified by μm equals one-millionth of a meter, or one-thousandth of a millimeter (0.00004 in.).

All forms of electromagnetic radiation, regardless of wavelength, travel through space at the speed of light, which is about 300,000 km (186,000 mi) per second. At that speed, it takes 8 minutes for energy from the Sun to reach Earth. The energy received from the other, more distant stars takes even longer to arrive at Earth. For instance, radiation from the next nearest star, Proxima Centauri, must travel through space for 4.3 years before reaching us. Though this

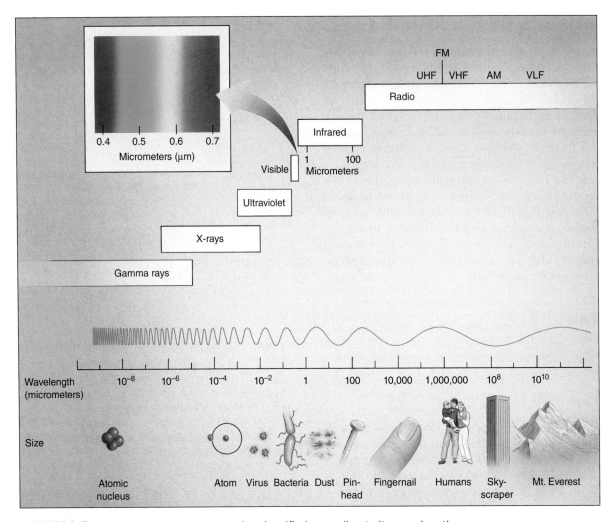

▲ **FIGURE 2–7** Electromagnetic energy can be classified according to its wavelength.

may seem like a long time, it is minuscule compared to the *billions* of years needed for light from a distant star to arrive at Earth.

Electromagnetic energy comes in an infinite number of wavelengths, but we can simplify things by categorizing wavelengths into just a few individual "bands," as indicated in Figure 2–7 and Table 2–1. The band with the shortest wavelengths consists of gamma rays, with a maximum wavelength of 0.0001 µm. Successively longer wavelength bands include X-rays, ultraviolet (UV), visible, near-infrared (NIR), thermal infrared (IR), microwave, and radio waves. Note that there is nothing unique or special about the visible portion of this electromagnetic spectrum other than the fact that our eyes and nervous systems have evolved to be able to sense this type of energy. Except for their wavelengths, visible rays are just like any other form of electromagnetic energy.

Checkpoint

1. What is electromagnetic radiation?
2. Explain how the electromagnetic spectrum is categorized into bands.

Intensity and Wavelengths of Emitted Radiation

All objects radiate energy, not merely at one single wavelength, but over a wide range of wavelengths. Figure 2–8a graphs the intensity of radiation emitted at all wavelengths every second by a square meter of the surface of the Sun and Earth. We can readily see that a unit of area on the Sun emits

TABLE 2–1

Wavelength Categorizations

Type of Energy	Wavelength (Micrometers)
Gamma	< 0.0001
X-ray	0.0001 to 0.01
Ultraviolet	0.01 to 0.4
Visible	0.4 to 0.7
Near Infrared (NIR)	0.7 to 4.0
Thermal Infrared (IR)	4 to 100
Microwave	1000 to 1,000,000 (1 meter)
Radio	> 1,000,000 (1 meter)

▶ **FIGURE 2–8** Energy radiated by substances occurs over a wide range of wavelengths. Because of its higher temperature, emission from a unit of area of the Sun (a) is 174,000 times more intense than that of the same area on Earth (b). (Notice that the units on the vertical axis of (a) are very much greater than those of (b)). The Sun also emits shorter wavelengths than does Earth.

(a)

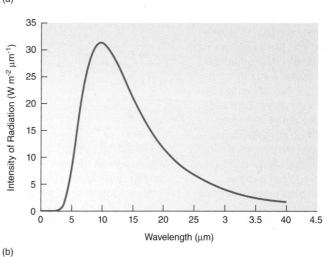

(b)

much more radiation (about 174,000 times more) than does the same amount of surface area on Earth. The shape of the curve showing the intensity of energy emitted by Earth at different wavelengths (Figure 2–8b) is similar to that of the Sun, but the total energy released is much less, and the peak of the curve corresponds to a longer wavelength.

Of course, the amount of radiation emitted and its wavelengths are not the result of mere chance; they obey some fundamental physical laws. Strictly speaking, these laws apply only to perfect emitters of radiation, so-called **blackbodies**. Blackbodies are purely hypothetical bodies—they do not exist in nature—that emit the maximum possible radiation at every wavelength. Earth and the Sun are close to blackbodies and therefore nearly follow the laws described shortly. Other materials may or may not approximate blackbodies. In particular, the atmosphere, composed mainly of gases, is especially far from a blackbody, so we will not treat it as one.

Stefan-Boltzmann Law The single factor that determines how much energy a blackbody radiates is its temperature. Hotter bodies emit more energy than do cooler ones; thus, not surprisingly, a glowing piece of hot iron radiates more energy than an ice cube. Interestingly, though, the amount of radiation emitted by

an object is more than proportional to its temperature. In other words, a doubling of temperature produces *more* than a doubling of the amount of radiation emitted. Specifically, the intensity of energy radiated by a blackbody increases according to the fourth power of its absolute temperature. This relationship, the blackbody version of the **Stefan-Boltzmann law**, is expressed as

$$I = \sigma T^4$$

where I denotes the intensity of radiation in watts per square meter, σ (Greek lowercase sigma) is the Stefan-Boltzmann constant (5.67×10^{-8} watts per square meter per K^4), and T is the temperature of the body in kelvins (see *Box 1–5, Physical Principles: The Three Temperature Scales*).

Because the intensity of radiation depends on the temperature raised to the fourth power, a doubling of temperature leads to a 16-fold increase in emission. Solving the Stefan-Boltzmann equation using the mean temperature of Earth's surface (about 288 K, 15 °C, or 59 °F) reveals that a square meter emits about 390 watts of power. In contrast, the surface of the Sun, with its temperature of about 5800 K (5500 °C, or 9900 °F), emits about 64 million watts per square meter.

Although true blackbodies do not exist in nature, they provide a useful model for understanding the maximum amount

The Nature of Radiation, Absorption, and Emission

We commonly describe electromagnetic energy traveling through space as a sequence of waves, but in some contexts it behaves as a stream of particles. The particle nature of radiation applies at the smallest scale of observation, as when visible light is emitted by a single atom or molecule. When light is emitted, there is a change in the orbital characteristics of the electrons in the emitter. As the orbit changes, a small bundle of energy, called a **photon**, is released.

Using the simplest example, imagine a hydrogen atom (with one proton and a single orbiting electron). The electron is not free to assume just any orbital distance from the nucleus. Instead, it is confined to fixed orbital distances, called *shells*. Each shell is associated with a given energy level; the greater the distance from the nucleus, the greater the energy level. As shown in Figure 1, when sufficient energy is absorbed by the hydrogen atom, it can become "excited," and its electron jumps from its "ground state" to a higher shell. Similarly, if the electron jumps back to its previous energy level, it gives off energy in the form of a photon. Because only a few discrete shells exist, only certain energy changes are possible. This means that a photon emitted by the atom can contain only certain discrete amounts of energy, corresponding to the atom's decrease in energy.

Similarly, an atom is restricted in what photons it can absorb, namely those with energies that push the atom into an allowable state. It so happens that the energy of a photon depends only on its wavelength: Photons at shorter wavelengths have more energy than photons at longer wavelengths. Consequently, if you know the wavelength of a photon, you can know its energy, and

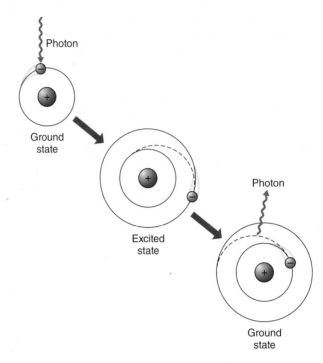

▲ **FIGURE 1** Electrons orbit the nucleus of atoms in prescribed zones called *shells*. This figure depicts a single electron orbiting the nucleus of a hydrogen atom. Upon receiving energy, the electron is in an excited state and jumps to its next shell. When the electron returns to its ground state, it releases energy in the form of a photon. Note that the energy emitted by such atoms must occur in discrete packets; at the atomic scale, units of energy are divided into individual parcels.

you can know whether or not the atom can emit or absorb that photon. This is a long-winded way of explaining that an atom will absorb and emit radiation only at certain wavelengths.

Of course, our atmosphere does not consist of single hydrogen atoms; it is mainly composed of molecules of gases. For these gases, changes in energy level are more complex than for simple hydrogen atoms, but the fact remains that emission and absorption involve a decrease or increase in energy level as a photon is released or absorbed. Furthermore, emission and absorption are again confined

to just those wavelengths that cause the molecule to move into an allowable energy state. That is, atmospheric gases, just like hydrogen, are selective absorbers and emitters. This is not true for liquids and solids, which tend to emit and absorb a wide range of wavelengths.

One very important consequence of all this is that the atmosphere and surface respond differently to radiation. In addition, the atmosphere responds quite differently to radiation of various wavelengths. As will be seen later, these basic principles go a long way toward explaining Earth's climate and climate change.

of radiation that can be emitted. Most liquids and solids can be treated as **graybodies**, meaning that they emit some percentage of the maximum amount of radiation possible at a given temperature. Whereas some substances (for example, water) are highly efficient at radiating energy, others (for example, aluminum) are less efficient. The percentage of energy radiated by a substance relative to that of a blackbody is referred to as its **emissivity**. Emissivities range from just above zero to just below 100 percent and are denoted by the Greek letter epsilon (ε). By incorporating the emissivity of any body, we derive the graybody version of the Stefan-Boltzmann law:

$$I = \varepsilon \sigma T^4$$

Including the emissivity factor means that the electromagnetic energy emitted by any graybody will be some fraction of what would be emitted by a blackbody. Note that even though the graybody form of the Stefan-Boltzmann law shows radiation intensity to be a function of both emissivity and temperature, most natural surfaces have emissivities above 0.9 (that is, above 90 percent of blackbody emission). In most cases, therefore, differences in emission are governed by temperature differences. The atmosphere is an exception to this because emission depends on a number of factors, such as the amount of water vapor in the air. Moreover, for a gas there is tremendous variability in emission with wavelength (see *Box 2–1, Physical Principles: The Nature of Radiation, Absorption, and Emission*); therefore, it is not accurate to think of atmospheric emission as just scaled-down blackbody emission. In other words, the concept of emissivity has less meaning when applied to the atmosphere, and use of the Stefan-Boltzmann equation is somewhat questionable. But regardless, we can say that the atmosphere is certainly not a perfect emitter of radiation and therefore emits less radiation at any particular temperature than would a blackbody.

Checkpoint

1. What does the Stefan-Boltzmann law tell us about the relationship between an object's temperature and the amount of radiation it emits?
2. How would an increase in Earth's temperature influence the amount of radiation emitted by Earth's surface?

Did You Know?

To be comfortable, humans need to maintain a skin temperature of about 306 K (33 °C or 91 °F). At this temperature, an average-sized person with surface area of 1.8 square meters emits 895 watts of power—about the same as that used by nine household lightbulbs. Of course, the wavelengths emitted are in the thermal infrared range, so we don't light up a dark room.

Wien's Law As we saw in Figure 2–8, the radiation emitted by the Sun and Earth (or any other body) is not of a single wavelength, nor are all wavelengths emitted in equal amounts. In the case of the Sun, the wavelength emitted more than any other is 0.5 μm, while energy radiated by Earth peaks at about 10 μm. For any radiating body, the wavelength of peak emission (in micrometers) is given by **Wien's** (pronounced "weens") **law**:

$$\lambda_{max} = \text{constant}/T$$

where λ_{max} refers to the wavelength of energy radiated with greatest intensity. More specifically, the constant in the preceding equation rounds off to the value 2898 for T in kelvins and λ_{max} in micrometers. Thus, we can determine the peak wavelength of radiation emitted as:

$$\lambda_{max} = 2898/T$$

Wien's law tells us that hotter objects radiate energy at shorter wavelengths than do cooler bodies. This is not surprising given that shorter wavelengths correspond to higher energies. Hot objects, possessing more thermal energy, necessarily radiate a higher proportion of energy at those shorter, more energetic wavelengths. Thus, for example, solar radiation is most intense in the visible portion of the spectrum, though it emits over a wide range of wavelengths. Most of the radiation has wavelengths less than 4 μm, which we generically refer to as **shortwave radiation**. Of the radiation emitted by the Sun, about 46.5 percent is near and thermal infrared, 46.8 percent is visible light, and 6.7 percent is ultraviolet.

Radiation emanating from Earth's surface and atmosphere consists mainly of that having wavelengths longer than 4 μm. This type of electromagnetic energy is called **longwave radiation**.

It is important to note that hotter bodies radiate more energy than do cooler bodies *at all wavelengths*. For example, the Sun radiates energy with wavelengths of about 0.5 μm most effectively and puts out far less energy at $\lambda = 10$ μm. Nevertheless, the Sun emits more radiation at those wavelengths than does Earth, despite the fact that Earth emits virtually nothing but longwave radiation (see *Box 2–2, Physical Principles: The Sun*).

The Stefan-Boltzmann law and Wien's law have some very useful and interesting applications. You have undoubtedly seen color-enhanced satellite images showing the distribution of clouds across North America, such as the one in Figure 2–9. This type of image depicts the height of cloud tops, which can be used as an indicator of the intensity of precipitation occurring below. The images are obtained by measuring the intensity of infrared radiation emitted by the cloud tops. Colder surfaces radiate less intense energy than do warm bodies. Weather satellites measure radiation intensity to determine the cloud-top temperatures across a target region. Because higher clouds tend to be colder than lower level clouds (remember that in the troposphere temperature tends to decrease with altitude), temperatures can be used to infer the height of the clouds, and thus their relative thickness. Thicker clouds, in turn, usually yield more intense precipitation. Infrared imagery can be obtained at night, as well as during the daytime, because it relies on energy radiated from the cloud tops rather than reflected light.

Checkpoint

1. What does Wien's law tell us about the relationship between the temperature of an object and the wavelength at which it radiates?
2. How has Wien's law been applied in the understanding of satellite images of clouds?

The Solar Constant

We all know that the Sun is extremely hot and we are protected from its great heat by our distance from the solar surface. But the electromagnetic energy moving through space

The Sun

The Sun may seem special to us, but compared to the 100 billion or more other stars in our galaxy, it is not particularly unique. Although stars vary considerably in size, temperature, brightness, and density, the Sun is about average in terms of these characteristics. Obviously, we have no first-hand observations of the solar interior, but based on physical principles we can infer the processes that go on within it. Using this information, we can divide the Sun into three sections (Figure 1).

The Interior

In the innermost portion of the Sun, the **core**, extremely high temperatures (about 15 million °C, or 27 million °F) and high densities lead to the energy-generating process of **nuclear fusion**. In this reaction, hydrogen atoms combine under tremendous heat and pressure to form a smaller number of heavier helium atoms. A certain amount of mass is lost in the process, and radiant energy is released—the same energy that works its way to the solar surface, travels through space, and ultimately warms Earth. The amount of this energy is staggering. Try to imagine the explosion of 100 billion one-megaton hydrogen bombs—that is equivalent to the amount of energy released in the core *every second!*

Energy initially travels outward from the core as electromagnetic energy through the **radiation zone** and into the base of the **convection zone**, where upwelling of the solar gases transfers the energy to the relatively thin solar surface.

The Photosphere

The layer of the Sun that radiates most of the energy away from the Sun is called

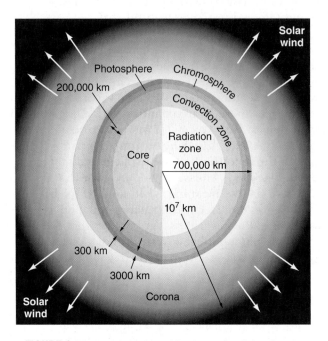

▲ **FIGURE 1** Energy is produced in the core of the Sun by nuclear fusion. Within the Sun, the energy is radiated to the base of the convection zone, where mixing transfers the energy upward to the base of the photosphere (the layer of the Sun visible from Earth).

the **photosphere**. It is the layer of the Sun we actually see as the **solar disk**. Although radiation travels from the photosphere to Earth in only about 8 minutes, the transfer of radiant energy within the Sun is incredibly slow. In fact, it takes about a million years for the energy unleashed in the core to travel to the base of the photosphere; thus, the energy reaching Earth is ancient.

Inspection of the photosphere with heavily filtered telescopes reveals that its outer layer is not a uniform, smooth surface. It is, instead, marked by a number of features of varying sizes and lifetimes.

Granules are the ever-present tops of convection cells that transport energy from the base of the photosphere to its surface. These features, analogous to bubbles in a pot of boiling water, are about 1000 km (600 mi) in diameter with lifespans on the order of 5 to 10 minutes. At any given time, there are literally millions of these on the surface of the photosphere.

Sunspots (each lasting a few weeks or months) are dark regions on the photosphere with diameters of about 10,000 km (6000 mi) and temperatures about 1500 °C cooler than the surrounding surface. They

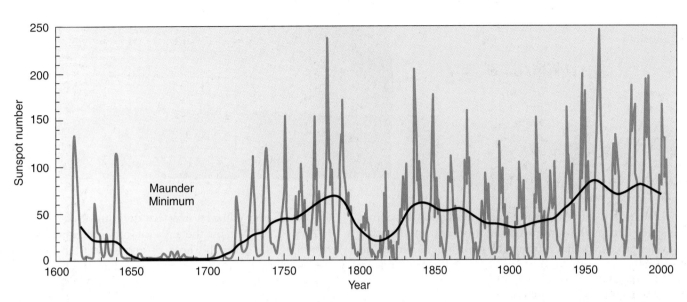

▲ **FIGURE 2** Sunspots appear in a somewhat regular manner, with peak occurrences normally observed about every 11 years. As seen in the smoothed curve (black), there are also long-term changes in average sunspot number. Much of the seventeenth century, for example, was marked by minimal sunspot activity. This period is known as the Maunder minimum.

form in response to locally strong magnetic fields, a thousand times more intense than those of the surrounding photosphere, which block the upwelling of heat from below.

Records of sunspot activity have been maintained since the days of Galileo Galilei (1564–1642), who observed their apparent movement across the solar surface. We now know that sunspots remain fixed in place and appear to move because of the rotation of the Sun (which takes about 24 days and 16 hours to complete one turn of its axis). The number of sunspots tends to peak every 11 years (Figure 2). Although the cycle is usually well defined, long episodes of minimal or unusually high sunspot activity have appeared during historic times. Figure 3, for example, shows a

long period of reduced sunspot activity during the seventeenth century. (This gap does not appear in all records.) Associated with these changes in sunspot number are very small changes in solar radiation. For example, the radiation change over an 11-year cycle is about 0.1 percent, and there has likely been change of about the same size over the last 2000 years. For many years, scientists have speculated about the possible role of sunspot activity in climate changes on Earth, but in light of such small radiation changes, it is not surprising that no definitive connections have been established. Because there is no known cause-and-effect mechanism between sunspots and climate, we should be very skeptical of attempts to link them.

Perhaps the most spectacular of solar disturbances are **flares**, intensely hot flashes (perhaps 100 million °C) across the photosphere surface due to magnetic instabilities. Temperatures within flares can achieve a staggering 100,000,000 K, and these features are so explosive that they have been likened to incredibly large bombs exploding on the solar surface. Although they exist for only a matter of minutes, they release a huge amount of energy, particularly in the form of X-ray and ultraviolet radiation.

▶ **FIGURE 2-9** A color-enhanced satellite image of North America. Cloud-top heights are determined by applying radiation laws; redder colors denote higher cloud tops.

is not depleted as it moves toward Earth. Radiation traveling through space carries the same amount of energy and has the same wavelength as when it left the solar surface. However, at greater distances from the Sun, it is distributed over a greater area, which reduces its intensity.

Consider a sphere completely surrounding the Sun whose radius is equal to the mean distance between Earth and the Sun, or 1.5×10^{11} m (Figure 2–10). As the distance from the Sun increases, the intensity of the radiation diminishes in proportion to the distance squared. This relationship is known as the **inverse square law**. By dividing total solar emission (3.865×10^{26} W) by the area of our imaginary sphere surrounding the Sun (the area of any sphere is given as $4\pi r^2$), we can determine the amount of solar energy received by a surface perpendicular to the incoming rays at the mean Earth–Sun distance. This incoming radiation is equal to

$$\frac{3.865 \times 10^{26}\,\text{W}}{4\pi \left(1.5 \times 10^{11}\,\text{m}\right)^2} = 1367\,\text{W/m}^2$$

▶ **FIGURE 2-10** The intensity of a beam of solar radiation does not weaken as it travels away from the Sun. However, as radiation travels farther from the Sun it is distributed over a greater area, thereby making it less effective at warming or illuminating any surface it hits. Imagine two spheres encompassing the Sun (such as the one with a radius equal to the mean Earth–Sun distance). All the radiation emitted from the Sun would be captured by this surrounding sphere. Now imagine that the surrounding sphere has a radius equal to the mean distance between the Sun and Mars. This sphere is larger than the previous one, so the energy emitted must be distributed over a greater area.

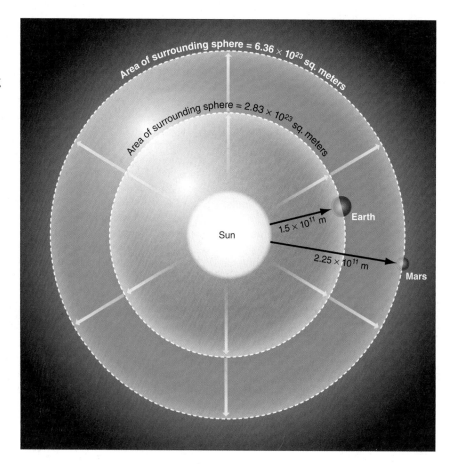

We refer to the value 1367 W/m² as the **solar constant** (although minor variations in solar output and other factors allow for some minor departures from this "constant"). For the sake of comparison, using the same procedure, we can determine that the solar constant for Mars (2.25×10^{11} m from the Sun) is 445 watts per square meter.

We should point out that when drawn at the compressed scale of Figure 2–10 the Sun's rays appear to diverge strongly as they pass Earth. In reality, the great distance between Earth and Sun means that the rays are almost parallel when they reach the planet, as is demonstrated in the solar radiation tutorial. It is much easier to understand variations in sunlight by assuming parallel rays, and that assumption is implicit in what follows.

The Causes of Earth's Seasons

Although the Sun emits a nearly constant amount of radiation, on Earth we experience significant changes in the amount of radiation received during the course of a year. These variations in energy manifest themselves as the seasons. We also know that the low latitudes (for example, the tropics and subtropics) receive more solar radiation per year at the top of the atmosphere than do regions at higher latitudes (for example, the Arctic and Antarctic). In this section, we will discuss how Earth's orbit around the Sun and its orientation with respect to incoming radiation influence the seasonal and latitudinal receipt of incoming solar radiation (often called **insolation**[1]).

TUTORIAL
EARTH–SUN GEOMETRY

Use the tutorial to explore how the tilt of Earth's axis and the position of Earth in its orbit influence the seasons and the receipt of solar radiation.

Earth's Revolution and Rotation

As we know, and as Figure 2–11 shows, Earth orbits the Sun once every 365¼ days as if it were riding along a flat plane. We refer to this imaginary surface as the **ecliptic plane** and to Earth's annual trip about the plane as its **revolution**.

The orbit is not quite circular but instead sweeps out an elliptical path, so that the distance between Earth and the Sun varies over the course of the year. Earth is nearest the Sun—at a point called **perihelion**—on or about January 3, when Earth–Sun distance is about 147,000,000 km (91,000,000 mi). Earth is farthest from the Sun—at a point called **aphelion**—on or about July 3, when Earth–Sun distance is about 152,000,000 km (94,000,000 mi). Thus, on perihelion Earth is 3 percent closer to the Sun than on aphelion. But because the intensity of incoming radiation varies inversely with the square of Earth–Sun distance (recall the inverse square law), the radiation is almost 7 percent more intense. (As mentioned at the outset of this chapter, however, this variation in intensity is not what causes the change of seasons.)

In addition to its revolution, Earth also undergoes a spinning motion called **rotation**. Rotation occurs every 24 hours (23 hours and 56 minutes, to be exact) around an imaginary line, called *Earth's axis*, connecting the North and South Poles. The axis is not perpendicular to the plane of the orbit of Earth around the Sun but is tilted 23.5° from it, as shown in Figure 2–12. Moreover, no matter what time of year it is, the axis is always tilted in the same direction and always points to a distant star called **Polaris** (the North Star). The constant direction of the tilt means that for half the year the Northern Hemisphere is oriented somewhat toward the Sun, and for half the year it is directed away from the Sun. The changing orientation of the hemispheres with regard to the Sun is the true cause of the seasons—not the varying distance between Earth and the Sun.

It is easier to visualize how the tilt of the axis influences the seasons if we consider a hypothetical situation in which the axis is tilted not 23.5°, but rather a full 90°, as depicted in Figure 2–13. Actually, this is the case for Uranus, so what we describe is not entirely hypothetical. Examine the situation when Earth is in position #1. The Northern Hemisphere is oriented directly toward the Sun so that it is fully illuminated over the entire 24-hour period of rotation. Meanwhile, the Southern Hemisphere undergoes 24 hours of continual darkness. This situation favors greater warmth in the Northern than in the Southern Hemisphere. Furthermore, a person standing at the North Pole would observe the Sun as

◄ **FIGURE 2–11** Earth's orbit around the Sun is not perfectly circular but is an ellipse. Earth is nearest to the Sun (perihelion) on about January 4 and farthest away (aphelion) around July 3.

Ecliptic plane

Perihelion

Sun

Aphelion

January 4

July 3

147 million km

152 million km

---

[1] It is sometimes reported that the term *insolation* is an acronym for incoming solar radiation. Such is not the case; it is from a Latin word and was first used in the 1600s.

▶ **FIGURE 2–12** When Earth is at the far left position the Northern Hemisphere has its maximum tilt toward the Sun (the summer solstice for that hemisphere). The Northern Hemisphere will then have a stronger orientation toward the Sun than will the Southern Hemisphere. The opposite situation exists when Earth is in the far right position. The situation at the bottom and top of the diagram represents the equinoxes, in which neither hemisphere has a stronger orientation toward the Sun.

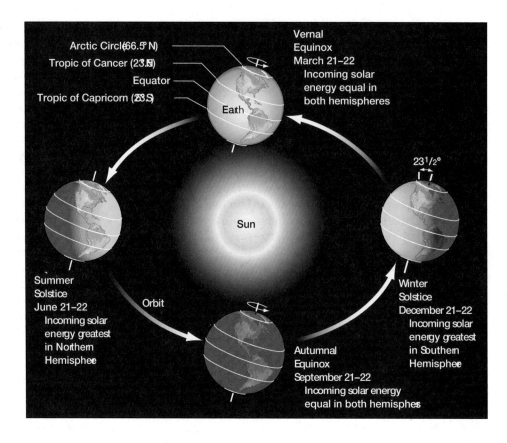

▶ **FIGURE 2–13** A hypothetical situation wherein Earth's axis is aligned along the ecliptic plane. In position #1, the Northern Hemisphere receives much energy from the Sun while the Southern Hemisphere is in constant darkness. The situation reverses six months later (position #3). In positions #2 and #4, both hemispheres receive equal amounts of solar energy.

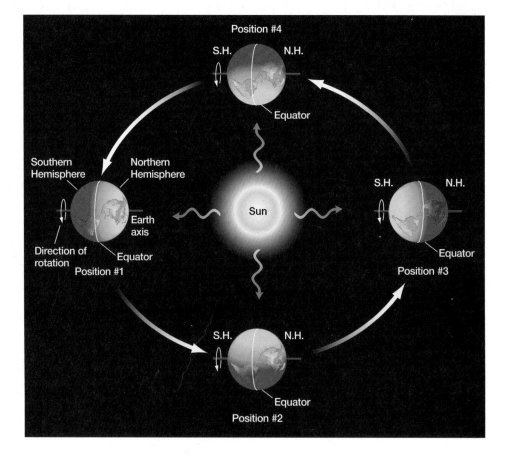

being directly overhead during the entire day. Moving away from the North Pole toward the equator, there is a gradual reduction in the angle of the Sun above the horizon, until at the equator the Sun appears to be right on the horizon. South of that line, the Sun is below the horizon and nighttime covers the Southern Hemisphere.

Now refer to position #3, which occurs 6 months after position #1. In this situation, the Southern Hemisphere is in continual sunlight while the Northern Hemisphere is subjected to 24 hours of darkness. Furthermore, someone standing at the South Pole would see the Sun directly overhead, and the apparent position of the Sun would shift toward the horizon for viewers located closer to the equator.

Finally, observe the intermediate positions, #2 and #4. In these two situations, the 90° tilt of the axis is neither toward nor away from the Sun, and the tilt becomes irrelevant to the receipt of insolation. Moreover, in positions #2 and #4 every place on Earth receives 12 hours of daylight and 12 hours of darkness because every latitude is half sunlit and half dark. Finally, note that at noon (when the longitude of any place in question is aligned directly toward the Sun), a person standing at the equator would observe the Sun to be directly overhead. Thus, Earth's revolution causes seasonal changes in the amount of heating of the surface. When either the Northern or Southern Hemisphere is oriented toward the Sun, that hemisphere receives a greater amount of insolation and therefore warms more effectively. Whichever hemisphere is oriented away from the Sun receives less radiation.

Solstices and Equinoxes Refer back to Figure 2–12, which shows the true seasonal change in orientation of Earth with respect to the Sun based on the actual 23.5° tilt of the axis. Although the axis is tilted only 23.5°, and not 90°, the principle just described still applies. During 6 months of the year, the Northern Hemisphere receives more sunlight than does the Southern Hemisphere; during the other 6 months, the Southern Hemisphere receives a greater amount of insolation. The four positions shown in the diagram represent 4 days that have particular significance.

TUTORIAL
EARTH–SUN GEOMETRY

Use the tutorial to observe how insolation varies over the course of a year as Earth moves through its orbit.

In the farthest left position in Figure 2–12, the Northern Hemisphere has its maximum tilt toward the Sun. This occurs on or about June 21, which we refer to as the **June solstice** (this is also called the *summer solstice* according to the corresponding Northern Hemisphere season). Although we designate this as the first day of summer, it actually represents the day on which the Northern Hemisphere has its greatest availability of insolation. Six months later (on or about December 21), the Northern Hemisphere has its minimum availability of solar radiation on the **December solstice** (*winter solstice* in the

Northern Hemisphere), which is called the first day of winter in the Northern Hemisphere and the first day of summer in the Southern Hemisphere. Intermediate between the two solstices are the **March equinox** (often called the *vernal* or *spring equinox* for the Northern Hemisphere) on or about March 21 and the **September equinox** (called the *autumnal equinox* in the Northern Hemisphere) on or about September 21. On the equinoxes, every place on Earth has 12 hours of day and night (the word equinox refers to "equal night"), and both hemispheres receive equal amounts of energy.

Of course, the transitions between the four positions shown in Figure 2–12 do not occur in sudden leaps; instead, a steady progression occurs from one position to the next. As shown in Figure 2–14, the 23.5° tilt of the Northern Hemisphere toward the Sun on the June solstice causes the *subsolar point* (the point on Earth where the Sun's rays meet the surface at a right angle—and where the Sun appears directly overhead) to be located at 23.5° N. This is the most northward latitude at which the subsolar point is located. The fact that the Sun never appears directly overhead poleward of 23.5° N gives that latitude special significance, and we call it the **Tropic of Cancer**.

Likewise, on the December solstice, the sun is directly overhead at 23.5° S, the **Tropic of Capricorn**. On the two equinoxes, the subsolar point is on the equator. Thus, the subsolar point migrates 47° (that is, between 23.5° N and 23.5° S) over a 6-month period, and on any particular day it is located somewhere between the Tropics of Cancer and Capricorn. This seasonal movement of the subsolar point is similar to what would happen if instead of Earth orbiting the Sun, its axis were to slowly rock back and forth, toward and away from the Sun.

▲ **FIGURE 2-14** Because Earth's axis is tilted 23.5°, the subsolar point is at 23.5° N during the summer solstice.

The latitudinal position of the subsolar point is the **solar declination**, which can be visualized as the latitude at which the noontime Sun appears directly overhead. Figure 2–15 plots the solar declination for several days of the year, while the arrows between the dates indicate the direction toward which the declination is moving at that time of year.

Now we are ready to see how the changing orientation of Earth with respect to the Sun directly affects the receipt of insolation through three mechanisms: (1) the length of the period of daylight during each 24-hour period, (2) the angle at which sunlight hits the surface, and (3) the amount of atmosphere that insolation must penetrate before it can reach Earth's surface.

Period of Daylight One way the tilt of the axis influences energy receipt on Earth is by its effect on the lengths of day and night. We have already seen that if the axis were tilted 90° from the plane of Earth's orbit there would be 1 day when the entire Northern Hemisphere underwent 24 hours of daylight and an equivalent period 6 months later of continuous darkness. But the axis is only tilted 23.5° from the plane of the orbit, so only the latitudes poleward of 66.5° (that is, 90° minus 23.5°) experience a 24-hour period of continuous daylight or night. These lines of latitude are the **Arctic Circle** (in the Northern Hemisphere) and the **Antarctic Circle** (in the Southern Hemisphere). This is illustrated in Figure 2–16. On the June solstice, any place north of the Arctic Circle has 24 hours of daylight. Just a short distance south of the Arctic Circle there is almost (but not quite) 24 hours of daylight. Moving toward the equator, the period of daylight decreases until reaching the equator,

where the day and night are both 12 hours long. Moving into the Southern Hemisphere, daylength shrinks until 66.5° S, where night is 24 hours long. The opposite pattern, of course, holds for the December solstice.

TUTORIAL

EARTH–SUN GEOMETRY

Use the tutorial to see how the period of daylight and solar angle are affected by latitude and the time of year.

Solar Angle Think back to the last time you spent a few hours lying out in the Sun. If you went outside early in the day when the Sun was low above the horizon, you probably did not feel a great deal of warming from its rays. But as the Sun got higher in the sky, it became more effective at warming your body. This change was largely due to a decrease in beam spreading. **Beam spreading** is the increase in the surface area over which radiation is distributed in response to a decrease of solar angle, as illustrated in Figure 2–17. The greater the spreading, the less intense the radiation is. In part (a) of the figure, the incoming light is received at a 90° angle, which concentrates it to a small area and increases its ability to heat the surface. In part (b), the rays hit the surface more obliquely and the energy is distributed over a greater area, leading to a less intense illumination (less energy per unit area). Thus, a beam of light is more effective at illuminating or warming a surface if it has a high angle of incidence (that is, the angle at which it hits the surface).

▶ **FIGURE 2–15** The solar declination gradually migrates north and south over the course of the year. On the June solstice the subsolar point marks its most northward extent, 23.5° N. Solar declination is 23.5° S on the December solstice.

(a)

(b)

▲ **FIGURE 2-17** The intensity of radiant energy hitting a surface is affected by its angle of incidence. More direct illumination causes a more intense amount of heating.

▲ **FIGURE 2-16** On the June solstice (a), every point north of 66.5° N has 24 hours of daylight and every point south of 66.5° S has continual night. During the December solstice (b), the situation is reversed. These latitudes are called the Arctic Circle and Antarctic Circle, respectively.

Noontime sun angles for any given latitude can be easily determined if the solar declination is known. To do this, you simply subtract the latitude of a given location from 90° and then add the solar declination. Thus, on either of the equinoxes (solar declination = 0°) Toronto, Ontario (latitude 44° N), has a noontime solar angle of 90° − 44° = 46°. During the June solstice, the noontime solar angle is 90° − 44° + 23.5° = 69.5°. Six months later the solar declination = −23.5°, with the negative sign indicating the sun is overhead in the Southern Hemisphere. At this time, the noontime solar angle = 90° − 44° − 23.5° = 22.5°.

The change in noontime solar angle over the course of a year can cause significant differences in the intensity of sunlight hitting the surface due to beam spreading. For example, beam spreading of the noontime sun at Toronto on the December solstice is almost two and a half times greater than on the June solstice. Thus, in the absence of other effects, the midday sun will be less than half as intense in the winter as in the summer.

Differences in the amount of beam spreading have a major role in causing the seasons. During the 6-month period between the March and September equinoxes, any latitude in the Northern Hemisphere has a more direct angle of incidence than does its Southern Hemisphere counterpart. Thus, insolation available to the Northern Hemisphere is subjected to less beam spreading, which promotes greater warming of the surface. During the following 6 months, the situation is reversed and the Southern Hemisphere has, on the whole, higher Sun angles.

Figure 2–18 shows the effects of noontime solar angle and length of daylight period for the solstices and equinoxes. On the June solstice, both factors work together to enhance warming in the Northern Hemisphere; on the December solstice, they combine for less effective heating in the north.

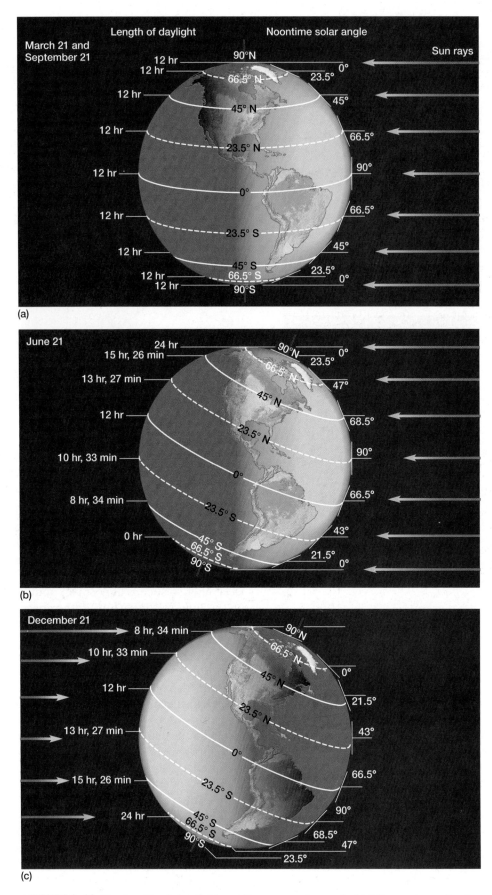

▲ **FIGURE 2–18** The length of day (left) and the noon solar angle (right) are shown for the equinoxes (a), June solstice (b), and December solstice (c).

Checkpoint

1. What is beam spreading?
2. How does the latitude of the point having the most intense midday solar heating change over the course of the year?

Atmospheric Beam Depletion The third way in which the tilt of the axis influences heating is in determining the amount of atmosphere that sunlight must penetrate before reaching the surface. As you can see in Figure 2–19a, insolation approaching the surface at a 90° angle passes through the atmosphere as directly as possible. Compare this to Figure 2–19b, in which sunlight approaches the surface at a low angle (as it does around sunrise or sunset). In this situation, a beam of sunlight must pass through a greater amount of atmosphere. Although the atmosphere is mostly transparent to incoming sunlight, some radiation is absorbed and even more is reflected back to space. The greater the thickness traveled, the more the beam is weakened. Because the solar altitude is lowest in the Northern

Hemisphere near the December solstice, at that time more energy is lost due to atmospheric effects than at any other time of the year.

Overall Effects of Period of Daylight, Solar Angle, and Beam Depletion We can now summarize some of the important points regarding controls on energy reaching Earth's surface. We must emphasize that we have not considered differences in atmospheric transparency or cloudiness at all; thus, the discussion assumes uniform optical properties. In other words, the patterns described below arise solely from geometrical considerations.

 TUTORIAL
EARTH–SUN GEOMETRY

Interact with the tutorial to examine the influence that solar angle has on the amount of beam spreading and atmospheric beam depletion.

(a)

(b)

◄ **FIGURE 2-19** A high solar angle (a) allows sunlight to pass through the atmosphere with a relatively short path. Lower sun angles such as those near sunrise and sunset (b) require that the energy pass through more of the atmosphere. This increase in atmospheric mass results in a greater depletion of energy than in (a).

1. **June solstice:**
   a. Solar declination = 23.5° N.
   b. Every latitude in the Northern Hemisphere receives more energy than the corresponding latitude in the Southern Hemisphere.
   c. Every place north of the Arctic Circle receives 24 hours of daylight, and every place south of the Antarctic Circle has 24 hours of night.
   d. The equator receives 12 hours of daylight.
   e. For the Northern Hemisphere as a whole, sunlight travels through less atmosphere than it does in the Southern Hemisphere.

2. **December solstice:**
   a. Solar declination = 23.5° N.
   b. Every latitude in the Southern Hemisphere receives more energy than the corresponding latitude in the Northern Hemisphere.
   c. Every place south of the Antarctic Circle receives 24 hours of daylight, and every place north of the Arctic Circle has 24 hours of night.
   d. The equator receives 12 hours of daylight.
   e. For the Southern Hemisphere as a whole, sunlight travels through less atmosphere than it does in the Northern Hemisphere.

3. **Equinoxes:**
   a. Solar declination is 0°.
   b. Every place on Earth has 12 hours of daylight.
   c. Both hemispheres receive equal amounts of insolation.

Changes in Energy Receipt with Latitude

We have seen that seasonal changes in the orientation of Earth with respect to the Sun determine the availability of solar radiation across the globe. Obviously, a large supply of energy is favorable for warm temperatures and a lack of energy leads to cold conditions. But the impact of radiation availability does not end with temperature. As we will see in later chapters, energy receipts also affect the distribution of pressure, which, in turn, affects winds, cloudiness, precipitation, and other aspects of weather and climate.

Having examined the effects of latitude and time of year separately, we can now look at how their combined effects set up differences in available insolation from one latitude

TABLE 2–2

Variations in Solar Angle and Daylength

	Solar Angle at Noon	Length of Day	Total Radiation for Day (Megajoules m^2)
December 21			
Winnipeg (50° N)	16.5°	7 hr, 50 min	7.1
Austin (30° N)	36.5°	10 hr, 04 min	18.6
June 21			
Winnipeg (50° N)	63.5°	16 hr, 10 min	44.5
Austin (30° N)	83.5°	13 hr, 56 min	43.9

to another. Table 2–2 presents the values of noontime solar angle and length of day for Winnipeg, Manitoba, Canada (50° N), and Austin, Texas (30° N), for the December and June solstices. (Remember that the values shown in the table represent the amount of energy available at the top of the atmosphere. Atmospheric conditions—especially the amount and type of cloud coverage—will reduce the amount of radiation that actually reaches the surface. The processes that affect the receipt of the energy at the surface are described in Chapter 3.) During the winter, Winnipeg has only 62 percent as much available solar radiation as does Austin because of its lower solar angle and the shorter period of daylight. During the June solstice, however, Winnipeg has a slightly greater amount of radiation because its greater period of daylight more than offsets its lower midday solar angle. Thus, during the winter, both factors lead to decreasing values of available insolation with latitude. In contrast, during the summer, the increasing daylength at the high latitudes offsets the effect of lower sun angles. The result is a weak latitudinal gradient in summer and strong north–south differences in winter. As will be seen in coming chapters, this has profound implications for seasonal temperature distributions and for seasonal changes in the vigor of large-scale atmospheric motion.

Did You Know?

On the June solstice, more solar radiation is available at the top of the atmosphere at the Arctic Circle than at either the Tropic of Cancer or the equator. In this particular instance the effect of the 24-hour period of sunlight at the Arctic Circle outweighs the lower Arctic Sun midday sun angles.

Summary

Virtually all the energy available for physical and biological processes originates within the Sun and travels through space as electromagnetic radiation. Such energy is propagated as a series of waves having characteristic wavelengths.

All matter emits electromagnetic energy in a continuum of different wavelengths, but not all substances or objects radiate the same amount of energy, nor do they emit it at the same dominant wavelengths. The Stefan-Boltzmann law tells us that hot objects emit radiation more intensely than do cooler objects, and Wien's law dictates that hotter objects emit radiation at shorter wavelengths. Together, these laws mandate that the Sun puts out hundreds of thousands of times more energy than does Earth and that the energy from the Sun is emitted at shorter wavelengths, with effectively no overlap.

Despite the fact that the distance between Earth and the Sun varies slightly through the year, these differences exert only a small impact on the seasonal heating of the planet. More important to the distribution of available radiation is the relative orientation of Earth to the Sun, specifically through the effects of solar angle, length of day, and the length of the path a beam of radiation must take through the atmosphere

before reaching the surface. The patterns described in this chapter deal with the availability of insolation before the atmosphere acts on the incoming radiation. In the next chapter, we describe the atmospheric processes that reduce the intensity of incoming radiation and the transfer mechanisms by which energy is exchanged between the atmosphere and the surface. These processes directly influence global temperature and pressure variations—variations responsible for all other weather phenomena.

Key Terms

energy *page 32*

joule *page 32*

power *page 32*

watt *page 32*

electromagnetic
 radiation *page 32*

kinetic energy *page 32*

potential
 energy *page 32*

conduction *page 33*

convection *page 33*

buoyancy *page 34*

radiation *page 34*

wavelength *page 35*

micrometers/
 microns *page 35*

blackbody *page 37*

Stefan-Boltzmann
 law *page 37*

graybody *page 38*

emissivity *page 38*

photon *page 38*

Wien's law *page 39*

shortwave radiation *page 39*

longwave radiation *page 39*

core *page 40*

nuclear fusion *page 40*

radiation zone *page 40*

convection zone *page 40*

photosphere *page 40*

solar disk *page 40*

granules *page 40*

sunspots *page 40*

flares *page 41*

inverse square law *page 42*

solar constant *page 43*

insolation *page 43*

ecliptic plane *page 43*

revolution *page 43*

perihelion *page 43*

aphelion *page 43*

rotation *page 43*

Polaris *page 43*

June solstice *page 45*

December solstice *page 45*

March equinox *page 45*

September equinox
 page 4 5

Tropics of Cancer and
 Capricorn *page 45*

solar declination *page 46*

Arctic and Antarctic
 Circles *page 46*

beam spreading *page 46*

Review Questions

1. Give several examples of kinetic and potential energy as they exist on Earth.

2. Conduction and convection are alike in that both transfer heat within a substance. What is the critical difference between them?

3. We have discussed sunlight and X-rays as two examples of electromagnetic radiation. Describe radiation as a wave phenomenon, and explain what is meant by "electromagnetic."

4. Why is wavelength important in radiation transfer? That is, when discussing radiation, why isn't it enough to specify the amount or rate of energy transfer?

5. Place the following wavelength bands in correct order of wavelength: visible, X-rays, ultraviolet, microwave, infrared.

6. Is there a temperature that has the same value on both the Fahrenheit and Celsius scales? If so, find that temperature. (*Hint:* Draw a graph of °C versus °F.)

7. Why is the Kelvin scale superior to the Fahrenheit and Celsius scales in many scientific applications?

8. Describe how the wavelengths and total energy emitted change as the temperature of an object increases.

9. The solar constant is about 1367 W/m^2. If the distance between Earth and Sun were to double, what would be the new value?

10. What is the most important factor responsible for seasons on Earth?

11. Describe the annual march of solar declination.

12. What is the significance of the Arctic and Antarctic Circles?

13. If the tilt of Earth's axis were 10°, where would we find the Arctic and Antarctic Circles? Would this cause a change in the dates of the solstices, equinoxes, and perihelion and aphelion?

14. Pick a day in the Northern Hemisphere winter. Describe the changes in daylength and solar position you would encounter if you were to travel from the North Pole to South Pole. Do the same for a day in the Northern Hemisphere summer.

15. Explain why the equator always has 12 hours of sunlight.

16. Explain how changes in solar position influence the intensity of radiation on a horizontal surface.

17. If you were to travel from the equator to the North Pole, on what day would variations in solar radiation be smallest? Why? Explain how daylength and solar angle change as you move poleward.

18. Burlington, Vermont, is located at 44.5° N. What is the angle of the noontime Sun on either of the equinoxes and on the solstices?

Critical Thinking

1. Goose down is composed of a large number of very small filaments. These separate the air within a parka or sleeping bag into many small packets that do not readily circulate. How does this feature make down such a good insulating material? How does this feature of goose down relate to the material in this chapter?

2. Some old buildings are warmed by radiators. And the latest trend in residential heating is "radiant heat" supplied by hot water running through tubing in floors. Is this a truly accurate descriptor of how rooms in those buildings are actually warmed?

3. Why is it not completely accurate to describe the energy coming from the Sun as visible radiation?

4. If Earth's speed of rotation were to change, would there be a corresponding change in the amount of energy the planet as a whole receives?

5. At noon the solar angle is always greater at Tucson, Arizona, than at Laramie, Wyoming. Is the same also true for 6 P.M.?

6. Locations near the equator typically have less seasonality than do locations farther away from the tropics. Explain why this is so.

7. Why is it that the solar angle cannot be considered the sole influence on the amount of radiation reaching Earth's surface? Is the situation different for the Moon?

8. How might the temperature change in the course of a day differ on east-facing vs. west-facing slopes?

9. At noon at 45° N latitude, the solar angle is 45° above the southern horizon. What would the angle of incidence be on a north-facing slope of 45°? Would the slope of the surface affect both beam spreading and atmospheric path length?

10. Describe the apparent path of the Sun to a person standing at the North Pole on June 22.

11. On the equinoxes a person at the equator would see the sun rise exactly to the east, pass directly overhead at noon, and set exactly in the west—all over a 12-hour period. How will this change on the solstices?

Problems and Exercises

1. An instrument measures the radiation emitted from an ocean surface as 365 watts per square meter. What law would you apply to determine the ocean surface temperature? (More advanced question: What would the ocean surface temperature actually be? *Hint*: You will need to rearrange one of the equations given in this chapter.)

2. Assume that a body has an emissivity of 0.9 and a temperature of 300 K. Which would have a greater impact on the intensity of radiation emitted: a 50 percent reduction in the emissivity or a 5 percent reduction in the absolute temperature?

3. Saturn is about 1.42×10^{12} m from the Sun, or about 9.5 times as far from the Sun as Earth is. Calculate the solar constant for Saturn. Do you suppose the distribution of wavelengths of the sunlight received at that distance is different from the distribution Earth receives?

4. One factor that influences the amount of insolation available is the varying Earth–Sun distance. Using the distances for perihelion and aphelion of 1.47×10^{11} m and 1.52×10^{11} m, respectively, determine the intensity of solar radiation at the top of Earth's atmosphere on those two days.

5. What is the difference in the noontime solar angle between the two solstices at a latitude of 10° N? How does this compare to the range of noontime solar angles at 30° N? Can you think of any significant outcomes of this difference?

6. Go to **http://aa.usno.navy.mil/data**. Check the first of the two hypertext options under the category "Positions of the Sun and Moon." This site will allow you to determine the solar angle throughout the course of any date you select. Plot the solar angle of the Sun throughout the daylight period for the solstices and the equinoxes and notice how the pattern changes through the year. Next, for the solstices and equinoxes, plot and compare the differing solar elevations over the course of the day for East Lansing, Michigan; Knoxville, Tennessee; and Gainesville, Florida. What generalizations can you make?

Quantitative Problems

This chapter has introduced some important laws and concepts describing the type and amount of radiation received by Earth. You can enhance your understanding of these laws by solving some quantitative problems on this book's Web site, **www.MyMeteorologyLab.com**. These brief problems should bolster your comprehension of the Stefan-Boltzmann and Wien's laws and the temperature scales discussed in this chapter.

Useful Web Sites

http://aa.usno.navy.mil/data/docs/AltAz.php
Gives altitude and azimuth of the Sun at 10-minute intervals for any day of year. Interactive format allows you to request data for particular cities or latitude–longitude coordinates. Includes precise sunrise/sunset data as well as solar angle data.

http://aa.usno.navy.mil/data/docs/EarthSeasons.php
Dates and times of perihelion, aphelion, equinoxes, and solstices.

www.srrb.noaa.gov/highlights/sunrise/gen.html
Fully interactive site that allows you to get time of sunrise/sunset for a choice of world cities or for precise latitude–longitude coordinates. Also gives solar declination and several other items of Earth–Sun position information, complete with glossary to explain their meanings.

MyMeteorologyLab™

The *Understanding Weather and Climate,* Sixth Edition, MyMeteorologyLab™ web site contains numerous multimedia resources to aid in your study of **Solar Radiation and the Seasons**.
Visit **www.MyMeteorologyLab.com** to:

- **Review** key chapter concepts.
- **Read** the Pearson eText, *In the News* RSS feeds, and other chapter-specific Web resources.
- **Visualize** challenging topics with Interactive Tutorials, Geoscience Animations, MapMaster interactive maps, and Weather in Motion movies and visualizations, and more.
- **Test Yourself** with self-study quizzes.

 ### TUTORIALS
RADIATION

EARTH–SUN GEOMETRY

Use the interactive animations and quizzes in these tutorials to review this chapter's key concepts.

WEATHER IN MOTION
View movies and visualizations of meteorology patterns and processes.

Solar Power
The Sun in Ultraviolet
Solar Eclipse
Global Variations in Insolation Through the Year
January and July Global Movies

3 Energy Balance and Temperature

After reading this chapter, you should be able to:

▶ Explain how incoming solar radiation is affected by the atmosphere.

▶ State the proportions of incoming solar energy that are reflected, scattered, or absorbed.

▶ Explain how energy is transferred between the surface and the atmosphere.

▶ Define the greenhouse effect.

▶ Describe the global distribution of temperatures on Earth's surface.

▶ List the factors that influence temperatures on Earth's surface.

▶ Describe daily and annual temperature patterns—including means, ranges, and temperature indices—and explain how temperatures are measured.

▶ Make use of thermodynamic diagrams and understand vertical temperature profiles.

▶ Summarize observations and predictions of climate change/global warming.

It was early morning on October 24, 1998, and Tropical Storm Mitch had just intensified into a true hurricane in the western Atlantic. While forecasters were certain the storm would strengthen further and move westward, nobody knew that within a few days Mitch would become the worst hurricane to hit Central America since 1780. Its ultimate devastation was hard to comprehend. The exact death toll from Hurricane Mitch will never be known but probably stands at somewhere between 9000 and 18,000 people. Wind gusts of more than 320 km/hr (200 mph) and rainfall amounts greater than 50 cm (20 in.) ravaged many villages. Worst hit was Honduras, where more than 20 percent of the population was suddenly homeless and about 70 percent of the crops were destroyed. The mountainous terrain of Nicaragua suffered devastating mudslides.

Hurricane Mitch lasted about three weeks as an organized storm. After ravaging Honduras and its neighbors, Mitch weakened and became a tropical storm that swept northeast across the Yucatan Peninsula and the Gulf of Mexico, crossing South Florida on November 5. From there, the storm headed across the Atlantic Ocean to Britain and Iceland, before dissipating in the North Atlantic.

Hurricane Mitch covered an area greater than 250,000 square kilometers, dumped millions of tons of rainwater, and brought winds capable of wiping out whole villages. Such activity requires an enormous amount of energy and, as we saw in Chapter 2, solar radiation provides virtually all of that energy. But there is more to the story, because most of the energy contained in the atmosphere does not accrue by the *direct* absorption of solar radiation. Instead, the majority of the energy comes *indirectly* from the Sun after first having been absorbed by Earth's surface. From there, several processes combine to transfer this absorbed energy to the atmosphere. But in the end, as much energy goes into the Earth surface–atmosphere system as goes out of it. We refer to this input and output as Earth's **energy balance**. In this chapter, we examine this energy transfer, which provides the fuel for everyday weather and for catastrophic events such as Hurricane Mitch.

◀ A satellite image of Hurricane Mitch, the deadliest hurricane to hit Central America in the last 200 years.

Atmospheric Influences on Insolation

Three processes—the absorption, scattering, and transmission of solar radiation—directly affect the distribution of temperature throughout the atmosphere. They also explain a number of atmospheric phenomena of everyday interest, such as the blue sky on a clear day or the redness of a sunset.

Solar radiation reaching the top of the atmosphere does not pass unimpeded through the atmosphere, but rather is attenuated by a variety of processes. The atmosphere absorbs some radiation directly and thereby gains heat. Another portion of radiation disperses as weaker rays going out in many different directions through a process we call *scattering*. Some of the scattered radiation is directed back to space; the remainder is scattered forward as the light we see from the portion of the sky away from the solar disk. In either case, the energy that is scattered is not absorbed by the atmosphere and therefore does not contribute to its heating. The end result of all this is what we see when we look upward, as in Figure 3–1.

The remaining insolation is neither absorbed nor scattered and passes through the atmosphere without modification, reaching the surface as direct radiation. But not all the energy reaching the surface is absorbed. Instead, a fraction is scattered back to space and, like the radiation scattered by the atmosphere, it does not contribute to the heating of the planet.

In this section, we explore the processes affecting incoming radiation.

Absorption

Atmospheric gases, particulates, and droplets all reduce the intensity of insolation by **absorption**. Absorption represents an energy transfer to the absorber. This transfer has two effects: the absorber gains energy and warms, while the amount of energy delivered to Earth's surface is reduced.

The gases of the atmosphere are not equally effective at absorbing sunlight, and different wavelengths of radiation are not equally subject to absorption. Ultraviolet radiation, for example, is almost totally absorbed by ozone in the stratosphere. Visible radiation, in contrast, passes through the atmosphere with only a minimal amount of absorption. This is of no minor consequence, because if the atmosphere *were* able to absorb all the incoming solar energy, the sky would appear completely dark. Artificial lights would be useless, because their radiation would likewise be absorbed. The very fact that we can see great distances suggests that the atmosphere is not particularly good at absorbing visible radiation, an impression that turns out to be correct.

Near-infrared radiation, which represents nearly half the radiation emitted by the Sun, is absorbed mainly by two gases in the atmosphere—water vapor and (to a lesser extent) carbon dioxide. This is why direct sunlight in the desert feels so hot and shade is so welcome, whereas in humid regions the apparent temperature difference between standing in direct sunlight and standing in shade is relatively small. When the humidity is high, water vapor absorbs a significant portion of near-infrared radiation, thereby reducing the amount of energy available to warm your skin. On dry days, the lack of water vapor allows a greater amount of near-infrared radiation to penetrate the atmosphere and raise your skin temperature.

Reflection and Scattering

The **reflection** of energy is a process whereby radiation making contact with some material is simply redirected away from the surface without being absorbed. The reason we are able to see is that the human eye can detect the receipt of visible radiation. Visible energy travels in all directions as it is reflected off objects in our field of view. Some of the reflected light comes into contact with our eyes, which in turn send

▲ **FIGURE 3–1** Reflection of sun and cloud off a building.

TABLE 3–1

Typical Surface Albedos

Surface	Albedo
Forest	0.14–0.20
Grassland	0.16–0.20
Green Crops	0.15–0.25
Sand	0.18–0.28
Snow (fresh)	0.75–0.95
Snow (old)	0.40–0.60
Soils	0.05–0.40

Source: Peixoto and Oort, *Physics of Climate*, 1992.

signals to optical centers in our brains. All substances reflect visible light, but with vastly differing effectiveness. Objects do not reflect all wavelengths equally. A shirt, for example, will appear green if it most effectively reflects wavelengths in the green portion of the spectrum. A fresh patch of snow very effectively reflects visible light, while a piece of coal reflects only a small portion of the visible radiation hitting its surface. The percentage of insolation reflected by an object or substance is called its **albedo**. Some typical albedo values are presented in Table 3–1.

Light can be reflected off a surface in a couple of different ways. When light strikes a mirror, it is reflected back as a beam of equal intensity in a manner known as **specular reflection**. In contrast, when a beam is reflected from an object as a larger number of weaker rays traveling in many different directions, it is called **diffuse reflection**, or **scattering**. When scattering occurs, you cannot see an image of yourself on the reflecting surface as you can in a mirror. Consequently, although a surface of fresh snow might reflect back most of the visible light incident on it, you would not be able to check out your appearance by looking at it. The vast majority of natural surfaces are diffuse rather than specular reflectors.

In addition to large solid surfaces, gas molecules, particulates, and small droplets scatter radiation. Furthermore, although much is scattered back to space, much is also redirected forward to the surface. The scattered energy reaching Earth's surface is thus **diffuse radiation**, which is in contrast to unscattered **direct radiation**. Figure 3–2 illustrates the process of scattering and the transformation of direct radiation to diffuse radiation. You can think of it this way: The blocking of direct radiation is what creates shadows, but a surface in the shadow of the direct radiation is not completely dark because it is illuminated by diffuse radiation. Notice that whether accomplished by a gas molecule, particulate, or droplet, this result is still a scattering process in which the radiation is redirected but not absorbed. The radiation can be scattered forward or backward (backscattering).

The characteristics of radiation scattering by the atmosphere depend on the size of the scattering agents (the air molecules or suspended particles) relative to the wavelength of the incident electromagnetic energy. Three very general categories of scattering exist: Rayleigh scattering, Mie scattering, and nonselective scattering.

▲ **FIGURE 3–2** Scattering is a process whereby a beam of radiation is broken down into many weaker rays redirected in other directions.

Rayleigh Scattering Scattering agents smaller than about one-tenth the wavelength of incoming radiation disperse radiation in a manner known as **Rayleigh scattering**. Rayleigh scattering is performed by individual gas molecules in the atmosphere. It primarily affects shorter wavelengths. Rayleigh scattering is particularly effective for visible light, especially those colors with the shortest wavelengths, so blue light is more effectively scattered by air molecules than is longer-wavelength red light. Furthermore, Rayleigh scattering disperses radiation both forward and backward. Combined with its greater effectiveness in scattering shorter wavelengths, this characteristic leads to three interesting phenomena: the blue sky on a clear day, the blue tint of the atmosphere when viewed from space, and the redness of sunsets and sunrises.

Figure 3–3 illustrates how Rayleigh scattering produces a blue sky. As parallel beams of radiation enter the atmosphere, a portion of the light is redirected away from its original direction. A person looking upward, away from the direction of the Sun, can see some of the scattered light that has been redirected toward the viewer. Because blue light is among the shortest (and therefore most readily scattered) of the visible wavelengths, the scattered radiation contains a higher proportion of blue light than yellow, green, or other longer-wavelength light. Rayleigh scattering occurs at every point in a clear atmosphere and diverts energy toward a viewer from all directions, so no matter where you look on a cloudless day, the sky is blue. Of course, not all the incoming radiation is scattered on a clear day. In fact, the amount of diffuse radiation received at the surface under cloudless skies is normally about one-tenth that of the direct radiation.

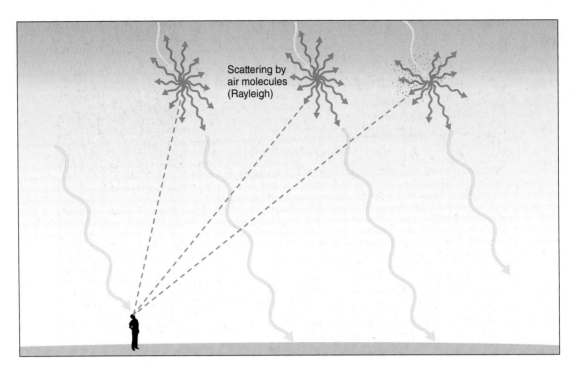

▲ **FIGURE 3–3** The sky appears blue because the gases and particles in the atmosphere scatter some of the incoming solar radiation in all directions. Air molecules scatter shorter wavelengths most effectively. Someone at the surface looking skyward perceives blue light, the shortest wavelength of the visible portion of the spectrum.

On the Moon, which has no atmosphere, the "sky" appears black (Figure 3–4). As a viewer looks toward the horizon on the Moon, there is no downward scattered light because of the absence of an atmosphere, and the sky appears little different from the way it does at night. All that can be seen is the energy reflected off the lunar surface and Earth.

The same process that leads to the blue sky as seen from the surface also produces the bluish tint of the atmosphere as viewed from space. Like forward scattering, backscattering is biased toward blue wavelengths, so diffuse radiation directed back to space appears blue.

Rayleigh scattering is also largely responsible for the redness of sunrises and sunsets. Figure 3–5 shows how this happens. When the Sun is barely over the horizon, sunlight must travel a greater distance through the atmosphere than it does during the middle of the day, and the longer path increases the amount of Rayleigh scattering. As the direct beam travels its long path, the shortest wavelengths of radiation are depleted, so the longer wavelengths constitute an increasing percentage of the direct sunlight. The sky in the general vicinity of the Sun thereby takes on a reddish tint due to the depletion of the green and blue (shorter-wavelength) light.

Mie Scattering Vertical motions in the atmosphere are sufficiently strong that the atmosphere always contains suspended aerosols. This is true not only in cities, which tend to have higher air pollution concentrations, but also in rural areas far removed from urban activities. The microscopic

aerosol particles are considerably larger than air molecules and scatter sunlight by a process known as **Mie** (pronounced "mee") **scattering**. Unlike Rayleigh scattering, Mie scattering is predominantly forward, diverting relatively little energy backward to space. Furthermore, Mie scattering does not have nearly the tendency to scatter shorter-wavelength radiation that Rayleigh scattering does. Thus, on hazy or polluted days (when there are high concentrations of aerosols) the sky appears gray, as the whole range of the visible part of the spectrum is effectively scattered toward the surface.

Mie scattering causes sunrises and sunsets to be redder than they would be due to Rayleigh scattering alone, so episodes of heavy air pollution often result in spectacular sunsets (Figure 3–6). Fires can also trigger enhanced Mie scattering. Residents of the western United States observed this phenomenon firsthand, when major fires burned across the region during the summer of 2002. If a fire is large enough, Mie scattering can be increased great distances downwind. In 1988, for example, fires in Yosemite National Park reddened the sky as far away as Minneapolis, Minnesota. Volcanic eruptions, such as the major eruption of Mount Pinatubo in 1991, can even enhance the color of sunrises and sunsets across an entire hemisphere, as stratospheric winds transport aerosols far from their source.

Nonselective Scattering The water droplets in clouds are considerably larger than suspended particulates; therefore, they scatter sunlight in yet another way, behaving more or less like lenses. An isolated water droplet affects various

▲ **FIGURE 3–4** An "Earthrise" from the Moon, as seen by the *Apollo 11* astronauts. Although this photo was taken during the day, the Moon has no blue sky. This is due to the absence of an atmosphere to scatter incoming solar radiation. Notice the blue tint of Earth, the result of Rayleigh scattering.

wavelengths of solar radiation differently. You see this whenever you witness a rainbow, which involves each wavelength being refracted (bent) a different amount, hence the bands of individual colors. In the aggregate, however, clouds reflect all wavelengths of incoming radiation about equally, which is why they appear white or gray. Because of the absence of preference for any particular wavelength, scattering by clouds is sometimes called **nonselective scattering**.

(a)

(b)

(c)

◄ **FIGURE 3–5** Sunrises and sunsets appear red because sunlight travels a longer path through the atmosphere when the Sun is low on the horizon. This causes a high amount of scattering to remove shorter wavelengths from the incoming beam radiation. The result is sunlight consisting almost entirely of longer (e.g., red) wavelengths.

▲ **FIGURE 3-6** The scattering of shorter wavelengths enhances the redness of sunrises and sunsets during episodes of heavy particulate concentrations.

Clouds are by far the most important agent in nonselective scattering and exert a tremendous impact on the global receipt of solar radiation by reflecting large amounts of energy back to space.

Checkpoint

1. What is the difference between specular reflection and scattering?
2. Compare and contrast Rayleigh, Mie, and nonselective scattering.

Transmission

When solar radiation travels through the vacuum of outer space, there is no modification of its intensity, direction, or wavelength. However, when it enters the atmosphere, only some of the radiation can pass unobstructed to the surface. The amount varies greatly, depending on atmospheric conditions. A clear, dry atmosphere might transmit as much as 80 percent of the incoming solar radiation as direct beam radiation without scattering or absorption. This is what you experience on a sunny, unpolluted day with sharp, distinct shadows. In contrast, when it is cloudy or hazy, only a small fraction of solar radiation will reach the surface as direct radiation. Under these conditions, there is both a reduction in the amount of radiation reaching the surface and a shift from direct radiation to diffuse, or scattered, radiation.

Spatial Distribution of Solar Radiation

The amount of solar radiation reaching the surface depends on two factors: the amount of insolation available at the

top of the atmosphere (**extraterrestrial radiation**) and the reduction in that amount due to absorption and backscattering by the atmosphere. Figure 3–7 shows the annual average solar radiation received by a horizontal surface over the United States for January (a), July (b), and annually (c). The pattern for January solar radiation receipt is strongly affected by latitude, with the northern tier states showing lower values than those to the south. This reflects the combined influence of shorter periods of daylight and lower solar angles at higher latitudes during the winter months.

During the summer months, as shown in (b), the reduction in midday solar angles with increasing latitude is partially offset by the longer period of daylight. Thus atmospheric conditions, especially cloud cover, dominate the pattern of solar radiation receipt. Highest insolation values occur over the western United States (except for the extreme northwest), where cloud cover is typically less than that found in the humid east. On an annual basis (c), the southwestern United States receives the greatest amount of insolation.

The Fate of Solar Radiation

Because Earth's orbit around the Sun is not perfectly circular, there are slight seasonal variations in the availability of insolation, with almost 7 percent more solar energy available on perihelion than on aphelion. Despite this variation, it is useful to think of a constant supply of radiation at the top of the atmosphere and to examine what happens, on average, to this energy. In other words, we need to account for the relative amount of radiation that is transmitted through the atmosphere, absorbed by the atmosphere and surface, and scattered back to space.

Such an exercise is more than a mere bookkeeping activity, because the amount of radiation absorbed by the atmosphere and surface will greatly influence their temperatures. For the sake of simplicity, we will assume that 100 units of insolation are available at the top of the atmosphere and then compare the amount of energy scattered back to space and absorbed by the atmosphere and surface to these 100 units. Bear in mind that the values presented in this discussion are annual and global averages; they do not necessarily apply to any particular place or time (see Figure 3–8). And though there is some degree of uncertainty, the numbers shown here represent the most recent values published in 2009.

As a global average, the atmosphere absorbs 23 of the 100 units available at the top of the atmosphere. Six of the 23 units are ultraviolet radiation absorbed in the stratosphere by ozone, with most of the remainder being near-infrared radiation absorbed in the troposphere by gases (mostly water vapor). Thus, most of the radiation absorbed by the atmosphere is not visible radiation—a situation that suits us well, because if visible radiation were strongly absorbed by the atmosphere, it would be harder for us to see. Also note

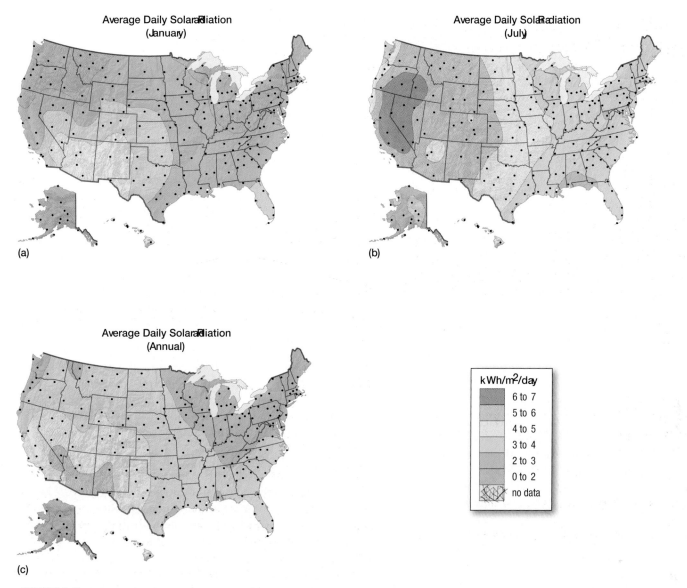

▲ **FIGURE 3-7** Average receipt of solar radiation at the surface in January (a), July (b), and annually (c). Values are in kilowatt hours per square meter per day.

that relatively little of the available shortwave radiation is absorbed by clouds; instead, clouds affect the incoming radiation primarily by scattering and reflection.

Though comparatively ineffective at absorbing shortwave radiation, cloud droplets scatter backward a large percentage of the incident energy. Their high albedo not only gives them their bright appearance, which makes them readily visible even from space, but also greatly reduces the amount of energy available for warming the atmosphere and surface. On average, the global cloud cover reflects 17 units of the incoming solar radiation back to space. But clouds are not the only agent of backscatter. Averaged across the globe, the backscattering of radiation by atmospheric gases and aerosols accounts for 6 of the 100 units of insolation at the top of the atmosphere, with Rayleigh scattering more

important than Mie scattering (largely because most Mie scattering is directed forward rather than back to space). Together, scattering by clouds and gases reflects back to space 23 units (that is, they yield an atmospheric albedo of 23 percent).

After atmospheric absorption and backscattering, 54 units of insolation are able to reach the surface. But not all the radiation reaching the surface is absorbed because Earth's surface is not completely black. Of the 54 units incident at the surface, 7 are reflected back to space. Overall, a total of 30 units of solar radiation (23 from the atmosphere and 7 from the surface) are scattered back to space, resulting in a **planetary albedo** of 30 percent. Note that the amount of insolation reflected from the surface is slightly smaller than that backscattered by atmospheric gases. In other words, when viewed from space,

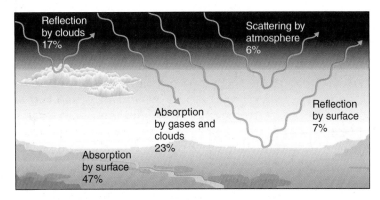

▲ **FIGURE 3–8** Incoming solar radiation available is subject to a number of processes as it passes through the atmosphere. The clouds and gases of the atmosphere reflect 17 and 6 units, respectively, of insolation back to space. The atmosphere absorbs another 23 units. Only 54% of the insolation available at the top of the atmosphere actually reaches the surface, where another 7 units are reflected back to space. The net solar radiation absorbed by the surface is 47 units.

the planet shines more from atmospheric reflection than from surface reflection.

The end result of these processes is that the atmosphere absorbs 23 units of energy, while the surface takes in 47 units. If this were the end of the story, we would all be in big trouble, because the constant supply of heat would cause continued warming of the planet. In fact, if this energy were stored in the upper few centimeters of Earth's surface, the surface would be heated at a rate of several hundred degrees Celsius each day!

Did You Know?

Because ultraviolet radiation (especially that having the shortest wavelengths in the UV band, known as UV-B) has very short wavelengths, Rayleigh scattering breaks down much of the energy into diffuse radiation. The fact that so much of the UV is diffuse and coming from different parts of the sky means that standing in the shade of a tree might not offer a great deal of protection against sunburn or skin cancer. Though you may be shielded from direct radiation by the tree, much of the diffuse radiation can still reach your skin.

Obviously, we do not observe the oceans boiling away, nor the land surface melting; thus, the surface must be continually losing energy. The same is true for the atmosphere, and for the Earth–atmosphere system as a whole. In other words, in the absence of rapid climate change, the surface, the atmosphere, and the planetary system must give up as much energy as they obtain. We now discuss the mechanisms involved in the maintenance of Earth's energy balance.

TUTORIAL
GLOBAL ENERGY BALANCE

Use the animations in Section 2.1 to follow solar radiation as it works its way down through the atmosphere.

Checkpoint

1. What happens to incoming solar radiation as it moves through the atmosphere?
2. Describe the geographic and seasonal variability of solar radiation reaching the surface over the United States.

Energy Transfer Processes Between the Surface and the Atmosphere

The atmosphere and surface continually exchange energy with each other. Much of this energy exchange is accomplished by the emission and absorption of radiation, but other mechanisms are also at work. This section describes the processes by which the energy transfer occurs.

Surface–Atmosphere Radiation Exchange

Like all other objects at terrestrial temperatures, Earth's surface and atmosphere radiate energy almost completely in the longwave (primarily thermal infrared) portion of the spectrum. Any discussion of longwave energy transfer is somewhat more complex than that of solar radiation because longwave energy has no obvious beginning or end point.

Longwave radiation emitted by Earth's surface is largely absorbed by the atmosphere. This increases the temperature of the atmosphere, which causes it to radiate more energy outward. The energy radiated by the atmosphere is transferred in all directions, including downward, and so the surface receives a considerable portion of this radiation. This causes further surface heating, which leads to an increase in longwave radiation emission from the surface, which again warms the atmosphere, and so on and so on. In other words, there is an infinite cycle of exchange, with energy constantly transferring back and forth.[1]

Figure 3–9 describes the globally averaged amount of longwave radiation exchanged between the atmosphere and

---

[1]Although tempting to do so, we should not think of longwave radiation being cycled back and forth, or "bouncing" between the surface and atmosphere. First, natural surfaces do not reflect much longwave radiation. Second, when radiation is absorbed, the energy no longer exists as radiation, so it can hardly bounce anywhere.

▲ **FIGURE 3-9** The disposition of longwave radiation (indicated as LW on the figure) between the surface and atmosphere is depicted, based on the same units of energy as in Figure 3–8. The surface radiates 116 units upward and receives 98 from the atmosphere, for a net longwave radiation loss of 18 units. The atmosphere radiates 156 units and receives 104 from the surface, for a net deficit of 52 units.

the surface. Beginning with the surface, we see that 116 units of longwave radiation are emitted up toward the atmosphere, with the greatest part (104 units) being absorbed by the atmosphere. The clear atmosphere absorbs longwave radiation far better than solar radiation, mainly due to the presence of water vapor and carbon dioxide. As shown in Figure 3–10, both of these gases are good absorbers of longwave radiation, with strong absorption bands in the longwave part of the spectrum (see *Box 3–1, Physical Principles: Selective Absorption by Water Vapor and Carbon Dioxide*).

Although water vapor, carbon dioxide, and other greenhouse gases are good at absorbing most wavelengths of longwave radiation, a portion of the longwave spectrum can pass through the atmosphere relatively unimpeded. Interestingly, wavelengths in this band, 8 to 12 μm, happen to match those radiated with greatest intensity by Earth's surface. This range of wavelengths, not readily absorbed by atmospheric gases, is called the **atmospheric window**. The atmospheric window must not be thought of as a place in the atmosphere or the absence of some gas; it represents just a certain range of wavelengths of special importance to the radiation balance.

Although the gases of the atmosphere are not effective at absorbing the wavelengths in the atmospheric window, clouds (even those of fairly modest thickness) readily absorb virtually all longwave radiation. This explains why cloudy nights do not cool off nearly as rapidly as do clear nights. When the evening sky is overcast, the cloud cover absorbs a large portion of the energy that otherwise would escape to space. Warmed by this energy, the clouds emit longwave radiation downward to the surface and lesser amounts upward to space. The clouds thus act something like a blanket, helping retain heat.

(a)

(b)

◀ **FIGURE 3-10** Earth's surface acts nearly as a blackbody in its emission of radiation (a), but the gases of the atmosphere absorb most of the energy with wavelengths outside of the range of 8 to 12 μm. The shaded area in (a) indicates the energy absorbed by the atmosphere. Figure (b) shows the effectiveness of individual gases in absorbing the energy. The percentage of area shaded indicates the percentage of longwave energy absorbed.

3–1 PHYSICAL PRINCIPLES

Selective Absorption by Water Vapor and Carbon Dioxide

Water vapor and carbon dioxide are the two gases that absorb the most longwave radiation emitted by the surface (see Figure 3–10). Why are these gases so selective, being nearly transparent for shortwave radiation but nearly opaque for longwave? Recall from Chapter 2 that isolated atoms have discrete energy states, with only certain energy states possible. As energy is absorbed and emitted by a gas molecule, its energy state rises and falls by discrete amounts from one allowable state to another. We have also seen that the energy associated with a photon of radiation is discrete and depends on its wavelength. Knowing the wavelength, we know the energy level of the photon.

It must be, therefore, that gas molecules absorb only certain photons, namely those that push the molecule into allowable energy states. Photons with higher or lower energy values will not be absorbed but will instead pass through the gas. Because a unique wavelength is associated with every energy level, this is equivalent to saying that only certain wavelengths can be absorbed by any particular gas. (The same is not true of liquids and solids, whose molecules interact to give much more continuous absorption.) Whether or not a particular wavelength can be absorbed depends on the molecular structure of the absorber (the configuration of electrons, etc.). As it happens, the gases in the atmosphere do not have strong absorption bands in the visible part of the spectrum. But some of them, including water vapor and carbon dioxide, do have molecular structures that permit absorption of longwave radiation. Combined, the various gases absorb most of the longwave energy passing through the atmosphere.

Refer again to Figure 3–9. The energy emitted by the atmosphere amounts to 156 units, of which 98 are directed down toward the surface and 58 are radiated to space. Notice that the longwave energy lost from the atmosphere exceeds the amount it absorbs from the surface. The difference between absorbed and emitted longwave radiation is referred to as the **net longwave radiation**. For the atmosphere, the net longwave radiation is a negative 52 (104–156) units.

Did You Know?

Recent examination of the solar radiation record across the globe suggests that between the late 1950s and early 1990s the amount of solar radiation reaching the surface may have decreased substantially—possibly by as much as 10 percent. Other studies have suggested smaller reductions (it is hard to determine the exact value, partly because of a paucity of stations across the globe), but the evidence of some amount of "dimming" is fairly strong. If the reduction in sunlight reaching the surface really has occurred, a likely contributor was an increase in aerosols emitted into the atmosphere by human activity. These aerosols are thought to have had a direct effect by absorbing and scattering solar radiation back to space, as well as an indirect effect by changing cloud properties. It now appears that environmental regulations instituted in many countries have led to a reduction in human-introduced aerosols, which has led to a recovery in the amount of solar radiation reaching the surface. While having many positive effects, the increase in incoming radiation could also contribute to future global warming.

Similarly, the surface receives 98 units of longwave radiation, but this amount is exceeded by the 116 units that are radiated, for a net longwave radiation deficit of 18 units.

Even though shortwave and longwave radiation undergo different amounts of absorption and reflection, they are not separate entities as far as the heating of the atmosphere and surface are concerned. When either is absorbed, the absorber is warmed. It is therefore natural to combine longwave and shortwave into **net all wave radiation**, or simply **net radiation**, defined as the difference between absorbed and emitted radiation, or equivalently, the net energy gained or lost by radiation.

Figure 3–11 summarizes the net radiation balance for Earth. The atmosphere absorbs 23 units of solar radiation but undergoes a net loss of 52 units of thermal radiation, for a net deficit of 29 units. The surface absorbs 47 units of solar radiation but has a longwave deficit of 18, resulting in a net radiative surplus of 29 units. In other words, the atmosphere has a net deficit of radiative energy exactly equal to the surplus attained by the surface.

If radiation were the only means of exchanging energy, the surplus of radiative energy obtained by the surface would result in a perpetual warming, while the deficit of the atmosphere would lead to a continual cooling. Eventually our feet would be scorched by a terrifically hot ground while the rest of our bodies would freeze, surrounded by a bitterly cold atmosphere. This, of course, is not about to happen, because energy is transferred from the surface to the atmosphere and within the atmosphere by two other forms of heat transfer: conduction and convection. The net transfer of energy by these two processes

▲ **FIGURE 3-11** Net radiation is the end result of the absorption of shortwave radiation and the absorption and radiation of longwave radiation. The surface has a net radiation surplus of 29 units, while the atmosphere has a deficit of 29 units.

allows the radiation surplus at the surface to be eliminated while at the same time offsetting the radiation deficit of the atmosphere.

Checkpoint

1. What is the atmospheric window?
2. Explain how the atmosphere and surface end up with net losses of longwave radiation.

 TUTORIAL

GLOBAL ENERGY BALANCE

Use the animations in Sections 2.2, 2.3, and 4 to follow energy transfers between surface and atmosphere.

Conduction

Conduction, described in general terms in Chapter 2, helps transfer energy near the surface. As radiant energy is absorbed by a solid Earth surface during the middle of the day, a temperature gradient (a rate of change of temperature over distance) develops in the upper few centimeters of the ground. In other words, temperatures near the surface become greater than those a few centimeters below. As a result, conduction transfers energy downward. At night the reverse process occurs; the topmost portion of the ground cools by the loss of longwave radiation and becomes colder than the underlying volume. Energy is then conducted upward.

Warming of the ground during the day also sets up a temperature gradient within a very thin, adjacent sliver of air called the **laminar boundary layer**. Although air is usually highly mobile and capable of being easily mixed, very thin layers on the order of a few millimeters in thickness resist mixing. During the middle of the day, very strong temperature gradients can therefore develop in the laminar boundary layer, through which a substantial amount of conduction can occur. Energy conducted through the laminar boundary layer is then distributed through the rest of the atmosphere by a mixing process called *convection*.

Convection

Convection is a process whereby heat is transferred by the bodily movement of a fluid—that is, a liquid or gas. In contrast to conduction, convection involves the actual displacement of molecules. Unlike conduction, which transfers energy from the surface to the atmosphere, convection circulates this heat between the very lowest and the remaining portions of the atmosphere. The direction of heat transfer is upward when the surface temperature exceeds the air temperature (the normal situation in the middle of the day). At night the surface typically cools more rapidly than the air, and energy is transferred downward. Convection can be generated by two processes in fluids: local heating (free convection) and mechanical stirring (forced convection).

Free Convection **Free convection** is the mixing process related to buoyancy, the tendency for a lighter fluid to float upward when surrounded by a denser fluid. Recall your days as a child, when you could annoy your parents by blowing bubbles through a straw in a glass of milk. As the air was injected into the milk, it would immediately rise upward because of its lesser density and cause turbulent mixing. This was free convection at work.

Column of air expanding over hot surface

Air flow replacing rising air

(a)

(b)

▲ **FIGURE 3-12** Convection (a) is a heat transfer mechanism involving the mixing of a fluid. In free convection, local heating can cause a parcel of air to rise and be replaced by adjacent air. Free convection can create updrafts able to keep a hawk airborne (b) without it having to flap its wings.

Free convection (shown in Figure 3–12a) often occurs when a localized parcel of air is heated more than the nearby air. Because warm air is less dense than cold, it is relatively buoyant and rises. On a warm summer day, we can see the effect of free convection by observing a circling hawk (Figure 3–12b) that stays airborne without flapping its wings. This flight is possible because the hawk's wings are designed to catch the rising parcels of buoyant air that carry it upward. Convection can have far more important impacts than helping to keep hawks in the air—it can lead to intense precipitation.

Forced Convection **Forced convection** (also called **mechanical turbulence**) occurs when a fluid breaks into disorganized swirling motions as it undergoes a large-scale flow. When water flows through a river channel, for example, it does not flow uniformly, as would a very thick syrup. Instead, the flow breaks down into numerous *eddies*. Forced convection in the atmosphere is shown in Figure 3–13. Horizontally moving air undergoes the same type of turbulence. Instead

of moving as a uniform mass, the air breaks up into numerous small parcels, each with its own speed and direction that are superimposed on the larger-scale flow. Because there is a strong vertical component to the eddy motions, the forced convection helps transport energy from the top of the laminar boundary layer upward during the day.

Generally speaking, higher wind speeds generate greater forced convection. Mechanical turbulence is also enhanced when air flows across rough surfaces (for example, forests and cities) rather than smooth ones such as glaciers. Both free and forced convection transfer two types of energy: sensible heat and latent heat.

Sensible Heat

The transfer of energy as **sensible heat** is simple. When energy is added to a substance, an increase in temperature can occur that we physically sense (hence the term *sensible*). This is what you experience when you rest outside on a warm, sunny day; the increase in your skin temperature results from a gain in sensible heat. The magnitude of temperature increase is related to two factors, the first of which is **specific heat**, defined as the amount of energy needed to produce a given temperature change per unit mass of the substance. In SI units, specific heat is expressed in joules per kilogram per kelvin. Everything else being equal, a substance with high specific heat warms slowly because much energy is required to produce a given temperature change. Likewise, it also takes longer for a substance with a high specific heat to cool off, assuming the same rate of energy loss.

The temperature increase resulting from a surplus of energy receipt also depends on the *mass* of a substance. Not surprisingly, a given input of heat results in a greater rise in temperature if it is applied to only a small amount of mass. For example, compare the amount of energy needed to boil

▲ **FIGURE 3-13** Forced convection. Air is forced to mix vertically because of its low viscosity (ability to be held together) and the deflection of wind by surface features.

water for a cup of tea to that needed to take a warm bath. In just a few minutes, a single burner on a kitchen stove can have the water ready for the tea, but your hot water heater must supply considerably more energy for the large amount of bath water. These relationships are shown in Figure 3–14.

Sensible heat travels by conduction through the laminar boundary layer and is then dispersed upward by convection. Through these mechanisms, 5 of the 29 units of net radiation surplus for the surface are transferred to the atmosphere, where they help offset the net radiation deficit. The remaining 24 units are transferred to the atmosphere by the convection of heat in another form.

Latent Heat

Latent heat is the energy required to change the phase of a substance (that is, its state as a solid, liquid, or gas). In meteorology we are concerned almost exclusively with the heat involved in the phase changes of water.

Recall that all physical processes require energy. The evaporation of water and the melting of ice are no exceptions to this rule—for either process, energy must be supplied. In the case of melting ice, the energy is called the *latent heat of fusion*. For the change of phase from liquid to gas, the energy is called the *latent heat of evaporation*. It takes seven and a half times more energy (2,500,000 joules) to evaporate a kilogram of liquid water than it does to melt the same amount of ice (335,000 joules). Although both forms of latent heat can be important locally, on a global scale the latent heat of evaporation is far more influential.

When radiation is received at the surface, it can raise the temperature of the land or the water. If water happens to exist at the surface (or can be brought up from below the surface through the root systems of plants), some of the energy that might have been used to increase the surface temperature is instead used to evaporate some of the water. This results in a smaller temperature increase than would occur for a dry surface. You have probably experienced this while walking

▲ **FIGURE 3–14** The heat content of a substance depends on several factors. In (a) the input of 4190 J of energy to a kilogram of water increases its temperature 1 °C, while a doubling of the energy input causes twice as much heating. The specific heat of a substance also influences the amount of temperature change resulting from an input of energy. In (b) the application of 4190 J to 1 kg of sandy soil produces more than five times the increase in temperature than it would for a kilogram of water. The amount of mass also affects the temperature change accruing for a given energy input. Note in (c) that the temperature increase for 2 kg of water is half as much as that for 1 kg.

barefoot on a hot pavement. If the pavement is watered down, the surface cools as energy is taken from the ground and used for evaporation. The amount of energy consumed can be quite large, as much as 90 percent of absorbed solar radiation for a completely wet surface.

Let's use another common example to illustrate the concept of latent heat. We all know that perspiration is a mechanism that lowers our body temperatures and keeps us from overheating. But how does it work? Clearly, it is not a matter of sweat being cold—it is as warm as the body producing it. The reason sweat cools a person is latent heat. As you exercise, the heat produced as a by-product causes your body temperature to rise. However, if your skin is covered with water and that water is free to evaporate, some of the energy produced by your body is used to evaporate the moisture rather than increase your body temperature.

Did You Know?

When you drop an ice cube in a glass of water, the cooling that results is not primarily due to the lower temperature of the ice cube. It is mostly due to the latent heat of fusion involved in melting the ice. The energy used to melt the ice is taken from the water, thereby lowering its temperature.

The energy needed to evaporate water or melt ice is said to be latent because it does not disappear. It is held, "latent" in the atmosphere, to be released later when the reverse process occurs—the condensation of water vapor into liquid cloud or fog droplets. In effect, then, the evaporation of water makes energy available to the atmosphere that otherwise would have warmed the surface. It thus acts as an energy transfer mechanism, taking heat from the surface to the atmosphere. On a global average basis, the amount of energy transferred to the atmosphere as latent heat amounts to 24 units, which makes latent heat considerably more important as a mode of heat transfer than sensible heat (5 units). Perhaps this is not surprising, given that the planet is mostly covered by ocean. The mean annual values of the net radiation components, latent and sensible heat, are depicted in Figure 3–15.

Checkpoint

1. What is the role of convection in the atmosphere?
2. How would Earth's average surface temperature be different if radiation were the only means of exchanging energy?

Net Radiation and Global Temperature

The balance between incoming and outgoing radiation is not merely fortuitous; physical laws dictate that it must be the case. To understand why, consider what would happen if the Sun were suddenly to increase its radiative output and raise the temperature of Earth. Because warm bodies radiate

▲ **FIGURE 3–15** Both the surface and atmosphere lose exactly as much energy as they gain. The surface has a surplus of 29 units of net radiation, which is offset by the transfer of sensible and latent heat to the atmosphere. The atmosphere offsets its 29 units of radiation deficit by the receipt of sensible and latent heat from the surface.

more energy than cooler ones, the planet would respond by emitting more longwave energy back to space. As long as the incoming energy from the Sun were to exceed that emitted by Earth, the global temperature would continue to rise, and emission to space would increase accordingly. Eventually the planetary temperature would increase to the point where outgoing energy equaled incoming energy, and a new equilibrium temperature would be established. The individual values in Figure 3–15 would all be changed, but the global inputs and outputs of radiation would still sum to zero.

Latitudinal Variations

The overall balance between incoming and outgoing radiation discussed in previous sections applies to the planet as a whole but not to any particular place. Figure 3–16 shows that the balance between incoming and outgoing radiation varies with latitude. On an annual basis, the Earth surface–atmosphere system gains more radiation than it loses between about 38° north and south latitudes, and it loses more than it gains poleward of these two parallels. The boundary between zones of radiant energy surpluses and deficits migrates seasonally. During the Northern Hemisphere summer, most of the area north of about 15° S gains more radiant energy than it loses. During the Northern Hemisphere winter, most areas south of about 15° N take in more radiation than they emit.

If no other processes were involved, the net gain of radiation between 38° north and south latitudes and the deficits outside this zone would cause the tropics and subtropics to undergo continual heating and the extratropical regions to constantly cool. This does not occur, however, because the energy surplus at low latitudes is offset by the horizontal movement, or **advection**, of heat poleward. This transfer is accomplished primarily by the global wind systems (75 percent), and secondarily by

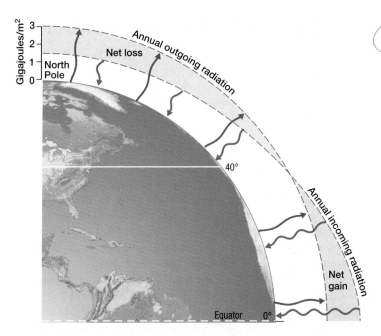

▲ **FIGURE 3-16** Annual average net radiation values for the atmosphere and surface combined. At latitudes between about 38° north and south, a radiant energy surplus exists. Poleward of these latitudes, the atmosphere and surface lose more radiant energy than is gained.

the oceanic currents (25 percent), as illustrated in Figure 3–17. As the wind and water currents move, they carry with them their internal heat, which is redistributed across the globe.

The Greenhouse Effect

The interactions that warm the atmosphere are often collectively referred to as the **greenhouse effect**, but the analogy to a greenhouse is not strictly accurate. Greenhouses, such as the one shown in Figure 3–18, are made primarily of glass, which is transparent to incoming shortwave radiation but opaque to outgoing longwave radiation. The glass therefore allows in more radiation than is allowed to escape, causing the temperature inside the structure to be warmer than outside. In that regard, a greenhouse is similar to the atmosphere, which also transmits most of the incoming solar energy but absorbs the vast majority of longwave radiation emitted upward by the surface.

The analogy breaks down, however, when we incorporate the effect of convection. A greenhouse not only reduces the loss of energy by longwave radiation but also prevents the loss of sensible and latent heat by convection. In contrast, the *greenhouse gases* (those that absorb longwave radiation) of the atmosphere do not impede the transfer of latent and sensible heat. Thus, it would be more accurate if the term "greenhouse effect" were replaced by "atmospheric effect."

If the atmosphere had none of the "greenhouse gases" that absorb outgoing longwave radiation, Earth would be considerably colder, on average, and the temperature would oscillate wildly from day to night. In fact, without greenhouse gases and clouds, Earth's surface would have an average temperature of −18 °C (0 °F), rather than the observed mean temperature of 15 °C (59 °F). The greenhouse effect keeps Earth warmer by absorbing most of the longwave radiation emitted

▲ **FIGURE 3-17** The circulation of ocean currents. Those moving warm water are depicted by red arrows, those moving cold water by blue arrows.

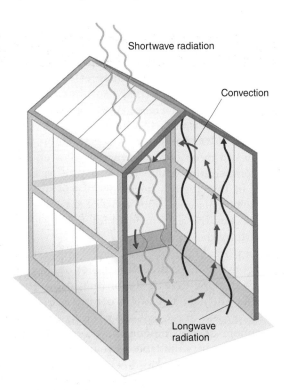

Shortwave radiation

Convection

Longwave radiation

▲ **FIGURE 3–18** The air inside a greenhouse is warmer than that outside because glass allows solar radiation to enter but is opaque to outgoing longwave radiation. It also precludes the movement of heat away from the surface by convection. This latter effect makes the action of a greenhouse different from that of the atmosphere. Therefore, the term "greenhouse effect" is not completely appropriate when applied to the atmosphere.

by the surface, thereby warming the lower atmosphere, which in turn emits radiation downward. (A more quantitative discussion of this topic is presented in *Box 3–2, Physical Principles: Earth's Equilibrium Temperature*.)

 TUTORIAL
GLOBAL ENERGY BALANCE

Use the animations in Section 6 to discover how greenhouse gases affect the temperature of the atmosphere.

These days you find considerable discussion in the media about the possibility of climatic warming due to the anthropogenic increase of greenhouse gases such as carbon dioxide and methane. We briefly discuss how human activities have contributed to a warming atmosphere later in this chapter. A more general discussion of climatic change appears in Chapter 16.

Checkpoint

> 1. What is one drawback of using a greenhouse as an analogy for how energy behaves in the atmosphere?
> 2. What role does the greenhouse effect play in keeping Earth a habitable planet?

Global Temperature Distributions

One of the most immediate and obvious outcomes of radiation gain or loss is a change in the air temperature. Figure 3–19 shows the mean air temperature distributions for January and July and the differences between the two.

Each line on the maps, called an **isotherm**, connects points of equal temperature. Several patterns of large-scale temperature are apparent in (a) and (b). First, as expected, temperatures tend to decrease poleward in both hemispheres. Second, the latitudinal temperature gradient is greatest in the hemisphere experiencing winter (that is, the Northern Hemisphere in January and the Southern Hemisphere in July). The strong gradients in the winter hemisphere occur because the midday sun angles and the length of day both decrease with latitude. During summer, the lower midday sun angles at the higher latitudes are offset by longer days, so the temperature gradients are relatively weak.

The third feature of the maps is that the isotherms shift poleward over land in the hemisphere experiencing summer and shift equatorward over land during the winter. In other words, temperatures over land tend to be warmer in the summer than over adjacent water bodies, and colder during winter. Finally, the Northern Hemisphere has a steeper temperature gradient in its winter (a) than the Southern Hemisphere does in its winter (b). This is because the Southern Hemisphere consists of a much greater proportion of ocean than land. As you can see from (c), landmasses have much greater annual ranges in temperature than do ocean bodies, for reasons discussed later in this chapter.

Influences on Temperature

Certain geographical factors combine to influence temperature patterns across the globe. These factors include latitude, altitude, atmospheric circulation patterns, continentality, ocean current characteristics along coastal locations, and local conditions.

Latitude

Most people know that outside the tropics, the annual mean temperature decreases with latitude—Santa Claus, living at the North Pole, must have a wardrobe suitable for extremely cold conditions. Not only does latitude influence the average temperature, it also affects seasonal patterns. As described in Chapter 2, the tilt of Earth's axis influences the amount of solar radiation available at any latitude on any particular day. Within the tropics, there is relatively little annual variation in the length of day and the midday solar angle, so energy receipts exhibit little change through the course of the year. Outside the tropics, the noontime solar angles exhibit a range of 47°, with the lowest solar angles coinciding with the periods of shorter days. As a result, the availability of incoming radiation (and therefore the temperature) is more variable as distance from the equator increases.

3–2 PHYSICAL PRINCIPLES

Earth's Equilibrium Temperature

If Earth had no atmosphere, and therefore no greenhouse effect (see page 69), the mean temperature of the planet would be much colder. Using the principles discussed thus far, we can easily estimate the magnitude of the greenhouse effect. To do so, we will compute the equilibrium temperature for a planet having no atmosphere. By comparing the computed and observed temperatures, we will see how the atmosphere influences Earth's temperature.

First, assume the planet acts as a blackbody with regard to longwave radiation, that the planetary albedo is 30 percent, and that the solar constant is 1367 watts per square meter. If Earth were a flat disk perpendicular to incoming radiation, each square meter would receive 1367 joules/second. But Earth is not a flat disk; it is a sphere, the surface area of which is four times larger than that of a disk of the same radius. Thus the intensity of radiation averaged over the sphere is one-fourth as large as for the imaginary disk. Because of this, each square meter of Earth receives 1367/4, or 342 watts/m². (These 342 watts per square meter are the

100 units of solar radiation discussed in the section, "The Fate of Solar Radiation," earlier in this chapter.) Given the planetary albedo of 30 percent, it must be that 70 percent of this incoming radiation is absorbed. In other words, total absorbed radiation is

$$1367 \text{ watts/m}^2 \times 0.25 \times 0.7 = 239.2 \text{ watts/m}^2$$

The planet must lose exactly as much energy as it gains, and the intensity of radiation for a blackbody is determined by rearranging and applying the Stefan-Boltzmann law. Recall that the Stefan-Boltzmann law for a blackbody states that

$$I = \sigma T^4$$

We know, however, that the intensity of radiation for the planet without an atmosphere must be 239.2 watts per square meter. We then rearrange the equation to solve for T, rather than I, to get

$$T^4 = I/\sigma$$

which can be reduced to

$$T = (I/\sigma)^{0.25}$$

Using the values, $\sigma = 5.67 \times 10^{-8}$ watts/(m²K⁴) and $I = 239.2$ watts/m²,

the equilibrium temperature works out to 254.9 K (−18.3 °C, 0 °F) Thus, the mean temperature of Earth would be far colder without an atmosphere.

Note that our calculation is highly simplified and somewhat questionable. For example, we used 30 percent for the planetary albedo, but that value arises in part from the albedo of the atmosphere, which our imaginary planet lacks. Should we therefore have used present-day surface albedo in the computation? Perhaps, but surface albedo on the real Earth is in part the result of temperature, the very thing we are trying to compute! Ideally, we would treat albedo as a variable, allowing it to respond to changes in temperature.

We see that even a beginning question about the global effect of greenhouse gases raises complications that are not easy to address with a simple model. Given this, it is not surprising that realistic computer models of atmospheric behavior are enormously complex, requiring huge computer resources. Nonetheless, computer models have become indispensable for daily weather forecasting (Chapter 13) and tell us much of what we know about potential climatic changes due to human activity (Chapter 16).

Altitude and Elevation

Any point within the atmosphere has some particular *altitude* (its height above mean sea level). Altitude is not synonymous with *elevation*, the distance above sea level for a land surface. For example, some particular city may have an elevation of 1000 m above sea level, while air that is 1000 m above the city would have an altitude of 2000 m. Altitude and elevation both deal with position relative to sea level, but altitude is related to the atmosphere and elevation refers to land.

As was shown in Figure 1–12, temperatures in the troposphere typically decrease with altitude above sea level. This occurs because the surface is the primary source of direct heating for the troposphere, and increased altitude implies a greater distance from the energy source. Figure 3–20 contrasts day and night temperatures at three locations. Position A is located a couple of meters (about 6 ft) above the surface at sea level. Position B is 3000 m (about 10,000 ft) directly above A, and Position C is 3000 m above sea level but just a couple of meters above the mountain surface.

During the middle of the day, Position A responds to the absorption of solar radiation at the surface and warms as the surface transfers energy upward by convection and the emission of longwave energy. Position B, a considerable distance from the surface, undergoes virtually no warming. Position C, although at the same altitude as B, is nearer to the primary source of warming, and its daytime temperature rises appreciably.

At night, the surface below Position A cools by the emission of longwave radiation. The air 2 m above the surface also undergoes a lowering of temperature in response to the cooling of the underlying surface. At Position B, the air undergoes little cooling because of its distance from the surface. Position C chills rapidly because the sparse atmosphere above does not effectively absorb the outgoing radiation from the surface. Cooling is often enhanced by rapid evaporation into the drier high-altitude air. Perhaps while on a mountain camping trip you've noticed how rapidly a hot meal becomes stone cold. This is a consequence of strong evaporation and weak

▲ **FIGURE 3-19** Distribution of mean January (a) and July (b) surface air temperatures. The difference in temperature between the two months is shown in (c).

(a)

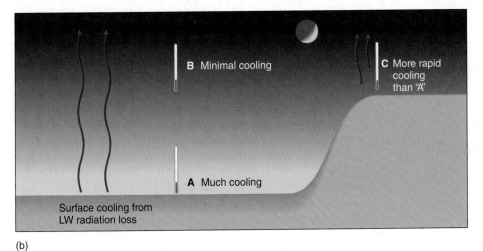

(b)

▲**FIGURE 3-20** Effects of elevation and altitude on daily temperature patterns.

atmospheric counterradiation, compared to lower elevations. The overall effect of elevation above sea level and altitude above the ground, therefore, is that cooling and warming cycles are minimal high above the surface, and the air just above a high-elevation surface will undergo greater cycles of cooling and warming than will air at the same altitude but farther above the surface. It is not uncommon that the nighttime loss of longwave radiation at the surface can lead to lower temperatures near the surface than aloft. This reversal of the normal pattern in the troposphere (lower temperatures with increasing distance from the surface) is known as an **inversion**. The characteristics of inversions are described in Chapter 6.

Atmospheric Circulation Patterns

As we will discuss in Chapter 8, an organized pattern of mean atmospheric pressure and air flow across the globe strongly influences the movement of warm and cold air, with a direct effect on temperature. These large-scale circulation patterns also influence the development of cloud cover, which has an indirect effect on temperature. Subtropical areas (latitudes 20° to 30° in both hemispheres), for example, tend to be regions of minimal cloud cover, and insolation passing through the atmosphere undergoes less attenuation on its way to the surface. In contrast, equatorial regions are often cloudy in the afternoon and experience a greater attenuation of incoming solar radiation. The result is that the highest temperatures on Earth tend to occur not at the equator but in the subtropics. Many other patterns of atmospheric circulation affect regional temperatures.

Contrasts Between Land and Water

Because the atmosphere is heated primarily from below, it should be no surprise that the type of surface influences air temperature. The greatest influence arises because of contrasts between land and water. Water bodies are far more conservative than land surfaces with regard to their temperature, meaning that they take longer to warm and cool when subjected to comparable energy gains and losses.

San Francisco, California, and St. Louis, Missouri, together provide an interesting example of the effect of **continentality**—the effect of an inland location that favors greater temperature extremes. They are at similar latitudes and elevations and both are subject to a predominantly west-to-east airflow. Yet San Francisco, situated along the Pacific Coast, has more moderate temperatures than does its inland counterpart. During July, for example, its mean temperature is 15 °C (59 °F), while St. Louis averages 26 °C (79 °F). The temperature difference is similar but runs in the opposite direction in January when San Francisco has an average temperature of 10 °C (50 °F) and St. Louis a mean of 0 °C (32 °F).

Four reasons cause water bodies to be more conservative than landmasses with regard to temperature:

1. The specific heat of water is about five times as great as that of land.

2. Radiation received at the surface of a water body can penetrate to several tens of meters deep and distribute its energy throughout a very large mass. In contrast, the insolation absorbed by land heats only a very thin, opaque surface layer.

3. The warming of a water surface can be reduced considerably because of the vast supply of water available for evaporation. Because much energy is used in the evaporative process, less warming occurs.

4. Unlike solid land surfaces, water can be easily mixed both vertically and horizontally, allowing energy surpluses from one area to flow to regions of lower temperature.

Warm and Cold Ocean Currents

Figure 3–17 shows warm and cold ocean currents. The warm currents typically move poleward in the western portion of the ocean basins near the east coasts of continents in the middle latitudes, carrying large amounts of energy with them. Similarly, along the eastern margins of oceans, cold ocean currents dominate in the middle latitudes. Where the water temperatures are high, heat is transferred to the atmosphere and promotes higher air temperatures. Thus, the existence of a warm ocean current offshore can cause a location along the east coast of a continent to have higher temperatures than would a cold current offshore along a west coast.

Compare, for example, Los Angeles, California, and Charleston, South Carolina, two coastal locations at similar latitudes and elevations. Summers are considerably warmer at Charleston than at Los Angeles, largely (but not entirely) because the temperature of the ocean off Charleston is higher than the temperature of the ocean off Los Angeles. The high water temperatures of the Gulf of Mexico and the western margin of the Atlantic Ocean allow the transfer of an enormous amount of heat to the atmosphere, and the average July temperature in Charleston of 31 °C (88 °F) is substantially warmer than that of Los Angeles (23 °C; 73 °F). In winter, temperatures are lower at Charleston than at Los Angeles because

the westerly winds blowing toward Los Angeles are subject to the moderating effects of the Pacific, whereas Charleston's winter temperatures are affected by colder prevailing winds passing over the continental interior. Thus, the influence of atmospheric circulation, which we noted previously, can interact with the position along the edge of a continent in influencing temperature patterns.

Local Conditions

A number of site-specific factors, such as slope orientation and steepness, can influence the temperature characteristics of an area. In the Northern Hemisphere, slopes that are south-facing receive midday sunlight at a more direct angle than do those oriented in other directions, thereby promoting greater energy receipt and higher surface temperatures. The heating of south-facing slopes often results in a greater amount of drying than on the opposite, north-facing slopes. Vegetation patterns often respond to the change in microclimate, with plants intolerant of dry conditions occupying north-facing slopes. Such a pattern is shown in Figure 3–21.

Densely wooded areas also have different temperature regimes than areas devoid of vegetation cover. In a region like that shown in Figure 3–22, a tall, dense vegetation cover reduces the amount of sunlight hitting the surface during the day and considerable evaporation of water from leaf surfaces occurs. At night the plant canopy reduces the net longwave radiation losses. These factors lead to lower daytime temperatures but warmer evenings.

The effects of vegetation on local climate can often be put to use in a manner that increases human comfort. For example, it is often a good idea to plant deciduous trees on the equatorward side of houses. During the warm summer months, trees can cast shade on houses, thereby keeping interior temperatures down or reducing air conditioning costs. When winter arrives, the loss of the leaves from the deciduous trees minimizes the reduction in sunlight, and this can help keep the building warm.

Checkpoint

1. How can an ocean current affect the temperature of nearby land?

2. Seattle, Washington, and Fargo, North Dakota, are both at similar latitudes. Which would you expect to have the lower temperatures in January? Why?

Daily and Annual Temperature Patterns

The principles of heat transfer discussed in Chapter 2 and earlier in this chapter directly affect the daily (often called *diurnal*) and seasonal temperature changes that occur at any location. Let's first consider typical daily patterns.

◄**FIGURE 3–21** Slope aspect is one of the local factors affecting temperature. In the Northern Hemisphere, north-facing slopes typically receive less intense daytime heating and therefore exhibit lower temperatures. This retards the rate of surface evaporation, making more water available for plant life. Looking east in this photo taken in San Diego County, California, you can see denser vegetation cover on the north-facing slopes (right) than on the south-facing slopes (left).

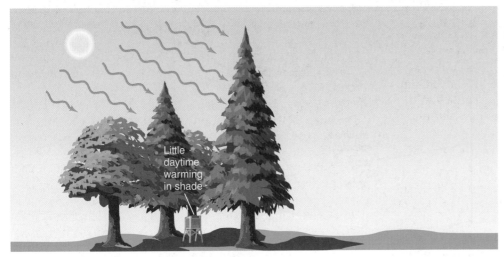

◄**FIGURE 3–22** A dense vegetation cover lowers daytime temperatures because of its shadowing effect on incoming solar radiation (a). At night (b) the forest canopy retards the loss of longwave radiation to space, resulting in higher nighttime temperatures than in the open.

Little daytime warming in shade

(a)

LW radiation loss

Little nighttime cooling by LW loss

(b)

Daytime Heating and Nighttime Cooling

First consider what happens over a 24-hour period on a cloudless day, beginning at sunrise (Figure 3–23a). As the Sun emerges above the horizon, the incoming energy does not effectively heat the surface due to the beam spreading and atmospheric depletion associated with the low solar angle. At sunrise and for some time period afterward, the receipt of solar radiation does not offset the loss of longwave radiation at the ground, leading to steady or slowly decreasing temperatures. (Though some people may assume that temperatures begin to rise as soon as the Sun comes up, such is not the case.) As the morning proceeds, the Sun sweeps a path from where it rose over the eastern sky toward the south (or north in the Southern Hemisphere), while at the same time rising upward from the horizon. Thus, as the outgoing longwave radiation remains roughly constant, the incoming solar radiation increases. At some point the Sun rises far enough above the horizon for its energy to offset the loss of longwave radiation, and warming occurs.

How long after sunrise does the ground begin to warm? That depends on several factors, including the rate at which the solar angle increases over the course of the morning, which in turn varies with latitude and time of year. Other factors include air temperature, wind speed, surface wetness, and the amount of heat conducted through the ground, to name a few.

At solar noon the Sun achieves its greatest angle above the horizon and the surface receives its greatest input of solar radiation. But experience tells us that noon is not the warmest time of the day, and that maximum air temperatures (as measured at the usual 5 ft above the surface) are more likely to occur at least a couple hours afterward. Recall that surface temperature increases as long as energy gained by the surface is greater than energy lost. Although solar radiation begins to decrease after noon, the net energy gain by all processes often exceeds the loss for some time afterward, and warming continues. Eventually the energy balance becomes negative, and the surface cools. Typically the cooling rate is much slower than the previous warming, largely because as the surface becomes colder than the ground beneath heat is conducted upward. This is not enough to prevent surface temperatures from falling, but it does slow the decrease.

▶ **FIGURE 3–23** Surface temperature increases whenever the energy gains exceed energy losses. On clear days (a) the availability of solar radiation in the middle of the day produces a large surplus that persists into the afternoon, but at night the longwave radiation loss results in substantial cooling. Overcast conditions (b) suppress the diurnal change in temperature mainly by reducing the incoming solar radiation gain during the day and by supplying more downward longwave radiation from the atmosphere at night.

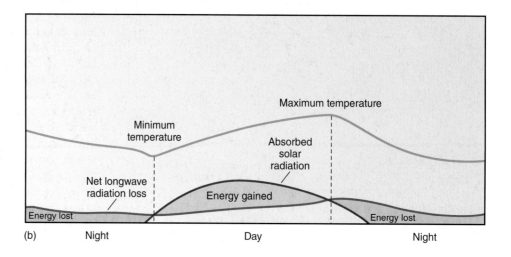

Ground heat flux to the surface continues through the night for as long as the surface temperature continues to fall. These trends of surface temperature are largely mirrored by air temperature near the ground, except that the amplitude of daily change is less, and the time of maximum air temperature lags farther behind solar noon.

Effects of Cloud Cover and Wind

The amplitude of the daily temperature pattern is reduced under overcast conditions. During the day the cloud cover can greatly reduce the daytime input of solar radiation and likewise reduce the magnitude of the net longwave radiation loss over the entire 24-hour period (Figure 3–23b). The overall effect is to lower the daytime maximum temperature while keeping nighttime temperatures somewhat higher than they would be on a clear night. Overall, the temperature difference over the 24-hour period is much less than in the previous example.

Time (CST)	Temperature °F (°C)	Wind (mph)
9 P.M.	41.0 (5.0)	W 16
10 P.M.	39.0 (3.9)	WNW 26
11 P.M.	39.0 (3.9)	W 26
12 A.M.	37.9 (3.3)	W 24
1 A.M.	37.9 (3.3)	W 21
2 A.M.	37.9 (3.3)	W 17
3 A.M.	37.9 (3.3)	W 20
4 A.M.	37.9 (3.3)	W 21
5 A.M.	37.9 (3.3)	W 20
6 A.M.	37.9 (3.3)	W 23
7 A.M.	37.9 (3.3)	W 20
8 A.M.	37.9 (3.3)	W 18
9 A.M.	37.9 (3.3)	WNW 23

▲ FIGURE 3-24 Cloudy, windy conditions can lead to very little cooling at night. This was illustrated at Chicago's Midway Airport on October 25–26, 2001, a day with total overcast. The combined effects of cloud cover and vigorous winds allowed only a 1.7 °C (3.1 °F) drop in temperature between 9 P.M. and 9 A.M.

Strong winds also moderate daily temperature ranges. Recall that higher wind speeds promote greater forced turbulence. When turbulence increases, the enhanced vertical movement causes any small parcel of air immediately in contact with the ground to be quickly displaced upward and replaced by another parcel. The result is that the sensible heat transferred from the ground is distributed to a greater mass of air, which reduces the increase in temperature near the surface. (Not only do strong winds reduce the rate of warming, they also make the air feel colder at any particular temperature, as discussed later in this chapter.) The same process works at night to inhibit cooling (Figure 3–24). If a lot of forced convection occurs, no parcel of air will remain in contact with the chilled surface long enough to undergo substantial cooling. In the absence of forced convection, a shallow layer of air is subject to cooling for an extended time period, and temperatures can drop rapidly.

Annual Cycle

The relative amounts of incoming and outgoing radiation also affect energy budgets and temperatures on an annual basis. The Northern Hemisphere experiences its greatest amount of solar radiation, but not its highest temperatures, on the June solstice. For about 4 to 6 weeks following the solstice, many places continue to have a positive energy balance, which allows warming to continue until sometime in July or August when maximum temperatures are achieved.

Measurement of Temperature

For everyday purposes, the measurement of temperature is a simple and routine procedure in which the expansion and contraction of the fluid in a thermometer is noted. The most accurate thermometers contain mercury, the only metal that exists as a liquid at normal Earth temperatures. Less expensive models are available using dyed alcohol.

It is often useful to know the daily maximum and minimum temperatures. A **maximum thermometer** (Figure 3–25a) is very similar to a regular thermometer, but with two differences. Unlike a regular thermometer, it must contain mercury—it cannot use dyed alcohol. The other difference is that in the tube just beyond the bulb is a very narrow constriction that allows the mercury to expand outward when the temperature increases but prevents it from contracting back into the bulb when the temperature decreases. The temperature shown on the maximum thermometer indicates the highest temperature experienced since the last time it was reset. Resetting the thermometer is easy; the mercury can be forced down simply by shaking the thermometer downward (note that an oral thermometer used at home or in a medical clinic is merely a type of maximum thermometer). If the instrument

Maximum Thermometer

(a)

Minimum Thermometer

(b)

▲ **FIGURE 3-25** Maximum thermometers (a) have a small constriction that only allows the mercury to move outward from the bulb. Minimum thermometers (b) have a sliding index indicating the lowest temperature since the instrument was last reset.

is mounted on a pivoting base, it can be reset by being spun so that centripetal force propels the mercury back into the bulb.

A **minimum thermometer** (see Figure 3–25b) is also similar to a regular thermometer, except that it can only contain dyed alcohol and has within it a small index shaped like a weightlifter's dumbbell. If the index is at the end of the alcohol and the temperature is decreasing, surface tension (the force that holds water molecules together) pulls the index toward the bulb. When the temperature increases, the index remains at its present position as the alcohol expands away from the bulb. Minimum thermometers are mounted horizontally with a latch that can be released to allow the instrument to be inverted. The instrument is reset by turning it upside down, allowing the index to slowly slide down to the end of the alcohol.

Another instrument for the measurement of temperature is the **bimetallic strip**, which consists of two thin strips of different metals bonded together. Because all metals have different rates of expansion and contraction with temperature, one undergoes a greater change in length than the other, causing the strip to bend. A pointer and scale are attached to the bimetallic strip, whose bending is amplified by a lever. When this mechanism is coupled with a rotating drum and a pen, the resulting **thermograph** (Figure 3–26) gives a continuous record of temperature. Thermostats for the heating and air conditioning units of many homes use bimetallic strips to determine the room temperature.

▲ **FIGURE 3-26** A thermograph.

The temperature sensors described above are no longer used at airports and other sites having electronic sensors, but they are still used widely at thousands of weather observing sites worldwide that report their data to national meteorological centers. Considerably more sophisticated instruments for measuring temperature are used at airports and other primary sites. Among these are **resistance thermometers**, instruments that send an electrical current through a very thin filament made of conductor or semiconductor material exposed to the air. The temperature of the filament is the same as that of the surrounding air, and its temperature influences its resistance to the electrical current. The instrument registers the amount of resistance and uses the reading to determine the air temperature. A **thermistor** is a particular type of resistance thermometer that uses a ceramic semiconductor instead of a metallic wire for a filament.

Resistance thermometers are fast-response instruments, meaning that they rapidly register changes in the ambient temperature. Consider what might happen if you were to put a thermometer (a slow-response device) in a refrigerator for an hour or so and then remove it. Although your kitchen is probably much warmer than the inside of your refrigerator, it would take several minutes before the instrument accurately measured the temperature of the room. When slow response is not a problem, thermometers can be perfectly acceptable instruments. Fast-response instruments must be used in situations where rapid temperature changes may be encountered. Experimental research stations that measure the transfer of sensible and latent heat in the atmosphere near the surface require such instruments. **Radiosondes**, packages of weather instruments carried by balloons, likewise require fast-response instruments as they rapidly ascend through air with highly varying temperature characteristics.

Instrument Shelters

You have no doubt heard statements such as this one: "It was 100 degrees in the shade." This is a common expression but one that might have originated from the Department of Redundancy Department because, to be meaningful, temperature must *always* be measured in the shade. A thermometer exposed to direct sunlight will be warmed by the absorption of insolation and assume a temperature greater than that of the air. But when we refer to "air" temperature, we are really concerned with the temperature of the air—not that of a thermometer. As a result, temperature-measuring instruments should always be kept in an instrument shelter like the one in Figure 3–27.

Because an instrument shelter is designed to reduce the influence of incoming radiation on the instruments, certain design criteria must be met. The shelter should be painted white so that its albedo will be maximized and reduce the absorption of radiation. It should also be paneled with slats rather than solid side walls to permit the free flow of air and the removal of any heat that might otherwise accumulate. The door must be mounted on the north side of the box (in the Northern Hemisphere) so that direct sunlight will not strike

▲ **FIGURE 3–27** Temperature instruments must be kept in shelters that protect them from the absorption of solar radiation. Note that the box is painted white and has slats to allow the movement of air. It also has a door that opens on the north side (in the Northern Hemisphere).

the instruments if the door is opened during the middle of the day. Finally, the shelter must conform to a standardized height, so that the thermometers will be mounted at 1.52 m (5 ft) above the ground.

Did You Know?

In August 2001 a heat wave broke temperature records across Europe. It is impossible to determine exact fatality numbers because one cannot know for sure whether high temperatures were the actual cause of death or an aggravating factor. Nonetheless, the World Meteorological Organization has estimated that some 21,000 people likely died as a result of the intense and prolonged heat. France may have been hit the worst; it is believed that nearly 15,000 people died from the heat wave there.

In the United States, temperature is observed hourly by automated systems at National Weather Service offices and Federal Aviation Agency (FAA) facilities at airports across the country. The automated systems (described in Chapter 13) use resistance thermometers for temperature readings. This network is supplemented by observations made at a large number of cooperative agencies (such as U.S. Forest Service stations) and by individual volunteers. At the cooperative stations, observations are made once or twice daily, normally in the early morning and/or midafternoon, with maximum and minimum thermometers housed in instrument shelters. Environment Canada is responsible for temperature data acquisition in Canada.

Temperature Means and Ranges

In just about all aspects of daily life, we use descriptive statistics to talk about the things around us. Although the concept of an average, or mean, value of a property is fairly straightforward, applying the concept involves occasional complications. For example, trying to determine a daily mean temperature poses a dilemma—exactly how many times during the day must we measure it to obtain a true mean? We could make observations every hour, every minute, or even every second, with each method giving us a separate value.

Did You Know?

Hawaii and Alaska share the same highest temperature ever recorded, 37.8 °C (100 °F). The situation is a little different with regard to the minimum temperature, however. While the lowest reading ever taken in Hawaii was −11 °C (12 °F), the record low for Alaska was −62 °C (−80 °F).

The standard procedure is simple—the *daily mean* is defined as the average of the maximum and minimum temperature for a day. The advantage of this method is that the daily mean can be obtained even at weather stations having just the most basic instrumentation; a minimum and maximum thermometer are all we need to compute daily mean temperatures. The disadvantage of using just the maximum and minimum temperature is that it introduces a bias. Observe the daily temperature pattern for a particular day, shown in Figure 3–28. Notice that the nighttime temperatures remain nearly equal to the minimum throughout most of the night, while afternoon temperatures are near the maximum for only a few hours. If we were to obtain a daily mean by taking 24 hourly spaced observations and dividing by 24, the mean value would be lower than that obtained by using just the maximum and minimum temperatures. Nonetheless, averaging the maximum and minimum temperatures is the accepted method for obtaining daily mean temperatures even though it introduces a bias.

The *daily temperature range* is obtained simply by subtracting the minimum temperature from the maximum.

Having obtained daily mean temperatures for an entire month, we calculate the *monthly mean temperature* simply by summing the daily means and dividing by the number of days in the month. Similarly, an *annual mean temperature* is obtained by summing the monthly means for a year and dividing by 12. The *annual range* is the difference between the highest and lowest monthly mean temperatures.

Global Extremes

More impressive than mean temperatures are the maximum and minimum values ever recorded. Not surprisingly, they tend to occur at continental locations. The highest temperature ever recorded in North America was at Death Valley, California. Death Valley is located only a couple of hundred kilometers from the Pacific Ocean, but the Sierra Nevada mountain range presents a barrier that eliminates any moderating influence of the water. In addition, Death Valley's position below sea level and its sparse vegetation cover further promote high temperatures. On July 10, 1913, an all-time high temperature of 57 °C (134 °F) was recorded. The world's highest measured temperature was 58 °C (136 °F) at Azizia, Libya.

The lowest temperature ever recorded occurred at the Vostok Research Station in Antarctica in 1983, with a reading of −89 °C (−129 °F). The research station is located atop thousands of meters of glacial ice, thereby combining the effects of high latitude, a continental locale, and a sparse atmosphere. In North America, the record low temperature of −63 °C (−81 °F) was observed at Snag, Yukon, in February 1947.

Checkpoint

1. How would you determine the daily temperature range for a location?
2. How are daily mean temperatures determined?

Did You Know?

Of the 50 U.S. states, Montana holds the record for the greatest spread between its all-time maximum and minimum temperatures. That state recorded a high temperature of 47 °C (117 °F) at Medicine Lake in 1937 and low temperature of −57 °C (−70 °F) at Rogers Pass in 1954, for a range of 104 °C (187 °F). The all-time maximum temperatures for each state and their dates and locations of occurrence can be found at http://ggweather.com/climate/extremes_us.htm.

▶ FIGURE 3–28 A continuous plot of temperature over a 24-hour period with clear skies. Note that the temperature is near that of the maximum for a relatively short time period. In contrast, the air temperature is near the daily minimum throughout most of the night.

3–3 FOCUS ON SEVERE WEATHER

Recent Severe Heat Waves

Summer heat waves are certainly no rarity in the United States and Canada. Sooner or later everybody endures an episode of unpleasantly high temperatures. But heat waves can cause much more than a few days of discomfort—they can kill. One of the most notable heat waves of the last few decades was the relatively brief but severe event in mid-July 1995 in the north-central United States.

Although extremely high temperatures occurred from the Great Plains to the Atlantic Coast, nowhere was the problem more acute than in Chicago, Illinois, where 525 people died from the heat. The heavy mortality resulted from a combination of high temperatures (Midway Airport recorded an all-time high temperature of 41.1 °C, or 106 °F) and unusually high humidities. The heat and humidity combined to make the "apparent temperature" equivalent to 47 °C (117 °F). Though the searing daytime heat

created plenty of misery on its own, it is believed that the major factor leading to the many deaths is the fact that the extreme heat went uninterrupted, with the apparent temperature exceeding 31.5 °C (89 °F) for nearly 48 consecutive hours. (Recent research suggests that such conditions pose a greater danger than do brief periods of more extreme heat.)

Four years later, in 1999, another major July heat wave occurred in the eastern two-thirds of the United States. Once again, Illinois was in the center of the action, with more than half of the 232 fatalities across the Midwest occurring in Chicago. Missouri was the second hardest hit state, with 61 fatalities. All across the region, power outages occurred from excessive demand, roads buckled, and crops wilted in the fields.

As July gave way to August, the heat moved eastward toward the Atlantic states, where it broke numerous weather records. Charleston, South Carolina, had an all-time high temperature of 40.5 °C (105 °F). Augusta, Georgia's high tempera-

ture of 39.4 °C (103 °F) came on the sixth consecutive day in which the record for the daily maximum temperature was tied or exceeded. And on August 8, Raleigh-Durham, North Carolina, broke the 100 °F mark (37.8 °C) for the 11th time that summer.

Severe heat has also impacted other parts of the world. Some 35,000 people died from heat-related causes during a heat wave across Europe in July and August of 2003. England recorded its all-time highest temperature on August 10, with a reading of 38.1 °C (100.6 °C) at Gravesend-Broadness. France, suffering through its hottest summer since World War II, was especially hard hit, with over 11,000 heat-related fatalities.

The decade of the 1990s and the first decade of the 2000s were remarkably warm relative to other periods in recorded history. This is particularly noteworthy because the topic of human-induced climatic warming has been a major issue for scientists and policy makers. This matter will be discussed further in later chapters of this book.

Undoubtedly, more extreme temperatures have occurred across the globe and North America at locations without temperature observation stations. (*See Box 3–3, Focus on Severe Weather, Recent Heat Waves.*)

Some Useful Temperature Indices

There are several situations in which basic temperature readings can be modified to provide a better understanding of the effects of temperature. In some cases the temperature data can be combined with other variables, such as wind and humidity, to help assess the effects on human comfort. In other cases average temperatures might be adjusted for planning purposes. Several of these indices are described in the following three sections. An additional temperature index that incorporates the effect of humidity will be discussed in Chapter 4, which addresses atmospheric moisture.

Wind Chill Temperatures

Temperature by itself exerts a major impact on human comfort, but the discomfort caused by high or low temperatures can be compounded by other weather factors. If low temperatures are accompanied by windy conditions, a person's

body loses heat much more rapidly than it would under calm conditions, due to an increase in sensible heat loss. Thus, a windy day with a temperature of −2 °C (28 °F) might feel colder than a calm day at −40 °C (−40 °F). As a result, when temperatures are low, it is common for weather reports to state both the actual temperature and how cold that temperature actually feels, the **wind chill temperature index** (or simply the *wind chill temperature*) (see Tables 3–2 and 3–3).

While helping to provide people a better idea of the danger imposed by the combination of windy conditions and low temperatures, the wind chill index has certain shortcomings. For example, calculations do not take into account the potential warming effect of sunlight on a person's body. Nonetheless, the index provides people with guidance in the way they should dress and the types of activities they should undertake under cold and windy conditions. This is especially important when one considers the fact that extreme cold is the number one cause of fatalities directly attributable to weather.

Just as windy conditions can make low temperatures feel even colder, high humidities can cause warm days to feel oppressively hot. As a result, a heat index has been calculated that incorporates the effect of high atmospheric moisture at high temperatures. This index is discussed in Chapter 5, Atmospheric Moisture.

TABLE 3–2

Wind Chill Temperature (°C)

Wind (km/hr)	Temperature (°C)									
	5	0	−5	−10	−15	−20	−25	−30	−35	−40
5	4	−2	−7	−13	−19	−24	−30	−36	−41	−47
10	3	−3	−9	−15	−21	−27	−33	−39	−45	−51
15	2	−4	−11	−17	−23	−29	−35	−41	−48	−54
20	1	−5	−12	−18	−24	−31	−37	−43	−49	−56
25	1	−6	−12	−19	−25	−32	−38	−45	−51	−57
30	0	−7	−13	−20	−26	−33	−39	−46	−52	−59
35	0	−7	−14	−20	−27	−33	−40	−47	−53	−60
40	−1	−7	−14	−21	−27	−34	−41	−48	−54	−61
45	−1	−8	−15	−21	−28	−35	−42	−48	−55	−62
50	−1	−8	−15	−22	−29	−35	−42	−49	−56	−63
55	−2	−9	−15	−22	−29	−36	−43	−50	−57	−63
60	−2	−9	−16	−23	−30	−37	−43	−50	−57	−64
65	−2	−9	−16	−23	−30	−37	−44	−51	−58	−65
70	−2	−9	−16	−23	−30	−37	−44	−51	−59	−66
75	−3	−10	−17	−24	−31	−38	−45	−52	−59	−66
80	−3	−10	−17	−24	−31	−38	−45	−52	−60	−67

TABLE 3–3

Wind Chill Temperature (°F)

Wind (mph)	Temperature (°F)																
	40	35	30	25	20	15	10	5	0	−5	−10	−15	−20	−25	−30	−35	−40
5	36	31	25	19	13	7	1	−5	−11	−16	−22	−28	−34	−40	−46	−52	−57
10	34	27	21	15	9	3	−4	−10	−16	−22	−28	−35	−41	−47	−53	−59	−66
15	32	25	19	13	6	0	−7	−13	−19	−26	−32	−39	−45	−51	−58	−64	−71
20	30	24	17	11	4	−2	−9	−15	−22	−29	−35	−42	−48	−55	−61	−68	−74
25	29	23	16	9	3	−4	−11	−17	−24	−31	−37	−44	−51	−58	−64	−71	−78
30	28	22	15	8	1	−5	−12	−19	−26	−33	−39	−46	−53	−60	−67	−73	−80
35	28	21	14	7	0	−7	−14	−21	−27	−34	−41	−48	−55	−62	−69	−76	−82
40	27	20	13	6	−1	−8	−15	−22	−29	−36	−43	−50	−57	−64	−71	−78	−84
45	26	19	12	5	−2	−9	−16	−23	−30	−37	−44	−51	−58	−65	−72	−79	−86
50	26	19	12	4	−3	−10	−17	−24	−31	−38	−45	−52	−60	−67	−74	−81	−88
55	25	18	11	4	−3	−11	−18	−25	−32	−39	−46	−54	−61	−68	−75	−82	−89
60	25	17	10	3	−4	−11	−19	−26	−33	−40	−48	−55	−62	−69	−76	−84	−91

Heating and Cooling Degree-Days

Because climate exerts a major impact on the energy demands of people across the world, it is useful to create an index that can help a planner anticipate energy needs more accurately. One such index, called **heating degree-days**, was derived by heating engineers. This index is based on the notion that buildings generally need artificial heating to bring their interior temperatures up to the desired temperature of 21 °C (70 °F) when the mean temperature for a day falls below 18 °C (65 °F). Each degree of temperature that the daily mean is lower than 18 °C (65 °F) constitutes one degree-day. Hence, a day with an average temperature of 13 °C (56 °F), for example, would account for 5 Celsius or 9 Fahrenheit heating degree-days. These values are then summed for each day of the year in which the temperature falls below the critical value, yielding the total number of heating degree-days for a location. Figure 3–29 (a) and (b) map the heating degree-days for the United States and Canada, respectively.

Cooling degree-days are the warm season analogy to heating degree-days. They are obtained by subtracting a base temperature—often but not always 18 °C (65 °F)—from the daily average temperature and summing those values. Like heating degree-days, these values have been calculated for a great many locations and are readily available from many sources. Figure 3–30 plots United States cooling degree-days using the base temperature of 65 °F. (Environment Canada has not produced an equivalent map for that country.)

Growing Degree-Days

Growing seasons for agricultural products are to a large extent determined by the length of time in which the temperature remains above some *base temperature* that varies by crop type. If a particular crop requires a daily average temperature of 10 °C (50 °F) to grow, the growing season will extend between the first day of spring and the last day of fall with mean temperatures above that base value. During the growing season each crop requires a particular amount of heat, reflected in the daily mean temperatures, to reach maturation for harvest. **Growing degree-days**, which are very similar to heating degree-days, provide a useful index for estimating when crops will have undergone the growth necessary to send them to market. Growing degree-days are calculated by subtracting a crop's base temperature from the daily mean temperature and summing those numbers over the growing season. While other factors, such as water availability, also help determine plant maturation, this index provides useful information for agriculturalists.

Thermodynamic Diagrams and Vertical Temperature Profiles

The distribution of temperatures across Earth's surface is a basic component of daily weather. But temperature variations do not only occur horizontally; temperatures are just as likely to vary vertically, and those variations can greatly affect atmospheric behavior. Rapid decreases with height, for example, increase the potential for cloud development and even severe thunderstorms. Temperature inversions (increasing temperature with altitude) have their own implications for the atmosphere, including the suppression of atmospheric mixing that can concentrate air pollutants near the surface.

Thermodynamic diagrams, which depict the vertical profiles of temperature and humidity with height above the surface, provide extremely important information to the forecaster. These charts enable forecasters to determine the height and thickness of existing clouds and the ease with which the air can be mixed vertically (a consideration in the near-term development of clouds and precipitation). The data on the charts are obtained from instrument packages called *radiosondes* that are carried aloft by weather balloons twice a day at weather stations across the globe.

Though extremely valuable, thermodynamic diagrams are not altogether simple to understand, and some of the information one can obtain from them is beyond the scope of this text. Consequently, rather than trying to present a complete guide to the use of these charts all at once, we will introduce various elements of thermodynamic diagrams as they pertain to the individual discussions in this text. As we progress, you will see just how crucial such charts can be to day-to-day forecasting. For now we will only discuss the plotting of vertical temperature profiles on these charts. This will provide a foundation for further understanding in later chapters.

Thermodynamic diagrams come in several varieties. The *skew-T graph* is commonly used by meteorologists but can be conceptually difficult for beginning students. The easiest to use is the **Stuve** (pronounced STU-vay) **diagram** (a simplified version is shown in Figure 3–31). The Stuve diagram starts with a rectangular grid, with temperature plotted on the horizontal axis and pressure on the vertical axis. Three things must be noted at the onset: First, the chart plots temperature as a function of the *pressure level*—not the height above the surface. The altitude of a particular pressure level—such as the 500 mb level—is not a constant, but varies over time and from one location to another). This convention allows a more direct application of meteorological laws than would plotting temperature against altitude.[2]

Second, note that the pressure decreases upward along the vertical axis. This is a response to the fact that atmospheric pressure invariably decreases with height from the surface. Finally, observe that the pressure is not plotted on a linear scale. Instead, the axis is plotted on a nearly logarithmic scale so that the 1000 mb and 900 mb lines are closer together than are the 300 mb and 200 mb lines. This mimics the way pressure actually decreases in the real atmosphere.

Let's look at the example of a simplified Stuve diagram in Figure 3–31. The temperature plotting (or *sounding*, as these plots are often called) was based on the ascent of a radiosonde launched at Slidell, Louisiana, June 9, 2002, at 12 noon, Greenwich mean time (7 A.M. CDT). In all, observations were recorded at 52 levels from the surface all the way up to the 100 mb level (well into the stratosphere). The data are made available in text form, which is useful for detailed observations, and then plotted automatically onto the Stuve diagram. At the surface, the air pressure was 1015 mb and the temperature was 22.8 °C (73.0 °F). As the radiosonde initially rose from the surface, it recorded increasing temperatures—an inversion—up to a pressure level of 988 mb (which in this case was 238 m—781 ft—above the surface). At the top of the inversion, the temperature was 25.0 °C (77.0 °F), or 2.2 °C (4.0 °F) warmer than at the surface. Ground-based inversions such as these are very common at night and during the early morning hours due to the loss of longwave radiation (inversions were briefly discussed earlier in this chapter—we will return to them in greater detail in Chapter 6). Above the inversion layer, the temperature decreases fairly consistently with altitude

---

[2]For reference, the height of the average altitude of several pressure levels is often indicated along the left, vertical axis of the diagram.

▶ **FIGURE 3-29** The distribution of heating degree-days in the United States in degrees Fahrenheit (a) and Canada in degrees Celsius (b).

(a)

(b)

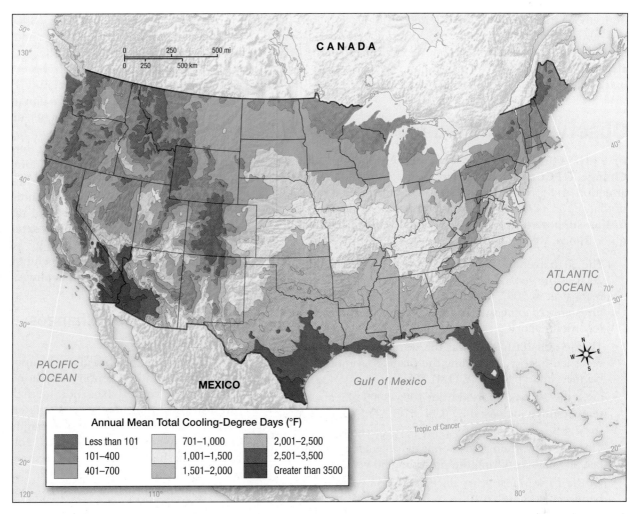

▲ **FIGURE 3–30** The distribution of cooling degree-days in the United States (°F).

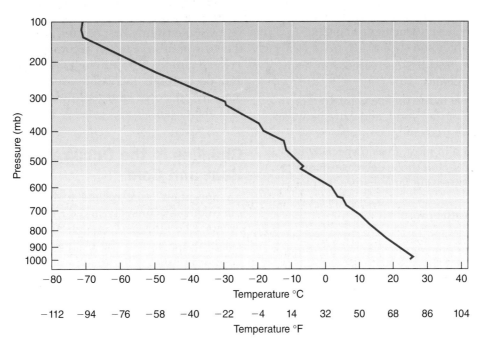

▲ **FIGURE 3–31** A simplified Stuve diagram showing the temperature profile for Slidell, Louisiana, on the morning of June 9, 2002.

until about the 140 mb level (located at about 14.7 km—9.1 mi—above sea level), where the air temperature becomes nearly constant with height at about −70 °C (−90 °F), the point at which the air temperature ceases to decrease with height marks the tropopause, which was described in Chapter 1.

Observed and Predicted Warming

In early 2007 the **Intergovernmental Panel on Climate Change (IPCC)** issued a report that garnered worldwide attention and a Nobel Prize for its members. The conclusions of that report pointing to a rapidly warming planet have been further supported by several additional years of data. Among other things, we know that

- The 2010 average global land surface temperature tied 2005 as the second warmest on record, following the record set in 2007. When land and ocean temperatures are combined, 2010 tied 2005 for the warmest in recorded history.

- The decade 2001–2010 was the warmest on Earth in recorded history, exceeding the planet's twentieth-century value by 0.56 °C (1.01 °F).

- Over the 100-year period from 1906–2005 global average temperatures increased by 0.74 °C (1.33 °F) ± 0.18 °C (0.32 °F).

- The rate of warming was about twice as great over the last 50 years as during the preceding 50 years.

- Of the 14 warmest years since 1850, 13 occurred between 1995 and 2010.

- Warming over the Arctic has been about twice as strong as that over the rest of the globe.

- Warming has been more marked over land areas than over the oceans, with greatest warming during the Northern Hemisphere winter and spring.

- The number of days with frost has decreased over many parts of the midlatitude regions.

- There has been a decrease in the number of extreme cold events across much of the world, and extreme warm events have become more frequent.

- Snow cover has decreased in most areas (especially in the Northern Hemisphere), and those decreases have mostly been driven by increasing temperature. In places where snow cover increased, increasing precipitation (rather than cooling) was the cause.

- The breakup date for river and lake ice occurred earlier by an average of 6.5 days per century, and the freeze-up date occurred later by an average of 5.8 days per century.

- Most of the observed global warming since the middle of the twentieth century is "very likely" (i.e., more than 90 percent probable) due to human activities that have increased greenhouse gas concentrations in the atmosphere.

In some regards all this was an old story; scientists have been warning about changing climate and the role of human activities for some time. But the news media and scientific community placed enormous importance on this report—and for good reason. The IPCC was created by the World Meteorological Organization to periodically summarize what the science community currently knows about climate change and its impacts (an overview of the organization and how it goes about its mission is presented in *Box 16-2, Focus on the Environment: Intergovernmental Panel on Climate Change*). It is the most authoritative organization assembled on the subject, and its findings are widely accepted by scientists and policy makers around the world. In this chapter we briefly summarize its findings specifically related to temperature. Other chapters will refer to the IPCC where appropriate. Chapter 16 presents a detailed analysis of climate change, including the processes by which human activities have altered the climate, the techniques by which climate scientists have arrived at their findings, and related impacts.

Recent Observed Changes in Temperature

There is no question that global temperatures have increased during the twentieth century, as shown in Figure 3–32. The values on the vertical axes indicate the difference in temperature for a given year relative to the climatic average for the period. Every year from 1880 through the late 1930s experienced below-normal temperatures. Following the late 1930s, most years had above-normal temperatures, with marked warming beginning in 1976. Over the entire period global surface temperatures increased by about 0.7 °C (1.26 °F) per century, but the rate of temperature increase since about 1976 has been about three times as great as that of the century as a whole.

The last decade of the twentieth century and the early years of the twenty-first century were remarkably warm, with the 10 warmest years on record having occurred during the 11-year period 1997 through 2007. Temperatures over land and ocean both increased over the period of record, but temperatures over land rose by about twice as much as those over the oceans.

Changes in the Occurrences of Extreme Heat and Cold

An important issue related to temperature change involves the frequency, intensity, and duration of extremely hot or cold events. The evidence suggests that extreme events may play a significant part in the observed warming. In other words, episodes of uncommonly high temperatures may account for a greater part of the recent warming than increases in more frequent day-to-day temperatures.

But because rare events are by definition infrequent, such a trend in unusual events may be hard to verify statistically. For example, what if a heat wave of a particular intensity occurs five times over a given 25-year period but then occurs seven times over the next 25-year period? Would that signify

▲ **FIGURE 3–32** Global temperatures from 1800–2007 plotted as departures from the 1901–2000 average.

a trend or might it just be the result of random variation? Despite this difficulty, numerous studies have looked at the change in the frequency of extreme heat and cold events in different parts of the world. These studies have shown no consistent pattern that applies for all regions, though, generally speaking, over the last half century there has been a greater decrease in extremely low temperatures than there has been an increase in occurrence of severe heat. Over the period 1979–2003, however, the increase in extreme maximum temperatures has outpaced the increase in anomalous minimum temperatures.

An increase in the severity and frequency of extreme heat can have a serious impact on human mortality (recall the discussion of recent severe heat outbreaks and the fact that severe heat annually kills more people in the United States than does lightning, tornadoes, or hurricanes). To make matters worse, the increase in nighttime temperature has been greater than the increase in daytime temperatures, and these high temperatures have been coupled with increasing humidity levels that create even higher "apparent temperatures" (discussed in Chapter 5). This is important because experts believe that the most dangerous health threat occurs not from very high daytime temperatures but from the persistence of high apparent temperatures for several days without intervening cool nights. On the other hand, the reduced frequency and intensity of major outbreaks of cold conditions can reduce loss of life due to that type of weather. Studies suggest that with continued global warming some countries will see, the decrease in cold-related deaths will more than offset warm-event deaths, whereas in other countries the reverse will be true.

Projected Changes in Temperature

Atmospheric scientists use complex mathematical programs called *general circulation models* (GCMs) to examine how the atmosphere might respond to increasing greenhouse gas concentrations in the future. We will discuss these models in

greater detail in Chapter 16, but for now we can say that they are based on scientific principles and provide the best estimates of how the climate will respond to continued emissions of greenhouse gases.

Figure 3–33 plots the global temperature changes from the 1980–1999 averages projected by different GCMs. In each of the plots the solid colored lines represent the output of a particular GCM, and the black dotted line is the average of the individual models. The panels, from top to bottom, illustrate modeled output from examples of high-, medium-, and low-emission scenarios, respectively (the exact scenarios involve more assumptions than just CO_2 emissions, and their complete description falls beyond the scope of this discussion). Not surprisingly, the model outputs differ from each other, and the differences become greater further into the future. But they all lead to the same conclusion that the warming experienced over the last half century or so will continue into the future.

Table 3–4 shows the average amount of warming predicted by the models for three 20-year time periods in the twenty-first century for each of the three emission scenarios. For 2011–2030 there is little difference in the warming predicted by the models under the three scenarios, ranging from between 0.64 °C (1.15 °F) and 0.69 °C (1.24 °F). The minimal difference is due to the fact that the expected warming is the result of the emissions that have already been put into the atmosphere rather than those that will be added between

TABLE 3–4

Global Mean Warming Projections for Future Time Periods

Scenario	2011–2030	2046–2065	2080–2099
A2	0.64 (1.15)	1.65 (2.97)	3.13 (5.63)
A1B	0.69 (1.24)	1.75 (3.15)	2.65 (4.77)
B1	0.66 (1.19)	1.29 (2.32)	1.79 (3.22)

Each scenario is based on different assumptions about future greenhouse gas emissions. Temperatures are in °C (°F).

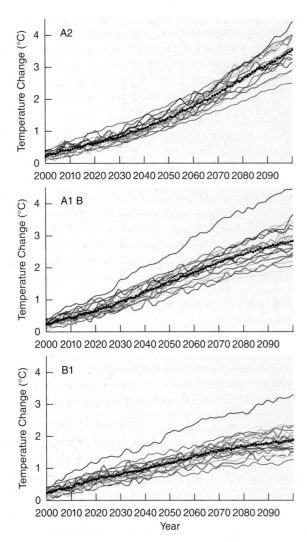

▲ **FIGURE 3–33** Projected global temperature changes as predicted by general circulation models, with the mean of the model output shown by the heavy dark line. From top to bottom the three graphs plot predicted temperatures for high-, medium-, and low-emission scenarios.

now and 2030. But by the time we get to the 2080–2099 period the three emission scenarios yield very different amounts of warming. The average model prediction under the high-emission situation yields 3.13 °C (5.63 °F) of warming, significantly higher than the 1.79 °C (3.22 °F) under the low-emission situation. In other words, the actions we take in the near future in terms of controlling emissions will not greatly affect the amount of warming experienced in the early part

of the twenty-first century, but they will have very significant impacts on conditions later in the century (and beyond).

Checkpoint

1. What is a general circulation model (GCM)?
2. Look at the graphs in Figure 3–33. What is the general trend in projected temperatures among the three models?

The Role of Human Activity

Perhaps the most significant conclusion of the Fourth Assessment Report of the IPCC can be summarized in the following two paragraphs from the Synthesis Report:

> Most of the observed increase in globally-averaged temperatures since the mid-20th century is very likely due to the observed increase in anthropogenic GHG (greenhouse gas) concentrations. This is an advance since the TAR's [Third Assessment Report's] conclusion that "most of the observed warming over the last 50 years is likely to have been due to the increase in GHG concentrations."

> The observed widespread warming of the atmosphere and ocean . . . support the conclusion that it is extremely unlikely that global climate change of the past 50 years can be explained without external forcing, and very likely that it is not due to known natural causes alone. During this period, the sum of solar and volcanic forcings would likely have produced cooling, not warming. . . . It is likely that increases in GHG concentrations alone would have caused more warming than observed because volcanic and anthropogenic aerosols have offset some warming that would otherwise have taken place.

The above statement indicates that the IPCC paid close attention to the influence of volcanic eruptions, solar radiation variability, and other factors that could influence global climates but still identified human greenhouse gas emissions as the primary cause of warming. Moreover, the panel concluded that these other factors not only do not account for the observed warming of the last half century but that those factors operating by themselves would probably have combined to cool the climate, rather than warm it. The effect of humans on climate is one of the foremost matters that societies will have to deal with for some considerable time to come.

Summary

We all know that the ultimate source of energy for all atmospheric processes is solar energy. But most of this energy is delivered to the atmosphere indirectly. This chapter first described the processes that take place as solar radiation (insolation) passes through the atmosphere. Most of the ultraviolet radiation is absorbed by ozone in the stratosphere, and much of the near-infrared radiation is absorbed in the troposphere by water vapor and other gases. Scattering, both forward and

backward, also reduces the intensity of the direct radiation reaching the surface. After atmospheric absorption and scattering back to space are accounted for, slightly more than half of the insolation available from the Sun reaches the surface, where some is again reflected back and the rest absorbed.

Having absorbed solar radiation, the surface and the atmosphere both radiate longwave energy. The end result of the transfer of solar and longwave radiation is that the surface has a net radiation surplus while the atmosphere has a deficit of equal value. The radiation deficit of the atmosphere and the surplus at the surface are offset by two other heat transfer mechanisms: conduction and convection. Conduction transfers energy across an extremely thin layer of air in contact with the surface, while convection distributes sensible and latent heat from near the surface to higher regions of the atmosphere.

The amount of energy obtained by the Earth system is not equal at all latitudes. Areas near the equator receive a surplus of energy, while more poleward regions have an energy deficit. This unequal energy distribution is offset by the latitudinal transfer of energy by wind movements and ocean currents. The most immediately recognizable outcome of spatial variations in absorbed energy is their influence on global temperatures.

The chapter also looked at various aspects of air temperature, including its average global distribution and the geographic factors that determine local temperature characteristics. We examined the ways that temperature is measured and statistically summarized and introduced the concept of heating and cooling days and the wind chill temperature index. In addition, the chapter included an introduction of the thermodynamic diagram, an extremely valuable tool for analyzing and predicting weather.

By many measures it is clear that the planet has undergone significant warming during the twentieth century and into the twenty-first century, and it very likely that this trend will continue for at least through the next few decades. The Intergovernmental Panel on Climate Change (IPCC), the leading organization studying the issue, has concluded that it is highly probable that human activities are responsible for most of this change. Human impacts on climate extend beyond the direct effects of temperature, as will be discussed in upcoming chapters—most extensively in Chapter 16.

Key Terms

energy balance *page 55*
absorption *page 56*
reflection *page 56*
albedo *page 57*
specular reflection *page 57*
diffuse reflection/ scattering *page 57*
diffuse radiation *page 57*
direct radiation *page 57*
Rayleigh scattering *page 57*
Mie scattering *page 58*
nonselective scattering *page 58*

extraterrestrial radiation *page 60*
planetary albedo *page 61*
atmospheric window *page 63*
net longwave radiation *page 64*
net all wave radiation *page 64*
net radiation *page 64*
laminar boundary layer *page 65*
free convection *page 65*
forced convection/mechanical turbulence *page 66*

sensible heat *page 66*
specific heat *page 66*
latent heat *page 67*
advection *page 68*
greenhouse effect *page 69*
isotherm *page 70*
inversion *page 73*
continentality *page 74*
maximum thermometer *page 77*
minimum thermometer *page 78*
bimetallic strip *page 78*
thermograph *page 78*

resistance thermometer *page 79*
thermistor *page 79*
radiosonde *page 79*
wind chill temperature index *page 81*
heating degree-days *page 82*
cooling degree-days *page 83*
growing degree-days *page 83*
thermodynamic diagram *page 83*
Stuve diagram *page 83*
Intergovernmental Panel on Climate Change (IPCC) *page 86*

Review Questions

1. Explain how the absorption and scattering of radiation in the atmosphere affect the receipt of solar radiation at the surface.

2. Which two gases are most effective at absorbing long-wave radiation?

3. How do specular reflection and diffuse reflection differ?

4. What does the term *albedo* mean?

5. What characteristics of Rayleigh scattering cause it to create a blue sky?

6. What properties of Mie scattering distinguish it from Rayleigh scattering?

7. Why are overcast days typically gray?

8. What is the numerical value of Earth's planetary albedo?

9. Which type of scattering accounts for the majority of Earth's planetary albedo?

10. Describe quantitatively how much solar radiation is absorbed and reflected by Earth's atmosphere and surface.

11. What is the atmospheric window?

12. Why is it incorrect to state that longwave radiation bounces back and forth between clouds and the surface?

13. Explain why the incoming and outgoing radiation for the Earth system (radiation entering and leaving the top of the atmosphere) must equal each other.

14. How do conduction and convection work together to transfer heat upward?

15. What is the difference between free convection and forced convection?

16. Describe sensible and latent heat.

17. How do the net input and output of radiation vary with latitude?

18. Which two processes transport energy from zones of radiation surplus to zones of radiation deficit?

19. Why does the term *greenhouse effect* inaccurately describe how the atmosphere is heated?

20. Discuss how geographic factors such as latitude and altitude influence the distribution of temperature across Earth's surface.

21. How do the various instruments that are used to observe temperature work?

22. Explain how daily, monthly, and annual mean temperatures are computed from observed temperatures. Discuss some of the factors that can bias resulting values.

23. Describe the horizontal and vertical scales on Stuve diagrams.

Critical Thinking

1. Shorter wavelength radiation is more subject to Rayleigh scattering than is longer wave radiation. Explain how this might affect the value of facing directly toward the Sun to pursue an even suntan.

2. Even on cloudy days, excessive exposure can lead to a danger of sunburn. What does this imply about clouds' effect on ultraviolet radiation?

3. Desert areas are often photographed with spectacular sunsets. Can you think of any reasons why they may be more inclined to have particularly flashy skies at dusk?

4. Our eyes are sensitive only to wavelengths between 0.4 and 0.7 micrometers. Would the sky appear any different on clear days if our eyes could also perceive wavelengths between 0.2 and 0.4 micrometers? How would the ratio of perceived diffuse to direct radiation change?

5. Would you expect both the Northern and Southern hemispheres to have the same average albedo? What factors might cause the two hemispheres to reflect different percentages of insolation back to space?

6. Snow often melts more rapidly in wooded areas immediately adjacent to trees than in nearby openings. What type of energy transfer processes could lead to this effect?

7. Net radiation values in the summer may be higher in forested areas than desert areas, despite the higher temperatures in the desert. How can this be?

8. The ratio of sensible to latent heat loss from a surface is called the *Bowen ratio*. How do you suppose Bowen ratios might differ among desert, wooded, and urban landscapes?

9. Clouds can reduce the amount of insolation reaching Earth's surface, but they can also reduce the amount of longwave radiation from the surface that escapes to space. How might this affect maximum and minimum temperatures? Do you think all types of clouds would produce similar effects?

10. Figure 3–17 shows the mean distribution of ocean currents. It is believed that climate change, through a variety of mechanisms, could cause a shift in the position of some currents. Can you identify any land regions whose climate could be vulnerable to shifts in nearby currents? Are there any localities whose climates could cool even if the average global temperature were to warm?

11. Instrument shelters protect thermometers from the heating effect of absorbed sunlight. Is it also true that shelters protect the thermometers from the chilling effect of wind?

12. An orchard farmer hears a weather forecast for overnight low temperatures to hover just above the freezing point of 0 °C (32 °F), but with wind chill temperatures expected to drop significantly lower. Will the wind chill increase the possibility of frost damage? Why or why not?

Problems and Exercises

1. Go to the Web site **http://rredc.nrel.gov/solar/old_data/nsrdb/redbook/atlas**. Select mean solar radiation as data type and horizontal flat plate for instrument orientation. Examine the maps of average solar radiation for January, April, July, and October. Describe each of the patterns. How much do you think Earth–Sun relationships affect the distribution relative to the effect of cloud cover and other weather elements?

2. Go to **http://earthobservatory.nasa.gov/Observatory/Datasets/netflux.erbe.html** and observe the seasonal shift in net radiation for the surface and atmosphere. What is the most obvious pattern? Are there significant differences between land bodies and adjacent oceans?

3. The latent heat for water is 2,500,000 joules per kilogram (kg), and the specific heat of water is 4190 joules per kg per degree Celsius of temperature change. Assume that a kg of water of water begins with a temperature of 20 °C (68 °F). Compare the amount of energy needed to bring the water to the boiling point to the amount of energy needed to evaporate the same amount of water.

4. View the maps of mean minimum temperatures in January and mean maximum temperatures in July at **www.climatesource.com/map_gallery.html**. Assess the relative importance of latitude, elevation, continentality, ocean current, and local conditions to these distributions.

Quantitative Problems

This chapter has described the energy balance of the atmosphere and the factors that influence global temperatures. The companion Web site for this book, **www.MyMeteorology Lab.com**, offers a brief quiz in which you can test your knowledge of the material by answering several quantitative problems. The brief exercise should help you better understand equilibrium temperatures, seasonal effects of cloud cover on energy receipt, and relative expenditures of surface energy on latent and sensible heat transfer. After entering the site, go to the bottom of the page and select Chapter 3. Then highlight the Quantitative Examples line on the lefthand panel.

Useful Web Sites

www.nrel.gov/gis/solar.html

Provides monthly and annual maps of solar radiation across the United States. For the material relevant to this chapter, be sure to check the box for *Horizontal Flat Plate* as instrument orientation mode.

itg1.meteor.wisc.edu/wxwise/museum/a2/a2net.html

Maps of average net radiation for each month and an animation showing the change in the net radiation distribution over an average year.

ncdc.noaa.gov/extremes/records

Access to U.S. temperature and precipitation records by date and state.

usairnet.com

Current distribution of temperatures across the 48 conterminous United States. Also provides links to maps of the current temperature and other meteorological values for each of the 48 conterminous United States.

www.srh.noaa.gov/ffc/?n=wci and
www.srh.noaa.gov/ffc/?n=hichart
Provide apparent temperature and wind chill converters.

MyMeteorologyLab™

The *Understanding Weather and Climate,* Sixth edition, MyMeteorologyLab™ Web site contains numerous multimedia resources and assessments to aid in your study of **Energy Balance and Temperature**. Visit **www.MyMeteorologyLab.com** to:

- **Review** key chapter concepts.
- **Read** the Pearson eText, *In the News* RSS feeds, and other chapter-specific Web resources.
- **Visualize** challenging topics with Interactive Tutorials, Geoscience Animations, MapMaster interactive maps, and Weather in Motion movies and visualizations, and more.
- **Test Yourself** with self-study quizzes.

 TUTORIAL
GLOBAL ENERGY BALANCE

Use the interactive animations and quizzes in this tutorial to review this chapter's key concepts.

WEATHER IN MOTION
View movies and visualizations of meteorology patterns and processes.

Temperatures and Agriculture
Seasonal Changes in Land Surface Temperatures
Global Warming Predictions
Heavy Convection over Florida
Global Sea-Surface Temperatures—Climatology

Atmospheric Pressure and Wind

After reading this chapter, you should be able to:

▶ Explain the concept of air pressure.

▶ Describe how pressure changes vertically and horizontally in the atmosphere.

▶ Apply the equation of state to calculate air density.

▶ Identify devices used to measure air pressure.

▶ Describe the distribution of air pressure across the globe, at sea level, and in the upper atmosphere.

▶ Explain the factors that affect the wind speed and direction.

▶ Describe winds in relation to pressure gradients in the upper atmosphere and near the surface.

▶ Explain the forces that produce anticyclones, cyclones, troughs, and ridges.

▶ Describe how meteorologists measure wind.

The coastline of the Pacific Northwest of the United States and the Canadian province of British Columbia are subject to extremely powerful storms coming in from the Gulf of Alaska. A particularly strong storm occurred on December 14, 2006, as winds peaking up to 180 km/hr (113 mph) were recorded at Mt. Rainier, Washington. Seattle-Tacoma International Airport received a record-breaking wind speed of 110 km/hr (69 mph), almost a quarter of a million residents went without electricity due to the downing of power lines, and at least 10 people died from weather-related causes.

As concerned as we sometimes are with wind conditions, few of us pay much attention to a closely associated component of weather—*atmospheric pressure*. After all, how many times have you canceled a picnic because the pressure was too low? Or how many people do you know who have special clothes they wear only on days of high pressure?

Although seldom considered in everyday life, air pressure deeply affects other weather variables that have much more immediate impact. For example, horizontal variations in atmospheric pressure are directly responsible for generating winds. The winds of the December 2006 storm in the Pacific Northwest were so strong because of just such a difference in air pressure. A region of low pressure moving in from the Pacific passed near a region of high pressure to the south. The considerable difference in pressure over a relatively short distance between the two pressure centers generated the powerful winds. Differences in air pressure also strongly influence the likelihood of cloud formation and precipitation. The reason for this is that air descends in areas of high surface pressure and rises in regions of low surface pressure.

This chapter introduces the basic concepts of atmospheric pressure and its vertical and horizontal distributions. We discuss the relationship between pressure and other atmospheric variables and the processes that create horizontal and vertical variations in pressure. With this foundation, we can go on to discuss storm patterns in later chapters.

◀ The high winds of the December 2006 storm in the Pacific Northwest toppled trees and power lines throughout the region, leaving many residents without electricity.

The Concept of Pressure

The atmosphere contains a tremendous number of gas molecules being pulled toward Earth by the force of gravity. These molecules exert a force on all surfaces with which they are in contact, and the amount of that force exerted per unit of surface area is **pressure** (see *Box 4–1, Physical Principles: Velocity, Acceleration, Force, and Pressure*). Of course, the concept of pressure is not confined to meteorology but rather is fundamental to all the physical sciences. In most physical science applications, the standard unit of pressure is the **pascal** (Pa), but in the United States meteorologists use the **millibar** (mb), which equals 100 Pa. Canadian meteorologists use yet another unit, the **kilopascal** (kPa), equal to 1000 Pa, or 10 mb. For purposes of comparison, air pressure at sea level is typically roughly 1000 mb (100 kPa)—or more precisely, 1013.2 mb.

To understand the characteristics of pressure, refer to Figure 4–1, which depicts a sealed container of air. The enclosed air molecules move about continually and exert a pressure on the interior walls of the container (a). The pressure of the air is proportional to the rate of collisions between the molecules and walls. We can increase the pressure in two ways. The first way is by increasing the density of the air either by pumping more air into the container or by decreasing the volume of the container (b). The second is by increasing the air temperature, in which case the molecules exert higher pressure because they are moving more rapidly (c). Thus, pressure reflects both the density and temperature of the gas.

If the air in the container is a mixture of gases (as it is in the atmosphere), each gas exerts its own specific amount of pressure, referred to as its *partial pressure*. The total pressure exerted is equal to the sum of the partial pressures. This relationship is known as **Dalton's law**.

On Earth the container in Figure 4–1 is surrounded by the atmosphere, which exerts pressure on the exterior walls. Consider what would happen if we were to remove the lid of the container or puncture a hole in its side. If the pressure *outside* the container were greater than that within, the outside air would be forced inward until the pressure equalized. (The force of this equalization is what causes the "whoosh" when you open a vacuum-packed container.) On the other hand, if the pressure were greater *inside* the container, air would be forced outward until the internal pressure decreased to match the surrounding air. In either case, within moments the air pressure exerted on the outside of the container would become exactly equal to that on the inside. This example introduces us to another characteristic of air: It constantly moves to establish an equilibrium between areas of high and low pressure.

Did You Know?

Many people know that air molecules are constantly moving about randomly, but people are surprised to learn just how fast they travel. On a day with a comfortable temperature, molecules near Earth's surface move at a rate of several hundred kilometers per second—comparable to the speed of a rifle bullet.

(a)

(b)

(c)

▲ **FIGURE 4-1** The movement of air molecules (indicated by the dots with the red arrows) within a sealed container exerts a pressure on the interior walls (a). The pressure can be increased by increasing the density of the molecules (b) or increasing the temperature (c). The speed of the molecules (and therefore the temperature) is indicated by the degree of redness and length of the arrows.

As described in Chapter 1, what we experience as atmospheric pressure is in fact the mass of the air above us being pulled downward by gravity. The pressure at any point reflects the mass of atmosphere above that point. We most commonly measure air pressure as it exists at the surface (**surface pressure**), but meteorologists are also concerned with air pressure at heights above the surface. As we go upward through the atmosphere, the mass of atmosphere above necessarily decreases, so pressure must also decrease. Note that pressure is unique among atmospheric variables in that it always decreases vertically. Other variables (such as temperature, moisture, and density) do not necessarily behave this way.

▲ **FIGURE 4-2** Because atmospheric pressure is a response to the weight of the overlying atmosphere, it always decreases with elevation. The pressure at the top of the mountain, p_1, is less than that at the base of the mountain, p_2, because of the greater amount of overlying air.

Despite the fact that the atmosphere is pulled downward by the force of gravity, air pressure is exerted equally in all directions—up, down, and sideways. Revisit the case of the sealed container we just described, with a greater pressure inside than outside. It doesn't matter whether the container is punctured along one of its sides, on its bottom, or on its top; the greater pressure within still forces the air outward.

Here is another way to visualize the fact that air pressure is exerted equally in all directions. Hold your arm directly out from your body. Air pushes on your arm, not only down but also along its side and almost equally on its underside.[1] If pressure were applied only downward, the weight of the air would be so great that even the strongest person would be unable to extend her arm outward. Under normal conditions at sea level, the force would equal about 14.7 pounds on every square inch—quite a load for even a short arm!

Vertical and Horizontal Changes in Pressure

As mentioned in Chapter 1, we need to measure and compare the differences in pressure that arise in different locations, since these differences produce horizontal movements of air. The job is complicated by the fact that elevation varies from place to place. (Recall that high elevations have lower pressure simply because there is less overlying air.) If we used just surface measurements for comparisons, it would be impossible to separate the effects of elevation from the

true pressure differences that lead to wind. To overcome this problem, meteorologists use the concept of sea level pressure.

Surface pressure is the pressure actually observed at a particular location, whereas **sea level pressure** is the pressure that would exist if the observation point were at sea level. Because most land surfaces are above sea level, surface pressure is almost always the lower of the two. Compare, for example, the surface pressures at the tall mountain peak and the nearby valley in Figure 4–2. Although the atmosphere is uniformly distributed over the area, surface pressure at the mountain location is considerably less than at the valley.

Sea level pressure allows us to compare pressure at different locations, taking into account differences in elevation. For locations not very much above sea level, we can get a good indication of sea level pressure by assuming a uniform change in pressure with elevation. At an elevation of 150 m (500 ft), for example, we add 14 mb to the surface pressure to obtain the sea level pressure (roughly a 1 mb increase for every 10 m). For high-elevation sites, however, this method is unreliable because we must account for compressibility of the atmosphere.

As shown in Figure 4–3 on page 99, pressure does not decrease with height at a uniform rate. Instead, it decreases most rapidly at low elevations and gradually tapers off at greater heights. For example, from sea level to 1 km (0.6 mi) in elevation, the average pressure decreases about 100 mb, but between 9 and 10 km it drops only half as much. This nonlinearity exists because, as we know, air is compressible. With atmospheric mass packed more densely at low levels, a small elevation change at low levels takes you through a large amount of atmosphere, resulting in a large pressure drop.

Though surface pressure also varies from place to place, horizontal pressure differences are very small compared

---

[1]The pressure on the bottom of your arm is very slightly greater than at the top. This is because pressure always decreases with altitude, and the top of your arm is a few centimeters higher than the bottom. This difference is extremely small, however, and can be disregarded in this example.

Velocity, Acceleration, Force, and Pressure

It is quite common in everyday conversation to hear the terms *force* and *pressure* used interchangeably, just as *velocity* and *speed* are often considered synonymous. In the language of science, however, intermixing these terms can lead to great confusion. Let us look briefly at how they differ.

Velocity and Acceleration

Any object that moves has a particular **speed**, defined as the distance traveled per unit of time. Speed is related to, but not the same as, velocity. **Velocity** incorporates direction as well as speed. Think, for example, of two cars traveling at 20 meters per second (44 mph) but moving in opposite directions. Though they have the same speed, their velocities are not equal because of their different directions. This distinction is crucial for understanding our next quantity, **acceleration**, the change in velocity (not speed) with respect to time.

Because velocity includes both speed and direction, a change in either speed or direction is an acceleration. Consider a car that at one moment in time travels at 20 m/sec; one second later, the same car has a speed of 19 m/sec; one second later, the speed is 18 m/sec, and so on. As each second goes by, the car's speed decreases 1 m/sec, (note that acceleration can be either positive or—as in this example—negative). An acceleration can also occur as a change in direction with respect to time, even for an object whose speed does not change. A car traveling at a constant speed but gradually turning undergoes an acceleration, just like a car whose speed is changing.

In meteorology there is one particular acceleration of utmost importance—**gravity** (*g*). This acceleration, 9.8 m/sec² (32.1 ft/sec²), is nearly constant across the globe. There is a slight decrease in *g* from equator to pole, and also a very small difference in *g* from the surface to the upper atmosphere. For most applications, however, these variations in *g* are so slight they can be ignored.

Force and Pressure

One of the most important tenets of physical science is Newton's Second Law, which relates the concept of **force** (denoted *F*) to mass *(m)* and acceleration *(a)*. Specifically, Newton's Second Law tells us that the acceleration of an object is proportional to the force acting on it and inversely proportional to its mass. Symbolically, this is expressed as

$$a = \frac{F}{m} \text{ or}$$

$$F = ma$$

Imagine that a fully loaded 18-wheeler truck is stopped at a traffic light next to a bicycle. As the light turns green, both begin to accelerate at the same rate. It is easy to see that if the two remain right next to each other, the much more massive truck requires a larger force (and more powerful "engine"). Likewise, if two bodies with equal mass are subjected to different forces, the one subjected to the greater force will undergo a greater acceleration.

Keep in mind that *F* in the equation above is the net force acting on the object. If various forces are acting simultaneously, they must all be considered together to determine the acceleration; we must account for both the magnitude and direction of each. As we will see with regard to falling raindrops (Chapter 7), forces acting in opposite directions reduce the net force and resulting acceleration, sometimes to zero.

Let's apply Newton's Second Law to our atmosphere. The atmosphere contains 5.14×10^{18} of mass. (To get an idea of what 5.14×10^{18} weighs, picture a million boxcars, each containing a billion elephants.) Multiplying the mass of the atmosphere by the acceleration of gravity, we determine that the force exerted on the atmosphere is about 5.0×10^{19} newtons (a newton [N] is the unit of force it takes to accelerate 1 kg one meter per second every second).

Force divided by the area on which it is exerted equals pressure. So dividing this force of 5.0×10^{19} newtons by the surface area of Earth gives us the average force per unit area, or average surface pressure of about 10.132 newtons per square centimeter. This is equivalent to 1013.2 mb, or about 14.7 pounds per square inch.

Having made the distinction between force and pressure, we now should address the question of how this distinction applies to the atmosphere. The answer is that despite the nearly constant total force of the atmosphere, its gases are not uniformly distributed

to vertical differences. For example, a sea level pressure of 1050 mb is considered very high, yet it is only 4 percent greater than the global average sea level pressure. Moreover, the difference between the highest and lowest sea level pressures over North America on a given day might amount to only about 25 mb—less than a 2.5 percent difference—and even this small percentage difference in pressure would normally be realized over a distance of many hundreds of kilometers.

In contrast, we need only go to the top of a modest hill or tall building to find an equivalent pressure change.

Checkpoint

1. What units are used in measuring pressure?
2. Why does air pressure decrease as you go upward through the atmosphere?

across the planet. Consider a particular area at Earth's surface with an imaginary column extending upward to the top of the atmosphere, as shown in the top right of Figure 1. Greater surface pressures exist at the bases of atmospheric columns that contain a greater number of molecules, and lower surface pressures are found where less air occupies the column. Just how these differences in pressure arise is considered later in this chapter; for the time being, the important point is that surface pressure reflects the mass of atmosphere within the column, as shown in Figure 1.

▶ **FIGURE 1** The downward force of the atmosphere is equal to the mass of the entire atmosphere times the acceleration of gravity. Because the amount of mass and the acceleration of gravity are constant through time, the force of the atmosphere does not change. Pressure is defined as the amount of force exerted per unit of area. Thus, the shaded area in the figure experiences a certain amount of pressure. Pressure varies because the mass of overlying air varies from place to place and time to time.

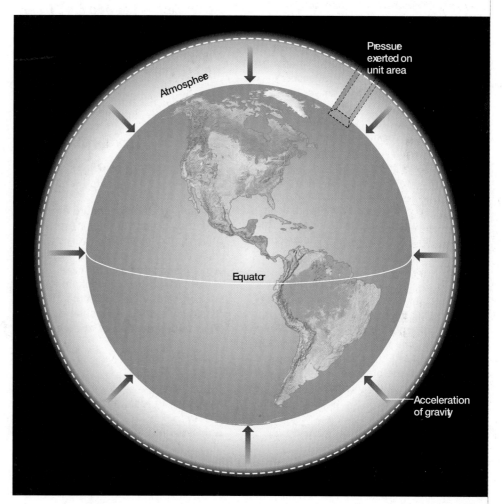

The Equation of State

Everyday experience indicates that gases tend to expand when heated and become denser when cooled. This suggests that temperature, density, and pressure are related to one another. As a matter of fact, their relationship is quite simple. It is described by the **equation of state** (also called the **ideal gas law**),

$$p = \rho RT$$

in which p is pressure expressed in pascals, ρ (the Greek letter rho) is density in kilograms per cubic meter, R is a constant equal to 287 joules per kilogram per kelvin, and T is temperature (in kelvins). To put the equation in words, it tells us that if the air density increases while *temperature is held constant*, the pressure will increase. Similarly, at constant density, an increase in temperature leads to an increase in pressure. (*Box 4–2, Physical Principles: Variations in Density* presents more information on the equation of state.)

4–2 PHYSICAL PRINCIPLES

Variations in Density

Perhaps you have wondered how much air weighs. The air around you has a particular density, and any volume of air contains a certain amount of mass. Changes in the density of air affect many everyday phenomena. For example, the density of the atmosphere influences how much lift a plane gets as it accelerates down a runway in preparation for takeoff. Likewise, automobile fuel injectors must account for variations in density to deliver the right mixture of gasoline and air into the car's engine. The density of air can even affect the amount of resistance the air exerts on a batted baseball, thereby influencing its distance traveled.

But are the variations in density really substantial? We can use the equation of state to see exactly how much variations in temperature affect the air's density. To do this, let's first rearrange the equation to the following:

$$\rho = p/RT$$

Let's now also compare the air density for two situations: a warm day with a temperature of 308 K (35 °C or 95 °F) and a cold one with a temperature of 278 K (5 °C or 41 °F). For consistency, we will assume that the pressure is 100,000 Pa (1000 mb; 100 kilopascals) in both instances.

Applying the equation of state for the warmer day, we find that the density of the air is

$$\rho = \frac{100,000\,\text{Pa}}{287\,\text{J kg}^{-1}\,\text{K}^{-1} \times 308\,\text{K}}$$

$$= 1.13\,\text{kg/m}^3 *$$

When we lower the air temperature to 278 K (5 °C or 41 °F), the equation yields an air density of

$$\rho = \frac{100,000\,\text{Pa}}{287\,\text{J kg}^{-1}\,\text{K}^{-1} \times 278\,\text{K}}$$

$$= 1.25\,\text{kg/m}^3$$

This is nearly 11 percent greater than the density the air had on the warmer day—a nontrivial amount.

In addition to temperature and pressure, the humidity of the air exerts an influence (although only a very minor one) on density. Let's see how. Molecular oxygen (O_2) and nitrogen (N_2) make up most of the mass of the atmosphere and exist in a constant proportion. Other, lesser constituents of the atmosphere are present in different amounts at different

---

*To get the units to balance, you must reduce the units of pascals (Pa) and joules (J) to their fundamental dimensions. Thus, Pa = kg m^{-1} sec^{-2} and J = kg m^2 sec^{-2}.

places and times, and because each has its own unique molecular weight (an expression of the relative amount of mass for molecules), their relative abundance can slightly affect the density of the atmosphere. Among these gases, water vapor usually accounts for about 1 percent of the atmospheric mass. Intuitively, we might assume that a greater humidity would favor a denser atmosphere. Actually, just the opposite is true.

Compare the amount of mass contained in individual molecules of water vapor and of the most abundant atmospheric gases. The molecular weights of nitrogen and oxygen are 28.01 and 32.00, respectively, and the mean molecular weight of the dry atmosphere is 28.5. Water vapor, on the other hand, has a molecular weight of only 18.01. Thus, as the proportion of the air occupied by water vapor increases, an accompanying reduction in the mean molecular weight of the atmosphere must occur. All other things being equal, humid air is less dense than dry air.

Incorporating the effect of varying moisture content requires only a small modification to the equation of state. Calculations using the revised formula show that at 15 °C (59 °F), air density declines by only 0.6 percent for a 1 percent increase in water vapor (from dry air to 1 percent water vapor).

Did You Know?

Throughout the troposphere the atmosphere consists of an extremely high concentration of molecules. At sea level, there are some 1 million billion molecules of air occupying each cubic millimeter of volume. But despite this large number of molecules, the atmosphere consists primarily of empty space. The air molecules are in fact spaced far apart relative to their size so that about only one-tenth of 1 percent of the volume is occupied by atoms. In contrast, liquids, which are much denser than gases, have much smaller intermolecular distances; about 70 percent of their volume is occupied by matter.

Measuring Pressure

Any instrument that measures pressure is called a *barometer*. Two types of barometers are most common for routine observations: one consisting of a tube partially filled with mercury and another that uses collapsible chambers.

Mercury Barometers

The standard instrument for the measurement of pressure is the **mercury barometer** (Figure 4–4), invented by Evangelista Torricelli in 1643. It is a simple device made by filling a long tube with mercury and then inverting the tube so that the mercury spills into a reservoir. Although the tube

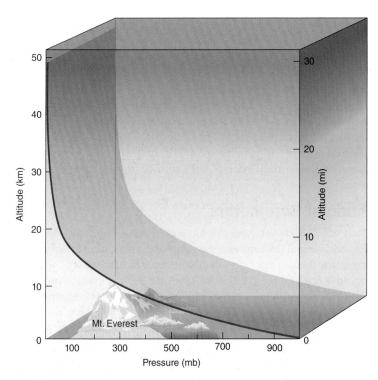

▲ **FIGURE 4–3** Pressure decreases with altitude by about half for each 5.5 km (3.3 mi).

is turned over completely, it does not empty. Instead, the air pushes downward on the pool of mercury and forces some of it up into the tube. The greater the air pressure, the higher the column of mercury.

Barometric pressure is often expressed as the height of the column of mercury in a barometer, which at sea level averages 76 cm (29.92 in.). This measure is inconsistent with the concept of pressure, however, because pressure does not have units of length. In other words, expressing barometric pressure in centimeters or inches is as incongruous as stating somebody's age as "30 miles per hour" or weight as "$1.99." The length measurements obtained from a barometer are a response to the atmospheric pressure but are not direct observations of pressure itself. Meteorologists prefer a unit that measures force per unit area, such as pounds per square inch or millibars. The simple conversion formulas for converting barometric heights to millibars are

$$1 \text{ centimeter} = 13.32 \text{ mb and}$$
$$1 \text{ inch} = 33.865 \text{ mb}$$

Mercury is an excellent fluid for use in a barometer because it is extremely heavy, with a density 13.6 times greater than that of water. This feature allows the instrument to be of a manageable size. Consider that if water were used instead of mercury, the column of water would need to be

▲ **FIGURE 4–4** A schematic showing how a mercury barometer works (left). An actual mercury barometer (right).

about 10 m (33 ft) tall to counterbalance the pressure of the atmosphere. On the other hand, although a three-story, water-filled barometer would be less than portable, it would make for very precise measurements because even small changes in pressure would translate into large height changes.

Corrections to Mercury Barometer Readings

One of the most important tools of a meteorologist is the weather map, which among other things plots the distribution of air pressure across the surface. Before barometer data can be used on the map, however, three corrections must be made to compensate for local factors that affect the readings.

The first correction compensates for the influence of elevation (described earlier in this chapter). If surface pressure values were plotted on weather maps, they would give a false representation of the distribution of the atmosphere. This happens because high elevations have lower surface pressures than do low elevations, even if the sea level pressures are the same. To standardize the observations, we must convert surface pressure readings to sea level values. For a station situated 100 m (328 ft) above sea level, about a centimeter (0.4 in.) is added, corresponding to about 13 mb. At higher elevations, a much greater adjustment might be needed. At Denver, Colorado (the "Mile High City"), for example, the correction is about 16 cm (6.24 in.), or 213 mb.

The second correction deals with the similarity between a mercury barometer and a thermometer. Just as the mercury in a thermometer expands with increasing temperature, so does the mercury in a barometer. The expansion reduces the density of the fluid and requires that it attain a greater height to offset the pressure of the atmosphere. In other words, on a hot day the height of the mercury column is greater than on a cold day, even if the atmospheric pressure is the same. For this reason, mercury barometers always have a thermometer attached to determine the temperature of the instrument, and a correction table tells us what height the mercury column

would be if the temperature were at the standard value of 0 °C (32 °F). At normal room temperature, this correction is small, requiring the subtraction of about 0.25 mm (0.01 in.).

The third correction accounts for the slightly greater acceleration of gravity at higher latitudes. To standardize the readings from all latitudes, we convert them to what they would be if the local gravity were equal to that at 45° north or south, or midway between the equator and poles. The latitudinal changes in gravity are small, however, and corrections are usually on the order of 0.25 mm (0.01 in.).

Aneroid Barometers

Mercury barometers are precise instruments, but they are also expensive and inconvenient to relocate. An alternative instrument for measuring pressure is the **aneroid** (meaning "without liquid") **barometer** (Figure 4–5a). Aneroid barometers are relatively inexpensive and can be quite accurate. They contain a collapsible chamber from which some of the air has been removed (b). The atmosphere presses on the chamber and compresses it by an amount proportional to the air pressure. A pointing device connected to a lever mechanism indicates the air pressure.

Aneroid barometers, which are often found in homes, must be calibrated when first installed. The user simply finds out the current sea level pressure and sets the instrument by turning a small screw on the back of the casing. Because there is no expandable fluid in an aneroid barometer, the instrument requires no temperature correction. Furthermore, the effects of altitude and latitude are already accounted for when the instrument is first calibrated. Thus, once calibrated, an aneroid barometer gives the sea level pressure without corrections or adjustments.

Sometimes it is useful to have a continual record of pressure through time. Aneroid devices that plot continuous values of pressure are called **barographs** (Figure 4–5c). A rotating drum (usually set to one rotation per week) turns a chart so that a pen traces a permanent record of the changing pressure.

(a)

(b)

(c)

▲ **FIGURE 4–5** An aneroid barometer (a) and its workings (b). A barograph (c).

Checkpoint

1. How does a mercury barometer work?
2. What corrections need to be made to readings from a mercury barometer and why are they necessary?

TUTORIAL
PRESSURE GRADIENTS

Use the animations to explore the processes that cause changes in air pressure from one place to another—both at the surface and aloft.

The Distribution of Pressure

The distribution of sea level pressure across the globe is a highly variable characteristic of the atmosphere. To visualize this distribution, meteorologists plot lines called *isobars* on weather maps.

Each **isobar** is drawn so that it connects points having exactly the same sea level pressure, and locations between any two isobars have pressures between those represented by the two lines. Isobars are drawn at intervals of 4 mb on U.S. surface weather maps, so the pressure difference between

adjacent isobars is the same everywhere on the map. The advantage of this is that the distance between adjacent isobars provides information about how rapidly pressure changes from one place to another. In other words, the spacing of the isobars indicates the strength of the **pressure gradient**, or rate of change in pressure, in the same way that spacing of isotherms reveals temperature gradients. A dense clustering of isobars indicates a steep pressure gradient (a rapid change in pressure with distance), while widely spaced isobars indicate a weak gradient.

By way of example, Figure 4–6 maps the sea level pressure distribution as it existed on March 4, 1994. The pressure over New England and southeast Canada was lower than over most of the West, and the strongest pressure gradient was over eastern North America.

Pressure Gradients

Pressure gradients provide the impetus for the movement of air we call *wind*. Imagine two people pushing against each other. The person who exerts the greater force pushes the other one back, and the greater the difference in force applied, the faster the pushed person will move. The same concept applies to air. If the air over one region exerts a greater pressure than the air over an adjacent region, the higher-pressure

▲ **FIGURE 4-6** A weather map showing the distribution of sea level air pressure on March 4, 1994. Note that the pressure is relatively low over the northeastern United States and eastern Canada. Also note that the highest and lowest pressure on the map are only within about 4 percent of each other.

air will spread out toward the zone of lower pressure as wind. The pressure gradient gives rise to a force called the **pressure gradient force** that sets the air in motion. For pressure gradients measured at constant altitude, we use the term *horizontal* pressure gradient force and call the resulting motion *wind*. Everything else being equal, the greater the pressure gradient force, the greater the wind speed.

Horizontal Pressure Gradients Meteorologists are very concerned with the distribution of pressure when examining weather maps. The map of sea level pressure shown in Figure 4–6 is fairly typical, having low- and high-pressure areas of average magnitude. Notice that the changes in pressure across the map are small. The lowest pressure observed is about 977 mb, while the highest is about 1021 mb. This 44 mb difference represents a mere 4 percent or so of the average pressure. Note also that the physical distance separating the areas of highest and lowest pressure is about 3000 km (1800 mi). In the most general sense, then, the pressure gradients across the map are on the order of 40 mb per 3000 km, or about 1 mb per 75 km. Clearly, on a continental scale at least, pressure gradients are usually small.

On a smaller scale, horizontal pressure gradients can be much greater. Hurricanes, for example, have steep gradients that produce violent and destructive winds. Yet even a hurricane may have a pressure in its interior only about 50 mb less than that just outside the storm, some 300 km (180 mi) away. Such a hurricane would have a pressure change of 1 mb per 6 km, yielding only a 5 percent difference in pressure over a considerable distance. This is in marked contrast to vertical pressure gradients, wherein a drop of 50 mb can occur within a vertical distance of only half a kilometer (0.3 mi).

Did You Know?

The difference in atmospheric pressure between the top and bottom of the Empire State Building is about the same as the pressure difference between the center and the outside of a strong hurricane.

Vertical Pressure Gradients Though we don't often think about vertical changes in pressure and usually aren't exposed to strong vertical air motions, both are important in the atmosphere. We've already seen that atmospheric pressure always decreases with altitude. Notice, for example, in Figure 4–3 that the mean sea level pressure of 1013.2 mb decreases to 500 mb at an altitude of 5640 m (about 18,000 ft). Thus, the average vertical pressure gradient in the lower half of the atmosphere is about 500 mb per 5640 m, or just less than 1 mb per 10 m. Compare that to the horizontal pressure gradient of an average hurricane, which we saw to be about 1 mb per 6000 m. The *average* vertical pressure gradient in this example is 600 times greater than the *extreme* horizontal pressure gradient associated with a hurricane! In sum, vertical pressure gradients are very much greater than changes in horizontal pressure and, as we will see in Chapter 8, strongly affect general atmospheric motions.

Hydrostatic Equilibrium

You already know that a pressure gradient force causes wind to flow from high to low pressure and that air pressure rapidly decreases with altitude. Given these two facts, you might infer that the wind must always blow upward. If this were the case, it would have troublesome implications for humans on the surface, who would suffocate as all the air around us literally exploded out to space in response to the vertical pressure gradient force.

Before you panic, however, consider a second relevant fact: Gravity pulls all mass, including the atmosphere, downward. Then why doesn't the atmosphere collapse all the way down to the point where we would be able to breathe only by getting on our hands and knees and sucking up the air that has fallen to the surface? Because the vertical pressure gradient force and the force of gravity are normally of nearly equal value and operate in opposite directions, a situation called **hydrostatic equilibrium**.

When the gravitational force exactly equals the vertical pressure gradient force in magnitude, no vertical acceleration occurs. When the gravitational force slightly exceeds the vertical pressure gradient force, downward motions result. Such downward motions are always very slow. On the other hand, the upward-directed pressure gradient force sometimes greatly exceeds the gravitational force, and updrafts in excess of 160 km/hr (100 mph) can develop. Such updrafts are associated with powerful thunderstorms. Although the gravitational and vertical pressure gradient forces are normally almost in balance, the exact value of each varies from place to place and time to time. The downward gravitational force on a volume of air is proportional to its mass (remember that force = mass × acceleration), so a dense atmosphere experiences a greater gravitational force than does a sparse atmosphere. Thus, if a dense atmosphere is to remain in hydrostatic equilibrium, it must have a greater vertical pressure gradient force to offset the gravitational force.

Examine the two identical columns of air in Figure 4–7a. Both have a surface pressure of 1000 mb that decreases to 500 mb at 5640 m above the surface. If the column on the right is heated, as shown in (b), it still contains as much mass as it did before, but the 500 mb level now lies at 5700 m above the surface. Thus, there is a greater distance required for the pressure to decrease 500 mb; in other words, there is a lessened vertical pressure gradient. At the same time, the density of the air decreases because the same amount of mass occupies a larger volume. So the weaker vertical pressure gradient and the decreased density are interrelated. This relationship has a major effect on horizontal motions in the upper atmosphere, which in turn greatly affect surface patterns. Therefore this concept should be thoroughly understood before continuing. (For mathematical details on the relation between density and the vertical pressure gradient, see *Box 4–3, Physical Principles: The Hydrostatic Equation*.)

The Hydrostatic Equation

The concept of hydrostatic equilibrium (in which the vertical pressure gradient force is equal and opposite to the gravitational force) can be succinctly summarized by the **hydrostatic equation**:

$$\frac{\Delta p}{\Delta z} = -\rho g$$

By convention, the Greek letter delta (Δ) stands for "change in." In this case, (Δp) refers to a change in pressure, while Δz refers to the change in altitude. Thus, $\Delta p / \Delta z$ on the left side of the equation refers to the change in pressure for a unit of increase in altitude.

We have met the symbols ρ and g before as density and the acceleration of gravity, respectively. The negative sign on the right-hand side accounts for the fact that

pressure decreases with height; that is, the left-hand side is always negative. For the two sides to balance, the right-hand side must also be negative.

Thus, the hydrostatic equation states that the rate at which pressure decreases with height equals the product of the air density and the acceleration of gravity. But because the acceleration of gravity is virtually constant, the rate at which pressure declines with altitude is determined almost completely by the density of the atmosphere. In particular, higher-density air has a greater vertical pressure gradient.

As an example, let us compare the two columns of air in Figure 4–7b, supposing that their temperatures are 0 °C and 40 °C. Using the surface pressure of 1000 mb, the equation of state gives the density of the warm air as 1.1 kg per cubic meter. At the same pressure, the cool air must have

higher density, in this case 1.3 kg per cubic meter. Assuming hydrostatic equilibrium, the corresponding vertical pressure gradients at the surface are

Warm Air Column

$$\frac{\Delta p}{\Delta z} = -1.1(9.8) = -10.8 \text{ Pa/m}$$

Cool Air Column

$$\frac{\Delta p}{\Delta z} = -1.3(9.8) = -12.8 \text{ Pa/m}$$

This confirms our earlier reasoning, where we concluded that pressure declines more rapidly in a cool, dense air column than in a warm air column. As we discuss in the body of the text, this sets up an upper-level horizontal pressure gradient between warm and cool air.

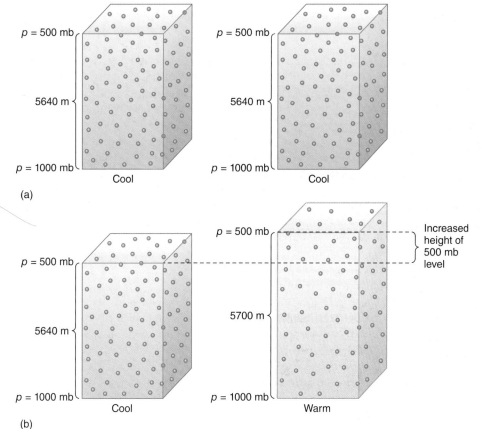

(a)

(b)

◀**FIGURE 4–7** Two columns of air with equal temperatures, pressures, and densities (a). Heating the column on the right (b) causes it to expand upward. It still contains the same amount of mass, but it has a lower density to compensate for its greater height. The pressure drops 500 mb over 5700 m within the warm air; it only take 5640 m of ascent for the pressure to drop the same amount in the cool air. Thus, the cool air has the greater vertical pressure gradient.

Checkpoint

1. What is the pressure gradient force?
2. Use the concept of hydrostatic equilibrium to explain why a cool, dense column of air has a greater vertical pressure gradient than a warmer, less dense column of air.

TUTORIAL

PRESSURE GRADIENTS

Use the animations in Section 6 to observe the changes that occur in the thickness of a column of air when heat is applied.

Horizontal Pressure Gradients in the Upper Atmosphere

Just as horizontal pressure gradients occur near the surface, they likewise occur in the upper atmosphere. As we have already seen, atmospheric pressure decreases more rapidly with height (that is, vertically) in a cold, dense air column than in a warm, less dense air column. Looking at Figure 4–7b, you can see that for the cool column at 5640 m above the surface, the air pressure is 500 mb; for the warm column the air pressure at 5640 m above the surface is greater than 500 mb. Thus, in the mid-troposphere around the 5640 m height there is a horizontal pressure gradient, with lower pressure over the cool column. Equivalently, the height of the 500 mb level is lower in the cool column.

If we look at the height of some given pressure level (such as the 500 mb level) on a map its height will often vary from one place to another. More specifically, the surface will slope downward in the direction of colder air, because cold dense air has stronger vertical pressure gradients and will attain a particular air pressure at a lower height (assuming for simplicity's sake that sea level pressure is uniform). Where a horizontal pressure gradient exists, there must also be a slope in the height of a particular pressure level, with heights decreasing toward colder air. It so happens that the horizontal pressure gradient force is proportional to the slope of the height of that pressure level. If we know the slope, we know the pressure gradient force. Figure 4–8 shows the distribution of the 500 mb level in an idealized atmosphere, with a gradually decreasing temperature toward the North Pole. Notice that the decrease in temperature toward the pole causes a decrease in the height of the 500 mb level, with the surface sloping downward toward colder air. On the "ground" of the diagram are contour lines showing the height of the 500 mb surface. The contour labels tell how high you must go to find a pressure of 500 mb. For example, if you stand on a line labeled 5500 m, the 500 mb surface is 5.5 km above you. The heights decrease toward the north; thus, the contour values also decrease northward.

TUTORIAL

PRESSURE GRADIENTS

Use the animations in Section 3 to change the horizontal pressure gradients and view the resulting changes in the slope of particular pressure surfaces.

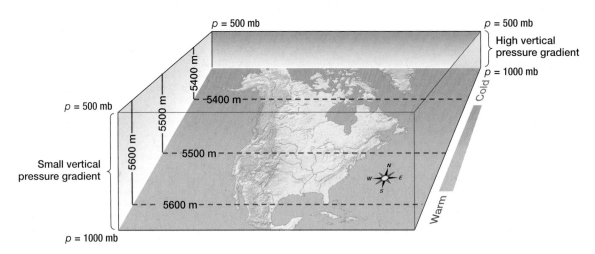

▲ **FIGURE 4–8** The gradual poleward decrease in mean temperature results in denser air occurring at high latitudes. As indicated by the hydrostatic equation, pressure drops more rapidly with height at high latitudes and lowers the height of the 500 mb level. The dashed lines depict the heights of the 500 mb level as they would be drawn on a 500 mb weather map.

Figure 4–9 shows a real 500 mb map for May 3, 1995. The height contours are labeled in decameters (units of 10 meters—so 564 decameters equals 5640 meters); thus, heights range from 5880 m in the south to 5220 m in the extreme northwest. Contours for 500 mb maps are drawn at 60 m intervals (every 6 decameters). Overall, the pattern is consistent with a decrease in height from south to north, following the temperature gradient from south to north.

Where the height contours are close together, the pressure gradient force is large. Thus, on that day, upper-level winds were strongest over Newfoundland, with speeds above 160 km/hr (100 mph). Over Nebraska, the pressure gradient force was much weaker, and winds were only 35 km/hr (22 mph).

Notice that the range in heights is only about 660 m from highest to lowest, a change of about 10 percent across a huge

distance. Like horizontal pressure gradients at the surface, these upper-atmosphere gradients are small and the pressure surfaces are nearly horizontal. Nevertheless, these weak gradients can produce significant accelerations and high winds, especially in the upper atmosphere where friction is nearly absent and density is low.

Upper-level maps are produced twice each day by the National Weather Service and similar agencies in other countries. They are extremely important for weather forecasting, as we will discuss in Chapter 13. In addition to the maps of the 500 mb level, similar maps are produced for the 850, 700, and 300 mb levels. These correspond to conditions at about 1500, 3000, 10,000, and 13,000 m above sea level (5000, 10,000, 33,000, and 43,000 ft), respectively. Keep in mind that these maps depict the varying heights of these individual pressure levels.

▲ **FIGURE 4–9** The distribution of the height of the 500 mb level on May 3, 1995. The height contours are labeled in units of decameters.

Checkpoint

1. How do horizontal changes in temperature in the lower atmosphere affect the height of the 500 mb level?
2. For the map in Figure 4–9, describe how the 500 mb level and the pressure gradient change along the Pacific Coast from Alaska to California.

Forces Affecting the Speed and Direction of the Wind

The unequal distribution of air across the globe establishes the horizontal pressure gradients that initiate the movement of air as wind. If no other forces were involved, the wind would always flow in the direction of the pressure gradient force. However, the situation is complicated somewhat by the effect of two other forces. The first arises from planetary rotation and alters the direction of the wind. The second force, friction, slows the wind.

TUTORIAL
ATMOSPHERIC FORCES AND WIND

Use the animations to observe the movement of air parcels in response to changing conditions.

The Coriolis Force

The pressure gradient force sets air in motion from higher pressure to lower pressure, and the magnitude of the pressure gradient force is most responsible for determining the strength of the wind. But to understand the direction the wind travels, you must consider the rotation of Earth, which gives rise to an apparent deflection (turning) of the wind. The phenomenon, called the **Coriolis** (pronounced Core-ee-OH-liss) **force**,[2] also causes an apparent deflection relative to Earth's surface in the flight of cannonballs, migrating birds, and jet aircraft. In fact, it has an impact on anything that moves in any direction. In all cases, however, the deflections are *apparent*, meaning they emerge only because we track motions across the rotating Earth. Therefore, this "force" is in one sense fictional. For that reason some texts prefer the term *Coriolis effect*.

To understand the effect Earth rotation has on moving objects, imagine a counterclockwise rotating platform with one person standing at the center and another person at one of the edges, as in Figure 4–10a. If the person at the center lobs a ball toward the person at the edge of the platform, the ball will travel in a straight direction toward the original target. But at the same time the rotating platform

moves the intended receiver away from the path of the ball. Figures 4–10b and 4–10c show the position of the ball and the movement of the platform after successive time increments. By the time the ball reaches the edge of the platform (d), the person who was supposed to have caught it has moved far enough away that he or she has no chance of catching it. To the people on the platform, the ball would appear to be turning to its right, even though it is in fact moving in a straight line. As it moves in a straight direction, it sweeps a curved path relative to the turning platform beneath it. (This concept is easier to visualize with the aid of animations. We *strongly* encourage you to refer to the Coriolis Force tutorial.)

Imagine the same phenomenon at Earth's North Pole. We know that Earth is nearly spherical, but to us it appears flat because of its large (relative to us) size. This means that at the North Pole, Earth is moving very much like the carousel just discussed—only rotating much slower, at once every 24 hours—and is therefore subject to the Coriolis force. The Coriolis force also applies to the wind, just as it does a projectile, so air moving in a straight line across the North Pole appears to be turning to its right. Does such a deflection occur at the equator? The answer is no, because the planet's 24-hour rotation imparts no twisting motion at the equator. Instead, any point on the surface of the equator travels a 40,000 km (24,000 mi) sweep around Earth's axis (as is shown in the Coriolis tutorial). In between the equator and the poles, there is a gradual increase in the strength of the Coriolis force with increasing latitude. The same relationship between latitude and the magnitude of the Coriolis force holds true in the Southern Hemisphere, with the only difference being that the apparent deflection is to the left, rather than to the right, because that hemisphere rotates in a clockwise direction.

The effect of the Coriolis force moving across different latitudes is illustrated in Figure 4–11. In part (a), we imagine no rotation of Earth and see that an aircraft traveling along a line of longitude from the North Pole southward would need no course adjustment to move southward. But the same plane directed southward undergoes a deflection to its right in the real (rotating) world as it moves equatorward, and the plane ends up west of the intended path if no adjustment is made.

The magnitude of the Coriolis force *increases* with wind speed (though this may seem counterintuitive at first). It is critically important to understand that the *Coriolis force involves the deflection that occurs over a given increment of time* (say, 1 second)—not the amount of deflection that occurs when an object moves from one particular location to another. Consider this: If an object has the minimum speed possible (zero), it would undergo no deflection. A very slow-moving object might be deflected slightly, but over any brief increment of time the deflection would be minimal. A fast-moving object would cover a considerable distance over the same time interval and therefore be subject to a greater deflection. (See *Box 4–4, Physical Principles: The Coriolis Force* for further information.)

---

[2]For G. G. de Coriolis (1792–1843), who gave the first mathematical explanation for this phenomenon.

The Coriolis Force

In describing wind and other motions, convention says we take the surface of Earth as a reference frame. For example, when we say there is a 10-meter-per-second wind, we mean the air is moving past the surface at 10 m/sec. Because the surface rotates, we are describing motions relative to a rotating reference frame. The result is that an object moving in a straight line with respect to the stars appears to follow a curved path on Earth's surface, as shown in Figure 4–10. In Figure 4–10a, the ball is at the center of a counterclockwise-rotating platform. As it moves toward the pin (b) and (c), the platform rotates beneath the ball. By the time the ball reaches the edge (d), the platform has turned considerably and so has the pin at which the ball was aimed. Using the platform as a reference, the ball appears to have been deflected to the right, although we know the ball traveled in a straight path.

The magnitude of the Coriolis force is determined by the rate at which the planet rotates (which is constant), the speed of the object (or wind) as it moves across the surface, and latitude. More precisely, it can be expressed as

$$F_c = 2\,\Omega\,v\,\sin\,\varphi$$

where F_c is the Coriolis force, Ω is Earth's rotation rate (1 revolution per day), v is wind speed, and φ is latitude (note that at the equator $\varphi = 0$, and there is no Coriolis force). As written here, the equation gives the Coriolis force per unit mass; that is, the force per kilogram of moving air. Combining the information with Newton's Second Law, which tells us that acceleration is force per unit mass, we see that F_c is an acceleration—specifically, the Coriolis acceleration. That said, from here forward we won't draw a distinction between the Coriolis acceleration and the Coriolis force, but instead will use the two interchangeably.

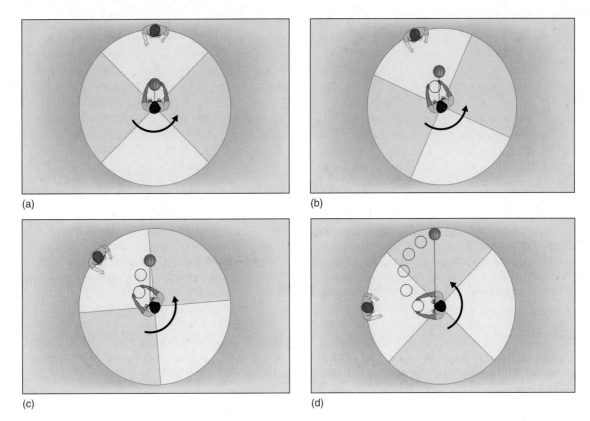

(a)　　　(b)　　　(c)　　　(d)

▲ **FIGURE 4–10** As an object moves along a rotating surface, its motion appears to curve away from the target. The ball in the center of the counterclockwise-rotating platform in (a) is about to move toward the target at the top of the figure. As the ball moves toward the target, the rotation of the platform causes the target to move away from its original position. This continues as the ball moves away from the center (b) and (c), so that by the time it nears the edge of the platform the ball appears to have curved to its right (d). Because all points on Earth (except along the equator) undergo some rotation, all moving objects experience this apparent displacement, which is ascribed to a force called the *Coriolis force*. This force not only acts upon projectiles; it acts on the movement of the atmosphere as well.

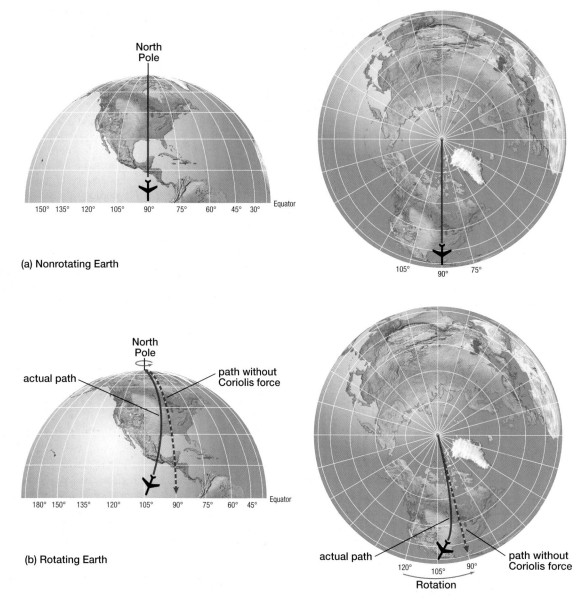

▲ **FIGURE 4-11** The Coriolis Force alters the movement of anything moving across Earth's surface (except at the equator). In (a) we see how an aircraft flying southward from the North Pole would travel in the absence of any Coriolis force; the flight maintains a steady path along a line of longitude. With the Coriolis force, however, the plane is deflected to its right and ends up west of where it was originally headed, unless adjustments are made by the flight crew.

After examining the tutorial on this topic, you should have a solid grasp of the following four fundamental characteristics of the Coriolis force:

1. The Coriolis force produces an apparent deflection in all moving objects, regardless of their direction of motion. The deflection is to the right in the Northern Hemisphere and to the left in the Southern Hemisphere.

2. The Coriolis force is zero at the equator and increases with increasing latitude, reaching a maximum level at the poles.

3. The Coriolis force acting on any moving object increases with the object's speed.

4. The Coriolis force changes only the direction of a moving object, never its speed.

The Coriolis force is not large relative to other commonly encountered forces. To produce detectable effects, the Coriolis force must act over relatively long periods of time. It is important mainly for the motion of objects traveling long distances, such as the air circulating around a hurricane. For motions that occur over short distances, its effect is negligible. Thus, although a basketball shot undergoes a minor deflection because of Earth's rotation, the deflection is so small that it cannot be used as an excuse for a miss. For the same reason, the

Coriolis force does not materially influence the motion of water spiraling into a bathtub or kitchen drain, contrary to what people commonly believe. Whether the water spirals clockwise or counterclockwise is usually determined by an asymmetry in the shape of the basin, or by an initial direction of spin imparted by the water supply. The Coriolis acceleration is present, but it cannot produce significant changes in direction because it has so little time to act before the water reaches the drain.

TUTORIAL
THE CORIOLIS FORCE

Use the animations to explore how the Coriolis force due to Earth's rotation affects air flow and depends on latitude and wind speed.

Friction

The other factor that influences the movement of air is **friction**, the force resisting the movement of a fluid or object as it passes along a surface or an adjacent gas or liquid. Air in contact with the surface experiences frictional drag, which decreases wind speed. Air just above, in contact with the slower-moving surface layer, likewise experiences frictional drag, but from the underlying air rather than from the surface. As this layer slows down, air at higher levels is similarly affected. The effects of friction therefore originate from the surface but are found throughout the lower atmosphere.

Generally speaking, friction is important within the lowest 1.5 km (1 mi) of the atmosphere—often called the **planetary boundary layer** or just the **boundary layer**. Friction lowers the wind speed for a given pressure gradient, which reduces the Coriolis force as well. In contrast, air in the **free atmosphere**, above about 1.5 km, experiences negligible friction. In the absence of friction, winds in the upper atmosphere are fundamentally simpler. We begin our discussion, therefore, with a description of upper-level wind patterns.

Checkpoint

1. What is the Coriolis force?
2. In what ways do latitude, wind speed, and direction affect the intensity of the Coriolis Force?

TUTORIAL
ATMOSPHERIC FORCES AND WIND

Use the animations in Section 6 to observe how friction affects air flow at different altitudes.

Winds in the Upper Atmosphere

The ground exerts a frictional force on the air moving along it. This frictional force gradually diminishes with distance from the surface, and at some height (often around 2 km—or

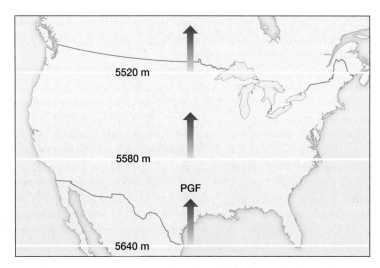

▲ **FIGURE 4-12** The pressure gradient force is directed from south to north in this hypothetical distribution of the height of the 500 mb level. On a nonrotating planet, this would cause air to move from south to north.

a little more than a mile) the frictional force becomes negligible. Thus when explaining winds in the upper atmosphere we deal with the interaction of only two forces: the pressure gradient and Coriolis forces.

Figure 4–12 illustrates the simplest type of pressure pattern that can exist in the free atmosphere. In this case the height contours are straight and parallel to one another, with a pressure gradient force directed northward. Assume that a parcel of air[3] contained in a balloon with a density equal to the surrounding air is tethered to a pole in the free atmosphere. Although the balloon is held in place, it nonetheless is susceptible to the pressure gradient force acting on it.

Figure 4–13 shows what happens when the connecting cord is severed. Initially, the balloon had no movement and therefore no Coriolis acceleration. But after the cord is cut (a), the pressure gradient force propels the balloon slowly toward the low-pressure area. As the pressure gradient force causes the balloon to move faster, there is an accompanying increase in the Coriolis force, and the balloon accelerates farther to its right (in the Northern Hemisphere) (b) and (c). Eventually, the air flows parallel to the height contour lines (d). At this point, the Coriolis force and pressure gradient balance one another, so no net force is acting on the balloon.

From here on, the balloon moves parallel to the height contours, which in this case is a straight line with a constant speed. In other words, the air flow becomes unaccelerated, with unchanging speed and direction. Such nonaccelerating flow is called the **geostrophic flow** (or **geostrophic wind**), and it occurs when the pressure gradient force equals the

---

[3]Meteorologists frequently refer to imaginary small units of air as air parcels.

Coriolis force. Geostrophic flow occurs only in the upper atmosphere where friction is absent and only the Coriolis and pressure gradient forces apply.

You may wonder why the airflow in Figure 4–13d does not turn all the way back toward the high-pressure area. The simplest answer is that if it were to move in that direction, it would have to flow against the pressure gradient force. This would slow down the air, reduce the Coriolis force, and thereby cause the flow to turn back to its left. Likewise, if it were to turn northward, the air would receive a boost from the pressure gradient force. This would lead to a stronger Coriolis acceleration, which would turn the air back to the south. We see that geostrophic flow is stable, meaning that once established, it is not easily disrupted. (For a more detailed look at this process, please refer to the Atmospheric Forces and Wind Tutorial.)

TUTORIAL
ATMOSPHERIC FORCES AND WIND

Use the animations in Section 4 to see how the pressure gradient and Coriolis forces interact to produce winds parallel to the pressure surfaces shown on upper-level maps.

Gradient Flow

In the simple situation in Figure 4–13, the pressure gradient force is uniform, with straight and parallel contours throughout the region. Such situations do occur in nature, but they are the exception rather than the rule. A more common pressure distribution is the type shown in Figure 4–14,

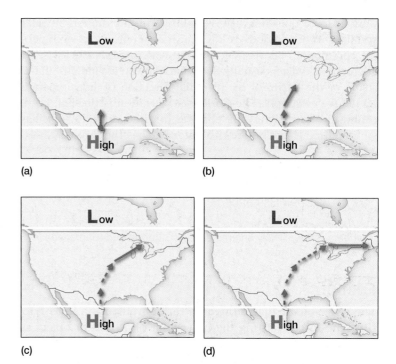

(a)   (b)

(c)   (d)

▲ **FIGURE 4–13** Geostrophic wind. Assume that there is a stationary parcel of air in the upper atmosphere subjected to a south-to-north pressure gradient force (a). If the parcel is tethered to an imaginary pole, no movement of the parcel can take place. Once the imaginary cord is cut, the horizontal pressure gradient accelerates the parcel northward (b). Initially, when the wind speed is low (as indicated by a short arrow), the Coriolis force is small. As the parcel speeds up (longer arrow), the strength of the Coriolis force increases and causes greater displacement to the right (c). Eventually, the wind speed increases the Coriolis force sufficiently to cause the air to flow perpendicular to the pressure gradient force.

5520 m

5580 m

5640 m

5700 m

▲ **FIGURE 4–14** Gradient wind. Gradient flow occurs in the upper atmosphere where the flow is unaffected by friction from the surface. Geostrophic flow is a special case of gradient flow where there is no acceleration.

in which the height contours curve and assume varying distances from one another. In the absence of friction, the air flows parallel to the contours, for the same reasons as in geostrophic flow. But this type of flow is not truly geostrophic because it is constantly changing direction and therefore undergoing an acceleration. In order for the air to follow the contours, there must be a continual mismatch between the pressure gradient and Coriolis forces. Meteorologists refer to this movement as **gradient flow** (or as **gradient wind**). Like geostrophic flow, gradient flow develops only in the absence of friction, and the wind flows perpendicular to the pressure gradient. In fact, geostrophic flow is simply a special case of gradient flow, arising if the contours happen to be straight and parallel.

Supergeostrophic and Subgeostrophic Flow

When airflow in the upper atmosphere flows around an area of high pressure, it must continually turn to its right in the Northern Hemisphere (to its left in the Southern Hemisphere). Similarly, air flowing around an upper level low in the Northern Hemisphere must constantly turn to its left (to its right in the Southern Hemisphere). This turning occurs because the Coriolis force is greater or less than what it would be for the same pressure gradient if the air were not forced to turn (i.e., geostrophic flow). Figure 4–15a examines gradient air flow around a circular region of high pressure in the upper atmosphere of the Northern Hemisphere. The pressure gradient force (PGF) is directed away from the center of high pressure while the Coriolis force is directed inward, as shown by the respective arrows. Here the air does not flow in a straight path; rather, it turns to the right and remains parallel to the pressure gradient. In order for it to turn right, the Coriolis force must exceed the pressure gradient force. However, for this to happen, the wind speed must be higher than what would occur under geostrophic conditions. Given the same pressure gradient force, the wind flows faster around a high-pressure region than it does where the height contours are straight because a larger Coriolis force is required to keep the flow turning to its right. This situation, in which the Coriolis force exceeds the pressure gradient force and causes the air to turn, is called **supergeostrophic flow**.

The opposite situation is illustrated in Figure 4–15b, where the air rotates counterclockwise around a zone of low pressure. The pressure gradient force directs the flow inward toward the low pressure, and the Coriolis force turns the air to the right. Once again, the air flows parallel to the height contours, but the forces do not balance one another—if that happened, the air would move in a straight line. If the air turns counterclockwise, it must be the case that the Coriolis force is weaker than the pressure gradient force. The existence of a weaker Coriolis force demands that the wind flow more slowly than it would if it were geostrophic. Such flow is said to be **subgeostrophic**.

(a)

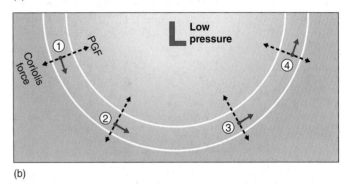
(b)

▲ **FIGURE 4–15** Supergeostropic flow (a) occurs in the upper atmosphere around high-pressure systems. As the air flows, it is constantly turning to its right. This turning motion occurs because the Coriolis force has a greater magnitude than the pressure gradient force (as represented by the length of the dashed arrows). Observe the changing direction of the four solid arrows 1 through 4. Subgeostrophic flow (b) occurs in the upper atmosphere around low-pressure systems. The pressure gradient force is greater than the Coriolis force and the air rotates counterclockwise in the Northern Hemisphere.

Checkpoint

1. How are geostrophic flow and gradient flow similar? How are they different?
2. How are the Coriolis force and the pressure gradient force related in supergeostrophic flow? In subgeostrophic flow?

Near-Surface Winds

Friction makes winds near the surface slower than those in the middle and upper atmosphere, given equal pressure gradients. The lower wind speeds reduce the Coriolis force and thereby prevent the flow from becoming gradient or geostrophic. Thus, the winds in the boundary layer do not flow parallel to the isobars; rather, they cross the isobars at an angle as they blow from high to low pressure. As always, there is deflection to the right in the Northern Hemisphere and to the left in the Southern Hemisphere (Figure 4–16). The angle of airflow relative to the pressure gradient is not constant, being greater at higher latitudes (because of the stronger Coriolis force) and over smooth surfaces where friction is minimized (such as oceans and large lakes).

(a)

(b)

▲ **FIGURE 4–16** Geostrophic flow cannot exist near the surface. Friction slows the wind, so that the Coriolis force is less than the pressure gradient force. Thus, the air flows at an angle to the right of the pressure gradient force in the Northern Hemisphere (a) and to the left in the Southern Hemisphere (b).

Anticyclones, Cyclones, Troughs, and Ridges

Experience tells us that the sea level pressure across the globe is not haphazardly distributed into disorganized, widely scattered zones of high and low pressure. Instead, it is usually organized into a small number of large high- and low-pressure regions. Thus, on a given day North America might have four or five major pressure centers at one time.

 TUTORIAL
ATMOSPHERIC FORCES AND WIND

Use the animations in Section 7 to observe how pressure gradient, Coriolis, and frictional forces affect wind flow and combine to create cyclones and anticyclones.

Anticyclones

Enclosed areas of high pressure marked by roughly circular isobars or height contours are called **anticyclones**. The wind rotates clockwise around anticyclones in the Northern Hemisphere, as the Coriolis force deflects the air to the right and the pressure gradient force directs it outward (Figure 4–17). In the boundary layer the air spirals out of anticyclones (a), while in the upper atmosphere it flows parallel to the height contours (b). In the Southern Hemisphere, the flow is counterclockwise (c) and (d).

Northern Hemisphere surface

(a)

Northern Hemisphere upper atmosphere

(b)

Southern Hemisphere surface

(c)

Southern Hemisphere upper atmosphere

(d)

▲ **FIGURE 4–17** Air spirals clockwise out of anticyclones in the Northern Hemisphere (a) and rotates clockwise around the high in the upper atmosphere (b). The flow is reversed in the Southern Hemisphere (c) and (d).

Cyclones

Closed low-pressure systems are called **cyclones**. As shown in Figure 4–18, air at the surface spirals counterclockwise into cyclones in the Northern Hemisphere and clockwise in the Southern. Now take a close look at Figure 4–18a. It may appear that the air turns to its *left* as it moves toward the low, but this is not the case. Figure 4–19 explains this apparent contradiction.

The air in position 1 has a pressure gradient force that directs it northward to the center of the low, but the Coriolis force deflects the air to the right so that it ends up in position 2. At position 2, the pressure gradient force still directs the air (now just to the west of due north) toward the center of the cyclone, and again the Coriolis force turns it to the right. This shift in the direction of the pressure gradient gives the trajectory the appearance of being deflected to the left. But at any moment in time, it is accelerated to the right.

The term *cyclone* sometimes causes confusion because people associate it with major tropical storms near southern Asia. In fact, any closed low-pressure system (even one that produces nothing other than gentle breezes and a few clouds) is a cyclone. Although the violent tropical storm goes by the same name, in this discussion we use the term in its most generic sense. We will discuss the more specific type of cyclone in Chapter 12.

Northern Hemisphere surface

(a)

Northern Hemisphere upper atmosphere

(b)

Southern Hemisphere surface

(c)

Southern Hemisphere upper atmosphere

(d)

▲ **FIGURE 4-18** Air spirals counterclockwise into surface cyclones in the Northern Hemisphere (a) and rotates counterclockwise around an upper-level low (b). The flow is reversed in the Southern Hemisphere (c) and (d).

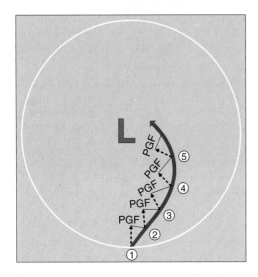

▲ **FIGURE 4-19** The counterclockwise flow of air into surface cyclones in the Northern Hemisphere turns to the left but is nevertheless deflected to the right by the Coriolis force. If there were no Coriolis force, it would cross the isobars at a right angle; instead, it is everywhere deflected to the right.

In addition to their characteristic horizontal winds, particular vertical motions are associated with cyclones and anticyclones. As it approaches the center of the cyclone, air moving into a low-pressure system at the surface has nowhere to go but up. The rising motions, as we will see in the next chapter, promote cloudy or stormy weather. In contrast, the air in anticyclones moving out from the center is replaced by sinking air. Anticyclones therefore typically have clear skies and fair weather. As a general rule, anticyclones are larger than cyclones and have weaker horizontal pressure gradients and weaker winds.

Figure 4–20 shows the distribution of sea level pressure across much of North America on October 18, 2007, with a large cyclone occupying much of the northcentral United States. Over the Dakotas the wind generally approaches the low pressure from the northwest (the wind barbs show the direction that the wind is blowing *from* and the number of tick marks indicate the approximate wind speed. See the left side of Figure 1–16 for further information on interpreting the wind barbs). Airflow approaches the low out of the southeast over Indiana, and generally comes out of the east over the Great Lakes region. A high-pressure area dominates much of the western United States. From western Montana to central California there are few isobars, indicating a weak pressure gradient. Not surprisingly, wind speeds are very low over most of the west, with the exception being over the extreme Pacific Northwest where a well-developed cyclone approaches from the west.

Troughs and Ridges

Pressure systems often occur as closed cells, which are fully enclosed in a somewhat circular fashion. However, many low- and high-pressure systems occur instead as elongated areas called **troughs** (low pressure) and **ridges** (high pressure). Examples are shown in Figure 4–21. There is a tendency for pressure to be distributed as cyclones and anticyclones at the surface and gradually give way to ridges and troughs in the upper atmosphere. This is depicted in Figure 4–22, which shows simultaneously obtained weather maps for the 850, 700, 500, and 300 mb levels.

Troughs and ridges also appear on surface maps. Figure 4–23 shows the distribution of sea level pressure on October 21, 2007, three days after the situation shown in Figure 4–20. In this map, a new anticyclone has developed over the west, but in this case the high pressure system has a much stronger pressure gradient than was found in Figure 4–20. Notice the northwestward bending of the isobars along the California coast (highlighted by the dashed brown line) extending from the region of low pressure off the southern California coast. This intrusion of lower pressure is a trough. This example is a noteworthy one because the strong pressure gradient over much of California fostered very strong winds, high temperatures, and dry air, which led to one of the worst outbreaks of wildfires in California history, with over 1500 homes destroyed.[4]

---

[4]This was a classic example of a Santa Ana wind, which is explained in Chapter 8.

Checkpoint

1. What is the difference between a cyclone and a trough?
2. In Figure 4–23, what resulted from the trough represented by the dashed brown line along the California coast? Explain.

▲ FIGURE 4–20 Distribution of sea level pressure on October 18, 2007. Air rotates counterclockwise into the large cyclone over the central United States. Winds are generally weak out of the anticyclone out west.

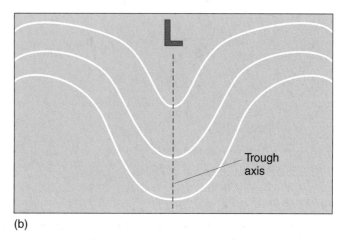

(a)   (b)

▲ FIGURE 4–21 Elongated zones of high and low pressure are called ridges (a) and troughs (b), respectively.

▲ FIGURE 4–22 Maps of the 850, 700, 500, and 300 mb levels for the same date and time. Observe how the pattern of cyclones and anticyclones at the 850 mb level gradually gives way to one of troughs and ridges.

▲ FIGURE 4–23 The surface weather map for October 21, 2007. An area of high pressure exists over much of the interior western United States and a trough is found just west of California. The tight packing of isobars produced very strong winds (though not reflected in the station models at this point in time) that fostered catastrophic wildfires in southern California.

Measuring Wind

Wind direction and velocity are measured at all major weather stations. Direction is always given as that from which the wind blows, so that a "westerly" wind is one from the west. It is often expressed by its azimuth. As shown in Figure 4–24, the **azimuth** is the degree of angle from due north, moving clockwise. Thus, due north is represented as 0° (or 360°), east has an azimuth of 90°, south of 180°, and west of 270°. A simple device for observing wind direction is the **wind vane**.

Wind speeds are measured with anemometers. **Anemometers** have rotating cups mounted on a moving shaft. Wind blowing into the cups turns the shaft and generates an electrical current. The strength of the current is proportional to the wind speed, which is then plotted on a strip chart or entered digitally into a computer.

FIGURE 4-24 The azimuth expresses direction based on the angle away from North (0°), moving clockwise. Thus, for example, due east is represented as 90°, south 180°, and west 270°.

Did You Know?

There is a famous dictum known as **Buys-Ballot's law** that states that if you stand with your back to the wind in the Northern Hemisphere, low pressure will be on your left (in the Southern Hemisphere the low pressure will be to your right). However, this is neither a true law in the physical sense nor was it originally postulated by Christoph Buys-Ballot (pronounced Bies Bal-LOW), for whom it is named. This relationship between pressure and wind direction was actually first postulated by William Ferrel, a meteorologist who did some early work on the general circulation of the atmosphere. Buys-Ballot was the first to offer empirical evidence for the idea in a paper published in 1857, and the "law" remains named for him to this day.

Looking like an airplane without wings, an **aerovane** (Figure 4–25) indicates both wind direction and speed. When the wind changes direction, it pushes against the tail and points the aerovane toward the wind. Like an anemometer, the aerovane has a propeller to generate an electrical current that indicates the wind speed.

▲ **FIGURE 4-25** An aerovane.

Winds in the middle and upper troposphere are every bit as important to weather analysis as surface winds and are especially significant to aviation. (See *Box 4–5, Focus on Aviation: Strong Winds and the Little Rock, Arkansas, Air Disaster,* for additional information on how strong winds can impact takeoffs and landings.) Upper-level wind measurements are obtained by **rawinsondes**, radiosondes (Chapter 3) whose movement is tracked by radar. These are launched twice a day at meteorological offices throughout the world.

4–5 FOCUS ON AVIATION

Strong Winds and the Little Rock, Arkansas, Air Disaster

Aviation in Canada and the United States has a stellar safety record, making air travel within the two countries the safest form of travel. In all of 1998, not a single commercial air traffic fatality occurred within the United States. Not only that, but planes travel millions of miles and undertake many flights in inclement weather. However, that does not mean that bad weather can be ignored, especially during takeoffs and landings. This was clearly the case at 11:50 P.M., CDT, on June 2, 1999, at Little Rock, Arkansas. Despite the fact

that a strong thunderstorm was under way and strong winds were being recorded, American Airlines flight 1420 from Dallas/Ft. Worth came in for a landing with 145 people on board.

As the plane touched down, it immediately went into a spin, skidded sideways down the runway for more than a mile, and crashed into a light tower. The plane then broke apart and burst into flames. The death toll, though regrettable, was surprisingly low, with 10 dead. More than 80 people were injured and, miraculously, 51 did not even require hospital treatment.

Investigators immediately focused their attention on the strong winds as the prob-

able cause of the crash. Little Rock was under a severe thunderstorm warning issued 34 minutes prior to the accident, stating that "the main severe weather threat in the warning area is strong and gusty winds." Just 1 minute before the crash, the automated weather sensor next to the runway recorded a wind gust of 140 km/hr (87 mph). Under these conditions, the wisdom of the attempted landing was truly questionable.

Nothing that occurred on June 2 should shake our confidence in our very safe air traffic system. Nonetheless, it would be foolish to become complacent or overconfident in the presence of severe weather.

Summary

The distribution of air pressure across Earth sets the winds in motion and determines whether air will rise or sink. Air pressure increases with both the density and temperature of the air, as dictated by the equation of state. Because pressure decreases with altitude, it is important to distinguish between surface air pressure and sea level pressure, the latter being a standard that allows comparisons between different locations. Mercury and aneroid barometers both measure air pressure, which is expressed in units of millibars, pascals, or kilopascals.

The nonuniform distribution of the atmosphere gives rise to horizontal pressure gradients. Although small compared to vertical gradients, horizontal differences in pressure can sometimes produce devastating winds. The vertical pressure gradient force is much larger but is usually offset by a nearly equal but opposite gravitational force, resulting in hydrostatic equilibrium. Under conditions of hydrostatic balance, the vertical pressure gradient is proportional to the density of the atmosphere, as shown by the hydrostatic equation. Spatial variations in the density of the lower atmosphere lead to horizontal pressure gradients in the upper atmosphere that initiate upper-level winds.

Horizontal winds respond to the combined effect of three forces: the pressure gradient force, the Coriolis force, and friction. These combine to form gradient and geostrophic flow in the upper atmosphere and flow across isobars near the surface. The distribution of pressure tends to organize into fairly large-scale patterns of anticyclones, cyclones, ridges, and troughs. Wind vanes, anemometers, and aerovanes each measure different components of the wind velocity.

Key Terms

pressure *page 94*

pascal *page 94*

millibar *page 94*

kilopascal *page 94*

Dalton's law *page 94*

surface pressure *page 94*

sea level pressure *page 95*

speed *page 96*

velocity *page 96*

acceleration *page 96*

gravity *page 96*

force *page 96*

equation of state/ideal gas law *page 97*

mercury barometer *page 98*

barometric pressure *page 99*

aneroid barometer *page 100*

barograph *page 100*

isobar *page 101*

pressure gradient *page 101*

pressure gradient force *page 102*

hydrostatic equilibrium *page 102*

hydrostatic equation *page 103*

Coriolis force *page 106*

friction *page 109*

planetary boundary layer *page 109*

free atmosphere *page 109*

geostrophic flow/geostrophic wind *page 109*

gradient flow/gradient wind *page 111*

supergeostrophic flow *page 111*

subgeostrophic flow *page 111*

anticyclones *page 112*

cyclones *page 112*

troughs *page 113*

ridges *page 113*

azimuth *page 116*

wind vane *page 116*

anemometer *page 116*

Buys-Ballot's law *page 116*

aerovane *page 116*

rawinsondes *page 117*

Review Questions

1. What is a partial pressure?

2. What is Dalton's law?

3. Why does pressure always decrease with altitude?

4. What is the difference between surface pressure and sea level pressure?

5. How do speed and velocity differ? How do force and pressure differ?

6. What are the equation of state and the hydrostatic equation, and what do they tell us?

7. What two variables determine air pressure?

8. Describe how mercury and aneroid barometers measure air pressure, and explain why corrections need to be used for the observations made from one but not the other.

9. In what way is it misleading to express pressure in inches of mercury?

10. Explain the concept of hydrostatic equilibrium.

11. Explain how air temperature affects the vertical pressure gradient.

12. Explain how the pressure gradient force, the Coriolis force, and friction determine the movement of air in the free atmosphere and in the planetary boundary layer.

13. Describe the roles (if any) that wind speed, latitude, and direction of motion have in determining the magnitude of the Coriolis force.

14. What are geostrophic and gradient flows? Why don't they occur near the surface?

15. What are supergeostrophic and subgeostrophic flows?

16. Define the terms *cyclone, anticyclone, trough,* and *ridge.*

17. Briefly describe the movement of air around cyclones and anticyclones in the Northern and Southern Hemispheres.

18. What do anemometers and aerovanes measure?

Critical Thinking

1. Pressurized cans of shaving cream advise users not to expose the product to excessive heat. What might happen if the advice is not followed? Will this potential problem remain throughout the life of the product?

2. On a particular day, the vertical pressure gradient at the surface is −11 pascals per meter. What is the vertical pressure gradient in units of millibars per kilometer? Would you be able to use this gradient to exactly determine the pressure at the top of a building 200 m tall?

3. If a low-pressure region were to instantaneously replace a high-pressure system (assuming normally encountered values of high and low pressure), do you think you would be able to notice the difference by the pressure in your ears? Why or why not?

4. Would a particular pressure gradient produce the same exact wind speed over an Arizona desert that it would over a dense forest of tall trees? Why or why not?

5. The pilot of a small plane wants to fly at a constant height above the surface. Can the pilot fly at a constant pressure level (such as 500 mb) to assure the constant height above the ground? Why or why not?

6. A rule of thumb is that the 850 mb level often marks the boundary between the free atmosphere and the boundary layer. Are there parts of North America where this relationship is likely not to be valid? If so, where?

7. The Coriolis force applies equally to objects moving in any horizontal direction. Do you think the Coriolis force also affects objects moving directly up or down? If so, how would latitude affect the magnitude of the force?

8. Consider a 90-story skyscraper with high-speed elevators. Would a person ascending from the 46th to the 90th floor undergo the same degree of ear popping as a person ascending from the 1st floor to the 45th? Why or why not?

Problems and Exercises

1. Refer to any Web site below that produces surface and 500 mb maps. Examine the current surface map and identify the major cyclones, anticyclones, troughs, and ridges at the surface. Then look at the 500 mb map. Does the same general pattern emerge? Do the troughs and ridges at the 500 mb level occur directly over the corresponding features on the surface map? (Relationships between the surface and 500 mb levels will be discussed further in Chapter 10.)

2. On a daily basis, go to the Weather Channel's Web site at **www.weather.com**. Read the narrative describing the general weather pattern across the United States and identify the most notable weather events occurring across the country. Then look at the surface and 500 mb weather maps. This process will help you to become more familiar with normal pressure distributions and the type of weather often associated with them. As you proceed

through this text, the pressure patterns and their association with daily weather will become more meaningful to you.

3. Examine today's 500 mb weather map. You will probably find a trend toward decreasing 500 mb heights with increasing latitude. Are there any exceptions on the map to that general pattern? If so, observe the surface temperatures across North America. Do the temperature

patterns have any association with the 500 mb pattern? If so, describe them.

4. Observe a surface weather map that plots isobars and station models (**www.atmo.arizona.edu/products** is a good source). Do the air flow patterns around cyclones and anticyclones shown by the station models completely correspond with the generalizations made in this chapter? If not, why not?

Quantitative Problems

Differences in atmospheric pressure across the globe affect all the other elements of weather. Several quantitative problems are presented in this book's Web site to help you understand the concept of pressure, how it decreases with height, and its sensitivity to changes in moisture. It also

provides problems to help illustrate the effect of latitude on the Coriolis force. To get to the problems, enter the Web site at **www.MyMeteorologyLab.com** and click on Chapter 4. Then click on the "Quantitative Problems" section on the bar at the left.

Useful Web Sites

www.atmo.arizona.edu/products
Offers numerous images, maps, and animations. Scroll down to the last two thumbnails to view the current surface and 500 mb weather maps for North America. These maps are plotted with the same conventions that weather forecasters have used for many decades. You can observe an animation of the 500 mb surface for North America (also showing the water vapor distribution) by clicking on the appropriate thumbnail under the heading *GOES Water Vapor Images*.

weather.uwyo.edu/upperair/uamap.html
Allows the user to produce upper air maps at different levels for any portion of the globe. Maps are archived for approximately 2 weeks.

www.unisys.com/nam/index.php
Maps of the predicted upper air patterns for standard pressure surfaces based on one of the primary forecast models. Output from other models is also available from this site, as are forecast maps for sea level pressure and precipitation.

www.princeton.edu/%7Eoa/safety/altitude.html
Provides interesting information on how low pressure associated with altitude affects humans.

MyMeteorologyLab™

The *Understanding Weather and Climate,* Sixth edition MyMeteorologyLab™ Web site contains numerous multimedia resources to aid in your study of **Atmospheric Pressure and Wind**.
Visit **www.MyMeteorologyLab.com** to:

- **Review** key chapter concepts.
- **Read** the Pearson eText, *In the News RSS feeds,* and other chapter-specific Web resources.
- **Visualize** challenging topics with Interactive Tutorials, Geoscience Animations, MapMaster interactive maps, and Weather in Motion movies and visualizations, and more.
- **Test Yourself** with self-study quizzes.

 TUTORIALS
PRESSURE GRADIENTS
ATMOSPHERIC FORCES AND WIND
THE CORIOLIS FORCE

Use the interactive animations and quizzes in these tutorials to review this chapter's key concepts.

WEATHER IN MOTION
View movies and visualizations of meteorology patterns and processes.

The Growth of Wind Power in the US
The Coriolis Effect on a Merry-Go-Round
Winds During a Drought
Hurricane Winds
Forecasting Wind Patterns

PART TWO

Water in the Atmosphere

5 Atmospheric Moisture

TUTORIAL Atmospheric Moisture and Condensation
What is saturation, and what role does air temperature play in determining the saturation point?

6 Cloud Development and Forms

TUTORIAL Atmospheric Stability
What factors affect atmospheric stability?

Water exists in the atmosphere in three distinct phases: vapor, liquid, and ice. This part of the book describes the conditions and processes involved in the transformation of water through these phases that lead to the creation of clouds. The discussion then turns to the processes by which liquid water and ice crystals in clouds grow sufficiently to fall as precipitation, and the processes that determine the type of precipitation to occur.

Fog above the Bulkley River, Telkwa, British Columbia

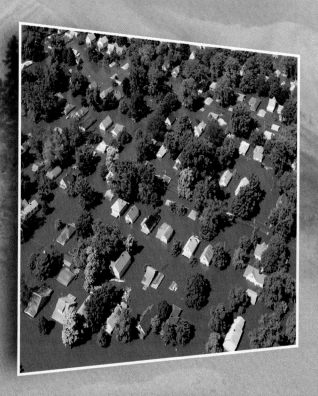

7 Precipitation Processes

TUTORIAL Precipitation

How do cloud droplets reach precipitation size in the middle latitudes?

5 Atmospheric Moisture

After reading this chapter, you should be able to:

▶ Describe the hydrologic cycle.

▶ Explain the concept of saturation.

▶ Identify the indices used in measuring the atmosphere's water vapor content.

▶ Describe how humidity is measured.

▶ Describe how water vapor is distributed in the atmosphere.

▶ Explain the processes that lead to saturation and the factors that affect condensation.

▶ Describe how diabatic and adiabatic processes produce cooling and condensation.

▶ List different forms of condensation and describe their distribution.

▶ Describe the formation and dissipation of cloud droplets.

▶ Describe how the effects of high humidity on human discomfort are rated.

▶ Describe possible effects of global warming and its effects on evaporation rates and atmospheric water vapor content.

At 10:17 P.M. EST, February 12, 2009, Colgan Air Flight 3407 was on approach to Buffalo Niagara Airport, New York, en route from Newark Liberty Airport in New Jersey. Flying through the type of wintry weather expected in the Northeast in winter, the crew had the anti-icing system in operation for most of the flight. As the plane approached for landing, a sequence of warning devices on the plane automatically activated to warn the flight crew of dangerously low air speed. The crew rapidly lost control of the aircraft, which slammed directly into a home 8 km (5 mi) ahead of the runway. The 49 passengers and crew on board all died in the crash, as did one person in the home.

The Federal Aviation Administration (FAA) concluded that the immediate problem was icing, but that the ultimate cause of the disaster was pilot error. Heavy accumulations of ice can alter a plane's aerodynamics and require greater air speeds to maintain an adequate amount of lift. The FAA determined that the captain of the aircraft did not undertake appropriate maneuvers in response to the situation and deemed the accident a result of pilot error.

Although other crashes have been caused by aircraft icing, such incidents are extremely rare. The fact remains that weather conditions can pose significant problems to aviation. Dangerous weather most often takes the form of extreme turbulence or rapidly changing wind conditions. Sometimes even relatively mild conditions can present significant risks. A mere fog bank or layer of overcast clouds can reduce visibility and pose a threat to safe ground and air travel. Despite the fact that clouds and fog are common to our everyday lives, many of us have a weak understanding of how they form.

This chapter opens Part Two, "Water in the Atmosphere," by describing the fundamentals of atmospheric moisture. It lays out the processes by which water can change from one phase to another and describes the common measures by which we express humidity. The chapter also describes the fundamentals involved in fog and cloud formation. Chapter 6 describes the processes of cloud development and the resultant cloud forms. Chapter 7 discusses how cloud droplets grow large enough to fall as precipitation. The topics discussed here are vital to understanding some of the most common weather phenomena, as well as those that sometimes have major human impacts.

◀ Ice accumulation on an aircraft.

The Hydrologic Cycle

Precipitation in all its forms—rain, snow, hail, and so on—is for many people the most notable feature of the atmosphere. Though the amount and timing of precipitation vary markedly from one region of Earth to another, the total amount of precipitation for the entire globe is relatively constant from one year to the next—at about 104 cm (41 in.) per year. And yet the atmosphere doesn't run out of water! Clearly, there is continual replenishment of water lost through precipitation. In other words, just as we have described for other gases, water vapor is constantly added to and removed from the atmosphere. As long as these balance, the atmospheric store will remain constant. The movement of water between and within the atmosphere and Earth is referred to as the **hydrologic cycle**. As it happens, the hydrologic cycle, depicted in Figure 5–1, is among the fastest of all geochemical cycles; atmospheric residence time for water vapor is only 10 days or so. The hydrologic cycle is a continuous series of processes that occur simultaneously. As with any cycle, the entire process has no real end or beginning.

Let's begin this discussion with the evaporation of water from Earth's surface into the atmosphere. Evaporation can occur directly from the oceans or from water bodies on land surfaces (such as lakes and rivers). It can also occur indirectly through plants via a process called *transpiration*. The water vapor that goes into the air eventually becomes water droplets or ice crystals in the form of clouds or fog. Many fog and cloud droplets or crystals will evaporate back into the air, but others precipitate down to the surface.

The shortest route the cycle can take is from the ocean to the atmosphere and back to the ocean. The situation is a bit more involved if water precipitates onto land. There, some precipitation might not reach the surface directly and may instead fall onto vegetation, accumulating as a coating of water or ice. The water that undergoes this process, called *interception*, might then drip or trickle down (after melting, if the precipitation fell as snow) the plant to the surface or evaporate back into the air. In some environments a sizable percentage—as much as 40 percent in some situations—of the precipitation that has fallen can be evaporated back into the atmosphere after interception.

Rainfall that does reach the land surface, either directly or after interception, might then flow above the surface into rivers, which then transport the water into a lake or ocean, or it can evaporate directly back into the atmosphere. If the precipitation falls as ice (the most obvious example being snow) it might temporarily remain on the ground before melting, or it might be locked away for eons as part of a glacier.

Liquid water at the ground does not always flow along the land surface but instead penetrates into the ground in the process called *infiltration*. Such water is pulled downward by gravity and can collect in the pores of underlying soil or rock as *groundwater*. Much of this groundwater eventually seeps into rivers for eventual transport toward the ocean, where the cycle continues. But much is almost immediately withdrawn by plants and transpired to the atmosphere. Still smaller quantities enter the animal kingdom as plants are consumed by browsers of all sizes, from elephants to microorganisms. And an ever-increasing fraction is drawn off by humans for agriculture (from which most is transpired), industry, and residential uses.

▲ **FIGURE 5–1** The hydrologic cycle.

Did You Know?

Just like the planet on which you live, you have your own hydrologic cycle. The human body is 50 to 60 percent water, on average, and none of the water molecules in your body reside there permanently. In Chapter 1 we saw that the residence time of water vapor in the atmosphere can be calculated by dividing its mass by the rate at which it enters and leaves the atmosphere. The same calculation can be made for the water in your body, and that calculation yields a residence time of about 14 days—not all that different from the residence time of water in the atmosphere (10 days). Also, most of the water molecules currently in your body are ancient, because water is a very stable molecule. Thus, those molecules in your body are well traveled and have probably been a part of the lives of many a famous person—at least temporarily.

This chapter deals with water in its various states and its transformation to and from the liquid and solid phases in the atmosphere. The two chapters following will examine clouds, the processes that form them, and precipitation. In reading those sections, keep in mind that the processes discussed represent individual components of the hydrologic cycle.

TUTORIAL

ATMOSPHERIC MOISTURE AND CONDENSATION

Use the tutorial to observe the processes of evaporation and condensation and see the effect of temperature on saturation.

Water Vapor and Liquid Water

Although matter in the gaseous phase is highly compressible, the density of a gas cannot be increased to an arbitrarily high level. At some point a limit is reached, forcing a change to liquid or solid state. For one atmospheric gas, water vapor, that limit is routinely achieved at temperatures and pressures found on Earth. (Other gases, such as nitrogen and oxygen, can be liquefied only at very low temperatures.) Air that contains as much water as possible is said to be *saturated*[1], and the introduction of additional water vapor results in formation of water droplets or ice crystals. The concept of saturation is fundamental to understanding the processes that form clouds and fogs. We begin our discussion with a hypothetical laboratory experiment that describes the general principles of evaporation and condensation. We then apply those principles to the processes that take place in the real atmosphere.

Evaporation and Condensation

Figure 5–2 depicts a hypothetical experiment, in which a tightly sealed container is partially filled with pure water (H_2O). Although it may seem obvious at this juncture, let's stipulate that the water in the jar has a perfectly flat surface. Furthermore, assume that at the onset of the experiment the water surface is covered by an impermeable coating, so no water vapor exists in the volume of the container above the water surface. Whether the volume above the water surface contains any air is entirely irrelevant to this experiment. The volume can contain normal air, pure hydrogen, methane, or fumes from French perfume—it can even be a complete vacuum. All that matters with respect to the evaporation/condensation process is that no water vapor be present initially.

---

[1]As we will see later in this chapter, the concept of saturated air becomes a little more complex than given here. Nonetheless, this description works well in the early stages of this discussion.

(a)

(b)

(c)

▲ **FIGURE 5–2** A hypothetical jar containing pure water with a flat surface and an overlying volume that initially contains no water vapor (a). When evaporation begins (b), water vapor accumulates in the volume above the liquid water. Initially, no condensation can occur because of the absence of water vapor above the liquid. But as evaporation contributes moisture to the overlying volume, some condensation can occur. Evaporation exceeds condensation for a while and thereby increases the water vapor content. Eventually enough water vapor is above the liquid for condensation to equal evaporation (c). At this point, saturation occurs.

Figure 5–2b shows what happens when we remove the covering on the liquid water surface. Without the covering, some of the molecules at the surface can escape into the overlying volume as water vapor. The process whereby molecules break free of the liquid volume is known as **evaporation**. The opposite process is **condensation**, wherein water vapor molecules randomly collide with the water surface and bond with adjacent molecules. At the beginning of our hypothetical experiment, no condensation could occur because no water vapor was present. As evaporation begins, however, water vapor starts to accumulate above the surface of the liquid.

At the early stages of evaporation, the low water vapor content prevents much condensation from occurring, and the rate of evaporation exceeds that of condensation. This leads to an increase in the amount of water vapor present. With increasing water vapor content, however, the condensation rate likewise increases. Eventually, the amount of water vapor above the surface is enough for the rates of condensation and evaporation to become equal, as shown in Figure 5–2c. A constant amount of water vapor now exists in the volume above the water surface due to offsetting gains and losses by evaporation and condensation. The resulting equilibrium state is called **saturation**. When this equilibrium exists in the atmosphere, the air is said to be saturated.

 The state of saturation described here can occur whether or not air (or other gases, for that matter) exists in the container. In other words, the water vapor is not "held" by the air (although this erroneous statement is frequently made). Water vapor is a gas, just like the other components of the air. Thus, it does not need to be "held" by air any more than the oxygen, nitrogen, argon, and other gases of the atmosphere need to be held by water vapor! When the air is saturated, there is simply an equilibrium between evaporation and condensation; the dry air plays no role in achieving this state. It is also important to realize that the exchange of water vapor and liquid described here applies as well to the change of phase between water vapor and ice. The change of phase directly from ice to water vapor, without passing into the liquid phase, is called **sublimation**. The reverse process (from water vapor to ice) is called **deposition**

(Figure 5–3). (Meteorologists sometimes use the word *sublimation* to apply to vapor-to-solid phase changes as well as solid-to-vapor. Because opposite processes should not have the same name, the use of the term *deposition* is preferred for vapor-to-ice changes.)

Checkpoint

1. What are evaporation and condensation?
2. In the hypothetical experiment described, how do rates of evaporation and condensation change until saturation is achieved?

Indices of Water Vapor Content

Humidity refers to the amount of water vapor in the air. Humidity can be expressed in a number of ways—in terms of the density of water vapor, the pressure exerted by the water vapor, the percentage of the amount of water vapor that can actually exist, or several other methods. There is no single "correct" measure, but, rather, each has its own advantages and disadvantages, depending on the intended use. All measures of humidity have one thing in common, however—they apply exclusively to water vapor, and not to liquid droplets or ice crystals suspended in or falling through the air. Let's now take a look at these measures.

TUTORIAL
ATMOSPHERIC MOISTURE AND CONDENSATION

Use the tutorial to explore how changes in temperature and the amount of water vapor in the air affect the indices of water vapor content.

Vapor Pressure

In Chapters 1 and 4, we saw that the air exerts pressure on all surfaces. Each gas that makes up the atmosphere contributes to the total air pressure, with the most abundant permanent gases accounting for most of the pressure. Because water vapor seldom accounts for more than 4 percent of the total atmospheric mass, it exerts only a small percentage of the total air pressure. The part of the total atmospheric pressure due to water vapor is referred to as the **vapor pressure**. Like the atmospheric pressure, vapor pressure is commonly expressed in units of millibars (mb) by U.S. meteorologists, and as kilopascals (kPa) by their Canadian counterparts, though in most scientific applications the pascal (Pa) is the preferred unit (100 Pa = 1 mb = 0.1 kPa)

The vapor pressure of a volume of air depends on both the temperature and the density of water vapor molecules (Figure 5–4). If the air temperature is high, water vapor molecules (along with all the other gaseous constituents of the

▲ **FIGURE 5–3** Deposition is the direct transfer of vapor to ice.

▲ **FIGURE 5-4** The movement of molecules exerts a pressure on surfaces, called *vapor pressure*. The vapor pressure increases with concentration and temperature.

atmosphere) move more rapidly and exert a greater pressure. Similarly, a greater concentration of water vapor molecules means that a greater amount of mass is available to exert pressure. In practice, temperature influences are small compared to density changes, so vapor pressure closely follows changes in the density or abundance of water molecules.

Because there is a maximum amount of water vapor that can exist, there is a corresponding maximum vapor pressure, called the **saturation vapor pressure**. The saturation vapor pressure does not represent the current amount of moisture in the air; rather, it is an expression of the maximum that *can* exist. The saturation vapor pressure depends on only one variable—temperature. Figure 5–5 shows the relationship between saturation vapor pressure and temperature, with

higher temperatures having higher saturation vapor pressures. For example, at 40 °C the saturation vapor pressure is 73.8 mb, while at 0 °C it is only 6.1 mb, less than one-tenth as much.

The increase in saturation vapor pressure with temperature is not linear. At low temperatures there is only a modest increase in saturation vapor pressure, but at high temperatures saturation vapor pressure grows rapidly. For example, a 2 °C increase in temperature, from 0 °C to 2 °C, increases the saturation vapor pressure from 6.1 mb to 7.1 mb, only a 1 mb difference. Raising the temperature the same amount from a higher starting point, from 40 °C to 42 °C, raises the saturation vapor pressure by 7.7 mb, from 73.8 to 81.5 mb. This nonlinear behavior is captured in a simple statement: At temperatures normally encountered near Earth's surface the saturation vapor pressure approximately doubles for every 10 °C (18 °F) increase in temperature.

Absolute Humidity

Another measure of water vapor content is the **absolute humidity**, which is simply the density of water vapor, expressed as the number of grams of water vapor contained in a cubic meter of air. Because absolute humidity represents the amount of moisture contained in a volume of air, its value changes whenever air expands or contracts. Thus, for example, if an air parcel expands (as it does when it is heated or lifted upward), its absolute humidity will fall, even though no water vapor is removed from the parcel. Because absolute humidity suffers from this drawback and has no strong advantage over any other index, it is not widely used.

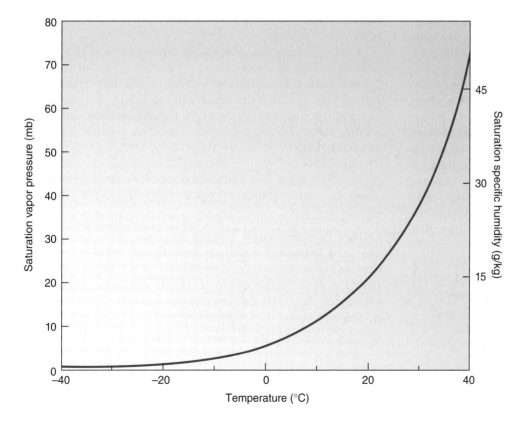

◄ **FIGURE 5-5** Saturation vapor pressure and saturation specific humidity as a function of temperature. The curve is steeper at higher temperatures, meaning that saturation vapor pressure is more sensitive to temperature changes when the air is warm.

Specific Humidity

Although not normally encountered outside scientific applications, **specific humidity** is a useful index for representing atmospheric moisture. Specific humidity expresses the mass of water vapor existing in a given mass of air. Consider, for example, a volume containing exactly 1 kg of air (at sea level such a volume would be about 0.8 cubic meters, or about 27 cu. ft). Of that kilogram, some number of grams would be water vapor. The proportion of the atmospheric mass accounted for by water vapor is the specific humidity. Most often, specific humidity is expressed as the number of grams of water vapor per kilogram of air. Because the water vapor outside the tropics usually is less than 2 percent of the mass of the air near the surface, specific humidities are normally less than 20 grams of water vapor per kilogram of air. Specific humidity, q, is expressed mathematically as

$$q = \frac{m_v}{m} = \frac{m_v}{m_v + m_d}$$

where m_v = the mass of water vapor, m = the mass of atmosphere, and m_d = the mass of dry air (all the atmospheric gases other than water vapor). Unlike vapor pressure, specific humidity is affected to a small degree by atmospheric pressure, because it depends in part on the total mass of the atmosphere, m.

Unlike absolute humidity, specific humidity has the advantage of not changing as air expands or contracts. When a kilogram of air expands, its mass is unchanged (it is still 1 kg), and the proportion that is water vapor is unchanged. As a result, the specific humidity is unaffected. Likewise, specific humidity is not temperature dependent. If a kilogram of air contains 1 g of water vapor, it still contains 1 g after heating. For this reason, specific humidity is a good indicator for comparing water vapor in the air at different locations whose air temperatures might be different from each other.

For example, if Toronto, Ontario, has a specific humidity of 10 grams of water vapor per kilogram of air on a given day and Albuquerque, New Mexico, has 5 g/kg, we can infer that Toronto has twice as much water vapor in the air as does Albuquerque, no matter what their temperatures are. This may not seem very profound, but the direct correspondence between specific humidity and water vapor content does not hold for the more frequently used index of moisture, relative humidity. Thus, specific humidity is a useful measure of water vapor whose only real drawback is the general public's unfamiliarity with the term.

Because there is a maximum amount of water vapor that can exist at a particular temperature, there is likewise a maximum specific humidity. This maximum is called the **saturation specific humidity**. This property is directly analogous to the saturation vapor pressure and increases in the nonlinear manner shown in Figure 5–5.

Mixing Ratio

The **mixing ratio** is very similar to specific humidity. In the case of specific humidity, we express the mass of water vapor in the air as a proportion of *all* the air. In contrast, the mixing ratio, r, is a measure of the mass of water vapor relative to the mass of the other gases of the atmosphere, or

$$r = m_v/m_d$$

(Note that the denominator denotes a mass of *dry* air as opposed to *all* air.) Numerically, the mixing ratio and specific humidity will always have nearly equal values. This is because the amount of water vapor in the air is always small, so that whether or not it is counted in the denominator hardly changes the ratio. The values of the mixing ratio and specific humidity are so similar that some meteorologists tend to use the two terms almost interchangeably.

A simple example should clarify the similarity between specific humidity and mixing ratio. If the specific humidity is 10 grams of water vapor per kilogram of air, the mixing ratio is 10 grams of water vapor per 990 grams of dry air. Note that 10 divided by 990 equals 10.011. In other words, if the specific humidity is 10.0 g/kg, the mixing ratio is only 1.1 percent higher, or 10.011 g/kg.

Using the mixing ratio as an index of moisture content offers the same advantages as using specific humidity. The maximum possible mixing ratio is called the **saturation mixing ratio**.

Relative Humidity

The most familiar measure of water vapor content is **relative humidity**, RH, which relates the amount of water vapor in the air to the maximum possible at the current temperature. The World Meteorological Organization defines the relative humidity as,

RH = (mixing ratio/saturation mixing ratio) × 100%[2]

To see how this works, let's refer to the left example in Figure 5–6, in which the actual mixing ratio is 6 g/kg of dry air, and the temperature of 14 °C (57 °F) yields a saturation mixing ratio of 10 g/kg of dry air. The relative humidity would thus be

$$RH = \frac{6}{10} \times 100\% = 0.6 \times 100\% = 60\%$$

The relative humidity is not uniquely determined by the amount of water vapor present. Because more water vapor can exist in warm air than in cold air, the relative humidity depends on both the actual moisture content and the air temperature. If the temperature of the air increases, more water vapor can exist and the ratio of the amount of water vapor in the air relative to saturation decreases. Thus, the relative humidity declines even if the moisture content is unchanged.

---

[2]The American Meteorological Society has a slightly different definition of relative humidity, wherin the vapor pressure is divided by the saturation vapor pressure. The two methods yield almost identical results.

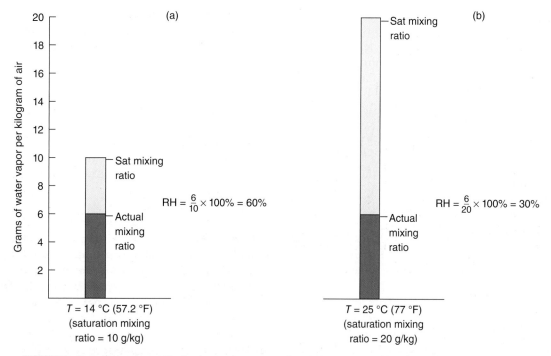

▲ **FIGURE 5–6** Relative humidity is dependent not only on the amount of water vapor in the air but also on the air temperature (which determines the saturation mixing ratio). Refer to the graphs showing various values of mixing ratio and saturation mixing ratio, and assume in both these examples that the mixing ratio is 6 grams of water vapor per kilogram of dry air. In example (a), we assume that the air temperature is 14 °C (57.2 °F), which has a saturation mixing ratio of 10 grams of water vapor per kilogram of dry air; hence the air contains 60 percent of what it could contain and the relative humidity is 60 percent. The same 6 grams of water vapor per kilogram of dry air air is plotted in (b), but the temperature is now 25 °C (77 °F). At this temperature, the saturation mixing ratio is 20 grams of water vapor per kilogram of dry air, so the relative humidity is only 30 percent—half of what it was in (a) even though the same amount of water vapor exists.

Again referring to the example in Figure 5–6, let's consider what would happen if the amount of water vapor remains constant but the temperature increases from its original 14 °C to 25 °C (77 °F). At the new temperature, the saturation mixing ratio increases to 20 grams of water vapor per kilogram of dry air, and the relative humidity becomes

$$RH = \frac{6}{20} \times 100\% = 0.3 \times 100\% = 30\%$$

The relative humidity decreased even though the amount of water vapor remained constant! This is a significant drawback to any index that is supposed to be a measure of humidity.

Because of its dependence on temperature, the relative humidity will change throughout the course of the day even if the amount of moisture in the air is unchanged. Relative humidity is usually highest in the early morning—not because of abundant water vapor, but simply because the temperature is lower. As the day warms up, the relative humidity typically declines because the saturation mixing ratio increases. Figure 5–7 shows a typical pattern of daily temperature and relative humidity values. Notice how relative humidity varies by a factor of 3 over the course of the day. Almost all of that variation arises from the change in air temperature.

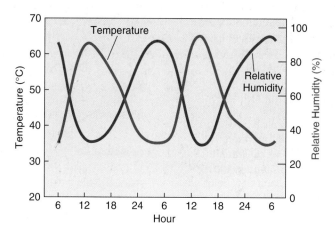

▲ **FIGURE 5–7** A typical plot of temperature and relative humidity over a 48-hour period. (The temperature scale is shown on the left vertical axis and the relative humidity on the right vertical axis.) Observe that as the temperature increases, the relative humidity decreases, and vice versa. In this example there is a substantial change in the relative humidity even though the actual water vapor content over the course of the day underwent only minimal changes. This shows the strong dependence of relative humidity on air temperature (and thus one of its serious limitations as an indicator of moisture content).

The influence of temperature on relative humidity creates another problem—it confounds direct comparisons of moisture contents at different places with unequal temperatures. Consider, for example, a cold morning in Montreal, Quebec, where the temperature is –20 °C (–4 °F) and the mixing ratio is 0.7 g/kg. At –20 °C, the saturation mixing ratio is 0.78 grams of water vapor per kilogram of dry air, and the resultant relative humidity is [(0.70 ÷ 0.78) × 100%], or 89.7%. Now compare that to the warmer situation at Atlanta, where the temperature is 10 °C (50 °F) and the mixing ratio 6.2 g/kg (nearly nine times greater than at Montreal!). At 10 °C, the saturation mixing ratio is 7.7 grams of water vapor per kilogram of dry air, so the relative humidity is [(6.2 ÷ 7.7) × 100%], or 79.9 %. Notice that the relative humidity is lower at Atlanta than at Montreal despite the fact that it contains much more water vapor. This illustrates why relative humidity is a poor choice for comparing the amount of water vapor in the air at one place to that at another.

Some people are confused about the true meaning of relative humidity. Some believe the term represents the percentage of the air that is water vapor. This is not correct. To see why, consider an instance in which the relative humidity is 100 percent. If the air were 100 percent water vapor, it would include no nitrogen or oxygen, and we would have a difficult time breathing, let alone discussing water vapor content! Another common misperception is about how high the relative humidity can be on a hot, humid day. Many people would estimate that on such a day the relative humidity would be about 99 percent. But in reality, very hot days never have relative humidities approaching that value. That is because at high temperatures the saturation mixing ratio is very much higher than the actual mixing ratios likely to be encountered. For example, if the temperature is 35 °C (95 °F), the saturation mixing ratio is 36.8 grams of water vapor per kilogram of dry air. But we have seen that outside the tropics it is unusual for the mixing ratio to exceed 20 g/kg—even when the air is humid. Thus, a 99 percent relative humidity is not a realistic possibility at that temperature. Indeed, warm days can be extremely uncomfortable even with relative humidities of only about 50 percent.

Dew Point

A useful moisture index that is free of the temperature relationship just described is the **dew point temperature** (or simply the **dew point**), the temperature at which saturation occurs. This quantity may seem confusing at first because it is expressed as a temperature, but it is a simple index to use and easy to interpret. And it is dependent almost exclusively on the amount of water vapor present.

Consider the parcel of unsaturated air in Figure 5–8. Initially the air temperature was 14 °C (58 °F), yielding a saturation mixing ratio of 10 grams of water vapor per kilogram of dry air. The initial mixing ratio was 8 g/kg. The relative humidity was therefore 80 percent. As the air cools, its relative humidity increases, and if the air is cooled sufficiently its relative humidity reaches 100 percent and it becomes saturated. Any further cooling leads to the removal of water vapor by condensation. In this example, the dew point is 10 °C (50 °F) because that is the temperature at

▶ **FIGURE 5-8** The dew point is an expression of water vapor content, although it is expressed as a temperature. In (a), the temperature exceeds the dew point and the air is unsaturated. When the air temperature is lowered so that the saturation mixing ratio is the same as the actual mixing ratio (b), the air temperature and dew point are equal. Further cooling (c) leads to an equal reduction in the air temperature and dew point so that they remain equal to each other.

5–1
FORECASTING

Dew Point and Nighttime Minimum Temperatures

Knowledge of the current dew point temperature is a useful tool to the forecaster for the prediction of the following morning's low temperature. If no major wind shifts or other weather changes are anticipated, the minimum temperature will often approximate the dew point. Consider a hypothetical evening with an air temperature of 15 °C (59 °F) and a dew point of 5 °C (41 °F). The spread between the air temperature and the dew point temperature is not very large, and a 10 °C (18 °F) lowering of the air temperature is feasible. If the air temperature does indeed drop to the dew point and there is little or no wind, a radiation fog has a good chance of forming. The fog would then inhibit further cooling, partly because latent heat is released and partly because water droplets are extremely effective at absorb-

ing longwave radiation from the surface. Without the loss of radiation, the surface temperature would remain almost constant, and the overnight low would equal the dew point temperature.

The relationship between dew point and minimum temperature will not hold under certain conditions. The first has to do with the changes in the big weather picture. Imagine, for example, that a mass of warmer air is moving into the forecast region. This large body of air can replace the one present at the time of the forecast and bring with it higher nighttime temperatures. Similarly, the passage of an advancing cold front (briefly described in Chapter 1 and discussed in more detail in Chapter 9) can lead to significant drops in temperature below the current dew point.

Both heavy cloud cover and strong winds inhibit a drop in air temperature, and their presence may keep minimum air temperatures above the dew point. Cloud cover

achieves this effect because of its absorption and downward reradiation of longwave energy. Strong winds prevent large temperature decreases at the surface by vertical mixing. A shallow layer of cold air that would otherwise develop is easily disrupted, leading to higher surface temperatures and more uniform temperatures with height.

Minimum temperatures won't go down to the dew point if the difference between the air temperature and the dew point temperature is very large. One can readily see how this might occur if a desert has a high temperature of 45 °C (113 °F) and a dew point of 0 °C (32 °F). Even with calm winds and no cloud cover, a cooling of 45 °C is unlikely over the course of a short summer night. Though the temperature won't always drop down as low as the dew point, it is always certain that unless a front passes through or the wind direction changes significantly, the minimum temperature will not fall much below the evening dew point.

which the saturation mixing ratio is 8 g/kg. Notice that even though the relative humidity increased as the temperature decreased, the dew point remained constant at 10 °C.

What would have happened if the specific humidity had remained constant and the temperature had increased from its initial 14 °C? The relative humidity would have decreased, yet the dew point would have remained constant. The dew point would not have changed because eventual cooling of the air to 8 °C still would have led to saturation.

The dew point is a valuable indicator of the moisture content; when the dew point is high, abundant water vapor is in the air. Moreover, when combined with air temperature, it is an indicator of the relative humidity. When the dew point is much lower than the air temperature, the relative humidity is very low. When the dew point is nearly equal to the air temperature, the relative humidity is high. Furthermore, when the air temperature and the dew point are equal, the air is saturated and the relative humidity is 100 percent.

Unlike relative humidity, the dew point does not change simply because air temperature changes. Moreover, if one location has a higher dew point than another, it will also have a greater amount of water vapor in the air, assuming the same air pressure. Once you are familiar with dew point, it is probably the most effective index of water vapor content. Dew points on very humid, hot days are typically in the low 20s on the Celsius scale (low 70s Fahrenheit). (When you see a dew point of 70 °F or higher, you can plan on a sleepless

night unless you have air conditioning.) On comfortable days that are neither humid nor dry, dew points may be in the low teens Celsius (mid-50s Fahrenheit); very dry days can have dew points in the minus 20s or lower on the Celsius scale (0 °F). The dew point temperature can sometimes serve as a predictor of overnight cooling, as explained in *Box 5–1, Forecasting: Dew Point and Nighttime Minimum Temperatures.*

The dew point is always equal to or less than the air temperature; under no circumstances does it ever exceed the temperature. So what happens if the air temperature is lowered to the dew point and then cooled further? In that case, the amount of water vapor exceeds the amount that can now exist, and the surplus is removed from the air. This happens by condensation of the water vapor to form a liquid or by the formation of ice crystals. In either case, the dew point decreases at the same rate as the air temperature, and the two remain equal to each other. This is illustrated in Figure 5–8b and c. When the temperature was lowered to 10 °C in Figure 5–8b, the air became saturated with 8 grams of water vapor per kilogram of dry air. As the air cooled further to 6 °C (Figure 5–8c), the saturation mixing ratio decreased to 6 g/kg. Because the mixing ratio, by definition, cannot exceed the saturation mixing ratio, 2 grams of water vapor (8 grams minus 6 grams) had to be removed from each kilogram of dry air by condensation. The removal of water vapor kept the specific humidity equal to that of the saturation specific humidity and also lowered the dew point. Note that when the temperature at which saturation would occur is below 0 °C (32 °F), we use the term **frost point** instead of dew point.

Checkpoint

> 1. What is the dew point?
> 2. What happens if air temperature drops below the dew point?

(a)

Measuring Humidity

Considering the fact that water vapor is an invisible gas mixed with all the other gases of the atmosphere, you might suspect that its measurement would entail some highly sophisticated instrumentation. Such is not the case. The simplest and most widely used instrument for measuring humidity, the **sling psychrometer** (Figure 5–9a), consists of a pair of thermometers, one of which has a cotton wick around the bulb that is saturated with water. The other thermometer has no such covering and simply measures the air temperature. The two thermometers, called the **wet bulb** and **dry bulb thermometers**, respectively, are mounted to a pivoting device that allows them to be circulated ("slung") through the surrounding air. If the air is unsaturated, water evaporates from the wet bulb, whose temperature falls as latent heat is consumed. After about a minute or so of circulating, the amount of heat lost by evaporation is offset by the input of sensible heat from the surrounding, warmer air, and the cooling ceases. Thereafter, the wet bulb maintains a constant temperature no matter how long the instrument is swung around.

The difference between the dry and wet bulb temperatures, called the **wet bulb depression**, depends on the moisture content of the air. If the air is completely saturated, no net evaporation occurs from the wet bulb thermometer, no latent heat is lost, and the wet bulb temperature equals the dry bulb temperature. On the other hand, if the humidity is low, plenty of evaporation will take place from the wet bulb, and its temperature will drop considerably before reaching an equilibrium value. To determine the moisture content, first you note the difference between the dry and wet bulb temperatures. Then, with the use of tables such as Tables 5–1 and 5–2, you obtain the dew point, relative humidity, or any other humidity measure by finding the value corresponding to the row for the air temperature and the column for the wet bulb depression.

Some psychrometers are equipped with fans that circulate air across the bulbs of the two thermometers. These **aspirated psychrometers** save the user the effort needed to sling the thermometers through the air (as well as the aggravation of cleaning up the mess after accidentally striking nearby objects). Another alternative to the sling psychrometer is the **hair hygrometer**, whose basic part is a band of human hair. Hair expands and contracts in response

(b)

▲ **FIGURE 5-9** (a) Sling psychrometer and (b) hygrothermograph.

to the relative humidity. By connecting strands of hair to a lever mechanism, we can easily determine the water vapor content. Often, the hygrometer is coupled with a bimetallic strip and rotating drum to give a continuous record of temperature and humidity. Such a **hygrothermograph** is shown in Figure 5–9b. Though more sophisticated instruments are used for keeping track of humidity at major weather platforms, such as those at airports, hygrothermographs are still often housed in instrument shelters (Chapter 3) at cooperative agencies. (The distribution of water vapor across the globe is routinely observed by water vapor imagery. This is described in *Box 5–2, Forecasting: Water Vapor and Infrared Satellite Imagery*.)

Checkpoint

> 1. Why does a sling psychrometer include two thermometers?
> 2. Why is the wet bulb reading lower than the dry bulb reading?

TABLE 5–1

Dew Points

Dry Bulb (°C)	Wet Bulb Depression, °C (Dry Bulb Temperature Minus Wet Bulb Temperature = Wet Bulb Depression)																					
	1	2	3	4	5	6	7	8	9	10	11	12	13	14	15	16	17	18	19	20	21	22
−20	−33																					
−18	−28																					
−16	−24																					
−14	−21	−36																				
−12	−18	−28																				
−10	−14	−22																				
−8	−12	−18	−29																			
−6	−10	−14	−22																			
−4	−7	−22	−17	−29																		
−2	−5	−8	−13	−20																		
0	−3	−6	−9	−15	−24																	
2	−1	−3	−6	−11	−17																	
4	1	−1	−4	−7	−11	−19																
6	4	1	−1	−4	−7	−13	−21															
8	6	3	1	−2	−5	−9	−14															
10	8	6	4	1	−2	−5	−9	−14	−28													
12	10	8	6	4	1	−2	−5	−9	−16													
14	12	11	9	6	4	1	−2	−5	−10	−17												
16	14	13	11	9	7	4	1	−1	−6	−10	−17											
18	16	15	13	11	9	7	4	2	−2	−5	−10	−19										
20	19	17	15	14	12	10	7	4	2	−2	−5	−10	−19									
22	21	19	17	16	14	12	10	8	5	3	−1	−5	−10	−19								
24	23	21	20	18	16	14	12	10	8	6	2	−1	−5	−10	−18							
26	25	23	22	20	18	17	15	13	11	9	6	3	0	−4	−9	−18						
28	27	25	24	22	21	19	17	16	14	11	9	7	4	1	−3	−9	−16					
30	29	27	26	24	23	21	19	18	16	14	12	10	8	5	1	−2	−8	−15				
32	31	29	28	27	25	24	22	21	19	17	15	13	11	8	5	2	−2	−7	−14			
34	33	31	30	29	27	26	24	23	21	20	18	16	14	12	9	6	3	−1	−5	−12	−29	
36	35	33	32	31	29	28	27	25	24	22	20	19	17	15	13	10	7	4	0	−4	−10	
38	37	35	34	33	32	30	29	28	26	25	23	21	19	17	15	13	11	8	5	1	−3	−9
40	39	37	36	35	34	32	31	30	28	27	25	24	22	20	18	16	14	12	9	6	2	−2

Dew point temperatures

Dry Bulb Temp. (°F)	Wet Bulb Depression (°F)																																		
	1	2	3	4	5	6	7	8	9	10	11	12	13	14	15	16	17	18	19	20	21	22	23	24	25	26	27	28	29	30	31	32	33	34	35
0	−7	−20																																	
5	−1	−9	−24																																
10	5	−2	−10	−27																															
15	11	6	0	−9	−26																														
20	16	12	8	2	−7	−21																													
25	22	19	15	10	5	−3	−15	−51																											
30	27	25	21	18	14	8	2	−7	−25																										
35	33	30	28	25	21	17	13	7	0	−11	−41																								
40	38	35	33	30	28	25	21	18	13	7	−1	−14																							
45	43	41	38	36	34	31	28	25	22	18	13	7	−1	−14																					
50	48	46	44	42	40	37	34	32	29	26	22	18	13	8	0	−13																			
55	53	51	50	48	45	43	41	38	36	33	30	27	24	20	15	9	1	−12	−59																
60	58	57	55	53	51	49	47	45	43	40	38	35	32	29	25	21	17	11	4	−8	−36														
65	63	62	60	59	57	55	53	51	49	47	45	42	40	37	34	31	27	24	19	14	7	−3	−22												
70	69	67	65	64	62	61	59	57	55	53	51	49	47	44	42	39	36	33	30	26	22	17	11	2	−11										
75	74	72	71	69	68	66	64	63	61	59	57	55	54	51	49	47	44	42	39	36	32	29	25	21	15	8	−2	−23							
80	79	77	76	74	73	72	70	68	67	65	63	62	60	58	56	54	52	50	47	44	42	39	36	32	28	24	20	13	6	−7	−53				
85	84	82	81	80	78	77	75	74	72	71	69	68	66	64	62	61	59	57	54	52	50	48	45	42	39	36	32	28	24	19	12	3	−12		
90	89	87	86	85	83	82	81	79	78	76	75	73	72	70	69	67	65	63	61	59	57	55	53	51	48	45	43	39	36	32	28	24	19	11	1
95	94	93	91	90	89	87	86	85	83	82	80	79	78	76	74	73	71	70	68	66	64	62	60	58	56	54	52	49	46	43	40	37	33	29	24
100	99	98	96	95	94	93	91	90	89	87	86	85	83	82	80	79	77	76	74	72	71	69	67	65	63	61	59	57	55	52	50	47	44	41	37
105	104	103	101	100	99	98	96	95	94	93	91	90	89	87	86	84	83	82	80	78	77	75	74	72	70	68	67	65	63	61	58	56	54	51	48
110	109	108	106	105	104	103	102	100	99	98	97	95	94	93	91	90	89	87	86	84	83	81	80	78	77	75	73	72	70	68	66	64	62	60	57
115	114	113	112	110	109	108	107	106	104	103	102	101	99	98	97	96	94	93	92	90	89	87	86	84	83	81	80	78	76	75	73	71	69	67	65
120	119	118	117	115	114	113	112	111	110	108	107	106	105	104	102	101	100	98	97	96	94	93	92	90	89	87	86	84	83	81	80	78	76	75	73
125	124	123	122	121	119	118	117	116	115	114	112	111	110	109	108	106	105	104	103	101	100	99	97	96	95	93	92	90	89	88	86	84	83	81	80
130	129	128	127	126	124	123	122	121	120	119	118	116	115	114	113	112	110	109	108	107	106	104	103	102	100	99	98	96	95	94	92	91	89	88	86

Dew point temperatures

TABLE 5–2

Relative Humidities

Wet Bulb Depression, °C
(Dry Bulb Temperature Minus Wet Bulb Temperature = Wet Bulb Depression)

Dry Bulb (°C)	1	2	3	4	5	6	7	8	9	10	11	12	13	14	15	16	17	18	19	20	21	22
−20	28																					
−18	40																					
−16	48	0																				
−14	55	11																				
−12	61	23																				
−10	66	33	0																			
−8	71	41	13																			
−6	73	48	20	0																		
−4	77	54	32	11																		
−2	79	58	37	20	1																	
0	81	63	45	28	11																	
2	83	67	51	36	20	6																
4	85	70	56	42	27	14																
6	86	72	59	46	35	22	10	0														
8	87	74	62	51	39	28	17	6														
10	88	76	65	54	43	33	24	13	4													
12	88	78	67	57	48	38	28	19	10	2												
14	89	79	69	60	50	41	33	25	16	8	1											
16	90	80	71	62	54	45	37	29	21	14	7	1										
18	91	81	72	64	56	48	40	33	26	19	12	6	0									
20	91	82	74	66	58	51	44	36	30	23	17	11	5									
22	92	83	75	68	60	53	46	40	33	27	21	15	10	4	0							
24	92	84	76	69	62	55	49	42	36	30	25	20	14	9	4	0						
26	92	85	77	70	64	57	51	45	39	34	28	23	18	13	9	5						
28	93	86	78	71	65	59	53	47	42	36	31	26	21	17	12	8	4					
30	93	86	79	72	66	61	55	49	44	39	34	29	25	20	16	12	8	4				
32	93	86	80	73	68	62	56	51	46	41	36	32	27	22	19	14	11	8	4			
34	93	86	81	74	69	63	58	52	48	43	38	34	30	26	22	18	14	11	8	5		
36	94	87	81	75	69	64	59	54	50	44	40	36	32	28	24	21	17	13	10	7	4	
38	94	87	82	76	70	66	60	55	51	46	42	38	34	30	26	23	20	16	13	10	7	5
40	94	89	82	76	71	67	61	57	52	48	44	40	36	33	29	25	22	19	16	13	10	7

Relative humidity values

Wet Bulb Depression (°F)

Dry Bulb Temp. (°F)	1	2	3	4	5	6	7	8	9	10	11	12	13	14	15	16	17	18	19	20	21	22	23	24	25	26	27	28	29	30	31	32	33	34	35
0	67	33	1																																
5	73	46	20																																
10	78	56	34	13																															
15	82	64	46	29	11																														
20	85	70	55	40	26	12																													
25	87	74	62	49	37	25	13	1																											
30	89	78	67	56	46	36	26	16	6																										
35	91	81	72	63	54	45	36	27	19	10	2																								
40	92	83	75	68	60	52	45	37	29	22	15	7																							
45	93	86	78	71	64	57	51	44	38	31	25	18	12	6																					
50	93	87	80	74	67	61	55	49	43	38	32	27	21	16	10	5																			
55	94	88	82	76	70	65	59	54	49	43	38	33	28	23	19	11	9	5																	
60	94	89	83	78	73	68	63	58	53	48	43	39	34	30	26	21	17	13	9	5	1														
65	95	90	85	80	75	70	66	61	56	52	48	44	39	35	31	27	24	20	16	12	9	5	2												
70	95	90	86	81	77	72	68	64	59	55	51	48	44	40	36	33	29	25	22	19	15	12	9	6	3										
75	96	91	86	82	78	74	70	66	62	58	54	51	47	44	40	37	34	30	27	24	21	18	15	12	9	7	4	1							
80	96	91	87	83	79	75	72	68	64	61	57	54	50	47	44	41	38	35	32	29	26	23	20	18	15	12	10	7	5	3					
85	96	92	88	84	81	77	73	70	66	63	59	57	53	50	47	44	41	38	36	33	30	27	25	22	20	17	15	13	10	8	6	4	2		
90	96	92	89	85	81	78	74	71	68	65	61	58	55	52	49	47	44	41	39	36	34	31	29	26	24	22	19	17	15	13	11	9	7	5	3
95	96	93	89	86	82	79	76	73	69	66	63	61	58	55	52	50	47	44	42	39	37	34	32	30	28	25	23	21	19	17	15	13	11	10	8
100	96	93	89	86	83	80	77	73	70	68	65	62	59	56	54	51	49	46	44	41	39	37	35	33	30	28	26	24	22	21	19	17	15	13	12
105	97	93	90	87	84	81	78	75	72	69	66	64	61	58	56	53	51	49	46	44	42	40	38	36	34	32	30	28	26	24	22	21	19	17	15
110	97	93	90	87	84	81	78	75	73	70	67	65	62	60	57	55	52	50	48	46	44	42	40	38	36	34	32	30	28	26	25	23	21	20	18
115	97	94	91	88	85	82	79	76	74	71	69	66	64	61	59	57	54	52	50	48	46	44	42	40	38	36	34	33	31	29	28	26	25	23	21
120	97	94	91	88	85	82	80	77	74	72	69	67	65	62	60	58	55	53	51	49	47	45	43	41	40	38	36	34	33	31	29	28	26	25	23
125	97	94	91	88	86	83	80	78	75	73	70	68	66	64	61	59	57	55	53	51	49	47	45	44	42	40	38	37	35	33	32	30	29	27	26
130	97	94	91	89	86	83	81	78	76	73	71	69	67	64	62	60	58	56	54	52	50	48	47	45	43	41	40	38	37	35	33	32	30	29	28

Relative humidity values

5-2
FORECASTING

Water Vapor and Infrared Satellite Imagery

We know that water vapor is an invisible gas, so it may seem strange that meteorologists look at satellite images showing the distribution of water vapor. These images are obtained by sensors on board weather satellites that observe the amount of radiation received at two particular wavelengths of electromagnetic energy, 6.7 and 7.3 μm (Figure 1). Water vapor is effective at emitting and absorbing this portion of the infrared band of wavelengths, so the energy emitted by the surface at these wavelengths is absorbed by water vapor before it can exit the atmosphere. At the same time, energy at 6.7 and 7.3 μm emitted by water vapor in the upper part of the troposphere does escape to space and is observed by the satellite. Thus the water vapor imagery actually depicts the distribution of water vapor (and clouds) in the upper atmosphere, making it extremely useful, in particular for discerning high-level moisture distributions. When a sequence of images taken at regular time intervals is combined to make a movie loop (such as the Weather in Motion loop), the resultant movie is extremely effective in showing upper atmospheric motions. These motions, which were introduced in Chapter 4, will be discussed more fully in later chapters.

▲ **FIGURE 1** A water vapor image of North America. These images are obtained from weather satellites with sensors that observe radiation emitted by water vapor in the upper troposphere. Regions shown in white have high water vapor contents; those that are dark are relatively dry.

Did You Know?

Averaged over the entire planet, Earth's atmosphere contains 13×10^{15} kilograms of water vapor (the equivalent of 3.4×10^{15} gallons when condensed to liquid). While this may sound like a lot, it represents a mere one-thousandth of 1 percent of all the free H_2O on the planet. If all of this water vapor were to condense and collect at Earth's surface, it would be the equivalent of about 2.5 cm (1 in.) of precipitation worldwide.

Distribution of Water Vapor

Water vapor at a point in the atmosphere is the result of either local evaporation or horizontal or vertical transport of moisture from other locations or altitudes. The effect of horizontal transport (advection) on the distribution of water vapor is clearly evident in Figure 5–10, which shows the spatial distribution of mean dew points (in °F and in °C) across the United States in January (top) and July (bottom). Looking at the eastern two-thirds of the country first, it is clear that for both months the amount of water vapor generally decreases with distance from the Gulf of Mexico. Because of the Gulf's high water temperatures, moisture is readily evaporated into the atmosphere year round, and this moisture can be transported northward. The decline in water vapor content is seen in a north–south direction and also moving westward from about the Mississippi River toward the Rocky Mountains during the summer. During the winter months, the amount of moisture extending into the Great Plains is low and only a minimal amount of east–west variation exists.

The effect of distance from the source of moisture is also evident in the West, with water vapor generally decreasing from the Pacific Coast to the Rocky Mountains. The most

▶ **FIGURE 5–10** The average distribution of dew points across the United States in January (top) and July (bottom).

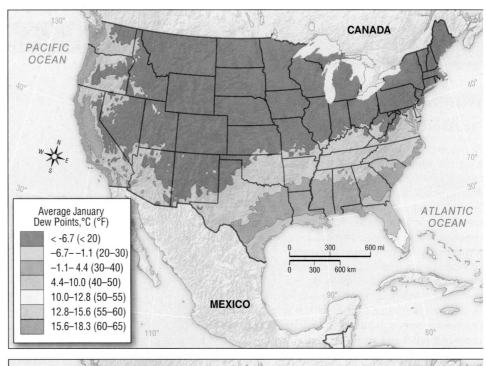

Average January
Dew Points,°C (°F)

	< -6.7 (< 20)
	−6.7– −1.1 (20–30)
	−1.1– 4.4 (30–40)
	4.4–10.0 (40–50)
	10.0–12.8 (50–55)
	12.8–15.6 (55–60)
	15.6–18.3 (60–65)

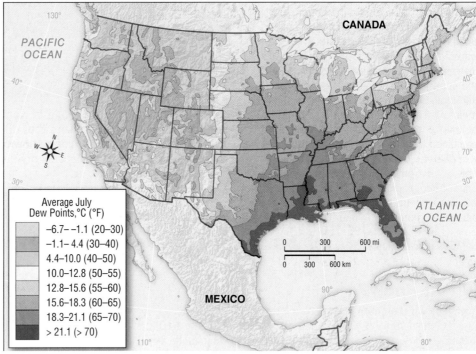

Average July
Dew Points,°C (°F)

	−6.7– −1.1 (20–30)
	−1.1– 4.4 (30–40)
	4.4–10.0 (40–50)
	10.0–12.8 (50–55)
	12.8–15.6 (55–60)
	15.6–18.3 (60–65)
	18.3–21.1 (65–70)
	> 21.1 (> 70)

rapid drop occurs very near the coast because local mountains block off substantial amounts of moisture from inland areas.

Comparing the two maps, you will note a substantial increase in the amount of water vapor in the air in July over that in January. This should not be surprising, because lower January temperatures preclude the existence of high water vapor contents. Thus, for example, along the Ohio River Valley the average dew point increases from about –7 °C (20 °F) in January to perhaps 17 °C (63 °F) in July. This is why residents of much of the country, especially those east

of the Rockies, are subject to uncomfortable dry skin in the winter—only to find themselves sweating profusely during the summer.

The general patterns described for the United States also apply to most of Canada. In this section we looked at the horizontal distribution of water vapor, but water vapor content also varies with distance from the surface, generally decreasing upward. *Box 5–3, Forecasting: Vertical Profiles of Moisture*, provides information on the analysis of vertical water vapor distributions.

5–3
FORECASTING

Vertical Profiles of Moisture

In Chapter 3, you saw how simplified thermodynamic diagrams can be used to plot the vertical profile of temperature. Because dew point values are likewise expressed as temperatures, they, too, can be plotted on thermodynamic charts. In fact, by plotting temperatures and dew points simultaneously, you can obtain considerable information on cloud conditions. Refer to the profiles (also called soundings) of temperature and dew point in Figure 1, taken from Stapleton Airport, near Denver, Colorado, at midnight, Greenwich mean time, on April 12, 2002.

The example in Figure 1 plots temperature (the curve on the right) and dew point (left). If you contrast this to the profile shown in Chapter 3 (page 85), taken at Slidell, Louisiana, you will notice that this sounding begins at a much lower pressure—at about the 840 mb level. The reason for this is very simple. Denver's elevation of 1625 m (5330 ft) causes its surface pressure to be much lower than that at Slidell's nearly sea level location (remember, pressure *always* decreases with elevation).

In Figure 1, the temperature at the surface is 17 °C (63 °F), and the dew point is –4 °C (25 °F). This large difference between the two values indicates that the relative humidity is low (calculated to be 23 percent). But as distance from the surface increases, temperature decreases more rapidly than dew point. At about the 560 mb level (at a height of about 4850 m, or 15,900 ft, above the surface), the two values become nearly equal to each other and the air is saturated (a slight measurement error accounts for the plotted temperature and dew point values not being exactly equal). The dew point and the air temperature then decrease at the same rate up to about the 460 mb level (about 6500 m, or 21,300 ft, above the surface), above which the temperature once again exceeds the dew point. Because the air is saturated in the layer of air between 4850 m and 6500 m above the surface, we can infer that a cloud occupies that 1650 m thick layer.

The thermodynamic diagram shown here is slightly more complex than the one in

▲ **FIGURE 1** A sounding of temperature and dew point. This chart plots temperature and dew points throughout the troposphere and much of the stratosphere. The slightly sloping red lines depict values of mixing ratio in grams of water vapor per kilogram of dry air (labeled just above the x-axis). Meteorologists use these lines along with the plots of temperature to determine the saturation mixing ratio; these lines and the dew point profiles are used to obtain the actual mixing ratio.

Chapter 3, because it includes an additional set of lines that provides one more type of moisture information. The red lines that slope gently to the left as they extend upward indicate the mixing ratio and the saturation mixing ratio at any level. The dew point profile is used to determine the mixing ratio at any pressure level; the temperature profile lets us obtain the saturation mixing ratio. Let's first see how the plot of the air temperature profile can be used to determine the saturation mixing ratio at a given pressure level. As an example, notice that at the 700 mb level the air temperature is 2 °C. It so happens that one of the sloping lines (labeled at the bottom of the diagram with the number 7) nearly intersects the temperature profile at the 700 mb level (highlighted by a red dot). This indicates that the air at the 700 mb level has a saturation mixing ratio of just under 7 grams of water vapor per kilogram of dry air.

We can follow a similar procedure to find out what the actual mixing

ratio is at the 700 mb level. To do this, we follow the dew point profile up to the 700 mb level. The point where the profile crosses the 700 mb line occurs right between the two sloping lines labeled 2 and 4. In other words, the actual mixing ratio at the 700 mb level is right between 2 grams of water vapor per kilogram of air and 4 g/kg—with 3 g/kg a very good approximation.

After estimating the actual mixing ratio and saturation mixing ratio at the 700 mb level, it is easy to use the values to obtain the relative humidity at the 700 mb level:

RH = (mixing ratio/saturation mixing ratio) × 100% = (3/7) × 100 = 43%

This procedure can be performed at the surface or any other level of the atmosphere.

Later in this chapter, we will see how thermodynamic diagrams can give us information in forecasting the likelihood of cloud development.

Processes that Cause Saturation

Air can become saturated by any one of three general processes: adding water vapor to the air; mixing cold air with warm, moist air; and lowering the temperature to the dew point. The first of these processes can be seen in your bathroom when you take a warm shower. The warm water from a showerhead evaporates moisture into the air in the room and brings it to the saturation point. Condensation first forms on your mirrors and other surfaces, and then a general fog develops. In the natural environment, the evaporation of water from falling raindrops can raise the dew point in the air beneath the cloud from which the rain falls. If enough vapor is added to the air to saturate it, a **precipitation fog** forms beneath the cloud.

Condensation from the second process, the mixing of cold with warm moist air, is illustrated in Figure 5–11. Consider parcels A and B of unsaturated air. Parcel A has a temperature of 0 °C (32 °F) and a specific humidity of 3 grams of water vapor per kilogram of dry air; Parcel B has a temperature of 30 °C (86 °F) and a specific humidity of 20 g/kg. If equal amounts of the two parcels are mixed together, the new parcel has a temperature of 15 °C (59 °F), exactly midway between the temperatures of the original parcels. However, such is not the case for the specific humidity.

Although the total H_2O content in the mixed parcel is exactly between the amounts of the original parcels, some

of the moisture occurs in the form of a liquid rather than as water vapor. Here is why: The midpoint between the original specific humidity values is 11.5 g/kg, but at 15 °C the saturation specific humidity is only 10.8 g/kg. In other words, the air contains 0.7 g/kg more water than can exist in the vapor form. The surplus therefore condenses to form fog droplets.

The preceding process is what causes contrails to form behind aircraft traveling at high altitudes. As the jet engines burn fuel, they put out a large amount of heat as well as water vapor. The air is extremely turbulent in the wake of the aircraft, so the hot, moist (but unsaturated) exhaust from the engines rapidly mixes with the cold surrounding air. At the subfreezing temperatures of the upper troposphere, the vapor directly forms into ice crystals or into liquid droplets that eventually freeze to form the contrail.

A similar but naturally occurring phenomenon is known as **steam fog**. As we learned in Chapter 3, water bodies are rather slow to change temperature. As a result, lakes can remain relatively warm well into the fall or early winter, even as air temperatures become low. Because of evaporation and the upward transfer of sensible heat, a thin, transitional layer of air exists just above the water surface that is warmer and moister than the air above. If a mass of cold air abruptly passes over the warm lake, the warm, moist transitional air mixes with the overlying layer of cold air to form a layer of fog a meter or two thick (Figure 5–12).

▶ **FIGURE 5–11** Saturation by the mixing of warm, moist air with cold air. Parcel A has a low temperature and can therefore contain only a small amount of water vapor. The warm parcel (Parcel B) has a higher moisture content. Parcel C results when Parcels A and B are mixed together and has a temperature in between those of the original parcels. The amount of moisture in the air is greater than that which can exist at the new temperature, so the excess water vapor condenses.

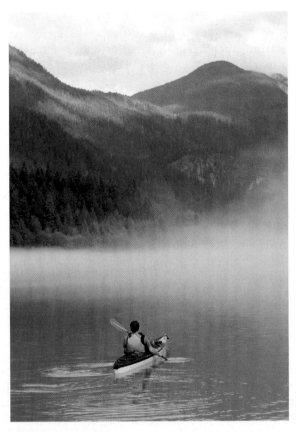

▲ **FIGURE 5–12** Steam fog.

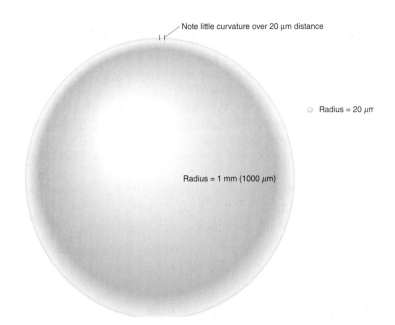

▲ **FIGURE 5–13** Large droplets are less curved than small droplets. The larger droplet on the left has virtually no apparent curvature across a 20 μm extent across its surface, unlike the much smaller droplet on the right. Droplets with highly curved surfaces evaporate more rapidly than droplets with less curved surfaces, and therefore require greater surrounding vapor pressures to remain in equilibrium.

Although we can cite several examples of how clouds form by the increase of moisture content or by mixing warm, moist air with cold, dry air, experience shows that most clouds form when the air temperature is lowered to the dew point. There are several ways this cooling can occur, requiring considerable explanation, as we will see later in this chapter. For now we simply note that this third mechanism, atmospheric cooling, is by far the most common process for cloud formation.

Checkpoint

1. What process that leads to saturated air causes the "fog" that forms in the bathroom when you take a shower?
2. How does steam fog form?

Factors affecting Saturation

This chapter opened with a hypothetical experiment in which water in a jar attained an equilibrium between condensation and evaporation. The experiment, assuming pure water with a flat surface, provided a foundation for understanding saturation. But meteorology studies the atmosphere—not what goes on in hypothetical jars. In the real atmosphere, we are concerned with the rates of evaporation and condensation across the surfaces of suspended cloud and fog droplets. Such droplets are neither flat nor made up of pure H_2O.

We therefore expand on our discussion of evaporation, condensation, and equilibrium to take into account the effects of curvature of cloud and fog droplets and the fact that they are not made of pure H_2O.

Effect of Curvature

Water droplets exist in nature not as tiny cubes with flat sides, but rather as microscopically small spheres with considerable curvature. Compare the two droplets shown in Figure 5–13. The one on the left is much larger than the one on the right and therefore has less pronounced curvature. We could even consider a more extreme example—Earth itself. Most of us are pretty certain by now that the planet is not flat but spherical. However, because of Earth's large size, its curvature is inconspicuous to anybody standing on it, and a centimeter of distance across its surface essentially forms a straight line. Moreover, a straight-edge ruler can lie flat against its surface. But hold a ruler over a tennis ball and only a small part of it is in contact with the ball's surface; it is more strongly curved than Earth.

Cloud droplets are much smaller than tennis balls, of course, and thus have even lesser semblance to a flat surface and curve markedly within a very short distance. But what does curvature have to do with saturation? The answer is that curvature has an effect on evaporation from cloud droplet surfaces and therefore on the vapor pressure necessary for saturation. Effects arising from surface tension lead to differences in the saturation point, as described shortly.

The graph of saturation vapor pressure versus temperature, shown as Figure 5–5, applies only to flat surfaces of pure H_2O. For curved water surfaces, the evaporation rate is greater. The enhanced rate of evaporation requires that condensation also be increased for the two to remain in balance. Thus, a highly curved droplet of pure water at any given temperature has a higher saturation vapor pressure than indicated in Figure 5–5. Stated another way, highly curved droplets of pure water require relative humidities in excess of 100 percent to keep them from evaporating away.

 TUTORIAL
ATMOSPHERIC MOISTURE AND CONDENSATION

Use the tutorial to observe the effect of water droplet curvature on saturation vapor pressure.

Figure 5–14 illustrates the effect of droplet size on the relative humidity needed to maintain an existing droplet of pure H_2O. For very small droplets (those with radii of about 0.1 µm), relative humidities in excess of 110 percent are needed to achieve an equilibrium between evaporation and condensation. In other words, those droplets require a *supersaturation* of 10 percent. The degree of supersaturation necessary to maintain a droplet rapidly decreases with increasing droplet size. Droplets with radii larger than 10 µm require supersaturations of less than 1 percent.

If the atmosphere were devoid of any aerosols, condensation would occur only by **homogeneous nucleation**, in which droplets form by the chance collision and bonding of water vapor molecules under supersaturated conditions. Such droplets would necessarily have only a small number of molecules and a high degree of curvature, and therefore would only exist at high levels of supersaturation. This process seldom if ever occurs, because certain **hygroscopic** (water-attracting) aerosols in the atmosphere assist the formation of droplets at relative humidities far below those necessary for homogeneous nucleation. The formation of water droplets onto hygroscopic particles is called **heterogeneous nucleation**, and the particles onto which the droplets form are called **condensation nuclei**. These particles enhance condensation through two processes, as described below.

Effect of Solution

Some condensation nuclei readily dissolve in water to form a solution. For any solution a certain number of the molecules at the surface are those of the *solute* (the material dissolved in the water) rather than H_2O molecules. With fewer H_2O molecules at the surface, the rate of evaporation is lower than for pure water. As a result, solutions require less water vapor above the surface to maintain an equilibrium between evaporation and condensation. We see that cloud droplets formed from soluble condensation nuclei will have a lower saturation point than droplets of pure water. Thus after formation, they will be less likely to evaporate than any formed by homogenous nucleation. The solute effect therefore promotes condensation, which is counter to the effect of curvature, and condensation normally occurs at relative humidities near or slightly below 100 percent.

Although the proportion of aerosols in the air that are hygroscopic is small, the atmosphere contains so many suspended particles that condensation nuclei are always abundant. Some materials are more hygroscopic than others, and large aerosols are generally more effective than smaller ones. Some are even capable of attracting water at relative humidities below 90 percent and forming extremely small droplets. We observe such droplets as **haze** (Figure 5–15).

Condensation nuclei originate from natural and human processes. Scientists once believed that most condensation nuclei in the atmosphere were natural, consisting mostly of continental dust, sea salt, and aerosols derived from volcanic eruptions, natural fires, and gases given off by marine phytoplankton. Research undertaken during the 1990s indicated that the role of human activity was previously understated and that anthropogenic sources may account for the major-

▶ **FIGURE 5–14** Small droplets of pure water require relative humidities above 100 percent to remain in equilibrium.

▲ **FIGURE 5-15** Haze.

ity of condensation nuclei over industrialized areas in the Northern Hemisphere. Still, there is considerable uncertainty about which materials are the most active cloud condensation nuclei.

TUTORIAL
ATMOSPHERIC MOISTURE AND CONDENSATION

Use the tutorial to explore how the presence of particles in the air can offset the effect of curvature on saturation.

Ice Nuclei

So far we have examined the formation of liquid water droplets when air becomes saturated. But saturation can occur at very low temperatures, which suggests that rather than liquid water droplets, ice crystals may form. On the other hand, many of us have walked through fog composed of liquid droplets even though the temperature was below 0 °C. So what really happens when saturation occurs at temperatures below the freezing point? Does this lead to the condensation of liquid droplets or to the deposition of ice crystals?

The answer is that it depends. Strangely, although ice always melts at 0 °C, water in the atmosphere does not normally freeze at 0 °C.

If saturation occurs at temperatures between 0 °C and −4 °C, the surplus water vapor invariably condenses to form **supercooled water** (water having a temperature below the melting point of ice but nonetheless existing in a liquid state). Ice does not form within this range of temperatures. Just as the formation of liquid droplets at relative humidities near 100 percent requires the presence of condensation nuclei, the formation of ice crystals at temperatures near 0 °C requires **ice nuclei**. Unlike condensation nuclei, which are always abundant, ice nuclei are rare in the atmosphere. This is because an ice nucleus must have a six-sided structure that mimics the alignment of molecules in an ice crystal (although exceptions to this rule exist).

A material's ability to act as an ice nucleus is temperature dependent, and no materials are effective ice nuclei at temperatures above −4 °C. Though ice can exist at temperatures between −4 °C and 0 °C, it does not form spontaneously in the atmosphere in this range. (In fact there is little ice in the atmosphere at temperatures above −10 °C.) Thus, between −4 °C and 0 °C, the removal of water vapor occurs only by the condensation of supercooled water.

As temperature decreases, the likelihood of ice formation increases, and at temperatures between about −10 °C (14 °F) and −40 °C (−40 °F) saturation can lead to the nucleation of ice crystals, supercooled droplets, or both. In clouds having temperatures within this range, liquid droplets and solid crystals will usually coexist, but with a greater proportion of ice at lower temperatures. At temperatures below −30 °C (−22 °F), the cloud will be mostly ice crystals. For temperatures below −40 °C, saturation leads to the formation of ice crystals only, with or without the presence of ice nuclei. As we will see in Chapter 7, the coexistence of ice crystals and liquid droplets in clouds is extremely important for the development of precipitation outside the tropics. (*Box 5–4, Focus on Aviation: Icing* describes how the formation of ice aloft creates a hazard for aircraft.)

Among the materials that serve as ice nuclei are components of natural soils called *clays*. Clay materials have a platy structure, microscopic sizes, and strong electrical attractions. These characteristics make them very difficult to dislodge from the surface and incorporate into the atmosphere, which largely explains their scarcity. Clays occur in a variety of compositions, of which the most effective seems to be *kaolinite* (aluminum silicate). Other types of clay serve as ice nuclei but are active only at lower temperatures.

Checkpoint

1. How does droplet size affect rates of evaporation and condensation?
2. Why are ice nuclei often required for the formation of ice crystals?

5–4 FOCUS ON AVIATION

Icing

The formation of ice onto aircraft, *icing*, can create a serious hazard to aircraft. In the period 1997 to 2006, icing was responsible for 10 percent of all weather-related accidents and 22 percent of those involving fatalities. Icing most commonly forms on the body of an aircraft (*structural icing*) but can also develop within the air intake system (*induction icing*) and cause the engine to shut down.

Even a thin film of relative ice can have a drastic effect on the aerodynamics of a plane, increasing drag (wind resistance), reducing the amount of lift provided by the wings, lowering the stall speed, and hindering aircraft maneuverability. Ice also adds additional weight to the aircraft, though this effect is secondary relative to that of the aerodynamic impacts.

Structural icing usually occurs in one of the following three forms:

Clear ice. Sometimes called *glazed ice*, this type of ice is translucent and has a smooth appearance. It is caused when a plane flies through a cloud containing supercooled drops that slowly freeze onto the wings and fuselage. This type of ice occurs at temperatures not far below the freezing point of water, so that the water spreads from the leading edge of the plane and migrates slowly backward, gradually forming a continuous film of ice. While generally maintaining a continuous coating the ice can also form lumps and undulations, therefore causing even greater impacts on the aerodynamics.

Rime. Rime ice usually forms at lower temperatures than does clear ice, generally within the range of –10 °C to –20 °C (14 °F to –4 °F). At these lower temperatures water can freeze rapidly and entrap small amounts of air, giving it a rough, milky appearance. Because the ice forms quickly the water does not flow across a wing or fuselage prior to freezing, and therefore accumulates at the front of the plane.

Mixed ice. This is the most common form of icing, consisting of a combination of clear ice and rime.

Aircraft icing is most threatening when clouds contain a large amount of liquid water. As shown in Figure 1, winter icing is more likely to occur near British Columbia–Washington State, the coast of Alaska, over the eastern Great Lakes, and over extreme eastern Canada. Pilots must know what to do if icing occurs, but it is absolutely essential that they anticipate its formation before taking off and while in flight. The National Weather Service Aviation Weather Center provides detailed information on current icing conditions at http://aviationweather.gov/adds/icing/ (Figure 2).

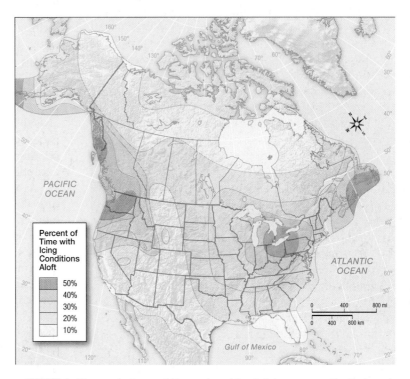

▲ **FIGURE 1** The occurrence of November–March icing across North America.

▲ **FIGURE 2** The National Weather Service Aviation Weather Center provides up-to-date information on icing conditions.

Cooling the Air to the Dew or Frost Point

Although condensation can occur from an increase in the amount of water vapor or from mixing cold air with warm, moist air, the most common mechanism for cloud formation is the lowering of the air temperature to the dew or frost point. We might expect that air temperature will change only in response to gains or losses of energy, but such is not the case. Air temperature changes can occur from two very general classes of processes: those that involve the removal or input of energy, and those that do not. These are referred to as *diabatic* (DIE-a-bat-ic) and *adiabatic* (A-dee-a-bat-ic), respectively.

Diabatic Processes

A **diabatic process** is one in which energy is added to or removed from a system. A pot of water placed over a stove warms diabatically, as does air that is warmed by conduction when in contact with a warm surface. Likewise, air that passes over a cool surface loses energy by conduction into the surface and therefore cools diabatically. Note that the direction of heat transfer is in accordance with the **second law of thermodynamics**, which dictates that energy moves from regions of higher to lower temperatures. Diabatic processes are frequently responsible for the formation of fog but are secondary to adiabatic processes for the development of clouds.

Adiabatic Processes

Processes in which temperature changes but no heat is added to or removed from a substance are said to be **adiabatic** (which literally means "not diabatic"). Although changes in temperature without the exchange of heat may at first seem counterintuitive, adiabatic processes are common in the atmosphere and provide the most important mechanism for the formation of clouds.

To adequately understand such processes, we refer to a version of the **first law of thermodynamics**, which states what happens when heat is added to or removed from gases. Specifically, if heat is added, there will be some combination of an expansion of the gas and an increase in its temperature. The law is given in numerical form as

$$\Delta H = p \cdot \Delta\alpha + c_v \cdot \Delta T$$

where ΔH = Heat added to the system, p is the air pressure, $\Delta\alpha$ is the change in volume (positive for expansion and negative for contraction), c_v is the specific heat for air (assuming a constant volume), and ΔT is the change in temperature. (Note that the Greek letter delta, Δ, preceding a symbol represents a change in the value of the quantity.) The first term on the right-hand side of the equation, $p \cdot \Delta\alpha$, is the work performed by the gas as expansion occurs. The second term, $c_v \cdot \Delta T$, refers to the change in internal energy. The important thing for us is that heat added to the air does not simply disappear but rather is apportioned between temperature and volume changes.

The first law of thermodynamics describes the underlying principle of what occurs in the cylinder of an internal combustion engine in an automobile, as shown in Figure 5–16. As the air–fuel mixture burns, it expands and pushes down on the piston (this is the work performed) and ultimately propels the car. In addition, there is an increase in the internal energy of the gas, which we observe as an increase in temperature. (Just ask anyone who has ever burned a hand on an exhaust manifold!) The energy unleashed with the combustion of the

▲ **FIGURE 5-16** A four-stroke automobile engine works on the principle invoked by the first law of thermodynamics. As the piston is pulled down by the crankshaft (a) and (b), a fuel-air mixture enters the cylinder. In the second stroke the mixture is compressed (c). The third stroke occurs when the spark plug fires causing combustion of the air-gas mixture (d). The energy released by the burning fuel is manifest as work done by the moving piston and an increase in internal energy (the increase in temperature). The burned fuel is expelled during the fourth stroke (e).

fuel is therefore manifested as work performed and an increase in temperature. Of course, good automotive design calls for the engine to convert most of the chemical energy to work performed, with little going toward an increase in internal energy. In other words, engine heat represents wasted energy, and a cold exhaust manifold would be the mark of good engineering. The same relationship among heat, temperature, and volume applies to our atmosphere.

An adiabatic process represents a special case of the first law of thermodynamics in which the left-hand side of the equation equals 0 (no heat is added or removed). Substituting 0 for ΔH yields

$$0 = p \cdot \Delta \alpha + c_v \cdot \Delta T$$

which can be rearranged as

$$p \cdot \Delta \alpha = -c_v \cdot \Delta T$$

or

$$-p \cdot \Delta \alpha = c_v \cdot \Delta T$$

Stated in words, the adiabatic form of the first law indicates that if no heat is added or removed from the system work performed *by* the air (the expansion of the gas) causes a decrease in internal energy (a decrease in temperature) and work performed *on* the gas (compression) leads to warming. Stated even more succinctly, expanding air cools and air undergoing compression warms.

This principle applies to what happens when you use a hand pump to inflate a bicycle tire. Compressing the air causes its temperature to increase even though no heat is added, as you can readily observe by touching the base of the pump. When air is allowed to escape from the tire valve, its temperature decreases because of expansion, and the jet of outgoing air feels cold. Likewise, blowing on a burned finger feels good, not because the air in your lungs is cool, but rather because the air cools as it expands past your lips. You can easily test this by blowing on a thermometer. The temperature will be around 29 °C (84 °F), a few degrees lower than skin temperature, and considerably less than its original temperature of 37 °C (99 °F).

Figure 5–17 applies this concept to a parcel of unsaturated air that is displaced upward. As the air rises, it encounters lower surrounding pressure, expands, and cools. The rate at which a rising parcel of unsaturated air cools, called the **dry adiabatic lapse rate** (DALR), is very nearly 1.0 °C/100 m (5.5 °F/1000 ft). Thus, a parcel of unsaturated air cools 1 degree Celsius for every 100 meters of ascent, despite the fact that no heat is removed. Likewise, the downward movement of unsaturated air leads to compression and warming at the same rate. It is important to note that the term *lapse* refers to a decrease in temperature with altitude. We therefore leave out the minus sign preceding the 1.0 °C/100 m.

If a parcel of air rises high enough and cools sufficiently, expansion lowers its temperature to the dew or frost point, and condensation or deposition commences. The altitude at which this occurs is known as the **lifting condensation level** (LCL). As the saturated air rises beyond the LCL, expansion continues to lower its temperature, but the cooling is partially offset by the release of latent heat from condensation (or the deposition to ice). Thus, the lifting of saturated air results in a less rapid cooling than the lifting of unsaturated air. The rate at which saturated air cools is the **saturated adiabatic lapse rate** (SALR), which is about 0.5 °C/100 m (3.3 ° F/1000 ft). The term *wet adiabatic lapse rate* is often used interchangeably with saturated adiabatic lapse rate. Unlike the dry adiabatic rate, the SALR is not a constant value but instead varies with temperature, as we explain in *Box 5–5, Physical Principles: The Varying Value of the Saturated Adiabatic Lapse Rate.*

Did You Know?

Water vapor is not the only familiar gas that can condense as a result of cooling due to expansion. Carbon dioxide in a bottle of beer or soda exists as a gas dissolved in the drink and is also found in the neck of the bottle between the top of the drink and the bottle cap. When you open the bottle, the carbon dioxide above the surface of the liquid expands out of the bottle neck and into the surrounding atmosphere. In doing so, its temperature drops adiabatically, and some condenses to form a visible miniature CO_2 cloud wafting out of the bottle top.

▶ **FIGURE 5-17** As a parcel of unsaturated air rises, its temperature decreases at the dry adiabatic lapse rate of 1 degree Celsius per 100 meters. The parcel shown here has an initial temperature of 10 °C but cools to 7 °C when lifted 300 meters.

(a)   (b)   (c)

5–5 PHYSICAL PRINCIPLES

The Varying Value of the Saturated Adiabatic Lapse Rate

The dry adiabatic lapse rate (DALR) has a constant value of 1 °C/100 m. The saturated adiabatic lapse rate (SALR) is usually about half that value, because the release of latent heat as saturated air rises partially offsets the cooling by expansion. But unlike the value of the dry adiabatic lapse rate, the SALR is not constant. Rather, it depends on the temperature of the saturated air parcel, with higher temperatures causing lower lapse rates. Figure 1 can help us see why this is the case. Recall that if air is saturated, its actual specific humidity is equal to the saturation specific humidity. As a rising parcel of saturated air is cooled adiabatically, the amount of water vapor that can exist decreases and the surplus water vapor is removed by condensation (or deposition).

Now observe what happens when the temperature of warm, saturated air decreases 5 °C (9 °F), from 30 °C to 25 °C. As the air cools, its specific humidity changes from 27.7 grams of vapor per kilogram of air to 20.4—a decrease of 7.3 grams of vapor per kilogram of air. The 7.3 g of water vapor do not simply vanish; rather, they are converted to an equal mass of liquid water. Upon condensation, each gram of water releases 2500 joules of latent heat, for a total of 18,250 J. Compare that to what happens if the temperature undergoes another 5 °C drop in temperature, but this time from

▲ **FIGURE 1** Unlike the DALR, the SALR is not a constant value. When warm, saturated air cools, it causes more condensation (and hence more latent heat release) than for cold, saturated air. For example, if saturated air cools from 30 °C to 25 °C (a 5° decrease), the specific humidity decreases from 27.7 grams of water vapor per kilogram of air to 20.4. A 5 °C drop in temperature from 5 °C to 0 °C lowers the specific humidity only 1.7 grams for each kilogram of air. This brings about less warming to offset the cooling by expansion, as well as a greater saturated adiabatic lapse rate.

5 °C to 0 °C. At this lower temperature, the 5° of cooling reduces the water vapor content from 5.5 to 3.8 g/kg—only a 1.7 g decrease—and only 4250 J of energy are released to the air. Thus, at low temperatures relatively little latent heat is released to offset the cooling due to expansion, and the SALR is nearly equal to the DALR. When warm, saturated air is lifted, a greater amount of latent heat is available to offset the cooling by expansion, and the SALR assumes a lower value.

The Environmental Lapse Rate

The adiabatic lapse rates are not to be confused with the **environmental** (or ambient) **lapse rate** (ELR), which applies to the vertical change in temperature through still air. A large mass of air is not likely to have a constant temperature; rather, its temperature usually decreases with altitude in the troposphere. The rate at which the ambient temperature decreases with height is called the *environmental lapse rate* (ELR). The ELR for the troposphere is highly variable. It changes from day to day, from place to place, and even from one altitude to another within the atmospheric column. An atmosphere in which temperature decreases rapidly with elevation is said to have a "steep" environmental lapse rate.

As an analogy, we can contrast the changes in temperature that would occur inside a rising balloon with that of the surrounding air. In Figure 5–18, a balloon rises through the atmosphere, which in this case has an ELR of 0.5 °C/100 m. A thermometer within the balloon would record the temperature change inside the expanding balloon, corresponding to the DALR as long as the air within remains unsaturated. A thermometer attached outside the balloon would record the temperature of the surrounding, nonmoving air, reflecting the environmental lapse rate. As the balloon rises, the temperature within the balloon will be lower than that of the air surrounding it, because in this instance the DALR exceeds the ELR.

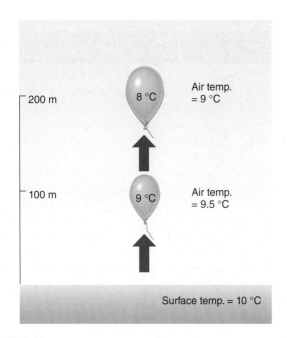

▲ **FIGURE 5–18** Do not confuse the adiabatic lapse rates with the environmental (or ambient) lapse rate. A balloon rising through air with an ELR of 0.5 °C/100 m passes through air whose temperature decreases from 10 °C at the surface, to 9.5 °C at 100 m, and 9.0 °C at 200 m. The air within the balloon cools at the dry adiabatic lapse rate of 1.0 °C/100 m, faster in this example than the ELR, and therefore attains a temperature of 8 °C at the 200 m level.

Checkpoint

1. What happens as a parcel of unsaturated dry air rises? Explain.
2. What is the difference between the environmental lapse rate and the adiabatic lapse rate?

Forms of Condensation

Saturation can lead to the formation of liquid water or ice crystals. The condensation or deposition can occur in the air as cloud or fog, or onto a surface as dew or frost. In this section we discuss the processes that give rise to these various types of condensation. Table 5–3 summarizes the general types of condensation and the processes that form them.

Dew

Dew (Figure 5–19a) is liquid condensation on a surface, often occurring during the early morning hours. At night the loss of longwave radiation can cause the surface to cool diabatically. Air immediately in contact with the cold surface cools by conduction, and if the temperature decreases all the way to the dew point, condensation forms. Dew is most likely to form on clear, windless nights, when the absence of clouds allows much longwave radiation to escape to space and the lack of wind precludes the mixing of warmer air from above. Together these conditions promote rapid cooling within a shallow layer of air immediately adjacent to the surface.

Frost

The formation of **frost** is similar to that of dew, except that saturation occurs when the temperature is below 0 °C. When the air temperature is lowered to the frost point, very small ice crystals are deposited onto solid surfaces, giving them a bright white appearance, as shown in Figure 5–19b. This type of deposition, sometimes referred to as *white frost* or *hoar frost*, occurs by the transformation of water vapor directly into ice, without going through the liquid phase.

Because it consists of a huge number of separate ice crystals rather than a solid, continuous coating of ice, frost

TABLE 5–3

General Types of Condensation

Condensation Form	Predominant Processes	Characteristics
Dew	Lowering of temperature to the dew point near the surface. Favored under clear skies and no wind. Diabatic process.	Appears as coating of liquid water on surfaces.
Frost	Lowering of air temperature to saturation point, when the saturation point is below 0 °C (32 °F). Diabatic process.	Appears as large number of small white crystals on surfaces.
Frozen Dew	Formation of dew at temperatures above 0 °C, followed by cooling to temperatures below 0 °C. Diabatic process.	Continuous layer of solid ice on surface.
Fog	Usually by cooling of layer of air with light winds. Sometimes by evaporating water from falling precipitation or by mixing warm, moist air with cold air. Diabatic or adiabatic process.	Large concentration of suspended droplets in layer of air near ground. Under extreme cold, can consist of suspended ice crystals.
Radiation fog	Cooling of air to dew point by longwave radiation loss. Diabatic process.	Same as above.
Advection fog	Cooling of air to dew point as it passes over cool surface. Diabatic process.	Same as above.
Upslope fog	Cooling of air as it flows upslope. Adiabatic process.	Same as above.
Precipitation fog	Increasing the water vapor content of the air by evaporation from falling droplets. Adiabatic process.	Same as above.
Steam fog	Mixing warm, moist air with cold air. Adiabatic process.	Same as above.
Clouds	Usually by lifting of air and adiabatic cooling.	Concentration of suspended droplets and/or ice crystals in air well above the surface.

on the windshield of a car is often easy to remove by a swift brushing with a credit card or window scraper. This is in contrast to a more troublesome type of condensation called *frozen dew*.

(a)

(b)

(c)

▲ **FIGURE 5–19** Dew (a), frost (b), and frozen dew (c).

Frozen Dew

Frozen dew differs from frost in both its structure and its manner of formation. It begins when saturation forms liquid dew at temperatures slightly above 0 °C. When further cooling brings its temperature below the freezing point, the liquid solidifies into a thin, continuous layer of ice. In contrast to frost, frozen dew is neither milky white nor easy to remove but instead bonds tightly to any surface on which it forms, as shown in Figure 5–19c.

Because it is a continuous coating of solid ice, a mere brushing does not come close to removing frozen dew. In addition to coating your car's windshield, the ice can make it difficult or impossible to get your key into the lock mechanism of your door. And even if you are lucky enough to be able to turn the key, the door can become frozen to the car frame as if it were welded shut. On the other hand, it may be just as well for you that you are not able to get into your car because of *black ice*, the smooth coating of frozen dew that forms on road surfaces. This is especially likely to form over bridges, causing dangerous, slippery driving conditions.

Fog

Fog is essentially a cloud whose base is at or near ground level (Figure 5–20). It can be extremely shallow, on the order of a meter or so in depth, or it can extend for tens of meters above the surface. Like any other form of condensation, fog can form by the lowering of the air temperature to the dew point, an increase in the water vapor content, or the mixing of cold air with warm, moist air.

Precipitation and Steam Fogs We have already described one example of condensation resulting from the addition of water vapor to the air, *precipitation fog*, which results from the evaporation of falling raindrops. Another, although very localized, type of fog occurs from adding water vapor to the air—the type you see right in front of you when you exhale on a cold day. We have also discussed the formation of *steam fogs*, which occur when cold, dry air mixes with warm, moist air above a water surface. All other types of fog result from a cooling of the air to the dew point. They include radiation fogs, advection fogs, and upslope fogs.

Radiation Fog **Radiation fogs** (sometimes called *ground fogs*) develop when the nighttime loss of longwave radiation causes cooling to the dew point. Like dew, a radiation fog is most likely to form on cloudless nights when longwave radiation from the surface easily escapes to space. Unlike dew, it is most likely to form with light winds of about 5 km/hr (3 mph) rather than in perfectly still air. Light breezes promote a gentle stirring of the lower atmosphere, which permits condensation to form throughout a layer of air. When the wind speed is much greater than 5 km/hr, the excess turbulence brings warm air down toward the surface and thereby inhibits cooling to the dew point.

Most radiation fogs begin to dissipate within a few hours of sunrise. Although we sometimes talk about a fog "lifting,"

▲ **FIGURE 5–20** Fog.

(a)

that is not what really happens. It is probably better (although still somewhat imprecise) to describe it as "burning off." When sunlight penetrates the fog, it warms the surface, which in turn warms the overlying air. As the air temperature increases, the fog droplets gradually evaporate. Because the evaporation of droplets is most rapid near the surface, the fog appears to lift, although it really undergoes no vertical displacement.

When radiation fogs are especially well developed, they can scatter backward the greater part of incoming solar radiation. Because the amount of energy reaching the surface is now reduced, the fog can persist throughout the day, especially in the winter when the days are short and the sun angles are low. Under the most extreme circumstances, fogs can persist for days on end.

A prime location for a persistent radiation fog is the Central Valley of California (Figure 5–21). To the west of the valley, the Coast Ranges block the moderating effects of the Pacific Ocean, while the Sierra Nevada isolates the valley from the east. In addition to the clear skies and light winds that often predominate during the winter, this heavily agricultural region has copious amounts of moisture evaporated into the air from the irrigated farmland. When the radiation fog (locally referred to as a *Tule fog*) covers the valley, visibility along the two major north–south highways can be dangerously reduced to near zero. This happened with fatal consequences on February 5, 2002, on California State Highway 99, south of Fresno. As the fog reduced visibility to as little as 15 m (50 ft), California Highway Patrol "pace cars" were dispatched to lead traffic at safe speeds. But, as has happened many times in the past, the effort was unsuccessful. Eighty-seven cars were involved in two chain reaction pileups that left two persons dead and many others injured. Just one month earlier, a similar set of accidents caused two other fatalities along Interstate 5 near the Sacramento airport.

On March 5, 2002—one month after the major pileup near Fresno—a thick, morning fog led to a major crash involving 125 vehicles on Interstate 75 in northwest Georgia, just south

(b)

▲ **FIGURE 5–21** The Central Valley of California often gets radiation fogs in winter that last for days on end (a). These persistent fogs are visible from space (b).

of Chattanooga, Tennessee. Four people died in the accident. This is another region where radiation fogs are common, and that morning the National Weather Service issued numerous advisories about hazardous driving conditions. Despite the warning, the near-zero visibility made any travel along the highway a risky endeavor. According to Sheriff Phil Summers, "It did not appear that there was any fault other than fog."

Radiation fogs—and the risks they impose on travelers—are frequent occurrences wherever air has the opportunity to cool at night with gentle stirring. Various state transportation agencies have tried to alleviate the threat of such accidents by painting reflective "fog lines" on at-risk highways and installing automated weather stations to provide real-time weather information, but measures such as these cannot overcome the inherent danger imposed by dense fog.

Radiation fogs are formed by diabatic cooling and are therefore associated with cold air. Because cold air is denser than warm air with otherwise similar characteristics, radiation fogs often settle into local areas of low elevation, where they are called *valley fogs*. It is tempting to attribute the fog's high density to the heavy droplets of liquid water, but they are not the cause. It is the low temperature—not the existence of water droplets—that causes the fog to be dense and settle into valleys. Because all the water droplets replace an equal mass of water vapor from which they formed, their presence does not make the air any heavier.

Advection Fog Advection fogs (Figure 5–22) form when relatively warm, moist air moves horizontally over a cooler surface (the term *advection* refers to horizontal movement). As the air passes over the cooler surface, it transfers heat downward; this causes it to cool diabatically. If sufficient cooling occurs, a fog forms. Such fogs can be advected for considerable distances and persist well downwind of the area over which they form. One of the most famous examples of this phenomenon occurs during the summer months over the San Francisco Bay area. As relatively warm Pacific air drifts eastward, it passes over the narrow, cold, southward-flowing California ocean current. The cooling of the air offshore forms the fog, which drifts eastward toward San Francisco. It is quite common for the fog to fully engulf San Francisco while Oakland and Berkeley, across the bay, remain warm and sunny.

Advection fogs can also form over water when warm and cold ocean currents are in proximity to each other. Off the coast of New England and the Maritime Provinces, the cold Labrador current flows just to the north of the Gulf Stream. When the moist air from the Gulf Stream region drifts over the Labrador current, it can be cooled to the dew point to form a persistent, dense fog that can last for weeks. Although most common in summer, it can occur any time during the year.

Another example of an advection fog is the fog that often covers London during the winter. Warm air passing over the warm Gulf Stream is advected over England, where it is chilled by the surface to form a thick fog.

The winds associated with advection fogs are often greater than those of radiation fogs. This promotes greater turbulence and allows the droplets to circulate to greater heights. Advection fogs can have thicknesses up to about a half kilometer (1500 ft).

Upslope Fog Of the three types of fog caused by the cooling of air, only **upslope fog** is formed by adiabatic cooling. When air flows along a gently sloping surface, it expands and cools as it moves upward. The western slope of the Great Plains of the United States provides an excellent setting for this type of condensation (Figure 5–23). Westward from the Mississippi River valley, the elevation gradually increases toward the foothills of the Rocky Mountains. Moist air from the Gulf of Mexico cools adiabatically as it ascends the slope of the plains to create widespread fog. Figure 5–24 presents a highly generalized description of the types of fog most prevalent over regions of the United States and Canada.

Distribution of Fog

Figure 5–25 depicts the number of heavy fog days (defined as limiting visibility to a quarter of a mile or less) across the 48 conterminous United States. The three significant centers

◄ **FIGURE 5-22** Advection fogs frequently form when air flows from a region of warmer water to over a cold ocean current. This is particularly common over the San Francisco Bay area.

▲ **FIGURE 5–23** Upslope fogs form by adiabatic cooling of air as it flows up a hill or mountainside.

of heavy fog are along the Pacific Northwest, New England, and the middle Appalachians. The Pacific Northwest and the coast of British Columbia experience numerous advection fogs, as westerly winds advect moist air over the cold California current. But it is not correct to attribute all of the fog formation to the cooling effect of the cold surface waters. As damp, cool air reaches the shore, fog formation is abetted by the *orographic* effect (the lifting effect caused as air crosses a mountain or similar barrier) as air approaches the steep Coast Ranges. Cape Disappointment, Washington, wins the prize for the foggiest U.S. location, being shrouded in heavy fog nearly one-third of the time. The zone of most persistent fog is confined primarily to the coastal region because the Coast Ranges block the flow of moisture inland.

New England and the Maritime Provinces experience a large number of heavy fog days. Along the coast, advection fogs dominate, with the coast of Maine having the highest incidence of dense cover. The advection fogs are most prevalent in summer. Inland, radiation fogs are very common at some of the higher elevation areas of New Hampshire and Vermont.

The southern Appalachians, the third major focus of fog, undergo a large number of radiation fogs, particularly in late summer and fall. Not surprisingly, fog is at a minimum in the desert Southwest.

Checkpoint

1. What type of fog is likely to form on a cool, cloudless night with light winds?
2. Compare and contrast the processes that form advection fogs and upslope fogs.

Formation and Dissipation of Cloud Droplets

Farther from the surface than fog, dew, or frost, clouds are usually the result of the adiabatic cooling associated with rising air parcels. In this section we take a closer look at such parcels as they rise, become saturated, and continue upward above the level at which condensation first occurs.

▶ **FIGURE 5–24** The different types of fog commonly found in North America.

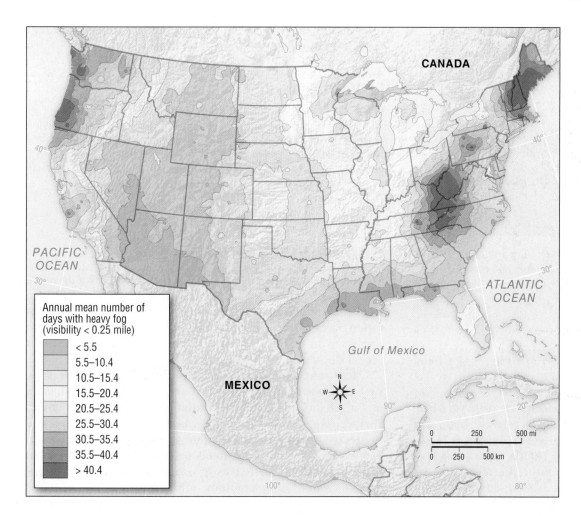

Annual mean number of
days with heavy fog
(visibility < 0.25 mile)

< 5.5
5.5–10.4
10.5–15.4
15.5–20.4
20.5–25.4
25.5–30.4
30.5–35.4
35.5–40.4
> 40.4

For reasons we will not discuss here, the dew point decreases as the air rises, at the rate of about 0.2 °C/100 m (1.1 °F/1000 ft). This decrease is called the **dew point lapse rate**. As unsaturated air is lifted, its temperature therefore approaches the dew point by 0.8 °C for every 100 m of ascent (i.e., 1.0 °C minus 0.2 °C). Thus, if the air temperature and dew point start out at 18 °C and 10 °C, respectively, an ascent of 1000 m is necessary to cause the air to be saturated.

Raising an air parcel above the lifting condensation level (LCL) initially leads to the formation of small cloud droplets. But at about 50 m or so above the LCL, all the condensation nuclei in the air will have attracted water, and further uplift leads only to the growth of existing water droplets. In other words, no new droplets form as the air continues to rise; instead, the existing droplets grow larger.

The processes that lead to the formation of a cloud do not continue forever. At some point in time, lifting will cease and there will be no further condensation. When this happens, further cloud development comes to an end.

Now consider what happens if a rising parcel of air reverses its movement and begins to subside. The parcel now warms at the *saturated* adiabatic lapse rate because the warming by compression is partially offset by the gradual evaporation of the droplets. The evaporation continues until the parcel has descended back to the original level at which condensation occurred. If the air then sinks below the original LCL, it warms at the DALR, because all the droplets will have evaporated. If brought down to the initial level at which uplift first began, the air will reassume its original temperature and dew point. The net effect of all this is that the cooling of the air and the condensation of liquid water are *reversible processes*. Note that we assume all the condensation products remain in the atmosphere and are thus available for evaporation during descent. Of course, the real atmosphere *loses* moisture by precipitation of rain and snow, which means that these processes are not strictly reversible. But a relatively small portion of condensed water falls out, so the concept remains generally valid.

In thinking about vertical motions and moisture, it is interesting that very small displacements can have such large consequences. After all, we hardly think about the effects of horizontal movements covering 100 km (62 mi). But in the vertical, movement of just 1 km (0.62 mi) can make the difference between a fine day and a ruined picnic, as rising air leads to clouds, and clouds give rise to precipitation. Like the formation of clouds, precipitation is no simple matter, as we will see in Chapter 7.

▲ **FIGURE 5-26** The combination of high temperature and high humidity can cause extreme discomfort and even medical emergencies.

High Humidity and Human Discomfort

Temperature is one of the most important weather variables with regard to human comfort, and excessively high and low temperatures account for more North American fatalities than do hurricanes, lightning, floods, and tornadoes combined. But the effects of temperature extremes can be compounded by other factors, such as humidity, the intensity of sunlight, and strength of wind (Figure 5–26). In Chapter 3 we saw how wind speeds can be combined with low temperatures to create a wind chill index. Similarly, the effect of humidity and high temperatures can be expressed in a **heat index**, sometimes referred to as the **apparent temperature**.

One of the human body's most effective mechanisms in guarding against excessive heat is perspiration. Sweat cools the body because, when released to the surface of the skin, it is free to evaporate into the atmosphere and consume latent heat. If the atmosphere has a high moisture content, however, the rate of evaporation is retarded and the loss of latent heat is reduced. In other words, the sweat is unable to effectively do what it is supposed to do. The heat index accounts for this effect (Table 5–4).

The apparent temperatures caused by the combination of heat and humidity provide useful guidelines for people. At values between 41 °C and 54 °C (105 °F and 129 °F), muscle cramps or heat exhaustion are likely for high-risk people, and even people who are not at high risk face the threat of heat stroke (a potentially fatal increase in the body's internal temperature). If the National Weather Service forecasts that the apparent temperature will exceed 41 °C (105 °F) for more than 3 hours, it issues a *heat advisory*. Apparent temperatures above 54 °C (129 °F) are considered extremely dangerous, and heat stroke is likely for at-risk people.

The apparent temperatures shown in Table 5–4 should be considered approximate guidelines. Some people react dif-ferently to heat, and the index does not account for variables such as exposure to bright sun or wind.

Did You Know?

Bangkok, Thailand, one of the hottest cities in the world, has such persistently high temperatures and humidity that it exceeds the U.S. National Weather Service threshold for issuing heat advisories more than half the days of a typical year.

Atmospheric Moisture and Climate Change

We have seen in this chapter that the amount of water vapor that can exist in the air increases with temperature. It is also clear that higher water temperatures lead to increased evaporation rates; and, in fact, ocean surface temperatures *have* increased in conjunction with higher air temperatures over recent decades.

Let's take the situation a step further. In Chapter 3 we saw that water vapor, like carbon dioxide, is an effective gas at absorbing longwave radiation emitted from Earth's surface. As such it is a greenhouse gas that keeps the planetary temperature at a value greater than it would otherwise be. We therefore need to consider whether global warming leads to increases in evaporation rates and atmospheric water vapor contents, and whether such increases would in turn contribute to further atmospheric warming—and by extension even greater water vapor contents. Self-perpetuating situations such as these are referred to as *positive feedbacks*.

The Intergovernmental Panel on Climate Change (IPCC) has made an extensive review of the published research on this phenomenon in its Fourth Assessment Report, issued in early 2007 (see Chapter 3). The panel cited studies around the world that have found increases in specific humidity near the surface have been associated with increasing temperatures since 1976. These increases in water vapor are not restricted to the surface, as increases in water vapor content even into the upper troposphere have likewise occurred. Observations suggest a 4 percent overall increase between 1970 and the early twenty-first century.

Despite the increase in moisture content, the average relative humidity across the globe has not increased. This is due to the effect of temperature on the saturation point. As the amount of water vapor in the air has increased, so, too, has the amount of water vapor that *can* exist because of higher temperatures. Over most oceanic areas the relative humidities have remained fairly constant, as increasing water vapor contents have been offset by increases in the saturation specific humidity. Over certain land areas the increases in specific humidity have been more than offset by increases in the saturation levels, leading to locally unchanged or slightly reduced relative humidities.

TABLE 5–4

Heat Index

Temp. °F (°C)	Relative Humidity (%)											
	40	45	50	55	60	65	70	75	80	85	90	95
110 (47)	136 (58)											
108 (43)	130 (54)	137 (58)										
106 (41)	124 (51)	130 (54)	137 (58)									
104 (40)	119 (48)	124 (51)	131 (55)	137 (58)								
102 (39)	114 (46)	119 (48)	124 (51)	130 (54)	137 (58)							
100 (38)	109 (43)	114 (46)	118 (48)	124 (51)	129 (54)	136 (58)						
98 (37)	105 (41)	109 (43)	113 (45)	117 (47)	123 (51)	128 (53)	134 (57)					
96 (36)	101 (38)	104 (40)	108 (42)	112 (44)	116 (47)	121 (49)	126 (52)	132 (56)				
94 (34)	97 (36)	100 (38)	103 (39)	106 (41)	110 (43)	114 (46)	119 (48)	124 (51)	129 (54)	135 (57)		
92 (33)	94 (34)	96 (36)	99 (37)	101 (38)	105 (41)	108 (42)	112 (44)	116 (47)	121 (49)	126 (52)	131 (55)	
90 (32)	91 (33)	93 (34)	95 (35)	97 (36)	100 (38)	103 (39)	106 (41)	109 (43)	113 (45)	117 (47)	122 (50)	127 (53)
88 (31)	88 (31)	89 (32)	91 (33)	93 (34)	95 (35)	98 (37)	100 (38)	103 (39)	106 (41)	110 (43)	113 (45)	117 (47)
86 (30)	85 (29)	87 (31)	88 (31)	89 (32)	91 (33)	93 (34)	95 (35)	97 (36)	100 (38)	102 (39)	105 (41)	108 (42)
84 (29)	83 (28)	84 (29)	85 (29)	86 (30)	88 (31)	89 (32)	90 (32)	92 (33)	94 (34)	96 (36)	98 (37)	100 (38)
82 (28)	81 (27)	82 (28)	83 (28)	84 (29)	84 (29)	85 (29)	86 (30)	88 (31)	89 (32)	90 (32)	91 (33)	93 (34)
80 (27)	80 (27)	80 (27)	81 (27)	81 (27)	82 (28)	82 (28)	83 (28)	84 (29)	84 (29)	85 (29)	86 (30)	86 (30)

Category	Heat Index	Possible heat disorders for people in high risk groups
Extreme Danger	130 °F or higher (54 °C or higher)	Heat stroke likely.
Danger	105 – 129 °F (41 – 54 °C)	Muscle cramps and/or heat exhaustion possible with prolonged exposure and/or physical activity.
Extreme Caution	90 – 105 °F (32 – 41 °C)	Muscle cramps and/or heat exhaustion possible with prolonged exposure and/or physical activity.
Caution	80 – 90 °F (27 – 32 °C)	Fatigue possible with prolonged exposure and/or physical activity.

Source: National Weather Service Office, Birmingham, AL.

Carbon dioxide contents will continue to increase at least for some time to come, which will likely lead to further atmospheric warming. With greater warming we can also anticipate further increases in specific humidity and more effective absorption of outgoing longwave radiation. So will this positive feedback ultimately result in cataclysmic temperature increases? Not necessarily, because strong negative feedbacks also operate. The most important of these is increasing emission of longwave radiation as temperature rises. You recall from Chapter 2 that for blackbodies, emission grows with the fourth power of temperature. Although Earth does not emit to space as a perfect blackbody, increasing longwave emission acts counter to the water vapor feedback.

Another feedback is related to clouds: It might be that increasing water vapor concentrations will lead to increases in global cloud cover. The effects of cloud cover on shortwave and longwave radiation depend on the type of cloud cover, the height of the clouds, and their location. Under some conditions an increase in cloud cover can lead to a substantial reduction in absorbed solar radiation that suppresses further warming. However, clouds are strong absorbers of longwave radiation, thus they also limit radiation emitted to space. This reduction in outgoing radiation can exceed the reduction in solar radiation absorption, thereby promoting further warming. Even on a global basis, it is an open question as to whether clouds will emerge as a negative or positive feedback in the future. This uncertainty contributes significantly to our lack of knowledge regarding the magnitude and regional distributions of future climate changes. This issue will be discussed in Chapter 16.

Summary

Water in its three phases—solid, liquid, and gas—constantly moves across the interface between the atmosphere and Earth's surface through what is known as the *hydrologic cycle.* Despite the fact that water accounts for only a small proportion of the mass of the air, in its three phases it is extremely important to the atmosphere—and to all life on Earth. To fully understand its role in the atmosphere, we need to comprehend the concept of saturation. At the surface of liquid water, molecules constantly move about. As they do so, some randomly break free of the surface to become water vapor. In a partially filled jar, this adds vapor to the volume above the water surface and thereby increases the water vapor content. As the water vapor concentration increases, so does the opposite process—condensation. As long as the rate of evaporation exceeds condensation, the amount of water vapor increases. The condensation rate in the jar eventually becomes equal to the evaporation rate, and saturation occurs.

The foregoing principle applies to the real atmosphere. In meteorology, we are particularly concerned with the exchange of water between a suspended cloud droplet and the air that surrounds it, because the equilibrium between evaporation and condensation is a prerequisite to the persistence of clouds and fog. When evaporation and condensation are in equilibrium, the air is saturated and small droplets can remain in the air without evaporating away.

There is no single best expression of humidity. One way of expressing the amount of water vapor is by the pressure it exerts (vapor pressure). The specific humidity and mixing ratio are similar ratios that relate the mass of water vapor to the air in which it is contained and to the mass of the other gases, respectively. Another measure, absolute humidity, is not widely used but is simply the density of water vapor. In contrast, relative humidity is a quantity we hear about almost daily, even though its partial dependence on temperature presents a serious drawback. Perhaps the most useful index is dew point, expressed as the temperature to which the air must be lowered for saturation to occur.

Although it might seem that condensation should occur at exactly 100 percent relative humidity, the situation in reality is slightly more complex. Condensation is controlled by two factors having opposite effects: curvature and the abundance of hygroscopic nuclei. Curvature retards condensation; without the presence of condensation nuclei, condensation would occur only at very high relative humidities. Natural and anthropogenic nuclei are sufficiently numerous that condensation usually occurs at relative humidities between 98 and 100.1 percent.

Humidity is an easy property to measure. It can be determined using paired thermometers (a psychrometer) that provide dry and wet bulb temperatures. The difference between the two temperatures is the wet bulb depression, which, when combined with dry bulb temperature, permits the use of simple tables to determine relative humidity and dew point. Humidity affects a human's susceptibility to heat-related dangers. This susceptibility is reflected in the heat index.

For any kind of condensation (such as dew, fog, or clouds) to form, the dew point must equal the air temperature. This can result from raising the vapor content of the air to the saturation level; mixing warm, moist air with cooler, dry air; or by lowering the air temperature to the dew point. The latter process is most important for cloud formation. Lowering the air temperature does not require that heat be removed from the air. In fact, most clouds form by adiabatic cooling—the lowering of the air temperature without the removal of heat. Adiabatic cooling results from the expansion of air that occurs when it is lifted and is a direct application of the first law of thermodynamics.

Noncloud forms of condensation—dew, frost, frozen dew, and fog—are distinguished from clouds by their proximity to or direct contact with the planet's surface. Though one type of fog (upslope fog) results from adiabatic processes, the others result from diabatic cooling of the air, involving the loss of energy. Because temperatures have risen over much of the globe, there has been an increase in specific humidity near the surface over most regions, though not all regions have experienced a concomitant increase in relative humidity. This is important not just in its immediate effect but also because it amplifies greenhouse warming and because of its potential to impact cloud formation and patterns.

We have discussed how rising air leads to adiabatic cooling and to saturation and cloud formation, but we have yet to provide any explanation for these vertical motions. As it happens, several mechanisms are capable of generating uplift, some of which are quite intricate and closely tied to the cloud form that results. The topic deserves extended discussion, which we provide in the next chapter.

Key Terms

wet bulb depression
page 132

aspirated psychrometer
page 132

hair hygrometer *page 132*

hygrothermograph
page 132

precipitation fog
page 138

steam fog *page 138*

homogeneous nucleation
page 140

hygroscopic *page 140*

heterogeneous nucleation
page 140

condensation nucleus
page 140

haze *page 140*

supercooled water
page 141

ice nucleus *page 141*

diabatic process *page 143*

second law of
thermodynamics *page 143*

adiabatic process
page 143

first law of thermodynam-
ics *page 143*

dry adiabatic lapse rate
page 144

lifting condensation level
page 144

saturated (wet) adiabatic
lapse rate *page 144*

environmental (ambient)
lapse rate *page 145*

dew *page 146*

frost *page 146*

frozen dew *page 147*

radiation fog *page 147*

advection fog *page 149*

upslope fog *page 149*

dew point lapse rate
page 151

heat index *page 152*

apparent temperature
page 152

Review Questions

1. What is the hydrologic cycle?
2. Why is it incorrect to refer to the air as "holding" water vapor?
3. What are deposition and sublimation?
4. What is vapor pressure? In what units of measure is it expressed?
5. Explain the concepts of equilibrium and saturation.
6. What units of measurement are used to describe mixing ratio and specific humidity? Why are the two values nearly equal?
7. Why is absolute humidity seldom used?
8. Define relative humidity.
9. Why is relative humidity a poor indicator of the amount of water vapor in the air?
10. Define dew point. What characteristics make this measure superior to relative humidity?
11. Why can't the dew point temperature exceed the air temperature? What happens if the air temperature is lowered to a value below the initial dew point?
12. Describe the distribution of average dew point across the United States in summer and winter.
13. What are the three general methods by which the air can become saturated?
14. What are the effects of droplet curvature and solution on the amount of water vapor needed for saturation?
15. Why doesn't homogeneous nucleation form water droplets in the atmosphere?
16. What are condensation nuclei and ice nuclei? Are they typically made of the same materials? Which is more abundant in the atmosphere?
17. What is supercooled water?
18. What are psychrometers? How do they work?
19. Define *dry bulb temperature, wet bulb temperature, and wet bulb depression.*
20. What is the heat index?
21. What is the first law of thermodynamics and how does it apply to cloud development?
22. Explain the difference between diabatic and adiabatic processes.
23. What are the numerical values of the dry and saturated adiabatic lapse rates? Under what circumstances are they applicable?
24. What does the environmental lapse rate refer to?
25. Describe the various processes that can lead to the formation of dew.
26. What is the difference between frozen dew and frost?
27. Describe the various processes that can lead to the formation of fog.

Critical Thinking

1. When rubbing alcohol is applied to a person's skin, it feels colder than the application of water would. Why?
2. A person sleeps through the night without waking up, but awakes in the morning weighing slightly less than the night before. What happened?
3. A crowded classroom is filled with students. In what way might the presence of the students affect the dew point and relative humidity in the room?

4. A person parks her car in the driveway on a warm afternoon and notices a small puddle of water beneath the car a few minutes later. Explain how using the car's air conditioning can account for the puddle.

5. At Wheeling, West Virginia, the evening temperature is 55 °F and the dew point is 48 °F. How would you assess the likelihood of fog forming overnight?

6. A map of North America shows the average distribution of vapor pressure across the continent. Will the distribution on the map be *only* a function of the amount of water vapor in the air, or will the distribution be affected by another factor as well? Explain.

7. The temperature within a forest is –2 °C (28 °F) and there is frost on the trees but no fog. Outside the woods there is a fog. Why wasn't this fog in the woods?

8. Is fog more likely to occur downwind or upwind of an oil refinery? Why?

9. All fogs are made of water droplets or ice crystals. Despite the fact that they have the same composition, how would you know if a particular fog is a radiation, advection, or upslope fog?

10. Diesel engines, like four-stroke engines, work because of burning fuel, but they do not require a spark plug or similar device for initiating the burning. Apply your knowledge of the first law of thermodynamics to explain how the fuel can be forced to burn.

Problems and Exercises

1. Assume that a kilogram of air consists of 995 g of dry air and 5 g of water vapor. Show that the specific humidity and mixing ratio are very nearly equal.

2. Assume that a kitchen measures 4 meters by 5 meters by 3 meters. If the air density is 1 kg/m^3 and the specific humidity is 10 g water vapor per kg of air, how much water vapor is in the room? If the doors and windows were sealed shut, would boiling 1 kg of water into the air make a substantial change in the humidity of the room?

3. The dry and wet bulb readings in Honolulu are 80 °F and 69 °F. At Charlottesville, Virginia, the readings are 50 °F and 45 °F.
   a. Use Tables 5–1 and 5–2 to determine the relative humidity and the dew point for both locations.
   b. Which of the two locations is more humid?

4. The dry and wet bulb temperatures are 70 °F and 54 °F. Use Tables 5–1 and 5–2 to answer the following questions:
   a. What are the dew point and the relative humidity?
   b. What will the dew point and relative humidity be if the air temperature increases to 80 °F? (*Hint:* Do not assume the same wet bulb depression.)
   c. What will the dew point and relative humidity be if the air temperature drops to 39 °F? (*Hint:* You don't need the tables for this one.)

d. What will the dew point and relative humidity be if the air temperature drops further, to 35 °F?

5. The numerical value of the specific heat for air in the first law of thermodynamics, c_v, is strictly valid only for air with no water vapor. Water vapor has a specific heat approximately twice as great as that of c_v. Will a humid mass of air undergoing expansion therefore undergo more or less cooling than would a dry mass of air?

6. Assume a parcel of air starts out at the surface with a temperature of 12 °C and a dew point of 9.6 °C. Then it is lifted.
   a. What will the air temperature be at the 100 m level?
   b. What will the dew point be at the 100 m level? (*Hint:* Don't forget the dew point lapse rate.)
   c. At what height will condensation occur?
   d. What will the temperature be when condensation occurs?
   e. What will the dew point be when condensation occurs?
   f. What will the temperature be 500 m above the surface?
   g. If the parcel of air is lowered back to the surface (assuming none of the condensed moisture was removed as rain), what will the temperature and dew point be?

Quantitative Problems

As this chapter has shown, several indices are used to express the moisture content of the air, with each having some advantages and disadvantages. In meteorology we are particularly concerned with how these indices change in response to the lifting of air adiabatically. One excellent way to reinforce your understanding of these measures is by solving the simple problem set given on the Chapter 5 page of this book's Web site (**www.MyMeteorologyLab.com**). These questions are particularly valuable as an aid to understanding how saturation is achieved in rising parcels.

Useful Web Sites

www.usatoday.com/weather/whumdef.htm
Much useful information on humidity and humidity measures.

weather.unisys.com/surface/contour.php
Select "Dewpoint" from the available plots to see the current dewpoint map or a map from the previous 12 hours.

www.wunderground.com/US/Region/US/HeatIndex.html
Map of the current heat index across the United States.

MyMeteorologyLab™

The *Understanding Weather and Climate,* Sixth Edition, MyMeteorologyLab™ web site contains numerous multimedia resources to aid in your study of **Atmospheric Moisture.**
Visit **www.MyMeteorologyLab.com** to:

- **Review** key chapter concepts.
- **Read** the Pearson eText, *In the News RSS feeds,* and other chapter-specific Web resources.
- **Visualize** challenging topics with Interactive Tutorials, Geoscience Animations, MapMaster interactive maps, and Weather in Motion movies and visualizations, and more.
- **Test Yourself** with self-study quizzes.

 TUTORIAL
ATMOSPHERIC MOISTURE AND CONDENSATION

Use the interactive animations and quizzes in this tutorial to review this chapter's key concepts.

WEATHER IN MOTION

View movies and visualizations of meteorology patterns and processes.

The Hydrologic Cycle
Forecasting Humidity
Global Evaporation Rates
A Satellite View of Fog
January and July Water Vapor

6 Cloud Development and Forms

LEARNING OUTCOMES

After reading this chapter, you should be able to:

▶ Describe four mechanisms that lift air.

▶ Explain the factors that determine the stability of parcels of air.

▶ Explain the factors that influence the environmental lapse rate.

▶ Explain what causes unstable air to stop rising.

▶ Explain what causes stable air and inversions to develop.

▶ Describe the different types of clouds.

▶ Describe how meteorologists measure cloud coverage.

All of us pay special attention to clouds at one time or another. Some are particularly beautiful, while others portend severe weather. But certain people do a lot more than passively observe clouds. *Storm chasers* set out in teams during the severe weather season with the goal of observing the formation and movement of tornadoes. Some storm chasers are nonprofessional weather enthusiasts, while others are trained meteorologists with advanced degrees or students working toward their credentials. The professionals and students go out not just for the thrill of witnessing severe weather firsthand (though the excitement accounts for much of the pursuit), but also to acquire information that helps us understand these phenomena.

During the afternoon of April 10, 1979, several teams of storm chasers outside of Seymore, Texas, observed a *wall cloud*, a large protrusion extending below the main base of a thunderstorm where tornadoes commonly develop. Howard Bluestein, now one of the leading experts on severe storms, photographed the wall cloud in Figure 6–1. Soon afterward a tornado formed. Though the observation teams tried to keep pace, the storm outran them as it tracked toward Wichita Falls. When the chasers arrived in town, they witnessed the horrible destruction that had occurred minutes before: 3000 homes were destroyed and 42 people killed. The tornado was still swirling in full force and moving northeast. As it continued out of Wichita Falls, the chasers saw the back side of the storm, which Bluestein describes as looking like an atomic bomb explosion.

Most clouds are far less dynamic than those that spawn tornadoes, yet even the most common cloud types can be quite beautiful. And clouds associated with everyday weather result from the same processes that cause condensation and deposition in clouds in severe conditions. From Chapter 5 we know that condensation or deposition can occur by (1) adding water vapor to the air; (2) mixing warm, moist air with cold air; or (3) lowering the air temperature to the dew point by adiabatic cooling of rising air. Although the first two processes can lead to cloud formation in many situations, lowering of air temperature to the dew point is the most common mechanism of cloud formation (especially for clouds that cause precipitation). This chapter discusses the processes and conditions associated with the formation of clouds due to upward motions. It also describes the cloud types that form as a result of those processes.

▲ **FIGURE 6–1** Wall cloud associated with severe thunderstorm near Seymore, Texas.

◀ Storm chasers in Wyoming observe a storm that may yield a tornado.

Mechanisms That Lift Air

Four mechanisms lift air so that condensation and cloud formation can occur:

1. Orographic lifting, forcing air above a mountain barrier
2. Frontal lifting, displacement of one air mass over another
3. Convergence, horizontal movement of air into an area at low levels
4. Localized convective lifting due to buoyancy

Orographic Uplift

As shown in Figure 6–2a, air flowing toward a hill or mountain will be deflected around and over the barrier. The upward displacement of air that leads to adiabatic cooling and possible cloud formation is called **orographic uplift** (or the **orographic effect**). The height to which these *orographic clouds* can rise is not limited to the height of the hill or mountain; their tops can extend many hundreds of meters higher and even into the lower stratosphere. The height of cloud tops is strongly related to characteristics of the air that vary from day to day, an issue we will describe in more detail later in this chapter.

Downwind of a mountain ridge, on its leeward side, air descends the slope and warms by compression to create a **rain shadow** effect, an area of lower precipitation. The Sierra Nevada mountain range provides a dramatic illustration of this effect. The ridge crest of the Sierra runs north to south and is essentially perpendicular to the predominantly

Air flow

(a)

(b)

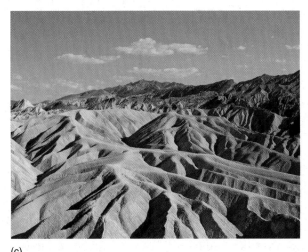

(c)

▲ **FIGURE 6-2** Orographic uplift. When air approaches a topographic barrier, it can be lifted upward or deflected around the barrier (a). The Sierra Nevada forms a major barrier to winds that generally blow from the west. This promotes enhanced precipitation on the windward side (b), and a rain shadow on the leeward side (c).

west-to-east airflow. With much of the range being higher than 3500 m (11,500 ft), precipitation on the western, windward side is greatly enhanced because of orographic lifting (refer to Figure 6–2b); in places, the mean annual precipitation exceeds 250 cm (100 in.). The eastern slope of the range is extremely steep and the valley floor is low, sometimes below sea level. Thus, the descending air on the leeward slope creates one of the strongest rain shadow effects on Earth. Death Valley, one of the driest places in North America, is just east of the range, whose windward slopes accumulate the majority of California's usable water (Figure 6–2c). A comparable rain shadow effect exists in South America, where the Andes Mountains form an abrupt barrier to the westerly winds. Figure 6–3 illustrates the rapid changes that can take place in orographic cloud development.

Frontal Lifting

Although temperature normally changes from place to place, experience tells us that such changes are usually quite gradual. In other words, if the temperature is 10 °C (50 °F) in Toronto, Ontario, chances are the temperature in Buffalo, New York, about 100 km (60 mi) away, will not be very much different. Sometimes, however, transition zones exist in which great temperature differences occur across relatively short distances. These transition zones, called *fronts*, are not like vertical walls separating warm and cold air but rather slope gently, as we will discuss in Chapter 9.

Airflow along frontal boundaries results in the widespread development of clouds in either of two ways. When cold air advances toward warmer air (a situation called a **cold front**), the denser cold air displaces the lighter warm air ahead of it, as shown in Figure 6–4a. When warm air flows toward a wedge of cold air (a **warm front**), the warm air is forced upward in much the same way that the orographic effect causes air to rise above a mountain barrier (Figure 6–4b).

Convergence

Because the mass of the atmosphere is not uniformly distributed across Earth's surface, large areas of high and low surface pressure exist. These pressure differences set the air in motion in the familiar effect we call the wind. Not surprisingly, the pattern of wind that results is very much related to the pattern of pressure.

In particular, when a low-pressure cell is near the surface, winds in the lower atmosphere tend to converge on the center of the low from all directions. Horizontal movement toward a common location is called **horizontal convergence**, or just **convergence** for short (Figure 6–5). Does convergence lead to increasing density near the center of the low, with imported air confined to its original altitude? No—instead, vertical motions near the center of the low carry away about as much

(a)

(b)

(c)

▲ **FIGURE 6-3** The development of cumulus clouds over the mountains east of San Diego, California. Notice the increased vertical development between (a) and (b). In (c), the cloud has decreased in depth due to precipitation. Photos were taken at about 10-minute intervals.

mass as is carried in. Air will rise. This will be explained in more detail later, but for now we can just make the connection between low-level convergence and rising air with adiabatic cooling.

Localized Convection

We saw in Chapter 3 that free convection is lifting that results from heating the air near the surface. It is often accompanied

► **FIGURE 6-4** Frontal boundaries. Cold fronts
(a) cause uplift as cold air advances on warmer,
less dense air. Uplift occurs along a warm front
(b) when warm air overruns the cold wedge of air
ahead of it.

(a)

(b)

by updrafts strong enough to form clouds and precipitation.
During the warm season, heating of Earth's surface can produce free convection over a fairly limited area and create the
brief afternoon thunderstorms that disrupt summer picnics.
In Canada and the United States east of the Rocky Mountains,
high moisture content of the air sometimes allows for tall
clouds with bases at relatively low altitudes. Such conditions favor vigorous precipitation over small regions (free
convection by its nature does not create updrafts more than
several tens of meters in diameter). Even in the deserts of
the Southwest, which are usually low in water vapor, intense
heating can lead to localized convection intense enough to
cause thundershowers.

Free convection arises from buoyancy, the tendency for a
lighter fluid to float upward through a denser one. By itself,
buoyancy can initiate uplift. But buoyancy can also speed or
slow the uplift begun by the orographic effect, frontal lifting,
and convergence. As we will see next, meteorology uses the
concept of static stability to summarize the effect of buoyancy
on uplift.

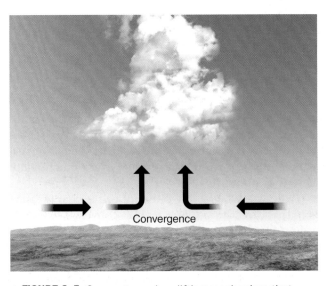

▲ **FIGURE 6-5** Convergence is a lifting mechanism that
occurs when air flows into a given area from different
directions.

Checkpoint

1. What are fronts?
2. How do fronts cause air to rise?

Static Stability and the Environmental Lapse Rate

Sometimes the atmosphere is easily displaced and an air parcel given an initial boost upward continues to rise, even after the original lifting process ceases. At other times, the atmosphere resists such lifting. The air's susceptibility to uplift is called its **static stability.** *Statically unstable* air becomes buoyant when lifted and continues to rise if given an initial upward push; *statically stable* air resists upward displacement and sinks back to its original level when the lifting mechanism ceases. *Statically neutral air* neither rises on its own following an initial lift nor sinks back to its original level; it simply comes to rest at the height to which it was displaced.

Static stability is closely related to buoyancy. When a parcel of air is less dense than the air around it, it has a positive buoyancy and floats upward. (In fact, buoyant parcels of air increase in speed as they move upward—even to the point of creating violent updrafts.) Air that is denser than its surroundings sinks if not subjected to continued lifting forces. Density differences between a parcel and its surroundings are determined by their temperatures. If the parcel is warmer than the surrounding air, it will be less dense and experience a lifting force. If it is colder, it will be more dense and have "negative" buoyancy.

TUTORIAL
ATMOSPHERIC STABILITY

Use the tutorial to lift parcels of air through the surrounding atmosphere to observe how varying temperature profiles affect the parcels' buoyancy.

If a rising parcel cools at a rate that makes it colder than the surrounding air, it will become relatively dense. This will tend to suppress uplift. If the lifted air cools more slowly than its surroundings, it will become warm relative to the surroundings and have positive buoyancy. This creates a tendency for a parcel to rise on its own, even in the absence of other lifting mechanisms. Thus, the buoyancy of a rising air parcel depends on its rate of cooling relative to the surrounding air. Temperatures in the parcel are governed by either the dry or saturated adiabatic lapse rate, whereas the surroundings are governed by the **environmental lapse rate** (ELR). (The adiabatic and environmental lapse rates were explained in Chapter 5.)

Consider a parcel of air near the surface that is lifted through the surrounding air. The lifted air cools at one of the adiabatic lapse rates, and the surrounding air maintains its original temperature profile. The relative density of the rising parcel thus depends on two conditions: (1) whether or not the parcel is saturated (which determines the applicable adiabatic lapse rate) and (2) the ELR. These factors combine to produce different types of air with regard to their static stability. These are *absolutely unstable, absolutely stable*, and *conditionally unstable.* Let's use the following examples to see how static stability can be determined by comparing the environmental lapse rate to the saturated and dry adiabatic lapse rates.

Absolutely Unstable Air

Figure 6–6a illustrates what happens when a parcel of unsaturated air is lifted and the ELR is greater than the dry adiabatic lapse rate (DALR). In other words, in Figure 6–6a, the rising air is cooling more slowly than its surroundings.

Let's suppose the air temperature at the surface is 10 °C and the ELR is 1.5 °C/100 m, which means that the air is cooling at the rate of 1.5 °C for every 100 m of height. As our parcel rises, it cools at the DALR (recall from Chapter 5 that the DALR is 1 °C/100 m). When it is lifted to the 100 m level, the rising parcel cools to 9 °C—half a degree warmer than the surrounding air. If the parcel is lifted to the 200 m level, its temperature becomes 8 °C, or one degree Celsius warmer than the surrounding air. Thus, the lifted parcel is becoming progressively warmer and more buoyant than the surrounding air.

The air in this instance is said to be *absolutely unstable* because once a parcel within is lifted, it continues to move upward. Not only does the parcel rise, but it does so at an ever-increasing speed. This occurs because the temperature difference between it and the surrounding air continually increases, leading to greater buoyancy, and also because it gathers momentum as it rises.

Figure 6–6b provides a second example of absolutely unstable air. In this case, the ELR is still 1.5 °C/100 m, but the air is now *saturated*. The lifted parcel of air therefore cools more slowly, at the saturated adiabatic lapse rate (SALR), and will be warmer than an unsaturated parcel. The temperature difference between the warm parcel and colder surrounding air is greater, giving rise to a stronger buoyant force. We conclude that the air is again unstable, even more so than in the previous example.

These two examples show that *whenever the environmental lapse rate exceeds the dry adiabatic lapse rate, the air is absolutely unstable and a parcel contained within it will continue to rise once lifted, regardless of whether or not it is saturated.* (Of course, upward motions cannot continue forever. But for now, we will put aside the issue of how far unstable air can rise so we can focus on the main concept associated with what happens to air in a zone of instability.)

Absolutely Stable Air

Figures 6–7a and 6–7b illustrate what happens when the ELR, in this case 0.2 °C/100 m, is less than the saturated

(a)

(b)

▲ **FIGURE 6-6** Absolutely unstable air. In both examples, the ELR of 1.5 °C/100 m exceeds the DALR. In (a) lifted unsaturated air cools at 1 °C/100 m (the DALR). The rising parcel does not cool as rapidly with height as does the air surrounding it. It therefore becomes warmer and more buoyant than the surrounding air. In (b) the lifted saturated air likewise cools less rapidly (0.5 °C/100 m) with height than the surrounding air. Thus, with an ELR greater than the DALR, air forced to rise becomes warmer and more buoyant than the surrounding air, whether it is unsaturated or saturated. Air with such temperature profiles is said to be absolutely unstable.

adiabatic lapse rate. As we see in (a), when a parcel of unsaturated air rises, its temperature drops more rapidly than the temperature of the air around it, making the parcel relatively heavier and less buoyant. Because of its negative buoyancy, the lifted air will sink back to its initial level if the lifting mechanism stops. Such air is *absolutely stable*. The same principle applies in part (b) of the figure. The saturated parcel becomes colder than the air around it. Like the unsaturated parcel in (a), it has a tendency to sink back to its original position.

From these two examples we can conclude that *whenever the environmental lapse rate is less than the saturated adiabatic lapse rate, the air will be absolutely stable and will resist lifting, regardless of whether or not it is saturated.*

Note that it is possible for the ELR to be such that the temperature does not change at all with height, or even for the temperature to increase with height, as will be discussed later in this chapter. Though we will not provide examples here, the logic described in this section applies to these situations as well. If the ELR = 0 °C/100m or if the temperature increases with height (a negative ELR), the air will be absolutely stable.

Conditionally Unstable Air

The preceding examples describe what happens when the ELR is less the SALR or greater than the DALR. But what happens when the ELR is between the dry and saturated

(a)

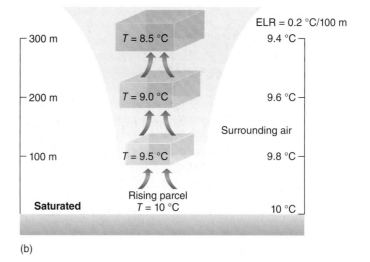

(b)

▲ **FIGURE 6-7** Absolutely stable air. In both examples, the ELR of 0.2 °C/100 m is less than the SALR of 0.5 °C. It does not matter whether the air is unsaturated (a) or saturated (b); rising air parcels become colder and less buoyant than the surrounding air. Regardless of whether the air is unsaturated or saturated, temperature profiles that exhibit ELRs less than the SALR indicate absolutely stable air.

adiabatic lapse rates? In this environment, the air is said to be *conditionally unstable*, and the tendency for a lifted parcel to sink or continue rising depends on whether or not it becomes saturated and how far it is lifted.

Let's suppose there is an ELR of 0.7 °C/100 m in the atmosphere through which an unsaturated parcel is rising (Figure 6–8a). Because the lifted parcel becomes colder (and therefore denser) than the surrounding air, it resists further uplift. The situation at this point is much the same as was described for absolutely stable air.

Figure 6–8b illustrates how the same environmental lapse rate can lead to a lifted parcel of air becoming buoyant relative to its surroundings if it is lifted to a great enough height. We again apply the ELR of 0.7 °C/100 m to a lifted parcel that eventually becomes saturated and forms a cloud. In this example, the air has an initial temperature of 10°C and a dew point, T_d,

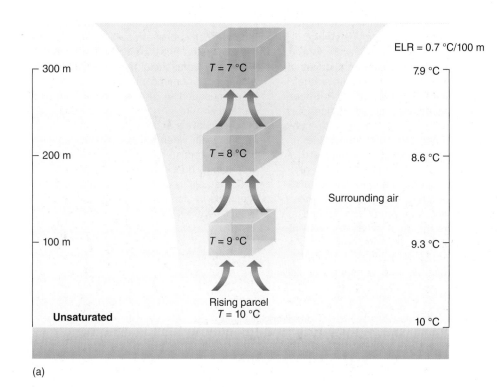

(a)

◄ **FIGURE 6–8** The atmosphere is conditionally unstable when the ELR is between the dry and saturated adiabatic lapse rates. In (a), the ELR is 0.7 °C/100 m and the air is unsaturated. As a parcel of air is lifted, its temperature is less than that of the surrounding air, so it has negative buoyancy. In (b), a parcel starts off unsaturated but becomes saturated at 100 m, where it is cooler than the surrounding air. Further lifting cools the parcel at the SALR. At the 200 m level, it is still cooler than the surrounding air, but if taken to 300 m, it is warmer and buoyant. Thus, if the air is lifted sufficiently, the parcel continues to rise by virtue of its buoyancy.

(b)

6–1
FORECASTING

Determining Stability from Thermodynamic Diagrams

The static stability of air can be determined numerically by comparing the environmental lapse rate to the saturated and dry adiabatic lapse rates. Thermodynamic diagrams can also be very useful in this regard. Figure 1 provides a simple schematic showing how this comparison is done. The figure compares three hypothetical temperature profiles against the SALR and the DALR on a portion of a simplified thermodynamic diagram. The lines labeled as *moist adiabat* and *dry adiabat* plot the change in temperature a saturated or unsaturated parcel of air would experience if it were lifted or lowered. The three hypothetical temperature profiles illustrate examples in which the air is absolutely unstable (Profile 1—the temperature decreases more rapidly than the DALR), conditionally unstable (Profile

2—ELR is between the DALR and SALR), and absolutely stable (Profile 3—ELR is less than the SALR).

In the real world, the ELR varies from the surface upward. For example, at one level the air might be absolutely stable, whereas at another level it might be conditionally or absolutely unstable. Thermodynamic diagrams allow the forecaster to observe the resultant changes in stability at different levels visually, rather than by having to compute the ELR repeatedly for comparison to the adiabatic lapse rates. Figure 2 shows a temperature (heavy line on right) and dew point (heavy line on left) sounding plotted on a complete thermodynamic diagram that includes dry and moist adiabats. The dry adiabats are shown in green and slope steeply to the left as they extend upward. The dashed blue lines are the moist adiabats. From the surface to 850 mb, the temperature profile is parallel to that of the adjacent dry adiabats, indicating the layer is nearly statically

unstable. Above that is a shallow layer, which is statically stable. From 800 mb to about 650 mb, the air is conditionally unstable. And just above that, there is a very shallow inversion.

The changes in stability at different levels may appear to make the use of the thermodynamic diagram a daunting task for the forecaster, but the situation has several remedies. Professional meteorologists have a number of numerical indexes calculated for every sounding. The indexes are based on temperature–dew point combinations at varying levels and are computed automatically when the soundings are plotted. Forecasters refer to these values for initial guidance in their interpretations of how stability conditions will influence the likelihood of cloud cover, precipitation, or violent weather.

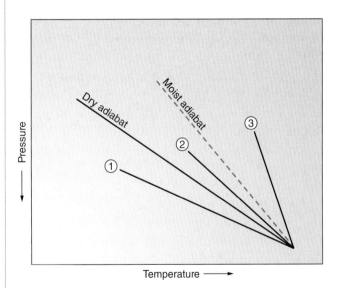

▲ **FIGURE 1** Stability can be determined by comparing temperature profiles with the slope of the dry and wet adiabats. In (a) Profile 1 is absolutely unstable, Profile 2 conditionally unstable, and Profile 3 absolutely stable.

▲ **FIGURE 2** A complete thermodynamic diagram plotting temperature (heavy line on right) and dew point (left) profiles for Detroit, Michigan, on June 27, 2002.

of 9.2 °C. It cools at the DALR until it reaches saturation (the lifting condensation level) at the 100 m level, where $T = T_d = 9.0$ °C. There the lifted parcel is colder and denser than the surrounding air; it is not buoyant and will not rise further unless something forces it to do so. If the parcel is lifted farther, it will cool at the saturated adiabatic rate, which is less than the ELR. At the 200 m level, the lifted parcel is still colder than the surrounding air, but the difference in temperature between it and the surrounding air is less than it was at the 100 m level. If the parcel is lifted to the 300 m level, it then becomes *warmer* than the ambient air. The lifted parcel thus becomes buoyant and will now rise on its own, even in the absence of external lifting. Thus, if the atmosphere is conditionally unstable, an air parcel becomes buoyant if lifted above some critical altitude. That altitude, called the **level of free convection**, is the height to which a parcel of air must be lifted for it to become buoyant and to rise on its own.

The condition in the term *conditionally unstable* refers to a parcel's ability to become buoyant only if it is lifted to some particular level. Air not lifted to that level does not become buoyant and will rise only if subjected to some other lifting mechanism (such as the passage of a front). When a parcel of conditionally unstable air rises above that level, it is common for clouds to rapidly increase in depth and yield precipitation. (*Box 6–1, Forecasting: Determining Stability from Thermodynamic Diagrams,* provides more information on the analysis of stability.)

Static and Potential Instability

The types of static stability discussed in this section pertain to a parcel of air's ability to rise when subjected to uplifting mechanisms. But in some situations, especially with regard to the potential for severe storm development, another type of stability is important: **potential instability** (sometimes called *convective instability*). While static stability describes what would happen to a small parcel of air that is lifted or lowered while the surrounding air is kept in place, potential stability describes what happens when entire layers of air are displaced upward (as in the case of a mass of warm air displaced upward by the movement of a cold front). When such layers are displaced upward, their environmental lapse rates may be altered so that statically stable conditions within the layer can change to statically unstable conditions, or vice versa. Potential instability is a feature that is particularly important with the forecasting of severe thunderstorms, as described in *Box 6–2, Forecasting: Potential Instability.*

Checkpoint

1. What is static stability?
2. How does air that is statically stable behave differently from air that is statically unstable?

Factors Influencing the Environmental Lapse Rate

ELRs are highly variable in space and time. Just as the surface air temperature at a location is subject to change, so is the vertical temperature profile. The following three factors can bring about changes in the ELR.

Heating or Cooling of the Lower Atmosphere

During the daytime, solar radiation heats Earth's surface, which in turn warms the atmosphere in contact with it. Because it is heated more rapidly than the air aloft, the lower atmosphere typically has a steeper ELR during the midday, as shown in Figure 6–9. The initial temperature profile indicated by the solid line changes through the course of the day, and steeper profiles (shown by the dashed lines) can result for several hundred meters above the surface. The effect of

◄ **FIGURE 6-9** The ELR can be changed by heating of the surface, as shown by the sequential changes in this temperature profile.

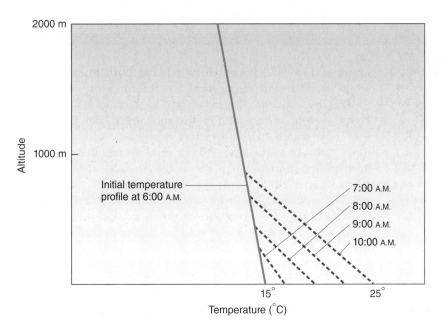

Initial temperature profile at 6:00 A.M.

7:00 A.M.

8:00 A.M.

9:00 A.M.

10:00 A.M.

2000 m

1000 m

Altitude

15°   25°

Temperature (°C)

Potential Instability

Air with strong temperature lapse rates is said to be statically unstable. Another type of instability that influences vertical air motions, called potential instability, arises when a layer of warm, dry air rests above one that is warm and humid. Lifting the two layers can cause the temperature lapse rate to increase, thus making the air statically unstable.

Consider the inversion situation shown in Figure 1. In Parcel 1, located in the lower layer just below the base of the inversion, the temperature (T) equals 26 °C and the dew point (T_d) is 22 °C. In Parcel 2, located in the layer above the base of the inversion, the temperature and dew point are 27 °C and 19 °C, respectively. Now consider what happens between Parcels 1 and 2 if some process lifts the two layers containing the parcels. Both parcels are initially unsaturated, so they cool at the dry adiabatic lapse rate (DALR) of 1 °C per 100 m, and their dew points decrease at 0.2 °C per 100 m. After 500 m of ascent, the temperature of both parcels has fallen by the same amount, so the temperature difference between them is unchanged. However, Parcel 1 is now saturated, so further lifting will cause its temperature to decline at the saturated adiabatic lapse rate (SALR). Meanwhile, Parcel 2 is still unsaturated, so further lifting leads it to cool at the DALR.

Now let's lift the two parcels another 500 m. Assuming an SALR of 0.5 °C per 100 m, the lower parcel cools 2.5 °C to

18.5 °C, while the upper parcel cools at the DALR to 17 °C. We can now see how uplift of the column of air containing the two parcels has affected its stability. Initially, the parcel in the upper layer was warmer than the one below, which meant the layer containing the two parcels was statically stable. After lifting both parcels 1000 m, however, the upper parcel became 1.5 °C cooler than the one below. This yields a temperature lapse rate between the two parcels of 1.5 °C per 100 m, making the air statically unstable. Thus, the air that is statically stable has the potential to become statically unstable, given sufficient uplift—hence the term *potential instability*.

Both theory and experience show that potential instability is an important factor in the development of severe thunderstorms. During spring and summer, the southern Great Plains region often has warm, humid air near the surface advected from the Gulf of Mexico. In the middle troposphere above the region, westerly winds bring dry air from the southern Rockies. This air in the middle troposphere sinks somewhat after passing the Rockies to form a subsidence inversion, which inhibits the development of

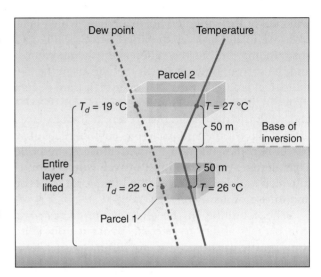

▲ **FIGURE 1** Potential instability occurs when a layer of warm, dry air overlies moist air. The temperature (solid line) and dew point (dashed line) profiles show such a situation. Examine what happens to the parcels of air 50 m above and below the base of the inversion. Because the lower parcel is nearly saturated (i.e., the dew point is close to the air temperature), lifting causes it to become saturated after 500 m of uplift. The parcel above the inversion does not become saturated until 1000 m of uplift. Thus, after both have been lifted 1000 m, the upper parcel has undergone more cooling than the lower one, and the temperature lapse rate becomes steep enough to produce statically unstable air.

air mass thunderstorms. But given sufficient uplift, the surface layer of air can become statically unstable and severe thunderstorms can develop.

---

solar radiation on the lapse rate is (of course) greatest on clear, sunny days, especially above unvegetated land surfaces, where abundant solar radiation is available and little energy is expended on evaporation.

Cooling of the surface, such as occurs at night, chills the lower atmosphere and decreases the ELR. With sufficient cooling, the air near the surface can become colder than the air above and create a situation in which air temperature *increases* with height. (This is known as a temperature inversion, a condition of extremely stable air that we discuss later in the chapter.)

Advection of Cold and Warm Air at Different Levels

Temperature profiles can be influenced by differences between wind directions at low and high levels. In Figure 6–10a, for

example, low- and high-level winds are both from the west, where surface and upper-level temperatures are 10 °C and 9.5 °C, respectively. The lapse rate is thus 0.5 °C/100 m. In Figure 6–10b, surface winds are unchanged, but upper-level flow now comes from the colder northeast, so that temperatures aloft are lower, only 9.0 °C. Cold air has been advected (transported horizontally) above the surface, resulting in a steeper lapse rate. Warm air can be similarly advected, if winds happen to blow from warmer toward colder locations.

Of course, the advection of warm or cold air can occur at any level. For example, if cold air is advected at low levels, the lapse rate declines, producing greater stability. Moreover, advection is not confined to a single altitude; we should not think of a moving slab of air at some height unconnected to the rest of the atmosphere. Most commonly, there is a gradual change in wind direction (and speed) with height.

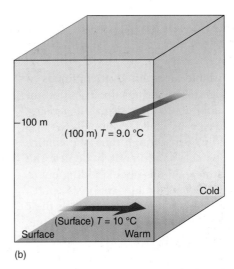

(a)   (b)

▲ **FIGURE 6-10** The ELR can be changed by the advection of air with a different temperature aloft. In (a), the winds at the surface and the 100 m level bring in air with temperatures of 10 °C and 9.5 °C, respectively, yielding an ELR of 0.5 °C/100 m. In (b), the surface winds still bring in air with a temperature of 10 °C. But the wind direction at the 100 m level has shifted to northeasterly, and the advected air has a temperature of 9.0 °C. This yields a steeper ELR of 1.0 °C/100 m.

Go outside on a cloudy, windy day and you will probably observe differences in the movement of clouds at different levels. Depending on how the winds are oriented relative to the temperature distribution, each altitude can have differing amounts of cold or warm advection. This does not mean advection is haphazard or random. As you will see in a later chapter, definite systematic relations exist between the wind and pressure fields. The point here is that the presence of advection is variable, both from day to day and from altitude to altitude within a column, and the effects on atmospheric stability are likewise variable.

Advection of an Air Mass with a Different ELR

The atmosphere has a strong tendency to be arranged into large areas distinguished by small horizontal differences in temperature and humidity. These so-called air masses maintain their temperature and moisture characteristics as they move from one place to another. When an air mass migrates to a particular area and replaces another, the initial ELR at that location gives way to that of the new air mass. In Figure 6–11, for example, the air at position A initially has a low temperature lapse rate, while an air mass approaching from the left of the diagram has a greater lapse rate (a). The new air mass with the steeper lapse rate reaches position A, bringing with it more unstable conditions (b).

Checkpoint

1. What are three factors that affect the environmental lapse rate?
2. How can differences in advection between different levels affect the ELR?

 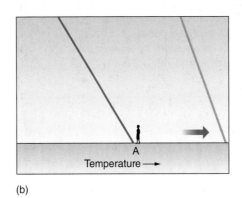

(a)   (b)

▲ **FIGURE 6-11** The ELR changes when a new air mass replaces one that has a different lapse rate. The green line represents the temperature profile originally encountered at position A. The blue line depicts the temperature profile replacing the earlier profile.

Limitations on the Lifting of Unstable Air

We now know that once unstable air is lifted, it continues to rise and even increases in speed. This brings up an important question: What causes unstable air to quit rising? If a rising parcel continued to rise forever it would eventually escape Earth, never to be seen again. Given enough time, the continued loss of the unstable air parcels would entirely deplete the atmosphere. Of course, our atmosphere is not exploding out to space, so something must occur to eventually suppress uplift.

The primary braking mechanism for rising parcels is their ascent into a layer of stable air. A second process that inhibits upward motions is called *entrainment*.

A Layer of Stable Air

The solid line in Figure 6–12 plots one of the infinite number of temperature profiles that can exist. From Earth's surface to the 500 m level, the air is unstable; above 500 m, it is stable. If a parcel of air rises from the surface (for simplicity, we assume it is unsaturated), it becomes buoyant throughout the lowest 500 m. Above 500 m, however, the rising parcel cools more rapidly than the ambient air and eventually becomes cooler than its surroundings. The parcel does not come to a screeching halt at that point, however, because it still has considerable upward momentum. Instead, the parcel gradually slows to a stop and then sinks back down because of its greater density relative to the surrounding air. The parcel may then bob up and down before coming to rest at some equilibrium level.

Is a stable layer always present at some altitude to contain uplift? The answer is yes, because, if nothing else, a rising parcel will eventually encounter the stratosphere, which is extremely stable. As a result, even the most rapidly ascending parcels of air must slow down and reach an equilibrium level above the tropopause. Although severe thunderstorms can have updrafts of more than 200 km/hr (120 mph), uplift seldom extends above the lowest kilometer or so of the stratosphere.

Entrainment

When we talk about a rising air parcel, we mean a small mass that undergoes motions distinct from the surrounding atmosphere. To some extent, we can imagine such a parcel as being like the air inside a balloon. But unlike a balloon, which has a rubber film to isolate the air within, an air parcel has no barrier to prevent it from mixing with its surroundings. In fact, as air rises, considerable turbulence causes ambient air to be drawn into the parcel. This process, called **entrainment**, is especially active along the edges of growing clouds. Entrainment suppresses the growth of clouds because it introduces unsaturated air into their margins and thus causes some of the liquid droplets to evaporate. The evaporation consumes latent heat and thereby cools the margin of the cloud, making it less buoyant.

Checkpoint

1. Why don't rising parcels of air eventually escape into space?
2. How does entrainment limit the ability of air to rise?

▶ **FIGURE 6–12** Air that is unstable at one level may be stable aloft. The solid line depicts a temperature profile that is unstable in the lowest 500 m but capped by an inversion. An unsaturated air parcel displaced upward would cool by the DALR (dashed line), making it initially warm and buoyant relative to the surrounding level. Some distance after penetrating the inversion layer, the rising air is no longer warmer than the surrounding air, and further lifting is suppressed. The parcel continues upward for some distance, however, because it has considerable upward momentum. As it does so, it cools more rapidly than the surrounding air and becomes relatively dense. After coming to a stop, the heavy parcel of air sinks back down and eventually comes to rest at some equilibrium level.

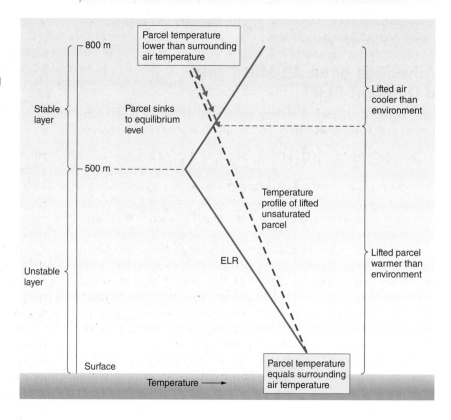

Radiation Inversions and Human Activities

On clear, still nights, air near the ground can cool rapidly. Of course, air higher up will not be subjected to the same cooling as air in contact with the surface, and temperature differences of several degrees Celsius can be observed over just a few meters in height. Thus, subfreezing temperatures can exist near the ground, while just a short distance above, temperatures may be safely above the freezing level. This has some important effects on agriculture in the southern United States.

When temperatures drop to the freezing point, winter crops are vulnerable to frost damage. To offset this problem, growers often set large "wind machines" atop masts, as shown in Figure 1a. When temperatures near the ground fall dangerously low, the machines are turned on to force the warm air in the upper part of the inversion down toward the surface. Thus, the potentially damaging air near the surface is replaced as warmer air circulates downward.

At even lower temperatures, growers may also activate *smudge pots* that burn heating oil, as shown in Figure 1b. Although the emission of longwave radiation by the smudge pots helps somewhat to protect the crops, their greater contribution is to produce free convection. Like the wind machines, the smudge pots produce a

(a)

(b)

▲ **FIGURE 1** Agricultural wind machines (a) blow air downward during times of potential frost damage to force warm air in the upper part of a radiation inversion toward the crops. During times of greater stress, growers also may employ smudge pots (b).

continual mixing of air between low and higher levels, so that surface air stays above the freezing point. Yet another tactic is to spray citrus crops with water. As water freezes, latent heat is released, which keeps the fruit within a few degrees of freezing, warm enough to prevent severe damage. Of course, on very cold nights inversions are so deep and cold that none of these remedies is effective. In such a case, frost damage can ruin an entire crop.

Problems associated with radiative cooling and the resultant inversions are not restricted to agriculture. The strong stability of an inversion can suppress the vertical motions that dilute the concentration of pollutants near the surface. It is common for urban dwellers to notice a low-lying layer of sooty haze near the surface on cold, clear mornings when radiation inversions are most likely to form.

Extremely Stable Air: Inversions

So far, we have been concerned with the mechanisms that cause air to rise, and the influence of instability on the effectiveness of those mechanisms. Now it is useful to examine the most extreme forms of stable air, those associated with inversions.

Although on average the temperature in the troposphere decreases with elevation, in some situations temperature *increases* with altitude. Layers of the atmosphere in which this situation exists are called **inversions**. Air parcels rising through inversions encounter ever-warmer surrounding air and therefore have strong negative buoyancy. Inversions are thus extremely stable and resist vertical mixing.

Several different processes can cause different types of inversions to develop. One of the most common is a *radiation inversion*, which results from diabatic cooling of the surface. On cloud-free nights with little or no wind, longwave radiation emitted by the surface easily escapes to space. This lowers the ground temperature, which in turn chills the air immediately in contact with it. Because the lower air chills more rapidly than that farther from the surface, an inversion develops at ground level.

If cooling is sufficient to lower the temperature to the dew point, a radiation fog forms. Inversions are associated with all radiation fogs, but if the cooling does not lower the temperature to the dew point, a radiation inversion can exist without the appearance of fog.

Radiation inversions occur throughout the world. Though they are usually restricted to fairly shallow depths above the surface, they can have important ramifications for agriculture and other activities. We discuss two examples in *Box 6–3, Special Interest: Radiation Inversions and Human Activities.*

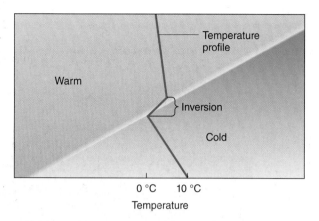

▲ **FIGURE 6–13** Frontal inversions. The temperature is plotted with the solid blue line.

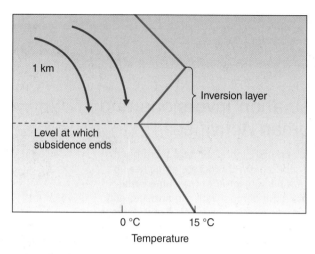

▲ **FIGURE 6–14** Subsidence inversions occur when air descends toward but not all the way to the surface. The base of the inversion marks the lowest point to which air has subsided.

Factors other than diabatic cooling of the surface can also produce inversions. When a cold or warm front is present, for example, a transition zone separates warm and cold air masses. The boundary is not horizontal but rather forms a wedge of cold air that underlies warmer air, as shown in Figure 6–13. The horizontal extent of these *frontal inversions* can be up to several hundred kilometers, and the height of the inversion increases with distance from surface position of the front. Rain falling into a very cold surface layer can freeze before reaching the ground (resulting in sleet), or on the ground, resulting in the much more dangerous freezing rain.

More extensive and meteorologically important than frontal inversions are the *subsidence inversions* that result from sinking (or *subsiding*) air. Recall that a layer of air is compressed and warmed during descent. As it is compressed, its thickness decreases, which means that the top of the layer descends a greater distance than does the bottom of the layer. The longer descent leads to greater temperature increases at the top of the layer than the bottom, and, if enough sinking occurs, an inversion forms.

Subsidence is common along the eastern margins of large areas of high pressure and downwind of major mountain ranges. Because the subsiding air does not descend all the way to the surface, the base of the inversion can be several hundred meters above the surface. As a result, subsidence inversions are clearly distinguishable from radiation inversions, the bases of which are at ground level. Figure 6–14 shows a typical temperature profile for a subsidence inversion. Notice that the top of the inversion layer is more than 10 °C (18 °F) warmer than its base. This is a large difference, but not at all uncommon.

A spectacular example of a subsidence inversion can be found every year between the months of April and September in the United States. A large high-pressure system called the Hawaiian High often forms over the middle latitudes of the North Pacific. As upper-level air rotates out of the high-pressure center, subsidence occurs over coastal southern California and forms an inversion. This provides a "cap" for the vertical dispersal of pollutants that helps to give Los Angeles some of the most polluted air in North America.

Checkpoint

1. What is an inversion?
2. How does a subsidence inversion form?

Cloud Types

Clouds can assume a variety of shapes and sizes and can occur near the surface or at high altitudes. Most cloud types occur in the troposphere, but some appear in the stratosphere and even in the mesosphere. Clouds can contain liquid droplets, ice crystals, or a mixture of the two. They can be thick or thin and have high or low liquid water or ice contents. It is easy to see why meteorologists would want a classification scheme to distinguish the many types of clouds from one another.

DID YOU KNOW?

Have you ever looked out the window of an airplane and found faint clouds formed right above the plane's wings? This is the result of adiabatic expansion of the air above the wing that can bring the immediate air temperature down to the dew point.

The first widely accepted system for cloud classification was devised by English naturalist Luke Howard in 1803. It divided clouds into four basic categories:

1. **Cirrus**—thin, wispy clouds of ice
2. **Stratus**—layered clouds
3. **Cumulus**—clouds having vertical development
4. **Nimbus**—rain-producing clouds

TABLE 6–1

Ten Principal Cloud Types

High Clouds (Heights Greater Than 6000 m, OR 19,000 ft)		
Cirrus (Ci)	(Figure 6-16)	Thin, white, wispy clouds resembling mares' tails.
Cirrostratus (Cs)	(Figure 6-19)	Extensive, shallow clouds somewhat transparent to sunlight, producing a halo around the Sun or Moon.
Cirrocumulus (Cc)	(Figure 6-20)	High, layered cloud with billows or parallel rolls.
Middle Clouds (Heights between 2000 m AND 6000 m, OR 6000 TO 19,000 ft)		
Altostratus (As)	(Figure 6-21)	Extensive, watery, layered cloud. Allows some penetration of sunlight but Moon or Sun appears as bright spot within cloud.
Altocumulus (Ac)	(Figure 6-22)	Shallow, mid-level cloud containing patches or rolls.
		Generally more opaque and having less distinct margins than cirrocumulus.
Low Clouds (Below 2000 m, OR 6000 ft)		
Stratus (St)	(Figure 6-23)	Uniform layer of low cloud ranging from whitish to gray.
Nimbostratus (Ns)	(Figure 6-24)	Low cloud producing light rain. Produces darker skies than altostratus.
Stratocumulus (Sc)	(Figure 6-25)	Low-level equivalent to altocumulus.
Clouds with Vertical Development (May Extend through Much of Atmosphere)		
Cumulus (Cu)	(Figures 6-26 and 6-27)	Detached billowy clouds with flat bases and moderate vertical development. Sharply defined boundaries.
Cumulonimbus (Cb)	(Figure 6-28)	Clouds with intense vertical development with characteristic anvil. May be tens of thousands of meters thick. Appear very dark when viewed from below.

Our current classification scheme is a modified version of Howard's typing that retains his four categories and also allows new combinations (for instance, *cirrostratus* clouds have the characteristics of cirrus clouds and stratus clouds). The ten principal types of clouds that result are then grouped according to their height and form:

1. **High clouds**—cirrus, cirrostratus, and cirrocumulus
2. **Middle clouds**—altostratus and altocumulus
3. **Low clouds**—stratus, stratocumulus, and nimbostratus
4. **Clouds with extensive vertical development**—cumulus and cumulonimbus

The ten principal cloud types based on this scheme are outlined in Table 6–1 and Figure 6–15. Several of the ten cloud types described in Table 6–1 can also be divided into subtypes, called species. These species are described in Table 6–2.

▲ **FIGURE 6-15** Generalized cloud chart.

TABLE 6-2

Cloud Species

Name	Types of Cloud	Description
Calvus	Cumulonimbus	Upper portion of cumulonimbus loses distinct outline as ice crystals form in lieu of water droplets.
Capillatus	Cumulonimbus	Fibrous upper portion consists of ice crystals. Further developed than cumulonimbus calvus.
Castellanus	Cirrus, cirrocumulus, altostratus, and stratocumulus	Towerlike vertical development on portion of cloud.
Congestus	Cumulus	Undertaking rapid and significant vertical development.
Fractus	Stratus and cumulus	Broken or ragged.
Humilis	Cumulus	Having slight vertical extent.
Incus	Cumulonimbus	Having anvil-shaped top.
Lentiucularis	Stratocumulus, altocumulus, cirrocumulus	Lens shaped.
Mammatus	Cumulonimbus	Pouchlike.
Pileus	Cumulus, cumulonimbus, stratocumulus	Caplike cloud immediately above larger cloud.
Translucidus	Altocumulus, altostratus, stratocumulus, and stratus	Somewhat transparent, so the Sun, Moon, or higher clouds may be discernible.

High Clouds

High clouds are generally above 6000 m (19,000 ft). They are almost universally composed of ice crystals instead of liquid droplets. Recall that within the troposphere the average temperature decreases from 15 °C (59 °F) at sea level at a rate of 6.5 °C/1000 m (3.6 °F/1000 ft). As a result, the average temperature for high clouds is usually no higher than –35 °C (–31 °F); cooling to the frost point causes the formation of ice crystals instead of supercooled droplets.

Where surface temperatures are very low, clouds composed exclusively of ice can occur at altitudes as low as 3000 m (10,000 ft). Thus, the definition of a high cloud (above 6000 m) is somewhat temperature dependent.

The simplest of the high clouds are *cirrus* (abbreviated Ci), which are wispy aggregations of ice crystals (Figure 6–16). The average thicknesses of cirrus clouds is about 1.5 km (1 mi), but they can be as thick as 8 km (5 mi). Given the very low temperatures at which they exist, they have little water vapor from which to form ice. So although they may be easily visible, the water content of these clouds is extremely low. In fact, the ice content of cirrus clouds is typically only about 0.025 grams per cubic meter, or about one-thousandth of an ounce per cubic yard.

Although the entire mass of ice contained in a cirrus cloud is small, the individual crystals can be as long as 8 mm (0.3 in.). These crystals fall at a speed of about 0.5 meters per second (about 1 mile per hour), which is sufficient for them to overcome updrafts and descend as *fall streaks* (Figure 6–17).

▲ **FIGURE 6-16** Cirrus clouds are wispy clouds of ice crystals.

Contrails (Figure 6–18) are a type of ice cloud frequently caused by jet aircraft. The very hot engine exhaust contains considerable water vapor as a result of fuel combustion, and turbulence in the wake of the aircraft rapidly mixes the exhaust with the cold, ambient air. As was explained in Chapter 5, the mixing of warm moist air with cold air can lead to saturation and, in this case, the rapid formation of ice crystals.

Cirrostratus (Cs) clouds (Figure 6–19), like cirrus, are composed entirely of ice but tend to be more extensive horizontally and have a lower concentration of crystals than cirrus. Though cirrostratus clouds reduce the amount of solar radiation reaching the surface, enough direct sunlight penetrates to allow objects at the surface to cast shadows. Furthermore, they do not fully obscure the Moon or Sun behind them. Instead, when viewed through a layer of cirrostratus, the Moon or Sun has a whitish, milky appearance but a clear outline.

A characteristic feature of cirrostratus clouds is the *halo*, a circular arc around the Sun or Moon formed by the refraction (bending) of light as it passes through the ice crystals. Ice crystals bend much of the passing sunlight or moonlight 22° away from its initial direction. So if you face toward the

▲ **FIGURE 6-17** Fall streaks result from falling ice crystals.

▲ **FIGURE 6-18** Aircraft contrails.

▲ **FIGURE 6-20** Cirrocumulus clouds. These frequently occur in rows of individual, puffy clouds.

Sun or Moon, refracted sunlight will be directed toward you from a ring that surrounds the Sun at a 22° angle.

Cirrocumulus (Cc) are often among the most beautiful of clouds. They are composed of ice crystals that are arranged into long rows of individual, puffy clouds (Figure 6–20). Cirrocumulus form during episodes of *wind shear*, a condition in which the wind speed and/or direction changes with height. Wind shear often occurs ahead of advancing storm systems, so cirrocumulus clouds are often a precursor to precipitation. Because of their resemblance to fish scales, cirrocumulus clouds are associated with the term "mackerel sky."

DID YOU KNOW?

For three days following the tragedy of September 11, 2001, all commercial aircraft were grounded in the United States. Formation of contrails stopped immediately; according to one study, this increased the daily temperature range across the country during that period. The evidence suggests that the temporary disappearance of contrails increased the amount of sunlight reaching the surface during the daytime (thus raising daytime surface temperatures), while increasing longwave radiation losses at night (thereby lowering nighttime temperatures). Researchers believe that the daily temperature ranges across the country averaged about 1.8 °C (3.2 °F) more than they normally would over the period. It is likely that over the Midwest, where contrails are most likely to form, the increase in the daily range may have been even greater.

Middle Clouds

Middle clouds occur between 2000 and 6000 m (6500 and 19,000 ft) above the surface and are usually composed of liquid droplets. The two major categories in this group are both prefixed by *alto*, which means "middle."

Altostratus (As) clouds (Figure 6–21) are the middle-level counterparts to cirrostratus. They differ from cirrostratus in that they are more extensive and composed primarily of liquid water. Altostratus scatter a large proportion of incoming sunlight back to space, thereby reducing the amount of sunlight that reaches the surface. The insolation that does make its way to the surface consists primarily or exclusively of diffuse radiation, so one way to distinguish altostratus from cirrostratus is by the absence of shadows. Furthermore, when viewing the Sun or Moon behind altostratus, one sees a bright spot behind the clouds instead of a halo.

Altocumulus (Ac) (Figure 6–22) are layered clouds that form long bands or contain a series of puffy clouds arranged in rows. They are often gray in color, although one part of

▲ **FIGURE 6-19** Cirrostratus clouds. Such clouds often create a halo around the Sun or Moon.

▲ **FIGURE 6–21** Altostratus clouds. These middle-level, layered clouds are composed of water droplets.

▲ **FIGURE 6–23** Stratus clouds. These are low-level, layered clouds.

the cloud may be darker than the rest. Consisting mainly of liquid droplets rather than ice crystals, they usually lack the beauty of cirrocumulus.

Low Clouds

Low clouds have bases below 2000 m. *Stratus* (St) (Figure 6–23) are layered clouds that form when extensive areas of stable air are lifted. They are normally between 0.5 and 1 km (0.3 to 0.5 mi) thick, in marked contrast to their horizontal extent, which can exceed that of several states. Usually the rate of uplift producing a stratus cloud is only a few tens of centimeters per second (less than 1 mph), and its water content is low, perhaps just a few tenths of a gram per cubic meter.

Stratus clouds do not necessarily form from the lifting of air but may also result from the turbulence associated with strong winds. Consider a situation in which the air is at rest and there is a moderate decrease in temperature and dew point with height. When the wind begins to blow, it stirs the

air by forced convection. This causes a slight decrease in dew point near Earth's surface and an increase in water vapor at the top of the layer, as moisture is redistributed vertically. At the same time, the air temperature decreases more rapidly with altitude after mixing because the uplifted air cools at the DALR rather than the original (more gradual) ELR. The reduction in air temperature and increase in moisture content in the upper portion of the layer causes the air to become saturated.

Low, layered clouds that yield precipitation are called **nimbostratus** (Ns) (Figure 6–24). Because nimbostratus clouds have low liquid water contents and weak updrafts to replenish moisture, they yield only light precipitation. Seen from below, these clouds look very much like stratus, except for the presence of precipitation.

Stratocumulus (Sc) (Figure 6–25) are low, layered clouds with some vertical development. Their darkness varies when seen from below because their thickness varies across the cloud. Thicker sections appear dark, and thinner areas appear as bright spots.

▲ **FIGURE 6–22** Altocumulus clouds. Layered, mid-level altocumulus clouds are often arranged in bands.

▲ **FIGURE 6–24** Nimbostratus clouds. This type of cloud produces light precipitation.

▲ **FIGURE 6-25** Stratocumulus clouds. These are low, layered clouds with some vertical development.

Checkpoint

> 1. What two criteria form the major basis of cloud classification?
> 2. How do cirrocumulus, altocumulus, and stratocumulus clouds differ from each other?

Clouds with Vertical Development

Cumuliform clouds are those that have substantial vertical development and occur when the air is absolutely or conditionally unstable. Vertical velocities within these clouds are commonly several meters per second, but they can achieve speeds well in excess of 50 meters per second (100 mph). In other words, updrafts in certain cumuliform clouds can be more rapid than the horizontal winds found in weak hurricanes! Liquid water contents are several times greater than those of stratiform clouds.

Cumulus (Cu) clouds fall into several subgroups distinguished by the extent of their vertical development. Fair-weather cumulus (Figure 6–26a), called *cumulus humilis*, have a single plume of rising air that often results from localized heating at the surface. They do not yield precipitation (hence the name, "fair-weather cumulus"), and they can evaporate away soon after their formation. Notice in Figure 6–26b that the clouds and the open areas between them form an invisible circulation system. The clouds mark the zone of rising air, and the cloud-free areas occur where the air sinks.

More intensely developed clouds are *cumulus congestus* (Figure 6–27). Unlike cumulus humilis, they consist of multiple towers, and each tower has several cells of uplift. This gives them a fortresslike appearance with numerous columns of varying heights. Their strong vertical development implies that these clouds form in unstable air.

The individual towers of a cumulus congestus have lifespans of only tens of minutes and are constantly being replaced by newly forming ones. Because of their vertical extent, cumulus congestus clouds can have large temperature differences from top to bottom, and even on hot days their upper portions can have subfreezing temperatures. Liquid droplets do not freeze into ice instantaneously as updrafts carry them into the cold portion of the cloud, however; instead, they can remain in their supercooled state for some time. Eventually the supercooled droplets do freeze, and the liquid cloud becomes dominated by ice crystals. We can actually observe the result of this process from the ground. When a portion of the cloud becomes *glaciated* (composed entirely of ice), it does not exhibit sharply defined edges like the portions consisting of water. Instead, it has a washed-out appearance that makes it readily distinguishable from the liquid portion of the cloud. The upper portion of the cloud in Figure 6–27 provides a good example of glaciation.

Zone of rising air   Zone of sinking air   Zone of rising air

▲ **FIGURE 6-26** Cumulus humilis, or fair-weather cumulus (a). Scattered fair-weather cumulus clouds form by rising air parcels, but the area between clouds has weak downdrafts (b).

▲ **FIGURE 6–27** Cumulus congestus clouds. Note the substantial vertical development.

Cumulonimbus (Cb) (Figure 6–28) are the most violent of all clouds and produce the most intense thunderstorms. In warm, humid, and unstable air, they can have bases just a few hundred meters above the surface and tops extending into the lower stratosphere. In other words, these clouds can occupy almost the entire depth of the troposphere and more!

A cumulonimbus is frequently distinguished by the presence of an **anvil** (so named for its resemblance to the blacksmith's tool). This feature, composed entirely of ice crystals, is formed by the high winds of the lower stratosphere that extend the cloud forward. The anvil appears as a wedge of ice at the top of the cloud that gradually thins out as it gets farther from the main body of the cloud. Strong winds can propel hailstones toward the anvil, where they are ejected and fall from the cloud. This is why airline pilots avoid flying near the anvil.

DID YOU KNOW?

The ice crystals that compose the anvil of a cumulonimbus can exist long after the rest of the cloud has completely dissipated. The resultant cirrus clouds can then be transported many hundreds of kilometers by upper-level winds before they sublimate away. Thus, the cirrus observed by residents of the eastern half of the United States and Canada may be the remnants of a major storm from a couple of days earlier and a considerable distance away.

Although cumulonimbus clouds have extremely strong updrafts, the updrafts vary across the cloud. The most rapidly rising air is found in about the upper third or so of the cloud, with weaker motions below. Furthermore, updrafts

▲ **FIGURE 6–28** Cumulonimbus cloud. Extending into the lower stratosphere, these clouds can create violent weather.

6–4 FOCUS ON AVIATION

Responding to Icing in Different Types of Clouds

In Chapter 5 we talked about the threat that icing presents to aviation. The first line of defense when having to fly through icing conditions is to activate the plane's anti-icing equipment. But pilots sometimes can also alter the course of the flight to lessen the threat, with the appropriate action depending on the cloud type.

When icing appears to occur within stratiform clouds it is a good idea to change elevation upward or downward by about 900 meters—3,000 feet (Figure 1). By definition stratiform clouds have relatively little vertical development, so changing altitude by that amount will likely get you out of the region of supercooled droplets.

Icing can occur rapidly and severely in cumuliform clouds, especially in updrafts and at temperatures ranging from –2 °C to –20 °C (28 °F to –4 °F). Because cumuliform clouds have very substantial vertical development, changes in altitude may be a less effective maneuver than in stratiform clouds. It is a good idea if possible to simply fly around such clouds. Icing in cumuliform clouds normally occurs at altitudes below about 8000 meters (27,000 ft).

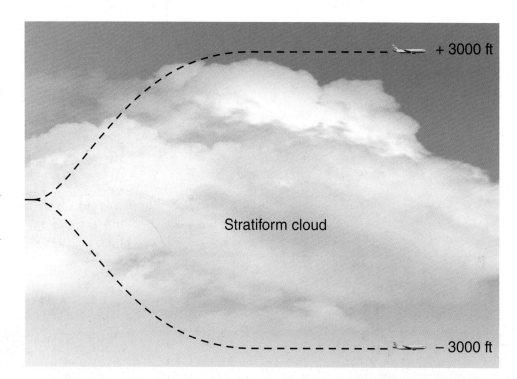

▲ **FIGURE 1** Pilots can avoid icing conditions in stratiform clouds by increasing or decreasing altitude.

along the margins of the cloud are generally less intense than those in the interior because of entrainment. Cumulonimbus even have regions where air descends. Although commercial aircraft can fly through such clouds, the rapid and dramatic shifts in vertical winds cause extreme turbulence that would violently jostle the plane and its occupants. Commercial pilots wisely fly around rather than through cumulonimbus clouds. Other aspects of aviation safety are discussed in *Box 6–4, Focus on Aviation: Responding to Icing in Different Types of Clouds.* In *Box 6–5, Special Interest: Why Clouds Have Clearly Defined Boundaries,* we examine an interesting aspect of cumulus clouds.

Checkpoint

1. Which type of cloud can extend into the lower stratosphere?
2. What are two reasons that airline pilots might want to avoid clouds of this type?

Unusual Clouds

Certain cloud types do not neatly fall into the categories mentioned above. **Lenticular clouds** form downwind of mountain barriers and have curved shapes like eyeglass lenses (Figure 6–29). They form when mountain ranges disrupt the flow of air to form a series of waves. As the air rises in each wave, adiabatic cooling leads to condensation; as the air descends, adiabatic warming causes the cloud droplets to gradually evaporate. Although new droplets constantly form on the upwind side of lenticular clouds and old droplets evaporate on the downwind side, the clouds remain in a fixed position. Thus, the flow of moisture into and out of lenticular clouds is similar to the movement of objects on a conveyor belt, with as much mass removed as is added. Usually no more than two or three lenticular clouds form downwind of the barrier, but if the conditions are just right, as many as six or seven may be observed. **Banner clouds** (Figure 6–30) are similar to lenticular clouds but are individual clouds located immediately above isolated peaks.

◄ **FIGURE 6–29** Waves formed by the passage of air over a topographic barrier can lead to the formation of lenticular clouds.

◄ **FIGURE 6–30** Banner clouds form atop isolated mountain peaks.

Sometimes portions of cumulonimbus clouds hang downward in sac-like shapes called **mammatus** (Figure 6–31). These features occur where downdrafts force water droplets below the cloud base or sloping edge of the cloud, usually near the anvil. Because of the high liquid water content of these clouds, the droplets contained in the downward-moving air require substantial descent (and resultant adiabatic warming) before they fully evaporate. Thus, the mammatus extend some distance below the droplets of the surrounding cloud base.

Why Clouds Have Clearly Defined Boundaries

The next time you see a cumulus cloud, notice its base and edges. You will see that the boundaries of the cloud are marked by flat, sharply defined bottoms and edges that are clearly distinct from the surrounding air. This clear definition of the cloud base is partly due to the rapid growth of droplets as they form just above the lifting condensation level.

Recall that liquid water at the base of a cloud initially forms onto condensation nuclei, of which there are a finite (but very large) number. Within a few tens of meters of the lifting condensation level, all the available condensation nuclei have attracted moisture, and further condensation occurs only onto existing droplets. When droplets first form, they are very small; but they quickly attain diameters of about a micrometer, which makes them effective at scattering visible light—hence the clearly visible base. (If instead the droplets grew slowly, there would be only a gradual increase in the number capable of scattering visible light from the base of the cloud. The cloud would therefore have a faintly visible base that would gradually become more discernible with height.)

Another factor is that cloud droplets evaporate within a very short distance of the cloud base. To see why, consider that if a droplet falls into unsaturated air below the cloud, it shrinks by evaporation.

Because they are so small, cloud droplets fall slowly to begin with and fall even more slowly as they evaporate. As a result, the distance a cloud droplet can fall without evaporating is minuscule, on the order of a centimeter. Viewed from a great distance, this will look like a flat surface. (By way of contrast, raindrops have survival distances measured in kilometers because they are so much larger.)

The sharp boundaries along the sides of cumulus clouds are the result of entrainment. When unsaturated air just outside the margins of the cloud is drawn into the cloud by turbulence, water droplets rapidly evaporate, leaving behind a distinct boundary between the unsaturated, ambient air and the large droplets remaining within the cloud.

▲ **FIGURE 6–31** Mammatus. These clouds are found on the margins of cumulonimbus clouds, are formed by downdrafts, and sometimes are distorted by complex motions.

Like most other weather features, the majority of clouds exist in the troposphere. However, two rare cloud types exist at higher levels and can be seen during the twilight hours of winter at high latitudes. **Nacreous clouds** (Figure 6–32) consist of supercooled droplets or ice crystals in the stratosphere at heights of about 30 km (20 mi). They have a soft, whitish appearance and sometimes are called *mother of pearl* clouds. Even higher are the **noctilucent clouds** (Figure 6–33), whose location in the mesosphere allows them to be illuminated after sunset (or before sunrise) when the surface and the lower atmosphere are in Earth's shadow.

DID YOU KNOW?

The sighting of noctilucent clouds in North America has historically been restricted to latitudes near the Canadian–United States border and northward. They have now recently been observed as far south as Colorado and Utah. This could reflect a decrease in the temperature of the stratosphere associated with climate warming in the lower atmosphere. If so, their increased appearance over the United States may have significant ramifications—beyond making the nighttime sky more dramatic.

Checkpoint

1. What is a lenticular cloud?
2. What role do adiabatic warming and cooling play in the formation of lenticular clouds?

▶ **FIGURE 6-32** Nacreous clouds. These stratospheric clouds are only observed at high latitudes.

▶ **FIGURE 6-33** Noctilucent clouds are in the mesosphere and can illuminate the sky at high latitudes during the twilight hours.

Cloud Coverage and Observation

In addition to their height and form, another characteristic of clouds is their breadth or **coverage**. Meteorologists use several terms to describe coverage, as shown in Table 6–3.

Clouds on any given day are not restricted to a single height above the surface. They can occur simultaneously at several different levels in the atmosphere, and each level can have different cloud types and a different amount of coverage.

Thus, for example, a detailed cloud report might describe the sky as having scattered cumulus at 1000 m, broken altostratus at 4000 m, and a layer of overcast cirrostratus at 7000 m.

Surface-Based Observation of Clouds

Cloud heights and coverages up to a height of 3650 m (12,000 feet) are now routinely determined by automated devices called laser *ceilometers* as part of automated sensing units installed at

6-6 PHYSICAL PRINCIPLES

The Surprising Composition of Clouds

We think of clouds as being liquid water and/or ice, but by far the greatest amount of the mass contained in a cloud is air. Although clouds contain a very large number of suspended droplets or particles—typically about 1000 per cubic centimeter—even the largest of these droplets are extremely small. Therefore, despite their large numbers they amount to relatively little mass. This might not be too surprising if you recall that water vapor only accounts for a small percentage of the mass of the atmosphere. Because the water vapor is so limited, it stands to reason that only a small amount of ice or liquid water can be deposited or condensed.

We can apply some simple arithmetic to determine the relative mass of air and liquid water in a cloud. The average cloud droplet has a radius of about 0.001 cm, and the volume of a sphere is equal to

$$\frac{4}{3} r^3$$

where r is the radius. Substituting 0.001 cm for r, we find that the average droplet has a volume of about 4×10^{-9} cubic centimeters.

Because the density of water is about 1 g/cm^3, it follows that the mass of each droplet is about 4×10^{-9} g. Multiplying this value by the 1000 droplets per cubic centimeter normally found in a cloud gives us a liquid water content of 0.000004 g/cm^3.

Now we can compare the 0.000004 g/cm^3 of water to the mass of the air. At an altitude of 5.5 km above sea level, for example, the density of the air is about half that at sea level, or roughly 0.0006 g/cm^3 Compare this to the mass of the liquid, and you will find about 150 times more air than water in the cloud.

You can think of this another way: If a cloud has a horizontal area of 1 square kilometer and a height of 1 km, it contains about 4 million kg (or more than 1 million gallons) of water. This is far less than the approximately 600 million kg of air in the same cloud.

---

many airports (Figure 6–34). The laser units emit a brief pulse of energy upward that gets reflected downward by cloud droplets or ice crystals. The laser beam travels at a known speed, so the amount of time it takes for the pulse to make its round trip can easily be translated to the height of the cloud base.

Ceilometers can also reveal the amount of coverage for up to three layers above the surface. This is done by evaluating repeated measurements over the previous 30 minutes and noting the amount of time that clouds were present at each level. To make the estimate more representative of the current time period, observations over the previous 10 minutes are given twice as much weight in the calculation as those of the other 20 minutes. See *Box 6–6, Physical Principles: The Surprising Composition of Clouds,* to get a better idea of what we are really looking at when we observe clouds.

Cloud Observation by Satellite

Virtually every television weather report shows the distribution and movement of cloud cover—both regionally and locally—as observed by satellite. In some cases, the satellites view the cloud cover by sensing reflected visible radiation in much the way that a digital camera would view it if fixed on an orbiting platform. On these *visible images*, areas of deep cloud cover appear as a brighter shade of white (Figure 6–35a). Such images have both advantages and shortcomings: They are useful for observing most clouds, but are not very good for distinguishing between high-, medium-, and low-level clouds; and, of course, they are unable to provide any information at night.

Better information can be obtained when visible imagery is coupled with *infrared images* (Figure 6–35b). Unlike visible imagery, which relies on solar radiation scattered back toward space by cloud tops, infrared imagery senses the amount of

TABLE 6–3

Cloud Coverage

Amount of Cloud Coverage	Condition
0	Clear
1/8 to 2/8	Few*
3/8 to 4/8	Scattered
5/8 to 7/8	Broken
8/8	Overcast

* Any cloud coverage at all up to 2/8 is classified as "few."

▲ **FIGURE 6–34** Ceilometers can determine the height of multiple layers of clouds. Laser beams are emitted upward and a portion of the energy is reflected back to the instrument from each cloud layer. Cloud heights are determined by the time required for the laser pulse to return to the ceilometer.

(a)

(b)

(c)

▲ **FIGURE 6–35** Visible (a), infrared (b), and color enhanced infrared (c) satellite images of western North America taken on the morning of June 15, 2005. The comma-shaped cloud pattern off the coast of Oregon and Washington is a typical mid-latitude storm system. The whitish area off the coast of southern and Baja California is a large zone of low clouds and fog, typical of the region in the spring. Notice that the storm system appears well in all three images, with the color enhanced image providing the best detail of the cloud structure. Notice also that the visible image in this instance provides the most distinct image of the low cloud and fog area off the southern coast.

electromagnetic energy *emitted* by clouds. (Infrared images are similar to the water vapor images discussed in Chapter 5, but they are based on the detection of different wavelengths, making them less sensitive to the presence of water vapor but better for tracking clouds.) As discussed in Chapter 2, all materials radiate energy, with the amount of energy and the wavelengths radiated dependent on temperature. Generally, cloud tops at higher elevations will have lower temperatures than those closer to the surface (recall that temperatures within clouds decrease with altitude), and thus emit less radiation than lower-altitude cloud tops. On the most basic types of infrared imagery, clouds of greater thickness appear in a brighter shade of white. Because thicker clouds are more likely to produce precipitation (or severe weather), such areas on satellite imagery are significant.

DID YOU KNOW?

Juneau, Alaska, is the cloudiest U.S. city, receiving only 30 percent of its possible sunshine. Rupert, British Columbia, is the cloudiest city in Canada, with overcast conditions occurring 70 percent of the time. In stark contrast, Yuma, Arizona, receives 90 percent of its possible sunshine on average, annually, making it the sunniest city in the United States and Canada.

Infrared imagery can be color-enhanced to provide more detailed information to the user. Areas on *color-enhanced infrared* images (Figure 6–35c) having the deepest convection are modified to appear as bright red or purple, while those of less-intense activity are assigned yellow and green hues. Lowest-level cloud tops are presented in white.

Images and movie loops such as those in Figure 6–35 are readily available on the Web from many government, commercial, and university sources. More information on using these images will be provided in Chapter 13.

Checkpoint

1. What is a ceilometer?
2. What are the advantages (and disadvantages, if any) of satellite images of clouds made using visible radiation and infrared radiation?

Summary

In the previous chapter we saw the role of lifting in the formation of clouds. Here we examine the processes by which this lifting occurs—frontal uplift, convergence, orographic uplift, and convection. The first three processes are enhanced or hindered by the static stability of the atmosphere; free convection occurs only when the air is unstable. Static instability implies that if a parcel is given an initial boost upward, it will become buoyant and continue to rise. On the other hand, if the air is stable, a parcel displaced vertically will tend to return to its original position.

Static stability or instability is determined by the air column's rate of temperature decrease with altitude. When the temperature lapse rate is less than the SALR, the air is statically stable; when it exceeds the DALR, it is unstable. Conditional instability arises when the lapse rate is between the two adiabatic rates. When the air is conditionally unstable, a lifted parcel will rise on its own accord only if it is lifted above a certain critical point called the *level of free convection*. The static stability of air can be altered by a condition called potential instability, which involves what happens when changing environmental lapse rates (and therefore static stability) lift whole layers of air.

Three processes modify the lapse rate: the inflow of warm and cold air at different altitudes, the advection of a different air mass, and heating or cooling of the surface. ELRs vary not only through time but also with elevation. Thus, a column of atmosphere might be unstable at one level but stable aloft. In fact, no matter what the condition of the troposphere, the stratosphere is always statically stable and thereby limits the maximum height of updrafts.

Inversions are a special case in which the temperature increases with altitude. Because of their strong static stability, inversions suppress the vertical motions necessary for cloud formation and for the dispersion of air pollution. Inversions are formed by subsidence (sinking air), the emission of long-wave radiation from the surface, and the presence of fronts.

Clouds have been categorized into ten distinct types according to their height and form. Layered high clouds are called *cirrostratus*, for example, while clouds aligned in rows or clusters with vertical development are called *cirrocumulus*.

Clouds with extensive vertical development are called *cumulus*. Many cumulus clouds are associated with fair weather, but others can lead to precipitation or even extremely violent thunderstorms. The most dramatic of these are the cumulonimbus clouds, whose updrafts can exceed tens of meters per second.

Given the importance of clouds to the overall weather, meteorologists routinely record information on the presence of clouds, noting their height, type, and degree of coverage. This information can be determined locally by the use of ceilometers, and an overall picture of the cloud pattern can be obtained using visible, infrared, and color-enhanced infrared satellite imagery.

By themselves, clouds are of fundamental interest to meteorology. But of even greater interest, perhaps, is the fact that many clouds precipitate. Why do some clouds cause precipitation and not others? The answer is that a large amount of growth must occur for cloud water droplets and ice crystals to become heavy enough to fall to the surface. The processes by which the droplets grow to precipitation size, and the resultant types of precipitation, are discussed in Chapter 7.

Key Terms

orographic uplift (orographic effect) *page 160*

rain shadow *page 160*

cold front *page 161*

warm front *page 161*

convergence (horizontal convergence) *page 161*

static stability *page 163*

environmental lapse rate *page 163*

level of free convection *page 167*

potential (convective) instability *page 167*

entrainment *page 170*

inversions *page 171*

cirrus *page 172*

stratus *page 172*

cumulus *page 172*

nimbus *page 172*

high, middle, and low clouds *page 173*

clouds with vertical development *page 173*

contrails *page 174*

cirrostratus *page 174*

cirrocumulus *page 175*

altostratus *page 175*

altocumulus *page 175*

nimbostratus *page 176*

Review Questions

1. Describe the four mechanisms that lift air and promote cloud formation.

2. Explain how buoyancy affects the air's susceptibility to uplift.

3. Describe the situations that can cause air to be absolutely stable, absolutely unstable, or conditionally unstable.

4. What two factors can ultimately stop rising parcels of air from continuing upward?

5. What will determine whether air that is conditionally unstable will become buoyant?

6. What is the level of free convection?

7. Describe the processes that bring about changes in the environmental lapse rates, and thus the stability of the atmosphere.

8. What is entrainment, and how does it affect the growth of clouds?

9. Define the term *inversion*, and describe the mechanisms that can cause the various types of inversions.

10. Describe the classification scheme for clouds based on their height and form.

11. List the major subtypes of high, middle, and low clouds.

12. What type of cloud produces a characteristic halo?

13. Other than height, what significant difference exists between altocumulus and cirrocumulus clouds?

14. How do cumulus humilis and cumulus congestus clouds differ from each other?

15. What distinctive feature characterizes a cumulonimbus cloud?

16. What conditions generate lenticular and banner clouds?

17. What are mammatus?

18. List and describe the types of clouds that exist above the troposphere.

Critical Thinking

1. Orographic uplift can cause cloud or fog to form. How might stability be a factor in determining which develops?

2. Except for the shallow zone near the surface, it is rare for the atmosphere to be absolutely unstable. Why is it difficult for a very steep ELR to develop in the middle and upper atmosphere?

3. Localized convection might be more vigorous over the desert in the summer than over a wooded region, but precipitation is more likely in the wooded environment. Why?

4. Is the stability of the air more likely to change rapidly near the surface or aloft? At what time of day are major changes in the ELR most likely?

5. What time of year will unstable conditions be most common over the continental United States and Canada?

6. Is it possible for radiation and subsidence inversions to occur simultaneously?

7. Unlike the Sierra Nevada range in the west, the Appalachian Mountains do not exhibit very strong rain shadow effects. Why not?

8. Some of the higher peaks in Hawaii have "cloud forests" at certain levels. What type of environmental situations would favor ubiquitous cloud cover?

9. In many regions, the orographic effect causes precipitation to increase with elevation. Can you think of any reason why this might not be true all the way up to the top of Mt. Everest?

Problems and Exercises

1. On a daily basis, examine the current surface weather maps and the satellite images depicting cloud cover from any of the Web sites offering those products. Do you notice any relationships between cloud cover and surface pressure distributions? How do clouds tend to appear when associated with cold or warm fronts? Are there any mountainous regions that seem to have a higher incidence of cloud cover? Describe the patterns you observe.

2. Assume that a parcel of air starts out with a temperature of 12 °C and a dew point of 10.4 °C and that the ELR = 0.7 °C/100 m. Determine the following:

a. Is the air stable, unstable, or conditionally unstable?

b. At what level will the air become saturated?

c. Where is the level of free convection?

3. Redo problem 2, but assume that the ELR = 0.4 °C. Explain why there will be no level of free convection.

Quantitative Problems

This chapter has demonstrated the role of temperature lapse rates in cloud formation. The change in ambient air temperature with height determines whether the local atmosphere is absolutely stable, absolutely unstable, or conditionally unstable. The quantitative problems on this book's Web site (**www.MyMeteorologyLab.com**) provide you with the opportunity to assess your knowledge of how these lapse rates determine the susceptibility of the air to upward motions. We suggest that you log on to the site and go to the Chapter 6 page, where you can work out a set of numerical problems to improve your understanding of the concept of stability.

Useful Web Sites

www.crh.noaa.gov/lmk/?n=cloud_classification
Examples of many types of clouds.

www.weather-photography.com/gallery.php?cat=louds
Extensive library of cloud photos and time lapse movies.

www.nsf.gov/news/special_reports/clouds
Interesting information on clouds and climate change.

www.cnn.com/EARTH/9608/27/cloud.harvest
A discussion of how people in arid regions "harvest" fog.

www.ssec.wisc.edu/data/geo.html
An excellent site for current visible, infrared, and water vapor satellite images. Storm chasers risk life and limb to track violent storms.

MyMeteorologyLab™

The *Understanding Weather and Climate,* Sixth Edition, MyMeteorologyLab™ web site contains numerous multimedia resources to aid in your study of **Cloud Development and Forms**.
Visit **www.MyMeteorologyLab.com** to:

- **Review** key chapter concepts.
- **Read** the Pearson eText, *In the News RSS feeds,* and other chapter-specific Web resources.
- **Visualize** challenging topics with Interactive Tutorials, Geoscience Animations, MapMaster interactive maps, and Weather in Motion movies and visualizations, and more.
- **Test Yourself** with self-study quizzes.

 TUTORIAL
ATMOSPHERIC STABILITY

Use the interactive animations and quizzes in this tutorial to review this chapter's key concepts.

WEATHER IN MOTION
View movies and visualizations of meteorology patterns and processes.
Clouds and Aviation
Gravity Wave Clouds
Identifying Clouds in Satellite Imagery
Clouds Developing over Florida
Is That a Cloud?

7

Precipitation Processes

After reading this chapter, you should be able to:

▶ Describe the processes involved in the growth of cloud droplets.

▶ Describe the distribution of precipitation and explain how different types of precipitation form.

▶ Describe how precipitation is measured.

▶ Summarize efforts to induce precipitation through cloud seeding.

During the spring of 1993 residents of the Midwestern United States witnessed flooding on an unprecedented scale. At St. Louis, Missouri, the Mississippi River crested at 6 m (19.5 ft) above flood stage. The river, normally about 800 m (0.5 mi) wide near St. Joseph, Missouri, stretched out to as much as 10 km (6 mi), putting nearly half of St. Charles County under water. At Kansas City, Missouri, the Missouri River rose 6.7 m (22 ft) above its banks. Across the Midwest, tens of thousands of homes were damaged or destroyed by the flooding, as entire neighborhoods and 77 small towns ended up under water. The flooding even brought its share of irony: Des Moines, Iowa, was without potable water for 12 days because of contaminated floodwaters. Forty-year-old Jacki Meek of suburban St. Louis probably spoke for all of the 85,000 people who had to evacuate their homes: "I feel about 65 right now. I see my house on the news, and I just cry."

But what many would have thought would be a once-in-a-lifetime event was repeated 15 years later, when another round of extensive flooding soaked the Midwest in 2008. A series of heavy rains hit the region in early June, including a few exceptionally strong ones that brought more than 22 cm (9 in.) of rain in a 2-day period. Gays Mills, Wisconsin, which had been inundated by flood waters from the Kickapoo River for the second time in 10 months, was forced to consider relocating the town on higher ground farther up the floodplain to avoid similar events in the future.

Indiana, Michigan, Illinois, and Missouri were beset by record-breaking floods from exceptionally heavy rain. On June 7, Edinburgh, Indiana, received 27.2 cm (10.71 in.) of rain, the highest amount ever recorded in a single day for the entire state. But no state was hit harder than Iowa, where 83 of its 99 counties were declared disaster areas, more than 8 percent of the corn and soybean acreage were under water (Figure 7–1), and damage was estimated at about $1.5 billion. Many towns and cities fought rising water with sandbags, but often unsuccessfully, as in Cedar Rapids (Figure 7–2).

◄ Record flooding inundates homes in Cedar Rapids, Iowa, in June 2008.

▲ **FIGURE 7–1** Flooding occurred over a large portion of the upper Midwest in June 2008. These satellite images show the contrast between the normal situation and that which existed during the peak of the flooding. Note the increased width of the rivers in 2008, especially the Cedar River west of Iowa City and Cedar Rapids, Iowa, and the amount of saturated ground.

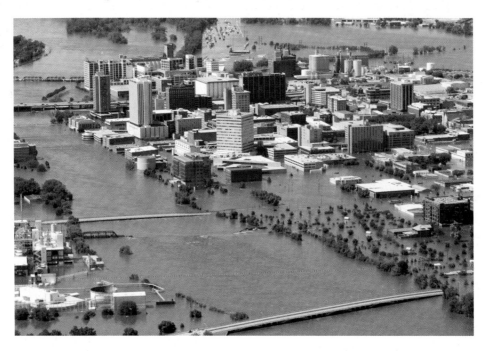

▲ **FIGURE 7–2** Flood damage at Cedar Rapids, Iowa, June 2008.

Overall, how did the 1993 and 2008 floods compare? They were similar in many respects. The 1993 flooding covered a larger area and maintained excessive river levels for a longer period of time. The 2008 floods were the result of more intense rainfall events that occurred over a shorter time period and had faster falling river levels after the peak of the flooding.

Rain and other forms of precipitation are a fact of life for everybody, although usually they are of far less consequence than the floods of 1993 and 2008. The desire to know what causes precipitation may have been one reason you picked up this book. In Chapter 6 you learned about the processes that lead to the formation of all clouds, precipitating or nonprecipitating. In this chapter we explain the processes by which nonprecipitating cloud droplets and ice crystals grow large enough to fall as precipitation.

Growth of Cloud Droplets

Acting alone, gravity would accelerate all objects toward the surface. But gravity is not the only force acting on a falling object; at the same time, the air exerts an opposing resistance or **drag**. As speed increases, so does resistance, until its force equals that of gravity and the acceleration ceases. The object falls, but at a constant speed, its **terminal velocity**. More than anything else, terminal velocity depends on size, with small objects falling much more slowly than large objects. (We examine details of the relationship between size and terminal velocity in *Box 7–1, Physical Principles: Why Cloud Droplets Don't Fall*.) Cloud droplets fall slowly because they are so tiny. Their small size is largely explained by the fact that condensation nuclei are very abundant; thus, cloud water is spread across numerous

small droplets rather than being concentrated in fewer large drops. With their small size, cloud droplets initially have extremely low terminal velocities, making it impossible for them to reach the surface.

This effect is apparent in Figure 7–3, which shows terminal velocities for various cloud constituents. The smallest are the condensation nuclei, on which liquid droplets form. (For the sake of simplicity, the figure applies only to clouds consisting of liquid water alone, without ice crystals.) Condensation nuclei are so small that they fall at an imperceptibly slow rate. Larger cloud droplets (but not falling as precipitation) typically range from about 10 μm to about 50 μm in radius (recall that 1 μm is one-millionth of a meter). These have fall speeds ranging from about 1 cm/sec (0.02 mph) to about 25 cm/sec (0.5 mph). By way of contrast, the much larger raindrops shown in the figure fall at 650 cm/sec, about 25 times faster.

Raindrops fall to the surface when they become large enough that gravity overcomes the effect of updrafts. How large is large enough? In terms of radius, raindrops are about 100 times bigger than typical cloud droplets. But in terms of volume or mass of water, raindrops are larger than cloud drops by a factor of a million, rather than just 100. The difference arises because volume for a sphere is proportional to the cube of the radius. If the radius is 100 times larger, the volume is 100 × 100 × 100 (1 million) times larger. Raindrops are not truly spherical, but the principle holds: Precipitation particles are vastly more massive than cloud drops. Although we do not think of clouds yielding massive falling objects, they certainly do, at least from the point of view of a cloud droplet. In the paragraphs that follow, we outline the processes that give rise to these "massive" falling objects.

7–1 PHYSICAL PRINCIPLES

Why Cloud Droplets Don't Fall

You are probably familiar with the legendary, late-sixteenth-century experiment of Galileo Galilei, who dropped two objects—a light one and a heavy one—off the Leaning Tower of Pisa. The objects, being subjected to the same gravitational acceleration, hit the ground at nearly the same time. Galileo's demonstration may seem inconsistent with our everyday experience, as an ant would surely take longer than a golf ball to fall from the top of a tall building. It is also at odds with our claim that small droplets fall slowly. The solution must be that a force besides gravity acts on falling objects: It is wind resistance, or drag. By examining how these two forces work together, we will gain some insight into why cloud droplets do not fall. To keep the discussion simple, we will assume spherical droplets throughout—using more realistic shapes would not change our conclusions.

Newton's second law tells us that if a net force is applied to a mass, it will undergo an acceleration (or change in velocity through time). For a given mass, the acceleration is directly proportional to the net force. In equation form, the law is given as

$$\text{net force} = \text{mass} \times \text{acceleration}$$

Notice that Newton's second law says that we must consider the net force, the result of all the forces acting on the object. As far as a falling droplet is concerned, there is the downward gravitational force, which is opposed by the force of wind resistance (drag). A droplet suddenly released in the atmosphere falls at increasing speed, but not indefinitely. Eventually the force of drag (F_d) balances the force of gravity (F_g), resulting in no net force:

$$\text{net force} = F_g - F_d = 0$$

With no net force, there is no acceleration, and the droplet falls at its terminal velocity. How fast does it fall? To answer that, we need to know something about the magnitude of the two forces.

Force of Gravity

The force of gravity is equal to mass times the acceleration of gravity, g. Whenever we step on a scale, we measure this force. For a liquid droplet presumed to be spherical, mass is the density of water, ρ, times the droplet volume, $4/3\pi r^3$, where r is the droplet radius. We therefore have F_g as

$$F_g = \rho \frac{4}{3} \pi r^3 g$$

Force of Drag

Drag between the droplet and surrounding air depends on the rate of fall and on the size of the droplet. Just like an automobile on a highway, a faster-moving droplet experiences greater resistance as it moves through the air. In fact, to a good approximation, the drag force increases with the square of wind speed, (v^2). So how does size influence drag?

It can be shown that the drag can be expressed as,

$$F_d = 0.5\, C_D \rho_a v^2\, \pi r^2$$

where C_D is a constant referred to as the drag coefficient, and ρ_a is the density of air (about one thousandth the density of water). The value of C_D is not important here; what matters is that F_d is proportional to the square of both the fall rate (v^2) and the radius (r^2).

Terminal Velocity

For a droplet falling at terminal velocity, we have said that gravity and drag are equal. If we use v_t for terminal velocity, we get

$$F_g = F_d$$

$$g\rho \frac{4}{3}\pi r^3 = 0.5 C_D \rho_a v_t^2\, \pi r^2$$

To find the terminal velocity, we rearrange and first solve for v_t^2

$$v_t^2 = (g\rho \frac{4}{3}\pi r^3)/(0.5 C_D \rho_a\, \pi r^2)$$

By consolidating the numerical values and constants in the above equation into a single constant, k_1, we get

$$v_t^2 = k_1 r$$

or

$$v_t = k_t \sqrt{r}$$

where $k_2 = \sqrt{k_1}$.

From this equation, we see that as droplet radius increases, so does terminal velocity. Large droplets fall faster than small droplets. What happens physically is that both F_g and F_d increase with radius, but the gravitational force increases more than drag, and a higher fall rate is therefore required to cancel F_g. Notice that as far as the droplet is concerned, falling through a still atmosphere at v_t is the same as remaining stationary in an updraft of speed v_t. Thus, the equation says that a strong updraft is needed to hold a large droplet aloft, whereas a small droplet is easily suspended.

Going back to the Leaning Tower of Pisa situation described at the beginning of this box, we can now understand why Galileo's objects fell at nearly the same speed. With large and therefore heavy objects, the gravitational forces far exceeded the drag forces throughout their short fall. With negligible drag, gravity accelerated both at nearly the same rate. If he had used objects of greatly different size, or if the objects had fallen far enough to reach their terminal velocities, differences in v_t would have emerged. The old, familiar story would be about wind resistance, and books like this would have no need for a feature on the topic.

Growth by Condensation

When cloud droplets begin to form by the adiabatic cooling of ascending air, they do so on condensation nuclei. But within a few dozen meters above the lifting condensation level, all the available condensation nuclei have attracted water, and any further condensation can only occur on existing droplets.

TUTORIAL
PRECIPITATION

Use the tutorial to explore the relationship between droplet radius and terminal velocity, and to see how collision and coalescence are related to droplet size.

▶ **FIGURE 7-3** The average characteristics of cloud constituents.

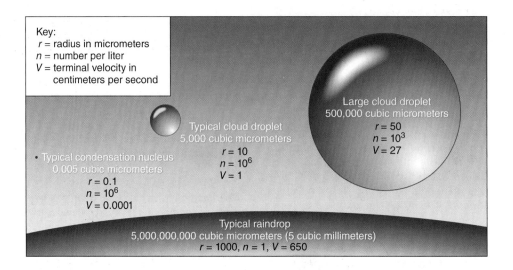

Key:
r = radius in micrometers
n = number per liter
V = terminal velocity in centimeters per second

Large cloud droplet
500,000 cubic micrometers
$r = 50$
$n = 10^3$
$V = 27$

Typical cloud droplet
5,000 cubic micrometers
$r = 10$
$n = 10^6$
$V = 1$

Typical condensation nucleus
0.005 cubic micrometers
$r = 0.1$
$n = 10^6$
$V = 0.0001$

Typical raindrop
5,000,000,000 cubic micrometers (5 cubic millimeters)
$r = 1000$, $n = 1$, $V = 650$

Condensation can lead to rapid growth for very small water droplets, but only until they achieve radii up to about 20 μm—far smaller than necessary to fall as precipitation. Beyond this point, further growth by condensation is minimal. To understand why, recall that relatively little water vapor is available for condensation. With so many droplets competing for a limited amount of water, none can grow very large. It is clear that if growth by condensation were the only process operating, we would experience little, if any, precipitation on Earth. We should therefore think of condensation as only the starting point for rain and snow. Two other processes are responsible for further droplet growth; their relative importance depends on the temperature of clouds.

Growth in Warm Clouds

Most precipitating clouds in the tropics, and some in the middle latitudes, are **warm clouds**, those having temperatures greater than 0 °C throughout. In warm clouds, the **collision–coalescence process** causes precipitation. This process depends on the differing fall speeds of different-sized droplets.

Cloud droplets come in different sizes, and therefore attain different terminal velocities. Refer to Figure 7–4 and consider what will happen when the largest droplet (the **collector drop**) falls through a warm cloud. As the collector drop falls, it overtakes some of the smaller droplets in its path because of its greater terminal velocity. This provides the opportunity for collisions and coalescence.

Collision As it falls, a collector drop collides with only some of the droplets in its path. The likelihood of a **collision** depends on both the absolute size of the collector and its size relative to the droplets below. If the collector drop is much larger than those below, the percentage of collisions (the *collision efficiency*) will be low. Figure 7–5 illustrates why. As the collector drop falls, it compresses the air in its path. The compressed air creates a small gust of wind that pushes the smaller droplets out of the way. The small gust of wind cannot push aside larger droplets, however, and the collector is able to collide with them. As a result, the collision

efficiency is lower for droplets that are very much smaller than the collector drops.

You have probably witnessed a similar phenomenon on a larger scale while driving down a country road in summer, with your windshield turning relatively large flying insects into "bug juice." Too heavy to get swept aside by the compressed air immediately ahead of the windshield, the bugs follow their own paths until the fateful moment of impact. Smaller bugs, in contrast, get blown out of harm's way.

Collision efficiencies are low for droplets nearly equal in size to the collector drop because their terminal velocities are so close to the collector's velocity that it is difficult for the collector to catch up to and collide with them. Continuing with the car analogy, collisions between vehicles are unlikely as long as all move at the same speed and direction.

Under certain situations, collision efficiencies can actually exceed 100 percent, and the collector can collide with more droplets than are in its path. A falling drop creates turbulence that can entrain small droplets outside its path and carry them back toward the top of the collector, where collision occurs.

Collector drop

▲ **FIGURE 7-4** Because of their greater mass, collector drops have greater terminal velocities (indicated by the length of the downward-pointing arrows) than do the smaller droplets in their path. Collector drops overtake and collide with the smaller ones.

▲ **FIGURE 7-5** As a collector drop falls (a), it compresses the air beneath it (b). This causes a pressure gradient to develop that pushes very small droplets out of its path (c). The small droplets get swept aside and avoid impact.

Recent research using mathematical models shows that turbulence in the form of whirling vortices greatly enhances collision efficiency. The vortices function like small centrifuges, separating droplets according to size as they spin around the center. The resulting variations in concentration significantly increase the average collision rate. In addition, rapid

spinning causes jets of droplets to detach from the air flow like rocks thrown from a sling. The ejected droplets have a high probability of colliding with other droplets, so this process enhances collision efficiency. Calculations show that only mild turbulence is required for the centrifuge and sling effects, which implies that the processes operate in most clouds.

Coalescence When a collector drop and a smaller drop collide, they can either combine to form a single, larger droplet or bounce apart. Most often the colliding droplets stick together. This process is called **coalescence**, and the percentage of colliding droplets that join together is the *coalescence efficiency.* Because most collisions result in coalescence, coalescence efficiencies are often assumed to be near 100 percent.

Collision and coalescence together form the primary mechanism for precipitation in the tropics, where warm clouds predominate. In the middle latitudes, most precipitating clouds have freezing temperatures, at least in their upper portions. This favors the growth of precipitation by another mechanism involving the coexistence of ice crystals and supercooled water droplets, the Bergeron process (also known as the Bergeron-Findeisen or ice crystal process) described in the next section.

Checkpoint

1. What are warm clouds and where are they most likely to be found?
2. What is the role of the collector drop in the formation of precipitation in a warm cloud?

Growth in Cool and Cold Clouds

Unlike their counterparts in the tropics, at least a portion of most midlatitude clouds have temperatures below the melting point of ice. Some, such as the one in Figure 7–6a, have temperatures below 0 °C throughout and consist entirely of ice crystals, supercooled droplets, or a mixture of the two. These are referred to as **cold clouds**.

Cool clouds (Figure 7–6b), on the other hand, have temperatures above 0 °C in the lower reaches and subfreezing conditions above. As we discussed in Chapter 5, saturation at temperatures between about −4 °C (25 °F) and −40 °C (−40 °F) can lead to the formation of ice crystals, if ice nuclei are present, or to the formation of supercooled liquid droplets, if ice nuclei are absent. Thus, a well-developed cumulus cloud might be composed entirely of water droplets in its lower portion,

(a)

(b)

◄ **FIGURE 7-6** Cold clouds (a) have temperatures below 0 °C from their base to their top. Cool clouds (b) have temperatures above 0 °C in their lower portions with subfreezing temperatures above.

▲ **FIGURE 7–7** A cumulonimbus cloud. The lower portion consists entirely of liquid droplets, the middle a mixture of ice and liquid, and the upper portion entirely of ice. Note the less sharply defined margins of the glaciated portion composed of ice.

a combination of supercooled droplets and ice crystals in its middle section, and exclusively ice crystals in its upper reaches (Figure 7–7). The processes described in this section operate in cold and cool clouds having a mixture of ice and liquid water.

As we will now see, the coexistence of ice and supercooled water droplets is essential to the development of most precipitation outside the tropics. The process by which droplets and crystals in midlatitude clouds grow to precipitation size was first described by one of the preeminent figures of modern meteorology, Tor Bergeron. This process is therefore often referred to as the Bergeron process.

The principle underlying the **Bergeron process** is that the saturation vapor pressure over ice (the amount of water vapor needed to keep it in equilibrium) is less than that over supercooled water at the same temperature. This is because molecules in an ice crystal bond to each other more tightly than molecules of liquid water. You should recall that saturation exists when the vapor pressure of the air is at the point where evaporation from a water droplet would be exactly offset by condensation back onto it, or if sublimation from an ice crystal would be offset by deposition. Within a certain range of temperatures below zero Celsius, both ice crystals and supercooled droplets can coexist in a cloud. Ice crystals, however, do not sublimate ice to vapor as rapidly as water droplets evaporate liquid to vapor. Thus, ice crystals do not require as high a surrounding vapor pressure as do water droplets to remain in equilibrium, and they are said to have a lower saturation vapor pressure. As a result, if there is just enough water vapor in the air to keep a supercooled droplet from evaporating away, then there is more than enough water vapor to maintain an ice crystal. Let us see how that leads to precipitation.

Refer to Figure 7–8 and consider the situation in which ice crystals and supercooled droplets coexist, and the vapor pressure is equal to that needed to keep the droplets in equilibrium. In (a) the rate of condensation onto the liquid droplet equals the rate of evaporation. But while vapor pressure in the cloud equals the saturation vapor pressure for the droplet, it exceeds that for the ice. This causes some of the water vapor in the air to be deposited directly on the ice. The vapor content of the air then falls, which in turn causes the liquid droplet to evaporate as it gives up water to restore equilibrium (b).

The process does not end there, because evaporation from the droplet increases the water vapor content of the air, which causes further deposition onto the ice crystals (c). This leads to a continuous transfer in which the liquid droplets surrender water vapor, which is subsequently deposited onto the ice crystals. In other words, the ice crystals continually grow at the expense of the supercooled droplets. Although Figure 7–8 suggests this process involves distinct steps, evaporation and deposition actually occur simultaneously.

The growth of ice crystals by the deposition of water vapor initiates precipitation. As the ice crystals grow, their increasing mass enables them to fall through the cloud and collide

(a)

(b)

(c)

▲ **FIGURE 7–8** The Bergeron process. If exactly enough water vapor is in the air to keep a supercooled water droplet in equilibrium, then more than enough moisture is present to keep an ice crystal in equilibrium. This causes deposition (the transfer of water vapor to ice) to exceed sublimation (the transfer of ice to water vapor), and the crystal grows in size (a). This, in turn, draws water vapor out of the air, causing the water droplet to undergo net evaporation (b). Evaporation from the droplet puts more water vapor into the air and facilitates further growth of the ice crystal (c). Although this is shown here as a sequence of discrete steps, the processes occur simultaneously.

with droplets and other ice crystals. The collisions cause two other important processes to occur that greatly accelerate the growth rate of the ice crystals: riming and aggregation.

Riming and Aggregation We have seen that the formation of ice crystals in the atmosphere usually requires the presence of ice nuclei, or particles that initiate freezing. It so happens that ice itself is a very effective ice nucleus. Thus, when ice crystals fall through a cloud and collide with supercooled droplets, the liquid water freezes onto them. This process, called **riming** (or accretion), causes rapid growth of the ice crystals, which further increases their fall speeds and promotes further riming.

Another process in the development of precipitation is **aggregation**, the joining of two ice crystals to form a single, larger one. Aggregation occurs most easily when the ice crystals have a thin coating of liquid water to make them more "adhesive." Water is more likely to be present when the cloud temperature is not much below 0 °C, so adhesion is more common at the warmer end of cold clouds. (Perhaps you have noticed that very large snowflakes are more common during warm, early season snows, as opposed to those that come in the dead of winter.)

The combination of riming and aggregation allows ice crystals to grow much faster than by the deposition of water vapor to ice alone. In fact, growth rates from the three processes combined allow the formation of precipitation-sized crystals within about half an hour from the initial formation

of the ice. When the ice crystals begin to fall, precipitation begins. What happens to these crystals as they fall determines the type of precipitation that occurs.

Checkpoint

1. What is the difference between a cool cloud and a cold cloud?
2. How does the fact that the saturation vapor pressure over ice is less than the saturation vapor pressure over supercooled water lead to precipitation?

Distribution and Forms of Precipitation

The processes described above are ultimately responsible for all the various kinds of precipitation. Figure 7–9 shows the distribution of mean annual precipitation based on worldwide weather records.

In the tropics, precipitation occurs primarily by the collision–coalescence process, and it can therefore occur only as rain. In the middle latitudes, where ice crystal processes dominate, precipitation occurs as a solid or a liquid, depending on the temperature profile of the air through which it falls. If precipitation reaches the surface without ever having melted, we recognize it as snow. If it melts on the way down,

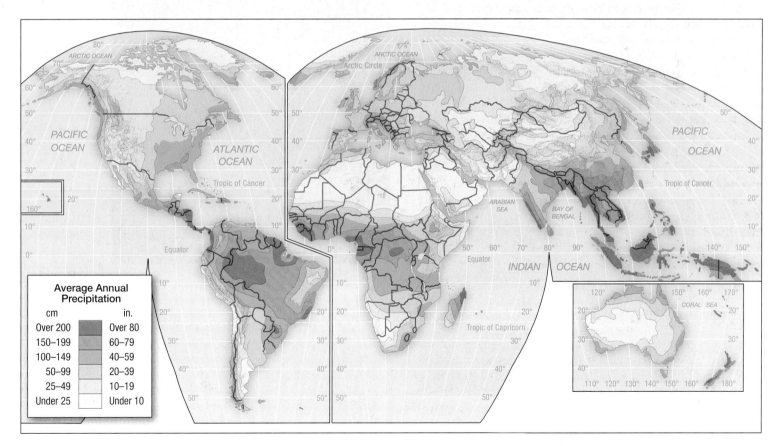

▲ **FIGURE 7-9** Average annual world precipitation.

TUTORIAL

PRECIPITATION

Use the tutorial to study saturation vapor pressure over ice and observe all three growth processes in cold clouds.

▲ **FIGURE 7–10** Snow is initiated by the Bergeron process and reaches the ground only if the crystals falling to the ground never encounter temperatures above zero degrees Celsius.

it might reach the surface as rain. But raindrops sometimes freeze again before, or immediately after, reaching the surface, and then a different type of precipitation results. We now discuss the various types of precipitation.

Snow

Precipitation as **snow** is initiated by the Bergeron process and the growth of crystals by riming and aggregation. Snow occurs if the falling crystals do not melt prior to reaching the ground. Therefore, cloud temperatures must be less than −4 °C (25 °F) and temperatures from the surface to the cloud base not much more than 0 °C (snow might not completely melt if it falls through a shallow layer of air not much warmer than 0 °C). Figure 7–10 illustrates a typical temperature profile required for snowfall.

Ice crystals in clouds can have a wide variety of shapes, including six-sided plates, columns, solid or hollow needles, and complex dendrites with numerous long, narrow extensions (Figure 7–11). The structure depends on the temperature and moisture conditions that exist when the crystal is formed.

If all of a crystal's growth occurs under similar conditions, its structure can be quite simple. If, on the other hand, the temperature and moisture conditions change during growth, a complex mixture of plate, needle, and dendrite can

▲ **FIGURE 7–11** Ice crystals can assume several general shapes, including dendrites (a), plates (b), and columns (c). Each is favored under certain conditions of moisture content and temperature.

develop. Consider, for example, a crystal that originates in the cold, upper reaches of a cloud and gradually falls through a warmer environment. Because each combination of moisture and temperature tends to favor a different type of structure, the crystal can have a particular form at its nucleus, with other forms superimposed.

Snowflakes exist in a wide range of shapes and sizes. They can be as small as about 50 μm or as large as 5 mm. Where riming is the dominant growth process (which is the case in relatively warm clouds), the crystals tend to form a dense "wet" snowpack, ideal for snowball fights but no friend to snowblowers. In contrast, very cold snow typically forms small snowflakes that accumulate on the ground with a lower density. Because of their low temperature, these crystals have less adhesion and are difficult to pack. Skiers know this type of snow as *powder*. There is a widely held misconception about snow. Some people believe it can be too cold to snow, but this is not the case. It can be too cold to snow *a lot*, but it is never too cold to snow *at all*. Because mass is conserved, any ice crystals that form must do so at the expense of the water vapor content of the air. At very low temperatures, only a small amount of water vapor can exist in the air. And without an ample supply of water vapor, cooling of the air can cause the deposition of only a limited supply of ice. It is still possible for some snow to occur, no matter how low the temperature.

North American Distribution Figure 7–12a maps the distribution of mean annual snowfall across Canada and the United States. In the western portion of North America, the distribution of snowfall is governed largely by the presence of north–south mountain ranges (the Coast Ranges, the Sierra Nevada, the Cascades, and the Rockies) that provide orographic uplift and enhance the precipitation from passing storm systems. At high elevations, these ranges have winter temperatures low enough that most precipitation occurs as snow. Over the eastern two-thirds of the continent, there is an increase in mean snowfall with latitude, mostly because the lower temperatures at higher latitudes favor snow rather than rain.

The distribution of annual snow is in marked contrast to the distribution of annual precipitation—rain plus the water equivalent of snow (Figure 7–12b). Total annual precipitation over the eastern two-thirds of North America decreases with latitude rather than increases, largely due to the fact that there is less water vapor in the air with increasing distance from the Gulf of Mexico. Furthermore, lower temperatures typically found at higher latitudes reduce the amount of water vapor that can exist in the air. Note, too, the decrease in precipitation westward across the Great Plains, revealing a rain shadow in the lee of the Rockies.

Lake-Effect Snow One feature of the snow distribution not shown in Figure 7–12 is the strong enhancement of snowfall that occurs downwind of the Great Lakes and other large bodies of water, referred to as **lake-effect snow** (Figure 7–13). This is most prevalent during the late fall and early winter, when lake temperatures are still moderately high but cold air can pass over from the north. As shown in Figure 7–14, the lake

▲ **FIGURE 7–12** Average annual snowfall in Canada and the United States (a), and average annual precipitation (b).

warms and evaporates moisture into the lower atmosphere. As the lower atmosphere warms, it can become unstable as the temperature lapse rate increases. Thus, air that was originally dry and stable becomes moist and statically unstable. When the air passes over land, the effects of topography, vegetation, and other features of the land surface slow the wind. The decrease in wind speed causes convergence, a mechanism for uplift and adiabatic cooling discussed in Chapter 6. Thus, the passage of cold air over the lakes and subsequent landfall provides the three mechanisms favorable for precipitation: unstable air, sufficient water vapor, and a mechanism for uplift.

▶ **FIGURE 7–13** Heavy lake-effect snow in Buffalo, New York.

This lake effect often produces snow showers restricted to a strip of land that can be anywhere from 1 to 80 km (0.6 to 50 mi) long and can extend more than 100 km (60 mi) inland. (It can also increase the amount of snowfall from storm systems passing over the lakes.) Lake-effect snow is most common along the northern part of the Upper Peninsula of Michigan, the western strip of the Lower Peninsula along Lake Michigan, and the eastern shores of Lake Erie and Ontario. Figure 7–15 maps the average yearly snowfall east of Lake Erie and Lake Ontario and illustrates the major impact that the lake effect has on that distribution. Figure 7–16 shows how intense the lake effect can be, plotting the distribution of snow over western

Did You Know?

The greatest recorded seasonal snowfall occurred at Mt. Baker Lodge, Washington, at 1500 m elevation (5000 ft) over the winter of 1998–99. The total of 2736 cm (1140 in., or 90 ft!) exceeded the previous record observed at Mt. Rainier, Washington, during the winter of 1971–72 (2693 cm, or 1122 in.).

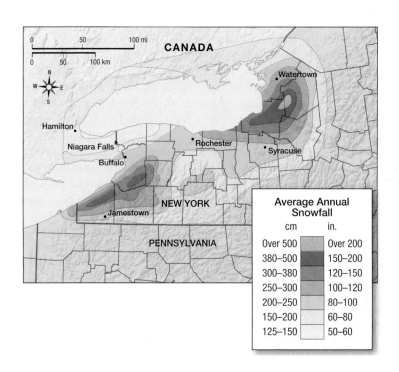

▲ **FIGURE 7–14** Lake-effect snow formation. As cold air moves over a warm lake, heat and moisture are brought into its base. The moist, unstable air mass is then subject to surface convergence as it slows down upon reaching the downwind shoreline, resulting in sometimes heavy snowfalls.

▲ **FIGURE 7–15** The average seasonal distribution of snow east of Lake Erie and Lake Ontario (inches). Note the increase in snow cover immediately downwind of the lakes.

▲ **FIGURE 7-16** Heavy lake-effect snow in February 2008. Snowfall locally exceeded 1 m (3.3 ft) in depth, while much of the general area received no snow.

and upstate New York over the 2-day period, February 19–20, 2008. Notice how large amounts of snow fell over two limited areas while nearby regions got no snow at all.

The winter of 1976–77 provided one of the most noteworthy seasons for lake effect snowfall in the Great Lakes region, when 51 days of lake-effect snow produced record accumulations in upstate New York. The 103 cm (40.5 in.) of snow that fell in a 4-day period from late November to early December in Buffalo provided only a preview of what was to come. By the end of January, Buffalo had received 3.6 m (12.5 ft) of snow for the 3-month period beginning November 1. Even greater accumulations occurred in northern New York, downwind of Lake Ontario, where up to 9.5 m (33 ft) of snow fell over the course of the winter.

Lake-effect snowfall can also occur on the northern side of Lake Ontario. In January 1999, for instance, Toronto experienced a series of storms that brought almost 120 cm (50 in.) of snowfall—an all-time monthly record for the city. During the first 2 weeks of the month, a normal year's worth of snow fell.

Checkpoint

> 1. What is lake-effect snow and where does it typically occur in the United States?
>
> 2. How does the movement of cold, dry air over a warm lake produce lake-effect snow?

Rain

As we have already seen, most precipitation in the tropics comes from warm clouds whose temperatures are somewhat above the melting point of ice. Furthermore, the air temperature

Did You Know?

The average precipitation for the entire world is 97 cm (38.8 in.) per year, and actual global precipitation occurring in any given year seldom departs much from this average. In fact, every year during the period 1900 to 2005 had global precipitation within ±5 cm (2 in.) of the average amount. So drier than normal years in some parts of the world are generally offset by wet conditions elsewhere.

below the clouds is well above freezing, so the rain does not freeze after leaving the base of the cloud. Thus, virtually all precipitation in this part of the world occurs as **rain**, except in high mountains such as Kilimanjaro in Tanzania. Figure 7–17a illustrates a typical temperature profile that could occur with rain of this type. In the middle latitudes, precipitation is usually initiated by the Bergeron process, so most rainfall results from the melting of falling snow (Figure 7–17b).

Rain Showers Convection can lead to the development of cumuliform clouds and precipitation within a few minutes. The episodic precipitation from these rapidly developing

▲ **FIGURE 7-17** Rain can fall when the collision–coalescence process occurs (a), or when falling ice crystals fall from a cloud and melt on the way to the ground (b).

(a) (b) (c) (d)

▶ **FIGURE 7-18** Raindrops are not teardrop shaped. They are initially spherical (a) but flatten out on the bottom as they fall (b). Wind resistance deforms the bottom of the drop, stretching it inward to resemble a parachute with a doughnut-shaped ring at the base (c). Eventually, the droplet breaks apart (d).

clouds is called **showers**. Showers can occur as either rain or snow, but because convection is usually most vigorous in the warm season, showers are more likely to occur as rain.

During a steady rain, droplets occur in a wide variety of sizes. In a shower, the first droplets all tend to be large and widely spaced, but within a short period of time the large droplets give way to a greater number of smaller ones. What happens is quite simple: Large and small droplets fall from the base of the cloud together, but the larger raindrops have greater terminal velocities and reach the surface while the smaller ones are still falling through the air.

Another factor favors the occurrence of large drops at the beginning of a rain shower. Because they take longer to fall through the unsaturated air, small raindrops are more likely to evaporate before reaching the surface. (Evaporation does decrease after a few minutes, however, when the first drops have sufficiently increased the moisture content of the air.)

Raindrop Shape One prevailing myth about weather is that raindrops are teardrop shaped. In reality, raindrops are initially spherical (Figure 7–18a). As they grow by collision and coalescence, their velocities increase, and the greater wind resistance flattens them along the bottom to give them a parachute or mushroom shape (b). As the bottom of the drops flattens out, the greater surface area increases the wind resistance. At this point, the drop begins to deform. The flat bottom becomes concave, somewhat in the shape of a parachute with the bottom surrounded by a relatively thick, doughnut-shaped rim (c). Eventually, wind resistance exceeds the surface tension that holds the droplets together, and the ring at the base of the drop and the bubble-shaped top break into multiple, smaller droplets

(d). The resulting small droplets then begin to grow again by collision and coalescence.

The breakup of falling drops explains why collision and coalescence do not produce enormously large droplets. If the drops were to grow continually on their way down, they could conceivably attain the size of basketballs! Under special conditions, droplets can have diameters of up to 5 mm or more, but they are seldom larger.

Graupel and Hail

When an ice crystal takes on additional mass by riming, its original six-sided structure becomes obscured and its sharp edges are smoothed out. The new ice may contain very small air bubbles that give it a spongy texture and milky-white appearance. This type of modified ice crystal is called **graupel** (pronounced GRAU-pull). Graupel pellets attain diameters up to 5 mm or so, giving them terminal velocities of about 2.5 m/sec (5 mph). Graupel pellets can fall to the ground as precipitation, but under other circumstances they can remain in the cloud and provide the nuclei upon which hailstones form.

Hail consists of multiple layers of ice that are usually no more than a few millimeters in thickness, but under extreme circumstances can achieve sizes comparable to softballs (Figure 7–19). Hailstones are often relatively concentric, though some have irregular shapes. No other type of precipitation is capable of producing drops or particles nearly as large as hail, so it should not be surprising that large amounts of water must be available in the cloud in which they grow and that very strong updrafts are required to allow precipitation this large to remain airborne. Hail therefore forms from cumulonimbus clouds.

(a) (b)

▶ **FIGURE 7-19** Though usually no larger than pea-sized, hail can be remarkably large (a). Its growth accrues by the repeated addition of new layers of ice (b).

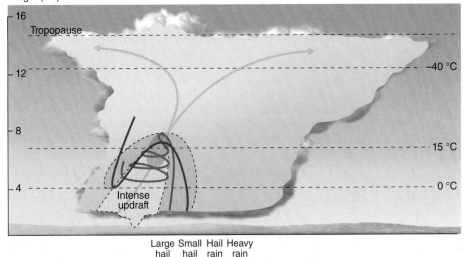

Height (km)

◄ **FIGURE 7-20** The formation of hail. Embryos
near zones of strong updrafts may be blown into
the updraft, rapidly lifted upward, and ejected
as small hail. Hail that forms as it rotates
around these updrafts can accumulate more ice
to become large and fall through a region called
a hail cascade.

Hail development begins with the formation of a small particle called an *embryo*, usually consisting of graupel, in a region of cloud where supercooled droplets coexist with ice crystals (Figure 7–20). The cumulonimbus clouds that produce hail do not have uniformly strong updrafts across their entire width; they have localized areas of extremely violent vertical motions surrounded by zones of relatively weak updrafts. Hail growth typically begins just outside these intense updrafts.

Some developing hailstones get blown from the edge of the strong updrafts into the core, where they are rapidly taken to the upper reaches of the cloud and ejected. When this happens the stones do not have time to accumulate large amounts of ice and fall from the cloud as relatively small hail.

Large hail can develop if the growing particles rise slowly around the more intense core of the updraft. This provides time for the hail to come into contact with supercooled liquid droplets that freeze onto the existing ice. This process can take place in one of two ways: the *dry growth* or *wet growth regimes*.

The dry growth regime occurs at temperatures well below 0 °C (32 °F), wherein supercooled water droplets collide with and freeze onto ice crystals. As the water freezes it releases latent heat that warms the crystal, but not sufficiently to bring its temperature to 0 °C.[1] The ice forms rapidly, allowing it to incorporate very small air bubbles that give it a whitish, opaque appearance.

The wet growth regime from the latent heat released when water freezes onto the ice is sufficient to bring its temperature to 0 °C. The water can remain as a liquid for some time, during which the air bubbles can be released. When the water does freeze, the lack of air bubbles gives it a much clearer, translucent appearance. The existence of supercooled water also allows the ambient wind conditions to deform the layer of water so that when it freezes it takes on a nonconcentric shape, as in Figure 7–19a.

Cumulonimbus clouds have extremely dynamic conditions, and things change readily from one part of the cloud to another and over very short time increments. As a result, growing hailstones can rapidly undergo changes in environment between the wet and dry regimes, causing multiple layers of ice having different appearances from their adjacent ones.

It was once believed that this layering was the result of hailstones making multiple passes upward and downward across the freezing level in a cloud. This is not the case, however, and we now know that the creation of multiple ice layers can result from growth during a single passage upward through a cloud, with the hailstone falling when it becomes too heavy to be maintained by the updrafts. Hailstones often fall in a localized area outside the zone of strongest updrafts, called a **hail cascade**.

Although most hailstones are pea sized, they can become as large as marbles, golf balls, or, under the most violent of conditions, even softballs!

Did You Know?

The largest hailstone ever recorded was found in Vivian, South Dakota, on July 23, 2010. It was officially measured at 19 cm (8 in.) in diameter and weighed 0.88 kg (1 pound, 15 ounces). A local ranch hand found the hailstone and immediately stored it in a freezer. Apparently the stone was even larger when initially recovered, but the storm knocked out power and the stone partially melted before it could be measured.

Hailstones consist mainly of ice with only a small amount of air mixed in. Because ice is relatively dense—90 percent as dense as liquid water—hailstones can become fairly heavy. Compare hailstones to snowflakes, whose volume is occupied mostly by air. Snowflakes have relatively little mass and low

[1]This latent heat of fusion is the heat released when liquid water freezes. It is analogous to the latent heat of evaporation released when water vapor condenses to form a liquid.

TABLE 7–1

Terminal Velocities of Hailstones

$V_t = \sqrt{\frac{1}{3}\rho/k}\sqrt{r} = 20\sqrt{r}$ (r in cm, V_t in m/sec)

Radius (cm)	Terminal Velocity (Meters per Second)
0.1	6 (13 mph)
1.0	20 (44 mph)
2.0	28 (62 mph)
3.0	35 (77 mph)

Did You Know?

Aviators are well advised to avoid cumulonimbus clouds for a variety of reasons, including the possibility of damaging hail. Hail can even be encountered near the anvil of cumulonimbus clouds as winds transport the stones large distances from the area in which they formed. But hail can even be a problem for aircraft on the ground. On May 24, 2011, major hailstorms caused damage to dozens of commercial aircraft on the ground at airports over the southern United States. American Airlines alone had to cancel some 700 flights so that planes could undergo inspections and repairs.

terminal velocities, so they flutter to the ground and make barely a sound. Hailstones, on the other hand, sound like a barrage of falling marbles as they hit the surface. Table 7–1 lists the terminal velocities of various sizes of hailstones.

Hailstones the size of baseballs (radius = 3.5 cm) contain about 160 g of ice (weighing about a third of a pound) and fall at about 40 m/sec (88 mph)! No wonder they are capable of producing tremendous damage. Hailstorms present a major threat to the Great Plains of the United States (Figure 7–21) and Canada (Alberta and interior British Columbia), where they are known to destroy entire fields of crops in a matter of minutes. Because they are most common in the spring and summer, it is often too late to replant the acreage with new seed. (See *Box 7–2, Physical Principles: The Effect of Hail Size on Damage,* for more information about the destructive potential of hail.)

Sleet

Sleet forms when raindrops freeze in the air while falling. Because most rain outside the tropics originates from the ice crystal process, sleet begins as falling ice crystals or snowflakes. As the ice falls through the air, it encounters warmer air and melts to form a raindrop. If the falling raindrop then encounters a lower layer of air whose temperature is below 0 °C, it can refreeze to form sleet. This process, shown in Figure 7–22, results in semitransparent pellets smaller than about 5 mm (0.2 in.) in diameter. Because the formation of sleet requires that a droplet fall through air that is cooler near the surface than aloft, it necessarily requires an inversion, usually one associated with a warm front (which we describe in Chapter 9).

Of course, a raindrop will not freeze instantaneously; sufficient cooling must take place as it falls through the

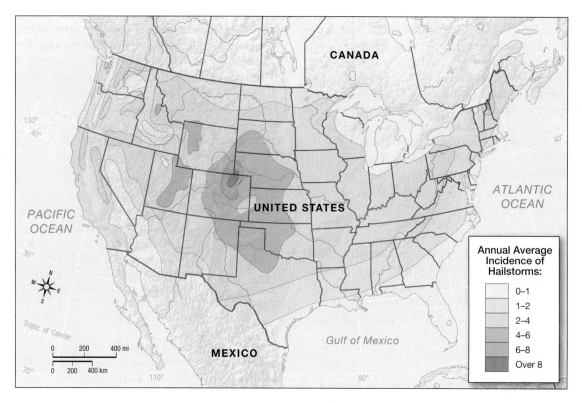

▲ **FIGURE 7–21** The annual average number of hailstorms over the United States.

7–2 FOCUS ON SEVERE WEATHER

The Effect of Hail Size on Damage

You might wonder why large hailstones are more damaging than small ones. After all, isn't the issue how much ice falls to the ground, and aren't many small stones roughly equivalent to a smaller number of large stones? As it happens, this is far from true. Damage done by hail increases very rapidly in a nonlinear fashion with increasing hailstone size. What matters is the amount of kinetic energy hail carries as it falls to the surface.

Kinetic energy depends on mass (m) and speed (v) according to

$$KE = \frac{1}{2}mv^2$$

The mass of a stone is its density (ρ) times its volume

$$\left(\frac{4}{3}\rho\pi r^3\right)$$

Velocity can be found using the formula in Table 7–1. We can thus write

$$KE = \frac{1}{2}\left(\rho\frac{4}{3}\pi r^3\right)(20^2 r)$$

We see that the kinetic energy of a hailstone is proportional to the fourth power of its radius. As a result, a hailstone with a radius of 1 cm (0.4 in.) packs not twice as much punch as a 0.5 cm (0.2 in.) hailstone, but 16 times as much. The threshold for automobile windshield damage due to falling hailstones is about 5 cm (2 in.) in diameter.

---

surrounding air. Thus, for sleet to develop, the layer of cold air beneath an inversion must be fairly deep. If it is too shallow, another type of frozen precipitation will occur, freezing rain.

Freezing Rain

Freezing rain (Figure 7–23) is one of the more deceptive weather events. It usually looks like a gentle rain—certainly nothing to cause major problems. In reality, widespread episodes of freezing rain (often referred to as **ice storms**) can literally paralyze transportation and communications for hundreds of square kilometers.

Freezing rain begins when a light rain or drizzle of supercooled drops falls through air with a temperature at or slightly below 0 °C. When the raindrops hit the surface, they form a thin film of water, but only for a moment. Soon afterward the water freezes to form a slick, continuous coating of ice.

When freezing rain hits roadways, the loss of friction that results leads to extremely dangerous conditions. The weight of accumulated ice also can cause tree limbs and telephone and power lines to snap and fall to the ground. When you imagine downed lines, impassable roads, and broken debris scattered about, it is easy to see why freezing rain can be so disruptive to human life. To make matters worse, freezing rain is often associated with slowly moving storm systems and may therefore persist for several days.

One of the most destructive ice storms in recent decades occurred from the Great Lakes through the southern Great Plains on December 9–11, 2007. Tens of thousands of homes and businesses lost power across Kansas, Missouri, Iowa, and Illinois, but the storm delivered its strongest punch to Oklahoma, where up to 3.5 cm (1.5 in.) of ice accumulated across the state. The entire state was declared a federal disaster area as 640,000 customers lost electrical power, some of them for up to a week. At least 27 people died from the storms, mostly from traffic accidents on the slick roads. If there were one favorable outcome from the event, it was the postponement of final exams across many of the colleges and universities in the Great Plains.

Checkpoint

1. Briefly define graupel, hail, sleet, and freezing rain.
2. How are the conditions under which sleet and freezing rain form similar? How are they different?

◄ **FIGURE 7–22** Sleet occurs as rain, falling from a cloud, passes through a cold layer and freezes into small pellets. This is most common along warm fronts.

▲ **FIGURE 7–23** Freezing rain.

Measuring Precipitation

Given the effects of precipitation on everyday activities, it is no surprise that we measure precipitation at many locations. Just as precipitation occurs in several forms, different types of gauges exist for measuring it. Each method has its own advantages and disadvantages.

Rain Gauges

Rainfall is usually measured with a **rain gauge** (Figure 7–24a). Standard gauges have collecting surfaces with diameters of 20.3 cm (8 in.). The precipitation funnels into a tube with one-tenth the surface area of the collector, so the depth of accumulated water undergoes a tenfold increase. This amplification lets us measure the precipitation level precisely by simply inserting a calibrated stick into the water, removing it, and noting the depth of the wet portion, rather like checking oil with a dipstick. Note that the measuring stick has a correspondingly graduated scale so the 1 cm mark is actually 10 cm from the base of the scale.

An automated collector known as a **tipping-bucket gauge** (Figure 7–24b) provides a record of the timing and intensity of precipitation. This instrument funnels precipitation from the top like a standard rain gauge, but as the water accumulates it is stored in one of two pivoting buckets. One of the buckets is initially upright, while the other, mounted on the opposite end of a pivoting lever, is tipped downward and away from the collector. When the upright bucket gathers rain equivalent to a certain depth (usually 0.01 in.), the weight of the water causes it to tip over, empty its contents, and bring the opposite bucket to the upright position. The tipping of the pivoting buckets triggers an electrical current to a computer that precisely notes the time of the event. The number of tips per unit of time indicates the precipitation intensity. Older recorders use a rotating drum and a printer to provide an analog record of the precipitation rate.

Weighing-bucket rain gauges are similar to tipping-bucket gauges insofar as new accumulations of rain are constantly recorded. A weighing mechanism in these devices translates the weight of the accumulated water in the gauge to a precipitation depth, and the information is stored automatically.

Rain gauges are found at virtually all weather-recording stations, which makes precipitation data plentiful in economically developed countries. Data are scarce in much of the rest of the world, especially over the more than 70 percent of Earth's surface covered by ocean. Furthermore, measurement accuracy is a concern (even with modern instruments), due to such problems as evaporation from the gauge and winds that can prevent rain from entering the gauge.

Rain Gauge Measurement Errors We tend to put unquestioned confidence in the readings we get from precipitation gauges, but unfortunately they have several inherent sources of error. First of all, they are *point measurements*, meaning that they represent the amount of precipitation that has fallen at only a single point or location—not across the street, 100 m away, or down the block. Compared to air temperature, precipitation shows wide variations across different locations, so we face some difficulty in trying to generalize from gauge readings. This not only frustrates attempts to estimate precipitation at points where measurements are unavailable, but it also greatly complicates mapping precipitation patterns.

(b)

(a)

▲ **FIGURE 7–24** A standard rain gauge with its component parts (a). Rain captured by the collector (bottom left) is funneled into the narrow tube (right). The calibrating stick is inserted into the collection tube, and the length of the wetted portion indicates the precipitation accumulation. The interior workings of a tipping-bucket gauge are shown in (b).

Precipitation gauges have other flaws. Wind-generated turbulence near the top of a gauge deflects precipitation away from the collecting surface, leading to an underestimate of the true value (especially for snow). Of the precipitation that does find its way into the gauge, a certain amount goes unrecorded—some because the water splashes out on impact, and some because water is retained as a thin film that does not accumulate at the base of the collector. On hot, windy days some of the collected water evaporates from the collector before measurement.

Other measurement errors have the opposite effect and cause an overestimate of precipitation. Just as water can splash out of a gauge, some can bounce off the surrounding ground and into the collector. Similarly, wind can cause snow to drift from nearby surfaces into the gauge. These errors can affect gauge amounts even under the most carefully monitored conditions. Other things can cause errors as well, such as failing to completely empty the gauge after the last measurement, placing it on a nonlevel surface, or spilling some of the accumulated water. At well-run weather stations where conditions can be controlled, errors such as these are usually very small. Imagine, however, the difficulty

in making reliable observations of precipitation on a ship or buoy. The collection surface of the gauge can be kept only as horizontal as a pitching and rolling ship deck will allow, and it is extremely difficult to prevent windblown seawater from entering the gauge. The moral here is that the distribution of precipitation is not as well recorded as we might suppose, nor as well recorded as befits its extreme value to agriculture and human welfare. Figure 7–25 shows estimated gauge errors for the world, based on a study that developed techniques for error estimation and correction. The most severe errors, which occur at high latitudes and over water, sometimes exceed 80 percent of the true values. Clearly, uncorrected gauge measurements must be used cautiously.

Precipitation Measurement by Weather Radar During recent years, the network of rain gauges has been augmented by radar measurements. Though radar will be discussed in more detail in Chapter 11, for now we can say that weather radars estimate the intensity of precipitation by emitting microwave radiation with wavelengths of several centimeters. Precipitating droplets, ice crystals, and hailstones scatter some of the emitted radiation back to the radar unit, which

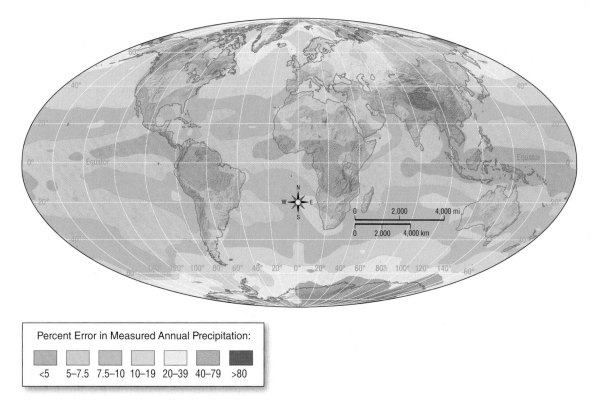

Percent Error in Measured Annual Precipitation:

<5 | 5–7.5 | 7.5–10 | 10–19 | 20–39 | 40–79 | >80

▲ **FIGURE 7–25** The global distribution of the percent error in measured annual precipitation.

records the intensity of the backscattered radiation. In general, the more intense the backscattered radiation, the more intense the precipitation. Meteorologists have developed schemes that relate the intensity of backscattered radiation to the rate of precipitation.

Radiation is not emitted continuously by the transmitter, but just for very brief periods that are interspersed with momentary pauses. Sufficient time is allowed for each pulse of radiation to echo back to the transmitter/receiver unit before the next beam is emitted. The closer the precipitation is to the radar, the quicker the pulse will return to the unit. By measuring the strength of the return radiation and the time taken for it to return to the unit, a profile can be taken showing how much precipitation is occurring and how far from the radar it is.

Did You Know?

On average, at any moment the atmosphere over the 48 conterminous United States contains the equivalent of 175 trillion liters (40 trillion gallons) of liquid water in the form of water vapor. Slightly more than one-tenth of this moisture condenses and falls to the surface daily as precipitation—enough to yield an average of 76 cm (30 in.) of precipitation annually across the country.

The transmitter/receiver slowly rotates as it emits and receives the radar beams, thus allowing a two-dimensional

depiction of the precipitation for several hundred kilometers from the radar. The information can be continually monitored and stored, so that the total amount of precipitation occurring over a fairly extensive region can be estimated (Figure 7–26). Such measurements have proven to be particularly useful in providing short-term forecasting for potential floods.

Snow Measurement

Rain gauges are particularly unreliable when precipitation occurs as snow, because captured snow can obstruct the inlet to the collecting tube. To estimate the precipitation in these environments, we measure the depth of accumulated snow. The *water equivalent* of the snow, which is the depth of water that would result if all the snow melted, can then be roughly estimated using a conversion ratio of 10:1. In reality, the ratio of snow depth to water equivalent can vary greatly—from about 4:1 to 50:1.

In remote mountainous areas, particularly in the western United States and Canada, observations of snow cover have been made for decades at hundreds of *snow courses*. Usually about ten observations are made at each snow course by pushing a collection tube into the snow, extracting the tube and its contents, and weighing them on a spring balance. The weight of the snow-filled tube is directly proportional to the water equivalent of the snow cover, and the average of the ten or so readings is used as the representative value.

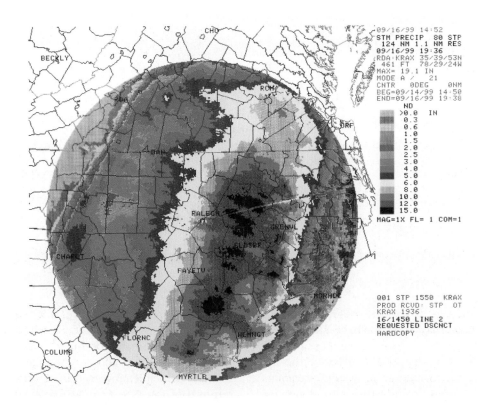

▲ **FIGURE 7–26** Precipitation estimates depicted by Doppler radar.

Manual snow course measurements are still obtained but are frequently augmented by automated snow pillows. **Snow pillows** are large air mattresses filled with an antifreeze liquid and connected to pressure recorders. As snow accumulates on a pillow, the increased weight is recorded and converted to a water equivalent. These instruments have radio devices that transmit the data to a centralized receiving station.

Checkpoint

> 1. Explain how each of the following devices works to measure precipitation: tipping-bucket gauge, weighing-bucket gauge, snow pillow.
> 2. What are some sources of error in measuring precipitation?

Cloud Seeding

Since the late 1940s, people have tried to induce precipitation from clouds, most often to alleviate droughts. This process, called **cloud seeding**, involves injecting one of two materials into nonprecipitating clouds. The objective is to convert some of the supercooled droplets in a cool cloud to ice and cause precipitation by the Bergeron process.

One of the materials, *dry ice* (frozen carbon dioxide), promotes freezing because of its very low temperature (below −78 °C or −108 °F). Very small shavings of dry ice can be ejected from a plane flying through the cloud. When introduced into a cloud, dry ice lowers the temperature of the droplets so that freezing can occur by homogeneous nucleation (see Chapter 5). (Recall that at temperatures below about −40 °C water droplets require no ice nuclei to freeze.)

The second agent for cloud seeding, *silver iodide*, initiates the Bergeron process by acting as an ice nucleus at temperatures as high as −5 °C (23 °F). Silver iodide owes its effectiveness as an ice nucleus to its six-sided structure. Like dry ice, silver iodide can be introduced directly into a cloud from aircraft. More often, it is mixed with a material that produces smoke when ignited in ground-based burners. Updrafts then carry the smoke and the silver iodide into the cloud. If a portion of the seeded cloud is cold enough, some of the supercooled droplets will freeze and begin the Bergeron process.

The cost-effectiveness of cloud seeding is widely debated. Under ideal circumstances, it can supplement water supplies somewhat. Take, for example, the case of the Sierra Nevada mountain range, which supplies much of the water for California and Nevada. Under the right wind, temperature, and moisture conditions, silver iodide released from the ground can enhance snowfall (which is later released as spring melt) by perhaps 10 percent. The right conditions are not usually met, however. Cloud seeding trials in the mountains of Colorado, Utah, and Montana provided disappointing results.

Strong theoretical reasons provide grounds to doubt the usefulness of cloud seeding, except in regions with a

7–3 FOCUS ON AVIATION

Fog Seeding at Airports

Seeding is sometimes used to clear fogs along airport runways. If a fog exists at temperatures below 0 °C (32 °F), introducing dry ice can instigate the Bergeron process, as it does in clouds. Some of the water droplets freeze into ice crystals, which grow at the expense of water droplets and fall out of the fog as snow, leaving a local area of clear air. Of course, this technique can only work for fogs containing supercooled droplets, which is the exception rather than the norm.

Missoula (Montana) International Airport has had good success seeding fog since 2006. Carbon dioxide is sprayed from a couple of pickup trucks into the air just upwind from the runways to stimulate localized snowfall out of the fog.

Several other methods have been tried for dispersing warm fogs, including flying helicopters over the runways to force down the warm air within a radiation inversion and introducing salt crystals into the fog with the intent of making some droplets larger—thereby accelerating growth by collision and coalescence.

---

continued uplift of air (such as where an orographic effect exists). Recall that water vapor accounts for only a small portion of the air, and that the liquid water content of a cloud is relatively small compared to that of the mass of the air contained within. Thus, the formation of heavy precipitation requires a constant resupply of moisture into clouds by updrafts. If such updrafts are already occurring, precipitation will probably occur with or without seeding. Consequently, many meteorologists believe that under most circumstances cloud seeding yields little or no additional precipitation.

Another question raises an ethical concern about cloud seeding. Let's assume that seeding a cloud produces rain. Would the cloud have yielded precipitation farther downwind if it had not been seeded? Residents downwind might argue they were deprived of precipitation that would have occurred naturally over their own fields and drainage basins. Such matters have in fact been litigated in civil court. In short, a number of open questions still exist about the value of cloud seeding for enhancing precipitation.

Seeding has also been attempted to reduce hail intensity. It was once believed that seeding hail-producing clouds could increase the number of growing ice pellets. Because clouds contain a limited amount of water that can freeze onto growing hailstones, increasing the number of hailstones would theoretically reduce their average size. The decrease in size would reduce the kinetic energy of the falling hailstones and thereby lower the likelihood of damage near the ground. However, a multiyear experiment in northeastern Colorado in the 1970s failed to support the usefulness of seeding as a hail suppression measure, and such attempts have been discontinued in the United States and Canada. Attempts to reduce the intensity of hurricanes by cloud seeding have likewise failed to yield convincing results.

The principles involved in cloud seeding also apply to the removal of fog. See *Box 7–3, Focus on Aviation: Fog Seeding at Airports,* to learn more about this.

Checkpoint

1. What are two substances that have been used in cloud seeding?
2. Suppose that you are a wheat farmer on the High Plains in Montana. Would cloud seeding intended to save your crop during a drought be cost effective? Explain why or why not.

Did You Know?

The viability of using dry ice to promote precipitation formation was discovered serendipitously by Vincent Schaefer in 1946. Working with a home freezer in his lab to test whether certain materials could work as ice nuclei, he introduced dry ice into the freezer with moist air to cool it further. He then saw some of the water droplets immediately turn to ice. Further tests by Schaefer and Irving Langmuir proved that introducing dry ice into real clouds could trigger the formation of ice crystals. A few years later, Bernard Vonnegut (brother of novelist Kurt Vonnegut) discovered that silver iodide could also promote ice crystal formation.

Summary

When you first picked up this book, you probably thought you would learn how precipitation occurs. That has been the focus of this chapter.

We can summarize the start of precipitation as the result of cloud droplets growing beyond a size that can remain suspended in the air. The amount of growth necessary for each droplet is tremendous. Some growth occurs through condensation onto existing droplets, but further growth depends on other processes. In the tropics, the primary mechanism for droplet growth is collision and coalescence. Outside the tropics, collision and coalescence are still important, but the Bergeron process dominates, wherein ice crystals grow at the expense of supercooled droplets. Once the Bergeron process has been set in motion, riming and aggregation promote even further growth of ice.

Precipitation occurs in several different forms. Outside the tropics, rain usually results from the melting of ice crystals or snowflakes as they fall to the surface. Ice crystals that do not melt before reaching the surface form snow. Graupel and hail form when supercooled water attaches to ice crystals and freezes. In the case of hail, growth occurs when water repeatedly freezes onto existing ice pellets as they rise above the freezing level on updrafts. Sleet and freezing rain both involve the freezing of raindrops. For sleet, freezing occurs before the drop reaches the surface; for freezing rain, it takes place on contact with the surface.

The standard instrument for measuring precipitation is the simple rain gauge, a collecting device that funnels precipitation into a narrow tube for measurement with a calibrated stick. The timing and intensity of precipitation can be recorded with a modified precipitation gauge called a *tipping-bucket gauge*. As direct and uncomplicated as precipitation measurement may seem, it is subject to a host of potential errors. As a result, our knowledge about the worldwide distribution of precipitation is subject to much uncertainty.

When precipitation is not sufficient to meet human needs, people sometimes resort to cloud seeding, which means introducing materials into clouds to stimulate precipitation by the Bergeron process. This can be done with dry ice, which causes supercooled droplets to freeze by homogeneous nucleation, or with silver iodide, which serves as an ice nucleus. At present, strong reasons exist to doubt the efficacy of cloud seeding in all but very limited circumstances.

Key Terms

drag *page 190*

terminal velocity *page 190*

warm cloud *page 192*

collision–coalescence
  process *page 192*

collector drop *page 192*

collision *page 192*

coalescence *page 193*

cold cloud *page 193*

cool cloud *page 193*

Bergeron process
  page 194

riming (accretion)
  page 195

aggregation *page 195*

snow *page 196*

lake-effect snow *page 197*

rain *page 199*

shower *page 200*

graupel *page 200*

hail *page 200*

hail cascade *page 201*

sleet *page 202*

freezing rain *page 203*

ice storms *page 203*

rain gauge *page 204*

tipping-bucket gauge
  page 204

weighing-bucket rain
  gauge *page 204*

snow pillow *page 207*

cloud seeding *page 207*

Review Questions

1. What determines the terminal velocity of falling droplets and raindrops?

2. Describe the characteristics that distinguish warm, cool, and cold clouds.

3. How do the growth processes of droplets in warm and cold clouds differ?

4. Why isn't growth by condensation able to create precipitation-sized droplets on its own?

5. How do collision and coalescence increase the size of cloud droplets?

6. Explain how variations in the saturation vapor pressure for ice crystals and supercooled water droplets affect the development of precipitation.

7. Why can't the Bergeron process take place in warm clouds?

8. What are riming and aggregation?

9. Why is precipitation greater in Mississippi than in Michigan?

10. How do lakes enhance precipitation downwind?

11. Why do rain showers start with only large drops?

12. Explain why the formation of sleet requires an inversion.

13. It is never too cold for snow to occur. Is that also true of sleet?

14. Why does hail consist of multiple layers of ice?

15. What are some inherent sources of error in rain gauges?

16. How do weighing-bucket and tipping-bucket gauges measure rainfall?

17. Explain how snow pillows measure snow accumulation.

18. What materials are used in cloud seeding, and how do they stimulate (or inhibit) precipitation?

Critical Thinking

1. Industrial activity can increase the number of cloud condensation nuclei. Would an increase in the number of such nuclei tend to promote the formation of rainfall or inhibit it? Why?

2. How might a warming of the atmosphere change how rainfall forms in the middle latitudes?

3. It is not possible for a cloud to precipitate all of its ice crystals or water droplets. Why not?

4. *Precipitable water* refers to the depth of water that would precipitate if all the water in a column of air above the surface were to rain out. Typically, precipitable water is on the order of about 2.5 cm (1 in.), but precipitation amounts can greatly exceed 2.5 cm. How is this possible?

5. Using weather radar to examine vertical profiles of clouds and precipitation shafts, it is easy to determine the height at which the temperature is 0 °C (32 °F). Why?

6. Aircraft icing is a serious threat to aviation at temperatures of about −4 °C to 0 °C (25 °F to 32 °F). Why is it less of a problem at lower temperatures?

7. Typical raindrops fall at a speed of about 6 m/sec—roughly 25 km/hr (15 mph). Snowflakes obviously fall more slowly. What does this imply about the depth of cloud required to yield precipitation through the collision–coalescence process compared to through the Bergeron process?

8. A shampoo company once advertised that its product was "pure as rainwater." Do you think this is true, and if so, does this speak well of the shampoo?

9. Hail sometimes shoots out of a cumulonimbus cloud near the anvil. How can large hailstones be found in this relatively shallow portion of the cloud?

10. Ice in the upper reaches of a cumulonimbus cloud over Colorado may be observed 2 days later over the eastern United States. How can these ice crystals manage to survive without having been precipitated out of the cloud or sublimated away?

Problems and Exercises

1. On a regular basis, examine the weather radar map at **weather.noaa.gov/radar/mosaic/DS.p19r0/ar.us .conus.sht\ml**. Then click on any region for a closer, regional view of the precipitation. Would you describe the precipitation intensity as uniform or spotty? This pattern is likely to vary by season and region. How might season, type of precipitation, and geography influence the spatial variability of precipitation within a storm?

2. A particular collector drop has a fall speed of 0.26 m/sec while a smaller, 10 μm droplet directly below falls at a rate of 0.01 m/sec. How long will it take for the two to collide? How far will each of them have fallen prior to collision?

3. The terminal velocities of spherical falling droplets and raindrops are proportional to the square root of their radii.
   a. If a cloud droplet with a radius of 10 μm falls at 0.01 m/sec, how fast would a droplet with a radius of 100 μm fall?
   b. If both droplets are within a cloud, 100 m above the cloud base, how long would it take the two of them to fall to the bottom of the cloud, assuming no growth or diminution in size?

Quantitative Problems

The Web site for this book (**www.MyMeteorologyLab.com**) offers several problems to help you obtain a quantitative understanding of precipitation. We suggest you enhance your comprehension of precipitation by spending a few minutes working on those problems.

Useful Web Sites

**www.ncdc.noaa.gov/oa/climate/research/2008/flood08
.html**
Detailed information on the Midwestern flood of 2008.

www.nws.noaa.gov/radar_tab.php
Composite map showing radar returns across the country.

Click on any region to get a more detailed view of the local
precipitation distribution.

weatheroffice.ec.gc.ca/radar/index_e.html
Canadian radar maps.

www.intellicast.com
Excellent source of information, including map of radar-
derived precipitation amounts for the last 24 hours or week.
To obtain the map, from the main page go to U.S. Weather,
and then click on RADAR. On the next page, click on the
pull-down window called HISTORIC, and then select DAILY
or WEEKLY PRECIPITATION.

lwf.ncdc.noaa.gov/oa/climate/severeweather/rainfall.html
Maps of rainfall records for the 50 states, as well as a number
of reports on significant precipitation.

www.nohrsc.noaa.gov/nsa
Map of U.S. snow cover depth, water equivalent, and other
snow-related parameters.

**weather.unisys.com/surface/sfc_daily
.php?plot=24p&inv=0&t=cur**
Daily U.S. precipitation totals.

www.cocorahs.org
Provides data from the Community Collaborative Rain, Hail
and Snow Network.

MyMeteorologyLab™

The *Understanding Weather and Climate,* Sixth Edition, MyMeteorologyLab™ web site contains numerous multimedia
resources to aid in your study of **Precipitation Processes.**
Visit **www.MyMeteorologyLab.com** to:

- **Review** key chapter concepts.
- **Read** the Pearson eText, *In the News RSS feeds,* and other chapter-specific Web resources.
- **Visualize** challenging topics with Interactive Tutorials, Geoscience Animations, MapMaster
  interactive maps, and Weather in Motion movies and visualizations, and more.
- **Test Yourself** with self-study quizzes.

 TUTORIAL
PRECIPITATION

Use the interactive animations and quizzes in this tutorial to review this chapter's key concepts.

WEATHER IN MOTION

View movies and visualizations of meteorology patterns and processes.

Rain and Flooding
The Importance of Wind Resistance
Global Precipitation
Lake-Effect Snow
Record-Breaking Hailstorm as Seen by Radar

Atmospheric pressure, wind, temperature and moisture patterns are not haphazardly distributed across Earth's surface. Certain patterns of pressure and wind occupy preferred positions across the globe at different times of the year and at different spatial scales. Likewise, it is common to find large areas of fairly uniform temperature and moisture separated from each other by well-defined boundaries called *fronts*. This section of the book examines the causes and distributions of the resultant pressure, wind, temperature, and humidity patterns that dominate much of our daily weather.

Katabatic winds from the East Antarctic ice sheet buffet a hiker near Wright Valley, Antarctica.

8 Atmospheric Circulation and Pressure Distributions

After reading this chapter, you should be able to:

▶ List the terms used to describe the spatial scales of weather phenomena.

▶ Describe the general circulation of the atmosphere in terms of the single-cell model and the three-cell model.

▶ Describe distribution and effects of semipermanent pressure cells.

▶ Explain the distribution of wind and pressure in the upper troposphere.

▶ Explain how the atmosphere affects the movement of ocean waters.

▶ Identify major wind systems such as monsoons; foehn, chinook, and Santa Ana winds; katabatic winds; and sea and land breezes.

▶ Explain sea–air interactions such as El Niño/La Niña, Walker circulation, Southern Oscillation, and Pacific Decadal Oscillation.

During the fall of 2007 the southeastern United States had settled into an unprecedented drought that led to water shortages and related problems, such as increased vulnerability to wildfires. By the end of the year, some parts of the region had accumulated rainfall deficits exceeding 40 cm (17 in.), and over a quarter of the region was classified as being in an area of "exceptional drought" (Figure 8–1). The water level at Lake Lanier, which supplies much of the water for Georgia, Florida, and Alabama, fell so low (Figure 8–2) that it brought urgency to a long-standing legal dispute among the three states. Officials throughout the region imposed mandatory conservation measures as water supplies ran dangerously low (Atlanta had a mere 90-day supply of water in its reservoirs at the peak of the drought). And for good measure, Governor Sonny Perdue of Georgia and some state legislators convened on the capitol steps to pray for an end to the drought.

Though the drought had eased somewhat by the spring of 2008, residents of the region (and elsewhere) had to deal with long- and short-term economic impacts. Water bills were set to increase to make up for lost revenue associated with decreased consumption during the peak of the drought, and the agricultural sector took a significant hit. Even the cost of Christmas trees increased that autumn due to the high cost of keeping them healthy during the drought.

Events such as these typically result from pressure patterns that, once established, persist for unusually long periods of time. Relief comes only when the pressure pattern evolves to permit wetter conditions. In Chapter 4 we saw that atmospheric pressure varies from one place to another, but its distribution is not haphazard. Instead, well-defined patterns dominate the distribution of pressure and winds across the global surface. The largest-scale patterns, called the **general circulation**, can be considered the background against which unusual events occur, such as the drought described above. Likewise, even mundane daily wind and pressure variations can be thought of as departures from the general circulation.

◀ The cracked lake bed of O.C. Fisher Lake in August 2011. A combination of repeated 100 plus degree days and persistent drought has dried up the lake that once spanned over 5400 acres.

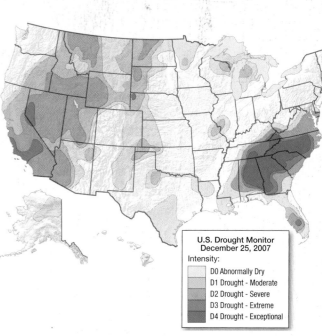

U.S. Drought Monitor
December 25, 2007
Intensity:

	D0 Abnormally Dry
	D1 Drought - Moderate
	D2 Drought - Severe
	D3 Drought - Extreme
	D4 Drought - Exceptional

▲ FIGURE 8-1 Map of conditions across the United States during the peak of the 2007–08 southeastern drought.

▶ **FIGURE 8-2** Dry conditions in 2007 caused a huge lowering of the water level at Lake Lanier, Georgia.

Our first goal in this chapter is to describe dominant planetary wind motions and look at the processes that generate them. In particular, we examine the interrelationships between the winds of the upper and lower atmosphere and the connections that occur at the boundary between the surface of the oceans and the lower atmosphere. We then consider wind and pressure patterns at sequentially smaller spatial and temporal scales. The chapter concludes with a discussion of air–sea interactions.

The Concept of Scale

Those who deal with atmospheric phenomena must be concerned with the notion of scale—in terms of both aerial coverage and time span. Some features of the atmosphere cover large portions of Earth and are maintained over an extensive time period (perhaps for weeks). Such features exist at what is called the **global scale**. Given the size and long duration of these features, you may not hear reference to them on local media weather reports. But as we examine large-scale wind and pressure patterns in this chapter you will see that global scale phenomena provide the background on which all weather and climate events take place.

The existence of high and low pressure patterns over large parts of continents occur at what is called the **synoptic scale**, meaning that they cover hundreds or thousands of square kilometers. Synoptic scale features persist for periods of days to as much as a couple of weeks. These are the most salient features that one sees on daily weather maps, such as the high and low pressure systems shown in Figure 8–3.

▶ **FIGURE 8-3** Sample weather map showing features at the synoptic scale (the high and low pressure systems) and meso scale (outflow boundaries).

Other elements of daily weather operate at the **mesoscale**, which may cover anywhere from just a few square kilometers to hundreds of square kilometers and for periods from as brief as half an hour to perhaps a large part of a day. A localized thunderstorm over western Oklahoma would be an excellent example of a mesoscale event, as would an organized cluster of storms covering several counties. By way of example, notice in Figure 8–3 the two areas marked as "outflow boundaries." Outflow boundaries (described in Chapter 11) are localized areas near regions of heavy precipitation where strong gusts of air flow outward and sometimes create other occurrences of severe weather.

The smallest exchanges of mass and energy operate at the **microscale**, such as those that might cause ripples to form on snow or a sandy beach—or the swirling of smoke emanating from an unattended fry pan. Microscale features are usually covered in more advanced texts.

This chapter examines typical patterns of pressure and wind, beginning with those at the largest scale and progressing to more localized situations.

Single-Cell Model

Scientists have sought to describe general circulation patterns for centuries. As early as 1735 a British physicist, George Hadley (1685–1768), proposed a simple circulation pattern called the **single-cell model** to describe the general movement of the atmosphere. One of his primary goals was to explain why sailors so often found winds blowing east to west in the lower latitudes. (Winds blowing east to west or west to east

are referred to as **zonal winds**; those moving north to south or south to north are called **meridional**[1]). Hadley's idealized scheme, shown in Figure 8–4a, assumed a planet covered by a single ocean and warmed by a fixed Sun that remained overhead at the equator. Hadley suggested that the strong heating at the equator caused a circulation pattern in which air expanded vertically into the upper atmosphere, diverged toward both poles, sank back to the surface, and returned to the equator. Hadley did not think winds would simply move north and south, however. He believed instead that the rotation of Earth would deflect air to the right in the Northern Hemisphere and to the left in the Southern Hemisphere, leading to the east–west surface winds shown in the figure.[2]

Hadley's main contributions were to show that differences in heating give rise to persistent large-scale motions (called *thermally direct circulations*) and that zonal winds can result from deflection of meridional winds. His idea of a single huge cell in each hemisphere was not so helpful, however.

A somewhat more elaborate model does a better—though still simplified—job of describing the general circulation. This **three-cell model** (Figure 8–4b) was proposed by U.S. meteorologist William Ferrel (1817–1891) in 1865.

---

[1]The wind is seldom purely zonal or purely meridional, but instead usually moves in some intermediate direction. In that case we will think of the wind as having both a zonal and a meridional component. It is possible for the two components to be equal (as in a southwesterly wind), but in general one component will be larger than the other.

[2]Hadley lived from 1685 to 1768, before Gaspard de Coriolis (1792–1843) quantitatively described the acceleration due to Earth's rotation. Nonetheless, Hadley had a qualitative knowledge of the Coriolis force and incorporated it into his model.

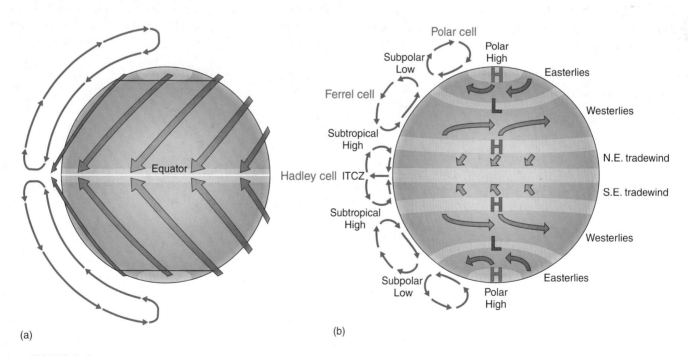

(a)   (b)

▲ **FIGURE 8-4** The single-cell (a) and three-cell (b) models of atmospheric pressure and wind. In the single-cell model, air expands upward, diverges toward the poles, descends, and flows back toward the equator near the surface. In the three-cell model, thermally driven Hadley circulation is confined only to the lower latitudes. Two other cells (more theoretical than real) exist in each hemisphere, the Ferrel and polar cells.

Checkpoint

1. What are zonal winds? Meridional winds?
2. What were two main contributions of Hadley's single-cell model?

The Three-Cell Model

The three-cell model divides the circulation of each hemisphere into three distinct cells: the heat-driven **Hadley cell** that circulates air between the tropics and subtropics, a **Ferrel cell** in the middle latitudes, and a **polar cell**. Each cell consists of one belt of rising air with low surface air pressure, a zone of sinking air with surface high pressure, a surface wind zone with air flowing generally from the high-pressure belt to the low-pressure belt, and an airflow in the upper atmosphere from the belt of rising air to the belt of sinking air. Though more realistic than the single-cell model, the three-cell model is so general that only fragments of it actually appear in the real world. Nonetheless, the names for many of its wind and pressure belts have become well established in our modern terminology, and it is important that we understand where these hypothesized belts are located.

The Hadley Cell

Along the equator, strong solar heating causes air to expand upward and diverge toward the poles. This creates a zone of low pressure at the equator called the **equatorial low**, or the **Intertropical Convergence Zone (ITCZ)**. The upward motions that dominate the region favor the formation of heavy rain showers, particularly in the afternoon. Heavy precipitation associated with the ITCZ is observable on weather maps and satellite images (Figure 8–5). Notice in the figure that the equatorial low exists not as a band of uniform cloud cover but rather as a zone containing many clusters of convectional storms. The ITCZ is the rainiest latitude zone in the entire world, with many locations accumulating more than 200 days of rain each year. Imagine how listless you might

Did You Know?

You might wonder why the general circulation of the atmosphere is approximated by three cells, rather than some other number. These wind and pressure belts that make up the cells arise because of interplay between Earth's rotation rate and the energy gradient between the equator and the poles. If Earth rotated faster, we would expect more belts. Thus, Jupiter, which turns on its axis every 11 hours, has many belts, not just three in each hemisphere.

▶ **FIGURE 8–5** The Intertropical Convergence Zone (ITCZ) is observable as the band of convective clouds and showers extending from northern South America into the Pacific on this satellite image.

feel in such an environment, with hot, humid afternoons giving way to heavy rain showers all year long. The area is also one in which winds can become light or nonexistent for extended time periods. This monotony is the basis for the old nautical term still used today, the *doldrums*.

Within the Hadley cell, air in the upper troposphere moves poleward to the subtropics, to about 20° to 30° latitude. As it travels, it acquires increasing west-to-east motion, primarily because of the conservation of angular momentum (see *Box 8–1, Physical Principles: Problems with the Single-Cell Model*). This westerly component is so strong that the air circles Earth a couple of times before reaching its ultimate destination in the subtropics. In other words, the upper-level air follows a great spiraling path out of the tropics, with zonal motion much stronger than the meridional component. Among other things, this explains why material ejected by tropical volcanic eruptions spreads quickly over a wide range of longitudes.

Upon reaching about 20° to 30° latitude, air in the Hadley cell sinks toward the surface to form the **subtropical highs**, large bands of high surface pressure. Because descending air warms adiabatically, cloud formation is greatly suppressed and desert conditions are common in the subtropics. The subtropical highs generally have weak pressure gradients and light winds. Such conditions exert minimal impact on long-distance travel today. But in preindustrial days when oceangoing vessels depended on the wind, its prolonged absence could be catastrophic. Ships crossing the Atlantic from Europe to the New World risked getting stranded in midocean while crossing the subtropics. Often among the cargo were cattle and horses to be brought to the New World, and legend has it that the crews of stalled ships threw their horses overboard. The jettisoned cargo has lent its name to what we colloquially call the **horse latitudes**.

In the Northern Hemisphere, as the pressure gradient force directs surface air from the subtropical highs to the ITCZ, the weak Coriolis force deflects the air slightly to the right to form the **northeast trade winds** (or simply the *northeast trades*). In the Southern Hemisphere, the north-ward-moving air from the subtropical high is deflected to the left to create the **southeast trade winds**. Notice that the trade winds are fairly shallow. Moving upward through the troposphere, the easterly motions weaken and are eventually replaced by westerly motion aloft. Together, the subtropical highs, the equatorial low, the trade winds, and upper-level westerly motions form the Hadley cells. Because it is produced thermally, the Hadley circulation is strongest in the winter season, when temperature gradients are strongest.

The Ferrel and Polar Cells

According to the three-cell model, the Hadley cell accounts for the movement and distribution of air over about half of Earth's surface. Immediately flanking the Hadley cell in each hemisphere is the Ferrel cell, which circulates air between the subtropical highs and the **subpolar lows**, or areas of low pressure. On the equatorial side of the Ferrel cell, air flowing poleward away from the Northern Hemisphere subtropical high undergoes a substantial deflection to the right, creating a wind belt called the **westerlies**. In the Southern Hemisphere, the pressure gradient force propels the air southward, but the Coriolis force deflects it to the left—thus producing a zone of westerlies in that hemisphere as well. Unlike the Hadley cell, the Ferrel cell is envisioned as an indirect cell, meaning that it does not arise from differences in heating, but is instead caused by turning of the two adjacent cells. Imagine three logs placed side by side, touching one another. If the two outer logs are turned in the same direction, friction will set the middle one in motion. Thus, the Ferrel cell shows the same kind of overturning as the other cells, but for different reasons.

In the polar cells of the three-cell model, surface air moves from the **polar highs** to the subpolar lows. Like the Hadley cells, the polar cells are considered thermally direct circulations. Compared to the poles, air at subpolar locations is slightly warmer, resulting in low surface pressure and rising air. Very cold conditions at the poles create high surface pressure and low-level motion toward the equator. In both hemispheres, the Coriolis force turns the air to form a zone of **polar easterlies** in the lower atmosphere.

The Three-Cell Model vs. Reality: The Bottom Line

Do the wind and pressure belts of the three-cell model adequately describe real-world patterns? The answer is: sort of. We have already seen that the ITCZ is real enough to be observed from space and that many deserts exist in their predicted locations. Furthermore, the trade winds are the most persistent on Earth. We would have to say that the Hadley circulation provides a good account of low-latitude motions. On the other hand, the Ferrel and polar cells are not quite as well represented in reality, though they do have some manifestation in the actual climate.

With regard to surface winds, much of the middle latitudes experience the strong westerly winds depicted by the model, especially in the Southern Hemisphere. Of course, local conditions often override this tendency (in fact, much of the central United States is dominated by a southerly flow during the summer). It is even more difficult to observe a persistent pattern of polar easterlies in the overall wind regime. They emerge in long-term averages, but are not a prevailing wind belt as the trades are.

With regard to upper-level motions, the three-cell model is not realistic at all. For example, where the Ferrel cell implies easterly motion in the upper troposphere, there is overwhelming westerly wind. Moreover, large overturning cells do not exist outside of the Hadley zones. Thus, the three-cell model mainly provides a starting point for a more detailed account. But perhaps its failures aren't surprising, given that it doesn't consider land–ocean contrasts or the influence of surface topography, two factors that surely ought to influence planetary winds and pressure.

Problems with the Single-Cell Model

Hadley's single-cell circulation model is a simple one based on the differential heating of Earth's surface. Contrary to Hadley's description, this thermally driven circulation occupies only the range of latitudes between about 30° N and 30° S. Why doesn't the single-cell model cover the entirety of each hemisphere instead of just a zone near the tropics? The answer boils down to the *conservation of angular momentum*.

Just as an object moving in a straight line has "linear" momentum, an object following a curved path has "angular" momentum. If we start a weight twirling at the end of a rope, as in Figure 1, we see the rope sweeping out a growing pie-shaped angle, hence the term *angular momentum*. The amount of angular momentum obviously increases with increasing mass (imagine trying to stop a very heavy weight). Angular momentum also increases with speed, because a larger area is swept out each second. Lastly, angular momentum increases with increasing radius, again because a larger area is swept out per unit time.

(a)

(b)

▲ **FIGURE 1** Twirling a weight at the end of a rope (a) illustrates the conservation of angular momentum. As the man pulls the rope closer to his body, shortening the length of the rope (b), the speed of rotation increases to maintain a constant angular momentum.

Putting this together, we can say that angular momentum is the product of mass, speed, and radius, or

$$A = mvr$$

where A is angular momentum, m is mass, v is speed, and r is the radius of rotation.

Checkpoint

1. How do rainfall and wind vary across a Hadley cell, from the equator to about 20° to 30° latitude? Explain.
2. What are the shortcomings of the three-cell model?

Semipermanent Pressure Cells

The three-cell model provides a good beginning for describing the general distribution of wind and pressure, but the real world is not covered by a series of belts that completely encircle the globe. Instead, we find a number of alternating **semipermanent cells** of high and low pressure, as shown in Figure 8–6. They are called *semipermanent* because they undergo seasonal changes in position and intensity over the course of the year. Some of these cells result from temperature differences, and others from dynamical processes

(meaning that they are related to the motions of the atmosphere). Among the most prominent features in the Northern Hemisphere during winter (a) are the **Aleutian** and **Icelandic lows**—over the Pacific and Atlantic Oceans, respectively—and the **Siberian high** over central Asia. In summer (b), the best-developed cells are the **Hawaiian** and **Bermuda–Azores highs** of the Pacific and Atlantic Oceans and the **Tibetan low** of southern Asia.

The size, strength, and locations of the semipermanent cells undergo considerable change from summer to winter. During the winter, a strong Icelandic low occupies a large portion of the North Atlantic, while the Bermuda–Azores high appears as a small, weak anticyclone. During summer, the Icelandic low weakens and diminishes in size, and the Bermuda–Azores high strengthens and expands. Even more dramatically, the Siberian high of the winter in interior Asia gives way to the Tibetan low of summer. As we will see later in this chapter, the seasonal shift in the distribution of semipermanent cells plays a major

If angular momentum is conserved, then *A* is constant. This simple statement has some serious consequences. Suppose, for example, that we decrease the radius by pulling on the rope. As *r* decreases, the speed must increase, even though we do not twirl harder. The same phenomenon allows an Olympic diver to twist or flip in the air. After leaving the diving board, the diver's angular momentum is fixed. When she tucks, bringing legs and arms close to her center of mass, she spins rapidly. At the end of the dive, she straightens out, slowing her spin rate to enter the water vertically without much rotation, and is rewarded with a high score (hopefully). It is remarkable to think that divers cannot adjust their angular momentum in the air but rather must leave the board with exactly what is needed for a particular dive.

Now consider the rotating Earth. At the equator, any fixed point on the surface travels a distance equal to the planetary circumference of 40,000 km (24,000 mi) over a 24-hour period. The same applies to a parcel of air that is stationary relative to the surface. At higher latitudes the circumference is smaller, so a parcel at rest travels a shorter distance in a day's time. For example, at 40° the speed is about 31,000 km per day.

Now let's see what happens as a fixed mass of air initially at rest relative to the surface moves northward from the equator. Traveling poleward, *r* decreases, so *v* must increase if angular momentum is conserved. At 40° latitude, *v* increases about 30 percent to 52,000 km/day. In other words, the parcel is moving eastward at 52,000 km/day as it travels around the axis of rotation. At that latitude, the surface is moving at only 31,000 km/day, so the parcel is moving over the surface. Standing on the ground, we would observe a wind speed of 21,000 km/day (52,000 minus 31,000). This is a huge value, equivalent to a wind speed of 243 m/sec, or 547 mph!

According to the single-cell model, upper-level air should flow all the way from the equatorial region to the poles. But if that were the case, the conservation of angular momentum would send the poleward-moving air eastward at unimaginably high speeds—far greater than what we experience in even the most severe tornadoes. Physical considerations prevent the air from maintaining such extreme flows, because even a slight random disruption would cause the wind to break down into numerous eddies with strong north–south motions.

The single-cell model involves another difficulty, also related to conservation of

angular momentum. To a large degree, the momentum possessed by the entire Earth–atmosphere system is unchanging—nothing stands outside the planet giving it a shove to increase its spin rate or to slow it down. (We are ignoring gravitational variations caused by roving planets and other small factors.) But angular momentum is transferred between Earth and atmosphere by friction whenever the air moves past the surface. Where there is westerly (from the west) wind, the atmosphere transfers momentum to the surface as it "pushes" in the direction of rotation. Similarly, easterly (from the east) winds imply a transfer of momentum from the surface to the atmosphere.

The single-cell model calls for easterly surface winds everywhere. If this were to occur, the planet would everywhere supply westerly momentum to the atmosphere, which would bring the atmosphere to rest within a week or two. To avoid this problem, easterly winds must be balanced by westerly winds somewhere else. More precisely, averaged over the whole Earth, momentum transferred from the surface to the atmosphere must equal momentum transferred from the atmosphere to the surface. Thus, a single-cell circulation would not be possible on Earth—or on any other rotating planet.

role in one of Earth's most important circulation patterns—the monsoon of southern and southeastern Asia.

We mentioned earlier that the Hadley cell is fairly easy to see in the real world, with the ITCZ appearing over much of the equatorial regions and the subtropical highs existing over much of the subtropics. But as you can also see in Figure 8–6, the subtropical highs exist primarily over the oceans (as the Hawaiian and Bermuda–Azores highs over the Northern Hemisphere) and not over land. Despite the absence of pronounced high sea level pressure over the subtropical land masses, the air in the middle troposphere does undergo sinking motions that inhibit cloud formation and promote desert conditions. The Sahara Desert of northern Africa, the interior desert of Australia, and the deserts of the southwestern United States and northwest Mexico clearly reflect this sinking process.

As the solar declination changes seasonally, so does the zone of most intense heating. Knowing that the Hadley cells

are thermal, we might expect the associated pressure and wind belts to move seasonally, and indeed they do. Although with a lag of several weeks, the ITCZ, subtropical highs, and trade winds all follow the "migrating Sun." This movement has a major impact on many of the world's climates—and on the people who inhabit them.

For example, many areas along the equator are dominated by the ITCZ year round and experience no dry season. Areas located near the poleward margins of the ITCZ, however, are subject to brief dry seasons as the zone shifts equatorward. Compare, for example, the average rainfall patterns for Iquitos, Peru (3° S), and San Jose, Costa Rica (9° N). Iquitos is located close enough to the equator so that it is perennially influenced by the ITCZ, but San Jose has a relatively dry period from January to March, when the low-pressure system is displaced to the south.

Similarly, some areas located on the equatorward edge of the subtropical highs are dry for most of the year, except

(a) January

(b) July

▲ FIGURE 8-6 Sea level pressure for January (a) and July (b).

briefly when the system shifts poleward during the summer. This condition exists in the Sahel of Africa, the region bordering the southern margin of the Sahara Desert (Figure 8–7a). Unlike the Sahara, which is dry all year, the Sahel normally experiences a brief rainy period each summer as the ITCZ enters the region (b). During the rest of the year, the descending air of the Hadley cell leads to dry conditions (c).

The migration of the Hadley cell has long supported a traditional lifestyle in which African herders followed the northward and southward shifting rains. During the 1960s and 1970s, the population of the region increased dramatically, which led to overgrazing and set the stage for catastrophe when a multiyear drought hit the area. Millions of head of livestock died from lack of food and water, which in turn led

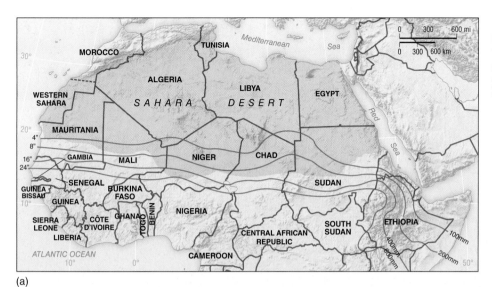

◄ **FIGURE 8-7** The Sahel is a region of Africa bordering the southern Sahara Desert (a). During the summer (b), the ITCZ usually shifts northward and brings rain to the region. For much of the year, the ITCZ is located south of the Sahel, and the region receives little or no precipitation (c).

to the deaths of tens of thousands of people. During the early 1990s, low rainfall again plagued the Sahel—this time along the eastern part of the region in Somalia. Coupled with the existing political and social instability that eventually gave rise to civil war in Somalia, the drought led to the starvation of an estimated 300,000 people. Thus, the existence of these cells—and variations in the way they develop during unusual years—are more than mere abstractions. They have real-world ramifications for millions of people.

The Upper Troposphere

In Chapter 4 we saw that pressure decreases more rapidly with altitude where the air is cold. We also know that temperature in the lower troposphere generally decreases from the subtropics to the polar regions. These two principles are critical in understanding the distribution of wind and pressure in the upper troposphere.

Figure 8–8 maps the global distributions of the mean height of the 500 mb surface (a convenient level representing conditions in the middle troposphere) for January (a) and July (b). In both months, the height of the 500 mb level exhibits a strong tendency to decrease toward the polar regions due to the lower temperatures found at higher latitudes. In January, the average height of the 500 mb surface over much of the southern United States is about 5670 m (18,600 ft), while over northern Canada it decreases to less than 5300 m (17,400 ft). A similar but less extreme change occurs in July as well.

Three features stand out from the maps in Figure 8–8. First, for both January and July the 500 mb heights are greatest over the tropics and decrease with latitude. Second, the gradient in height is greater in the hemisphere experiencing winter (the Southern Hemisphere in July and Northern Hemisphere in January). Third, at all latitudes the height of the 500 mb level is greater in the summer than during the

(a) January

(b) July

| 5970 | Height contours |

▲ **FIGURE 8–8** The mean heights of the 500 mb levels (in meters) for January (a) and July (b). The pattern is mostly zonal with decreasing heights toward the poles.

winter. All three of these features result from the general distribution of temperature in the lower-middle atmosphere; areas of warm air have greater 500 mb heights.

TUTORIAL
UPPER-LEVEL PRESSURE AND WINDS

Use the tutorial to observe the influence of temperature on upper-level pressure patterns, change the amplitude of troughs and ridges, and observe air flow within stationary and moving trough and ridge patterns.

Westerly Winds in the Upper Atmosphere

Recall from Chapter 4 that height differences correspond to pressure differences and that when the 500 mb surface slopes steeply a strong pressure gradient force exists. We can therefore infer from the 500 mb maps that there is always a pressure gradient force across the middle latitudes trying to push the air toward the poles. Of course, in the absence of friction, the winds do not blow poleward, but rather blow parallel to the height contours, from west to east. The pressure gradient force is strongest in winter (the height contours are closely spaced), so the upper-level westerlies are strongest in winter. What does this mean to you? For one thing, it explains why most midlatitude weather systems migrate from west to east. In other words, it tells us why a storm over Chicago might find its way over the East Coast a day or two later, but seldom if ever does such a storm make the reverse trip.

The predominance of westerly winds in the upper troposphere also affects aviation. For example, a commercial aircraft going from Chicago to London has an estimated flight time of about 7.5 hours, while the return trip normally takes an hour longer because it must overcome headwinds. The difference in flight time would be even greater were it not for the fact that airlines route their planes to take advantage of tailwinds and avoid headwinds.

Wind speeds generally increase with height between the surface and the tropopause. Partly this is because of decreasing friction, but more importantly, the pressure gradient force is typically stronger at high altitudes. As illustrated in Figure 8–9, the surfaces representing the 900, 800, and 700 mb levels all slant downward to the north, but not by the same amount. Higher surfaces slope more steeply, which means that the pressure gradient force is greater. You may recall from Chapter 4 that the pressure gradient force is directly proportional to slope, without regard to density.

But why do those higher surfaces slope more steeply? The air is warmer at point B, so the layer of air from 900 mb to 1000 mb is thicker at point B than at point A. Similarly, the thickness from 800 mb to 900 mb is greater at point B. In other words, the height change from B' to B'' is greater than the change from A' to A'' It must be, therefore, that the 800 mb surface slopes downward more than the 900 mb surface ($B'' - A''$ is greater than $B' - A'$). The difference in heights between successive surfaces continues to increase upward, leading to stronger winds.

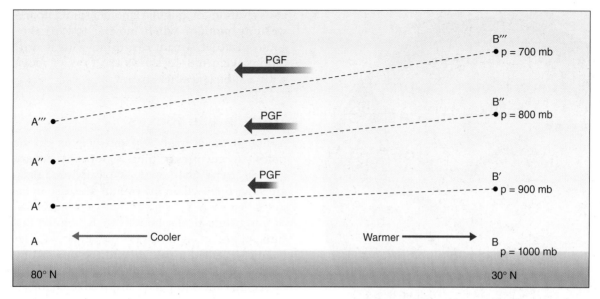

▲ **FIGURE 8-9** Latitudinal temperature gradients cause pressure surfaces to slant poleward. In this example, we assume a constant gradual decrease in temperature with latitude. The intensity of the pressure gradient force remains constant from one latitude to another but increases with altitude.

The Polar Front and Jet Streams

The gradual change in temperature with latitude depicted in Figure 8–9 does not always occur in reality. Instead, areas of gradual temperature change often give way to narrow, strongly sloping boundaries between warm and cold air. One such boundary, the **polar front**, is shown in Figure 8–10.

Outside of the frontal zone, the changes in temperature with latitude are gradual (as they were in Figure 8–9), and the slopes of the 900, 800, and 700 mb levels respond accordingly. But within the front, the slope of the pressure surfaces increases greatly because of the abrupt horizontal change in temperature. With steeply sloping pressure surfaces there is a strong pressure gradient force, resulting in the **polar jet stream**. Thus, we see the jet stream as a consequence of the polar front, arising because of the strong temperature gradient. At the same time, the jet stream reinforces the polar front. In Chapter 10, we will see that a jet stream is necessary to maintain the temperature contrast across the front.

Jet streams can be thought of as meandering "rivers" of air, usually 9 to 12 km (30,000 to 40,000 ft) above sea level (Figure 8–11). Their wind speeds average about 180 km/hr (110 mph) in winter and about half that in summer, though peak winds can exceed twice these values. Like rivers on land, jet streams are highly turbulent, and their speeds vary considerably as they flow. Unlike rivers on land, they have no precisely defined banks. Furthermore, a single jet stream will often diverge at some point and fork off into two distinct jets. Thus, the locations and boundaries of these features on weather maps are often difficult to accurately pinpoint.

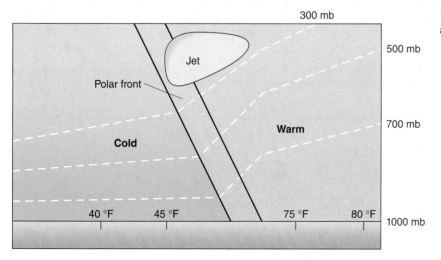

◄ **FIGURE 8-10** The polar jet stream is situated above the polar front near the tropopause.

conservation of angular momentum imparts ever-growing westerly motion. When moving toward the northeast, the subtropical jet stream can bring with it warm, humid conditions. Figure 8–12 shows the flow of moisture associated with a subtropical jet stream.

Troughs and Ridges

Although on average 500 mb heights decrease toward the poles, at any given time significant departures from the general trend will exist. Typically undulations, or waves, are superimposed on the overall decrease in height toward the poles. Figure 8–13 is a cartoon view of this, showing an axis of low height in the middle of the United States, flanked by "mountains" of high heights on either side. The valley of low heights is called a *trough*, and the upward bulges are called *ridges*. Also shown on the diagram are height contours—notice that they, too, have a wavelike character. The air flows parallel to the contours, so there is wavelike motion to the airstream as well.

▲ **FIGURE 8–11** Jet streams are localized areas of intense winds in the upper troposphere. We often see both a polar jet stream and subtropical jet stream.

TUTORIAL
UPPER-LEVEL PRESSURE AND WINDS

Use the tutorial to explore the concept of upper-level troughs and ridges and their representation on weather maps.

The polar jet stream greatly affects daily weather in the middle latitudes. Nearer the equator is another prominent jet, the **subtropical jet stream**, associated with the Hadley cell. As the upper-level air flows away from the ITCZ, the

Figure 8–14 shows simplified contour maps depicting the relationship between 500 mb heights and ridges and troughs. No east–west height changes (no trough, no ridge) occur in (a), and the flow is completely zonal. In (b), on the other

▶ **FIGURE 8–12** The subtropical jet stream appears in this infrared satellite image as the band of cloudiness extending from Mexico through Florida. The flow in this subtropical jet stream is from southwest to northeast. Note that redder colors generally indicate thicker cloud cover.

▲ **FIGURE 8–13** A hypothetical drawing of the 500 mb surface. Heights decrease from south to north but also rise and fall through the ridges and trough. Vertical changes are highly exaggerated in the figure. Actual height changes are very small compared to the size of the continent.

hand, a trough is in the midsection of the country. Going from point 1 to 2, there is a height decrease from 5610 m to 5580 m. Going from 2 to 3, heights increase again. In other words, going from 1 to 3 requires us to cross a valley (trough) of low heights. We see that height contours are displaced toward the equator in troughs and toward the pole in ridges. We also see that air winding its way poleward around ridges and equatorward around troughs will have a meridional component as well as a zonal component. In fact, when waves are pronounced, we say the flow is "meridional," as opposed to "zonal" when the flow is nearly all westerly.

Rossby Waves

We have seen that ridges and troughs give rise to wavelike flow in the upper atmosphere of the middle latitudes. The largest of these are called *long waves*, or **Rossby waves**.[3] Usually, there are anywhere from three to seven Rossby waves circling the globe. Like other waves, each has a particular *wavelength* (the distance separating successive ridges or troughs) and *amplitude* (its north–south extent). Though Rossby waves often remain in fixed positions, they also migrate west to east (Figure 8–15), or on rare occasions from east to west.

Rossby waves undergo distinct seasonal changes from summer to winter. They tend to be fewer in number, have longer wavelengths, and contain their strongest winds during winter. The latter two characteristics—wind speed and wavelength—affect the rate at which Rossby waves migrate downwind (see *Box 8–2, Physical Principles: The Movement of Rossby Waves*).

---

[3]Named for Carl Gustaf Rossby, who contributed much pioneering research on upper-level air flow in the early 1900s.

▲ **FIGURE 8–14** Troughs occur in the middle troposphere where the 500 mb height contours dip equatorward. In (a), positions 1–3 have the 500 mb level at 5610 m. Farther to the south, at positions 4–6, the 500 mb level is at 5640 m. In (b), the contour lines are in the same position over the East and West coasts as they were in (a), but they shift equatorward over the central portion of the continent. Thus, positions 2 and 5 have lower pressure than the areas east and west of them. The zone of lower pressure over the central part of the continent is a trough.

 TUTORIAL
UPPER-LEVEL PRESSURE AND WINDS

Use the tutorial to observe air flow relative to stationary and moving Rossby waves and see how these waves can migrate and change their form.

Rossby waves exert a tremendous impact on day-to-day weather, especially when they have large amplitudes. They are capable of transporting warm air from subtropical regions to high latitudes or cold polar air to low latitudes. Because upper-level air tends to change temperature only slowly in the absence of strong vertical motions, Rossby waves can bring anomalous temperatures to just about any place within the middle to high latitudes. This is illustrated by Figure 8–16, which shows a strong Rossby wave over North America on

8–2 PHYSICAL PRINCIPLES

The Movement of Rossby Waves

Like all atmospheric phenomena, the movement of Rossby waves results not from mere chance, but from the combined actions of numerous physical forces. Three factors determine the rate at which a Rossby wave propagates: (1) the westerly component of its internal wind speed, (2) its latitudinal position, and (3) its wavelength. In particular, Rossby waves with vigorous winds and shorter wavelengths move most rapidly, as indicated by the formula

$$C = U - bL^2/4\pi^2$$

where C is the speed at which the wave propagates downwind (m/sec); u is the westerly component of the wind speed within the wave (m/sec); b is a function of latitude equal to 1.6×10^{-11} m^{-1} sec^{-1} at 45°, and L is the wavelength (m).

The rates of downwind migration for Rossby waves at 45° latitude with various wavelengths and wind speeds are presented in Table 1. As you can see, a wave having a westerly wind speed of 20 m/sec (45 mph) and a wavelength of 3000 km advances 60 percent faster than does one with the same speed but a wavelength of 5000 km (16 m/sec versus 10 m/sec).

For a given wind speed and latitude, there is some particular wavelength, L_{crit} at which the waves do not migrate at all. Waves longer than this critical value actually migrate from east to west and are said to be *retrograding waves*. By rearranging the equation and setting $C = 0$ (that is, by assuming the waves are stationary), we can determine the critical wavelength:

$$L_{crit} = 2\pi \sqrt{u/b}$$

TABLE 1

Rates of Downwind Rossby Wave Migration at Latitude 45° N or 45° S

Wavelength	20 m/sec Wind Speed	40 m/sec Wind Speed
3000 km	16 m/sec wave speed	36 m/sec wave speed
5000 km	10 m/sec wave speed	30 m/sec wave speed

For example, at 45° a Rossby wave with a jet-stream speed of 40 m/sec will migrate upwind if its wavelength is more than about 10,000 km. This, however, is a particularly long wavelength. At high wind speeds, retrograding motion occurs only for exceptionally long waves, which are quite rare. For lower wind speeds, the critical wavelength is less (only 5000 km for wind at 10 m/sec). As it happens, shorter waves (with lower wind speeds) are more common. Thus, when westward movement appears, it tends to occur when upper-level winds are weak, not strong.

▲ **FIGURE 8–15** A sequence of 500 mb maps showing the migration of Rossby waves at 24-hour intervals.

(a)

(b) Minimum temperatures (°F)

▲ **FIGURE 8–16** Rossby waves in the upper atmosphere can advect cold or warm air from one location to another (a). On September 22, 1995, such a wave brought very mild conditions to interior Alaska, while southward-moving air brought record-breaking low temperatures to much of the central United States (b).

September 22, 1995. Record-breaking low temperatures for the date were observed over much of the central United States as the wave brought cold air from the far north. Farther upwind, a southwesterly flow brought mild air to the extreme northwest of North America, with Fairbanks, Alaska, basking in temperatures in the mid-20s Celsius (mid-70s Fahrenheit).

In addition to redistributing cold or warm air, Rossby wave patterns have a less obvious impact on local weather. Changes

The Dishpan Experiment

Although Rossby waves are the largest of the atmospheric waves, other swirling motions of varying sizes likewise exist. Why this complexity? At the simplest level, the behavior of the upper atmosphere is the inevitable result of three factors: (1) the unequal heating of the atmosphere from the equator to the poles, (2) the rotation of the planet, and (3) the inherently turbulent nature of the atmosphere.

To illustrate the interaction of these three, we can reproduce the migrating waves and eddy motions of the upper atmosphere with a relatively simple piece of hardware—a pan of water that rotates at a constant speed with a cooling of the fluid near the center and warming along the edge (Figure 1). The "dishpan experiment" simulates the rotating Earth with a surplus of net incoming radiation at low latitudes, and a net deficit closer to the poles. Even this very simple exercise yields motions of the fluid that in many ways resemble those of the upper troposphere.

Long waves form in the pan, resembling atmospheric Rossby waves. Superimposed on the long waves are smaller-scale eddies similar to smaller flows on Earth. Changes in the speed of rotation or the differential heating between the edge and center of the pan cause observable changes in the waves and eddies, with more extreme differences in heating and slower rotation rates leading to an increase in the amplitude of large waves at the expense of smaller-scale eddies. This implies that the

▲ **FIGURE 1** The pattern of eddies of different size of a "dishpan experiment."

Direction of rotation

oscillations in the atmosphere represent an inherent characteristic of any fluid (liquid or gaseous) on a rotating surface with spatially varying inputs of heat. Such observations are not restricted to simple dishpan experiments; elaborate computer models that simulate the motions of the atmosphere reveal similar patterns.

---

in the flow along the wave lead to *divergence* and *convergence*. When air in the upper atmosphere diverges (or spreads out), it draws air upward from below, causing adiabatic cooling. Thus, divergence in the upper atmosphere can serve as a mechanism for cloud development and precipitation. Convergence in the upper atmosphere has the opposite effect of forcing air downward and inhibiting cloud formation. We will return to this concept in Chapter 10. (*Box 8–3, Physical Principles: The Dishpan Experiment,* elaborates on the nature of Rossby waves.)

Checkpoint

1. What are jet streams?
2. How do jet streams form, and how can they affect the weather?

The Oceans

As we have seen, the movement of the atmosphere is strongly influenced by the input of heat from the surface. We have also seen that land and water undergo differential rates of heating and cooling; in part this is because of the vertical and horizontal motions of the water's surface. In this section, we take a close look at how the atmosphere affects the movement of oceanic waters. Later in this chapter, we look at some of the mutual interactions in the ocean–atmosphere systems.

Ocean Currents

Ocean currents are horizontal movements of surface water that are often found along the rims of the major basins. These currents, discussed briefly in Chapter 3, have a great impact on the exchange of energy and moisture between the oceans and the lower atmosphere. In many instances their effect on climate is conspicuous, with warm ocean currents, for example, favoring the existence of warm, humid air.

Ocean currents are driven by winds in the lower atmosphere that exert a drag on the water. Contrary to what you might expect, the surface water moves not in the same direction as the wind, but at an angle of 45° to the right of the air flow in the Northern Hemisphere (to the left in the Southern Hemisphere). Furthermore, neither the direction nor the speed of the current is uniform with depth. The current turns increasingly to the right (in the Northern Hemisphere) and decreases in speed at greater depths. At about 100 m below the surface, the direction of the current approaches 180° to the direction of the wind, and the current dies out. Figure 8–17 illustrates this pattern, called the **Ekman spiral**.

The movement of the current at the surface corresponds well to the global wind patterns described earlier in this chapter, as shown in Figure 8–18. Let's have a look at the surface currents of the North Atlantic, whose pattern is quite typical of those in the other major oceans. Just north of the equator, the easterly trade winds drag the surface water westward as the **North Equatorial Current**. Upon reaching South America, most of the westward moving water turns north, but some is deflected to the south toward the equator. A similar pattern appears in the Southern Hemisphere, where a portion of the **South Equatorial Current** is deflected northward to the equator. The water from the North and South Equatorial Currents converges and piles up in the western equatorial Atlantic and creates the eastward-moving **Equatorial Countercurrent**.

In the Northern Hemisphere, most of the North Equatorial Current reaching the South American coast turns northward to form the warm **Gulf Stream** (Figure 8–19). Near 40° N, the westerlies force the current to the east, where it becomes the **North Atlantic Drift**. The current remains warm as it flows

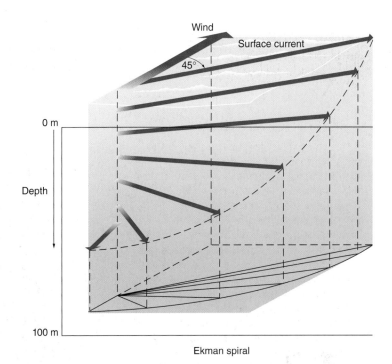

▲ **FIGURE 8–17** The Ekman spiral. Surface currents flow at an angle 45° to the right (in the Northern Hemisphere) of the winds that drive them and continue to shift clockwise as their speed decreases. At a depth of about 100 m, the current approaches the opposite direction of the surface current and begins to die out.

toward northern Europe, which makes winter conditions there unusually mild for those latitudes. Even the Scandinavian countries have surprisingly warm winter temperatures, considering their far northerly position. The North Atlantic Drift gradually cools and becomes the cold **Canary Current** as it turns southward.

At corresponding middle latitudes, temperatures are considerably warmer over the western part of the Atlantic

▲ **FIGURE 8–18** The ocean currents of the world with the approximate location of some of the major global wind systems.

▶ **FIGURE 8-19** The Gulf Stream flows in a complex pattern with eddies of different size superimposed, as shown by the infrared satellite image.

than they are over the east. Thus, for example, the average temperature off the New Jersey shore is about 8 °C (14 °F) warmer than off northern Portugal. The effect of a cold ocean current is also seen along the beaches of California, where even during the summer the water temperature seldom rises much above 22 °C (72 °F).

Finally, the cold **Labrador Current** that flows southward along the Maritime Provinces of Canada is fed by the **East** and **West Greenland Drift**.

Upwelling

In addition to the major ocean currents that circulate large masses of water horizontally, localized vertical motions of ocean water have a bearing on weather and climate. Because solar radiation is mostly absorbed in the uppermost layer of the ocean, the surface is usually warmer than the waters below. But strong *offshore* (blowing from land to ocean) winds along a coastal region sometimes drag the warmer surface waters seaward, which draws up cooler waters from below to take their place. This process, called

Did You Know?

The Gulf Stream, the largest of all the ocean currents, transports about 100 times as much water per unit of time as do all the world's rivers put together.

upwelling, greatly influences sea surface temperatures (and thus weather and climate) over the eastern portions of the major oceans.

Nowhere is upwelling better illustrated than along the western coast of South America. The equatorial coast of Colombia is one of the rainiest places in the world, with average annual precipitation approaching 700 cm (290 in.). But nearby, between the latitudes of about 7° and 30° S, lies the world's driest desert—the Atacama. At the heart of this desert is Arica, Chile (18° S), which does not have a single month in the year that averages as much as 1 mm (0.04 in.) of precipitation—and most years receives no precipitation at all! This dryness is due largely to the upwelling of cold water along the coast, because cold water chills the lower atmosphere and favors the development of stable air.

The same process occurs along the central California coast during the summer (Figure 8–20). Air circulating out of the Hawaiian high sets up an Ekman spiral, with a net flow of water away from the coast. Upwelling is particularly strong just north of the San Francisco Bay region, making the water there particularly cool. Beachgoers in southern California are also familiar with the occasional effects of upwelling. Episodes of hot, dry Santa Ana winds (discussed later in this chapter) bring huge crowds to the beaches. But although the high air temperatures and clear skies are great for sunbathing, the offshore winds lead to strong upwelling that creates low water temperatures and keeps many people out of the water.

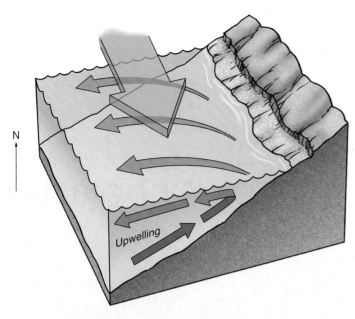

▲ **FIGURE 8–20** Upwelling can occur if a wind blows parallel to a coast. In this example, a wind blowing from north to south in the Northern Hemisphere propels surface water offshore. Water from below rises to replace the displaced surface water.

Checkpoint

1. How are winds and ocean currents related?
2. What are the Gulf Stream and the North Atlantic Drift, and how do they affect northwestern Europe?

Major Wind Systems

Monsoons

We know that several large regions of the world undergo changes in mean pressure between summer and winter. The seasonal oscillation between high- and low-pressure cells is nowhere more evident—or more dramatic in its effects—than over Earth's largest continent, Asia. The great size of this continent by itself would foster strong continentality (described in Chapter 3), but the presence of the Himalaya Mountains (Figure 8–21) enhances the effect in two ways: It imposes a barrier that blocks the northward and southward flow of moisture, and it alters the flow of upper-level winds that influence surface conditions.

Figure 8–22 illustrates the seasonal reversal in surface winds that characterizes the **monsoon** (from the Arabic word meaning "season") of southern Asia. Although the term is often erroneously associated only with the heavy summer rains that occur over southern Asia, *monsoon* refers to the climatic pattern in which heavy precipitation alternates with hot, dry conditions on an annual basis. During January (a), the winds generally flow southwestward toward the Indian Ocean from the southern Himalayas. The descending air is compressed and warmed, leading to dry conditions over most

of India and Southeast Asia. But offshore flow does not arise from the Tibetan high, as might be surmised from the sea level map in Figure 8–6, because the Tibetan high is quite shallow and the air cannot cross the Himalayas into southern India. Air flows down the southern slope of the Himalayan Mountains due to more rapid cooling over land than over the Indian Ocean. This flow is enhanced by a portion of the jet stream flowing along the southern flank of the mountain range. Interactions between the jet stream and mountains cause convergence in the upper troposphere, which leads to sinking air that gets forced southward.

The situation changes abruptly during late spring or early summer, when heating of the continent contributes to a reversal in the wind direction at both low and high levels. Aloft is an easterly jet stream, with often divergent motion promoting uplift. At low levels, onshore flow occurs, bringing warm, moist, and unstable air from the Indian Ocean to the southern part of the continent, where it rises orographically and by convection as it passes over the hot surface. Cloud formation is further enhanced by a much stronger orographic effect as the air ascends the southern slopes of the Himalayas.

The combination of moist air and strong uplift produces precipitation in amounts unimaginable to inhabitants of more moderate climates. The heavy rainfall in northeastern India is enhanced even further by the movement of **monsoon depressions** (or monsoon lows), areas of low pressure superimposed in the southeasterly air flow out of the Bay of Bengal. Consider, for example, the average distribution of monthly precipitation at Cherapunji, India (Figure 8–23). Rainfall amounts are generally low during the winter months, but by early summer mean values approach 300 cm (125 in.) *per month!* Keep in mind that this extreme climate exists in a very heavily populated area of the world, affecting the lives of millions of people.

The southeastern part of North America is geographically similar to Asia. Like Asia, it lies to the north of a warm water body, the Gulf of Mexico, and it undergoes a seasonal reversal of the surface winds due to warming and cooling of the land mass. But despite this similarity, the southern United States does not have a strong monsoon climate. Part of the reason for this is that the smaller size of North America results in weaker oscillations in pressure than those of Asia. More important, however, is the fact that the southern United States does not have a major east–west mountain chain comparable to the Himalayas. In fact, the area is essentially flat, except for the relatively low southern Appalachians that extend southwest to northeast. The Appalachians do not create a strong orographic barrier to the southerly wind flow of summer, and therefore they do not promote the extreme rainfall seen in south Asia. They also do not block the passage of winter storms from the north, as do the Himalayas, nor do they promote persistent upper-level convergence. With all of that, it is easy to see why winter is hardly a dry season in eastern North America.

The desert of the southwestern United States experiences what residents often call the *Arizona monsoon.*[4] In spite of

---

[4]Also called the *Southwest monsoon* or the *Mexican monsoon.*

▶ **FIGURE 8–21** Computer-generated images of the Himalayas and Asia. The Himalayan mountains—the tallest range in the world—play a major role in the formation of the monsoon climate.

(a)

(b)

its name, the Arizona monsoon bears little resemblance to the real thing. It occurs each summer when the desert southwest heats considerably, creating low pressure over Arizona and extreme southeastern California (Figure 8–24). Warm, moist air flowing in from the south and southwest and strong surface heating can lead to heavy convection to trigger scattered thundershowers. Unlike the intense precipitation of the Asian

monsoon, that of the Arizona monsoon does little to alleviate the desert conditions of the American Southwest.

Foehn, Chinook, and Santa Ana Winds

Foehn, Chinook and Santa Ana winds are those that flow downslope in response to the distribution of high and low pressure systems over and near large mountain areas. Compression of the descending air leads to adiabatic warming. We discuss each of these individually.

Foehn Winds Foehn (pronounced "fern" like the common plant) is the generic name for synoptic-scale winds that flow down mountain slopes, warm by compression, and introduce

Checkpoint

> 1. Why does the summer monsoon produce heavy rainfall?
> 2. How do continentality and the Himalayas help cause the monsoon in southern Asia?

(a)

(b)

◄ **FIGURE 8–22** The monsoon of Asia results from a reversal of the winds between the winter and summer. During winter (a), dry air flows southward from the Himalayas. When summer arrives (b), moist air is drawn northward from the equatorial oceans. Surface heating, convergence, and a strong orographic effect cause heavy rains over the southern part of the continent. The dashed lines represent areas of convergence, where winds from different directions tend to come together.

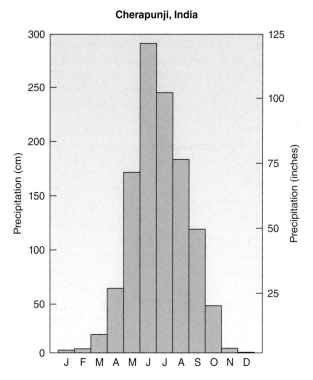

Cherapunji, India

▲ **FIGURE 8-23** Monthly mean precipitation at Cherapunji, India, highlights the sudden increase in precipitation that occurs when the south winds of the monsoon begin. Over much of the monsoon region, there is an abrupt increase in precipitation in May or June.

hot, dry, and clear conditions to the adjacent lowlands. Although the term *foehn* strictly applies to winds coming from the Alps of Europe, we generally use it to describe this type of wind anywhere in the world. In Europe, foehns develop when midlatitude cyclones approach the Alps from the southwest. The air rotates counterclockwise toward the center of low pressure and descends the northern slopes. These winds bring unseasonably warm conditions to much of northern Europe during the winter, when they are most prevalent.

Chinook Winds When winds warmed by compression descend the eastern slopes of the Rocky Mountains in North America, they are called **chinooks**. Low-pressure systems east of the mountains cause these strong winds to descend the eastern slopes at speeds that can exceed 150 km/hr (90 mph) when funneled through steep canyons. Like their European counterparts, chinooks are most common during the winter when midlatitude cyclones routinely pass over the region.

Sometimes the presence of a large mass of cold, dense air near the base of a mountain range prevents a chinook from flowing all the way down the slope. The hot air then overrides the cold air, and no warming is observed near the surface. If the chinook strengthens sufficiently, however, it can push the cold air out of its path, and the foothill region undergoes a rapid temperature increase. But if the hot winds weaken even momentarily, the cold air can return to the foothills and bring another sudden change in temperature—this time a nearly instantaneous cooling. Such reversals can take place repeatedly over a short period of time, with each bringing

◄ **FIGURE 8-24** The "Arizona monsoon" occurs each summer as intense heating creates low pressure at the surface. Moist air flows toward the low pressure from the south and southwest, and localized convection can trigger heavy thundershowers.

another rapid and sometimes extreme temperature change. Rapid City, South Dakota, for example, once experienced three such cycles over a 3-hour period, with temperature changes as large as 22 °C (40 °F) occurring with each shift.

Chinook winds can be a blessing to ranchers in the western Great Plains who rely on them to melt the snow that covers their rangelands. To others, the rapid temperature oscillations can be a source of misery. Imagine leaving your house in the morning when the temperature is 0 °C (32 °F), getting out of your car when it is 38 °C (100 °F), going for lunch when the temperature is down to –10 °C (14 °F), and returning from lunch when it is up to 35 °C (95 °F). Such changes have been known to occur from chinooks, and there is anecdotal evidence (although nothing has been proven) suggesting an increase in violent crime, depression, and suicide during such episodes. Other problems not related to human discomfort can also arise from chinooks. During the 1988 Winter Olympics at Calgary in Alberta, Canada, chinook winds melted the snow cover and forced a postponement of the ski competition for several days.

In parts of the western Great Plains, chinooks are frequent enough and strong enough to increase the average winter temperatures, as exemplified by Rapid City and Sioux Falls, South Dakota. Rapid City is situated in the foothills of the Black Hills and commonly experiences chinooks, whereas Sioux Falls is several hundred kilometers to the east and well out of their range. But despite the fact that its elevation is 500 m

(1650 ft) greater, Rapid City has an average January temperature 4 °C (7 °F) greater than that of Sioux Falls, thanks to frequent chinook winds.

Santa Ana Winds The **Santa Ana winds** of California are similar to foehns and chinooks, but they arise from a somewhat different synoptic pattern. These winds, common in the fall and to a lesser extent in the spring, occur when high pressure develops over the Great Basin (Figure 8–25). Air flowing away from the high pressure descends to lower elevations and warms by compression, just as air flowing along the eastern slopes does for the chinooks. The difference is that the Santa Ana winds occur in response to a large area of high pressure causing air to flow out of the Rockies, whereas the chinook forms in response to air flowing across the range.

During Santa Ana conditions, the sinking air can warm by 30 °C (54 °F) or more and attain temperatures in excess of 40 °C (104 °F) near the coast. Contrary to what some people believe, Santa Anas are *not* warm because they pass over hot desert surfaces—it is compression, and compression only, that causes their high temperatures. In fact, during a well-developed Santa Ana, the coastal areas of California are usually hotter than interior desert locations such as Las Vegas, Nevada.

Santa Ana winds often contribute to the spread of tremendously destructive fires in California (see *Box 8–4, Focus on the Environment: Wildfires*). The natural vegetation of the

▲ **FIGURE 8-25** A weather map showing a high-pressure system over the Rockies, causing a Santa Ana wind over southern California.

8–4 FOCUS ON SEVERE WEATHER

Wildfires

In the summer and fall of 2002, the landscape of the western United States was particularly dry due to an extended drought. That situation, coupled with the accumulation of potential fuel that had built up over the years since previous fires, set the potential for devastating wildfires. Many ecologists believe that fire suppression, though it may save homes and large areas of woodland in the short run, creates the conditions for uncontrollable fires. According to this argument, allowing fires to burn simply allows the ecosystem to maintain its normal conditions and precludes the buildup of highly volatile material that ultimately sets off massive conflagrations.

In 2002 the worst fears were indeed realized. By the first day of summer, hundreds of thousands of acres of land had already burned across the West. Unusually high temperatures and strong winds set the stage for the first of the most destructive fires in June in western Colorado. A couple weeks later, two major fires in Arizona joined together to form a massive wall of flames. And before summer was over, enormous, long-lived fires had also broken out over Oregon. All three fires were the worst ever in the history of the respective states.

Part of what makes massive fires such as these so devastating is that they create their own weather in a way that fosters more rapid spreading. Intense flames create strong thermal updrafts that lower the surface air pressure. This sets up strong pressure gradients, and the resulting strong winds help transport burning embers over great distances. It is little wonder that such fires are so difficult to contain and are often able to jump fire lines.

In one regard, fires sometimes set up weather conditions that actually assist fire crews. The same convection that can set up strong pressure gradients can also trigger rain showers that help extinguish the flames.

It is ironic, but not surprising, that while the major fires of the spring and summer of 2002 were burning in the West, other regions were coping with severe flooding. Rivers topped their banks in the upper Midwest in the spring. In the summer it was Texas that endured extremely high floodwaters, especially near San Antonio. Why is it not surprising that floods occurred in the central United States while the West was coping with drought? The answer is related to the Rossby waves, discussed earlier in this chapter. When these waves assume favored positions for extended periods of time, some portions of the waves favor high precipitation while others promote dry conditions. This is discussed further in Chapter 10.

region is dominated by an assemblage of species collectively referred to as *chaparral* (Figure 8–26), which is dry and highly flammable. When Santa Anas develop, the combination of hot, dry winds, low humidity, and an abundant source of fuel can set the stage for a major conflagration.

In October 2007 three large fires spawned by Santa Ana winds ravaged San Diego County. The fires forced the evacuation of more than half a million people as about 1500 square kilometers (560 sq. miles) of land and nearly 1600 homes burned. Seven people died as a result of the fires. Smoke was so thick in places that the Sun appeared as a subdued red spot in the sky (Figure 8–27). And yet the residents of the area could say they had been through worse—a mere 4 years earlier in October 2003, when the same dry and windy conditions spread fires that ravaged an even larger area, destroyed more than 2400 homes, and killed 16 people (Figure 8–28). At the same time, another 1100 homes were destroyed and 24 other people perished in San Bernardino County, east of Los Angeles.

Did You Know?

Several explanations, some rather implausible, have been offered for the origin of the name *Santa Ana*. The most widely accepted explanation is that during the early part of the twentieth century a local newspaper in Orange County reported on such a wind blowing out of the nearby Santa Ana Canyon. Eventually the name was used to describe the type of wind rather than its location.

Such fires are not restricted to southern California. In October 1991, Santa Ana–like winds fanned a wildfire through the Oakland Hills, east of the San Francisco Bay area. It, too, destroyed more than a thousand homes and killed 24 people.

A recent paper published by the California Climate Change Center combined observations with numerical models to conclude that the frequency of Santa Ana winds has declined over the last half of the twentieth century and will likely continue to decline through the current century due to global climate change. This is associated with complex interactions at both the synoptic and mesoscales.

▲ **FIGURE 8-26** Much of coastal California is covered by chaparral, a vegetation type adapted to the summer-drought conditions of the region.

◄ **FIGURE 8–27** Smoke was so thick during the San Diego County fires that the sun barely appeared in the sky as a dull, red ball.

◄ **FIGURE 8–28** A satellite image of southern California during the October 2003 fires. Santa Ana winds caused the fires to spread rapidly westward during the onset of the fires and created the large plumes of smoke that were blown offshore. Few clouds appear in this image except over a couple of bays along Baja California and over the extreme bottom portion.

▶ **FIGURE 8-29** Katabatic winds form over high plateaus, especially over large ice sheets such as over Greenland and Antarctica. Radiative cooling causes air over the ice sheet to become dense and flow downwards toward lower elevations.

Katabatic Winds

Like foehn and Santa Ana winds, **katabatic winds** warm by compression as they flow down slopes (Figure 8–29). Unlike foehns and Santa Anas, however, katabatic winds do not result from the migration of surface and upper-level weather systems. Rather, they originate when air is locally chilled over a high-elevation plateau. The air becomes dense because of its low temperature and flows downslope. The two best locations for such winds are along the margins of the Antarctic and Greenland ice sheets.

Katabatic winds usually occur as light breezes, but when funneled through narrow, steep canyons they can attain speeds in excess of 100 km/hr (60 mph). They happen sporadically because it can take some time for the air over the plateau to chill sufficiently. As soon as a mass of cold air forms over a plateau, gravity pulls it downslope and sets the wind in motion. When the cold air has been depleted, the wind ceases.

Much of coastal Antarctica is characterized by these alternating gusts and lulls of wind. And they can be very strong; one site, Cape Denison, occasionally experiences gusts up to 200 km/hr (120 mph) and holds the distinction of having the greatest average recorded wind speed on Earth.

Katabatic winds are not restricted to Greenland and Antarctica. They also flow out of the Balkan Mountains toward the Adriatic coast, where they are called *boras*. In France they are called *mistrals* as they flow out of the Alps and into the Rhone River Valley.

Checkpoint

> 1. Under what conditions do foehns, chinooks, and Santa Ana winds form? Why are these winds warm?
> 2. How do katabatic winds differ from foehns, chinooks, and Santa Ana winds? Explain.

Sea and Land Breezes

Near coastal regions or along the shores of large lakes, the differential heating and cooling rates for land and water form a diurnal (daily) pattern of reversing winds. During the daytime

(especially in summer), land surfaces warm more rapidly than the adjacent water, which causes the air column overlying the land to expand and rise upward (Figure 8–30). At a height of about 1 km, the rising air spreads outward, which causes an overall reduction in the surface air pressure. Over the adjacent water less warming takes place, so the air pressure is greater

(a)

(b)

(c)

▲ **FIGURE 8-30** The development of a sea breeze. Heating over the inland area causes air to expand upwards and diverge at higher altitudes. This creates a surface low pressure area, and the sea breeze flows inland from the sea.

◄ **FIGURE 8-31** This image of Hawaii shows the effect of a sea breeze. Heating of the land causes the air to expand upward. Coastal air flowing toward the interior is lifted as it passes over the mountains, causing orographic cloud cover.

than that over land. The air over the water moves toward the low-pressure area over the land, which sets up the daytime **sea breeze** (remember, winds are always named for the direction *from* which they blow). (See Figure 8–31.)

As a sea breeze encroaches landward, a distinct boundary exists between the cooler maritime air and the continental air it displaces. This boundary, called the **sea breeze front**, usually produces a small but abrupt drop in temperature as it passes. This does not mean that the temperatures will not continue to rise after the sea breeze front moves on—only that there is a temporary lull in the rate of warming. Farther inland more heating can occur before the sea breeze passes, and temperatures become higher than those along the coastal strip.

Note that though the rising air in the heated column creates low pressure at the surface, it also creates higher pressure in the middle atmosphere. But we know that pressure always decreases with height. So how can this be? Remember that the terms *high pressure* and *low pressure* are relative—that is, they mean that the pressure is higher or lower than surrounding air *at the same level.* Thus, in this example the surface pressure over land is less than that over the adjacent ocean because the outflow of air above the 1 km level reduces the total amount of air over the surface. In the middle atmosphere, however, the rising of air from below increases the amount of mass above a particular level, and thereby increases the pressure relative to that of the surrounding air (though the pressure is still less than that below).

At night, when the land surface cools more rapidly than the water, the air over the land becomes dense and generates a **land breeze**. That is, lower land temperatures make for higher surface pressure and offshore flow. Compared to land breezes, sea breezes are usually more intense and last for a longer period of time each day. Sea breezes tend to be strongest in the spring and summer when the greatest daytime temperature contrasts occur between land and sea. A typical sea/land breeze pattern is described in Table 8–1, which shows average wind characteristics for Los Angeles, California.

TABLE 8-1

Average Wind Speed and Direction—Los Angeles, CA

Time (PST)	Winter Dir	Winter Speed (m/s)	Spring Dir	Spring Speed (m/s)	Summer Dir	Summer Speed (m/s)	Fall Dir	Fall Speed (m/s)
4 A.M.	ENE	1.0	E	0.5	WSW	0.4	ENE	0.6
10 A.M.	ENE	1.4	WSW	1.9	WSW	3.2	WSW	1.0
1 P.M.	WSW	2.5	WSW	5.3	WSW	5.5	WSW	4.5
4 P.M.	WSW	3.5	WSW	5.5	WSW	5.8	W	5.1
10 P.M.	NE	0.5	W	1.6	WSW	2.4	W	0.7
1 A.M.	ENE	1.0	—	0	WSW	1.0	NNE	0.2

Source: California Air Resources Board.

The sea/land breeze type of circulation is not confined to coastlines but also occurs along the shores of large lakes, in which case it creates daytime **lake breezes**. Such systems occur along the margins of the Great Lakes in the United States and Canada, but there they occupy a narrower zone than do those along the Pacific or Atlantic coasts.

Valley and Mountain Breezes

A diurnal pattern of reversing winds similar to the land/sea breeze system also exists among mountains and valleys. During the day, mountain slopes oriented toward the sun heat most intensely. The air over these sunny slopes warms, expands upward, and diverges outward at higher altitudes in much the same way as it does over inland areas when a sea breeze develops. The **valley breeze** occurs when air flows up from the valleys to replace it (Figure 8–32a).

At night, the mountains cool more rapidly than do low-lying areas, so the air becomes denser and sinks toward the valleys to produce a **mountain breeze** (Figure 8–32b).

(a)

(b)

▲ **FIGURE 8–32** A valley breeze (a) forms when daytime heating causes the mountain surface to become warmer than nearby air at the same altitude. The air expands upward and the air flows from the valley to replace it. Nocturnal cooling makes the air dense over the mountain and initiates a mountain breeze (b).

Mountain breezes are usually just about as intense as valley breezes but tend to be somewhat gustier. Whereas valley breezes blow fairly continuously at speeds below 15 km/hr (10 mph), the nighttime air may be still for several minutes and then suddenly flow downslope.

Air–Sea Interactions

Earlier in this chapter we described how atmospheric circulations propel ocean currents. In a similar manner, the oceans exert an important influence on the input of heat and moisture to the atmosphere. Warm surface waters heat the overlying atmosphere by the transfer of sensible and latent heat. The addition of this heat, in turn, affects atmospheric pressure. Thus, the atmosphere and the ocean are linked as a complex system.

Compared to those in the atmosphere, oceanic motions and temperature changes are exceedingly slow. Temperatures in the lower troposphere can change tens of degrees in a matter of hours, while oceanic temperatures are very stable. Because the ocean surface changes so slowly, information about the current distribution of temperature can be a useful tool for atmospheric scientists in making long-term weather forecasts. In this section we examine some of the important interrelationships between the ocean and the atmosphere.

 TUTORIAL
ENSO

Use the tutorial to study the evolution of air and sea patterns associated with neutral, El Niño, and La Niña conditions as well as maps showing average conditions and movies of changes in sea-surface temperatures.

El Niño, La Niña, and the Walker Circulation

El Niño and La Niña events are patterns of tropical Pacific sea surface temperatures that may persist for months to perhaps a year. Though they are oceanographic phenomena, they are closely linked with the atmosphere through what is referred to as the Walker circulation.

El Niño Before its dramatic reappearance in 1983, the phenomenon known as **El Niño** was largely unknown to the public. But that changed in the early part of the year when the unusually warm waters in the eastern Pacific Ocean that mark an El Niño helped spawn a series of powerful storms in southern California. The storms not only caused severe flooding but also generated heavy surf that caused extensive coastal property damage and completely washed the sand away from many beaches. Another recent (1997–98) El Niño was also unusually strong, again leading to episodes of heavy surf, landslides, and flooding in southern California. In addition, precipitation across the southern tier of states was well above normal and severe storms were more frequent; some storms spawned very damaging tornadoes. Residents of the northern United States and eastern Canada also experienced anomalous weather conditions with unusually mild temperatures during the winter. During the fall and winter of 2002–2003, another weaker El Niño developed over the eastern Pacific.

High water temperatures promote two conditions favorable for major storm activity: increased evaporation into the air and reduced air pressure. There is therefore little doubt that the unusual episodes of 1983 and 1997–98, in which the surface waters were as much as 6 °C (11 °F) warmer than normal, played a role in the formation and passage of the storms. So what exactly is El Niño and how does it form?

Walker Circulation At 2- to 7-year intervals (every 40 months on average), the surface waters of the eastern Pacific, especially near the coast of Peru, become unusually warm. (The tendency for this warming to occur near the Christmas season led to the name El Niño, a reference to the Christ child.) But El Niño is much more than a simple oceanic warming; it results from a complex interaction with the atmosphere in what is called the **Walker circulation**. The Walker circulation is an atmospheric phenomenon in the tropical Pacific wherein air normally rises over the western ocean and descends over the eastern Pacific, as shown in Figure 8–33.

Under normal conditions, the trade winds move warm surface waters near the equator westward, causing higher temperatures and even a difference in sea level—about half a meter greater in the western Pacific compared to the eastern Pacific. At the same time, the upwelling of colder water from below replaces the warm water migrating westward in the eastern Pacific. Warmer water in the western Pacific leads to higher air temperatures, lower surface air pressure, and more convective precipitation. An El Niño develops when the trade winds weaken or even reverse and flow eastward. The

warm water normally found in the western Pacific gradually "sloshes" eastward as a slowly moving wave, in part due to the higher sea level, and eventually makes its way to the coast of North and South America. The eastern Pacific warms, upwelling weakens, and the warm surface layer deepens both there and throughout the tropical Pacific basin. The change in sea surface conditions linked with the change in the atmospheric pressure distribution is called the **Southern Oscillation**. The El Niño and Southern Oscillation are closely intertwined, and atmospheric scientists refer to their combined occurrences as **ENSO events**.

Did You Know?

The Atlantic Ocean has its own version of El Niño events. The processes responsible for Pacific El Niños also operates in the Atlantic and similarly gives rise to anomalously warm surface water in the eastern part of the ocean. But Atlantic El Niños are smaller, occur more frequently, and are not correlated with their Pacific cousins.

Figure 8–34 shows the sea surface temperature (SST) anomalies associated with the very strong El Niño of 1997–98. The colors on the figure depict temperature anomalies, the differences in temperature from normal conditions. Of course, no two ENSO events look exactly alike. The shape and exact locations of the pools of warm water vary between events, as do the magnitudes of the SST anomalies. Figure 8–35 shows the overall average distribution of November–March SST anomalies for eight average-to-large El Niño events. Figure 8–36 illustrates the variability in El Niño situations by plotting temperature anomalies for six different events.

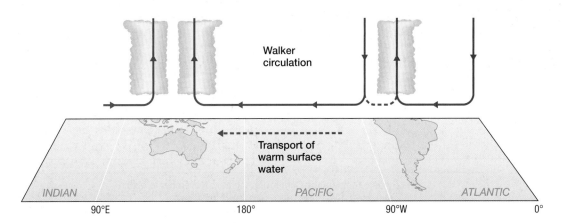

▲ **FIGURE 8–33** The Walker circulation involves a westward flow of surface air over the equatorial Pacific. (Although many texts show return flow in closed loops, recent evidence indicates this is wrong. Sinking air is supplied by convergence in the upper atmosphere, not eastward flow in the upper atmosphere across the Pacific basin.) The surface flow drags warm surface waters into the western Pacific. When the westerly flow weakens or reverses, the warm waters to the west migrate eastward and cause an El Niño.

▲ **FIGURE 8-34** The 1997–98 El Niño event contained a very large area of much above-normal sea surface temperatures over the tropical east Pacific (shown in bright red). The image is based on satellite data obtained on November 3, 1997.

Sea Surface Temperature Composite Anomaly, November to March
1958, 1966, 1973, 1983, 1987, 1992, 1996, 1998

−2.0 °C   −1.5 °C   −1.0 °C   −0.5 °C   0.0 °C   0.5 °C   1.0 °C   1.5 °C   2.0 °C

▲ **FIGURE 8-35** The average sea surface temperature differences from normal observed during the November through March period for eight El Niño episodes. The most prominent feature is the +2 °C (4 °F) increase in temperatures in the equatorial tropical east Pacific.

▲ FIGURE 8–36 Every El Niño event looks somewhat different from all others; some are larger than others and some have greater temperature anomalies than others. This figure depicts the observed sea surface temperature anomalies for six El Niño events.

La Niña When an El Niño dissipates, it can be followed either by a return to normal sea surface conditions, or by further cooling of the tropical eastern Pacific. If the waters cool to below normal temperatures, we have the reverse of an El Niño, called a **La Niña** (Figure 8–37). See *Box 8–5, Physical Principles: What Causes El Niños and the Southern Oscillation?* for a discussion of the factors that lead to changing ENSO conditions. Also, animations depicting the onset and demise of theoretical and observed El Niño and La Niña events are provided in the ENSO tutorial.

To study ENSO events and their effects on conditions elsewhere, a quantitative measure of ENSO activity is needed—an index that captures the occurrence and magnitude of an event. Several such indices have been used over the years. Some are based on sea surface temperatures in defined regions of the eastern equatorial Pacific. Another, the **Southern Oscillation Index (SOI)**, is defined as the monthly sea level pressure departure from normal at Tahiti minus the departure from normal at Darwin, Australia. Thus, if the sea level pressure for a particular month at Tahiti is 1 mb above normal for that month, and the pressure at Darwin is 1 mb

below the monthly norm, the SOI for that month is 2. Since Tahiti is east of Darwin, a positive SOI indicates a stronger than average pressure gradient from east to west, and a La Niña situation exists. A negative SOI indicates an El Niño. Note that there is no minimum SOI value, positive or negative, that absolutely delineates the existence of El Niño or La Niña from neutral conditions.

Though the SOI is a simple numeric value, it can be used to provide a historical perspective on ENSO occurrences through time. The fact that it is based on sea level pressure data (which have been recorded at Darwin and Tahiti for well over a century), instead of the shorter record of sea surface temperatures, allows for an analysis of ENSO for a longer time period than can be performed using any index based on water temperature. Figure 8–38 plots monthly values of the SOI since the late 1860s, with positive areas (in blue) indicating La Niña conditions and negative areas (in red) representing El Niños. Though there have been some periods in which El Niños and La Niñas were more frequent or less frequent than others, it is clear that both have occurred with substantial regularity over the last century and a half. Moreover, close

What Causes El Niños and the Southern Oscillation?

There are actually two parts to this question: (1) Why do these anomalies arise? That is, what causes an El Niño or La Niña to become established? and (2) What causes the reversals, or oscillations, between the two extremes? An answer to the first question was proposed more than 35 years ago in a pioneering study by Jacob Bjerknes, and has been essentially confirmed by more recent analysis and data. Unlike some other types of climate change, both anomalies arise without any external forcing. That is, they do not result from a nonclimatic event such as an asteroid impact, a volcanic eruption, or a change in the Sun's radiation. Rather, they result from processes internal to the atmosphere–ocean system. They are, in fact, classic examples of a *positive feedback* mechanism, in which relatively small changes are continually reinforced and amplified.

Consider, for example, the La Niña, where the strengthening trades push more warm water westward. In the western Pacific, the growing pool of warm water promotes more uplift, precipitation, and lower pressure. Lower pressure in the west intensifies the pressure gradient and strengthens the trades, which in turn leads to further transport of warm water. At the same time, export of warm water cools the eastern Pacific, enhances upwelling, and raises the thermocline. All of these processes promote further growth of the anomaly. Similar reasoning, but in reverse, explains why El Niños develop. In this case warmer water in the eastern Pacific weakens the trades, which leads to less heat export by westward-moving water. With less heat exported, warmth in the eastern Pacific intensifies, further reducing the pressure gradient and reducing export of heat. Thus we see that once the system begins to

drift toward either condition, there will be a tendency for the anomaly to intensify and become firmly established. This line of reasoning favors two very different states. In the El Niño state there is a thick pool of anomalously warm water in the eastern Pacific lying above a relatively deep thermocline. At the other extreme the eastern Pacific is anomalously cold, the thermocline is shallow, and the trades are strong.

If positive feedback explains the appearance of El Niños and La Niñas, what leads to their breakdown? That is, once established, why don't they persist indefinitely? Why is there oscillation from one state to the other? Much less is known with certainty about this phenomenon, and the topic remains a matter of intense interest within the scientific community. However, there is little doubt that for oscillations to arise naturally in the system, there must be delayed negative feedback processes that are out of phase with the positive feedbacks described above. That is, there must be one or more restorative processes that lag behind processes driving the system toward the extreme states. As an El Niño or La Niña builds, the delayed restoring processes also grow and eventually become large enough to overwhelm the amplifying processes. When that happens, the system moves away from the extreme state (El Niño or La Niña) toward more normal conditions. But the positive feedbacks are still in play, and if they are large enough, the system moves to the other extreme. So, in general terms, an oscillation between the extremes arises from the joint effects of positive and lagged negative feedbacks.

A number of candidate processes have been proposed as delayed negative feedbacks. One possibility is that during an El Niño Rossby waves generated in the east-central Pacific travel westward and are reflected backward from the western ocean boundary as small waves on the ocean surface. These so-called Kelvin waves bring a deeper thermocline to the

eastern Pacific, counteracting the El Niño. Similar waves form with La Niña, but they bring a shallower thermocline to the eastern Pacific, and thus similarly move the system toward more normal conditions. Another possibility, termed the *recharge oscillator*, contends that the buildup and discharge of warm water from the tropical Pacific gives rise to the oscillation. The idea is that while an El Niño builds, the entire tropical Pacific experiences a gradual "recharge" of heat as the thermocline deepens. Some time during the El Niño proper the excess warmth is discharged to higher latitudes, the thermocline becomes shallower, and the system moves toward the other extreme. This institutes another cycle of recharge, followed by another flushing of heat to the extratropics (the middle and high latitudes) with the ensuing El Niño.

Another possibility is that during an El Niño the central Pacific atmosphere warms because of enhanced convection and condensation, and this in turn generates low-pressure cells on either side of the equator. Initially these cyclones amplify the El Niño, but eventually they pump cold water eastward and thereby destroy the event they helped construct. Still other mechanisms have been suggested as important in the oscillations, including forcing by disturbances unrelated to the ENSO. No single process has been shown as responsible for ENSO events, and many experts believe that multiple processes are involved to varying degrees from one episode to the next.

Support for all these ideas comes from computer models that are able to reproduce ENSO-like cycles. But there is no agreement about what controls the period of oscillation, why it averages about 4 years instead of some other period, or even why the period of oscillation is variable, with ENSOs appearing every 2 to 7 years. Answers to these and other questions about ENSO are important both for understanding our present climate and for anticipating the nature and impact of future climate changes.

analysis of the data suggests that ENSO events have been more frequent in recent years than in the mid-1800s.

Some useful observations have been made regarding the impact of El Niño events on seasonal weather. The central

coast of California, for example, appears to have a greater likelihood of unusually high amounts of precipitation when El Niño is present, while the northwestern United States and Canadian Pacific coast regions tend to be unusually dry. The

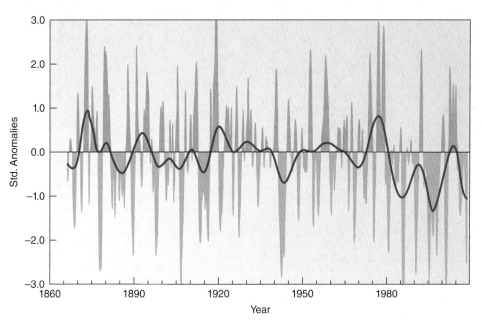

▲ FIGURE 8-37 The average sea surface temperature differences from normal observed during the November through March period for eight La Niña episodes. Note the large area of cooler-than-normal temperatures over much of the eastern Pacific, extending poleward along the North and South American coasts. Cold waters are also found in the Indian Ocean.

effects of El Niño are not restricted to the western coast of North America, however. As you have seen, the large-scale patterns of the atmosphere are strongly influenced by the position of Rossby waves. When high-pressure or low-pressure systems exist in some locations, they affect not only local weather conditions but also the overall size, shape, and position of the entire Rossby wave pattern. The establishment of

an upper-level trough over the Pacific Coast, for example, will promote the development of a ridge farther to the east. For this reason, certain weather conditions in the eastern United States are set up through **teleconnections**, the relationships between weather or climate patterns at two widely separated locations. El Niños favor the formation of ridge-trough patterns that often bring rainy conditions to the southeastern

◄ FIGURE 8-38 Time series plot of the Southern Oscillation Index (SOI). Positive values indicate La Niña conditions; negative values are El Niños. Several prominent El Niños have occurred since the 1980s.

United States and mild, dry conditions to the northeastern United States and eastern Canada.

These teleconnections also occur outside North America; El Niños tend to promote enhanced precipitation over coastal Ecuador and Peru, the central Pacific, parts of the Indian Ocean, and eastern equatorial Africa. The western tropical Pacific, Australia, India, southeastern Africa, and northeast-ern South America usually undergo decreased precipitation with the occurrence of El Niño events. Figure 8–39 depicts some of the worldwide climatic patterns generally associated with El Niños and La Niñas, and Figure 8–40 shows the average temperature and precipitation conditions that have occurred over the United States during recent ENSO episodes.

(a)

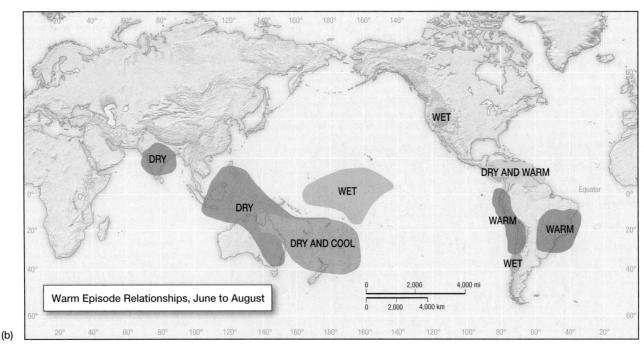

(b)

▲ **FIGURE 8–39** Worldwide climatic conditions associated with the occurrence of El Niño—(a) and (b)—and La Niña—(c) and (d)—events, for the periods of December–February and June–August. While the indicated conditions often appear during ENSO events, they do not occur with all El Niños or La Niñas.

(c)

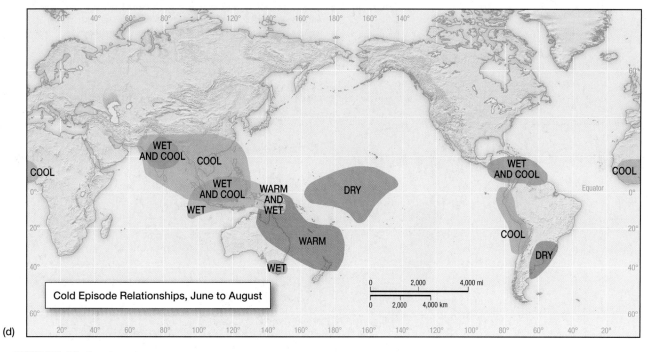

(d)

▲ **FIGURE 8–39** *Continued*

Did You Know?

Coral reefs provide information on ocean water temperatures that can be used to analyze the frequency and intensity of El Niño events of the distant past. One such study has suggested that the last century or so has witnessed the most intense El Niños of the last 130,000 years, with those of 1982–83 and 1997–98 possibly being the greatest to have occurred over that time span. It is possible, though not conclusively proven, that this change could be a response to global warming over the last century.

Like its warm-water counterpart, a La Niña tends to favor distinct (but different) regional and teleconnection patterns, including dry conditions along the California coast, the southern United States, and Peru. Enhanced precipitation tends to occur in the Pacific Northwest and western Canada, Indonesia, the Caribbean, and south Asia.

The fact that certain types of weather tend to occur in the presence of El Niños or La Niñas should not be taken to mean that some particular type of weather will always accompany the event. If the relationship between sea surface temperatures

▶ **FIGURE 8–40** El Niño and La Niña events tend to promote temperature and precipitation responses across the conterminous United States. During eight recent El Niño events, November–March temperatures have tended to be higher than normal in the north-central United States and lower than normal across the southern tier of states (a). At the same time, coastal California and the southeast generally have wet conditions, while relative drought tends to occur in the Pacific Northwest and over some of the Ohio River Valley (b). La Niñas tend to promote warm winter conditions over much of the southeast (especially coastal Texas) and parts of the northwest and the north-central United States (c), along with wetness in the northwest and relative drought in the southeast (d).

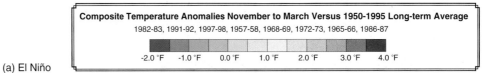

Composite Temperature Anomalies November to March Versus 1950-1995 Long-term Average
1982-83, 1991-92, 1997-98, 1957-58, 1968-69, 1972-73, 1965-66, 1986-87

| -2.0 °F | -1.0 °F | 0.0 °F | 1.0 °F | 2.0 °F | 3.0 °F | 4.0 °F |

(a) El Niño

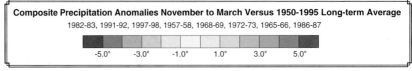

Composite Precipitation Anomalies November to March Versus 1950-1995 Long-term Average
1982-83, 1991-92, 1997-98, 1957-58, 1968-69, 1972-73, 1965-66, 1986-87

| -5.0" | -3.0" | -1.0" | 1.0" | 3.0" | 5.0" |

(b) El Niño

◄ **FIGURE 8–40** *Continued*

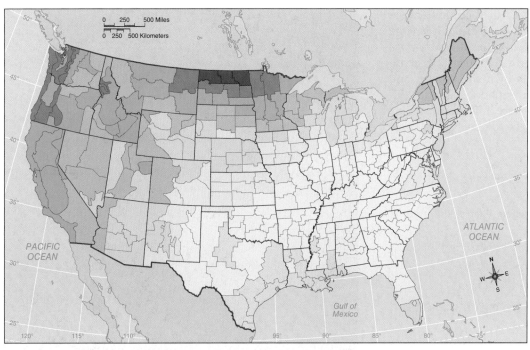

Composite Temperature Anomalies November to March Versus 1950-1995 Long-term Average

1954-55, 1955-56, 1964-65, 1970-71, 1973-74, 1975-76, 1988-89, 1998-99

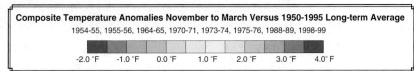

-2.0 °F   -1.0 °F   0.0 °F   1.0 °F   2.0 °F   3.0 °F   4.0 °F

(c) La Niña

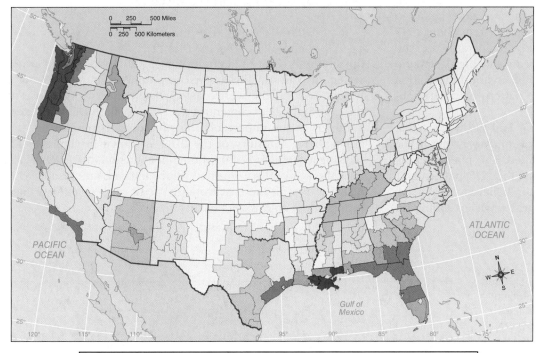

Composite Precipitation Anomalies November to March Versus 1950-1995 Long-term Average

1954-55, 1955-56, 1964-65, 1970-71, 1973-74, 1975-76, 1988-89, 1998-99

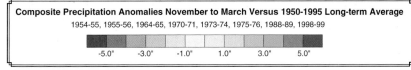

-5.0"   -3.0"   -1.0"   1.0"   3.0"   5.0"

(d) La Niña

and atmospheric patterns were that well defined, long-range weather forecasting would be very much easier than it is. One of the reasons individual El Niños and La Niñas do not always produce the same kind of atmospheric response is that no two are exactly the same. Each has its own size, shape, and location, and therefore influences atmospheric patterns differently. Not only that, but the atmosphere is far too complex to be governed entirely by any single factor, such as oceanic water temperatures.

Checkpoint

1. How do El Niño conditions in the eastern Pacific differ from "normal" conditions?
2. How can El Niño affect the weather? Include teleconnections in your answer.

Pacific Decadal Oscillation

The ENSO is not the only oscillation pattern across the Pacific Ocean. A much larger and longer-lived reversal pattern exists, known as the **Pacific Decadal Oscillation** (PDO). The PDO represents a pattern in which two primary nodes of sea surface temperature exist, a large one in the northern and western part of the basin, and a smaller one in the eastern tropical Pacific. At periods that range from about 20 to 30 years, the sea surface temperatures in the two zones undergo fairly abrupt shifts. For example, the period from about 1947 to 1976–77 was marked by generally low temperatures in the northern and western part of the ocean and high temperatures in the eastern tropical Pacific. Then an abrupt transition occurred and the reverse temperature pattern became established until the late 1990s.

In the late 1990s TOPEX/Poseidon and Jason 1 satellite data suggested that another reversal was underway, but in the early 2000s this shift seems to have halted. This phenomenon, along with other evidence, suggests that the PDO pattern may in fact be more complex than originally believed. The image in Figure 8–41 shows a pattern of warm water extending off

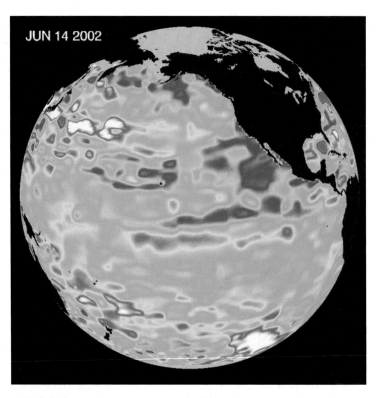

▲ **FIGURE 8–41** Pacific sea surface temperatures obtained in June 2002 suggest that a reversal in the Pacific Decadal Oscillation has begun. If so, this represents a major shift from the pattern in existence between 1976 and the late 1990s.

the southern coast of Japan in June 2002. A wedge of warmer waters also lies in the eastern tropical Pacific. Mostly cooler waters exist off the west coast of North America. Because of the enormous heat content of the oceans, the PDO may exert a direct, major impact on the pressure distribution of the atmosphere. A history of the PDO is shown in Figure 8–42, plotting a useful indicator of the phenomenon, the *PDO index*, over the last century. Positive values indicate an episode with warm waters in the east tropical Pacific, and negative values representing cold waters in that area.

▶ **FIGURE 8–42** The PDO Index is a commonly used indicator of the strength and mode of the Pacific Decadal Oscillation. Positive values of the index, plotted for the years 1900–2004, indicate warm waters in the eastern tropical Pacific. In the late 1990s it appeared that a shift from positive to negative was occurring, but this shift seems to have abruptly ended in the early 2000s.

▲ **FIGURE 8-43** The North Atlantic Oscillation relates to changes in the intensity of the high and low pressure systems over the North Atlantic. When the two pressure cells are more intense than normal (a) the jet stream becomes more intense and storms vigorously track toward northern Europe. Negative NAO phases mean that the pressure cells are weaker than normal, leading to a less vigorous jet stream than normal. Northern Europe becomes cold and dry while southern Europe and the Mediterranean tend to have increased winter storm activity.

Recent research has also suggested that the phase of the PDO influences the impact that El Niño events have on climate. When the PDO is in a "warm phase" (high temperatures in the eastern tropical Pacific), El Niño's impacts on weather are more pronounced than when the PDO is in a cold phase (low temperatures in the eastern tropical Pacific).

Arctic Oscillation and North Atlantic Oscillation

Multiyear oscillations are not restricted to the Pacific Ocean. Such patters also exist in the Atlantic. One is called the **Arctic Oscillation** (AO), which is closely related to another observed pattern called the **North Atlantic Oscillation** (NAO). Refer back to Figure 8–6 and you will see that the atmosphere over the North Atlantic Ocean is dominated by the Bermuda–Azores High and the Icelandic Low. Figure 8–6 depicts the average pressure pattern for summer and winter, but of course this pattern is variable and some years are marked by enhanced or reduced differences in pressure between the two semipermanent cells. The NAO is said to be in a *positive phase* when the pressure gradient is greater than normal (Figure 8–43a), and negative when it is less than normal (Figure 8–43b).

When the NAO is in a positive phase there is an enhanced pressure gradient from south to north that causes both an intensification and a northward shift of the polar jet stream. The more vigorous jet stream brings increased storminess to Northern Europe in the winter months along with generally mild temperatures, as the storms pass over the warm ocean current in the North Atlantic. The eastern United States likewise experiences mild, wet conditions with a positive NAO, but eastern Canada and Greenland, north of the jet stream, usually have cooler and drier winter conditions.

Negative NAO phases occur when the high and low pressure systems of the North Atlantic are less well developed, leading to a reduced pressure gradient. This weakens the strength of the jet stream, which is also directed more toward southern Europe and the Mediterranean, This usually leads to wet winters over southern Europe while the northern part of the continent is cold and relatively dry. The eastern United States experiences more frequent outbreaks of cold air surging southward from interior Canada. Figure 8–44 plots a numerical index of the NAO. The 1990s were dominated by positive NAO conditions; a shift to negative NAO dominance took place just before the beginning of the new millennium and lasting into the beginning of its second decade. The extreme winter of 2010–2011 over the eastern United States occurred during a negative NAO regime.

Checkpoint

1. What is the PDO index?
2. How do changes in the NAO influence weather conditions across Europe and eastern North America?

▲ **FIGURE 8–44** The NAO tends to remain dominated by a positive or negative phase for years at a time. The positive phase occurred over much of the 1990s followed by a shift prior to 2000.

Summary

Although air pressure is not the first thing that comes to mind when we think of weather, it is the driving force for atmospheric motions. Horizontal changes in air pressure distribution not only initiate the wind but also influence the vertical motions that promote or inhibit cloud formation.

While the three-cell model of wind and pressure belts, or bands, provides a good foundation for understanding the general circulation, in reality the atmosphere is dominated by a number of semipermanent cells of pressure.

The circulation of the upper troposphere is largely the result of latitudinal differences in temperature, with a strong tendency for westerly winds outside of equatorial latitudes. The westerlies dominate, but other atmospheric movements occur in the upper troposphere. The most prominent are the Rossby waves, the large, looping motions of air associated with the jet streams. Adding further complexity to the upper-level winds are short waves and other eddying motions.

Although this chapter (and indeed this book) deals primarily with the atmosphere, weather and climate phenomena interact with the oceans. The large-scale winds of the atmosphere initiate the large, slow-moving currents of the ocean that likewise influence the input of energy and water vapor into the air. The most famous interaction between the oceans and the atmosphere is the ENSO event (El Niño–Southern Oscillation).

One of the most significant patterns of wind flow is the monsoon of Asia, the seasonal reversal of winds that are dry in the winter and extremely wet in the summer. Smaller-scale winds can also affect weather and climate. Foehn winds occur near many mountain ranges and bring hot, dry conditions. In North America, foehn winds called *chinooks* and *Santa Anas* warm the Great Plains and the West Coast, respectively. Daily sea/land breeze patterns are prominent features of coastal zones, while mountain and valley breezes produce similar patterns inland.

The circulation of the atmosphere is strongly intertwined with sea surface conditions across the oceans. We discussed three important patterns of atmospheric/oceanic behavior: the ENSO phenomenon, the Pacific Decadal Oscillation, and the Arctic Oscillation.

Key Terms

general circulation *page 215*
global scale *page 216*
synoptic scale *page 216*
mesoscale *page 217*
microscale *page 217*
single-cell model *page 217*
zonal wind *page 217*
meridional winds *page 217*
three-cell model *page 217*
Hadley cell *page 218*
Ferrel cell *page 218*
polar cell *page 218*

Intertropical Convergence Zone (ITCZ; equatorial low) *page 218*
subtropical highs *page 219*
horse latitudes *page 219*
northeast trade winds *page 219*
southeast trade winds *page 219*
subpolar lows *page 219*
westerlies *page 219*
polar highs *page 219*

polar easterlies *page 219*
semipermanent cells *page 220*
Aleutian low *page 220*
Icelandic low *page 220*
Siberian high *page 220*
Hawaiian high *page 220*
Bermuda–Azores high *page 220*
Tibetan low *page 220*
polar front *page 225*
polar jet stream *page 225*

subtropical jet stream *page 226*
Rossby wave *page 227*
ocean currents *page 230*
Ekman spiral *page 231*
North Equatorial Current *page 231*
South Equatorial Current *page 231*
Equatorial Countercurrent *page 231*
Gulf Stream *page 231*

Review Questions

1. Describe the single-cell and three-cell models of the general circulation.

2. What is the Hadley cell and where is it found?

3. Of the pressure and wind belts described in the three-cell model, which have the strongest basis in reality?

4. Why do the trade winds flow from the northeast and southeast instead of directly from the east?

5. What are the Ferrel and polar cells?

6. Describe the distribution of semipermanent cells and their seasonal changes in location and size.

7. What is the Sahel? Describe its seasonal cycle of rainfall and explain its origin.

8. Describe the average patterns of the 500 mb level for January and July. What causes the patterns?

9. Explain why upper-atmospheric winds outside the tropics have a strong westerly component, on average.

10. Why does the equatorward bending of height contours for the 500 mb level imply the presence of a trough?

11. Explain how temperature patterns lead to the development of the polar jet stream.

12. Describe the distribution of Rossby waves and their impact on daily weather.

13. What is the Ekman spiral?

14. What is upwelling? How is it caused?

15. Describe the scope of global, synoptic, mesoscale, and microscale wind systems.

16. Describe the wind patterns associated with the monsoon of southern Asia.

17. Describe foehn winds.

18. How do katabatic winds differ in origin from foehn winds?

19. What causes sea/land and mountain/valley breezes to develop?

20. What is an El Niño and how is it related to the Walker circulation?

21. Describe the Pacific Decadal Oscillation and the Arctic Oscillation.

Critical Thinking

1. Figure 8–4a depicts the classic description of the single-cell model. Can you think of any reason the arrows should not be directed in a straight line toward the equator?

2. The three-cell model of circulation places the center of the equatorial low right along the equator. Do you think that the varying solar declination through the course of the year would be able to shift the center of the low all the way to the tropics of Cancer and Capricorn?

3. Which of the belts depicted in the three-cell model is likely to exhibit the greatest temperature gradients?

4. Examine Figure 8–6 and determine which of the semipermanent cells have the highest and lowest surface air pressures. How do the strengths of the winter and summer cells in the Northern and Southern Hemispheres compare to each other?

5. Figure 8–6 shows that pressure gradients, and therefore wind speeds, are strong in the middle-high latitudes of the Southern Hemisphere (such as at the southern tip of South America). Why doesn't a similar feature exist in the Northern Hemisphere?

6. If the western Great Plains were to have very low temperatures and the East Coast relatively warm conditions, what inferences would you make about the position and amplitude of the Rossby wave pattern?

7. Weather forecasters often use a type of weather map depicting the thickness of the 1000–500 mb layer of the atmosphere. What information would this map convey to the meteorologist?

8. Why isn't upwelling an important process of the Gulf Stream?

9. Why do ocean surface temperature patterns change so slowly, especially when compared to atmospheric patterns?

10. El Niño and La Niña conditions help meteorologists make seasonal forecasts for the southeastern United States—thousands of kilometers away. What atmospheric mechanism permits such extrapolations?

Problems and Exercises

1. Go to **weather.noaa.gov/fax/nwsfax.html**. In the first group of maps (labeled "standard barotropic levels"), select 500 mb. Then click on the link under *4a*, called *Height/temp*. The map that will appear is a map of the 500 mb pattern, with solid lines depicting the height of the 300 mb level (in tens of meters).

   a. Locate the position of the major troughs and ridges.

   b. Is the current pattern a zonal or meridional one?

   c. Where is the jet stream most prominent?

   d. Does today's jet stream appear across the entire region mapped?

   e. How does today's compare to the mean distributions shown in Figure 8–8? Why is the map for today (in all likelihood) less zonal than those in Figure 8–8?

2. If you have a high-speed connection to the Web, go to **weather.unisys.com/nam/nam .php?plot=500&inv=0&t=1** to see an animation depicting the change in the 500 mb pattern as predicted by a numerical model.

   a. Describe the predicted movement of the Rossby waves.

   b. How does the movement of the Rossby waves respond to their wavelengths and internal wind speeds (as implied by the spacing of the contours)?

Quantitative Problems

Rossby wave patterns and other phenomena discussed in this chapter are best understood when actual numbers are applied to them. We suggest you log on to this book's Web site (**www.MyMeteorologyLab.com**) and work out the quantitative problems for Chapter 8.

Useful Web Sites

weather.noaa.gov/fax/nwsfax.html
The National Weather Service site offers many current weather maps. The first group includes standard maps for various pressure levels, including maps for North America and the entire Northern Hemisphere.

www.ssec.wisc.edu/data/sst/latest_sst.gif
Satellite image depicting current ocean temperatures.

www.cpc.noaa.gov/products/analysis_monitoring/enso_update/sstanim.shtml
Weekly update on ENSO with several animations showing sea surface temperature distributions.

topex-www.jpl.nasa.gov/science/el-nino.html
Information on El Niño and La Niña, along with numerous links (including one on the PDO).

virga.sfsu.edu/gif/jetsat_00.gif
Satellite image of North America with superimposed arrows depicting strength and position of current jet stream.

iri.columbia.edu/climate/ENSO
An excellent overview of ENSO.

MyMeteorologyLab™

The *Understanding Weather and Climate,* Sixth Edition, MyMeteorologyLab™ web site contains numerous multimedia resources to aid in your study of **Atmospheric Circulation and Pressure Distributions**. Visit **www.MyMeteorologyLab.com** to:

- **Review** key chapter concepts.
- **Read** the Pearson eText, *In the News RSS feeds,* and other chapter-specific Web resources.
- **Visualize** challenging topics with Interactive Tutorials, Geoscience Animations, MapMaster interactive maps, and Weather in Motion movies and visualizations, and more.
- **Test Yourself** with self-study quizzes.

 ## TUTORIALS
UPPER-LEVEL PRESSURE AND WINDS
EL NIÑO-SOUTHERN OSCILLATION

Use the interactive animations and quizzes in these tutorials to review this chapter's key concepts.

WEATHER IN MOTION
View movies and visualizations of meteorology patterns and processes.

Floods and Droughts
Global Fire Patterns
Black Carbon Aerosols Trace Global Winds
El Niño
La Niña

9 Air Masses and Fronts

After reading this chapter, you should be able to:

▶ Describe the types of air masses and explain how they form.

▶ Describe the importance of source regions in air mass formation.

▶ Identify the different types of fronts and their characteristics.

▶ Describe the processes that lead to different types of cloud conditions for different types of fronts.

Nobody ever expects warm winters in Alaska, but sometimes conditions are even more severe than usual. This was particularly true in late January and early February of 1999, when a prolonged period of extreme cold produced the worst cold snap in a decade. Galena, in the northern part of the state, just missed an all-time low temperature when the reading dipped to –53 °C (–64 °F) in February. Conditions were so extreme that the city government officially closed down, maintaining only emergency services. If that was not cold enough, consider the –75 °C (–103 °F) wind chill factor at Kotzebue along the northwestern coast! Things were not exactly balmy in Fairbanks, either, when on February 15 the low temperature dipped below –37 °C (–35 °F) for a record-breaking nineteenth consecutive day. Needless to say, the air was also extremely dry, as is always the case under cold conditions.

Contrast those conditions with those experienced in Texas in the summer of 2011. College Station, by way of example, had its warmest May, June, and July average temperatures ever, as maximum temperatures exceeded 38 °C (100 °F) 51 times between May 31 and September 3. The heat was further compounded by high humidities and unusually sparse rainfall throughout the period.

Have you ever wondered about episodes such as the two described above, in which large areas experience more or less similar weather? At times like these, broadcasters use phrases such as "throughout the Midwest," or "across the eastern seaboard," or "Today the Pacific Northwest experienced." By way of contrast, it often happens that places within an hour's drive of each other have very different weather, with essentially nothing in common. What is the explanation for this behavior? Why does the atmosphere sometimes organize itself into broad uniform patches, and other times show extreme variation over short distances? This chapter addresses these and related questions using some very simple, powerful concepts.

◀ Extremely cold air causes ice to form on orange trees near Tampa Bay, as temperatures dropped to 26 degrees Fahrenheit on the morning of January 3, 2008.

The Alaska and Texas situations represent two extreme instances in which large regions are covered by a body of air having more or less uniform temperature and moisture. These large volumes of air are called **air masses**. Often an area the size of North America will be covered by several air masses at the same time so that, for example, the northeastern United States and southeastern Canada may experience cold, dry conditions, while the southern United States is dominated by warm, moist air. Consequently, a person might board an airplane in Nashville feeling perfectly comfortable in shirt sleeves, only to end up shivering in Boston. Moreover, these air masses are commonly separated from each other by fairly narrow boundary regions, called **fronts**, across which conditions change rapidly. The passage of a front is a significant weather event, because fronts often bring abrupt changes in temperature, humidity, and wind. They also provide a lifting mechanism that can lead to the formation of clouds and precipitation.

In this chapter we describe the formation and nature of air masses, the fronts that separate them, and their influences on local weather.

Formation of Air Masses

The temperature, pressure, and moisture characteristics of the atmosphere arise in large part from the continuous exchange of energy and water vapor near the surface. When energy inputs exceed energy losses, the temperature of the air increases. In the same way, when more evaporation than precipitation takes place, the moisture content of the atmosphere increases. Because heat and water are not uniformly distributed across the globe, the cooling and warming of the atmosphere vary from place to place, as does the net input of water vapor. Thus, air over the tropical Pacific, for example, takes on different characteristics from air over northern Canada.

Source Regions

The areas of the globe where air masses form are called **source regions**. Heating or cooling large bodies of air requires many days, as do changes in the moisture content, so air must remain over a source region for a substantial length of time for an air mass to form. Air mass source regions occur only in the high or low latitudes; middle latitudes are too variable and do not have the quiet periods necessary for an air mass to take on the characteristics of the underlying surface. Also, an area must be quite large—many tens of thousands of square kilometers—to act as a source region. Iceland is too small, for example, to allow the formation of air masses.

Although air masses have fairly uniform temperature and moisture content in a horizontal direction, they are not uniform from the surface to the upper atmosphere. In fact, large vertical gradients in temperature can easily occur in an air mass. These vertical differences in temperature affect the stability of the atmosphere (Chapter 6), which greatly affects the likelihood of precipitation. Some air masses, by their very nature, are more likely than others to yield precipitation.

Air masses are classified according to the temperature and moisture characteristics of their source regions. Based on moisture content, air masses can be considered either *continental* (dry) or *maritime*[1] (moist). According to their temperature, they are either *tropical* (warm), *polar* (cold), or *arctic* (extremely cold). Meteorologists use a two-letter shorthand scheme for categorizing air masses. A small letter *c* or *m* indicates the moisture conditions, followed by a capital letter *T*, *P*, or *A* to represent temperature. Continental polar air, for example, is designated cP, and maritime tropical air is mT. While this combination of letters could theoretically yield six possible types of air masses, maritime arctic (mA) air masses do not occur in nature because water bodies do not get cold enough to foster arctic air (they freeze at arctic temperatures, which largely removes their maritime character). Thus, there are five types of air masses, as described in Table 9–1. The major source regions of North America are shown in Figure 9–1.

Given the wide continuum of temperature and moisture contents that can exist, arbitrarily categorizing air masses into only five types might seem somewhat limiting. How, for example, would you classify a day with a temperature of 20 °C (68 °F) and a dew point of 10 °C (50 °F)? There really is no answer in that case, as it would hardly feel either tropical or polar. Or looking at the question another way, why even try to classify such bodies of air? The answer is that the air mass concept is useful when we wish to identify the air on either side of a frontal boundary or when a very simple description of the air will suffice.

Air masses are not permanently confined to their source regions; they are able to migrate to regions marked by less extreme weather conditions. The movement of an air mass away from its source causes two things to happen: (1) The region to which the air mass moves undergoes a major change in temperature and humidity, and (2) the air mass becomes more moderate. We will now examine the various types of air masses, the weather they bring, and the transformations they experience as they travel.

TABLE 9–1

Air Masses

Type	Source Regions	Properties at Source
Continental Arctic (cA)	Highest latitudes of Asia, North America, Greenland, and Antarctica	Extremely cold and very dry. Extremely stable. Minimal cloud cover.
Continental Polar (cP)	High-latitude continental interiors	Cold and dry. Very stable. Minimal cloud cover.
Maritime Polar (mP)	High-latitude oceans	Cold, damp, and cloudy. Somewhat unstable.
Continental Tropical (cT)	Low-latitude deserts	Hot and dry. Very unstable.
Maritime Tropical (mT)	Subtropical oceans	Warm and humid.

---

[1]Sometimes called *marine*.

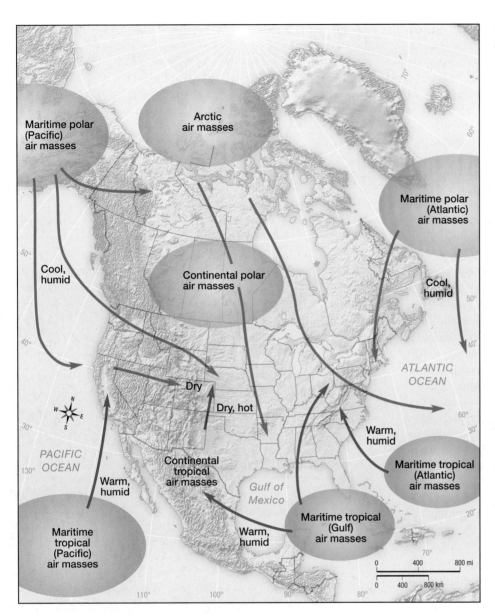

◄ **FIGURE 9-1** The source regions for North American air masses.

Continental Polar (cP) and Continental Arctic (cA) Air Masses

Continental polar (cP) air masses form over large, high-latitude land masses, such as northern Canada and Siberia. In winter, these regions have short days and low solar angles. They also are usually snow covered during the winter and therefore reflect much of whatever solar radiation does reach the surface. This combination of circumstances virtually guarantees that the air will lose more radiant energy in the winter than it receives. The cooling of the air from below leads not only to low temperatures but also to radiation inversions and highly stable conditions.

In addition to having very low temperatures, winter cP air masses are extremely dry. Recall from Chapter 5 that little water vapor can exist in cold air and that the dew point (or frost point) temperature is always equal to or less than the air temperature. If the air temperature is very low—say

−30 °C (−22 °F)—a kilogram of air at sea level can contain a maximum of only 0.24 g of water vapor. If the air is not saturated, the actual amount of moisture will be even lower.

The combination of dry air and stable conditions ensures that few if any clouds will form over a cP source region. Furthermore, the lack of water vapor reduces the absorption of incoming solar radiation by the atmosphere. Thus, despite their low temperatures, cP air masses over their source regions are usually bright and sunny. On the other hand, the stability of the atmosphere inhibits vertical mixing, so pollutants introduced at the surface remain concentrated near the ground. Because cold conditions lead to increased consumption of heating fuel (often coal and fuel oil), it isn't surprising that cP air is often associated with poor air quality over urban areas.

Summer cP air masses are similar, but much less extreme; they are both warmer and more humid than the winter version. They tend to remain at higher latitudes than winter cP

▶ **FIGURE 9–2** The arctic front separates a shallow layer of extremely cold arctic air from cold polar air. As you can see, it is much shallower than the polar front.

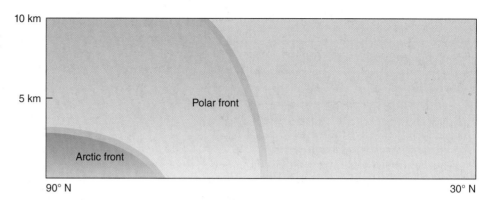

air masses, so they do not influence as much of the globe as their winter counterparts. Daytime radiation inversions do not form because the air mass develops over a snow-free surface when the days are long. In fact, it's not uncommon for some convective uplift to occur, resulting in fair-weather cumulus clouds (scattered puffy clouds in an otherwise blue sky). Still, these air masses form over a continent, so they do not contain enough moisture to sustain significant precipitation.

Continental arctic (cA) air is colder than continental polar, but the distinction between the two is more than just a matter of degree. cA and cP air are separated by a transition zone, similar to the polar front (Chapter 8), called the *arctic front*. While the polar front can extend upward several kilometers from the surface, the arctic front is shallow and does not usually extend beyond a kilometer (0.6 mi) or two above the surface (Figure 9–2). Thus, we feel the change in temperature associated with the passage of an arctic front, but the relatively shallow front does not produce a great deal of uplift that might promote snowfall. On rare occasions, cA air can extend as far southward as the Canadian–United States border region. Some meteorologists believe that the distinction between cA and cP is rather minor, and therefore restrict their temperature categorization to just polar and tropical air masses.

Did You Know?

On February 28, 2002, continental polar air surged unusually far south and with lower than normal temperatures across much of the southeastern United States. An unusually low temperature of −10 °C (14 °F) occurred in Crestview, Florida, the lowest temperature ever observed for that state on a February 28. Though temperature anomalies were greatest in the Southeast, virtually all of Canada and the continental United States were under the influence of cold conditions on that day. Such occurrences demonstrate the impact that the movement of air masses can have far away from their source regions.

Modification of cP Air Masses Leaving their source region, cP air masses bring cold weather to more temperate latitudes. Figure 9–3 illustrates the movement of a hypothetical mass of continental polar air. In (a) the frontal boundary is north

(a)

(b)

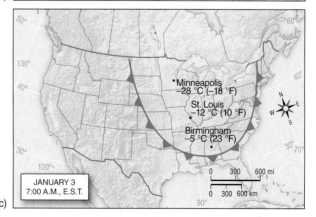

(c)

▲ **FIGURE 9–3** A sequence of surface weather maps showing the southward movement of continental polar air.

of Minneapolis, and moderate temperatures exist throughout the central United States. Twenty-four hours later (b), the boundary of the continental polar air has passed over Minneapolis, causing a 20 °C (36 °F) drop in temperature. Farther to the south, St. Louis and Birmingham experience little temperature change from the previous day because the cold air has not yet extended that far southward. By the third day (c), the cold air has penetrated to the Gulf Coast and caused noticeable temperature declines at St. Louis and Birmingham, while Minneapolis remains under the influence of extremely cold air.

Notice that at each successive location the decline in temperature associated with the passage of the front is less pronounced than at the next most northerly site. Minneapolis experienced the most severe drop in temperature, while St. Louis and Birmingham underwent less extreme cooling. This occurs because of the gradual moderation of the cP air as it leaves its source region.

Checkpoint

1. What is the source region of a continental polar (cP) air mass? What are its properties?
2. Would you expect to find cP air masses in the southern hemisphere? Explain.

Maritime Polar (mp) Air Masses

Maritime polar (mP) air masses are similar to continental polar air masses but are more moderate in both temperature and dryness. Maritime polar air forms over the North Pacific as cP air moves out from the interior of Asia. The warm Japan current adds heat and moisture to the cold, dry air and converts it from cP to mP. The developing air masses migrate eastward across the Pacific; most of them pass across the Gulf of Alaska before reaching the west coast of North America. They approach the northern west coast of North America throughout the year but influence the California coast mainly in the winter.

Maritime polar air also affects much of the East Coast, but the manner in which it approaches is different. When midlatitude cyclones (rotating counterclockwise) pass over a region, the rotating air sweeps around the low-pressure system and approaches the coast from the northeast. The resultant winds are the famous **northeasters** (or *nor'easters*) that can bring cold winds and heavy snowfall (see *Box 9–1, Special Interest: Maritime Air Masses Invade Eastern North America*).

Continental Tropical (cT) Air Masses

Continental tropical (cT) air forms during the summer over hot, low-latitude areas, such as the southwestern United States and northern Mexico. The desert areas where cT air masses form have little if any available surface water and a minimal amount of vegetation to extract the water below the surface. Solar radiation inputs are extremely high during the summer. That, coupled with the lack of moisture, makes for very high ground temperatures that warm the overlying air by the transfer of sensible heat. The result is that these air masses are extremely hot and dry and, often, cloud free.

The extremely high surface temperatures cause intense heating of the air nearest the ground, which brings a steep temperature lapse rate and unstable conditions. Despite the instability, cT air masses often remain cloud free because of their inherent dryness. Consider, for example, a hot, dry air mass near the surface with a temperature of 45 °C (113 °F) and a dew point of 0 °C (32 °F). Because the temperature of a rising parcel of unsaturated air approaches the dew point at a rate of 0.8 °C/100 m (5 °F/1000 ft) of ascent, the air must be lifted 5.6 km (3.4 mi) before it can become saturated.[2] The unstable layer tends to be much lower than this height, which means that parcels at the surface may not be lifted sufficiently for clouds to develop. On the other hand, if the unstable layer is deep or if some moisture exists in the air, intense thunderstorms can develop. Thunderstorms can also develop near mountain peaks where converging valley breezes promote uplift.

Maritime Tropical (mT) Air Masses

Maritime tropical (mT) air masses develop over warm tropical waters. They are warm (though not as hot as cT), moist, and unstable near the surface—ideal for the development of clouds and precipitation.

Maritime tropical air masses have an enormous influence on the southeastern United States, especially during the summer. These air masses form over the Atlantic and the Gulf of Mexico and migrate into North America. As the air flows inland, heating from the warm surface increases the environmental lapse rate and makes the air even more unstable. The combination of high moisture content and increased instability favors the development of heavy but short-lived precipitation, often thunderstorms over relatively small areas.

The passage of a midlatitude cyclone can also trigger precipitation in mT air. This kind of precipitation covers larger areas and lasts longer than precipitation initiated by localized uplift. An mT air mass moving poleward gradually loses moisture by precipitation. Thus, there is a shift toward lower dew points northward. This is not to say that the air becomes dry as it reaches Chicago or Toronto, but it is unmistakably less oppressive than over New Orleans or Miami.

The southwestern United States occasionally experiences the effects of mT air advected inland from the eastern tropical Pacific. Off the coast of southern California, the cold ocean current moderates the overlying air so that it does not have either the extreme moisture or the high temperature associated with air over the southeastern United States. In the late summer, however, moisture sometimes moves northeastward in the form of high-level outflow from tropical storms or

---

[2]Recall that the DALR = 1.0 °C/100m, but at the same time the dew point lapse rate = 0.2 °C/100m. Thus, a rising parcel of unsaturated air approaches the dew point at a rate of 0.8 °C/100 m.

9–1 FOCUS ON SEVERE WEATHER

Maritime Air Masses Invade Eastern North America

The movement of air masses produces many abrupt changes in the weather. Examples of such transitions are almost limitless, and new ones occur regularly. Let's take a look at one very notable example.

The Storm of January 6–8, 1996

Few American storms can rival the one that hit the East Coast in January 1996. It began on January 6 as a trough of low pressure from Louisiana to western Maryland and eastern Kentucky (Figure 1a and b). Within 24 hours the storm intensified and moved eastward to the Georgia–South Carolina coast (Figure 1c and d), and moderate to heavy snow fell over a large portion of the eastern United States.

By the morning of January 8 (Figure 1e and f), up to a meter (3 ft) of snow had accumulated over Virginia and North Carolina. But a large area of the Northeast was still experiencing the worst the storm had to offer. Airports across the eastern United States had shut down, and people in cities as far away from the storm as St. Louis spent uncomfortable nights at airports after their connecting flights to the East Coast were canceled. In all, 7600 flights—a third of those scheduled to depart in the United States—were canceled as a result of the storm.

Snow depths were particularly great in part because the snow was unusually light and fluffy—in some places the snow–water ratio exceeded 18:1. In other words, 18 cm of snow corresponded to 1 cm of actual precipitation—well in excess of the average ratio of about 10:1. By the time it was over, Philadelphia had experienced its greatest recorded snowfall ever, with 73 cm (30 in.) on the ground. Providence, Rhode Island, suffered

(a)

(b)

(c)

(d)

(e)

(f)

▲ **FIGURE 1** The northeaster of January 1996. The maps on the left panels and the infrared satellite images on the right show the position of the storm at 7 A.M. on sequential days.

through its second-greatest snowfall, and Washington, D.C., and New York City experienced their third-deepest accumulations.

At least 86 people died as a result of the storm. But things could have been much worse. Fortunately, the National Weather Service forecast for the region had been extremely accurate in predicting the storm's timing, location, and intensity. This gave local citizens and disaster authorities ample time to prepare for the onslaught of maritime air.

hurricanes off the west coast of Mexico. This can lead to a layer of high clouds and increased humidity over southern California. It can also produce local thundershowers. These usually occur over the inland mountains and deserts, but occasionally interrupt the normal summer drought over the coastal region as well. Farther to the east in Arizona and eastern California, the introduction of mT air over the deserts causes what is locally (but not very accurately) called the *Arizona monsoon.* Another type of maritime air flow that can affect the western United States is the Pineapple Express, which is described in *Box 9–2, Special Interest: The Pineapple Express.*

As we emphasized at the beginning of this chapter, precise categorization is often difficult. The concept of an air mass is most meaningful when applied to large bodies of air separated by boundary zones with horizontal extents of tens (or perhaps a couple hundred) of kilometers. These boundary zones are categorized into four types of fronts.

Checkpoint

1. What is the source region of a maritime tropical (mT) air mass? What are its properties?
2. Suppose an mT air mass moves inland from the Gulf of Mexico north to Chicago. How would it change?

Fronts

Fronts are defined as boundaries that separate air masses with differing temperature and other characteristics.[3] Often they represent the boundaries between polar and tropical air. They are important not only for the temperature changes they

---

[3]Though this is essentially correct, one should understand that they are actually transition zones between two air masses—often with rapid changes in air temperature and humidity over very short distances.

9–2 FOCUS ON SEVERE WEATHER

The Pineapple Express

Most precipitation along the west coast of North America results from storms that cross the Pacific Ocean between November and April, and most of these storms pass through the Gulf of Alaska. Unlike the winter storms that affect the central part of Canada and the United States, the Pacific storms are dominated by maritime polar air, which seldom brings extremely low temperatures to the region. Thus, the precipitation associated with them almost always occurs as rain along the immediate coastal region and as rain or snow in inland (especially high-elevation) locations.

The snow line (the elevation marking the change from rain to snow) for these storms depends on the temperature. It usually varies between several hundred and a few thousand meters above sea level. The colder the storm, the lower the snow line. Mountain snowpack is critical to every economic aspect of California. Water is stored in the watersheds as snow melts in the springtime and flows into the large reservoir system that provides drinking water and the irrigation supply. Precipitation that falls as rain evaporates more quickly than snow and is more likely to be lost before it can replenish the reservoirs.

When air temperatures remain above 0 °C long enough, the snow begins to melt. In southern California, this can be a matter of hours or days after snowfall. But in the upper reaches of the Sierra Nevada, snow can remain on the ground for months before it begins to melt in the spring.

In some years, the majority of Pacific storms do not approach the West Coast from the northwest but instead travel eastward from the vicinity of the Hawaiian Islands along a plume of moist, humid air. This storm track is referred to locally as the *Pineapple Express* (Figure 1). When storms take this path repeatedly, as they did during the El Niño winters of 1982–1983 and 1992–1993, the greatest amounts of precipitation are concentrated farther to the south than during more normal years. Instead of a general pattern of increasing precipitation with latitude, the southern part of the coast might receive greater amounts than the northern coast. Furthermore, because the Pineapple

▲ **FIGURE 1** A satellite image depicting a Pineapple express. These plumes of warm moist air and associated storms follow a path generally from the Hawaiian Islands toward the west coast of North America.

Express passes over warmer waters than do storms from the Gulf of Alaska, the air tends to be warmer and more humid, and the height of the snow line increases considerably. As a result, even though this type of storm track can lead to heavy precipitation in the mountains, a smaller percentage of it accumulates as snow—much to the chagrin of local residents who depend on the snow for skiing and other recreational activities.

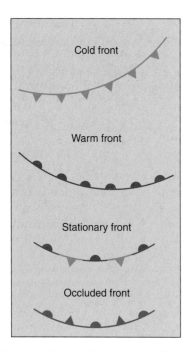

▲ **FIGURE 9-4** The symbols used to represent the four types of fronts on weather maps. Note that the direction that the triangles and semicircles point toward indicate the direction of movement of the fronts.

▲ **FIGURE 9-5** A typical midlatitude cyclone. Cold and warm fronts separated by a wedge of warm air meet at the center of low pressure. Cold air dominates the region outside the wedge separating the cold and warm fronts.

bring but also for the uplift they cause. Cold air is typically more dense than warm air, so when one air mass encroaches on another, the two do not mix together. Instead, the denser air remains near the surface and forces the warmer air upward. These upward motions lead to adiabatic cooling and sometimes to the formation of clouds and precipitation.

A cold front occurs when a wedge of cold air advances toward the warm air ahead of it. A warm front, on the other hand, represents the boundary of a warm air mass moving toward a cold one. A stationary front is usually similar to a cold front in structure but has not recently undergone substantial movement (in other words, it remains stationary). Unlike the other three types of fronts, occluded fronts do not separate tropical from polar air masses. Instead, they appear at the surface as the boundary between two polar air masses, with a colder polar air mass usually advancing on a slightly warmer air mass ahead of it. Figure 9–4 includes the symbols used to show the location of these fronts on surface weather maps.

Though some fronts are named for the temperature contrasts associated with them, distinct changes in other weather elements also occur along these boundaries. People experiencing the passage of a front not only notice a rapid change in temperature; they may also note a shift in wind speed and direction, a change in moisture content, and an increase in cloud cover.

Figure 9–5 illustrates the overall structure of typical Northern Hemisphere midlatitude cyclones, of which cold, warm, and (sometimes) occluded fronts are an important part (we discuss midlatitude cyclones in detail in Chapter 10).

The pressure distribution is somewhat circular in the larger cold sector on the poleward side of the cyclone, leading to a counterclockwise rotation of the air (in the Northern Hemisphere). The cold front separates the cold air that typically comes out of the northwest from the warmer air ahead. Between the cold front and the warm front ahead of it, pressure decreases toward the apex where the two fronts join together, and the wind is typically southwesterly. Let's now look at the structure of fronts in more detail.

Cold Fronts

We are all familiar with the dramatic shifts in weather that can occur quickly with the approach and passage of a **cold front**. During the winter, they can disrupt life for millions of people as they cause intense flooding from heavy rain or whiteout conditions from blowing snow.

TUTORIAL

MIDLATITUDE CYCLONES

Use the tutorial to observe the movement of air relative to fronts as the fronts move across the landscape.

Figure 9–6 shows the structure of a typical cold front, with its upper boundary sloping back in the direction of the cold air. As polar air surges away from its source region it does not do so as a vertical wall of cold air; in fact, the boundary of the front slopes strongly backward with distance from the ground. You should note that the vertical scale in this figure is highly exaggerated and makes the slope of the front appear far steeper than it really is. In reality, the typical cold front has a surface slope of about 1:100, meaning that its surface rises only 1 m for every 100 m of horizontal extent.

Observe also that the frontal boundary is not flat but curved; close to the ground the slope of the front is steeper than it is aloft. This is due to the varying effect of friction.

◄ **FIGURE 9-6** Cold fronts typically move more rapidly and in a slightly different direction from the warm air ahead of them. This causes convergence ahead of the front and uplift of the warm air that can lead to cumuliform cloud development and precipitation. In this example, the cold air (in blue) advances from west to east (notice that the wind speed depicted by the thin arrows increases with height). The warm air (in red) is blowing toward the northeast. The cold air wedges beneath the warm air and lifts it up.

Friction is strongest at the ground but decreases upward. Close to the ground there is a rapid decrease in friction with height. This allows the air farther from the ground to advance more rapidly than the air below, advancing the air farther from the ground to the position of the surface front, and giving the lower portion of the frontal boundary a very steep profile. Farther away from the ground the effect of friction is fairly negligible, so there cannot be a large difference in friction with height. The result is that the wedge of cold air moves forward more uniformly and the highest portion of the cold air mass does not surge forward. The steep slope near the ground and the gentler slope aloft results in the convex profile. Cold fronts move at widely varying speeds, ranging from a virtual standstill to about 50 km/hr (30 mph).

The cloud cover we usually observe with the passage of a cold front results from the convergence of the two opposing air masses. Differences in wind speed and direction allow cold northwesterly wind to converge on the warm air ahead of it and displace it upward. Furthermore, the air ahead of a cold front tends to be unstable and therefore easily lifted. This promotes development of cumuliform clouds along these boundaries. With their large vertical extent, cumuliform clouds can often produce intense precipitation. However, because of the limited horizontal extent and rapid movement of the frontal wedge, such precipitation is often of short duration.

The production of surface weather maps is now almost entirely automated and performed by computer. But meteorologists still plot the location of fronts manually because identifying them is often a subjective process. Meteorologists will look for the following five features to determine cold front positions:

1. Significant temperature differences between adjacent regions, with lower temperatures behind the cold front

2. Dew point differences, with drier air in the cold sector

3. Bands of cloud cover and precipitation, which nearly coincide with the position of the front

4. Narrow zones where wind direction changes, typically northwesterly in the cold sector to southwesterly in the adjacent warm region

5. Boundaries separating regions where the atmospheric pressure has decreased over the 3-hour period (typically ahead of the cold front) and those where the pressure has increased over the previous 3 hours (behind the cold front)

While one might logically expect that the best indicator of the position of a cold front is the shift in temperatures, that is not always true. When a cold front advances rapidly, the temperature does not decrease as rapidly behind the front as does the humidity. Therefore, the distribution of dew points is sometimes the better indicator of frontal position.

Though the position of cold fronts is plotted as a line on surface maps, they are three-dimensional and usually extend upward past the 500 mb level. Nonetheless, they are not plotted on upper-level weather maps, in part because there are not enough data available aloft to adequately determine their location and in part because their surface position is more significant than their position aloft.

When seen from space, the cloud bands ahead of cold fronts often appear to contain a fairly uniform distribution of cloud cover. Closer examination, however, reveals that the cloud bands often contain pockets of thicker cloud cover and more intense precipitation. For example, Figure 9–7 shows a frontal zone extending across the central United States on March 31, 1998. The portion of the front that extends from northeastern Missouri to central Texas is the cold front. The satellite view (Figure 9–7a) provides a compelling example of how the clouds are aligned parallel to the front, with a sharp transition between the warm, moist air ahead of the front and much drier, colder air behind. But as is often the case, this image gives a false impression of uniform precipitation.

The radar composite map (Figure 9–7b) gives us a much better description of the intensity of precipitation along the front. Weather radar imagery is obtained by emitting energy with wavelengths of about 10 cm (4 in.). Clouds composed

(a)

(b)

▲ **FIGURE 9-7** A midlatitude cyclone. The cloud cover in the satellite image (a) appears to be a continuous, uniform band. The radar composite map (b) reveals that the cloud cover is, in fact, marked by areas of varying precipitation intensities.

of large droplets scatter some of the electromagnetic energy back toward the radar unit, which displays the position and other characteristics of the cloud. The color on the radar display indicates the intensity of the returned energy, with bright red portions indicating large droplets and heavy precipitation.

In this instance, four pockets of very heavy precipitation are located in central Missouri, eastern Kansas, eastern Oklahoma, and central Texas. Notice that at the time of the map, no precipitation is indicated along the front in northern Texas and central Oklahoma. However, those areas were hardly dry for the duration of the frontal passage. In fact, central Oklahoma received more rain than any other location, with recorded amounts above 4 cm (1.6 in.). This tendency for spotty, short-lived precipitation is typical of strong cold fronts, a feature that places them in distinct contrast to warm fronts, which we discuss next. An example of a recent cold front is provided in *Box 9-3, Special Interest: An Example of a Cold Front*.

Checkpoint

1. Describe the slope of the boundary along a cold front and the typical position of clouds relative to the boundary.
2. What types of information do meteorologists consider when plotting the location of a cold front?

Warm Fronts

Warm fronts separate advancing masses of warm air from the colder air ahead. As with cold fronts, the differing densities of the two air masses discourage mixing, so the warm air flows upward along the boundary. This process is called **overrunning**.

Figure 9–8 illustrates the typical sequence of clouds that form along the frontal surface. The warm air flows up along the frontal boundary in much the same way that air rises above a mountain slope. But the gradual slope of the frontal surface leads to a very gradual progression of cloud types. As the air rises along the boundary, adiabatic cooling first leads to formation of low-level stratus clouds. As the air continues to rise, a sequence of higher clouds develops, with nimbostratus, altostratus, cirrostratus, and, finally, cirrus occurring in that order. As the front moves eastward or northeastward, the leading segment of the cloud sequence (the cirrus) is seen first, followed by the continually thickening and lowering cloud cover. Thus, even an amateur forecaster can predict an episode of continuous light rain from a warm front a day or two in advance, simply by observing the sequence of clouds as they pass overhead.

Warm fronts are about half as steep as cold fronts (their slopes are about 1:200), which causes the lifting of the warm air to extend for greater horizontal distances than for cold fronts. Although the clouds above the warm front cover a greater horizontal extent than the clouds above a cold front, they usually consist of smaller droplets and have a lower liquid water content. This results in less intense precipitation than is usually associated with cold fronts. For this reason, warm fronts in the summer are considered a friend to farmers; they provide a light, steady rain that nourishes crops but does not produce flash floods or severe storms.

The overrunning air above a warm front is usually stable, which normally leads to the formation of wide bands of stratiform cloud cover. Though the rising air may sometimes be unstable and lead to formation of cumuliform clouds, this is the exception rather than the rule. Precipitation along the front tends to be light, but the wide horizontal extent and typically slow movement of warm fronts (usually about

▲ **FIGURE 9-8** A warm front. Overrunning leads to extensive cloud cover along the gently sloping surface of cool air.

20 km/hr—12 mph) allows rain to persist over an area for up to several days.

The clouds along a warm front exist in the warm air above the wedge of dense, cold air. Thus, when rain falls from the base of these clouds, the falling droplets pass through the cold air below on their way to the surface. If the falling droplets are substantially warmer than the air they fall through, they can evaporate rapidly and form a frontal fog.

If the air is cold enough during the passage of winter warm fronts, the drops freeze on their way down to form sleet, or they solidify on contact with the surface to form freezing rain. So summer warm fronts are generally beneficial, while in winter they can cause widespread problems.

Figure 9–9 shows the typical change in the slope of a warm front over time. Because wind speeds are greater aloft than near the surface ahead of the front, the upper portion of the wedge advances more rapidly than does the part near the surface, and the slope of the frontal boundary becomes less steep.

Meteorologists use much the same rules for identifying warm front positions that they do for cold fronts, but with some differences. One must first look for a zone where warmer air advances toward cooler air (the opposite of cold fronts). Dew points typically increase behind the position of the warm front; winds commonly shift from southwesterly ahead of the front to southeasterly behind it; and cloud cover and precipitation bands are common. Also, the zone ahead of the warm front generally undergoes decreasing air pressures while the area immediately behind the front typically has stable air pressure.

Checkpoint

1. What is overrunning?
2. What cloud types typically form at different altitudes along the leading surface of a warm front? Explain.

Stationary Fronts

More often than not, the boundaries separating air masses move gradually across the landscape; at other times, they stall and remain in place for extended time periods. Nonmoving boundaries are called **stationary fronts.** They are identical to cold or warm fronts in terms of the relationship between their air masses. As always, the frontal surface is inclined, sloping over the cold air.

(a)

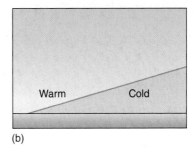

(b)

▲ **FIGURE 9-9** Warm fronts have gentler sloping surfaces and do not have the convex-upward profile of cold fronts. Surface friction decreases with distance from the ground, as indicated by the longer wind vectors away from the surface (a). This causes the surface of the front to become less steep through time (b).

9–3 FOCUS ON SEVERE WEATHER

An Example of a Cold Front

The winter of 2007–2008 was a cold and snowy one for much of the United States and Canada, with a seemingly endless series of snowstorms and cold spells. One of the season's notable events occurred in mid-February as a cold front pushed southward from Canada and the north central United States, bringing extremely cold conditions to the Midwest and colder than normal air to the Deep South.

Figure 1 shows the surface weather conditions across the 48 conterminous United States on the morning of February 17. A couple of relatively weak cold fronts extended southward out of Missouri, but neither was marked by extreme temperature contrasts, as morning low temperatures over the Southeast were at least 10 °C (50 °F), and in places well higher (Figure 2). But the cold front situated along the United States–Canadian border was set to substantially alter things further to the south.

By the morning of the 18th (Figures 3 and 4), the front located along the border the day before had advanced to the south-east, extending to eastern Missouri–western Illinois, and much of the north central United States recorded considerable drops in temperature compared to the day before. Fargo, North Dakota, which had a very mild morning on the 17th (at least by Fargo standards for mid-February) with a minimum temperature of −6 °C (21 °F) experienced significant cooling that took the temperature down to −23 °C (−9 °F) the next day, for a

▲ **FIGURE 1** Surface weather map for February 17, 2008, 7 A.M. EST. The cold front near the United States–Canadian border expanded southward and eastward to bring extreme cold to much of the central and eastern United States.

▲ **FIGURE 3** United States surface weather map, February 18, 2008.

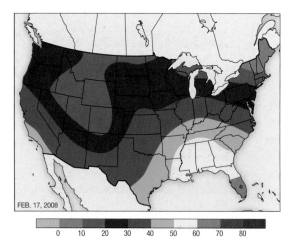

▲ **FIGURE 2** Minimum temperatures in °F recorded over the previous 24-hour period as of 7 A.M., February 17.

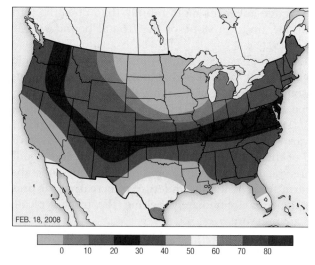

▲ **FIGURE 4** Minimum temperatures in °F recorded over the previous 24-hour period as of 7 A.M., February 18.

17 °C (30 °F) decrease. Des Moines, Iowa, which achieved a low of 3 °C (38 °F) on the 17th, cooled down to –12 °C (11 °F) on the 18th.

The front moved southward and eastward by the morning of February 19 (Figures 5 and 6), and cities such as St. Louis, Missouri experienced drops in minimum temperatures from the previous day on the order of some 8–11 °C (15–20 °F). At the same time, areas that had been hit by the cold a day earlier were subjected to even further cooling. Des Moines recorded a minimum temperature of –20 °C (–4 °F).

By the next day (Figures 7 and 8), the southern margin of the continental polar air mass had not moved much, and in some places along the frontal boundary the temperatures began to moderate. But deep within the core of the cold air mass (from the Dakotas to east to Wisconsin) the combination of snow-covered ground (with its high albedo) and clear skies (allowing a rapid loss of longwave radiation at night) brought temperatures down to severe levels. Des Moines registered an overnight low of –23 °C (–10 °F)—a full 17 °C (30 °F) colder than normal, and Fargo's temperatures plummeted to –35 °C (–31 °F).

This surge of continental polar air behind a cold front is hardly unique, but it illustrates the type of rapid temperature changes that occur during many winters with the passage of cold fronts and the change in air mass regime.

▲ **FIGURE 5** United States surface weather map, February 19, 2008.

▲ **FIGURE 7** United States surface weather map, February 20, 2008.

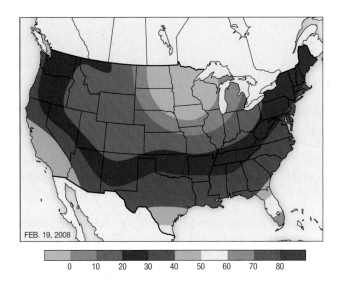

▲ **FIGURE 6** Minimum temperatures in °F recorded over the previous 24-hour period as of 7 A.M., February 19.

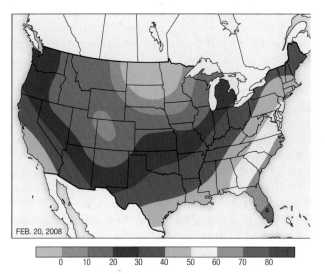

▲ **FIGURE 8** Minimum temperatures in °F recorded over the previous 24-hour period as of 7 A.M., February 20.

Determining whether a front is stationary is somewhat subjective. For example, if the front advances at a rate of 1 km/hr (0.6 mph), does that constitute enough movement for it to be nonstationary? If not, what about 2 km/hr? In practice, a meteorologist will make the designation by looking at the previous one or two surface weather maps (compiled at 3-hour intervals). If there has been no movement during that period, the front is considered stationary.

Did You Know?

Though the concepts behind the location of fronts on weather maps may seem simple, there are occasional ambiguities, and sometimes what appears to be a frontal boundary may in fact be something else. For example, local temperature contrasts may arise due to elevation differences, with high plateaus having lower temperatures (especially at night) than adjacent areas at lower elevations. Ranges such as the Appalachian Mountains sometimes form a barrier between warm Atlantic air and colder cT air to the west. Snow-covered areas might also have lower temperatures than nearby areas, producing temperature contrasts not associated with fronts.

If this sounds simple, keep in mind that the exact position of a front at any point in time is difficult to pinpoint. This is in part because fronts are zones of transition rather than abrupt boundaries; they are located at no precise line. Furthermore, maps are compiled from a finite number of weather stations. A front may undergo some movement that is not detected because it does not move across an observing station.

Occlude Fronts

The most complex type of front is an **occluded front** (sometimes called an *occlusion*). The term *occlusion* refers to closure—in this case, cutting off a warm air mass from the surface by the meeting of two cold fronts. Occluded fronts often form when a faster-moving cold front "catches up" to the warm front ahead. The warm air rises. At the surface, one mass of cold air merges with another, so there is a smaller difference in temperature (and likewise dew point) from one side of an occluded front to the other. A warm air mass is present, but it is aloft, pinched off from the surface, and therefore not reflected in surface temperature.

Figure 9–10a shows a typical midlatitude cyclone prior to occlusion, with the cold and warm fronts intersecting at the center of low pressure. When the cold front meets the warm front ahead of it, the segments of the two fronts closest to each other become occluded, as shown in (b). The warm air does not disappear but instead gets lifted upward, away from the surface. The occluded front becomes longer as more and more of the cold front converges with the warm front. Eventually, the cold front completely overtakes the warm front (c), and the entire system is occluded.

In this occlusion, the air behind the original cold front was colder than that ahead of the warm front. This is an example of a *cold-type occlusion*. This is the more common type of occluded front over the interior of North America because the cP air behind the cold front migrates southward from higher latitudes and is usually colder than the cold air ahead of the warm front.

In British Columbia and the Pacific Northwest of the United States, *warm-type occlusions* are more common, especially in winter (Figure 9–11a). In these, the relatively mild mP air behind a Pacific cold front is warmer than the frigid cP air ahead of the warm front. When the cold front catches up with the warm front, the less dense air behind the cold front rides up the slope of the air ahead of the warm front (Figure 9–11b and c).

Occluded fronts can also form without having a cold front catching up to a warm front. Some occlusions occur when the circular core of low pressure near the junction of the cold and warm fronts changes shape and stretches backward, away from its original position (Figure 9–12). In (a), the cold and warm fronts are joined at the dashed line. At some later time (b), the cold and warm fronts have the same orientation with respect to each other as they did in (a), but both have been pulled back beyond the dashed line. The circular isobar pattern of (a) becomes elongated to form a trough over the occluded region. Thus, the structure of the occluded front is the same as that proposed in the traditional model, but the process responsible for its formation is somewhat different. In other cases, the cold front moves eastward relative to the warm front, so that the point where the fronts intersect moves progressively farther east (Figure 9–13). The occluded front appears as a relic portion of the warm front, situated to the west of the new intersection of the cold and warm fronts.

TUTORIAL
MIDLATITUDE CYCLONES

Use the tutorial to study the typical characteristics, movement, and evolution of fronts and their overall setting.

Yet another way occluded fronts can form is by a combination of distinct surface and upper-level processes. At low levels, a cold front overtakes and merges with a warm front, while aloft a rapidly moving upper-level frontal zone overtakes and joins with the surface front. Again the result is an occlusion, but not for reasons outlined in the classical description.

The types of fronts described above, like just about every other kind of natural feature on Earth, are subject to much variability. For example, temperatures might drop more strongly across some cold fronts than other cold fronts, and some warm fronts might move more rapidly than other warm fronts. Experienced weather watchers tend to observe these variations whenever they look at weather maps.

(a) Mature midlatitude cyclone

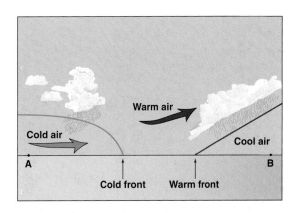

◄ **FIGURE 9-10** The traditional explanation of the occlusion process, with a cold front overtaking a warm front. The panels on the left show the placement of the fronts relative to the surface, along with the location of a sample cross section (the dashed line with the letters *A* and *B* at the end of the line). The panels on the right show the profiles of the fronts across the transect shown on the corresponding panel on the left.

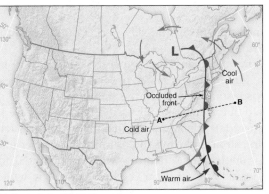

(b) Partially occluded midlatitude cyclone

(c) Occluded midlatitude cyclone

Drylines

Though fronts are the most common type of boundary separating air masses, we find another type of boundary in the spring and summer in Texas and the southern Great Plains. **Drylines** are areas where mT and cT air masses reside next to each other. Though the temperatures on either side of the boundary may not differ enough to create a front, the changes in water vapor content across the dryline can be substantial.

The dryline forms as warm moist air advects westward and meets up with cT air moving in from the west or southwest. Figure 9–14 shows a dryline moving across Texas at

noon CST on March 30, 2002. The temperatures and dew points (in °F)[4] at the selected weather stations are shown on the top left and bottom left of the station models, respectively. The dryline clearly demarcates the moist air from the Gulf and the drier air to the west.

Drylines tend to have greater temperature differences across their boundaries during the day than at night. The difference in temperature usually gets smaller overnight as the lower water vapor content of the cT air promotes more

---

[4]We are using °F rather than °C in this instance because U.S. surface weather maps still rely on the Fahrenheit scale.

(a)

(b)

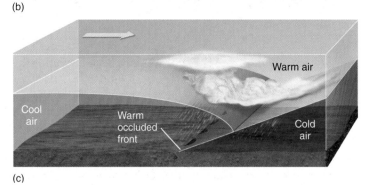

(c)

▲ **FIGURE 9–11** A warm-type occlusion. These often occur over British Columbia and the Pacific Northwest (a). The air behind the cold front is not as cold (and therefore not as dense) as the air ahead of the warm front (b), so when the cold front catches up to the warm front (c) it is displaced upward.

rapid cooling than occurs for the mT air to the east. By the early morning the greater cooling rate allows the cT air temperatures to approach those of the mT air, reducing the

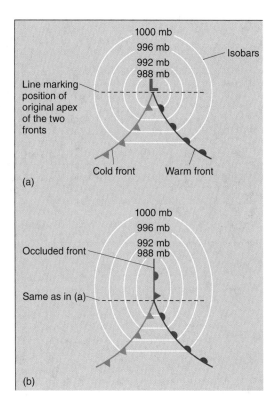

▲ **FIGURE 9–12** Some occluded fronts form when the surface low elongates and moves away from the junction of the cold and warm fronts.

temperature contrast. During the day, more rapid warming occurs west of the line.

Table 9–2 shows the change in temperature and dew point observed at an automated weather-observing system near Austin, Texas. A substantial drop in the humidity occurred as the dryline passed between 2 and 3 P.M., though no increase in temperature occurred. It is noteworthy that thunderstorms had been going on in the area during the morning hours because drylines often promote thunderstorms with the potential for tornadoes or other severe weather. The mechanism by which this happens will be discussed further in Chapter 11.

TABLE 9–2

Temperatures and Dew Points Prior to and Following the Passage of a Dryline

Time (CST)	Temperature °F (°C)	Dew Point °F (°C)
10 A.M.	66 (19)	64 (18)
11 A.M.	66 (19)	62 (17)
12 P.M.	71 (22)	64 (18)
1 P.M.	77 (25)	66 (19)
2 P.M.	78 (26)	62 (17)
3 P.M.	78 (26)	46 (8)
4 P.M.	80 (27)	46 (8)
5 P.M.	75 (24)	39 (4)

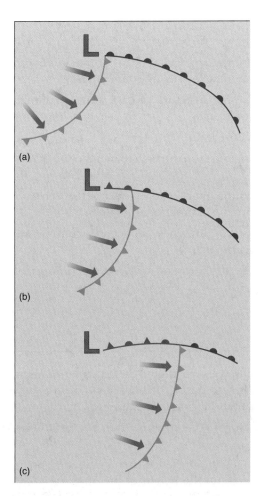

▲ **FIGURE 9-13** Some occlusions occur when the intersection of the cold and warm fronts slides along the warm front.

Checkpoint

1. How are cold-type and warm-type occlusions similar? How are they different?
2. What is one way an occluded front can form that differs from the processes in which a cold front catches up to a warm front?
3. What is a dryline?
4. How do drylines affect local weather? Develop a hypothesis to explain why this is so.

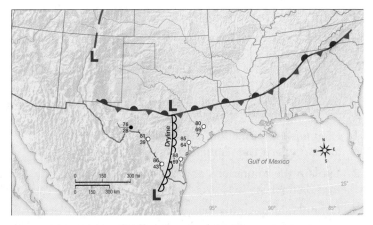

▲ **FIGURE 9-14** A dryline over Texas. West of the line the humidity is low, with dew points ranging from the upper 20s to the 40s (°F). East of the line the humidity is greater, owing to the flow of air from the Gulf of Mexico.

Summary

We have seen that the atmosphere tends to arrange itself into large masses with relatively little horizontal change in temperature and humidity. Air masses form over particular source regions—large areas of high or low latitudes of either continental or maritime locations. The five major types of air masses are continental polar, continental arctic, maritime polar, continental tropical, and maritime tropical.

The boundaries of these masses, called *fronts*, are particularly important to meteorology—not only because

their passage causes abrupt changes in the temperature and humidity, but because they promote uplift and cloud formation. The four types of fronts are cold, warm, stationary, and occluded. They are components of weather systems called *midlatitude cyclones*, which we discuss in the next chapter. Drylines are similar to fronts in that they separate air masses, though the dryline marks a humidity boundary rather than a temperature boundary. Drylines are preferred regions for severe storm activity, a topic discussed in Chapter 11.

Key Terms

air masses *page 260*
fronts *page 260*
source regions *page 260*
continental polar (cP) air masses *page 261*

continental arctic (cA) air masses *page 262*
maritime polar (mP) air masses *page 263*
northeasters *page 263*

continental tropical (cT) air masses *page 263*
maritime tropical (mT) air masses *page 263*
cold front *page 266*

warm front *page 268*
overrunning *page 268*
stationary front *page 269*
occluded front *page 272*
drylines *page 273*

Review Questions

1. What are the requirements for an area to serve as a source region?

2. Where are the primary air mass source regions in North America?

3. Describe the characteristics of cA, cP, cT, mT, and mP air masses.

4. Of the five types of air masses, which are the hottest, driest, coldest, and most damp?

5. What is the primary difference between arctic and polar air masses?

6. Which of the air mass types are likely to be stable or unstable?

7. Describe the changes that occur when a continental air mass migrates out of its source region.

8. Describe the structure of cold, warm, stationary, and occluded fronts.

9. What is overrunning?

10. Why do cold fronts have steeper slopes than warm fronts?

11. How do the alternative models of the occlusion process differ from the traditional model?

12. What is the difference between warm-type and cold-type occlusions?

13. What are drylines and why are they important?

Critical Thinking

1. Contrast the formation of air masses in the Northern and Southern Hemispheres.

2. Explain the limitations and benefits of classifying air into distinct masses.

3. Continental polar air masses can migrate into Florida during the winter but not into northern India. Why not?

4. What parts of North America are likely to experience the most frequent changes in air mass during the summer and winter?

5. Distinct temperature changes can be detected across narrow regions that are not associated with fronts. What can cause such conditions to exist?

6. The southwestern United States experiences what is locally referred to as a monsoon. Can we say the same thing about Florida or Texas?

7. Warm fronts are extremely rare over southern California. Why?

8. What does the presence of a continental polar air mass tell you about the height of the 500 or 300 mb levels?

9. Where in North America might you expect to see the collision of mT and cT air masses? What time of year would this be most likely?

10. Which types of fronts are most and least likely to have inversions?

Problems and Exercises

1. The following temperatures and dew points are observed. What are the likely types of air masses present for each?

   a. T = 29 °C (85 °F)   Dew Pt = 19 °C (66 °F)
   b. T = –18 °C (0 °F)   Dew Pt = –21 °C (–5 °F)
   c. T = 3 °C (38 °F)   Dew Pt = 0 °C (32 °F)
   d. T = 38 °C (100 °F)   Dew Pt = –4 °C (25 °F)

2. Assume that a cold front has a slope of 1:100 and that the height of the 700 mb level is 1500 m. How far behind the surface position of the front will the 700 mb position be?

Quantitative Problems

The Web site for this book (**www.MyMeteorologyLab.com**) offers some quantitative problems that can help reinforce the concept of air masses and fronts. We encourage you to go to the Chapter 9 page of the Web site and answer the problems.

Useful Web Sites

www.hpc.ncep.noaa.gov/html/sfc_archive.shtml
Archive of current and past maps depicting the positions of fronts across North America, including the most recent 7-day loop.

weather.unisys.com/surface/sfc_front.html
Map of current position of surface fronts across North America.

MyMeteorologyLab™

The *Understanding Weather and Climate,* Sixth Edition, MyMeteorologyLab™ web site contains numerous multimedia resources to aid in your study of **Air Masses and Fronts.**
Visit **www.MyMeteorologyLab.com** to:

- **Review** key chapter concepts.
- **Read** the Pearson eText, *In the News RSS feeds,* and other chapter-specific Web resources.
- **Visualize** challenging topics with Interactive Tutorials, Geoscience Animations, MapMaster interactive maps, and Weather in Motion movies and visualizations, and more.
- **Test Yourself** with self-study quizzes.

 TUTORIAL
MIDLATITUDE CYCLONES

Use the interactive animations and quizzes in this tutorial to review this chapter's key concepts.

WEATHER IN MOTION

View movies and visualizations of meteorology patterns and processes.

Effects of the 2011 Groundhog Day Blizzard
Radar Reflectivity and Air Masses
Tornadoes Ahead of a Cold Front
An Infrared View of the 2011 Groundhog Day Blizzard
Hurricanes and Air Masses

PART FOUR

Disturbances

10 Midlatitude Cyclones

TUTORIAL Midlatitude Cyclones

What is the relationship between midlatitude cyclones and flow in the middle troposphere?

11 Lightning, Thunder, and Tornadoes

TUTORIAL Doppler Radar

How can Doppler radar detect motions at various levels in a thunderstorm cloud?

We are subject to a wide variety of weather conditions from calm to severe—or even capable of bringing widespread death and destruction. The settings for these storms may be the passage of midlatitude cyclones, localized thunderstorm activities, organized clusters of thunderstorms, or tropical storms and hurricanes. This section of the book looks at the conditions associated with such activity and the processes involved in their formation and development.

Dust storm near Yuma, Arizona

12 Tropical Storms and Hurricanes

How does air circulate inside a hurricane?

Polar Front Theory

From 1914 to 1918, the world experienced one of history's great catastrophes, World War I. The advent of new weapons, including the machine gun and mustard gas, made it nearly impossible for opposing armies to gain large tracts of ground from their opponents. Until then, an army attacking with rifles and bayonets could charge its opponents with some reasonable chance of success. However, against a foe dug into trenches and armed with the latest weapons, such maneuvers were almost doomed to failure. Thus, the war zone remained stagnant for long periods, since neither army could advance across the *front*, or battle line.

While the war was going on in western Europe, Vilhelm Bjerknes (pronounced *bee-YURK-ness*) established the Norwegian Geophysical Institute in the city of Bergen. With several colleagues,[1] including his son Jacob, Bjerknes developed a modern theory of the formation, growth, and dissipation of midlatitude cyclones, storms that form along a front in middle and high latitudes. Bjerknes observed the systems forming along a boundary separating polar air from warmer air to the south. Comparing that boundary to the one separating the opposing armies in western Europe, he called his model the **polar front theory** (also called the *Norwegian cyclone model*). This theory has stood the test of time remarkably well. Though we now have far more observational information available than did Bjerknes (especially for the middle and upper troposphere), we still describe the life cycle of a midlatitude cyclone in much the same way that the scientists of the Bergen school did decades ago.

Midlatitude cyclones are large systems that travel great distances and often bring precipitation—and sometimes severe weather—to wide areas. Lasting a week or more and covering large portions of a continent, they are familiar as the systems that bring abrupt changes in wind, temperature, and sky conditions. Indeed, all of us who live outside the tropics are well acquainted with the effects of these common events.

The Life Cycle of a Midlatitude Cyclone

The Bergen meteorologists were perfectly placed to witness the atmosphere along the polar front, and they used their observations to describe the formation of midlatitude cyclones, a process called **cyclogenesis**, along this boundary. Although many cyclones do originate along the polar front, they also form in other areas, especially downwind of major mountain barriers. We first discuss the formation of midlatitude cyclones at the polar front, as described in Bjerknes's classical model. Later in this chapter we incorporate more recent insights into cyclogenesis.

<hr />

[1]Tor Bergeron, who discovered the ice crystal process for the formation of precipitation (Chapter 7), was among the scientists in this group.

TUTORIAL

MIDLATITUDE CYCLONES

Use the tutorial to observe the structure, development, movement, and evolution of midlatitude cyclones.

Cyclogenesis

Figure 10–1 illustrates the classical description of the life cycle of a typical midlatitude cyclone. Initially, the polar front separates the cold easterlies and the warmer westerlies (a). As cyclogenesis begins, a minor "kink" (b) develops along the boundary. The cold air north of the front begins to push southward behind the cold front, and air behind the warm front advances northward. This creates a counterclockwise rotation (in the Northern Hemisphere) around a weakly developed low-pressure system. With further intensification (c), the low pressure deepens and distinct warm and cold fronts emerge from the original polar front. Convergence associated with the low pressure can lead to uplift and cloud formation, while linear bands of deeper cloud cover develop along the frontal boundaries, as described in Chapter 9. Cyclogenesis along the polar front represents the initial stage in the development of a system that may affect millions of people thousands of kilometers away several days later as it progresses toward the occluded stage (d) and (e).

The Bergen scientists could not explain *why* cyclogenesis occurs, but they did observe that it commonly happened near zones of thermal contrasts (such as along coastal regions or at the boundaries between warm and cold ocean currents) or where topographic features (such as major mountain chains) disrupt the normal air flow.

Mature Cyclones

Figure 10–2a illustrates the cloud patterns, wind, uplift processes, and precipitation patterns associated with a **mature cyclone**. (The precipitation probabilities, listed as percentages in the figure, should not be interpreted too literally; they merely give a general picture.) A band of mostly cumuliform cloud cover runs along and ahead of the cold front, caused by denser, cold air displacing warm air. The likelihood of precipitation along the front increases toward the center of the low pressure, where large-scale convergence adds to the uplift caused by the meeting of the two air masses. Because of the high moisture content and generally unstable conditions typically found ahead of a cold front, precipitation—in the form of rain, snow, or even sleet or hail—can be intense. But the band of cloud cover and precipitation is relatively narrow, so precipitation may last for only a brief period before the frontal zone moves on.

A wider band of mostly stratiform clouds lies ahead of the warm front. As we found with the cold front, the likelihood of precipitation increases toward the center of low pressure. Precipitation tends to be light along the warm front because its more gradual slope leads to slower uplift. But the warm front's greater horizontal extent and generally slower forward motion

Polar Front Theory

From 1914 to 1918, the world experienced one of history's great catastrophes, World War I. The advent of new weapons, including the machine gun and mustard gas, made it nearly impossible for opposing armies to gain large tracts of ground from their opponents. Until then, an army attacking with rifles and bayonets could charge its opponents with some reasonable chance of success. However, against a foe dug into trenches and armed with the latest weapons, such maneuvers were almost doomed to failure. Thus, the war zone remained stagnant for long periods, since neither army could advance across the *front*, or battle line.

While the war was going on in western Europe, Vilhelm Bjerknes (pronounced *bee-YURK-ness*) established the Norwegian Geophysical Institute in the city of Bergen. With several colleagues,[1] including his son Jacob, Bjerknes developed a modern theory of the formation, growth, and dissipation of midlatitude cyclones, storms that form along a front in middle and high latitudes. Bjerknes observed the systems forming along a boundary separating polar air from warmer air to the south. Comparing that boundary to the one separating the opposing armies in western Europe, he called his model the **polar front theory** (also called the *Norwegian cyclone model*). This theory has stood the test of time remarkably well. Though we now have far more observational information available than did Bjerknes (especially for the middle and upper troposphere), we still describe the life cycle of a midlatitude cyclone in much the same way that the scientists of the Bergen school did decades ago.

Midlatitude cyclones are large systems that travel great distances and often bring precipitation—and sometimes severe weather—to wide areas. Lasting a week or more and covering large portions of a continent, they are familiar as the systems that bring abrupt changes in wind, temperature, and sky conditions. Indeed, all of us who live outside the tropics are well acquainted with the effects of these common events.

The Life Cycle of a Midlatitude Cyclone

The Bergen meteorologists were perfectly placed to witness the atmosphere along the polar front, and they used their observations to describe the formation of midlatitude cyclones, a process called **cyclogenesis**, along this boundary. Although many cyclones do originate along the polar front, they also form in other areas, especially downwind of major mountain barriers. We first discuss the formation of midlatitude cyclones at the polar front, as described in Bjerknes's classical model. Later in this chapter we incorporate more recent insights into cyclogenesis.

---

[1]Tor Bergeron, who discovered the ice crystal process for the formation of precipitation (Chapter 7), was among the scientists in this group.

TUTORIAL
MIDLATITUDE CYCLONES

Use the tutorial to observe the structure, development, movement, and evolution of midlatitude cyclones.

Cyclogenesis

Figure 10–1 illustrates the classical description of the life cycle of a typical midlatitude cyclone. Initially, the polar front separates the cold easterlies and the warmer westerlies (a). As cyclogenesis begins, a minor "kink" (b) develops along the boundary. The cold air north of the front begins to push southward behind the cold front, and air behind the warm front advances northward. This creates a counterclockwise rotation (in the Northern Hemisphere) around a weakly developed low-pressure system. With further intensification (c), the low pressure deepens and distinct warm and cold fronts emerge from the original polar front. Convergence associated with the low pressure can lead to uplift and cloud formation, while linear bands of deeper cloud cover develop along the frontal boundaries, as described in Chapter 9. Cyclogenesis along the polar front represents the initial stage in the development of a system that may affect millions of people thousands of kilometers away several days later as it progresses toward the occluded stage (d) and (e).

The Bergen scientists could not explain *why* cyclogenesis occurs, but they did observe that it commonly happened near zones of thermal contrasts (such as along coastal regions or at the boundaries between warm and cold ocean currents) or where topographic features (such as major mountain chains) disrupt the normal air flow.

Mature Cyclones

Figure 10–2a illustrates the cloud patterns, wind, uplift processes, and precipitation patterns associated with a **mature cyclone**. (The precipitation probabilities, listed as percentages in the figure, should not be interpreted too literally; they merely give a general picture.) A band of mostly cumuliform cloud cover runs along and ahead of the cold front, caused by denser, cold air displacing warm air. The likelihood of precipitation along the front increases toward the center of the low pressure, where large-scale convergence adds to the uplift caused by the meeting of the two air masses. Because of the high moisture content and generally unstable conditions typically found ahead of a cold front, precipitation—in the form of rain, snow, or even sleet or hail—can be intense. But the band of cloud cover and precipitation is relatively narrow, so precipitation may last for only a brief period before the frontal zone moves on.

A wider band of mostly stratiform clouds lies ahead of the warm front. As we found with the cold front, the likelihood of precipitation increases toward the center of low pressure. Precipitation tends to be light along the warm front because its more gradual slope leads to slower uplift. But the warm front's greater horizontal extent and generally slower forward motion

After reading this chapter, you should be able to:

▶ Identify the polar front theory.

▶ Outline the life cycle of a midlatitude cyclone as described in Bjerknes's model.

▶ Explain processes in the middle and upper troposphere that relate to midlatitude cyclones.

▶ Explain how surface fronts and upper-level patterns are related.

▶ Describe the behavior of a typical midlatitude cyclone as it crosses North America.

▶ Explain how flow patterns and large-scale weather patterns affect the development, steering, and dissipation of midlatitude cyclones.

▶ Explain the modern, conveyor-belt model of midlatitude cyclones.

▶ Describe anticyclones and the weather associated with them.

▶ Describe how scientists think climate change may affect midlatitude cyclones.

In late 1999 much of Europe was shaken by the most powerful storm in 50 years, with winds that topped 190 km/hr (110 mph). It brought power outages to nearly two million households, shut down three nuclear power plants, and crippled much of France's air and land transportation systems. Worse still, 97 people from across western Europe died because of the storm, mostly from being hit by falling or flying debris.

Among the European countries, France was the worst hit—both in terms of loss of life and material damage. Many of Paris's most famous landmarks were badly damaged, and 10,000 trees were toppled in Versailles. Germany, Switzerland, and Great Britain also sustained major damage and numerous fatalities.

Although European forecasters saw the storm approaching from the Atlantic and predicted that it would reach the continent, they did not expect its winds to be as severe as they were by the time it made landfall. As is the case with the west coast of North America, weather forecasting in Europe is complicated by the paucity of weather stations over the expanse of ocean to the west of the continent—a fact sometimes overlooked by the public. No sooner had the criticism started to mount over the understated forecast—two days after the storm—than another, even stronger, storm arrived, killing at least 22 more people. Thus, one very rare event was followed very shortly by a second. This rare combination of events should remind us that regardless of the technological sophistication of the twenty-first century, we are subject to the random whims of nature.

The two storm systems brought hurricane-force winds but were distinctly different from hurricanes. Unlike hurricanes, which originate over tropical waters, these *midlatitude cyclones* originate in the middle or high latitudes and are marked by well-defined fronts separating two dissimilar air masses. This chapter describes the midlatitude cyclones that are a common feature of the weather outside the tropics.

◄ Waves pummel a lighthouse on Lake Michigan on December 12, 2010, in Milwaukee, as a powerful storm hit the upper Midwest. The storm closed major highways, forced cancellation of more than 1600 flights in Chicago and caused the collapse of the roof of the Minnesota Vikings' stadium.

10 Midlatitude Cyclones

We are subject to a wide variety of weather conditions from calm to severe—or even capable of bringing widespread death and destruction. The settings for these storms may be the passage of midlatitude cyclones, localized thunderstorm activities, organized clusters of thunderstorms, or tropical storms and hurricanes. This section of the book looks at the conditions associated with such activity and the processes involved in their formation and development.

Dust storm near Yuma, Arizona

12 Tropical Storms and Hurricanes

How does air circulate inside a hurricane?

▲ **FIGURE 10–1** The life cycle of a midlatitude cyclone. (a) The stationary polar front separates opposing masses of cold and warm air. (b) Cyclogenesis first appears as a disruption of the linear frontal boundary. (c) The cyclone becomes mature; distinct warm and cold fronts extend from a low-pressure center. (d) Occlusion begins as the cold front catches up to the warm front. (e) Occlusion intensifies as more of the cold front has caught up to the warm front.

allow clouds and precipitation to last longer. Clear skies characteristically occur over the warm sector between the cold and warm fronts, although squall lines and other disturbances (discussed in Chapter 11) develop under certain conditions.

The isobar pattern, depicting the distribution of pressure within the cyclone (Figure 10–2b), is interrupted along the two fronts. This causes abrupt transitions in wind direction along the boundaries. The isobars are nearly straight in the

(a)   (b)

▲ **FIGURE 10–2** The typical structure of a mature midlatitude cyclone and the processes causing uplift. Shaded areas represent the presence of cloud cover. The numbers in (a) represent an approximation of the precipitation probability. The isobar pattern is shown in (b).

warm sector but become curved in the larger, cold region. Looking at the warm front, the winds shift from southeasterly on the cold side to southwesterly in the warm sector. Across the cold front, the winds shift from southwesterly in the warm sector to northwesterly on the cold side.

Though common, the pattern shown in Figure 10–2—with the fronts coming together in an inverted "V" shape and extending toward the southwest and southeast—does not apply to all midlatitude cyclones.[2] Like people, few midlatitude cyclones look exactly alike, and the orientation and position of the fronts can differ considerably. Figure 10–3 shows two examples of midlatitude cyclones. The midlatitude cyclone in (a) has a warm front that stretches eastward from the center of the low and a cold front extending to the south. In (b) the low-pressure system over the southern Great Lakes area has a well-defined cold front that sweeps southwestward into eastern Texas and a stationary front that extends northeastward into eastern Canada. Although the exact orientation of fronts can vary between storms, the one consistent characteristic midlatitude cyclones share is that the warm front will be located ahead of the cold front relative to the direction of movement of the system.

Occlusion

Occlusion represents the end of the cyclone's life cycle, with an **occluded front** having replaced the warm and cold fronts of the mature phase. Refer back to Figure 10–1 and observe the latter stages of a midlatitude cyclone shown in (d) and (e). Although a temperature contrast exists across the occluded front, temperature differences here are not as great as those

[2]In the Southern Hemisphere, typical midlatitude cyclones have a similar shape, but the "V" opens to the north. Thus, in both hemispheres warmer air is on the equatorial side of the storm.

along the original cold or warm fronts. West of the frontal boundary, the air flows out of the northwest and is extremely cold. Slightly warmer air approaches the occluded front from the east, but this air originates in the cold sector of the cyclone. Thus, the temperature difference is less than where the fronts separate warm, tropical air from cold, polar air.

The transitions from cyclogenesis to maturity, and from the mature phase to occlusion, are gradual, so no obviously identifiable points in time exist when the cyclone changes from one stage to another. Also, the evolution of the system coincides with a generally eastward migration of the midlatitude cyclone, though it may also have a northward or southward component.

Movement of Cyclones

Let's look at a hypothetical but realistic scenario to illustrate how the development and movement of the cyclone affects the weather. A weak disturbance in the airflow may have little noticeable effect as cyclogenesis begins off the east coast of Japan. But when the system develops into maturity and moves eastward, it can bring rain to the coastal portions of western North America and snow to the coastal mountains. If this storm occurs in the winter, upper-level winds may guide the storm southward into central and southern California and then eastward into the Rocky Mountain states. Passing over the lee of the mountains, the midlatitude cyclone may intensify and then barrel northeastward to bring blizzard conditions to the northeastern United States and southeastern Canada. As it moves offshore into the western Atlantic, a week or two after its formation in the western Pacific, the storm may undergo complete occlusion.

At a particular place (say, Kansas City, Missouri), the passage of the system brings predictable effects. As a warm front approaches, cloud cover usually deepens and increases and light to moderate precipitation is possible. The rain or snow eventually gives way to warmer, sunny conditions with the passage of the warm front, and the wind shifts from a southerly to southwesterly direction. Clear, warm conditions may then persist for a day or so. But as the cold front approaches, a strong, fast-moving band of heavy clouds and precipitation can cause major snowfall or rainfall. Afterward, the cold air behind the front brings cold, clear conditions.

Did You Know?

Scientists at the Goddard Institute for Space Studies (GISS) have compiled an online atlas of midlatitude cyclones across the globe. On average, 1071 midlatitude cyclones form each year, 578.6 in the Northern Hemisphere and 492.1 in the Southern Hemisphere. Thus, in each hemisphere an average of about one and a half new midlatitude cyclones form each day. Midlatitude cyclones of the Southern Hemisphere average lower minimum pressures at their most intense stage (972 mb) than do those of the Northern Hemisphere (988 mb). Monthly and seasonal maps of cyclone tracks for almost 40 years are available for download at http://data.giss.nasa.gov/stormtracks/.

▶ **FIGURE 10-3** Two examples of midlatitude cyclones. On June 24, 1994 (a), a fairly typical system was centered just south of the Great Lakes. A cold front extends southwestward into north Texas, and a warm front stretches eastward to the Atlantic Coast. On October 17, 1993 (b), another low is centered south of the Great Lakes, with a stationary front extending northeastward and a cold front oriented toward the southwest. Shaded areas show where precipitation is occurring. Dashed brown lines indicate the presence of a trough.

Interestingly, it was not known until the eighteenth century that storms move at all—previously they were thought to form and die in the same location. By an unfortunate turn of luck, Ben Franklin made this discovery in Philadelphia in 1743. Franklin had hoped to witness an eclipse of the moon but was disappointed when a midlatitude cyclone brought overcast conditions that completely obscured the event. Later he was told by a friend that the sky was clear in Boston at the time of the eclipse, but that storm system (the one that had ruined Franklin's observation of the event) arrived some time later. From this, Franklin rightly concluded that the system responsible for the clouds had migrated to the northeast. What he was unable to resolve, however, was how the cloud cover could move to the northeast while the winds at the time of the eclipse were from the northwest. We now know, of course, that the answer is related to the counterclockwise spiral within the midlatitude cyclone. That is, regardless of the direction in which a system is moving, winds at different points within the midlatitude cyclone flow in different directions.

Checkpoint

1. What initial conditions are required for the formation of a midlatitude cyclone?
2. In a mature cyclone, which area of precipitation might deliver the greatest rainfall totals: the area ahead of the cold front or the area ahead of the warm front? Explain.

Processes of the Middle and Upper Troposphere

The life cycle of midlatitude cyclones described above represents the state of knowledge that existed up to the 1940s. Another leap in understanding occurred during World War II, when British and U.S. pilots flying missions over Europe and Japan observed winds with speeds up to 400 km/hr (250 mph). Among meteorologists, this finding stirred an interest in the upper-tropospheric flow and how it might relate to weather conditions on the surface. As we have seen, Bjerknes and his colleagues had no information about upper-air patterns when they developed their polar front theory, and therefore they were unable to identify the causes of midlatitude cyclone development and occlusion. The next major breakthrough in the theory of midlatitude cyclones came about largely through the work of Carl Gustaf Rossby (who first described what are now known as **Rossby waves**). Rossby explained mathematically many of the linkages among upper- and middle-tropospheric winds, cyclogenesis, and the maintenance of midlatitude cyclones.

Rossby Waves and Vorticity

In Chapter 8 we described the large Rossby waves of the upper troposphere. Figure 10–4 illustrates how the air turns left and right as it flows through these waves. Moving from points 1 to 3, the air rotates counterclockwise (as indicated at the bottom left of the figure). Between points 3 and 5, it rotates clockwise. The rotation of a fluid (such as air) is referred to

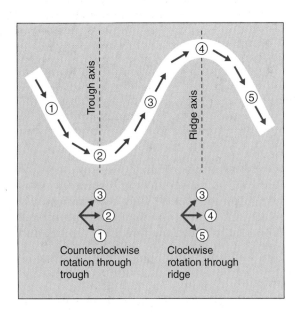

▲ **FIGURE 10-4** Vorticity around a Rossby wave. As air flows from positions 1 to 3, it undergoes a counterclockwise rotation. Along the ridge, the air turns clockwise from positions 3 to 5. The bottom of the figure shows the wind vectors representing the flow at the five positions.

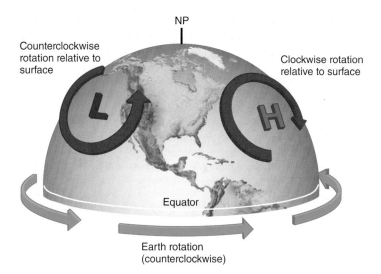

▲ **FIGURE 10-5** Earth vorticity and relative vorticity of air. As the Northern Hemisphere rotates counterclockwise, it generates Earth vorticity. Relative vorticity is the rotation of air relative to the surface, without regard to the planet's rotation. Absolute vorticity is the sum of the two.

as its **vorticity**.[3] The figure shows vorticity changes in the moving air, relative to the surface. Viewed from space, there is an additional component of vorticity arising from Earth's rotation on its axis. The overall rotation of air, or its **absolute vorticity**, thus has two components: **relative vorticity**, or the vorticity relative to Earth's surface, and **Earth vorticity**, which is due to Earth's daily rotation about its axis.

🌐 **TUTORIAL**
UPPER-LEVEL WINDS AND PRESSURE

Use the tutorial to see how midlatitude cyclones are maintained by regions of divergence in the upper-level flow. Be sure to toggle between side-view and top-view perspectives.

Relative vorticity depends on air motions with respect to Earth's surface, while Earth vorticity is a function solely of latitude—the higher the latitude, the greater the vorticity—with zero vorticity at the equator.[4] If the flow of air relative to the surface is in the same direction as the rotation of Earth itself (counterclockwise in the Northern Hemisphere), relative and Earth vorticity complement each other and increase the total or absolute vorticity (Figure 10–5). For this reason, counterclockwise rotation in the Northern Hemisphere is said to have *positive vorticity*, to be consistent with the convention used for the Coriolis force (see Chapter 4). Air rotating clockwise possesses *negative vorticity*.

---

[3]For this discussion, we are only concerned with rotation with respect to the local vertical (such as on a merry-go-round). *Vorticity* can also refer to rotation around a horizontal axis (such as with a horizontal roll of paper towels).

[4]Earth vorticity is proportional to the sine of the latitude. Thus, Earth vorticity is only 16 percent greater at latitude 55° (sin 55° = 0.819), than at 45° (sin 45° = 0.707). This means that over the midlatitudes, where Rossby waves are most likely to occur, the Earth vorticity changes with latitude are fairly small.

Figure 10–6 shows the trough from Figure 10–4 in greater detail so we can examine the air's vorticity. In segment 1, the air flows southeastward with no change in direction or speed. Because it undergoes no rotation, the air has zero relative vorticity. In segment 2, the air continuously turns to its left to yield positive relative vorticity. In segment 3, the air flows continuously toward the northeast and has no relative vorticity. Thus, the trough has three distinct regions: two with zero relative vorticity and one with positive relative vorticity. Two transition zones separate the regions of maximum and minimum (zero) relative vorticity. Across transition zone A, vorticity increases as the air flows, whereas across transition zone B the relative vorticity decreases. (The Rossby wave shown here covers a limited range of latitude. As a result, Earth vorticity changes only slightly in Figure 10–6, and changes in absolute vorticity correspond closely to changes in relative vorticity.)

At this point you might logically ask, "So what?" The answer is that vorticity changes in the upper troposphere lead to pressure changes near the surface. Let us see how. As you learned in Chapter 8, angular momentum is conserved in the absence of any outside forces. As a cowboy's twirling rope is pulled in, the reduction in the area swept by the rope causes it to twirl faster. The same thing happens when the vorticity or rotation of a parcel of air changes. That is, as the horizontal area occupied by an air parcel decreases by convergence, its vorticity or spin must increase, as in transition zone A. Decreasing vorticity, as in transition zone B, leads to divergence. This very important relationship can be summarized in the simple equation

$$-\frac{1}{\zeta}\frac{\Delta\zeta}{\Delta t} = div$$

where $-\frac{1}{\zeta}\frac{\Delta\zeta}{\Delta t}$ is the standardized change (here, the decrease) in absolute vorticity (ζ) with respect to time (t), and div = divergence. Conversely, an increase in absolute vorticity with respect to time leads to convergence. (For simplicity, we limit

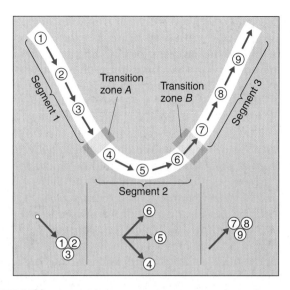

▲ **FIGURE 10-6** The change in vorticity along a Rossby wave trough. As the air flows from positions 1 to 3, it undergoes little change in direction and thus has no relative vorticity. From positions 4 to 6, it turns counterclockwise and thus has positive relative vorticity. The air flows in a constant direction from positions 7 to 9. Thus, the trough has three segments based on vorticity, separated by two transition zones.

our discussion of divergence and convergence to horizontal changes in area.)

Divergence in the upper atmosphere, caused by decreasing vorticity, draws air upward from the surface and provides a lifting mechanism for the intervening column of air. This, in turn, can initiate and maintain low-pressure systems at the surface (Figure 10–7). Conversely, increasing upper-level vorticity leads to convergence and the sinking of air, which creates high pressure at the surface. Surface low-pressure systems resulting from upper-tropospheric motions are referred to as **dynamic lows**

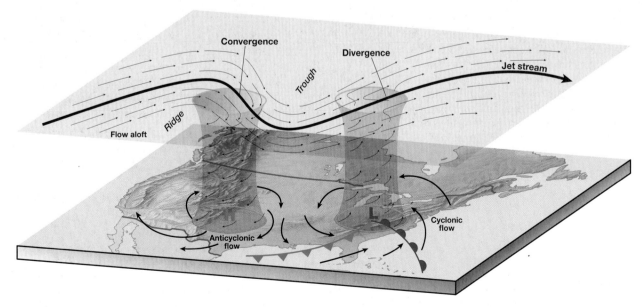

▲ **FIGURE 10-7** Upper-level convergence and divergence along favored positions on a Rossby wave create high and low pressure at the surface.

(also called *cold core lows*)—distinct from the **thermal** (*warm core*) **lows** caused by localized heating of the air from below. Cold core lows at the surface typically exist beneath regions of decreasing vorticity in the upper atmosphere, just downwind of trough axes. They are therefore associated with low pressure aloft, though the center of the low is not located directly above the surface low. In contrast, warm core lows have high pressure aloft. This arises because they have a relatively weak decline in pressure with altitude due to their higher temperatures. Thus, at some height in the middle troposphere their pressure becomes greater than that of the surrounding air, and a zone of high pressure exists. (Hurricanes, discussed in Chapter 12, are classic examples of warm core lows.)

Figure 10–8 shows a typical relationship between the distribution of 500 mb heights (representative of the pressure pattern in the middle troposphere) and absolute vorticity. The areas of greatest vorticity (shaded purple) occur along the two trough axes (in this case, over northern California and the lower Mississippi Valley). Downwind of these zones, vorticity decreases very rapidly. Thus, as air flows away from the vorticity maxima, upper-level divergence occurs, which in turn promotes low pressure at the surface. The region of lowest absolute vorticity (shaded in red) occurs near the ridge axis, centered over the Dakotas. An area of increasing vorticity that is a likely center of high pressure at the surface exists just downwind of this region. (See *Box 10–1, Physical Principles: Vorticity and the Maintenance of Rossby Waves,* for further information on vorticity patterns.)

Checkpoint

1. What are positive vorticity and negative vorticity?
2. Discuss the role of vorticity in the formation of the cyclone shown in Figure 10-7.

Surface Fronts and Upper-Level Patterns

At this point, we know that upper-level divergence causes the formation and intensification of surface midlatitude cyclones, whereas upper-level convergence causes high pressure at the surface. We have also seen that the airflow along Rossby waves can generate upper-level divergence and convergence. (Further information on divergence and convergence is presented in *Box 10–2, Physical Principles: A Closer Look at Divergence and Convergence.*) Thus, the airflow in the middle and upper troposphere has a significant effect on surface patterns. But the causality is not one way, because patterns at the surface, in particular the presence of cold and warm fronts, have their own effects on the middle and upper troposphere.

As you saw in Chapter 4, the hydrostatic equation states that the decrease in pressure with altitude (the vertical pressure gradient) is determined by the density of the air—cold, dense air has a greater vertical pressure gradient than does warm, light air. It follows that in a cold air column, the greater decrease in pressure with altitude should lead to lower pressure aloft compared to warm air. The differences in temperature on either side of a cold front must therefore lead to significant differences in upper-level pressure there (the same reasoning applies to warm fronts). We develop this idea more fully in the following section.

Cold Fronts and the Formation of Upper-Level Troughs

Rossby waves consist of large, alternating troughs and ridges that establish patterns of upper-level divergence and convergence. The troughs in the waves normally develop behind the position of surface cold fronts, not by some grand coincidence

▶ **FIGURE 10–8** Values of absolute vorticity (shown in dashed lines) on a hypothetical 500 mb map. Notice that the greatest values appear near trough axes (vorticity in units of 10^{-4} sec^{-1}). As air flows away from the areas of maximum vorticity, the decrease in that property results in upper air divergence. In this instance, that would occur near the Northern California–Nevada border and Louisiana–Arkansas regions.

Vorticity and the Maintenance of Rossby Waves

We have seen that the vorticity associated with airflow has two components. The first, called *Earth vorticity*, arises from the planet's 24-hour rotation. A person sitting on a chair at the North Pole undergoes one complete rotation each day and thus has maximum Earth vorticity. Like the Coriolis force, Earth vorticity increases with latitude so that it is at its maximum at either pole and is nonexistent at the equator.

We define the second source of rotation, *relative vorticity*, in terms of motions of air relative to the surface. Relative vorticity itself has two sources. The first is the *shear* that occurs when the speed of a fluid varies across the direction of flow. Figure 1a illustrates this process as water flows through a channel (the same applies to air, but we use water as an example simply because it is easier to visualize).

The flow is faster along the right bank. If a paddlewheel was fixed in the middle of this stream, the faster-moving water to the right would exert a greater force on the wheel than would the slower-moving flow on the left. This would cause the wheel to rotate counterclockwise and thereby undergo vorticity due to shear.

The second source of relative vorticity depends on the *curvature* of the flow, as shown in Figure 1b, where the curved channel forces a fluid to turn counterclockwise. As air flows through a Rossby wave, it undergoes this type of curvature. As we said, relative vorticity is the sum of curvature and shear; both are signed quantities. When they have the same sign, they act in the same direction. When the signs are different, they offset one another, perhaps completely. In general, shear is much less important to the occurrence of relative vorticity than curvature.

Absolute vorticity, the overall rotation of a fluid, like angular momentum, is conserved—that is, in the absence of intervening forces,

it remains constant. Figure 2 shows a Rossby wave in the Northern Hemisphere. As the air flows southward, west of the trough axis, its Earth vorticity decreases (recall that Earth vorticity decreases toward the equator). But because absolute vorticity is conserved, an increase in relative vorticity compensates for the decrease in Earth vorticity, causing the air to turn counterclockwise. Then, as it starts to flow poleward, the Earth vorticity increases. Thus, the air turns back to its right and once again exhibits negative relative vorticity. (For the sake of clarity, we are assuming no shear exists.)

In short, there is a constant trade-off between Earth and relative vorticity. As the air moves poleward, it assumes a greater clockwise rotation relative to the surface; as it moves toward the equator, it turns in a counterclockwise manner. Such reversals in relative vorticity, along with the conservation of angular momentum, help maintain Rossby waves.

(a)

(b)

▲ **FIGURE 1** Relative vorticity by shear (a) occurs when a fluid moves at a differential speed across the direction of flow. A paddle wheel fixed across the fluid rotates counterclockwise as a greater forward stress is exerted to the right of the direction of flow. Curvature in the direction of flow can also produce relative vorticity (b).

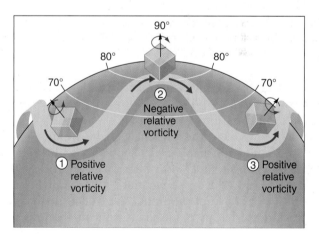

▲ **FIGURE 2** The maintenance of a Rossby wave. As the air moves poleward from position 1, the gain in Earth vorticity is compensated for by a decrease in relative vorticity. As the parcel approaches position 2, the relative vorticity becomes negative and the parcel turns back equatorward. At position 3, the reduction of Earth vorticity causes an increase in relative vorticity, and the air again turns poleward.

A Closer Look at Divergence and Convergence

Upper-level divergence and convergence are changes in the horizontal area occupied by an air parcel, but their real importance lies in their effects on vertical motions. In its simplest form, divergence is nothing more than air spreading out over a greater horizontal area (convergence, of course, is the exact opposite). Divergence and convergence can occur in two ways. The first is by an increase or a decrease in the speed of air as it flows. The second is by a stretching out or pinching inward of the air, in a direction perpendicular to the direction in which it is moving. The divergence and convergence described earlier in this chapter can take either form.

Speed Divergence and Speed Convergence

Speed divergence and **speed convergence** occur when air moving in a constant direction either speeds up or slows down. Consider the two parcels of air, A and B, in Figure 1a. Both parcels are moving in the same direction, but parcel B moves faster, as indicated by the length of the arrows.

▲ **FIGURE 1** Speed divergence and convergence. Note that the values shown on the lines in (c) and (d) represent the height of the 500 mb level in meters.

Because the leading parcel has greater speed than the one behind it, the distance between the two increases with time (b). This is an example of speed divergence.

This form of divergence is analogous to what might happen in a race with many entrants at the starting line. Initially, the runners cluster together, with little space between them. When the starting gun goes off, the people at the front of the pack dash away from those farther back, who shuffle along as they wait for the crowd to move forward. The cluster of people gradually thins out as the faster runners pull away from the slower ones, and the same number of people now occupy a greater area.

Because wind speed on an upper-level weather map is directly proportional to the

---

but in response to the presence of the fronts. Figure 10–9 illustrates how this happens by showing the temperature and pressure changes in a 1 km thick layer of air on either side of a cold front. The air above point A lies entirely within the warm sector ahead of the front. The frontal boundary slopes backward so that at B the air is cold in the lowest 500 m and warm in the upper 500 m. Cold air occupies the entire kilometer-thick layer above C.

Now let's assume (for simplicity's sake) that the surface pressure is 1000 mb at all three locations and compare the vertical pressure distributions. Above A, the pressure drops 55 mb in the lowest 500 m and another 53 mb in the next 500 m, to yield a pressure of 892 mb at the 1 km level. Above B, the pressure drops 58 mb in the cold, lowest 500 m (3 mb more than that above A). But, as in the upper 500 m above A, the pressure drops 53 mb to the 1 km level. Thus, the pressure at that height is 889 mb—3 mb less than at A. Over C, the pressure drops 58 and 56 mb, respectively, in the lower and upper

500 m layers, so that the pressure at 1 km is 886 mb. Thus, the pressure at 1 km decreases from 892 to 889 to 886 mb across the frontal boundary. This is how the differing temperature characteristics cause a horizontal pressure gradient in the middle atmosphere. The sloping boundary of a warm front exerts the same sort of effect.

The cold front shown in Figure 10–9 marks a zone of strong horizontal temperature contrasts in the lower half of the troposphere. Because the near-surface temperature patterns strongly influence the upper-level pressure distribution, the polar jet stream lies right above the frontal boundary. This is exactly what was described in Chapter 8, but now we see that the polar jet stream can often be part of a larger Rossby wave pattern. In fact, the polar jet stream usually demarcates the boundary of the Rossby waves. We can also see that this Rossby wave pattern and the polar jet stream are strongly linked with a midlatitude cyclone's cold and warm fronts.

spacing of height contours, we can use these maps to identify regions of speed divergence. Specifically, speed divergence occurs where contour lines come closer together in the downwind direction. In Figure 1c, the wind speed, indicated by the length of the arrows, increases in the direction of flow and causes speed divergence.

Speed convergence occurs when faster-moving air approaches the slower-moving air ahead. In the example of runners in the race, convergence might occur behind a muddy part of the track that slows the runners. The fastest runners, who have pulled ahead of the others, are the first to encounter the muddy spot. As they slog through the muck, the trailing runners have the opportunity to catch up. The entire pack bunches up in a smaller area and convergence occurs. A similar phenomenon occurs in Figure 1d.

Diffluence and Confluence

A second type of divergence and convergence, **diffluence** and **confluence**, occurs when air stretches out or converges horizontally due to variations in wind direction. In Figure 2a, a certain amount of air is contained in the shaded area between points 1 and 3. As it passes to the region

between points 2 and 4, the same amount of air occupies a greater horizontal area. This is diffluence, a pattern that commonly appears wherever vorticity changes cause divergence. Confluence is shown in Figure 2b.

Close inspection of Figure 2 reveals an interesting relationship between the different types of convergence and divergence. Diffluence occurs where height contours on an upper-level map spread apart in the upwind direction (confluence occurs where they converge). But where height contours are spread farther apart there is a lesser horizontal pressure gradient, and therefore weaker winds. So diffluence (a type of divergence) in the upper figure occurs at the same time that speed convergence is occurring; two opposite processes are occurring at once. We know that divergence occurs downwind of a trough axis. How does divergence actually occur downwind of a trough axis when diffluence and speed convergence occur simultaneously? The answer is that in most instances the diffluence is greater in magnitude than the accompanying speed convergence, so the sum of the two yields a net divergence.

(a)

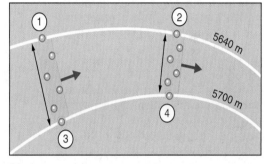

(b)

▲ **FIGURE 2** Diffluence and confluence.

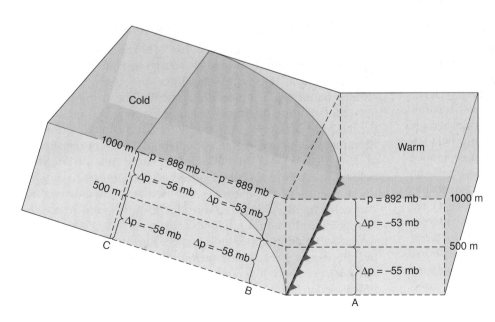

◄ **FIGURE 10-9** Temperature distributions in the lower atmosphere lead to variations in upper-level pressure. Above A, the entire column of air in the lower atmosphere is warm, so the pressure drops relatively slowly with height. At B, cold air occupies the lowest 500 m, with warmer air aloft. This leads to a slightly lower pressure at the 1 km level. At C, cold air occupying the lowest 1000 km causes a greater rate of pressure decrease with altitude and, as a result, a lower pressure at the 1 km level. In this way, the existence of sloping frontal boundaries establishes horizontal pressure gradients in the upper and middle atmosphere.

10–3
FORECASTING

Short Waves in the Upper Atmosphere and Their Effect on Surface Conditions

While Rossby waves play a role in establishing regions of upper-level divergence and convergence, they are not the only waves in the atmosphere that do so. The atmosphere also contains smaller eddies. Some of these, called **short waves**, are smaller ripples superimposed on the larger Rossby waves. These eddies migrate downwind within the Rossby waves and exert their own impact on the life cycle of midlatitude cyclones. Depending on where they are located within the Rossby waves, they can either enhance or reduce the local divergence or convergence.

The formation of short waves depends on **temperature advection**, the horizontal transport of warm or cold air by the wind. Because air in the upper troposphere is well removed from the direct source of atmospheric heating (the surface), the temperature changes we experience from day to night are barely perceptible in the upper atmosphere. Upper-level air changes temperature very slowly as it moves from one region to another, and air flowing

▲ **FIGURE 1** A barotropic atmosphere (a) exists where the isotherms (the dashed lines showing the temperature distribution) and height contours (solid lines) are aligned in the same direction. No temperature advection occurs when the atmosphere is barotropic. A baroclinic atmosphere occurs where the isotherms intersect the height contours. Cold air advection is occurring in (b), warm air advection in (c).

horizontally from a warm region can retain its high temperature as it moves into a region otherwise occupied by cold air. We refer to the horizontal movement of relatively warm air as **warm air advection**. The opposite, of course, is called **cold**

air advection. Both types of temperature advection appear on Rossby waves and, as we will see, affect the development of surface cyclones.

A useful method for detecting warm and cold air advection on a map of

---

Interaction of Surface Upper-Level Conditions

Despite the fact that we commonly refer to the "surface level" and the "upper level" of the atmosphere, they are *not* separate entities; they are simply different parts of a single atmosphere that are fully connected and intertwined with each other. The upper-level winds influence surface conditions by generating divergence and convergence that lead to the formation of surface cyclones and anticyclones (Chapter 4). At the same time, the temperature patterns in the lower atmosphere affect the rate at which pressure decreases with altitude and thereby influence the upper-level wind flow. More specifically, upper-level patterns with strong north–south components (meridional flow) cause the formation of midlatitude cyclones downwind of trough axes (where vorticity decreases). At the same time, the presence of fronts can cause Rossby waves in the upper

atmosphere. The usual juxtaposition of surface cold fronts and upper-level troughs and ridges is shown in Figure 10–10.

The bottom line is that the interconnectedness between surface patterns and those aloft provides the true foundation for understanding the life cycle of midlatitude cyclones. While the model of cyclogenesis, maturity, and occlusion described by Bjerknes and his colleagues provided an excellent *description* of the life cycle of midlatitude cyclones, the *explanation* behind the processes eluded the early scientists. Today we know that upper-level patterns and their associated divergence and convergence affect pressure distributions (and hence the changes in midlatitude cyclones) at the surface. Furthermore, we know that the upper-level patterns are influenced in turn by temperature conditions near the surface. (See *Box 10–3, Forecasting: Short Waves in the Upper Atmosphere and Their Effect on Surface Conditions,* for a more detailed analysis of upper-level and surface phenomena.)

the upper atmosphere is to compare the orientation of height contours and isotherms. Figure 1 illustrates three possible patterns of 500 mb height levels and temperature advection. In Figure 1a, parallel height contours (solid lines) are aligned in a west-to-east direction so that a geostrophic wind flows from west to east. The isotherms (the dashed lines showing the temperature distribution at the 500 mb level) run parallel to the height contours and indicate a northward decrease in temperature. Because the airflow is parallel to the height contours (and in this case the isotherms), the temperature is the same (just less than –25 °C) at positions 1 and 2, and there is neither cold nor warm air advection. When the height contours and isotherms are in alignment, the atmosphere is said to be **barotropic**.

In Figure 1b, the height contours are parallel to each other, as are the isotherms. But in this instance the height contours and isotherms intersect each other, and the temperature increases from position 3 (below –25 °C) to position 4 (above –20 °C). This is an example of cold air advection, wherein a parcel of colder air is transported from 3 to 4. The opposite situation, warm air advection, occurs in Figure 1c, where the temperature

decreases in the direction of airflow. When the height contours and the isotherms intersect, as in both (b) and (c), the atmosphere is said to be **baroclinic**.

Refer to Figure 2 and observe the two baroclinic zones at positions 1 (cold advection) and 2 (warm advection). Where cold advection exists, the entering air is denser than the air ahead of it because of its lower temperature. This gives it a negative buoyancy that causes it to sink downward, bringing cold air toward the surface. Cold air advection typically occurs behind a cold front, thereby enhancing the temperature contrast found on either side of the front. Where warm advection occurs, entering air is warmer and more buoyant than the air ahead of it and therefore rises. The warm and cold air advection thus cause vertical motions similar to those associated with statically unstable air (Chapter 6). This situation is called **baroclinic instability**.

In addition to undergoing rising or sinking motions, the air in areas of warm or cold air advection undergo a slight turning—to

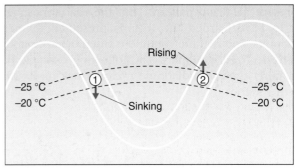

▲ **FIGURE 2** Warm and cold air advection around a Rossby wave. The air at position 1 flows from colder to warmer air, resulting in cold air advection. Along a zone of cold air advection, sinking motions and a turning of the air to the right tend to take place. Warm air advection occurs at position 2 along with a rising of the air.

the right in areas of cold air advection and to the left in regions of warm air advection. These motions are what cause the ripples (short waves) to form on the Rossby waves. When a short wave is located downwind of a Rossby wave trough axis, the divergence is enhanced and surface cyclones intensify.

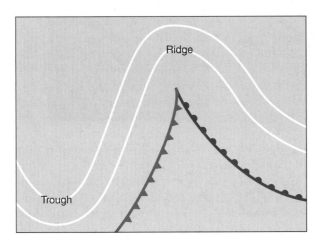

▲ **FIGURE 10-10** The effect of differing vertical pressure gradients on either side of warm and cold fronts leads to upper-tropospheric troughs and ridges.

Checkpoint

1. Assuming the same pressure at the surface, which would have higher pressure at an altitude of 1 km: a cold air column or a warm air column? Explain.
2. Why does the polar jet stream lie above the frontal boundary?

Did You Know?

People are often surprised to learn that the National Oceanic and Atmospheric Administration (NOAA) routinely flies planes into active hurricanes as part of their tracking and forecasting procedures. This sounds risky indeed. Yet the maximum winds in those hurricanes are often only half as strong as those of the polar jet stream that lies above the surface fronts—and commercial aircraft fly through these jet streams every day!

An Example of a Midlatitude Cyclone

The descriptions and examples of troughs, ridges, cyclones, anti-cyclones (each described in Chapter 4) and fronts (Chapter 9) represent idealized patterns. But the real atmosphere seems not to have learned the lessons presented in meteorology textbooks, and real-life conditions often depart markedly from these idealized examples. The following example from 1994 is a fairly typical midlatitude cyclone making its trek across North America.

April 15

Figure 10–11 shows the surface and 500 mb weather maps and a satellite image of North America on April 15. Clear skies, low humidities, and light winds dominate the western United States and southwest Canada. In contrast, the midlatitude cyclone over the north-central part of the United States has brought strong winds, overcast skies, and heavy rain showers. The surface wind in the warm sector flows northward out of the southern states and turns somewhat to the northwest as it approaches the center of the low pressure. North and west of the system, the air rotates counterclockwise around the low. Temperatures in the

(a)

(b)

(c)

▲ **FIGURE 10–11** Weather maps of the surface (a) and 500 mb level (b), and a satellite image (c) for April 15, 1994, 7 A.M. EST. Note that the positions of the surface fronts have been superimposed on the 500 mb map. Precipitation occurs in shaded areas.

warm sector are typically in the 60s to low 70s °F, considerably greater than those to the west of the cold front.[5]

At the 500 mb level (b), a well-defined trough exists to the west of where the surface cold front is located. Strong westerly winds in excess of 100 km/hr (60 mph) occur over the western portion of the Canadian–United States border

---

[5]Observed temperatures, dew points, wind velocities, and sea level pressures are depicted on the surface map for a few stations. Note that the temperatures and dew points (°F) are shown on the upper and bottom left, respectively, of the so-called *station models*. A line coming out of the circle shows the direction from which the wind is blowing, and the number of short and long tickmarks attached near the end of the line represents the wind speed. *Appendix C: Weather Map Symbols* provides more detailed information.

and then become northwesterly as they enter the trough. Downwind of the trough axis, the air again flows eastward toward the North Atlantic.

Just downwind of the trough axis, the vorticity decreases over northern Missouri, Iowa, and Minnesota. As we would expect, diffluence in this region leads to net divergence aloft and low pressure at the surface. Out west, downwind from the ridge axis, upper-level convergence leads to high pressure at the surface centered over central Idaho.

April 16

Over the next 24 hours (Figure 10–12), the midlatitude cyclone migrated some 800 km (500 mi) to the northeast. It

(a)

(b)

(c)

▲ **FIGURE 10–12** Weather maps and a satellite image as in Figure 10–11, for April 16.

also intensified, with about a 5 mb decrease in sea level pressure from the previous morning. For the first time, a portion of the cyclone becomes occluded north of the Great Lakes and rainshowers cover southeast Canada and the northeastern United States. The anticyclone over the western United States expanded southeastward and intensified slightly to a maximum sea level pressure of about 1027 mb.

The 500 mb pattern (Figure 10–12b) shows considerable change from the day before. The trough moved east, and its axis changed from the north–south orientation of the day before to a northwest–southeast orientation. The trough also intensified, with peak winds now reaching about 150 km/hr (90 mph) over the east-central United States. Another interesting change over the last 24 hours is the closing of the 5280 m contour over southern Ontario to form what is called a **cutoff low**. Although the main flow of air loops around the low and eventually flows off the map, a circular rotation of the air is also embedded in the trough.

April 17

During the 24 hours preceding the morning of April 17, the surface low-pressure center migrated a short distance to the northeast, with no change in central pressure (Figure 10–13).

(a)

(b)

(c)

▲ **FIGURE 10–13** Weather maps and a satellite image as in Figure 10–11, for April 17.

▶ **FIGURE 10-14** Weather maps and a satellite image as in Figure 10–11, for April 18.

(a)

(b)

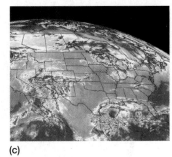

(c)

At the same time, the occluded front swept northward so that it is situated in a nearly west-to-east direction over southeast Canada. Precipitation, mainly as snow, is scattered along the frontal boundary and concentrated near the low-pressure center east of Hudson Bay.

The upper-level low (Figure 10–13b) has continued to deepen from the day before, as indicated by the lower heights of the 500 mb level. Not only has the height of the 500 mb level at the center of the cutoff low decreased by about 90 m, but the number of closed height contours has increased to four (from the previous one).

April 18

By April 18 (Figure 10–14), the center of the low-pressure system and most of the frontal boundaries have migrated off the surface map, but the system still is evident on the 500 mb map as the large trough extending to the southeast. As far as most of the population of eastern Canada and the United States is concerned, the midlatitude cyclone no longer exerts any direct influence on the weather.

Checkpoint

1. Referring to Figure 10–11, describe the situation that existed on the morning of April 15 at the surface and at the 500 mb level.

2. How closely does the polar jet stream correspond to the location of the major fronts of the storm over the entire period depicted in Figures 10–11 through 10–14?

Flow Patterns and Large-Scale Weather

We have seen that changes in upper-level vorticity create divergence and convergence patterns that influence the formation, intensification, and dissipation of surface cyclones and anticyclones. Thus, strong looping motions of upper-level air are likely to create distinct regions of high and low pressure at the surface. Conversely, air over a large area such as North America flowing in a straight westerly direction will have uniform vorticity, with little chance for significant upper-level divergence or convergence. Without upper-level divergence or convergence, no major areas of high or low pressure develop at the surface.

Compare the 500 mb maps in Figure 10–15. In (a), the height contours exhibit a zonal pattern with a minimum of north–south displacement. In contrast, the pattern in (b) shows a strong meridional component.

(a)

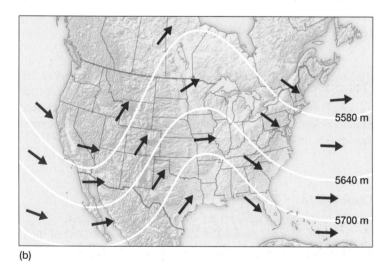

(b)

▲ **FIGURE 10–15** Zonal (a) and meridional (b) flow patterns. Meridional air flow has a greater likelihood of generating large-scale precipitation.

Because they have no pronounced vorticity changes, zonal patterns hamper the development of intense cyclones and anticyclones. They are therefore more often associated with a large-scale pattern of light winds, calm conditions, and no areas of widespread precipitation. Certainly there may be areas of localized precipitation—and the precipitation may even be quite heavy, as when orographically produced—but this activity will be spotty and widely scattered. Meridional flow, in contrast, can lead to the formation of major cyclones and anticyclones. If you look at an upper-level weather map and see strongly meridional flow, you can expect that some areas are experiencing cloudy and wet conditions while others are calm and dry. If you see a zonal pattern, it is less likely that large temperature contrasts exist from place to place or that there are large areas of heavy precipitation.

Experience shows that large-scale wind patterns in the upper atmosphere often persist, with one general type of pattern dominating for weeks or longer at a time. Such persistence of a zonal or meridional pattern can lead to droughts or episodes of heavy precipitation. A persistent zonal pattern can cause very widespread droughts due to the lack of vorticity. Regional droughts can also occur if a meridional pattern remains in place, with the zone of upper-level convergence downwind of a ridge axis persisting over a particular region.

The Steering of Midlatitude Cyclones

Upper-level winds have another important effect on surface conditions by governing the direction and speed at which the surface systems move. Outside of the tropics, the upper atmosphere includes a strong component of west-to-east flow. Likewise, experience tells us that both cyclones and anticyclones outside of the tropics typically migrate eastward. These two facts are not mere coincidence. In fact, the movement of surface systems can be predicted by the 500 mb pattern, with the surface systems moving in about the same direction as the 500 mb flow, at about one-half the speed. Keep in mind, however, that the 500 mb level wind pattern changes through time, so predicting the track of a cyclone involves more than just examining the current upper-level flow and assuming a constant movement parallel to the current height contour pattern; one must also predict the change in the 500 mb pattern.

Many midlatitude cyclones have their origin over the north Pacific off the coast of Japan. Upon reaching the Aleutian Islands of Alaska, the systems can die out, migrate toward the southeast, or continue on an eastward path across British Columbia. The most likely path the cyclones take upon reaching North America varies with the season, with northern treks favored in the summer, and movement toward the southeast more likely in the winter.

Upper-level winds are about twice as vigorous on average in the winter than in the summer. During the winter, net radiation decreases rapidly with increasing latitude, giving rise to a stronger latitudinal temperature gradient than in summer (Chapter 3). This results in greater pressure gradients (and winds) in the upper atmosphere. It is no surprise that midlatitude cyclones generally move faster in the winter.

Did You Know?

At one point the famous "storm of the century" that hit the eastern United States and Canada in March 1993 recorded a central barometric pressure of 960 mb—lower than that associated with 80 percent of the hurricanes and tropical storms that make landfall in the United States. It was also an unusually large midlatitude cyclone that affected 26 of the 50 United States, spawning blizzards that shut down virtually every airport on the eastern seaboard.

Though a winter midlatitude cyclone can take many different paths across North America, two are particularly common: the Alberta Clipper and the Colorado low (Figure 10–16). The Alberta Clipper is associated with zonal flow and a polar jet stream that sweeps across southern Canada and the northern United States. Though it can bring frigid conditions, snowfall is usually light. In contrast, some midlatitude cyclones passing farther to the south over western North America spawn new centers of low pressure as they pass over the central Rocky Mountains. They then follow a path from the southern Plains toward the northeastern United States and eastern Canada. These storms, usually warmer and containing greater amounts of water vapor in the air, often produce extremely heavy snowfall.

Storms occasionally have their genesis well to the south of the polar front (in contrast to the original model of the early Norwegian meteorologists) and can track northward along the eastern United States. Such storms often have strong uplift and high water vapor contents—conditions favorable for the development of extremely heavy snowfalls. Such conditions led to the "storm of the century" in March of 1993, which produced strong winds and record-breaking snow accumulations over the eastern third of the United States.

Migration of Surface Cyclones Relative to Rossby Waves

For a midlatitude cyclone to form, there must be upper-level divergence. If there is more divergence aloft than convergence near the surface, the surface low deepens and a cyclone forms. If the convergence at the surface exceeds the divergence aloft, more air flows into the low than exits it, and the low weakens and dies out,

Although the optimal place for midlatitude cyclones to develop is just beneath the zone of decreasing vorticity aloft, they don't usually remain in a fixed position relative to the upper-level trough. Instead, they are usually pushed along so that they migrate in the same direction (and at about half the speed) as the winds at the 500 mb level. Figure 10–17 shows the movement of a surface cyclone relative to an upper-level trough and ridge pattern (for simplicity, we assume that the position of the Rossby wave containing the trough and ridge remains fixed in time, though this is not usually the case for extended periods). In (a), the surface low associated with a cyclone exists under the zone of upper-level divergence. The divergence aloft provides the uplift that maintains the surface low. At the same time, however, the upper-level winds are moving the surface system northeastward relative to the Rossby wave. Thus the cyclone moves away from the region of maximum divergence aloft and eventually locates beneath the zone of upper-level convergence (b). At this point air descends toward the center of the low, the requisite uplift that maintained the surface low ceases, and the subsiding air causes the surface pressure to increase. The cyclone eventually dies out as its air pressure becomes nearly the same as that surrounding it. Note that this process occurs simultaneously with the occlusion process.

Checkpoint

1. How does an Alberta Clipper differ from a Colorado low? Explain.
2. What role do convergence and divergence in Rossby waves play in the eventual dissipation of a midlatitude cyclone?

The Modern View: Midlatitude Cyclones and Conveyor Belts

We have examined the structure of midlatitude cyclones at the surface and in the middle troposphere, but we should not overlook the fact that cyclones are three-dimensional

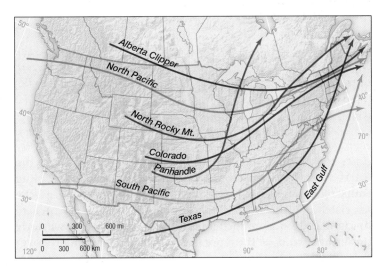

▲ **FIGURE 10–16** Typical winter storm paths across North America. These midlatitude cyclones are steered by the jet stream.

▲ **FIGURE 10–17** Surface high- and low-pressure centers can migrate relative to the Rossby wave aloft. In (a), the surface low exists below the area of upper-level divergence, and the resultant uplift maintains or strengthens the cyclone. Surface systems are generally guided by upper-level winds, so eventually the center of the low might be displaced to the zone where upper-level convergence occurs (b). The sinking air fills into the low, causing its demise. This process occurs as the cyclone undergoes the transition from its mature to occluding stages.

entities. In other words, the 500 mb level (or any other level) is not separate from the surface portion—it is simply a different part of the same system. The **conveyor belt model** (Figure 10–18), which we look at next, provides a better depiction of the three-dimensional nature of midlatitude cyclones.

This model describes the midlatitude cyclone in terms of three major flows. The first flow, the *warm conveyor belt*, originates near the surface in the warm sector and

flows toward and over the wedge of the warm front. As the warm belt flows up the frontal surface, adiabatic cooling leads to condensation and precipitation. Moreover, as the air rises into the middle troposphere, it begins to turn to its right and become incorporated into the general westerly flow downwind of the upper-level trough. The cloud cover associated with the rising warm conveyor belt appears prominently as the bright, wide band extending from south to north in (b).

(a)

(b)

▲ **FIGURE 10–18** The conveyor belt model of midlatitude cyclones (a). The warm conveyor belt between the cold and warm fronts is responsible for much of the large band of cloud cover evident in the satellite image (b). The cold conveyor belt flows toward the center of the low pressure system at the surface and rises as it does so, producing the cluster of clouds over the Dakotas and Minnesota. The dry conveyor belt flows in the upper troposphere, with some of the flow sinking as it approaches the cold front. The rest of the conveyor belt remains in the upper atmosphere and remains dry and relatively cloud free. That is evident in (b) as the gap over Wisconsin separating the major cloud regions.

The *cold conveyor belt* lies ahead (north) of the warm front. It enters the storm at low levels as an easterly belt flowing westward toward the surface cyclone. But like the warm conveyor belt, it too ascends as it flows, turns anticyclonically (clockwise in the Northern Hemisphere), and becomes incorporated into the general westerly flow aloft. Although it originates as cold (and therefore relatively dry) air, the cold conveyor belt gains moisture from the evaporation of raindrops falling from the warm conveyor belt above. The cold conveyor belt extends in (b) from northern Michigan to eastern South Dakota.

The final component of the three-dimensional circulation is the *dry conveyor belt* that originates in the upper troposphere as part of the generally westerly flow. This broad current brings the coldest air into the cyclone, and it is important in maintaining the strong temperature contrast across the cold front. The upper-level air sinks slightly as it approaches the cold front from the west, but then it rises and merges with the general upper-level flow. The dry conveyor belt separates the cloud bands from the warm and cold conveyor belts, creating the relatively cloud-free area seen between the brightly colored area of clouds associated with the warm and cold conveyor belts in (b).

 TUTORIAL
MIDLATITUDE CYCLONES

Use the tutorial to observe the airflow patterns in and around midlatitude cyclones. Be sure to toggle between side-view and top-view perspectives.

Anticyclones

So far we have said little about anticyclones, but they are as much influenced by upper-level conditions as are cyclones, and they also have an impact on weather. (Recall from Chapter 4 that these phenomena are areas of high pressure around which the wind blows clockwise in the Northern Hemisphere.) While cyclones can bring heavy precipitation and strong winds, anticyclones foster clear skies and calm conditions because the cool air within them slowly sinks toward the surface.

That should not be taken to mean that anticyclones are always associated with wonderful weather. Indeed, outbreaks of continental polar (cP) air over the eastern United States are associated with anticyclones behind southward- or southeastward-moving cold fronts. Furthermore, anticyclones often tend to remain over a region for a long time, which can lead to droughts. Finally, anticyclones over the Rocky Mountains can lead to Santa Ana wind conditions over the West Coast.

This chapter has presented the characteristics of midlatitude cyclones and anticyclones that influence weather outside the tropics. We know, however, that the atmosphere often undergoes violent types of weather, usually on smaller time and space scales than those associated with midlatitude cyclones. In Chapter 11 we will meet thunderstorms and tornadoes, phenomena that can cause considerable damage and loss of life.

Climate Change and Midlatitude Cyclones

Midlatitude cyclones are intricately associated with the polar front and polar jet stream, which situate farther poleward in the summer months and migrate to lower latitudes during the cold season. This is particularly evident over the North Pacific Ocean, where summer storms routinely approach North America along western Canada and Alaska, but commonly pass southward to southern California and Baja California during the winter.

We have seen in earlier chapters that increasing greenhouse gas concentrations has affected global temperatures, cloud conditions, and precipitation, and will very likely continue to do so. We also need to consider whether climate change might also cause a shift in the average tracks of midlatitude cyclones; the evidence suggests this is indeed the case.

Recent studies have shown that since the middle of the twentieth century, the average path of wintertime midlatitude cyclone tracks has shifted poleward. This change has been strongest in the Southern Hemisphere but also shows up in the Northern Hemisphere. This could have dire consequences for some regions of the globe. Take the case of the arid and semiarid western United States, where some very extensive and severe droughts have contributed to extremely destructive and widespread fires in recent years. If this northward shift in storm tracks becomes permanent, the region will witness fewer storms each year and be forced to deal with reduced water supplies and an increasing vulnerability to major burns.

In 2007 the Intergovernmental Panel on Climate Change (IPCC) concluded that continued warming will likely lead to a further poleward displacement of storm tracks. Some of the regional patterns identified included a reduction of Atlantic midlatitude cyclones approaching the Mediterranean as storms become more frequent in extreme Northern Europe, increasing aridity over southern Australia, and further drying over much of the western United States.

Summary

Although much of our knowledge of midlatitude cyclones comes directly from the work of Norwegian meteorologists in the early twentieth century, recent insights into upper-level winds have greatly increased our understanding. Midlatitude cyclones and anticyclones both depend on a close interaction between processes occurring in the upper and lower troposphere. Counterclockwise rotation in the Northern Hemisphere has positive vorticity, while clockwise rotation has negative vorticity. The greatest positive vorticity occurs around the trough axis of a Rossby wave. The decreasing vorticity immediately downwind of the axis causes divergence, which leads to the formation of cyclones at the surface. At the same time, the fronts associated with midlatitude cyclones help form the Rossby waves that create upper-level convergence and divergence. Thus, a constant interaction takes place between the upper and lower atmosphere.

Traditionally, the distribution of clouds along frontal boundaries has been linked to convergence and overrunning on either side of the fronts. The modern approach explains the cloud distributions as the result of three separate air flows, or conveyor belts. A cold conveyor belt flows toward the center of low pressure ahead of the warm front. As it approaches the low pressure, it rises, and adiabatic cooling produces the wide band of cloud cover. Similarly, a rising conveyor belt of warm air flows ahead of the cold front to provide another band of cloud cover. Both the warm and cold conveyor belts turn anticyclonically near the center of low pressure and join the upper-level westerly flow. A dry conveyor belt in the middle atmosphere flows above cold fronts.

Surface cyclones and anticyclones migrate in the direction of the mid-tropospheric (700 mb) winds of Rossby waves. Cyclones thus move from regions of upper-level divergence (which help maintain or intensify the surface low pressure) to regions of upper-level convergence (which weaken the cyclones). The mean position of storm tracks has apparently shifted poleward in recent decades and is likely to continue this shift. This will alter the precipitation distribution in some parts of the world that may have very significant social impacts, such as increased water shortage and greater vulnerability to fire.

Key Terms

polar front theory *page 282*

cyclogenesis *page 282*

mature cyclone *page 282*

occlusion *page 284*

occluded front *page 284*

Rossby wave *page 286*

vorticity *page 286*

absolute vorticity *page 286*

relative vorticity *page 286*

Earth vorticity *page 286*

dynamic low *page 287*

thermal low *page 288*

speed divergence *page 290*

speed convergence *page 290*

diffluence *page 291*

confluence *page 291*

short waves *page 292*

temperature advection *page 292*

warm air advection *page 292*

cold air advection *page 292*

barotropic *page 293*

baroclinic *page 293*

baroclinic instability *page 293*

cutoff low *page 296*

conveyor belt model *page 300*

Review Questions

1. Define *cyclogenesis*. Where does it most commonly occur, according to the original polar front theory?

2. Describe the isobar and wind patterns associated with mature midlatitude cyclones.

3. Where is precipitation most likely to be found within midlatitude cyclones?

4. Where within a midlatitude cyclone are overrunning, convergence, and instability likely to serve as precipitation-inducing processes?

5. What are Earth, relative, and absolute vorticity?

6. Why is counterclockwise rotation in the Northern Hemisphere said to have positive vorticity?

7. Where are the zones of positive, negative, and zero vorticity in a typical ridge and trough pattern?

8. In what part of a ridge and trough system do you find the areas of decreasing and increasing vorticity? Why is their existence important?

9. How do dynamic and thermal lows differ from each other?

10. Where are upper-level divergence and convergence most likely to occur?

11. What type of upper-level condition typically lies above and behind a cold front?

12. Where are upper-level ridges generally located relative to midlatitude cyclones?

13. What are diffluence and speed divergence? How do they differ from confluence and speed convergence?

14. If the upper-level air flow over North America is zonal, what can you infer with regard to widespread precipitation conditions?

15. Which type of general upper-level air flow is more likely to occur if minimal temperature variations exist across North America?

16. Describe the three conveyor belts of the conveyor belt model of midlatitude cyclones. How does the conveyor belt model differ from the description of airflow presented in the original polar front model?

17. Explain how the movement of midlatitude cyclones relative to the upper-level airflow contributes to the demise of cyclones.

Critical Thinking

1. Why is the term *polar front theory* probably a misnomer in view of current knowledge about cyclogenesis?

2. A commercial aircraft is flying at the 300 mb level and goes across a midlatitude cyclone over both the cold and warm fronts. What kind of weather changes might the aircraft encounter?

3. After a front is fully occluded, its demise is imminent. Why can't the occluded front persist for several more weeks?

4. Why can't systems similar to midlatitude cyclones develop over the tropics?

5. Vorticity is usually discussed with reference to a vertical axis. What types of Earth, relative, and absolute vorticity conditions would you expect to exist with regard to a horizontal axis?

6. Why don't thermal lows migrate like dynamic lows do?

7. Why do forecasters take particular interest in the distribution of vorticity on 500 mb weather maps?

8. Clear skies often portend warm conditions during the summer but are often associated with very cold conditions in the winter. Explain why this is true.

9. Are Rossby waves likely to have greater representation in the Southern Hemisphere or the Northern Hemisphere? Why?

Problems and Exercises

1. Go to **weather.uwyo.edu/upperair/uamap.html** and click the Get Map button to view the current 500 mb map for North America. Without referring to a surface map, make an educated guess about the position of surface midlatitude cyclones, cold fronts, and warm fronts. Then go to **weather.uwyo.edu/surface/front.html** to see how well your educated guess worked out (be sure to highlight Analysis for image type).

2. Using the same Web site as for problem 1, print the maps of the surface, 300 mb, 500 mb, 700 mb, and 850 mb levels. What patterns emerge as you move upward?

3. Visit **weather.uwyo.edu/surface/front.html** on a daily basis and keep track of existing midlatitude cyclones and their associated fronts (click the most recent map available). Do the systems correspond well to the life cycle described in this chapter?

Quantitative Problems

This chapter has discussed the importance of temperature differences in the establishment of upper-level troughs and the critical roles of vorticity and divergence in the life cycle of midlatitude cyclones. The Chapter 10 page of this book's Web site (**www.MyMeteorologyLab.com**) gives you the opportunity to perform some calculations to help you better understand the material. We recommend you access the site and perform some of the simple but revealing problems.

Useful Web Sites

www.giss.nasa.gov/data/stormtracks
An online atlas of storm tracks from 1961 through 1998. Maps are available for each year's seasonal or monthly storm track frequency, intensity, and individual paths.

www.hpc.ncep.noaa.gov/sfc/namfntsfcwbg.gif
Simplified, current surface weather map highlighting frontal systems.

weather.unisys.com/archive/index.html
An excellent source of archived surface and upper-air weather maps and satellite images.

MyMeteorologyLab™

The *Understanding Weather and Climate,* Sixth Edition, MyMeteorologyLab™ web site contains numerous multimedia resources to aid in your study of **Midlatitude Cyclones**.
Visit **www.MyMeteorologyLab.com** to:

- **Review** key chapter concepts.
- **Read** the Pearson eText, *In the News RSS feeds,* and other chapter-specific Web resources.
- **Visualize** challenging topics with Interactive Tutorials, Geoscience Animations, MapMaster interactive maps, and Weather in Motion movies and visualizations, and more.
- **Test Yourself** with self-study quizzes.

 ## TUTORIALS
MIDLATITUDE CYCLONES

Use the interactive animations and quizzes in this tutorial to review this chapter's key concepts.

WEATHER IN MOTION

View movies and visualizations of meteorology patterns and processes.

A Midlatitude Cyclone's Effects on Society
Water Vapor Transport by Midlatitude Cyclones
Short Waves and Long Waves
Winds During the Floods of 1993
A Midlatitude Cyclone's Dry Slot

Lightning, Thunder, and Tornadoes

LEARNING OUTCOMES

After reading this chapter, you should be able to:

▶ Describe the processes involved in lightning formation.

▶ State measures that can be taken to ensure safety from lightning.

▶ Explain how the different types of thunderstorms form.

▶ Describe the geographic and temporal distribution of thunderstorms.

▶ Explain how tornadoes form.

▶ Describe aspects of tornadoes as a natural hazard, including areas most at risk for tornadoes, damage, and fatalities.

▶ Explain the enhanced Fujita scale.

▶ Describe tornado watches and warnings.

▶ Explain tornado outbreaks.

▶ Explain how a waterspout forms.

It is hard to imagine anybody failing to be impressed by the beauty—as well as the danger—brought about by a thunderstorm. Spectacular though they may be, such storms occur about 40,000 times each day. Their frequency varies greatly from place to place, yet virtually every location on Earth is vulnerable to thunder and lightning from time to time.

Lightning can create inconveniences—such as blowing out all the electrical appliances in a house. It can also do considerable damage, such as starting forest fires. And, of course, it can kill; during an average year, about 70 people are killed by lightning in the United States and Canada. But considering that the population of these two countries exceeds 300 million people, it is easy to see that your chances of being struck are extremely remote.

Consider, however, the experience of the McQuilken family on their trip to Sequoia National Park, California, in August 1975. As the sky began to darken, Sean, Michael, and their sister Mary noticed their hair standing on end. Recognizing the apparent comedy of the situation, the boys posed for the photograph shown in Figure 11–1. Hail followed almost immediately. Then lightning struck—literally—and Sean was knocked unconscious. Michael quickly administered artificial respiration, which probably saved Sean's life. Another victim was less fortunate, however. The lightning had apparently forked off, with another branch hitting two nearby people, one of whom was killed.

The effects of lightning and thunder are eclipsed by an even greater menace—tornadoes. We will now examine how, where, and why violent weather occurs, and we will look at the situations that cause some storms to be weak and others to become destructive and deadly.

▲ FIGURE 11–1 Sean and Michael McQuilken in a strong electric field just prior to lightning.

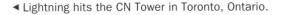

◀ Lightning hits the CN Tower in Toronto, Ontario.

Processes of Lightning Formation

About 80 percent of all **lightning** results from the discharge of electricity *within* clouds, as opposed to discharge from cloud to surface. This **cloud-to-cloud lightning** occurs when the voltage gradient within a cloud, or between clouds, overcomes the electrical resistance of the air. The result is a very large and powerful spark that partially equalizes the charge separation. Cloud-to-cloud lightning causes the sky to light up more or less uniformly. Because the flash is obscured by the cloud itself, it is commonly called **sheet lightning**.

The remaining 20 percent of lightning strokes are the more dramatic events in which the electrical discharge travels between the base of the cloud and the surface. Most of this **cloud-to-ground lightning** occurs when the negative charges accumulate in the lower portions of the cloud. Positive charges are attracted to a relatively small area in the ground directly beneath the cloud. This establishes a large voltage difference between the ground and the cloud base. The positive charge at the surface is a local phenomenon; it arises because the negative charge at the base of the cloud repels electrons on the ground below. Although the term *cloud-to-ground* is used, the same effect occurs in water—and lightning often strikes lakes, rivers, and oceans.

Although a stroke of lightning may come and go in just a few moments, a regular sequence of events must occur for the event to take place. Electrification of a cloud is the initial stage in all lightning. After that, a path must develop through which electrons can flow. Only then is electricity actually discharged to produce a lightning stroke.

Charge Separation

All lightning requires the initial separation of positive and negative charges into different regions of a cloud. Most often the positive charges accumulate in the upper reaches of the cloud, negative charges in lower portions. Small pockets of positive charges may also gather near the cloud base (Figure 11–2a). Now the question is: How does this **charge separation** occur in the first place? Nobody knows for sure, because clouds that produce lightning and thunder happen to be particularly inhospitable laboratories. But we do know several facts from which we can get some idea of how charges separate. Lightning occurs only in clouds that extend above the freezing level, and it is also restricted to precipitating clouds. Thus, the ice crystal processes responsible for precipitation must also influence charge separation. Laboratory experiments with artificial clouds and numerical calculations suggest that electrification results from collisions between ice crystals and graupel surrounded by cloud droplets. It seems likely that charges are transferred across thin films of water present on ice crystals and soft hail.

Did You Know?

Worldwide, there are about 4 million lightning discharges per day, resulting in about 2000 injuries and 600 deaths per year.

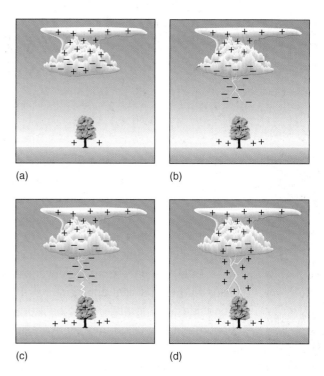

▲ **FIGURE 11–2** After charge separation is established in a cloud (a), the first step in the formation of lightning is the development of a stepped leader. The leader approaches the surface as a very rapid sequence of steps (b) and (c), until contact is made with an object at the ground. The flow of electricity produces the lightning stroke (d).

Though we normally don't notice it, solids are often coated with a liquid surface layer just a few molecules thick. This layer consists of molecules only weakly bound to the solid below. It is present even at temperatures well below the freezing point. (Among other things, the presence of this layer explains why ice is so slippery at temperatures well below zero degrees Celsius.[1]) In a cloud, when an ice crystal and soft hail collide, some of the liquid-water molecules on the hailstone's surface migrate to the ice. In fact, evidence exists suggesting that the collision actually increases the tendency for liquid to exist, thereby increasing the likelihood of liquid-water migration to the ice crystal. There is usually a transfer of positive charge that accompanies the movement of water from the hailstone to the crystal. Or equivalently, there is a transfer of negative charge from the crystal to the hailstone. In this way ice crystals surrender negative ions to the much larger hailstones, which then fall downward toward the cloud base.

Depending on the liquid-water content and temperature of the cloud, a *positive* charge is occasionally transferred to the hailstone, leading to an accumulation of negative charges aloft. Very recent research suggests that surface defects also regulate the charge transfer between particles. (See *Box 11–1, Physical Principles: Electricity in the Atmosphere,* for more information on electrical charges in the air.)

---

[1]You may have heard that pressure from an ice skate's blade melts enough ice to create a slippery film of water, but that's not correct. The ice is slippery whether you press hard or not.

11–1 PHYSICAL PRINCIPLES

Electricity in the Atmosphere

Lightning is, of course, an electrical disturbance, much of which can be explained by the basic principles of atmospheric electricity. You know from Chapter 1 that ions (charged particles) are most abundant high in the atmosphere (in the ionosphere, from about 80 to 500 km, or 50 to 300 mi). The upper atmosphere has a positive charge, just as we find near the positive pole of a battery. In the same way that a battery stores energy, electrical charges in the atmosphere represent stored energy and have the potential to do work. For both batteries and the atmosphere, this electrical potential is expressed by voltage, which is simply the energy per unit charge. For example, if a battery is rated at 1.5 volts (V), it means that 1.5 joules are available per coulomb of charge (1.5 J/C). A coulomb (C) is equivalent to the charge carried by about 6×10^{19} electrons. The higher the voltage, the greater the energy release for each coulomb transferred.

In the case of Earth, a huge voltage difference exists between the surface and the ionosphere—about 400,000 volts! This voltage gradient sets up what we call the **fair-weather electric field**. The fair-weather field is always present, even in bad weather, so a better name might be the **mean electric field**. The fair-weather field can be thought of as the background situation, on which extreme events such as lightning are superimposed.

Does electricity flow in response to the voltage gradient of the fair-weather field? Yes, but because air is a good insulator, the current is weak, about 2000 coulombs per second (2000 A) for the entire planet. In North America, individual houses are typically wired for 200 A service, so we see that the atmospheric current is truly very small. Nevertheless, it does represent a continuous leakage, whereby electrons are transferred from the surface, or (equivalently) positive charges are transferred from the atmosphere. This implies that for the mean electric field to be maintained, it must be continuously replenished. As a matter of fact, lightning discharges in thunderstorms are thought to be the primary recharge mechanism. In other words, cloud-to-ground lightning discharges transfer electrons to the surface, maintaining the voltage difference and the resulting electric field.

In the lower atmosphere, the fair-weather electric field gradient is on the order of 100 V per meter. (Although this might sound impressive, remember that few ions are present, so the total available energy is very low.) Of course, for lightning to occur, the field strength must be greatly intensified above the background value. How this happens can be only partly explained today, nearly 250 years after Ben Franklin performed his famous kite experiment.

Runaway Discharges

For many years scientists thought of lightning as a huge spark of static electricity similar to what one sometimes sees just before touching a doorknob after walking across a carpet. In that case, the voltage gradient becomes so large that the insulating properties of the air are overcome, and a conducting channel forms. It is now believed that lightning does not operate that way. For one thing, the voltage gradients necessary to produce such a discharge are not observed in clouds. Secondly, recent measurements show that X-rays are routinely emitted as part of lightning discharges. There is no known source for these X-rays within the traditional "big spark" view. Many physicists have therefore returned to an idea first proposed in 1925. Lightning is a consequence of electrons that have been accelerated to very high speeds (near the speed of light). Electrons moving faster than a few percent of the speed of light do not experience the resistance felt by ordinary slow-moving electrons. In fact, resistance actually decreases as their speed increases. So, if an electric field accelerates a fast electron, the resistance falls and the electron moves even faster, the resistance decreases further, and so on. Eventually the electron approaches the speed of light and gains a huge amount of energy. Fast-moving electrons create others by colliding with atoms, and this can result in an avalanche of runaway electrons. When a large number of runaway electrons accumulate in a small volume, the energy is released in a so-called **runaway breakdown**. There are still many uncertainties about all this that need to be explained, particularly how initial fast electrons are accelerated (cosmic rays have been suggested). But the fast electron model does explain the existence of X-rays just before lightning flashes, which is something the traditional model cannot do.

Leaders, Strokes, and Flashes

In cloud-to-ground lightning, the actual lightning event is preceded by the rapid and staggered advance of a shaft of negatively charged air called a **stepped leader** (Figure 11–2b) from the base of the cloud. The leader is not a single column of ionized air; it branches off from a main trunk in several places. Only about 10 cm (4 in.) in diameter, each section of the column first surges downward about 50 m (165 ft) from the base of the cloud in about a millionth of a second (or microsecond). This invisible leader pauses for about 50 microseconds, as electrons pile up at the tip and generate a strong electric field in the surrounding area. The field generates more runaway electrons that surge downward another 50 or so meters in the next step. The downward movement in a rapid sequence of individual steps gives the stepped leader its name. In each step, the newly created runaway electrons collide with air molecules and trigger a flare of X-rays.

▲ **FIGURE 11-3** Lightning.

▲ **FIGURE 11-4** Positive lightning stroke.

When the leader approaches the ground (Figure 11–2c), a spark surges upward from the ground toward the leader. When the leader and the spark connect, they create a pathway for the flow of electrons that initiates the first in a sequence of brightly illuminated **strokes**, or **return strokes** (Figure 11–3). The electrical current, flowing at about 20,000 amperes (A), appears to work its way downward from the base of the cloud, but the stroke actually propagates upward (Figure 11–2d). The conducting path is completed at the surface, from which there is a surge of positive charge upward toward the cloud. The current heats the air in the conducting channel to temperatures up to 30,000 K (54,000 °F), or five times that of the surface of the Sun!

The electrical discharge of the first stroke neutralizes some, but not all, of the negatively charged ions near the base of the cloud. As a result, another leader (called a **dart leader**) forms within about a tenth of a second, and a subsequent stroke emerges from it. We call the combination of strokes a lightning **flash**, the net effect of which is to transfer electrons from the cloud to the ground. While most flashes consist of only 2 or 3 strokes, some consist of as many as 20 individual strokes. Because they occur in such rapid succession, they appear to be a single stroke that flickers and dances about.

The total transfer of electrons is not large, only about as many electrons as we use in burning a 100-watt lightbulb for half a minute or so. So how is lightning able to split trees and perform other dramatic work? For one thing, in lightning the charge transfer is rapid, and so the electric current is discharged many thousands of times faster than in a household current. (Think of a lightbulb as having low current flowing for a relatively long time. The total charge transferred is the same as in lightning, where a huge current flows for just an instant.) Another factor is that the voltage is much larger than in a household circuit, so the energy release is much larger for each electron transferred. Taken together, these facts mean that a huge amount of energy gets released over a very brief period of time, making each stroke extremely powerful. A lightbulb would need a month or two to release the same amount of energy.

Positive charges found at the top of thunderstorm clouds can also lead to lightning. When high-level winds are strong, thunderstorm clouds become tilted, with positive charges carried ahead of the storm (Figure 11–4). These positive charges induce negative charges at the surface, resulting in a lightning strike that shoots positive charges to the surface. As a result, it often happens that the first of a storm's lightning strikes are positive, and measurements reveal that they can be many times stronger than the negative strokes that follow. This positive form of lightning is therefore particularly dangerous. It can occur several miles away from the storm, where people do not feel threatened; it tends to have larger peak electrical currents; and it typically lasts longer, making fires more likely. Though an estimated 9 percent of all cloud-to-ground flashes in the United States and Canada originate this way, there is considerable variability across North America. Strokes originating from positive polarity are more common in the upper Midwest of the United States and throughout much of Canada, where they represent perhaps 20 percent of all occurrences.

Types of Lightning

Far less common than strokes and leaders is the bizarre type of electrification called **ball lightning** (Figure 11–5). Ball lightning appears as a round, glowing mass of electrified air, up to the size of a basketball, that seems to roll through the air or along a surface for 15 seconds or so before either dissipating or exploding. One form is a free-floating, reddish mass that tends to avoid good electrical conductors and flows into closed spaces or through doorways and windows. Another form is considerably brighter and is attracted to electrical conductors (including people). Various explanations for ball lightning have been offered for at least 150 years, but until recently none could account for all aspects of the phenomenon, and most had glaring weaknesses. The situation improved significantly early in 2000, with the report of experiments involving artificial lightning strikes on soil. It was seen that lightning reduces silicon compounds in the soil to tiny nanoparticles of silicon carbide (SiC), silicon monoxide (SiO), and metallic silicon (Si). Unlike the original silicon compounds, these contain significant chemical energy and are unstable in an oxygen environment. They are ejected into the air, where they cool

11-2 SPECIAL INTEREST

A Personal Account of Ball Lightning

I saw ball lightning during a thunderstorm in the summer of 1960. I was 16 years old. It was about 9:00 P.M., very dark, and I was sitting with my girlfriend at a picnic table in a pavilion at a public park in upstate New York. The structure was open on three sides and we were sitting with our backs to the closed side. It was raining quite hard. A whitish-yellowish ball, about the size of a tennis ball, appeared on our left, 30 yards away, and

its appearance was not directly associated with a lightning strike. The wind was light. The ball was 8 feet off the ground and drifting slowly toward the pavilion. As it entered, it dropped abruptly to the wet wood plank floor, passing within 3 feet of our heads on the way down. It skittered along the floor with a jerky motion (stick-slip), passed out of the structure on the right, rose to a height of 6 feet, drifted 10 yards further, dropped to the ground and extinguished nonexplosively. As it passed my head, I felt no heat. Its acoustic mission I liken to that of a freshly

struck match. As it skittered on the floor it displayed elastic properties (a physicist would call them resonant vibrating modes). Its luminosity was such that it was not blinding. I estimate it was like staring at a less than 10-watt light bulb. The whole encounter lasted for about 15 seconds. I remember it vividly even today, as all eyewitnesses do, because it was so extraordinary. Not until 10 years later, at a seminar on ball lightning, did I realize what I had witnessed.

Source: Graham K. Hubler, reprinted by permission from *Nature*, Copyright 2000, Macmillan Magazines Ltd.

▲ **FIGURE 11-5** Ball lightning depicted in a 19th century engraving.

rapidly and condense into filmy chains and networks. The networks are light, so they float easily in the atmosphere. Most important, they burn brightly as they oxidize, releasing the stored energy in the form of visible light. (See *Box 11-2, Special Interest: A Personal Account of Ball Lightning,* for a firsthand description of this phenomenon.)

Did You Know?

Lightning strokes can extend more than 16 km (10 mi) from the side of clouds. Thus, lightning can hit in an area where clear skies prevail.

St. Elmo's fire (Figure 11-6) is another rare and peculiar type of electrical event. Ionization in the air—often just before the formation of cloud-to-ground lightning—can cause

tall objects such as church steeples or ships' masts to glow as they emit a continuous barrage of sparks. This often produces a blue-green tint to the air, accompanied by a hissing sound.

Recent observations and photographs from space shuttle missions have revealed the existence of previously unknown electrical phenomena at the tops of thunderstorms. **Sprites** (Figure 11-7) are very large but short-lived electrical bursts that rise from cloud tops as lightning occurs below. A sprite looks somewhat like a giant red jellyfish, extending up to 95 km (57 mi) above the clouds, with blue or green tentacles dangling from the reddish blob. Sprites accompany only about 1 percent of all lightning events. (Interestingly, military and commercial pilots now admit to having seen sprites before they were observed from shuttle missions, but they did not often report them lest they be accused of having hallucinations.)

▲ **FIGURE 11-6** St. Elmo's Fire viewed from the cockpit of a military aircraft.

▲ **FIGURE 11-7** A sprite.

Blue jets (Figure 11–8) are upward-moving electrical ejections from the tops of the most active regions of thunderstorms. They shoot upward at about 100 km/sec (60 mph) and attain heights of up to 50 km (30 mi) above the surface. In all likelihood, other types of electrical activity above thunderstorms remain to be discovered.

Thunder

The tremendous increase in temperature during a lightning stroke causes the air to expand explosively and produce the familiar sound of **thunder**. Although sound travels rapidly—about 0.3 km (0.2 mi) per second—it is much slower than the speed of light (300,000 km, or 186,000 mi, per second). This

▲ **FIGURE 11-8** A blue jet.

difference creates a lag between the flash of light and the sound of thunder; the farther away the lightning, the longer the time lag. You probably know the familiar rule of thumb for estimating the distance of a lightning stroke: Simply count the number of seconds between the stroke and the thunder and divide by three to determine the distance in kilometers (divide by five for the distance in miles).

This method does not work for very distant strokes, those more than about 20 km (12 mi) away. The decrease in the density of air with height causes sound waves to bend upward. At relatively short distances, the amount of bending is negligible. But beyond about 20 km, it is sufficient to displace the sound waves so that they cannot be heard at ground level. Lightning that seems to occur without thunder is sometimes called **heat lightning**, though this term is misleading in that it implies there is something unusual about it. The only oddity is that the sound of distant thunder does not reach the listener.

You have probably noticed that nearby thunder sounds like a loud, brief clap, while more distant thunder often occurs as a continuous rumble. A lightning stroke producing thunder may be several kilometers in length, so one part of it may be significantly farther from a listener than other parts. Thus, thunder makes a continuous sound as it takes longer for the sounds of more distant parts of the stroke to reach the listener. At greater distances, the echoing of sound waves off buildings and hills can cause the thunder to make a rumbling sound.

Did You Know?

On June 29, 2005, Jeff Johnson was struck by lightning as he worked in his office near Des Moines, Iowa. The lightning appears to have been carried into the office via electrical cabling, disabling the computer equipment and hitting Johnson. What made this event noteworthy was that Johnson was working at his job as a meteorologist at a National Weather Service forecast office and was following the progress of a lightning storm—the same one that hit him. After a visit to the hospital, Johnson took the rest of the day off at home—in good physical condition.

Lightning Safety

Despite its splendor, we must not forget that lightning can be lethal, killing an average of 69 people each year in the United States and 7 in Canada. Fortunately, our current understanding of lightning suggests some safety rules.

First and foremost, in the presence of lightning, always take cover in a building, being careful not to make contact with any electrical appliances or telephones. Do not stand under a tree or other tall object that is likely to serve as a natural lightning rod. Lightning hitting a tree can easily flow through it and electrocute or burn those near this particularly vulnerable location. Avoid standing on rooftops, hill crests, or other high areas where lightning requires a shorter path from the cloud base. And of course, stay out of the water. Do not watch the lightning storm from a pool, lake, or hot tub!

11-3 FOCUS ON AVIATION

Lightning and Aircraft

Lightning strikes on aircraft are not uncommon (Figure 1). Fortunately, they seldom result in serious damage, injury, or fatalities. Over the 57-year period prior to 2009 there have been 18 reported air disasters due to lightning worldwide—less than one every 3 years. The last confirmed U.S. civilian plane crash attributed to lightning was reported in 1967, the result of a fuel tank explosion.

Aircraft are protected in part by their highly conductive skins that conduct the electrical current around the fuselage and extremities of the plane. In the United States, the Federal Aviation Administration mandates high levels of additional protection against the potential direct and indirect consequences of lightning hits. For example, all electronic equipment must be shielded against power surges. The nose cones of commercial aircraft that house the radar (the radome) must be surrounded by lightning diverter strips that act much like lightning rods do on homes. The fuel tank and lines transporting fuel to the engines must be protected against the possibility of any sparks that can ignite the fuel, and the fuel itself is now designed to produce less explosive vapors.

Sometimes an existing lightning stroke will make contact with a plane, and passengers and crew will be greeted by a bright flash and a loud noise. Pilots may observe a flickering of lights in the cockpit. At other times the plane itself will cause a lightning event as it passes through a heavily ionized part of a cloud.

Scientists have recently begun to examine the incidence of ball lightning in and near planes. A recent analysis of 87 such events from 1938 to 2007 indicated that about half of these occurred inside the aircraft and the other half outside the airframe. Sometimes the ball lightning appeared to be a side effect of a regular lightning strike, but it can also occur independently. Though no damage or minimal damage is usually incurred, there have been reports of major damage and even three aircraft downings due to ball lightning.

▲ **FIGURE 1** Lightning strikes a commercial jet aircraft over Florida.

Automobiles (other than convertibles) are relatively safe, but not because the rubber tires provide insulation against grounding (as believed by many people). The real reason is that if a car is hit, the electricity will flow around the car body rather than through the interior (or its occupants). The same fact explains why lightning seldom brings down airplanes, even though any particular commercial aircraft is hit on an average of once a year (see *Box 11–3, Focus on Aviation: Lightning and Aircraft*).

We often associate deadly lightning strikes with golfing during a thunderstorm and other foolish behaviors, but the danger is not always easy to avoid. On January 1, 2000, for instance, a single lightning bolt killed a family of six near Mount Darwin, Zimbabwe, in a tragedy eerily similar to one that had occurred a few months earlier in Zimbabwe, when a single strike killed six persons near the city of Gokwe. Lightning deaths are becoming ever more frequent throughout that region as forests are cleared, leaving villages more exposed in open areas. The problem is greatly exacerbated by the use of dry thatch as a roofing material. Soot from cooking fires impregnates the thatch with carbon, making the roof highly conductive and attractive for lightning.

Did You Know?

Only about 10 percent of lightning victims die after being hit. By far the greatest number of those who are killed die from cardiac arrest; victims only rarely suffer external burns. Though most people are not killed when struck by lightning, survivors often suffer long-term psychological problems, including irritability, personality changes, chronic fatigue, and depression.

In the United States a substantial decline (greater than 50 percent) has occurred in the number of lightning fatalities since the 1920s. This fact is even more impressive when one considers that since that time the country's population has increased from about 150 million to well over 300 million. The decline in fatalities is believed to be due in part to a population decline in rural areas, where the risk of being struck is greater. Better public education, improved warnings, advances in medical aid, and modernized electrical systems in homes and other buildings are also believed to have contributed to the decline.

Checkpoint

1. Describe the various types of lightning and associated electrical events.
2. What processes lead to the formation of cloud-to-ground lightning strokes?

Thunderstorms: Air Mass, Multicell, and Supercell

Most lightning events are associated with localized, short-lived storms that dissipate within tens of minutes after forming. These storms, called **air mass thunderstorms**, actually extinguish themselves by creating downdrafts that cut off the supply of moisture into the precipitating clouds. For that reason, they do not normally produce severe weather. On other occasions, however, downdrafts from heavy precipitation do not impede the replenishment of moisture into the storm from updrafts. Those storms can become severe and produce serious damage and loss of life.

Air Mass Thunderstorms

Air mass thunderstorms are the most common and least destructive of thunderstorms. They also have very limited lifespans, usually lasting for less than an hour. Despite the name, which implies that these thunderstorms might occupy entire air masses (which are very large), air mass thunderstorms are very localized. But the term does make sense when you consider that air mass thunderstorms occur within individual air masses and are well removed from frontal boundaries. Think of it this way: Air mass thunderstorms are contained *within* uniform air masses, but they do not occupy the *entire* air mass.

Air mass thunderstorms will not occur unless certain conditions exist: The air must contain sufficient moisture with the temperature being not too much greater than the dew point, as that permits condensation to occur fairly close to the ground. The air also must be somewhat unstable so that uplift can be sufficient to produce a deep cumulus cloud. And of course there must be some uplift mechanism, such as heating of the surface. Another characteristic of air mass thunderstorms is that they form in an environment where there is little *wind shear*, a change in wind velocity with altitude.

Our current understanding of air mass thunderstorms is based on the Thunderstorm Project, which examined such events in Ohio and Florida during the late 1940s. An air mass thunderstorm normally consists of a number of individual updrafts (*cells*), each undergoing a sequence of three distinct stages—cumulus, mature, and dissipative (Figure 11–9).

Cumulus Stage The first stage of an air mass thunderstorm begins when unstable air begins to rise, often by the localized convection that occurs as some surfaces undergo more rapid heating than others. Because air mass thunderstorms frequently occur at late evening when the air is cooling, we know that other lifting processes can also trigger uplift. Regardless of which process causes uplift, the rising air cools adiabatically to form fair-weather cumulus clouds. These initial clouds may exist for just a matter of minutes before evaporating. Although they do not directly lead to any precipitation, the initial clouds play an important role in thunderstorm development by moving water vapor from the surface to the middle troposphere. Ultimately, the atmosphere becomes humid enough that newly formed clouds do not evaporate but instead undergo considerable vertical growth. This growth represents the **cumulus stage** in the air mass thunderstorm.

Clouds in the cumulus stage grow upward at 5 to 20 m/sec (10 to 45 mph). Within the growing clouds, the temperature decreases with height at roughly the saturated adiabatic lapse rate, and a portion of the cloud extends above the freezing level. Ice crystals begin to form and grow by the Bergeron process. The sky rapidly darkens under the thickening cloud; when precipitation begins to fall, the storm enters its next stage of development.

Mature Stage The **mature stage** of the air mass thunderstorm begins when precipitation—as heavy rain or possibly hail—starts to fall. As the falling rain or hail drags air toward the surface, downdrafts form in the areas of most intense precipitation. You can observe this process in your own yard.

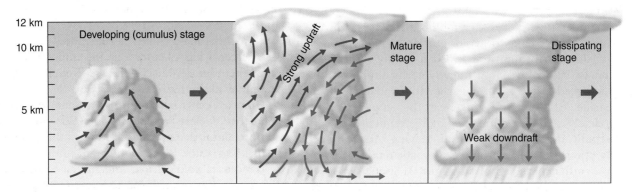

▲ **FIGURE 11–9** The cumulus (a), mature (b), and dissipative (c) stages of an air mass thunderstorm.

Simply turn on a garden hose full blast and put your hand just outside the stream of water; you will notice a breeze in the direction in which the hose is pointed. The downdrafts are strengthened by the cooling of the air—by as much as 10 °C (18 °F)—that occurs as the precipitation evaporates.

The mature stage marks the most vigorous episode of the thunderstorm, when precipitation, lightning, and thunder are most intense. The top of the cloud extends to an altitude where stable conditions suppress further uplift. Strong winds at the top of the cloud push ice crystals forward and create the familiar anvil shape extending outward from the main part of the cloud.

During the cumulus and mature phases of the storm, an abrupt transition exists between the edge of the cloud and the surrounding unsaturated air. Updrafts dominate the interior of the cloud, while downdrafts occur just outside it. This sets up a highly turbulent situation that encourages entrainment (Chapter 6). The entrainment of unsaturated air causes the droplets along the cloud margin to shrink and cool the cloud by evaporation. The outer part of the cloud becomes more dense and less buoyant, thus suppressing further uplift.

Dissipative Stage As more and more of the cloud yields heavy precipitation, downdrafts occupy an increasing portion of the cloud base. When they occupy the entire base, the supply of additional water vapor is cut off and the storm enters its **dissipative stage**. Precipitation diminishes and the sky begins to clear as the remaining droplets evaporate. Only a small portion—perhaps 20 percent—of the moisture that condenses within an air mass thunderstorm actually falls as precipitation. The greatest amount simply evaporates from the cloud.

Figure 11–10 shows an air mass thunderstorm. As is typical for thunderstorms in the mature phase, each tower consists of an individual cell and is in a different part of its life cycle. Notice in particular that some of the storm cloud appears washed-out and less well-defined than the rest. Such areas consist entirely of ice crystals, with no liquid droplets, and are said to be *glaciated*. They are not necessarily colder than other parts of the cloud; they are merely old enough so that all the supercooled droplets have had a chance to freeze.

Checkpoint

1. What are three stages in the formation of an air mass thunderstorm?
2. What is the series of steps that leads to the formation of a mature airmass thunderstorm?

Multicell and Supercell Storms

Sometimes thunderstorms develop into clusters referred to as **multicell thunderstorms**. These organized groups of thunderstorms are generally referred to as **mesoscale convective systems (MCSs)**. In some cases, MCSs occur as linear

▲ **FIGURE 11-10** An air mass thunderstorm. The part of the cloud on the right of the photo that has a washed-out appearance has become glaciated.

bands called **squall lines**. At other times, they appear as oval or roughly circular clusters called **mesoscale convective complexes (MCCs)**. Regardless of how they are arranged, the individual storm cells of an MCS form as part of a single system: They are not just a grouping of individual storms that happen to be near each other. The storm cells develop from a common origin or exist in a situation in which some cells directly lead to the formation of others.

MCSs can bring intense weather conditions to areas covering several counties. They often have lifespans of up to 12 hours, but in some cases they can exist for as long as several days. They are fairly common in North America, and in some parts of the central United States and Canada they account for as much as 60 percent of the annual rainfall. Because the surrounding circulation supports an MCS, they lead to much stronger winds and heavier precipitation than is normally found in an air mass thunderstorm.

Even more severe at times are **supercells**, intensely powerful storms that contain a single updraft zone. Though supercells often appear in isolation, they can also occur as a part of an MCS.

Mesoscale convective systems and supercells are capable of producing **severe thunderstorms**. By definition, severe thunderstorms have wind speeds that exceed 93 km/hr

(58 mph),[2] hailstones larger than 2.4 cm (1 in.) in diameter, or spawn tornadoes. The downdrafts and updrafts in severe storms reinforce one another and thereby intensify the storm. This reinforcement usually requires suitable conditions over an area from 10 to 1000 km (6 to 600 mi) across. In other words, most severe thunderstorms get a boost from a mesoscale atmospheric pattern that allows the wind, temperature, and moisture fields to "cooperate" and thereby create very strong storms.

Certain conditions are necessary for the development of all severe thunderstorms. Among these are wind shear, high water vapor content in the lower troposphere, some mechanism to trigger uplift, and a situation called *potential instability,* described earlier in *Box 6–2, Forecasting: Potential Instability.* Note that the requisite wind shear for these storms is a major factor that allows the development of multicell and supercell storms, as opposed to the less intense air mass thunderstorms.

We will now briefly describe MCCs, squall lines, and supercells. Keep in mind that the descriptions of these storm types are quite general, and individual systems might not be easy to categorize. It is also common for storm systems to evolve from one type of system to another.

Mesoscale Convective Complexes In the United States and Canada, severe weather often arises from mesoscale convective complexes (MCCs) (Figure 11–11). In the most general sense, MCCs are defined as oval or roughly circular organized systems containing several thunderstorms.[3]

Although not all MCCs create severe weather, they are self-propagating in that their individual cells often create

---

[2]This seemingly odd value was originally designated as 50 nautical miles per hour, or knots.

[3]Meteorologists have some precise criteria for classifying a system as an MCC, based on its signature on satellite imagery. We will simply apply the term to organized systems of thunderstorms clustered in a pattern that is closer to circular than linear.

downdrafts, leading to the formation of new, powerful cells nearby. To see how this occurs, imagine a large cluster of thunderstorms. At the surface a flow of warm, humid air comes from the south, and in the middle troposphere the wind flows from the southwest. This setting provides the wind shear necessary for a severe thunderstorm. As we have already seen, the precipitation from each thunderstorm cell creates its own downdraft. The downdraft is enhanced by the cooling of the air as the rain evaporates and consumes latent heat. Upon hitting the ground, the downdraft spreads outward and converges with the warmer surrounding air to form an **outflow boundary** (Figure 11–12).

Figure 11–13 illustrates the progressive movement of cells in an MCC. Initially, at time $t = 0$, five cells labeled $A,$ $B, C, D,$ and E are moving toward the northeast. Somewhat later, at $t = 1$, cells B through E have migrated to the northeast. Cell $A,$ located farthest to the north, has died out, while along the southern margin of the complex a new cell, $F,$ has formed. At $t = 2$, cell B has died out, to be replaced by cell G along the south. In this manner, the complex of thunderstorms moves eastward, though each individual cell moves to the northeast.

Near the southern side of the MCC, the cold outflow collides with the large-scale southerly surface flow and lifts it upward. The warm, humid air is drawn into the southern edge of the MCC, where it forms new cells. At the same time, the older cells on the northern side of the MCC dissipate because they lack the updrafts needed for replenishment.

Squall Line Thunderstorms Squall line thunderstorms consist of a large number of individual violent storm cells arranged in a linear band, typically about 500 km (300 mi) in length

▲ **FIGURE 11–11** Satellite image showing a mesoscale convective complex over eastern South Dakota.

▲ **FIGURE 11–12** A radar image highlighting two outflow boundaries.

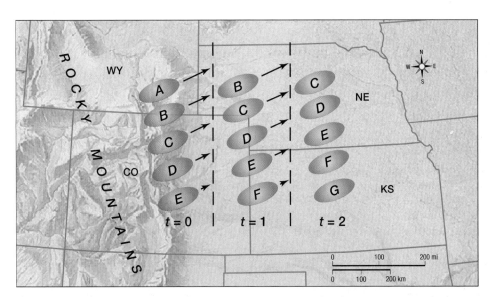

◄ **FIGURE 11-13** The movement of thunder-storm cells in a mesoscale convective complex. Initially (at *t* = 0), all the cells are moving toward the northeast. Cell A is the oldest, and cell E is the most recently formed. Later (*t* = 1), cell A has dissipated, but a new cell, F, has formed along the southern margin of the complex. At *t* = 2, cell B has dissipated while a new cell (G) has formed.

(Figure 11–14). While squall lines may be as much as 100 km (60 mi) in width, the embedded band of intense showers is much narrower, usually about 5 km (3 mi), and located near the leading edge of the advancing band of cloud cover and precipitation. A wide area of continuous precipitation normally exists in the central and western portions of the squall line, immediately behind the band of heavy showers.

Squall lines usually form in the warm sector of a midlatitude cyclone immediately ahead of the cold front but then advance more rapidly than the front itself. They may eventually reside about 300 to 500 km (180 to 300 mi) ahead of the front. Squall lines are most common over the southern United States during spring and summer. The average squall line has a lifespan on the order of 10 hours, though some have lasted up to four days.

As is the case for all intense thunderstorms, strong vertical wind shear is an important component of squall line thunderstorms. As shown in Figure 11–15a, wind velocities in the direction of storm movement typically increase with height. The strong winds aloft push the updrafts ahead of the downdrafts and allow the rising air to feed additional moisture into the storm. As the downdrafts reach the ground, they surge forward as a wedge of cold, dense air, called a **gust front** (Figure 11–15b). Gust fronts act in much the same way as advancing cold fronts by displacing warm air upward and are often discernible by the cloud of dust picked up from the ground transported by the heavy winds. They can also lift the warm air ahead to form a **shelf cloud** (sometimes called an *arcus*) just above the gust front and ahead of the main portion of the thunderstorm (Figures 11–16 and 11–17). Beneath the leading edge of the gust front, horizontally rotating air can produce a **roll cloud** (see Figure 11–17).

Supercell Storms Few weather systems are as awesome as a supercell storm (Figure 11–18). With diameters that range from about 20 to 50 km (12 to 30 mi), they are smaller than either squall lines or MCCs. On the other hand, they are usually more violent and provide the setting for most very large tornadoes. Unlike MCCs and squall lines, a supercell storm consists of a single, extremely powerful cell rather than a number of individual cells.[4] Supercell storms also undergo a large-scale rotation absent from squall lines and MCCs. The typical lifespan of a supercell is 2 to 4 hours.

Despite their single-cell structure, supercell storms are remarkably complex, with the updraft and downdraft bending and wrapping around each other due to strong wind shear (Figure 11–19). As in any other weather system that spawns severe weather, the downdrafts serve to amplify the adjacent updrafts.

▲ **FIGURE 11-14** Radar image of a squall line.

---

[4]Supercells usually occur as isolated storms, though "squall lines" of supercells have been observed.

► **FIGURE 11-15** Squall line thunderstorms require the presence of wind shear. The arrows in (a) represent the wind speeds with respect to the movement of the storm. The movement of the air within the cumulonimbus is shown in (b). The upper part of the cloud is pushed forward more rapidly than the lower part, which helps to draw in warm, moist air. Note the gust front near the ground ahead of the rain shaft.

(a)

(b)

▲ **FIGURE 11-16** A dramatic example of a shelf cloud.

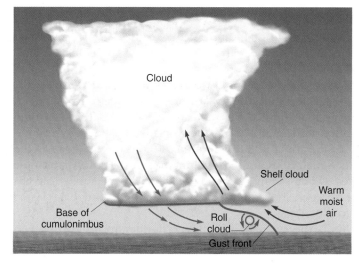

▲ **FIGURE 11-17** A shelf cloud and roll cloud associated with a gust front.

▲ **FIGURE 11-18** A supercell. Note the tornado at the far left.

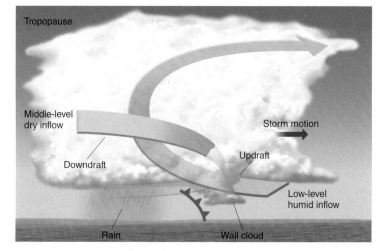

▲ **FIGURE 11-19** The internal structure of a supercell.

(a)

(b)

▲ **FIGURE 11–20** Radar images of supercells typically display certain features, such as a hook echo and vault (a). The vault is a zone of substantial cloud cover that appears on a radar image as a cloud-free zone because of the small size of the cloud droplets. The hook echo extends around the vault. An actual radar image showing a hook is seen in (b).

Meteorologists follow supercells with tremendous interest. Fortunately, they have at their disposal an extremely useful tool in the form of weather radar. Radar can reveal one of the most noteworthy features of a supercell, called a **hook** (or **hook echo**), which looks like a small appendage attached to the main body of the storm on the radar image (Figure 11–20).[5]

TUTORIAL

DOPPLER RADAR

Use the tutorial to see how Doppler radar detects the position and movement of clouds.

The zone with no radar return between the main part of the supercell and the hook echo, known as a **vault**, is where the inflow of warm surface air enters the supercell. The air entering the vault rises, and water vapor condenses to form a dense concentration of water droplets. But the newly formed droplets in the vault are too small to effectively reflect radar waves. Thus, this zone does not show up on the radar image despite the dense concentration of water droplets.

The appearance of hook echoes has long been significant to forecasters, as they often are followed by tornadoes. Today **Doppler radar** can allow us to observe rotation within clouds that provide an even better prediction of tornadoes (see *Box 11–4, Forecasting: Doppler Radar*).

Drylines Drylines were briefly described in Chapter 9 as boundaries separating mT and cT air masses. They are most

likely to form in spring and early summer over Texas and Oklahoma, where they can be the site of severe weather. Drylines develop as moist air near the surface from the Gulf of Mexico flows northward, separated from the dry air to the west. Terrain effects are important, as the elevation of the surface gradually increases westward from east Texas and Oklahoma toward the Rocky Mountains. The higher elevation prevents a westward incursion of the moist air. Dry air flowing eastward out of the high plains overrides the moist air below, creating a situation called *potential instability* (described in Chapter 6). Potential instability can lead to severe storm activity if enough lifting occurs below the boundary of the warm and moist air. This can occur due to surface convergence as the mT air from the Gulf of Mexico and the westerly flow of cT air collide along the dryline.

Downbursts, Derechos, Microbursts, and Haboobs

We have seen how downdrafts are an important feature of thunderstorms, especially in the maintenance of severe thunderstorms. Strong downdrafts may also create **downbursts**, potentially deadly gusts of wind that can reach speeds in excess of 270 km/hr (165 mph). When strong downdrafts reach the surface, they can spread outward in all directions to form intense horizontal winds capable of causing severe damage at the surface. In fact, damage attributed to tornadoes may in some cases be the result of downbursts.

Strong downdrafts associated with mesoscale convective systems can produce very powerful, larger-scale horizontal winds called **derechos** (the Spanish word meaning "straight ahead").

---

[5]Echos refer to the way radar waves reflect off cloud constituents, as sound waves echo off canyon walls.

Doppler Radar

Just as we are able to distinguish different colors of light by their wavelengths, so can we differentiate sounds by the length of their sound waves. If an object making a sound is moving away from a listener, the sound waves are stretched out and assume a lower pitch. Sound waves are compressed when an object moves toward the listener, making them higher pitched. Unconsciously, we use this principle, called the **Doppler effect**, to determine whether an ambulance siren is coming closer or moving away. If the pitch of the siren seems to become higher, we know the ambulance is getting nearer (of course, the siren would also sound louder). A similar process occurs when electromagnetic waves are reflected by a moving object: The light shifts to shorter wavelengths when reflected by an object moving toward the receiver

and to longer wavelengths as it bounces off an object moving away from the receiver.

Doppler radar is a type of radar system that takes advantage of this principle. It allows the user to observe the movement of raindrops and ice particles (and thus determine wind speed and direction) from the shift in wavelength of the radar waves, as well as the intensity of precipitation. Like any other type of radar, Doppler radar has a transmitter that emits pulses of electromagnetic energy with wavelengths on the order of several centimeters. Depending on the wavelength used, water droplets and snow crystals above certain critical sizes reflect a portion of the radar's electromagnetic energy back to the transmitter/receiver. Doppler radar is special in its ability to observe the motion of the cloud constituents. If a cloud droplet is moving away from the radar unit, the wavelength of the beam is slightly elongated as it bounces off the

reflector. Such reflections are normally indicated on the display monitor as reddish to yellow. Likewise, a droplet moving toward the radar unit undergoes a shortening of the wavelength. Echoes from these constituents are displayed as blue or green on the radar screen.

A radar unit must rotate 360 degrees to get a complete picture of the weather situation surrounding the transmitter/receiver unit. When the transmitter makes one complete rotation at a fixed angle, it is said to have completed a **sweep**. The angle can then be increased as a second sweep is taken that depicts a higher cloud level. This can be repeated several times so that the radar can peer into multiple levels of the cloud. The compilation of all the individual sweeps takes approximately 5 to 10 minutes and produces a **volume sweep**.

Figure 1 shows a pair of Doppler radar images of a major storm near Dallas–Fort

(a)

(b)

▲ **FIGURE 1** Doppler radar images of a storm near Dallas–Fort Worth, Texas, on March 29, 2000. Part (a) depicts the intensity of precipitation; part (b) shows the storm radial velocity (SRV) pattern, which is the movement of different parts of the storm toward or away from the radar unit.

Worth, Texas, on March 29, 2000. Figure 1a shows the reflectivity of the storm, with redder regions indicating intense cloud cover and green areas representing less intense cloud cover. The white arrows point toward a hook echo (described in the main text of this chapter). Figure 1b displays the *storm radial velocity (SRV) pattern*, which describes the motions taking place within the cloud. SRV displays use redder colors to represent winds blowing away from the radar and green to indicate movement toward the radar. The yellow arrows on this image highlight a region of counterclockwise rotation. As we discuss later in the chapter, this pattern, called a *mesocyclone,* often precedes the formation of a tornado. After the onset of rotation, it takes only 30 minutes or so for the tornado to form, which allows meteorologists to give warning in advance. In this particular case, a tornado did hit the city of Fort Worth.

During the early 1990s, the National Weather Service began replacing the old system of conventional radar with a modern Doppler network called NEXRAD (for NEXt Generation Weather RADar). The first unit, installed at Norman, Oklahoma, became operational in early 1991 and on its very first day in service tracked a tornado that destroyed two houses. Fortunately, the radar allowed forecasters to issue a warning that may have contributed to no one being killed or injured.

A study published in 2005 concluded that the implementation of Doppler radar units across the United States has prevented 79 fatalities and 1050 injuries per year. Doppler radar has increased the average warning time for all tornadoes from 5.3 minutes to 9.5 minutes. Results are even more impressive for EF-5 tornadoes, the most deadly of all. Warning time for those tornadoes has increased to 16.23 minutes from the previous 11.7 minutes.

Today about 160 Doppler sites are scattered across the United States (Figure 2). The National Weather Service operates 113 of these sites; the rest are owned by the Federal Aviation Administration and the Department of Defense. In part because of budgetary cutbacks, the Atmospheric Environment Service of Canada has just a handful of Doppler radar installations. Because both sides of the border area tend to be heavily populated, Doppler radar from the United States provides extensive cover-

▲ **FIGURE 2** Doppler radar sites in the United States.

age of severe storms that could affect many large Canadian urban centers.

NEXRAD is also useful for flood forecasting, providing continual precipitation estimates over large areas. Doppler radar can sometimes observe wind movements even when no clouds exist, as large clusters of flying bugs or heavy dust concentration scatter radar waves back toward the transmitter. The resultant echoes are called *clear air echoes*.

The NEXRAD network of Doppler radar, installed primarily for forecasting purposes, provides information on tornadoes that has proven useful to researchers. But NEXRAD radar units are spaced too far apart to provide close scrutiny of most passing tornadoes, so researchers have looked to

transportable radar to observe tornadoes at close range. For this reason, tornado researchers have come to rely on portable Doppler radar units called Doppler on Wheels that rely on units mounted to a flatbed truck (Figure 3). These units have been instrumental in acquiring new information on tornado dynamics.

▲ **FIGURE 3** Doppler on wheels.

(a) May–August

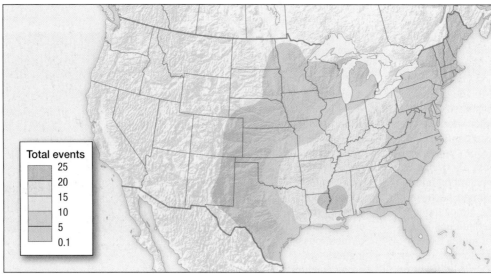

(b) September–April

▲ FIGURE 11-21 Frequency of derechos across the United States (a) annually, (b) between May–August, and (c) between September–April.

Microbursts

Microbursts can pose a serious threat to aircraft, especially during take-offs and landings (Figure 1). The horizontal spreading of a microburst creates strong wind shear when it reaches the surface. For example, air may flow westward on one side of the microburst while spreading eastward on the opposite side. Imagine what this might do to an aircraft attempting to land in a microburst. As the plane enters the microburst, a headwind provides lift, to which the pilot might respond by turning the aircraft downward. As soon as the plane passes the core of the downdraft, however, the headwind not only disappears, it is replaced by a tailwind, decreasing lift. Coming after the pilot's earlier downward adjustment, this causes the plane to abruptly drop in altitude. Because the plane is not far above the ground when these events occur, the pilot may not have time to compensate before a deadly crash occurs. Fortunately, such disasters are rare. They are also becoming less likely because the installation of Doppler radar at about 40 U.S. airports has proven highly effective at detecting microbursts, with a detection rate of about 95 percent.

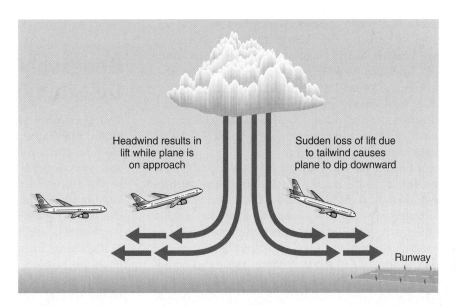

▲ **FIGURE 1** Microbursts can make aircraft landing perilous. A plane flying into the headwinds of a microburst gets a sudden increase in lift. This lift suddenly disappears and is replaced by a tailwind as it exits the downdraft, thereby reducing the lift. If the pilot overcompensates and guides the plane downward while entering the downdraft, a dangerous drop in altitude may occur.

Such winds may last for hours at a time and achieve speeds higher than 200 km/hr (120 mph)—comparable to those of many tornadoes. As already described, these downdrafts spread outward upon reaching the ground. But the winds can be especially powerful if the descending upper-level air has high wind speeds prior to being brought downward. This momentum does not disappear as the air sinks, so when it reaches the surface, it is capable of bringing destructive winds. Between 1986–2003, an average of nearly 21 derechos per year occurred in the United States, resulting in an annual average of 8.5 fatalities. Derechos are most common over eastern Oklahoma (Figure 11–21a), with the area of highest incidence extending northeastward to the southern Great Lakes. Figures 11–21 (b) and (c) show the seasonal variation in the distribution of derechos, being more common in the lower Mississippi River valley between September and April. Over the entire United States, derechos are most prevalent between May and July (Figure 11–22).

One of the most notable recent derecho events occurred shortly after midnight on Labor Day in 1998, when a derecho

▲ **FIGURE 11–22** The average number of derechos over the United States by month.

some 50 km (30 mi) in length brought wind gusts of up to 180 km per hour (115 mph) that killed two people, injured eight others, downed thousands of trees, and damaged scores of structures. The storm proceeded eastward through the night, passing through southern Vermont and New Hampshire prior to moving across the entire state of Massachusetts. The following July saw a derecho that lasted 22 hours, sweeping across extreme eastern North Dakota, across much of south central Canada, and into New England. That windstorm also killed two people and caused widespread damage.

Downbursts with diameters of less than 4 km (2.5 mi), called **microbursts**, can produce a particularly dangerous problem when they occur near airports (see *Box 11–5, Focus on Aviation: Microbursts*).

Very strong horizontal winds created by downdrafts over desert regions can dislodge sand and dust from the surface and incorporate them into the wind gusts. These winds can be remarkably turbulent and disburse the sand and dust to heights as great as 3 kilometers (10,000 feet) above the ground. These winds can persist for hours at a time and typically advance at speeds of about 50 km/hr (30 mph), creating an ominous wall of loose material that appears like a very dense, dark cloud.

These sandstorms are called **haboobs**, from the Arabic word meaning "wind." They are most common in northern and central Sudan, occurring perhaps two dozen times each year. A very dramatic haboob occurred over central Arizona in July 2011 (shown on the Part IV opening photo on page 278) and attracted considerable media attention.

Checkpoint

1. Briefly define each of the following terms in your own words: MCCs, squall line, supercell.
2. How are downbursts, derechos, microbursts and haboobs similar? How are they different?

Geographic and Temporal Distribution of Thunderstorms

Thunderstorms are extremely common across much of the globe, numbering some 14.5 million per year. They are most likely to develop where moist air is subject to sustained uplift, and, not surprisingly, such conditions occur most commonly in the tropics. Until recently the occurrence of lightning in low-populated or economically underadvantaged areas precluded the gathering of reliable statistics across the globe. Fortunately, satellites have provided high-quality observations of lightning incidence since the mid-1990s (Figure 11–23). Lightning strikes most frequently over the Congo basin of central Africa and occurs frequently over many other regions of the world. Colder and less humid regions typically have a lower incidence of lightning. Outside of the tropical regions, lightning is most common during the summer months when midday sun angles are greatest, while equatorial regions usually have their peak activity near the equinoxes (Chapter 2), due to abundant solar radiation.

▲ **FIGURE 11-23** Data from space-based optical sensors reveal the uneven distribution of worldwide lightning flashes. Units: flashes/km²/yr.

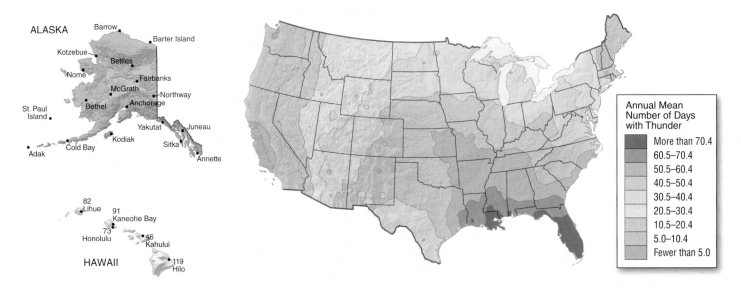

▲ **FIGURE 11-24** The average number of days annually in which thunder is heard.

Did You Know?

Thunderstorms account for a large part of the total rainfall that occurs over much of North America. Thunderstorms provide almost half the total rainfall that occurs over the Mississippi River Valley, which covers 41 percent of the conterminous 48-state land area. In the south-central United States they account for as much as 70 percent of total precipitation. Across the western portion of North America the percentages are far lower; generally less than 10 percent of the rainfall in that area falls from thunderstorms.

Detailed data on cloud-to-ground lightning flashes have been assembled using ground-based data from the Canadian Lightning Detection Network (CLDN) and the National Lightning Detection Network (NLDN) of the United States. The two networks' sensors detect electromagnetic radiation emitted during lightning strikes and automatically relay information on the location, timing, and polarity (positive or negative) of each stroke to a central facility. The networks have recorded an average of 28 million cloud-to-ground lightning flashes and 100,000 thunderstorms (280 flashes per thunderstorm) annually across the continental United States and Canada.

Across the eastern two-thirds of North America (Figure 11–24) there is a general pattern of decreasing thunderstorm activity northward. By far the state with the greatest incidence is Florida, where thunderstorms occur on average more than 100 days per year. Much of the necessary lifting of air results from strong solar heating of the surface. But the situation over the Florida peninsula is unique within the continental United States, because the area is almost completely surrounded by warm water. Thus, air that flows into the interior to replace lifted air has a very high moisture content, which in turn supports heavy precipitation and the development of thunderstorms. The area of the United States

and Canada west of the Rocky Mountains (except for Hawaii) experiences considerably fewer thunderstorms than the eastern two-thirds of the continent.

Figure 11–25 illustrates the areas that have the maximum number of lightning strikes during the daytime (8 A.M. to 8 P.M. local standard time) and those occurring during the 12 hours following 8 P.M. Much of North America exhibits a daytime maximum in lightning flashes, though much of the mid-continent has a greater likelihood of nighttime events.

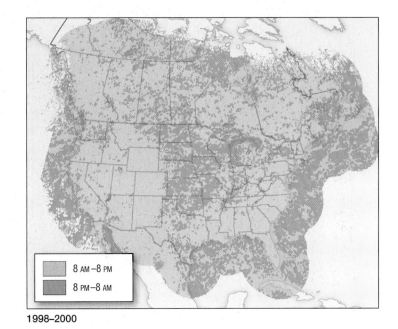

1998–2000

▲ **FIGURE 11-25** The spatial distribution of areas where the majority of lightning takes place from 8 A.M. to 8 P.M. and 8 P.M. to 8 A.M.

Checkpoint

> 1. From a global perspective, where are the conditions necessary for the formation of a thunderstorm most commonly found? Explain.
>
> 2. Look at Figure 11–24. What can you infer about climate in the United States from the distribution of annual mean number of days with thunder?

Tornadoes

The large hail and strong winds of a severe thunderstorm can bring widespread destruction, but even hailstorms are relatively tame compared to **tornadoes** (Figure 11–26). Tornadoes are zones of extremely rapid, rotating winds beneath the base of cumulonimbus clouds. Though the overwhelming majority of tornadoes rotate cyclonically (counterclockwise in the Northern Hemisphere), a few spin in the opposite direction. Some appear as very thin, rope-shaped columns, while others have the characteristic funnel shape that narrows from the cloud base to the ground. Regardless of their shape or spin, tornadoes are extremely dangerous.

Strong tornadic winds result from extraordinarily large differences in atmospheric pressure over short distances. Over just a few tenths of a kilometer, the pressure difference between the core of a tornado and the area immediately outside the funnel can be as great as 100 mb. To put this in perspective, on a typical day the highest and lowest sea level pressure across all of North America may differ by only about 35 mb—and this difference exists over horizontal distances of up to thousands of kilometers.

Tornado Characteristics and Dimensions

It is difficult to generalize about tornadoes because they occur in a wide variety of shapes and sizes. While most have diameters about the length of a football field (100 yards or so), some are 15 times as large. Usually they last no longer than a few minutes, but some have lasted for several hours. Tornadoes normally travel northeastward at speeds comparable to a car driving down a city street—about 50 km/hr (30 mph). A typical tornado covers about 3 or 4 km (2 to 2.5 mi) from the time it touches the ground to when it dies out.

Estimates of wind speeds within tornadoes are based primarily on the damage they have produced. The weakest have wind speeds as low as 105 km/hr (65 mph); the most severe are in excess of about 450 km/hr (280 mph).

Did You Know?

A tornado that touched down near Hallam, Nebraska, on May 22, 2004, set the record for being the largest known tornado, having a diameter of almost 4 km (2.5 mi). It surpassed the previous record holder that touched down near Gruver, in the Texas panhandle, in June 1971.

Tornado Formation

Tornadoes can develop in any situation that produces severe weather—frontal boundaries, squall lines, mesoscale

▶ **FIGURE 11–26** Tornadoes come in a wide range of shapes and sizes.

(a)

(b)

convective complexes, supercells, and tropical cyclones (see Chapter 12). The processes that lead to their formation are not very well understood. Typically, the most intense and destructive tornadoes arise from supercells.

Supercell Tornado Development In a supercell storm, the first observable step in tornado formation is the slow, horizontal rotation of a large segment of the cloud (up to 10 km—6 mi—in diameter). Such rotation begins deep

within the cloud interior, several kilometers above the surface. The resulting large vortices, called **mesocyclones**, often precede the formation of the actual tornado by some 30 minutes or so.

The formation of a mesocyclone depends on the presence of vertical wind shear. Moving upward from the surface, the wind shifts direction and its speed increases. This wind shear causes a rolling motion about a horizontal axis, as shown in Figure 11–27a. Under the right conditions, strong updrafts in

► **FIGURE 11–27** Mesocyclones can form when a horizontal vortex of air (a) becomes tilted upward (b), yielding vertical counterclockwise and clockwise rotating vortices.

(a)

(b)

▲ **FIGURE 11-28** (a) A wall cloud protrudes below the main body of a supercell. (b) A closer look at a wall cloud.

the storm tilt the horizontally rotating air so that the axis of rotation becomes approximately vertical (Figure 11–27b). This provides the initial rotation within the cloud interior.

Intensification of the mesocyclone requires that the area of rotation decrease, which leads to an increase in wind speed.[6] The narrowing column of rotating air stretches downward, and a portion of the cloud base protrudes downward to form a **wall cloud** (Figure 11–28). Wall clouds form where cool, humid air from zones of precipitation is drawn into the updraft feeding the main cloud. The cool, humid air condenses at a lower height than does the air feeding into the rest of the cloud. Wall clouds most often occur on the southern or southwestern portions of supercells, near areas of large hail

and heavy rainfall (Figure 11–29). They are particularly noteworthy because the most significant tornadoes associated with supercells usually form within or near wall clouds. For this reason professional storm chasers[7] often try to position themselves ahead and to the right of the position of the wall cloud, relative to the storm's motion. From this vantage point they maximize their chances of not having the tornado obscured by rainfall while minimizing their chances of being hit.

Funnel clouds form when a narrow, rapidly rotating vortex emerges from the base of the wall cloud. As air is drawn upward into the zone of rotation, condensation and the importation of dust and debris from the surface may give

---

[6]This is another application of the conservation of angular momentum, initially described in Chapter 8.

[7]Storm chasing can be extremely dangerous if undertaken without the supervision of a professional. Do not ever attempt it on your own—*always* treat tornadoes with the utmost caution.

▶ **FIGURE 11-29** An idealized supercell, showing a wall cloud on the southwestern portion.

the funnel a dark, ominous appearance. A funnel cloud has all the characteristics and intensity of a true tornado; the only difference between the two is that a funnel cloud has yet to touch the ground.

Because Doppler radar enables forecasters to observe the rotating winds of mesocyclones, the network has greatly increased lead times in issuing tornado warnings (refer again to *Box 11–4, Forecasting: Doppler Radar*). Only about 20 percent of all mesocyclones actually spawn tornadoes, however. Exactly why some mesocyclones produce tornadoes and others do not is unknown, and, for that reason, forecasters cannot tell in advance which mesocyclones will produce tornadoes. Despite these uncertainties, Doppler radar has proven itself to be the most important tool yet developed in "nowcasting" (making short-term predictions about) the formation of tornadoes.

Nonsupercell Tornado Development The exact mechanisms that lead to nonsupercell tornadoes are also poorly understood, though recent research indicates that these tornadoes have their origins nearer to the surface than do those that begin as mesocyclones. Figure 11–30 illustrates one situation that may lead to nonsupercell tornadoes. The arrows show the outflow of air from two thunderstorm regions: at the top left and bottom right of the figure. From the bottom left to the top right of the figure is a zone of convergence between the two masses of air. At certain areas along the convergence zone (the circled areas), strong rotation develops. Another possible mechanism is shown in Figure 11–31, where strong convection along the convergence zone causes uplift and the formation of a cumulus cloud. In Figure 11–31c, the cloud develops into a cumulonimbus, and the strong rotation stretches down from the cloud base to form a funnel cloud.

(a)

(b)

(c)

▲ **FIGURE 11–30** Some nonsupercell tornadoes appear to develop where the outflow from separate storm downdrafts causes convergence. The arrows depict the wind flow over a region of about 400 sq km. The circled areas, organized from the bottom left to upper right, represent zones of rapid rotation where tornadoes appeared.

▲ **FIGURE 11–31** The evolution of a tornado along a convergent boundary. Spinning motions along the boundary (a) can be carried upward if there is sufficient convection (b). Once the cumulonimbus develops (c), the downward movement of the strong rotation can lead to tornadoes.

▶ **FIGURE 11-32** Tornadoes around the globe. The areas of greatest dot concentration correspond to those of greatest tornado frequency.

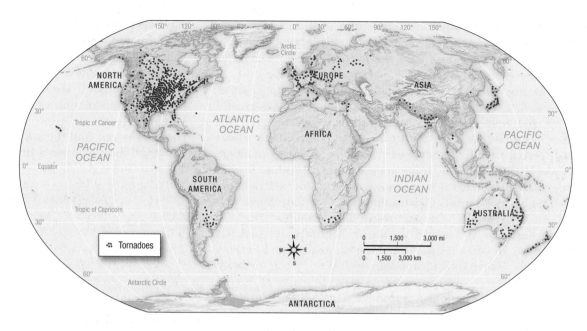

▶ **FIGURE 11-32** Tornadoes around the globe. The areas of greatest dot concentration correspond to those of greatest tornado frequency.

The Location and Timing of Tornadoes

Tornadoes are a very American phenomenon; no other country in the world has nearly as many as does the United States (Figure 11–32). Several factors combine to make North America a haven for tornadoes. The continent covers a wide range of latitudes: Its southeastern portion borders the warm Gulf of Mexico, while the northernmost portion extends into the Arctic. Furthermore, much of the eastern portion of the continent is relatively flat and, in particular, no major mountain range extends in an east–west direction. Together, these features allow for a collision of northward-moving maritime tropical air from the Gulf of Mexico with southward-moving continental polar air along the polar front. This setting, coupled with the frequent presence of potential instability (Chapter 6), provides a favorable situation for tornado development. The frequent occurrence of drylines also contributes to the high incidence of tornadoes across much of the southern Great Plains states (such as Oklahoma and much of Texas).

Tornadoes occur at least occasionally in almost all 50 states. Figure 11–33 maps the average incidence and

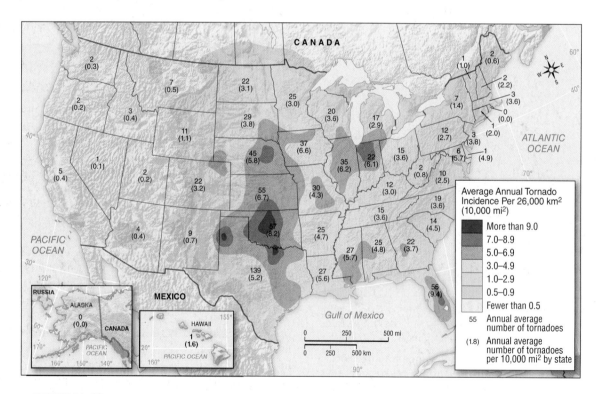

▲ **FIGURE 11-33** Annual average number of tornadoes between 1953–2004 (top number) and the annual average number of tornadoes per 10,000 square miles (lower numbers in parentheses).

concentration of tornadoes. A great many tornadoes touch down along a wide strip extending from east Texas to eastern Kansas and Nebraska, commonly called *Tornado Alley* (though the incidence of tornadoes is also high up through the Mississippi River Valley and the lower Great Lakes region).

Ever since Dorothy was swept up by a tornado and dropped into the Land of Oz, many people have believed that Kansas leads the nation in tornado incidence. In fact, that state ranks only third. Texas easily leads the rest of the states in total number of tornadoes, with considerably more than 100 annually. But when the large size of that state is taken into account, it ranks only ninth in number of tornadoes per unit area. Florida (which happens to be well removed from Tornado Alley) has the highest tornado density (number of tornadoes per unit area) of all the states, followed by Oklahoma. Unlike the twisters in Tornado Alley, many of Florida's tornadoes are embedded in passing hurricanes and tropical storms, and many occur offshore as relatively weak waterspouts (described later in this chapter). Oklahoma is especially dominant with regard to what are considered strong tornadoes—those categorized as EF-2 or higher on the Fujita scale, which is discussed later in this chapter—so that state is the overall leader with regard to the density and severity of tornadoes.

Tornadoes can occur at any time of the year. But, as shown in Figure 11–34, a strong concentration of tornadoes occurs during the spring, when air mass contrasts are especially strong. May has the greatest number of tornadoes, with

June a close second. Tornadoes are most likely to occur in the afternoon and early evening between 3 and 8 P.M., but many also occur well into the night.

Tornadoes are far less common in Canada, with an annual average of only about 100. The greatest concentration is in the extreme southern part of Ontario, between Lake Huron and Lake Erie. Most Canadian tornadoes outside Ontario occur in the southern region of the Prairie provinces and southwestern Quebec. The Canadian tornado season extends from April through October, with the greatest frequency in June and July. Despite the fact that the greatest concentration of tornadoes in general and of very strong tornadoes in particular occurs in Ontario, it is interesting to note that the worst Canadian tornado outbreak in recent decades occurred in 1987 in Edmonton, Alberta. The Edmonton tornadoes of July 31 left 27 people dead, more than 300 injured, and thousands homeless. Total damages were estimated at $300 million.

Did You Know?

Between 1950 and 2005 only 27 percent of U.S. tornadoes occurred at night, but these nocturnal tornadoes were responsible for 39 percent of the fatalities. This is likely due to their being harder to see at night and the fact that many potential victims are asleep and therefore unable to take protective measures.

Checkpoint

1. What type of weather situation most often leads to the formation of a large, destructive tornado?
2. Starting with a mesocyclone, what are the steps in the development of a funnel cloud?
3. What factors account for the distribution of tornadoes in Figure 11–32?

Trends in U.S. Tornado Occurrence

A plot of tornado occurrences in the United States, based on a log of events collected by the National Severe Storms Laboratory of NOAA, shows a doubling of the tornado frequency from 1950 to the early 2000s (Figure 11–35). Taken at face value, this would be a very disturbing trend. But there are reasons to believe that this increase might only partially reflect an actual increase in tornado activity, and that the trend may be primarily due to a greater likelihood that a given tornado will actually be observed. One possible explanation relates to population increases. As population centers have expanded out into formerly rural areas, there is a greater probability that a tornado will hit a structure or be observed directly. Furthermore, installation of the Doppler radar network has undoubtedly contributed to improved detection and an increase in the number of participants in the national storm spotter network. Changes in the way we classify tornadoes may also have played a role in the apparent increase. Numerous researchers are currently working to explain the relative effects of these factors on the apparent large increase in tornado activity and whether there really has been a significant increase.

▲ **FIGURE 11-34** U.S. tornadoes occur with greatest frequency in May and June. Values represent monthly averages for the years 1950 to 1995.

▶ **FIGURE 11-35** The number of observed tornadoes in the United States has shown an apparent doubling since the 1950s, but this increase may be mostly due to better observation rather than to a genuine increase in tornado activity.

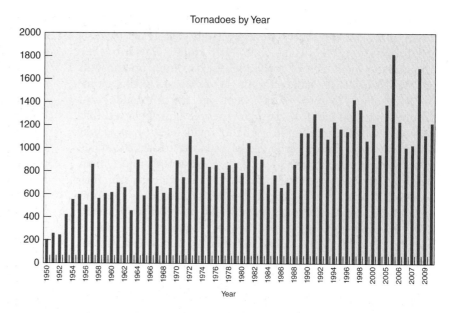

The debate on recent global warming and the importance of human activities to that warming has led to concern about a future rise in tornado frequency. At this point in time, however, these ideas are still speculative.

Tornado Damage

Most structural damage from tornadoes results from their extreme winds. People once believed that homes were destroyed mostly by the pressure differences associated with a tornado's passage, which supposedly caused the interior air to push outward against the walls so violently that the house would explode. For this reason, people were advised to open their windows if they saw an approaching tornado, so that the pressure within the house could be reduced.

We now know that this was not good advice, in part because few homes actually explode. Moreover, though winds are the major factor in tornado damage, flying debris is the primary cause of tornado injuries, and opening a window increases the risk of personal injury from flying debris. (We must also suspect that opening the windows is useless in any case, because they are likely to be "opened" anyway by flying objects.)

Although most tornadoes rotate around a single, central core, some of the most violent ones have several relatively small zones of intense rotations (about 10 m—30 ft—in diameter) called **suction vortices** (Figure 11–36). It is these small vortices that probably cause the familiar phenomenon of one home being totally destroyed while the next one remains relatively unscathed. Sometimes the path of a tornado is remarkably well defined after a sweep across the landscape (Figure 11–37).

Except for those rare times when tornado chasers make firsthand observations of passing tornadoes, it is impossible to get a precise reading on their pressure changes and wind

▲ **FIGURE 11-36** Suction vortices sweep around a tornado center and often cause major destruction (a). A multiple vortex tornado over Friendship, Oklahoma, on May 11, 1982 (b).

◄ **FIGURE 11–37** Swath of damage from the tornado that hit near Stoughton, Wisconsin, in 2005.

speeds. But it is possible to classify them according to the magnitude of the damage they cause. The **enhanced Fujita scale** (named for the late eminent tornado specialist Theodore Fujita) provides a widely used system for ranking tornado intensity. As shown in Table 11–1, tornadoes fall into six levels of intensity, with each assigned a particular EF-value (called an F-value on the original Fujita scale) ranging from 0 to 5. Fujita values are determined by trained observers who assess a tornado's impact on any of 28 categories of natural objects or built structures (such as softwood trees, service station canopies, motels, small barns, single-wide mobile homes, and so on) called *damage indicators*. The observers then assign a numerical value to represent the *degree of damage* to each damage indicator. A certain minimum wind gust speed lasting 3 seconds or longer is then assumed to create any particular degree of damage to a damage indicator, and this is translated to an EF value based on the values shown in Table 11–1.[8]

In the United States, the majority (69 percent) fall into the *weak* category, which includes EF-0 and EF-1 tornadoes. Twenty-nine percent of tornadoes are classified as *strong* (EF-2 and EF-3), which makes them capable of causing major structural damage even to well-constructed homes. Fortunately, only 2 percent of tornadoes are *violent* (EF-4 and EF-5). Those tornadoes are capable of wreaking incredible destruction. Cars can be picked up and carried tens of meters, pieces of straw can be driven into wooden beams, and freight cars can be carried off their tracks. Indeed, these storms are the true stars in all the movies and videos about tornadoes.

---

[8]Those familiar with the original Fujita scale will note that the wind speeds for the larger categories have been reduced in the enhanced version, based on better knowledge of the wind speeds necessary to incur certain types of damage. The damage that would have resulted in the classification of, say, an F4 on the original scale is the same as that which would be considered an EF-4 on the new scale.

TABLE 11–1

Enhanced Fujita Intensity Scale

Intensity	Maximum 3-second Wind Gust (km/hr)	Maximum 3-second Wind Gust (mph)	Typical Amount of Damage
EF-0	105–137	65–85	Light: Broken branches, shallow trees uprooted, damaged signs and chimneys.
EF-1	138–177	86–110	Moderate: Damage to roofs, moving autos swept off road, mobile homes overturned.
EF-2	178–217	111–135	Considerable: Roofs torn off homes, mobile homes completely destroyed, large trees uprooted.
EF-3	218–266	136–165	Severe: Trains overturned, roofs and walls torn off well-constructed houses.
EF-4	267–322	166–200	Devastating: Frame houses completely destroyed, cars picked up and blown downwind.
EF-5	Over 322	Over 200	Incredible: Steel-reinforced concrete structures badly damaged.

Note: EF-0 and EF-1 tornadoes are collectively called weak, EF-2 and EF-3 strong, and EF-4 and EF-5 violent.

Deadly 2011 Tornado Season

By the end of June (as this is written) the 2011 season has already turned out to be remarkably deadly. The anomalous number of fatalities resulted from the combination of a high number of tornadoes, including very powerful ones, and the bad luck of several very large ones passing over populated areas.

April 2011

Though April is on average the third most active month for United States tornadoes (see Figure 11–34), this particular April was anything but normal. The year 2011 proved not only to have the most in the United States tornadoes ever recorded for an April (875), it in fact turned out to eclipse the record for the number of tornadoes recorded for *any* month (542 in May 2003). The number of fatalities was also brutally high, at 369. To put this in perspective, over the previous 11 years the average annual number of fatalities was 55.

There were six significant multiday tornado outbreaks over the United States that month. One of them swept a path from Oklahoma to North Carolina on April 14–16, producing about 155 tornadoes—more than is typically encountered in an entire April—and killing 38 people. On April 22 an EF-4 tornado passed through parts of St. Louis, Missouri, and did substantial damage to Lambert St. Louis International Airport.

But the worst outbreak that month was that of April 25–28, which unleashed tornadoes—about 305 of them—all the way from Texas to New York State. Many of the tornadoes were particularly violent, with three of the classified as EF-5, 12 as EF-4, and 21 as EF-3. The deadliest tornado of the outbreak was an EF-4 that hit the cities of Tuscaloosa and

▲ **FIGURE 1** Dozens of homes lay in ruin in Pratt City, near Birmingham, Alabama, following a devastating April 28, 2011 tornado.

Birmingham, Alabama (Figure 1) on the 27th. With winds that peaked at 306 km/hr (190 mph) and a width of 2.4 km (1.5 miles), the tornado cut a path of 129 km (80 miles), killed at least 71 people and injured at least another 1000. Over the entire region on the 27th alone there were 317 fatalities, making it the worst single day for tornado fatalities since 1925.

Between 1950 and 1999, 51 EF-5 tornadoes occurred in the United States and none in Canada. Thus, they happen about once a year in all of North America. Texas holds the lead for the greatest number of EF-5 tornadoes (6) during that period, while Alabama, Iowa, Kansas, and Oklahoma each had 5.

Fatalities

Because they are small and last for such a short time, the overwhelming majority of tornadoes kill no one. A summary of data compiled by the University of Nebraska shows that the more than 48,000 tornadoes reported in the United States from 1950–2005 resulted in some 4700 fatalites. Thus, well over 43,000 (90 percent) of those reported tornadoes killed no one. And the actual proportion of fatality-free tornadoes must have been even higher, for at least two reasons. First, tornadoes in which no one dies are preferentially undercounted, because a storm that kills is almost certain to be reported. Second, most fatalities result from a few very large storms that kill up to dozens of people. According to the National Severe Storms Laboratory (NSSL), fewer than 5 percent of all

U.S. tornado deaths are associated with weak (EF-0 or EF-1) tornadoes, nearly 30 percent with strong tornadoes (EF-2 and EF-3), and about 70 percent with violent tornadoes (EF-4 and EF-5). Thus, only 2 percent of all tornadoes are responsible for more than two-thirds of all fatalities. Although the number of reported tornadoes has increased over the years, the number of resulting fatalities has dropped markedly since the 1930s, with much of that reduction occurring since the 1970s (Figure 11–38). There are several reasons for this drop, including better building construction, vastly improved technology for predicting and tracking tornadoes, and a better network for broadcasting emergency information to the public.

While Figure 11–33 shows that the region of maximum tornado activity in the United States is the southern Great Plains, Figure 11–39 indicates that the distribution of tornado fatalities is surprisingly different. For the period of January 2000 through June 2011, Alabama easily wins the dubious distinction of leading the country in tornado fatalities, followed by Missouri. It is noteworthy that both these states suffered extremely devastating killer tornadoes in the first half of 2011 that greatly inflated their fatality

May 2011

May of 2011 had more tornado activity than normal, with at least 370 tornadoes recorded. Though the number of tornadoes was less than there were the month before, it was still a deadly month with an estimated total of 171 fatalities. The month would not have been an abnormally deadly one were it not for the supercell that generated an EF-5 tornado that hit the heavily populated, southern part of Joplin, Missouri, during the afternoon of May 22 (Figure 2). The tornado, which had a maximum diameter of 1.2 km (0.75 mi), cut a nearly 10 km (6 mi) path through the town with maximum winds exceeding 320 km/hr (200 mph). About 171 people died in Joplin, making the tornado the deadliest during the modern period of record keeping that began in 1950. The Joplin tornado ranks as the seventh deadliest in U.S. history.

The Role of Tornado Watches and Warnings

As horrific as the death and injury statistics were for the spring of 2011, they almost certainly would have been far worse without the alerts given by the National Weather Service. When a roof was partially blown off one of the terminals at Lambert International in St. Louis there were no fatalities, as much of the airport had already been evacuated following a tornado warning.

The National Weather Service

▲ **FIGURE 2** Damage from an EF-5 tornado that hit Joplin, Missouri, on May 22, 2011.

played a life-saving role throughout the outbreaks of April and May, with tornado warnings issued on average 24 minutes prior to tornado strikes. This type of advanced warning was given to the residents of Joplin, Missouri, at 5:17 P.M. local time by the forecast office in Springfield. The first report of tornadoes touching ground occurred at 5:41 P.M. A tornado watch had also been issued some 4 hours prior to the tornado, giving the public additional advanced safety information.

totals (see *Box 11–6, Focus on Severe Weather: Deadly 2011 Tornado Season*). Even excluding the anomalous events of 2011, however, Missouri and Alabama still ranked two and three in the country, following Tennessee, for fatalities. None of those three states rank in the top ten for total tornado incidence.

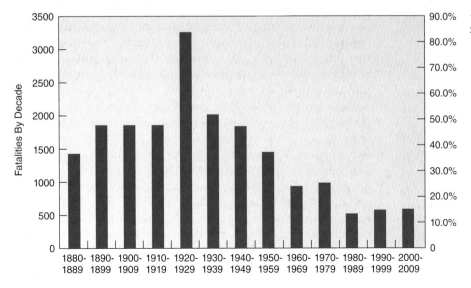

◄ **FIGURE 11–38** The number of reported United States tornado fatalities by decade.

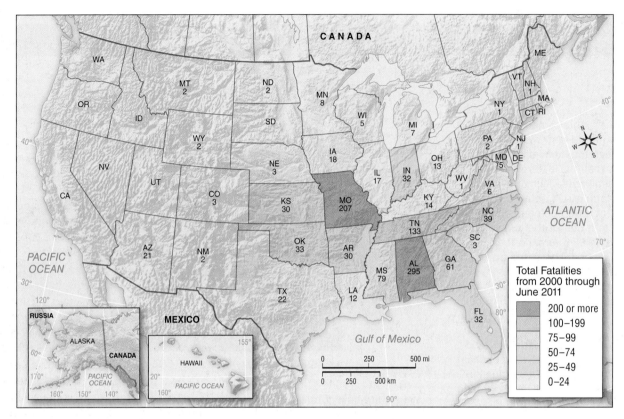

▲ **FIGURE 11-39** Total number of tornado fatalities by state, 2000–2011.

Recent research has tried to identify the reasons for the anomalously large number of fatalities over those states, and several possible explanations have emerged. One possible factor is that tornadoes in the region have a greater likelihood of occurring during the night when they are more difficult to spot and when people may be sleeping and unable to hear emergency warnings. It has also been hypothesized that residents of the area may be more complacent with regard to tornadoes because they are not as frequent and they are more dispersed throughout the year than they are in the southern Plains. Thus there is less of a public awareness of tornadoes in the Deep South than there is in Tornado Alley, where residents are particularly mindful of the threat, especially as they enter into the most dangerous spring months. But there is little question that the unusually high percentage of people living in mobile homes in the Deep South is a major factor in the high fatality rate.

Did You Know?

Items picked up by tornadoes can be lifted to great heights and transported long distances. Following the F3 tornado on August 18, 2005, in Stoughton, Wisconsin (Figure 11–37), birth certificates and other personal papers were found in Milwaukee, more than 100 km (60 mi) away.

If it seems that news reports of killer tornadoes tend to focus on mobile home parks, it is for a reason. A disproportionately large percentage of tornado-related deaths do in fact occur in mobile homes. Trailers offer little protection to their occupants because they are easily blown off their foundations and tossed about by strong winds.

According to the National Weather Service's Storm Prediction Center, between 1997 and 2007, 49 of the 705 United States tornado fatalities (7 percent) were in motor vehicles. These victims were often those who panicked and made the fatal mistake of leaving the relative safety of their homes to outrun the storm in their vehicles. By far the safest place you can be when a tornado threatens is inside a well-constructed building, preferably in the basement and away from the windows.

A famous video shot in 1991 by a television news crew has led many people to believe that highway overpasses provide good shelter from a tornado. The video shows a film crew along a Kansas highway fleeing an approaching tornado. The reporters stopped at an overpass, where they advised a terrified father and his daughter to take shelter under the girders. Huddled with the man and child, the news crew continued to film the tornado as it passed directly over them—with nobody being hurt. The video then proceeded to show a truck driver who luckily survived after his rig had been blown off the highway, and reported the death of another nearby motorist. The film provided a remarkable view of a tornado from the inside and naturally was repeatedly played on television stations across the country. Unfortunately, it did not issue a disclaimer about the relative lack of safety provided by overpasses in most instances. Though the people in the video survived their ordeal and may have saved their lives by moving under the overpass, this is not usually the best course of action. In this case, the center of the tornado passed just to the south of the overpass so that the road above happened to provide considerable shelter. However, had the tornado taken a slightly different path, the outcome would not have been so fortuitous. Also, this particular tornado was a weak one—either an EF-0 or EF-1—and the overpass was a strong one with a narrow opening that allowed the people to hold on tightly as they were buffeted by winds. And finally, the fact that the incident was in a very rural area led to a small amount of debris being thrown toward the people—a situation that would be far different in a more heavily populated area.

People should always obey the following safety rules when a tornado threatens:[9]

1. Stay indoors and seek shelter in a basement.

2. If you are in a building with no basement, move to an interior portion of the lowest floor and crouch to the floor. If possible, cover yourself with a mattress or some other form of padding to protect yourself from falling or flying debris. Despite what is sometimes said, the southwest corner of a building does not offer additional protection.

Did You Know?

Between 1985 and 2003, 40 percent of all tornado fatalities in the United States occurred in mobile homes. This is greater than the 31 percent of fatalities that occurred in permanent homes, and it is very noteworthy in view of the relatively small proportion of people who live in mobile homes.

3. If you are in a mobile home, you should immediately vacate it for a nearby shelter or sturdy structure.

4. If you are caught outdoors and are unable to quickly get to a shelter or sturdy building, you should get into your vehicle, fasten your seat belt, and drive to the nearest available place of shelter. If you encounter flying debris while in your vehicle, you should do either of the following: (a) pull over with your seat belt still fastened and lower your head to below window level (covering your head with your hands or other covering), or (b) leave the vehicle for a noticeably lower-lying area, lie down, and cover your head.

Checkpoint

1. In your own words, describe the level of damage associated with each level of the enhanced Fujita scale.

2. Look at the map in Figure 11–39. What factors might account for the state-to-state differences in tornado fatalities per year? Explain.

Watches and Warnings

Without question, one of the key responsibilities of the National Weather Service is to issue severe weather advisories. These take two forms, watches and warnings, either of which can be issued for severe storms or tornadoes.[10] The declaration of a watch does not mean that severe weather has developed or is imminent; it simply tells the public that the weather situation is conducive to the formation of such activity. Most watches are issued for a period of 4 to 6 hours, for an area that normally encompasses several counties—about 50,000 to 100,000 sq km (20,000 to 40,000 sq mi).

The Storm Prediction Center (SPC) of the U.S. Weather Service in Norman, Oklahoma, has responsibility for putting out **severe storm and tornado watches** for the entire country. Operating 24 hours a day, every day, the center constantly monitors surface weather station data, information from weather balloons and commercial aircraft, and satellite data for all of the United States. If any particular part of the country appears vulnerable to impending severe storm activity, SPC issues a severe storm or tornado watch. The advisory then goes to the local office of the National Weather Service, which notifies local television and radio stations. The broadcast media then relay the information to the public. Watches for the entire country are also available from a number of sources, many of which are on the World Wide Web. Here is a portion of the text of the tornado watch issued hours before the devastating Joplin, Missouri, tornado of May 22, 2011:

URGENT - IMMEDIATE BROADCAST REQUESTED
TORNADO WATCH NUMBER 325
NWS STORM PREDICTION CENTER NORMAN OK
130 PM CDT SUN MAY 22 2011

THE NWS STORM PREDICTION CENTER HAS ISSUED A
TORNADO WATCH FOR PORTIONS OF

    NORTHWEST ARKANSAS
    SOUTHEAST KANSAS
    SOUTHWEST AND CENTRAL MISSOURI
    EASTERN OKLAHOMA

EFFECTIVE THIS SUNDAY AFTERNOON AND EVENING FROM 130 PM UNTIL 900 PM CDT.

TORNADOES . . . HAIL TO 4 INCHES IN DIAMETER...THUNDERSTORM WIND GUSTS TO 70 MPH . . . AND DANGEROUS LIGHTNING ARE POSSIBLE IN THESE AREAS.

THE TORNADO WATCH AREA IS APPROXIMATELY ALONG AND 75 STATUTE MILES EAST AND WEST OF A LINE FROM 30 MILES WEST NORTHWEST OF JEFFERSON CITY MISSOURI TO 30 MILES SOUTH OF MUSKOGEE OKLAHOMA. FOR A COMPLETE DEPICTION OF THE WATCH SEE THE ASSOCIATED WATCH OUTLINE UPDATE (WOUS64 KWNS WOU5).

REMEMBER . . . A TORNADO WATCH MEANS CONDITIONS ARE FAVORABLE FOR TORNADOES AND SEVERE THUNDERSTORMS IN AND CLOSE TO THE

---

[9]Updated by the National Weather Service and American Red Cross in June 2009.

[10]Warnings and watches can be issued for other types of threatening weather, such as hurricanes and flash floods.

WATCH AREA. PERSONS IN THESE AREAS SHOULD BE ON THE LOOKOUT FOR THREATENING WEATHER CONDITIONS AND LISTEN FOR LATER STATEMENTS AND POSSIBLE WARNINGS.

AVIATION . . . TORNADOES AND A FEW SEVERE THUNDERSTORMS WITH HAIL SURFACE AND ALOFT TO 4 INCHES. EXTREME TURBULENCE AND SURFACE WIND GUSTS TO 60 KNOTS. A FEW CUMULONIMBI WITH MAXIMUM TOPS TO 600. MEAN STORM MOTION VECTOR 27025.

If a severe thunderstorm has already developed, the public is notified by a **severe thunderstorm warning**. Likewise, **tornado warnings** alert the public to the observation of an actual tornado (usually by a trained weather spotter) or the detection of tornado precursors on Doppler radar. Unlike watches, which are issued by the SPC, warnings are given by local weather forecast offices. They warn the public to take immediate safety precautions, such as finding shelter in a basement. The information is broadcast immediately by television and radio stations, and civil defense sirens are sounded.

Sometimes Doppler radar enables meteorologists to give warning of an impending tornado about half an hour before it actually forms. Prior to the advent of Doppler radar, warnings were usually issued only after a funnel cloud had been seen. Many tornado warnings are still based on *in situ* observations, but the use of Doppler radar provides better lead time in many instances. When tornado warnings are based on visual sightings, observers must report their findings to the local Weather Service office, where the meteorologist in charge then passes on the information to the broadcast media. The procedure normally takes several minutes. But most tornadoes have lifespans of about 3 to 4 minutes, so in many instances the warnings are issued to the public only after the tornado has died out. Though this might make the procedure seem pointless, that really is not the case. First of all, one tornado is often followed by others within the same storm system. Consequently, the precautions taken for the defunct tornado might still save lives when subsequent ones appear. Second, rare killer tornadoes tend to have considerably longer lifetimes and can remain in existence long after the warning has been issued. Thus the warning goes out while the tornado still presents a serious risk to people.

Convective Outlooks

The Storm Prediction Center issues **convective outlook** maps on its Web site (listed at the end of this chapter). These maps are issued several times each day, showing the probability of severe weather hitting parts of the United States for the current day, the next day, or the day after. Figure 11–40 shows such a series of maps for May 15, 2003, with (a) plotting the threat of severe weather of any kind across the country. Here the greatest possibility of severe weather occurs around the Texas and Oklahoma panhandles, with much of the Ohio River Valley and Southeast having a slight risk. Panels (b),

(c), and (d) give more specific probabilities for the occurrence of EF-2 or higher tornadoes, damaging winds, and large hail, respectively, occurring within 25 miles of any point within the outlined areas. So, for example, a resident of north Texas or western Oklahoma has about a 25 percent probability that a tornado or dangerous winds will occur within 25 miles of home on that day, with an even greater probability (35 percent) of large hail. Residents of Tennessee and Kentucky, on the other hand, have less than a 2 percent probability of tornadoes hitting, but a much greater likelihood (25 percent) of strong winds or large hail. In addition to these maps, the SPC accompanies the probability maps with narratives further describing the threats.

Despite the excellent warning system, disasters do indeed happen. Sometimes, sheer luck is all that stands in the way of major loss of life. On September 2, 2002, at 4:20 P.M. CDT, an F3 tornado hit the town of Ladysmith, Wisconsin, destroying 26 businesses and 17 homes. This time no tornado warning was issued until 17 minutes *after* the tornado hit the town, although a severe thunderstorm warning had been in effect for more than half an hour. (The fact that the nearby Doppler radar unit at La Crosse, Wisconsin, had been temporarily out of order may have contributed to the absence of a more timely warning.) Although most residents received no warning about the impending tornado, somehow the town escaped with no fatalities.

Tornado Outbreaks

The most devastating of all outbreaks in the United States was the Tri-State tornado that rolled through eastern Missouri, southern Illinois, and southwest Indiana on March 18, 1925. The tornado killed about 700 people (roughly 600 of them in Illinois) and leveled several towns. To this day we do not know for sure if the Tri-State tornado was a single, incredibly large tornado, or part of a larger **tornado outbreak**—an event in which a single weather system produces at least six tornadoes.

Though the Tri-State was the deadliest of all tornado events, it was not the largest. That honor goes to the outbreak of April 3–4, 1974, in which at least 148 tornadoes swept across a wide region extending from northern Alabama and Mississippi to as far north as Windsor, Ontario, in Canada. Thirty of those tornadoes had EF-4 or EF-5 intensity and accounted for many of the more than 300 fatalities incurred over a 16-hour period.

Checkpoint

> 1. In terms of assessing the risk of a tornado, what is the difference between a severe thunderstorm watch and a tornado warning?
> 2. What is the value of convective outlook maps in predicting the risk of tornadoes for an area on a given day? Explain.

▲ **FIGURE 11–40** Examples of convective outlook maps mapping the threat of any type of severe weather (a), tornadoes (b), strong winds (c), and large hail (d).

Waterspouts

So far we have discussed tornadoes over land. Similar features, called **waterspouts** (Figure 11–41), occur over warm-water bodies. Waterspouts are typically smaller than tornadoes, having diameters between about 5 and 100 m (17 to 330 ft). Though they are generally weaker than tornadoes, they can have wind speeds of up to 150 km/hr (90 mph), which makes them strong enough to damage boats.

Some waterspouts originate when land-based tornadoes move offshore. Most, however, form over the water itself.

These "fair-weather" waterspouts develop as the warm water heats the air from below and causes it to become unstable. As air rises within the unstable atmosphere, adiabatic cooling lowers the air temperature to the dew point, and the resultant condensation gives the waterspout its ropelike appearance. Waterspouts form in conjunction with cumulus congestus clouds, those having strong vertical development but not enough to form the anvil that characterizes cumulonimbus.

Contrary to what we might assume, the visible water in the waterspout is not sucked up from the ocean below; it actually

▶ **FIGURE 11–41** A waterspout.

comes from the water vapor in the air. Waterspouts are particularly common in the area around the Florida Keys, where they can occur several times each day during the summer.

Although tornadoes produce the most extreme winds on Earth, they are relatively small and short-lived. Tropical storms and hurricanes, on the other hand, usually produce less intense winds but can wreak more extensive destruction because of their longer lifespans and larger extent. Chapter 12 examines the processes that produce these devastating storms and their consequences.

Summary

Some of the most dramatic of all storms are those that produce lightning and thunder. Lightning begins when negative electrical charges build up near the base of a cloud and positive charges gather at the top. In the case of cloud-to-ground lightning, rapidly growing leaders extend downward from the base of the cloud. When they connect with some object at the surface, a visible stroke develops. Most often a rapid sequence of multiple strokes follows the initial one to produce a lightning flash. Extreme heating of the air within the stroke causes the air to expand explosively and create the sound of thunder.

Thunderstorms create downdrafts that can cause the storm to die out or intensify. In air mass thunderstorms, downdrafts eventually destroy the storm. They undergo a sequence from the cumulus to the mature to the dissipative stages in tens of minutes. Severe storms—those that produce large hail, damaging winds, or tornadoes—develop when downdrafts reinforce the storm. Such storms occur within squall lines, in multicellular mesoscale convective complexes, or as supercells.

Tornadoes, among the most fearsome of all natural phenomena, occur more often in the United States than in any other country of the world—and nearly all the states and Canadian provinces experience them sooner or later. They are most likely to form in the spring or early summer, but they can occur at any time of the year. Many tornadoes (especially those that emerge from supercells) follow the formation of large rotating areas within storm clouds called *mesocyclones*. However, other unexplained processes can also lead to their formation.

The majority of tornadoes are classified as weak and cause no fatalities, but relatively rare, extremely large tornadoes can cause tremendous devastation and loss of life. Fortunately, the U.S. Weather Service has a system of tornado watches and warnings that alert the public to threatening weather. In recent years, Doppler radar has shown tremendous value as a research and forecast tool. This tool has also become valuable in the tracking of another weather phenomenon—hurricanes—that we will examine in Chapter 12.

Key Terms

lightning *page 308*

cloud-to-cloud lightning *page 308*

sheet lightning *page 308*

cloud-to-ground lightning *page 308*

charge separation *page 308*

fair-weather electric field *page 309*

mean electric field *page 309*

runaway breakdown *page 309*

stepped leader *page 309*

stroke, return stroke *page 310*

dart leader *page 310*

flashes *page 310*

ball lightning *page 310*

St. Elmo's fire *page 311*

sprites *page 311*

blue jets *page 312*

thunder *page 312*

heat lightning *page 312*

air mass thunderstorms *page 314*

cumulus stage *page 314*

mature stage *page 314*

dissipative stage *page 315*

multicell thunderstorms *page 315*

mesoscale convective systems (MCSs) *page 315*

squall lines *page 315*

mesoscale convective complexes (MCCs) *page 315*

supercells *page 315*

severe thunderstorms *page 315*

outflow boundary *page 316*

gust front *page 317*

shelf cloud *page 317*

roll cloud *page 317*

Doppler radar *page 318*

hook or hook echo *page 319*

vault *page 319*

downbursts *page 319*

derechos *page 319*

Doppler effect *page 320*

sweep *page 321*

volume sweep *page 321*

microbursts *page 324*

haboob *page 324*

tornadoes *page 326*

mesocyclones *page 327*

wall clouds *page 328*

funnel clouds *page 328*

suction vortices *page 332*

enhanced Fujita scale *page 333*

severe storm and tornado watches *page 337*

severe thunderstorm warning *page 338*

tornado warning *page 338*

convective outlooks *page 338*

tornado outbreak *page 338*

waterspouts *page 339*

Review Questions

1. How common is cloud-to-ground lightning relative to cloud-to-cloud lightning?

2. Describe the current theories regarding the formation of charge separation.

3. What is the difference between a lightning stroke and a lightning flash?

4. Describe the sequence by which electrical imbalances lead to lightning strokes.

5. Briefly describe the following phenomena:
   a. ball lightning
   b. St. Elmo's fire
   c. sprites
   d. blue jets

6. What causes thunder?

7. Why is the term *heat lightning* misleading?

8. What are the three stages of an air mass thunderstorm?

9. How big are air mass thunderstorms and how long do they usually persist?

10. Explain why some thunderstorms have short lifespans and yield little damage and others are able to develop into severe thunderstorms.

11. Describe the following types of storm systems:
    a. mesoscale convective systems
    b. squall lines
    c. mesoscale covective complexes
    d. supercells

12. How are outflow boundaries formed, and what effect do they have?

13. Describe the processes that lead to tornado development in supercell and nonsupercell storms.

14. What features of Doppler radar make it an effective tool for severe storm forecasting?

15. What are hook echoes and vaults, and why are they important?

16. Explain how microbursts form and why they present a serious threat to aviation.

17. Describe the location and timing of tornadoes in North America.

18. Describe the process of tornado formation from supercell storms.

19. What are wall clouds, and why is their appearance a cause for concern?

20. What is the leading threat to human safety when tornadoes hit?

21. Describe the enhanced Fujita scale for classifying tornadoes. Which category is most common? What is the highest F-value that can actually occur in nature?

22. What is the difference between a tornado watch and a tornado warning?

23. How do waterspouts compare to tornadoes, on average, in terms of intensity?

Critical Thinking

1. Is charge separation necessary for sheet lightning and/or ball lightning?

2. Why does the environmental lapse rate affect the distance at which "heat lightning" can be observed?

3. You are outside on a sunny afternoon and observe a thunderstorm far to the west. An hour later, the storm passes over you. Is this more likely to have been an air mass thunderstorm or some sort of mesoscale convective system?

4. In what fundamental ways are gust fronts different from passing cold fronts?

5. Why is the incidence of thunderstorms much lower near the Pacific Coast than at the Atlantic Coast?

6. What conditions east of the Rocky Mountains promote a much greater incidence of tornadoes than exists in western North America?

7. Why is it not possible for a mesocyclone to occur within an air mass thunderstorm?

8. Why is it extremely unlikely that a tornado will move from east to west?

9. Other than their location with respect to land vs. water, how do waterspouts differ from tornadoes?

Problems and Exercises

1. Compare the map of annual hailstorms (Figure 7–21) to the map depicting the distribution of lightning flashes (Figure 11–25). Identify the regions where both are frequent, and those (if any) where only hail or thunderstorms frequently occur.

2. A tornado has a ring of uniform winds of 200 km/hr (120 mph) around the vortex 20 m (66 ft) away from its center. If the tornado moves to the northeast at 50 km/hr (30 mph), what is the effective wind speed on the northwestern and southeastern portions of the tornado?

3. On a regular basis, go to the Storm Prediction Center (SPC) Web site at **www.spc.noaa.gov** and note the location of all current severe thunderstorm or tornado watches. Then go to one of the Web sites mentioned in previous chapters to observe surface weather maps, satellite images, and radar return maps. Describe the position of the watch area relative to what the maps and images depict.

Quantitative Problems

The Web site for this book, **www.MyMeteorologyLab.com**, offers quantitative problems that can help reinforce your understanding of several of the concepts discussed in this chapter. The problems deal with the timing of the arrival of thunder following a lightning strike, the characteristics of tornadic winds and pressure gradients, and potential instability. We suggest you go to the Web page for Chapter 11 and work out these straightforward problems.

Useful Web Sites

www.npr.org/2011/05/26/136655052/before-and-after-a-birds-eye-view-of-joplin?sc=emaf
A very dramatic before and after photo of Joplin, Missouri, showing the effects of the May 2011 tornado.

www.lightningstorm.com
Click on the lightning map icon to see where lightning has occurred over the United States during the last 30 minutes. A similar map for Canada exists at **www.weatheroffice.gc.ca/lightning/index_e.html**.

www.nssl.noaa.gov
This site provides interesting lightning information from the National Severe Storms Laboratory.

www.uic.edu/labs/lightninginjury
Information on the medical aspects of lightning strikes.

www.nssl.noaa.gov/headlines/outbreak.shtml and
An excellent site for comprehensive information on the deadly tornado outbreak of May 1999.

www.lightningsafety.noaa.gov
Comprehensive information from the National Weather Service includes medical effects of being struck by lightning, myths, and indoor and outdoor safety tips.

www.srh.noaa.gov/oun/?n=safety-overpass-slide01
This is an outstanding presentation on the dangerous use of highway overpasses as tornado shelters.

www.crh.noaa.gov/sgf/?n=event_2011may22_summary
National Weather Service summary of 2011 Joplin, Missouri, tornado.

www.srh.noaa.gov/bmx/?n=event_04272011tuscbirm
National Weather Service summary of 2011 Tuscaloosa, Alabama, tornado.

www.spc.noaa.gov/climo/online/monthly/newm .html#latestmts Updated statistics on United States tornadoes.

MyMeteorologyLab™

The *Understanding Weather and Climate,* Sixth Edition, MyMeteorologyLab™ web site contains numerous multimedia resources to aid in your study of **Lightning, Thunder, and Tornadoes.**
Visit **www.MyMeteorologyLab.com** to:

- **Review** key chapter concepts.
- **Read** the Pearson eText, *In the News RSS feeds,* and other chapter-specific Web resources.
- **Visualize** challenging topics with Interactive Tutorials, Geoscience Animations, MapMaster interactive maps, and Weather in Motion movies and visualizations, and more.
- **Test Yourself** with self-study quizzes.

 TUTORIAL
DOPPLER RADAR

Use the interactive animations and quizzes in this tutorial to review this chapter's key concepts.

WEATHER IN MOTION

View movies and visualizations of meteorology patterns and processes.

The Deadliest Tornado Since Modern Recordkeeping Began
Thundersleet and Thundersnow
Identifying Tornadic Thunderstorms Using Radar Reflectivity Data
Identifying Tornadic Thunderstorms Using Radar Velocity Data
A Satellite View of the Joplin, Missouri Tornado

12

Tropical Storms and Hurricanes

LEAR

346 Understanding Weath

In this chapter
canes and tropic
characteristics,
movement, c
ing system

In just about any other year, Hurricane Wilma (Figure 12–1) would have been the most remarkable of hurricanes. But 2005 was no ordinary hurricane season. By mid-October, when Wilma was threatening Mexico, the Caribbean, and Florida, there had already been 11 hurricanes in the western Atlantic–Gulf of Mexico region, about twice the normal number. Many of these storms were unusually powerful, as well, including the infamous Hurricane Katrina. But Wilma was still noteworthy—not only because of the damage it brought to the Yucatan Peninsula and Florida, but also because of its extremely low sea level pressure of 884 mb (a record for the North Atlantic), and because of its explosive growth from a tropical storm to a Category 5 hurricane (the highest possible) in a mere 24 hours.

The Miami and Ft. Lauderdale areas experienced considerable damage, including the shattering of windows in large buildings and the downing of large trees. Roads were impassible soon after the storm, millions of people lost electric power for extended time periods, and residents were advised to boil their tap water before drinking it. Earliest reports identified ten fatalities.

Hurricane Wilma then proceeded northward in the western Atlantic. Though its strong winds did not hit coastal areas, it fed additional moisture into a nor'easter, hitting Canada and the northeastern United States and Canada. This increased the amount of snow and rainfall for an area that had already experienced extreme precipitation amounts for the month. Across the United States, total insured damages caused by Wilma were estimated to be between $6 billion and $10 billion. Hurricanes do not restrict their fury to coastal and inland regions; they have been the nemesis of mariners for centuries. They have sunk an untold number of ships and even played a role in World War II when a single typhoon (the equivalent of a hurricane over the western Pacific) sank or heavily damaged several U.S. ships in the Philippine Sea, killing nearly 800 sailors. The death toll exceeded that of most naval battles during the war.

▲ **FIGURE 12–1** Satellite image of Hurricane Wilma, October 2005.

◄ Heavy wind and rain from Hurricane Katrina hit Pensacola, Florida.

345

we first describe the setting for hurri-
storms. We then describe their general
stages of development, and typical patterns of
ncluding with hurricane monitoring and warn-
s.

urricanes Around the Globe: The Tropical Setting

Extremely strong tropical storms go by a number of different names, depending on where they occur. Over the Atlantic and the eastern Pacific they are known as **hurricanes**. Those over the extreme western Pacific are called **typhoons**; those over the Indian Ocean and Australia, simply **cyclones**. In structure, the three kinds of storms are essentially the same, although typhoons tend to be larger and stronger than the others. We will use the term *hurricane* for the general class of storm, regardless of location.

Most U.S. residents associate hurricanes with storms that form in the Atlantic Ocean or the Gulf of Mexico. Yet other parts of the world have a much greater incidence of hurricanes (Table 12–1 and Figure 12–2). The Atlantic and Gulf of Mexico receive an average of 6.4 hurricanes each year, while the eastern North Pacific off the coast of Mexico has an average of 8.9. Most tropical storms in the east Pacific move westward, away from population centers, and so they receive little public attention. Sometimes, however, they migrate to the northeast and bring severe flooding and loss of life to western Mexico.

The region having the greatest number of these events—by far—is the western part of the North Pacific. In a typical year, 16.5 typhoons hit the region. At the other extreme, no hurricanes form in the Southern Hemisphere Atlantic, even at tropical latitudes (except for a very unusual event off the coast of Brazil in 2004). As you will see later, hurricanes depend on a large pool of warm water, a condition that does not arise in the relatively small South Atlantic basin.

In Chapter 8 we saw that during much of the year air spirals out of massive high-pressure cells that occupy large parts of the Atlantic and Pacific Oceans. Middle- and upper-level air along the eastern side of these anticyclones sinks as it approaches the west coasts of the adjacent continents. Because the air does not descend all the way to the surface, a subsidence inversion (see Chapter 6) forms above the surface. This particular subsidence inversion is called the **trade wind inversion**. The air below the inversion, called the **marine layer**, is cool and relatively moist.

The thickness of the marine layer and the height of the inversion base vary across the tropical oceans. The inversion is lowest along the eastern margins of the oceans, where upwelling and cold ocean currents maintain a relatively cool marine layer. Here the inversion may be only a few hundred meters above the surface. Farther to the west, the warmer surface waters heat the marine layer and cause it to expand to a greater height. Over the eastern part of the oceans, the low inversion inhibits vertical cloud growth, and low stratus clouds often occupy the region. Farther to the west, the greater height of the inversion (or even its total disappearance) allows for more convection, and deep cumulus clouds are more likely to form. For this reason, more hurricanes occur along the western portion of the ocean basins.

Checkpoint

1. What are hurricanes called in the Pacific Ocean? The Indian Ocean?
2. What is the role of the trade wind inversion in the development of hurricanes along the western parts of ocean basins? Explain.

TABLE 12–1

Number of Storms by Basin. Calculated for 1981–2010 in Northern Hemisphere, 1981–1982 to 2010–2011 in Southern Hemisphere

Basin	Tropical Storm (greater than 17 m/s sustained winds)			Hurricane/Typhoon/Severe Tropical Cyclone (greater than 33 m/s sustained winds)		
	Most	Least	Average	Most	Least	Average
Atlantic	28	4	12.1	15	2	6.4
Northeast Pacific	28	8	16.6	16	3	8.9
Northwest Pacific	35	14	26.0	23	7	16.5
Northern Indian Ocean	10	2	4.8	5	0	1.5
Southwest Indian Ocean	14	4	9.3	8	1	5.0
Australia/Southeastern Indian Ocean	16	3	7.5	8	1	3.6
Australia/Southwestern Pacific	20	4	9.9	12	1	5.2
Globally	102	69	86.0	59	34	46.9

Source: Atlantic Oceanographic and Meteorological Laboratory, NOAA.

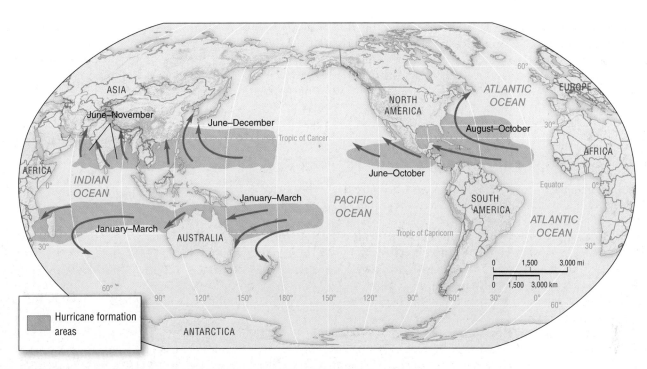

▲ **FIGURE 12-2** Hurricane formation regions around the globe.

Hurricane Characteristics

Hurricanes are the most powerful of all storms. As such, they have fascinated meteorologists, who have exhaustively studied hurricanes' powerful winds, source of energy, structure, process of formation, and movements. The energy unleashed by just a single hurricane can exceed the annual electrical consumption of the United States and Canada. By definition, hurricanes have sustained wind speeds of 120 km/hr (74 mph) or greater. Though their wind speeds are less than those of tornadoes, hurricanes are very much larger and have far longer lifespans. Sea level pressure near the center of a typical hurricane is around 950 mb, but pressures as low as 870 mb have been observed for extremely powerful hurricanes. The weakest hurricanes have central pressures of about 990 mb.

In contrast to tornadoes, whose diameters are typically measured in tens of meters, hurricanes are typically about 600 *kilo*meters (350 mi) wide. Thus, the typical hurricane has a diameter thousands of times greater than that of a tornado. Remembering that the area of a circle is proportional to the square of its radius, and knowing that tornadoes and hurricanes are roughly circular, we see that the area covered by a hurricane is likely to be millions of times greater than the area covered by a tornado. Furthermore, a tornado exists only for a couple of hours at most, while a hurricane can have a lifetime of several days or even a week or more.

Though hurricanes are usually about one-third the size of midlatitude cyclones, the pressure difference across a hurricane is about twice as great. They therefore have extreme horizontal pressure gradients that generate powerful winds: Average hurricanes have peak winds of about 150 km/hr

(90 mph), and the most intense hurricanes have winds up to 350 km/hr (210 mph). In addition to being smaller and more powerful than midlatitude cyclones, hurricanes differ by not having the fronts that characterize cyclones outside the tropics.

Because hurricanes obtain most of their energy from the latent heat released by condensation, they are most common where a deep layer of warm water fuels them. Given that tropical oceans have their highest surface temperatures and evaporation rates in late summer and early fall, it is not surprising that August and September are the prime hurricane months in the Northern Hemisphere, with January to March the main season in the Southern Hemisphere.[1]

Checkpoint

1. What is a hurricane's source of energy?
2. What two factors explain the spatial and temporal distribution of hurricanes?

Hurricane Structure

Hurricanes do not consist of only one uniform convective cell. Instead they contain a large number of thunderstorms arranged in a pinwheel formation, with bands of thick clouds and heavy thundershowers spiraling counterclockwise (in

---

[1]The United States National Hurricane Center defines the hurricane season as the period from June 1 to November 30. Tropical storms in other months are rare events—from 1871 to 1996, only six storms formed in December.

▶ **FIGURE 12-3** A cross-section of a typical hurricane. Looking into the hurricane there are parallel bands of cloud cover that spiral in toward the hurricane center. Cloud bands tend to be deeper toward the center of the hurricane until they meet the *eye wall*, which separates the intense part of the storm with strong winds and heavy showers from the relatively calm *eye*.

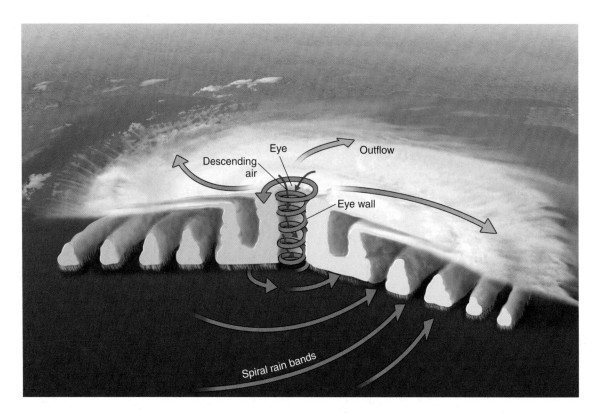

the Northern Hemisphere) around the storm center. Observe in Figure 12–3 the cloud bands separated by distinct gaps that give them a pinwheel like appearance. The bands of heavy convection are separated by areas of weaker uplift and even descending air and less intense precipitation. The wind speed and the intensity of precipitation both increase toward the center of the system (its *eye*), reaching a maximum 10 to 20 km (6 to 12 mi) away from the center, at what is called the *eye wall* (the eye and eye wall are described in the next section).

Figure 12–4 depicts generalized cross sections through a hurricane. The thickness of the cloud bands (a) corresponds well to the intensity of rainfall (b). On the other hand, the distributions of pressure (c) and wind speed (d) do not exhibit a similar banding. Both the pressure gradient and wind speed increase gradually toward the center of the storm and then increase rapidly in the vicinity of the eye wall.

Hurricanes also differ from midlatitude cyclones in that they are warm-core cyclones. As air flows inward toward lower pressure, the warm ocean surface supplies large amounts of latent and sensible heat to the overlying air. Because pressure within the moving air decreases as it flows toward the low, adiabatic expansion keeps the temperature from increasing dramatically, with the result that there is little temperature difference across the base of the storm. Nevertheless, much thermal energy is added, resulting in a "warm" central core. Aloft, after condensation and the release of latent heat, the warmth is reflected in temperature, so that temperatures near the center are much higher than those of the surrounding air (Figure 12–5).

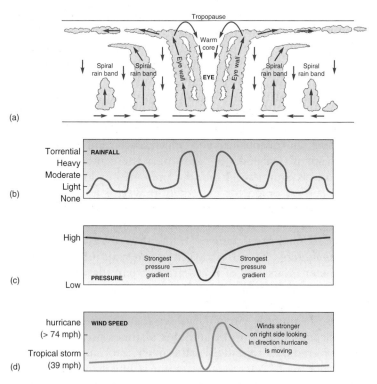

▲ **FIGURE 12-4** Typical cross section of a hurricane. Clouds bands become deeper toward the center of the hurricane (a). Rainfall becomes more intense under each cloud band (b). Air pressure decreases gradually in the outer portions of the hurricane but intensifies rapidly toward the center of the hurricane (c), as does wind speed (d).

◄ **FIGURE 12-5** Temperature differences across a hurricane relative to the surrounding air (°C). Near the surface, temperatures increase only slightly toward the eye. Aloft, however, temperatures exceed those of the surroundings by about 10 °C (18 °F).

As a warm-core low, pressure within a hurricane decreases relatively slowly with increasing altitude (see Chapter 4). Thus, the horizontal pressure gradient within the storm gradually decreases with altitude. At about 7.5 km (25,000 ft)—about the 400 mb level—the air pressure is the same as that immediately outside the storm. From this height to the lower stratosphere, the hurricane has relatively high pressure. So unlike the lower part of the hurricane in which the air rotates cyclonically, the air in its upper portion spirals anticyclonically from its center (clockwise in the Northern Hemisphere). Figure 12–6 shows a schematic of how surface winds typically flow inward, then rapidly rotate around the eye wall as they ascend to higher levels, and then spiral outward.

In the upper reaches of the storm, the low temperatures cause water droplets to freeze into ice crystals. As the ice crystals spiral out of the storm center, they create a blanket of cirrostratus clouds that overlies and obscures the pinwheel-like structure of the storm. That explains why hurricanes on satellite images often appear to have a uniform thickness and intensity, when in fact they are strongly banded.

The Eye and the Eye Wall

One of the most distinctive characteristics of a hurricane is its **eye**, a region of relatively clear skies, slowly descending air, and light winds. Hurricane eyes average about 30 km (20 mi) in diameter, with most ranging from 20 to 50 km (15 to 30 mi). Eye diameters vary considerably among individual storms, with some as small as 6 km (3.5 mi) and others almost as large as 100 km (60 mi). The change in the size of an eye through time gives some indication of whether the hurricane

is intensifying or weakening. Generally, a shrinking eye indicates an intensifying hurricane.

Along the margin of the eye lies the **eye wall**, the zone of most intense storm activity. The eye wall contains the strongest winds, thickest cloud cover, and most intense

▲ **FIGURE 12-6** Air trajectories initially rise gradually as they approach a hurricane eye wall. Most rapid ascent occurs in the eye wall until the air reaches the upper troposphere and flows outward anticyclonically.

precipitation of the entire hurricane. Directly beneath the eye wall, rainfall rates of 2500 mm/day (100 in/day) are not uncommon. The abrupt transition from an eye wall to the eye causes a strong and rapid change in weather. Imagine a hurricane about to make a direct hit on a small island. As the hurricane center approaches the island, the intensity of the wind and rain steadily increase, becoming most intense as the eye wall arrives. But when the eye reaches the island, the storm seems to suddenly dissipate, as blue skies and calm conditions take hold. Of course, the storm has not dissipated at all. Rather, there is just a brief lull until the opposite side of the eye wall covers the island and storm conditions resume. Because the average hurricane eye is about 30 km (18 mi) in diameter and travels at about 20 km/hr, (12 mph) the calm associated with passage of the eye lasts about an hour or two. Clearly, if the eye just grazes the island, the break in the storm will be even shorter.

Did You Know?

Although hurricane wind speeds are, by definition, extremely fast, it still takes a parcel of air an average of 8 days to flow into and out of a hurricane. This is due to the great distance traversed by the air as it spirals upward, and the relatively slow speed the air has as it exits as outflow.

▲ **FIGURE 12-7** Some hurricanes develop double eye walls, such as Hurricane Emily in 2005. Usually occurring as the hurricane achieves maximum strength, both eye walls contract, with the innermost eye wall eventually dissipating and giving way to the outer eye wall.

Some intense hurricanes develop **double eye walls**, as surrounding rain bands contract and intensify. As the interior eye wall contracts, it can begin to dissipate and the surrounding band can take its place. Figure 12–7 shows this process occurring in Hurricane Emily in 2005, with the outer band of rainfall (shown as green arcs embedded with red areas of heavier activity) surrounding the small eye wall in the center.

Sometimes, especially with very strong hurricanes, an inner eye wall will contract in size so that the eye's diameter becomes relatively small. If that happens, there can be a major reduction in the inflow of moisture to the inner eye wall and it can begin to lose its intensity and eventually die out. At the same time, what had been an outer wall can intensify and contract toward the center of the hurricane and take over as the single hurricane eye wall. This process, called **eye wall replacement**, often leads to an initial weakening of the hurricane winds (as the inner eye wall begins to die out), followed by renewed strengthening as the outer wall contracts and intensifies. Some hurricanes undergo this process more than once, with each cycle occurring on the order of about a half a day to two or three days. This is illustrated by Figure 12–8, showing satellite images of Hurricane Wilma over a 48-hour period, October 19–21, 2005.

NASA scientists have recently uncovered the existence of **hot towers** (Figure 12–9) embedded in some eye walls that last between 30 minutes and 2 hours. Hot towers are localized portions of eye walls that rise to greater heights (up to 12 km,

or 7 mi) than the rest of the eye wall. The researchers found that development of hot towers indicates a greater likelihood that the hurricane will intensify within the next 6 hours.

The air temperature at the storm's surface within an eye is several degrees warmer than outside the eye because compression of the sinking air causes it to warm adiabatically. The air is also drier, because warming the unsaturated air lowers its relative humidity. Contrary to common belief, however, the air is not entirely cloud free within the eye; instead, fair-weather cumulus clouds are scattered throughout the otherwise blue sky.

Checkpoint

1. Where in a hurricane would you find ascending air? Descending air?
2. What are some of the major changes that can occur in the hurricane associated with weakening or intensification?

Steps in the Formation of Hurricanes

Hurricanes do not suddenly appear out of nowhere; they begin as much weaker systems that often migrate large distances before turning into hurricanes. In this section we examine hurricane development, with particular emphasis on how it occurs in the Atlantic.

▲ **FIGURE 12-8** Hurricane Wilma underwent eye wall replacement between October 19 and 21, 2005. Initially a very tight eye wall is evident in (a). As the inner wall weakened, the eye became less distinct and the maximum wind speeds weakened (b). A day later the outer wall has completely replaced the original eye wall (c).

▶ **FIGURE 12-9** This hot tower in Hurricane Rita in 2005 was observed by NASA's Tropical Rainfall Measuring Mission (TRMM) spacecraft. These localized areas of deep cloud cover and intense precipitation often precede hurricane intensification.

Tropical Disturbances

Although most tropical storms attain hurricane status in the western portions of the oceans, their earliest origins often lie far to the east as small clusters of small thunderstorms called **tropical disturbances**. Tropical disturbances are disorganized groups of thunderstorms having weak pressure gradients and little or no rotation.

Tropical disturbances can form in several different environments. Some form when midlatitude troughs migrate into the tropics; others develop as part of the normal convection associated with the ITCZ. But most tropical disturbances that enter the western Atlantic and become hurricanes originate in **easterly waves**, large undulations or ripples in the normal trade wind pattern. Figure 12–10 illustrates a sequence of easterly waves. Because pressure gradients in the tropics are normally weaker than those of the extratropical regions, the easterly waves are better shown by plotting lines of wind direction (called *streamlines*) rather than isobars. The air in the wave initially flows westward, turns poleward, and then

flows back toward the equator, with the entire wave pattern extending 2000 to 3000 km (1200 to 1800 mi) in length. On the upwind (eastern) side of the axis, the streamlines become progressively closer together (convergence). With convergence there is low pressure and rising motion (see Chapter 6); thus the tropical disturbance is located upwind of the wave axis (the dashed line) of the easterly wave. Surface divergence downwind of the wave axis leads to clear skies. (An explanation for why the streamlines are convergent and divergent is somewhat complicated; the main factor involves changes in relative vorticity that occur as the air moves poleward and equatorward.)

The tropical disturbances that affect the Atlantic Ocean, Caribbean, and the Americas mostly form over western Africa, south of the Saharan desert. Being in the zone of the trade winds, these storms tend to migrate westward. When they reach the west coast of Africa, they weaken as they pass over the cold Canary current over the eastern Atlantic. There the low water temperatures chill the air near the water surface and cause the air to become statically stable. If the

▶ **FIGURE 12-10** Easterly waves have surface convergence and cloud cover east of the axis and divergence to the west. They represent troughs of low pressure, embedded in the trade winds, that migrate westward. They sometimes develop into tropical depressions, storms, or hurricanes.

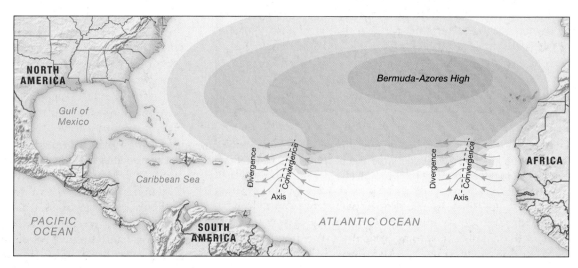

disturbances migrate beyond the coastal zone of surface upwelling, however, warmer waters farther offshore raise the temperature and humidity of the lower atmosphere and cause the air to become unstable. Then, as the storms continue westward, a small percentage develop into more intense and organized thunderstorm systems. Easterly waves move westward at about 15 to 35 km/hr (10 to 20 mph), and so it typically takes about a week to 10 days for an embedded tropical disturbance to migrate across the Atlantic.

Tropical Depressions and Tropical Storms

The vast majority (probably more than 90 percent) of tropical disturbances die out without ever organizing into more powerful systems. But some undergo a lowering of pressure and begin to rotate cyclonically. When a tropical disturbance develops to the point where there is at least one closed isobar on a weather map, the disturbance is classified as a **tropical depression**. If the depression intensifies further and maintains wind speeds above 60 km/hr (39 mph), it becomes a **tropical storm**. (At this point the system is named. See *Box 12–1, Special Interest: Naming Hurricanes,* for more information on this practice.) A further increase in sustained wind speeds to 120 km/hr (74 mph) creates a true hurricane. While only a small fraction of tropical disturbances ever become tropical depressions, a larger proportion of depressions become tropical storms, and an even greater percentage of tropical storms ultimately become hurricanes.

Hurricanes

The location at which hurricanes are most likely to form varies seasonally. Early in the Atlantic season, weak fronts in the western ocean extend southward over warm tropical water. Wind shear across the fronts provides the circulation necessary for cyclone development. Later in the season, fronts are confined to higher latitudes and no longer play a role in cyclogenesis. Instead, warm waters are found progressively farther to the east, so that disturbances leaving the African continent can grow into full-scale cyclones. (Systems that become tropical storms in the tropical waters just off western Africa and become hurricanes before reaching the Caribbean are often referred to as *Cape Verde hurricanes,* so named for the islands near which they originate.) The net effect is that the birthplace of tropical cyclones moves from west to east across the tropical ocean during the first half of the season. In the late fall, the breeding ground moves westward as frontal activity again emerges as a primary agent of cyclone genesis.

As with their Atlantic counterparts, Pacific hurricanes move westward during their formative stages. Many come near Hawaii, but most bypass the islands or die out before reaching them. Unfortunately, this is not always the case. In September 1992, Hurricane Iniki battered the island of Kauai with wind gusts up to 258 km/hr (160 mph) and brought heavy flooding to the beach resort areas. The hurricane destroyed or severely damaged half of the homes on the island and devastated most of the tourist industry.

Conditions Necessary for Hurricane Formation

Although the dynamics of hurricanes are extremely complex, meteorologists have long recognized the conditions that favor their development. Great amounts of heat are needed to fuel hurricanes, and the primary source of this energy is the release of latent heat supplied by evaporation from the ocean surface. Because high evaporation rates depend on the presence of warm water, hurricanes form only where the ocean has a deep surface layer (several tens of meters in depth) with temperatures above 27 °C (81 °F). The need for warm water precludes hurricane formation poleward of about 20 degrees; sea surface temperatures are usually too low there. Hurricanes develop most often in late summer and early fall, when tropical waters are warmest.

Also related to the presence of deep layers of warm water is the need for high moisture contents in the air. As water evaporates from the sea surface putting latent heat into the atmosphere, it also feeds enough water vapor into the air to help it maintain a deep layer of air with high humidity. This provides the necessary water vapor for the growth of deep cumulus clouds.

Hurricane formation also depends on the Coriolis force, which must be strong enough to prevent filling of the central low pressure. The absence of a Coriolis effect at the equator prohibits hurricane formation between 0° and 5° latitude. This factor and the need for high water temperatures explain the pattern shown in Figure 12–2, in which tropical storms attain hurricane status between the latitudes of 5° and 20°.

Stability is also important in hurricane development, with unstable conditions throughout the troposphere an absolute necessity. Along the eastern margins of the oceans, cold currents and upwelling cause the lower troposphere to be statically stable, inhibiting uplift. Moreover, the trade wind inversion puts a cap on any mixing that might otherwise occur. Moving westward, water temperatures typically increase and the trade wind inversion increases with height or disappears altogether, and so hurricanes become more prevalent. Finally, hurricane formation requires an absence of strong vertical wind shear, which disrupts the vertical transport of latent heat.

Once formed, hurricanes are self-propagating (just as severe storms outside the tropics are self-maintaining). That is, the release of latent heat within the cumulus clouds causes the air to warm and expand upward. The expansion of the air supports upper-level divergence, which draws air upward and promotes low pressure and convergence at the surface. This leads to continued uplift, condensation, and the release of latent heat.

So if hurricanes are self-propagating, can they intensify indefinitely, until they attain supersonic speeds? No, because they are ultimately limited by the supply of latent heat, which in turn is constrained by the temperature of the ocean below and by the processes underlying evaporation and convection. The importance of ocean temperature suggests that if the oceans were to become warmer, hurricanes would theoretically become more intense. This topic has received considerable attention lately because of climatic warming, which could be accompanied by higher ocean temperatures.

12-1 SPECIAL INTEREST

Naming Hurricanes

During hurricane season, several tropical storms or hurricanes can arise simultaneously over various oceans. Meteorologists identify these systems by assigning names when they reach tropical storm status. The World Meteorological Organization (WMO) has created several lists of names for tropical storms over each ocean. The names on each list are ordered alphabetically, starting with the letter *A* and continuing up to the letter *W*. When a depression attains tropical storm status, it is assigned the next unused name on that year's list. At the beginning of the following season, names are taken from the next list, regardless of how many names were unused in the previous season. Six lists have been compiled for the Atlantic Ocean, and the names on each list are used again at the end of each 6-year cycle (Table 1). English, Spanish, and French names are used for Atlantic hurricanes. If all the available names on a season's list are used, subsequent storms will be given the names of the letters of the Greek alphabet. Thus, the 22nd named storm for a season would be Alpha. This is exactly what happened in October 2005 when Tropical Storm Alpha appeared in the western Atlantic—the first time ever that the list of names was

TABLE 1

Western Atlantic Tropical Storm and Hurricane Names

2012	2013	2014	2015	2016	2017
Alberto	Andrea	Arthur	Ana	Alex	Arlene
Beryl	Barry	Bertha	Bill	Bonnie	Bret
Chris	Chantal	Cristobal	Claudette	Colin	Cindy
Debby	Dorian	Dolly	Danny	Danielle	Don
Ernesto	Erin	Edouard	Erika	Earl	Emily
Florence	Fernand	Fay	Fred	Fiona	Franklin
Gordon	Gabrielle	Gonzalo	Grace	Gaston	Gert
Helene	Humberto	Hanna	Henri	Hermine	Harvey
Isaac	Ingrid	Isaias	Ida	Ian	Irene
Joyce	Jerry	Josephine	Joaquin	Julia	Jose
Kirk	Karen	Kyle	Kate	Karl	Katia
Leslie	Lorenzo	Laura	Larry	Lisa	Lee
Michael	Melissa	Marco	Mindy	Matthew	Maria
Nadine	Nestor	Nana	Nicholas	Nicole	Nate
Oscar	Olga	Omar	Odette	Otto	Ophelia
Patty	Pablo	Paulette	Peter	Paula	Philippe
Rafael	Rebekah	Rene	Rose	Richard	Rina
Sandy	Sebastien	Sally	Sam	Shary	Sean
Tony	Tanya	Teddy	Teresa	Tobias	Tammy
Valerie	Van	Vicky	Victor	Virginie	Vince
William	Wendy	Wilfred	Wanda	Walter	Whitney

Checkpoint

1. What initial conditions are necessary to begin the process of hurricane formation?
2. What makes a hurricane self-propagating?
3. What factors could cut off or limit the intensification of a developing hurricane? Explain.

Hurricane Movement and Dissipation

The movement of tropical systems is related to the stage in their development. Tropical disturbances and depressions are guided mainly by the trade winds and, therefore, tend to migrate westward. The influence of the trade winds often diminishes after the depressions intensify into tropical storms. Then the upper-level winds and the spatial distribution of water temperature more strongly determine their speed and direction (with the storms tending to move toward warmer seas).

Hurricane Paths

Once fully developed, tropical storms become more likely to move poleward, as shown in Figures 12–11 and 12–12. Figure 12–12 shows the locations of all hurricane landfalls from 1950 through 2010. (*Box 12–2, Focus on Severe Weather: 2004 and 2005: Two Historic Hurricane Seasons*), describes the paths and damage done by several notable hurricanes of these two remarkable hurricane years. Hurricanes and tropical storms often move in wildly erratic ways—for example, moving in a constant direction for a time, then suddenly changing speed and direction, and even backtracking along its previous path. Figure 12–13 plots some particularly erratic paths along the east coast of North America.

unable to accommodate all of the tropical storms in a single season.

Particularly notable hurricanes can have their names "retired" by the WMO if an affected nation requests the removal of that name from the list. All replacements are made with names of the same gender and language. As of 2011, 75 names had been retired from the Atlantic hurricane list (Table 2). If a hurricane with a Greek letter merits special designation, it would do so with its year appended to the letter. Thus if in 2013 Alpha is very destructive, it can be noted as Alpha 2013.

The practice of naming tropical storms and hurricanes appears to have begun during World War II when meteorologists in the Pacific assigned female names (possibly after wives and girlfriends) to tropical storms and typhoons. This practice was adopted by the U.S. National Weather Service (then called the Weather Bureau) in 1953 and maintained until 1979, when male names were added to the lists.

TABLE 2

Retired Atlantic Names by Year

				1954	1955	1956	1957	1958	1959
				Carol Hazel	Connie Diane Ione Janet		Audrey		
1960 Donna	**1961** Carla Hattie	**1962**	**1963** Flora	**1964** Cleo Dora Hilda	**1965** Betsy	**1966** Inez	**1967** Beulah	**1968** Edna	**1969** Camille
1970 Celia	**1971**	**1972** Agnes	**1973**	**1974** Carmen Fifi	**1975** Eloise	**1976**	**1977** Anita	**1978**	**1979** David Frederic
1980 Allen	**1981**	**1982**	**1983** Alicia	**1984**	**1985** Elena Gloria	**1986**	**1987**	**1988** Gilbert Joan	**1989** Hugo
1990 Diana Klaus	**1991** Bob	**1992** Andrew	**1993**	**1994**	**1995** Luis Marilyn Opal Roxanne	**1996** Cesar Fran Hortense	**1997**	**1998** Georges Mitch	**1999** Floyd Lenny
2000 Keith	**2001** Allison Iris Michelle	**2002** Isidore Lili	**2003** Fabian Isabel Juan	**2004** Charley Frances Ivan Jeanne	**2005** Dennis Katrina Rita Stan Wilma	**2006**	**2007** Dean Felix Noel	**2008** Gustav Ike Paloma	**2009**
2010 Igor Tomas									

While hurricanes can hit any part of the Gulf of Mexico or Atlantic coasts at any time during hurricane season, they have a greater likelihood of taking particular paths during different months. Figure 12–14 shows the likelihood of hurricane passage for August (a), September (b), and October (c). In August, the most likely path of hurricanes tracks over the West Indies. From there, hurricanes are about equally likely to track toward the Texas coast or along the Atlantic coast from Florida to North Carolina. Two prominent paths dominate in September. One goes from between the Yucatan Peninsula and western Cuba, northward toward the central Gulf of Mexico coast; the other moves northward from around Haiti, the Dominican Republic, and Puerto Rico into the western Atlantic. Hurricanes taking the more easterly track are most likely to hit the middle Atlantic states if they make landfall. October paths exhibit a greater tendency to track from eastern Mexico northward to Florida and the rest of the southeastern United States.

Atlantic tropical storms and hurricanes can travel great distances along the North American east coast, but they usually weaken considerably as they approach the northeastern United States and the Maritime Provinces of Canada. These storms usually lack the strong winds that characterize hurricanes in the low latitudes but can still bring intense rainfall and flooding. On rare occasions, the storms can maintain their strong winds even as they move considerable distances from the subtropics. For example, an intense September 1938 hurricane brought 200 km/hr (120 mph) winds to Long Island, New York, as it moved toward New England. Its estimated 600 fatalities made it the fourth deadliest of all U.S. hurricanes. More recently, in September 1985, Hurricane Gloria brought considerable wind and flood damage to Long Island and Connecticut.

Although hurricanes and tropical storms can move into the northeastern United States, along the West Coast they do not migrate nearly as far north without weakening to tropical depressions. The reason for this is the difference in

▲ **FIGURE 12–11** The tracks of all western Atlantic and eastern Pacific major hurricanes (Category 3 or higher), from 1851–2007 for the Atlantic and 1949–2007 in the Pacific.

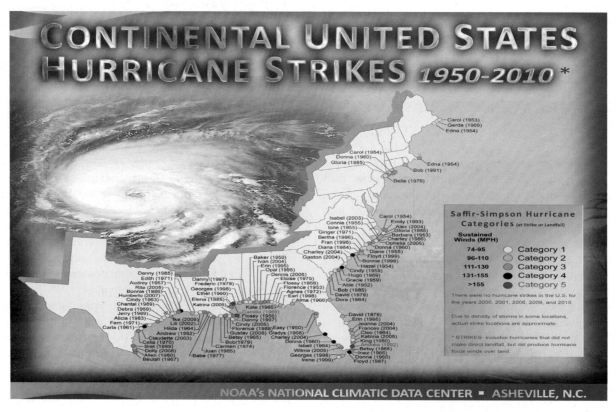

▲ **FIGURE 12–12** The locations of all hurricane landfalls over the continental United States from 1950 through 2007. Saffir-Simpson categories are those at the time of landfall.

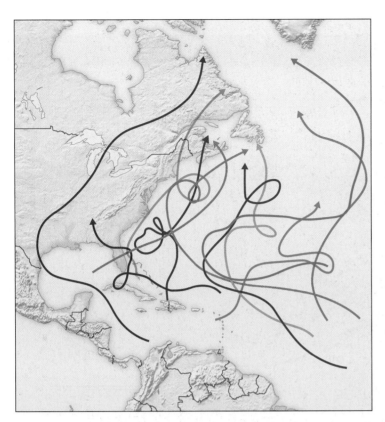

▲ **FIGURE 12–13** Tropical storms and hurricanes have a tendency to move north or northeast out of the tropics along the southeast coast of North America. Their paths are often erratic, as seen in these examples.

water temperatures along the two coasts. The Pacific Coast is dominated by upwelling and the cold California current, while the warm Gulf Stream flowing along the East Coast provides a greater supply of latent heat. Sometimes, storms off the coast of Mexico move to the northeast across Baja California and into southern California. These storms lose their supply of latent heat and lose their intensity as they move inland, but they can still bring heavy rains and flooding. In 1976 Hurricane Kathleen caused massive flooding in the desert of southern California that wiped out part of Interstate Highway 8.

Did You Know?

A quick glimpse of Figure 12–2 reveals that hurricanes never occur in the South Atlantic. Well, almost never. On March 27, 2004, for the first time in recorded history a hurricane hit the coast of Brazil. Tropical cyclones have hit the Brazilian coast twice before, but "Catarina" (so named for the Brazilian state of Santa Catarina, which was hit hardest) packed 147 km/hour (90 mph) winds—making it a true hurricane. This storm did not develop in exactly the same manner as do most of those in the North Atlantic. It appears to have been a hybrid between a tropical system and a midlatitude cyclone, having initially formed along a cold front in the South Atlantic. Considerable damage occurred along the coastal region, but effective warnings to the public minimized the loss of life.

Effect of Landfall

After making landfall, a tropical storm may die out completely within a few days. Even as the storm weakens, though, it can still import huge amounts of water vapor and bring very heavy rainfall hundreds of kilometers inland. This is especially true when the remnant of a hurricane moving poleward joins with a midlatitude cyclone moving eastward. Exactly this happened in 1969 when one of North America's most notorious hurricanes, Camille, moved northward from the coast of Mississippi (Figure 12–15). After its high winds and tidal flooding brought extreme damage to the Gulf Coast, the storm moved northeastward toward the western slopes of the Appalachians. There, orographic uplift coupled with low pressure and the high water vapor content of the remnant hurricane could easily have produced serious flooding. But to make matters worse, an eastward-moving cold front reached the mountains at the same time as the former hurricane. The combination of moist air, low pressure, an approaching front, and the orographic effect set the stage for intense rains and flash flooding that killed more than 150 persons.

Hurricane Destruction and Fatalities

Hurricanes can bring death and devastation in several ways, through any combination of strong winds, heavy rain, hurricane-spawned tornadoes, and the elevation of coastal waters combined with heavy surf.

Wind

By definition, hurricane winds exceed 120 km/hr (74 mph), and many are much faster. It is not surprising, then, that hurricane-force winds can cause extensive damage and even destroy well-built homes. An example of this happened on August 24, 1992, when Hurricane Andrew devastated much of southern Florida. It approached southern Florida almost directly from the east and cut westward across the state, with its eye passing about 40 km (25 mi) south of Miami Beach. There was relatively little damage due to coastal flooding or heavy rainfall, but winds completely leveled the town of Homestead, killed 24 people, and left 180,000 homeless in the state before crossing into the Gulf of Mexico.

Heavy Rain

Hurricanes produce staggeringly intense rainfall, with rates on the order of meters per day found beneath the eye wall. The rate for a location that remains stationary beneath a passing storm is smaller, but still huge—on the order of 25 cm/day (10 in./day).

Heavy rain made Hurricane Mitch the deadliest hurricane to hit the Western Hemisphere in the last 200 years, killing thousands of people in Central America in October 1998 (some effects of Mitch were described at the beginning of Chapter 3). Mitch hit Central America from the east. Although

2004 and 2005: Two Historic Hurricane Seasons

The period between 1995 and 2005 was marked by an unusually high number of Atlantic tropical storms and hurricanes, with many making landfall. Of the ten seasons during this time span, all but two had at least eight hurricanes—well above the annual average of 5.9. But 2004 and 2005 were particularly noteworthy. The year 2004 was the costliest hurricane season to hit the United States up to that time, bringing $42 billion in losses. Hurricanes damaged one-fifth of all the homes in Florida and killed 117 people that year. Incredibly, that devastation was dwarfed by the hurricane season that hit the following year. The year 2005 produced one of the greatest natural disasters in U.S. history—Hurricane Katrina—and several other major hurricanes that made landfall; in addition, it proved to be the most active tropical storm–hurricane season in American history, with 27 named storms (breaking a record set in 1933). Here we present a brief review of three hurricanes from 2004 and Hurricanes Rita and Wilma in 2005 (Figures 1 and 2). Given its enormous death toll and destruction, we discuss Hurricane Katrina separately later in this chapter.

Hurricane Charley, August 9–14, 2004

Hurricane Charley was the strongest hurricane to hit the United States since Hurricane Andrew 12 years earlier. At the time it was also the second costliest hurricane to strike the United States, bringing an estimated $14 billion in damages along with 10 fatalities that were directly attributable to the storm.

Hurricane Ivan, September 3–24, 2004

Hurricane Ivan became a truly remarkable system for both its intensity—achieving Category 5 status on three separate occasions—as well as for its highly unusual path. Ivan was the southernmost hurricane on record in the Atlantic. Ivan's Category 3 winds first hit the southern portion of Grenada on September 7, destroying

▲ **FIGURE 1** The paths of four 2004 hurricanes.

14,000 homes and killing 39 people. On September 10, Ivan passed south of Jamaica as a Category 4 hurricane, destroyed 5600 homes, damaged 47,000 others, and killed 17 people. Grand Cayman Island was next, with near-Category 5 winds damaging or destroying 95 percent of the island's homes.

Ivan then headed for the United States and made landfall on September 16 near Gulf Shores, Alabama, as a Category 3 hurricane.

Most storms would have done all their damage by this point, but this was no ordinary storm. On September 22, Ivan moved to the northwest and made landfall once again as a tropical depression over southeastern Louisiana, where it rapidly died out.

Ivan was responsible for 92 deaths—25 of them in the United States. U.S. damages were estimated at well over $14 billion—about the same as Hurricane Charley had brought a month and a half earlier.

Hurricane Jeanne, September 14–28, 2004

By far the deadliest of the 2004 Atlantic hurricanes was Jeanne, which killed more

than 3000 people. Most of its fatalities were associated with flooding in the Caribbean, especially in Haiti, as high rainfall rates coupled with slow storm movement led to very large precipitation totals. Although Puerto Rico had fewer fatalities than did Haiti, it suffered record-breaking floods.

Hurricane Rita, September 18–25, 2005

Following the devastation of Hurricane Katrina (see *Box 12–3, Special Interest: Hurricane Katrina*) only a few weeks earlier, Hurricane Rita became the second Category 5 hurricane to develop in the Gulf of Mexico that year—the first time in recorded history that two such powerful storms had ever occurred in the Gulf in the same season (Hurricane Wilma also achieved Category 5 status in the Caribbean Sea that year, but it lost Category 5 intensity before moving into the Gulf). At its worst, Rita was a monster storm. Its maximum wind speeds topped out at 280 km/hr (175 mph), hurricane-force winds extended 110 km (70 mi) away from the center, and tropical storm-force winds reached out to 295 km (185 mi). Its minimum sea level air pressure of 897 mb was the third lowest ever observed in the Atlantic Ocean.

Hurricane Rita's first brush with land occurred as it passed south of the Florida Keys on September 20 as a Category 2 hurricane. Though Rita's path took it too far to the south to deliver its worst punch, it still did considerable damage to the Keys, downing trees and creating a 1.5 m (5 ft) storm surge that topped U.S. Highway 1 and caused some flooding to business and homes. As it moved westward, the hurricane rapidly intensified and reached Category 5 status on the afternoon of September 21. With the public very much aware of the devastation wrought by Hurricane Katrina, citizens along much of the Gulf Coast took the call for evacuation very seriously, causing massive traffic problems in east Texas as an estimated 3 million people headed inland.

Rita struck the Texas coast just west of Louisiana as a Category 3 storm on September 23. It delivered hurricane-force winds to areas as far inland as 240 km (150 mi) and tropical storm-force winds as far north as southern Arkansas. Storm surges as high as 4.6 m (15 ft), though smaller than originally feared, caused very serious flooding and the total destruction or major damage of several communities. In addition, a 2.4 m (8 ft) storm surge in New Orleans reopened several breaches in the levees that had been temporarily repaired following Hurricane Katrina. Overall damage from Hurricane Rita was very large but far short of that brought about by Hurricane Katrina. Initial reports indicated a total loss of life of about 119 people. Only 6 of the deaths appear to have resulted directly from the hurricane; the rest were indirectly related, such as those caused by a fire that engulfed a bus full of elderly patients evacuating Houston.

Hurricane Wilma, October 17–25, 2005

At 11 A.M. EDT Tuesday, October 18, Tropical storm Wilma became Hurricane Wilma—the 12th of the season—as a Category 1 hurricane. By 5 A.M. EDT Wednesday, the hurricane packed sustained winds of 280 km/hr (175 mph)—

▲ **FIGURE 2** The paths of Hurricanes Katrina, Rita, and Wilma in 2005.

completing an explosive transition from tropical storm to Category 5 hurricane in less than 24 hours! Hurricane reconnaissance aircraft observed a sea level air pressure of 884 mb—the lowest ever observed in the Atlantic.

Wilma then headed for the Yucatan Peninsula. It made a direct hit on Cozumel Island, southeast of the mainland, on Friday morning, The storm remained over the area through much of Friday and Saturday before turning into the Gulf of Mexico and heading toward the Florida coast. An estimated 22,000 tourists and many more residents of the resort area had to find shelter from the 200 km/hr (125 mph) winds and heavy downpours. Meanwhile, 700,000 people had been evacuated from the west side of Cuba in anticipation of the heavy flooding that hit that portion of the island. At least 19 people were killed in the Caribbean area by the hurricane.

Wilma moved rapidly across the Gulf of Mexico as it headed toward the southwestern coast of Florida. It hit the western Gulf Coast on Monday morning, October 24, 2005, as a Category 3 hurricane. Storm surges up to 2.75 m (9 ft) caused

extensive flooding, especially in Key West. Widespread wind damage also occurred across much of the Gulf Coast. The hurricane moved very rapidly inland toward the Northeast. Away from the Gulf Coast, high winds brought the most severe damage, especially on the right-hand side of the hurricane. In fact, the rapid movement of the storm contributed to the surprisingly high wind speeds encountered over much of southern Florida. The Miami and Ft. Lauderdale areas experienced considerable damage, including the shattering of windows in large buildings and the downing of large trees. Roads were impassable soon after the storm, millions of people lost electric power for extended time periods, and residents were advised to boil their tap water before drinking it. Earliest reports identified ten fatalities.

Hurricane Wilma then proceeded northward in the western Atlantic. Across the United States, total insured damages caused by Wilma were estimated to be between $6 billion and $10 billion.

▶ **FIGURE 12–14** The predominant paths of Atlantic hurricanes in different months. In August (a) hurricanes may enter the Gulf of Mexico or sweep up the East Coast, but are more likely to remain in the Caribbean. By September the likelihood of hurricanes making landfall along the Gulf of Mexico or East Coast increases considerably (b). Late season hurricanes occurring in October are more confined to the far western Atlantic and Caribbean, but it is important to note that the overall frequency of hurricanes is less than it is in earlier months (c).

(a)

(b)

(c)

▲ **FIGURE 12–15** The movement of Hurricane Camille. Camille hit the Gulf coast as a major hurricane. Winds diminished considerably after advancing inland but heavy rains continued, especially over the Appalachians (c). Very serious flooding occurred as the passage of a cold front intensified rainfall (c and d). Eventually the remnants of the storm passed eastward into the Atlantic (e).

Mitch's winds weakened substantially, the remnant system brought heavy rainfall across the region as it tracked northward toward the Gulf of Mexico. Intense rains lasted for several days, and parts of Honduras and Nicaragua received estimated precipitation totals of 85 cm (35 in.), causing extensive flooding and mudslides in this mountainous region.

In 2002 Tropical Storm Allison demonstrated that a tropical storm need not attain hurricane status to become a major disaster. Allison hit the south Texas coast on June 5 and hovered over the area for nearly a week. Its heavy rainfall (as much as 96 cm, or 40 in.) caused major flooding across Texas and Louisiana, killing 24 people and flooding more than 46,000 homes and businesses.

Tornadoes

Many hurricanes also contain clusters of tornadoes, most often in the right-forward quadrant (Figure 12–16). They usually occur far enough away from the center that they are surrounded by relatively weak winds. It appears that slowing of the wind by friction at landfall contributes to tornado formation. Hurricane-spawned tornadoes tend to have shorter lifespans than tornadoes in the central United States.

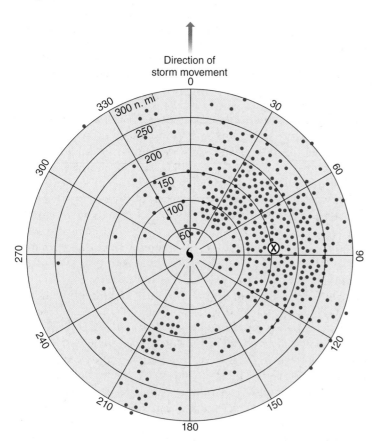

▲ **FIGURE 12–16** Tornadoes most often form in the right-forward quadrant of hurricanes (based on the direction in which the storm is moving). The figure is based on data from 373 hurricane-embedded tornadoes between 1948 and 1972 in the Northern Hemisphere. Each dot represents a tornado. The circled X indicates the mean position of tornadoes relative to the storm center.

Storm Surges

In addition to the threat of heavy rain, strong winds, and tornadoes, coastal regions are vulnerable to a special problem called the **storm surge**, a rise in water level induced by the hurricane (Figure 12–17). Two processes create a storm surge, the major one being the piling up of water as heavy winds drag surface waters forward. Strong winds blowing toward a coast force surface waters landward and thereby elevate sea level, while also bringing heavy surf. The low atmospheric pressure in a hurricane also contributes to the storm surge, in the same way that the height of a column of mercury in a barometer responds to variations in atmospheric pressure. For every millibar the pressure decreases, the water level rises 1 cm (0.4 in.). For most hurricanes along a coastal zone, the storm surge elevates the water level only a meter or two. But in extreme circumstances, storm surges can increase the water level by as much as 7 m (23 ft), as was the case for Hurricane Camille along the coast of Mississippi in 1969.

Storm surges along low-lying coastal plains can be extremely devastating where the rise in sea level brings waters far inland. Furthermore, the heavy waves generated by the strong winds pound away at structures, with debris

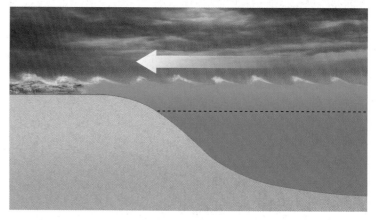

▲ **FIGURE 12–17** A storm surge is created primarily by strong winds blowing ocean water ashore, the effect of which is compounded by heavy wave activity. The reduction in air pressure associated with a hurricane contributes to the rise in sea level but is not the primary factor.

Did You Know?

Hurricane damage is not all water and wind. For example, in Navarre Beach, Florida, Hurricane Katrina left behind meters of sand along a 20-mile (12 km) stretch of road (Figure 12–18). Months of work were needed to clear these storm-produced drifts.

▲ **FIGURE 12–18** A stretch of road at Navarre Beach, Florida, cut off by blowing sand from Hurricane Katrina.

carried by the waves adding to the problem. Figure 12–19 illustrates the type of devastation that can result from a storm surge, in this case the result of Hurricane Katrina. Storm surges are most destructive when they coincide with high tides, especially over bays and inlets that have an extreme range of height between high and low tide.

▲ **FIGURE 12–19** Devastation from Hurricane Katrina's storm surge.

▲ **FIGURE 12–20** The varying intensities of the wind on the left and right sides of hurricanes. This figure shows the movement of the storm itself (indicated by blue arrow), 25 mph to the northeast. The wind speed on the right-hand side of the storm was 150 mph (125 mph + 25 mph); the wind speed on the left-hand side was 100 mph (125 mph – 25 mph). The red arrows indicate the net wind speed on the left and right sides of the hurricane.

Hurricane winds and storm surges are most intense on the right-hand side of the storm relative to the direction it is moving. Figure 12–20 shows Hurricane Charley as it approaches the west coast of Florida. The storm moved to the northeast at 25 mph, while winds relative to the eye were moving at about 125 mph. To the right of the eye relative to its direction of motion, the 125 mph wind speeds were supplemented by the 25 mph movement of the storm in the same direction. To the left of the eye, the counterclockwise rotating air was toward the southwest—the opposite direction of the storm itself. Thus the net speed there was 100 mph.

Though storm surges present the greatest potential for catastrophic coastal destruction and have claimed thousands of American lives over the last few centuries, they do not account for the majority of American hurricane fatalities. A study published by a researcher at the National Hurricane Center revealed that between 1970 and 2002, more than half the fatalities from tropical storms and hurricanes in North America resulted not from storm surges, but rather from freshwater flooding from heavy rain. Only about one-quarter of the fatalities associated with tropical storms and hurricanes (or their remnants) occurred in coastal counties. For several decades prior to 2005 and Hurricane Katrina, there had been a decrease in the incidence of storm surge fatalities. (See *Box 12–3, Focus on Severe Weather: Hurricane Katrina*, for more information on this historic event.) The reduction in deaths associated with

12–3 FOCUS ON SEVERE WEATHER

Hurricane Katrina

At the end of August 2005 we witnessed one of the major natural catastrophes in American history: Hurricane Katrina. Katrina was the first of three Category 5 hurricanes to form in the Gulf of Mexico or the Caribbean in 2005, bringing with it substantial flooding in southern Florida, the inundation of New Orleans, and a devastating storm surge in coastal Louisiana, Mississippi, and Alabama. Two months later, the storm's death toll was complete: More than 1300 people had perished in the United States. Damage estimates exceed $100 billion. In this section we present a chronology of what happened meteorologically. In a book such as this, we cannot even begin to analyze comprehensively the human impact of the disaster.

Wednesday, August 24, 2005: Tropical Storm East of South Florida

At 11 P.M. EDT, Tropical Storm Katrina was located to the east of south Florida (Figure 1a shows the location of the storm, as well as the official NHC predictions for future movements and status). Hurricane warnings had recently been issued for the southeast Florida coastline, with anticipated landfall near Miami. The storm was forecast to be a Category 1 hurricane by 8 P.M. the following day very near the shoreline. Though the cone depicted a fairly extensive range of possible positions, the forecast models the center used were in close agreement, and the storm moved much as predicted over the short term.

Thursday, August 25, 2005: Hurricane Katrina Landfall

At 5 P.M. EDT on August 25 (Figure 1b), Katrina had developed into a hurricane and was near landfall at the position forecast the night before (though it had moved somewhat faster than expected). Landfall occurred at about 6:30 P.M. near the Broward/Miami–Dade County state line (Figure 2). At this point, forecasters officially called for the cyclone to move directly westward and, upon entering the Gulf Coast the following afternoon, to turn toward the northwest and follow a track somewhere in the eastern Gulf along the west coast of Florida.

What is not revealed by Figure 1b is that the forecast models showed highly differing forecasts. Several suggested the path similar to the official NHC forecast, but three others indicated a southwestward movement across southern Florida. The storm did indeed move to the southwest, reaching the western Florida coast 7 hours after initial landfall. The hurricane moved rapidly across the state, which reduced the amount of weakening normally undertaken as a hurricane passes over land. So even though Katrina traveled across southern Florida mainly as a tropical storm, it was able to reintensify into a hurricane when it reached the Gulf.

Katrina did considerable damage to Florida directly and indirectly due to heavy rain that exceeded 26 cm (16 in.) in places. The heavy rainfall led to major flooding and trees toppled by the combination of saturated soils and strong winds. Six people died in Florida from Katrina, which also caused $100 million in damages and $423 million in agricultural losses. But there was much, much more to follow.

Friday, August 26, 2005: In the Eastern Gulf of Mexico

Shortly after midnight, Katrina entered the Gulf of Mexico. By dawn it had become clear to forecasters that the hurricane had the potential to become a huge danger to parts of the Gulf Coast. Early morning forecasts called for the storm to initially move to the west and then begin to arc northwestward toward the coast anywhere from the Florida panhandle to extreme eastern Louisiana (Figure 1c). But by late evening, the system had moved farther to the southwest than anticipated, and by 11 P.M. a very different track was predicted—one that would put Katrina on a collision course with New Orleans and the Mississippi coast (Figure 1d). The hurricane was now set to pass over a region of very warm Gulf waters, which the NHC described as ". . . like adding high octane fuel to the fire." All the forecast models predicted further intensification of Katrina, one of them calculating that wind speeds would top out at more than 243 km/hr (151 mph)—a strong Category 4. Also noteworthy was the fact that all the computer models were giving the same guidance in terms of the storm's path, creating little doubt of what was to come.

Saturday, August 27, 2005: On Course

By Saturday night (Figure 1e) the previous day's forecasts had proven very accurate, and there was little change in the expected path of the storm. The NHC issued a hurricane warning that included some of the following text:

. . . POTENTIALLY CATASTROPHIC HURRICANE KATRINA MENACING THE NORTHERN GULF COAST . . .

A HURRICANE WARNING IS IN EFFECT FOR THE NORTH CENTRAL GULF COAST FROM MORGAN CITY LOUISIANA EASTWARD TO THE ALABAMA/FLORIDA BORDER . . . INCLUDING THE CITY OF NEW ORLEANS AND LAKE PONTCHARTRAIN.

MAXIMUM SUSTAINED WINDS ARE NEAR 175 MPH . . . WITH HIGHER GUSTS. KATRINA IS A POTENTIALLY CATASTROPHIC CATEGORY FIVE HURRICANE ON THE SAFFIR-SIMPSON SCALE. SOME FLUCTUATIONS IN STRENGTH ARE LIKELY DURING THE NEXT 24 HOURS.

HURRICANE FORCE WINDS EXTEND OUTWARD UP TO 105 MILES FROM THE CENTER . . . AND TROPICAL STORM FORCE WINDS EXTEND OUTWARD UP TO 205 MILES.

COASTAL STORM SURGE FLOODING OF 18 TO 22 FEET ABOVE NORMAL TIDE LEVELS . . . LOCALLY AS HIGH AS 28 FEET ALONG WITH LARGE AND DANGEROUS BATTERING WAVES . . . CAN BE EXPECTED NEAR AND TO THE EAST OF WHERE THE CENTER MAKES LANDFALL. SIGNIFICANT STORM SURGE FLOODING WILL OCCUR ELSEWHERE ALONG THE CENTRAL AND NORTHEASTERN GULF OF MEXICO COAST.

The most vulnerable city in the United States, New Orleans, Louisiana, was about to be hit by a catastrophic storm. Bounded by Lake Pontchartrain to the north and surrounded by the winding Mississippi River, the city—much of which is below sea level—had long been known to be extremely vulnerable. The levees protecting much of the city were believed to be able to withstand a direct hit from a Category 3 hurricane, but they had never been seriously tested, as they would be soon. Meanwhile, coastal Mississippi and Alabama were lined up on the right-hand side of an enormously powerful hurricane track.

Sunday, August 28: Landfall Imminent

By late Sunday night the hurricane was just offshore (Figure 1f). Those who had

▲ **FIGURE 1** The position and forecasted movement of Hurricane Katrina at various times. The area in white shows the range of possible locations for the hurricane center at the forecast times shown.

(continued)

set out to evacuate were long gone. Those unable to leave were directed to shelters or hunkered down to take their chances at home. That morning, winds had easily exceeded the threshold for a Category 5 hurricane, with sustained winds of 282 km/hr (175 mph). The hurricane had become as intense as Hurricane Camille, which devastated coastal Mississippi in 1969, and was even larger (Figure 3). Katrina's minimum air pressure of 902 mb was the fourth lowest ever observed for an Atlantic storm.

▲ **FIGURE 2** Radar image of Hurricane Katrina as it approached Miami. This is shown as a movie on this book's CD.

▲ **FIGURE 3** Radar image of Hurricane Katrina as it approached the Louisiana coast. This is shown as a movie on this book's CD.

storm surges is partially the result of an episodic decrease in the number of strong hurricanes hitting populated coastal regions during the 30-year period, along with a better ability to predict the movement of hurricanes and improved evacuation procedures. (We discuss trends and cycles in hurricane activity later in this chapter.)

Hurricane Forecasts and Advisories

Responsibility for tracking and predicting Atlantic and east Pacific hurricanes lies with the National Hurricane Center (NHC) in Miami, Florida. During hurricane season, this office of the National Weather Service obtains constantly updated surface reports and satellite data to determine current storm conditions. Sophisticated numerical models on a supercomputer predict the formation, growth, and movement of tropical storms and hurricanes. When active hurricanes approach land, specially equipped aircraft fly into the storms and provide reconnaissance data from airborne radar and *dropsondes*, packages containing temperature, pressure, and moisture sensors and transmitters released from the plane into the storm.

The NHC uses the standard computer models for conventional weather forecasting (discussed in Chapter 13), as well as others developed specifically for hurricanes. The latter fall into three categories: *statistical*, *dynamical*, and *hybrid*. Statistical models apply information on past hurricane tracks and use those tracks as predictors for current storms. Dynamical models take information on current atmospheric and sea surface conditions and apply the governing laws of physics to current data. Hybrid models combine elements of statistical and dynamical models. The models repeatedly forecast the movement and internal changes of hurricanes for short time increments and then print information on projected storm positions, air pressure, and wind at 6-hour intervals. Not surprisingly, model forecasts become less accurate as lead time increases and are unreliable for more than about 72 hours.

Television news networks that had been following the storm nonstop relayed warnings from the NHC: There was absolutely no doubt that bad things were in store for the Gulf Coast.

Monday, August 29: Landfall

Hurricane Katrina made landfall early Monday morning. Coastal Louisiana was battered first as a huge storm surge overtook the area. Though the storm had weakened enough to be a strong Category 4, everybody knew that a historic disaster was occurring. The eye tracked just to the northeast of New Orleans, putting the city on the less-threatening "left side" of the storm. This initially gave the false impression that the city had narrowly escaped a disaster. This, of course, proved entirely incorrect. As New Orleans was subjected to wind speeds on the order of 160 km/hr (100 mph), Lake Pontchartrain waters rose along the levees, which were unable to hold them back. Eighty percent of the city came underwater (Figure 4).

For several weeks after the disaster, the U.S. Army Corps of Engineers argued that the levees were overtopped in places and that turbulent floodwaters undermined the base of the concrete walls atop the levees.

Later assessments (still subject to further analysis at the time this book was going to press) suggested that this may have occurred in some locations, but in other places the walls gave way without having been overtopped by the storm surge.

In some areas of coastal Mississippi, the flood surge wreaked total devastation. Farther inland, wind destroyed all of or most of many communities. Figure 2 in Box 12–2 plots the entire track of Hurricane Katrina, and this book's CD includes a movie compiled from satellite images that shows the entire movement of Katrina from the time it approached Florida to landfall in Louisiana and Mississippi. An enormous amount of information is also available on the Web. Check out the Web sites at the back of this chapter for links to valuable resources.

▲ **FIGURE 4** Downtown New Orleans under water.

Did You Know?

Meteorologists once believed that seeding clouds with silver iodide (see Chapter 7) could be an effective way to reduce hurricane strength. The idea was to seed parts of the cloud outside the eye wall. If the seeding successfully led to enhanced convection outside the eye wall, the new zone of heavy activity could compete with and thereby weaken the strength of the eye wall. Seeding was undertaken sporadically between 1962 and 1983 on 8 days in four different hurricanes. Although the results once appeared promising, the method ultimately proved unviable.

Hurricane forecasting requires a tremendous amount of data and places a great demand on computer hardware. The *National Oceanic and Atmospheric Administration (NOAA)* has deployed geostationary orbiting GOES satellites (meaning they remain over fixed locations on Earth) that provide improved data acquisition from space. NOAA also flies aircraft into hurricanes and observes conditions from on-board radar systems and the use of GPS dropsondes (Figure 12–21). Dropsondes are instrument packages that observe pressure, temperature, and moisture conditions as they descend through a hurricane. They are dropped from the aircraft, and they have parachutes that allow them to slowly descend through the hurricane and transmit their observations. NOAA relies heavily on supercomputers to run complex forecast models designed specifically for hurricanes.

The improvement in hurricane forecasting in recent years has been substantial, with a 50 percent reduction in the tracking error for the projected position of hurricanes between the mid-1990s and 2010. Still, the movement of hurricanes is particularly difficult to predict.

Hurricane Watches and Warnings

Improvements in forecasting capabilities and the need to give people as much notice as possible in anticipation of a hurricane led the National Hurricane Center (NHC) to change

(a)

Square-cone parachute increases stability of dropsonde

Vents fill chute within 10 seconds after release from aircraft

Shock cord reduces stress when chute opens

GPS antenna

Microprocessor controls the transmitter and digitizes data from the sensors

GPS receiver collects data from GPS satellites used tocalculate wind speed and direction

Battery pack provides power for at least one hour

Radio transmitter sends temperature, humidity, pressure, and GPS (wind) data to the aircraft every 0.5 second

Pressure sensor

Humidity sensors and temperature sensor

Fall speed ranges from 36 mph at 20,000 feet to 24 mph at sea level. A drop from 20,000 feet lasts 7 minutes.

(b)

▲ **FIGURE 12–21** A photo of a typical dropsonde used for hurricane surveillance (a). These packages are dropped into hurricanes and record storm characteristics as they descend via parachute. A schematic of a dropsonde instrument package (b).

their criteria for issuing hurricane watches and warnings. A **hurricane watch** is issued when hurricane conditions are *possible* for a particular coastal region. These watches are issued 48 hours ahead of the predicted onset of tropical storm–force winds for that location. A **hurricane warning** means that hurricane conditions are *expected* somewhere within the identified area, and are issued 36 hours ahead of the expected onset of tropical storm–force winds. Note that hurricane watches and warnings are issued in an anticipation of the arrival of only tropical storm–force winds rather than hurricane force winds. This is done because evacuation and other measures are very difficult to undertake once tropical storm–force winds have already arrived. If the winds are not expected to achieve hurricane status, then the NHC issues **tropical storm watches** and **tropical storm warnings** with the same anticipated lead time.

Figure 12–22 provides an example of how watches and warnings evolve as a hurricane approaches land. In Figure 12–22a, Hurricane Earl's position is marked by the bull's-eye near the Bahamas at 8 A.M. EDT on September 1, 2010. The map shows the forecasted movement of the hurricane along with a cone indicating the range of positions at which the storm may be located at indicated times. A hurricane watch has been issued for much of the North Carolina coast up to eastern Maryland and Virginia. Twenty-four hours later (Figure 12–22b) the hurricane has come

closer to land and most of the North Carolina coast has come under a hurricane warning. In addition, a short stretch of coast to the southwest of the warning area and a large area extending northeast to southern New England is under a tropical storm warning, with a small hurricane watch area farther up the coast. In the end, the Mid-Atlantic coast of the United States experienced only tropical storm winds from Earl. However, the storm, whose intensity had been weakening as it passed east of the Mid-Atlantic states, reintensified back to a hurricane and hit the coast of southern Nova Scotia on the morning of September 4.

The erratic nature of hurricanes makes them notoriously difficult to predict. When predicting hurricane movements, forecasters must weigh the effects of issuing watches or warnings for hurricanes that never make landfall versus the consequences of failing to issue a watch or warning for a storm that ultimately does hit. Obviously, the failure to warn people to evacuate may lead to unnecessary loss of life and property. On the other hand, false warnings have serious ramifications, especially if they occur repeatedly. Repeated false warnings can make the public so complacent that people will eventually disregard warnings that prove accurate.

Evacuations based on these advisories have immense economic costs for the general public, government agencies, and industry. Local residents and small businesses have their lives

▲ **FIGURE 12–22** Hurricane warnings and watches were issued as a result of Hurricane Earl on September 1 and 2, 2011. On September 1 (a) Earl was expected to move from its position near the Bahamas to the North Carolina coast, for which a hurricane watch was issued. By the next morning (b) the storm had advanced toward the coast and a hurricane warning had been issued for much of the North Carolina coast, with an extensive band of shoreline under a tropical storm warning.

Did You Know?

Between 1992 and 2001 the average forecast error for the 24-hour hurricane landfall position was 149 km (92 mi). If you live near a threatened coastal zone, you must keep in mind that the endangered area extends far beyond the point of landfall. Hurricane-force winds can occur for distances well exceeding 160 km (100 mi) from the eye in any direction. So even if the point of expected landfall is far away or if the eye of the storm is still offshore, you may be subject to extremely hazardous hurricane conditions.

thoroughly disrupted as they board up windows and prepare to evacuate or take shelter, while large industries (such as petroleum mining and processing) incur costs measured in tens of millions of dollars from having to shut down and reopen their plants.

Hurricane Intensity Scale

In addition to alerting the public to the location and projected movement of hurricanes, meteorologists use a simple scale to categorize their intensity. The **Saffir-Simpson scale** (Table 12–2) classifies hurricanes into five categories based on the highest current 1-minute average winds in the hurricane. Generally, higher-category hurricanes have lower central pressures and larger storm surges. Figure 12–23 gives a general idea of how high water levels would be for various categories of landfall hurricanes. Of course, the water would not rise up gently like a bath tub, but would hit the area with violent wave activity, intense rainfall, and extreme wind conditions. Though storm surges are usually the more destructive element of hurricanes upon landfall, the scale is based on wind speeds because storm surges are affected by nonmeteorological factors such as coastal configuration and the steepness of the offshore continental shelf. Extremely violent hurricanes are rare, with only three Category 5 and 16 Category 4 hurricanes having hit the mainland United States between 1900 and 2010.

Of course, Category 4 and 5 hurricanes are far more deadly and devastating than lower-category hurricanes. The effects of Hurricane Camille (Category 5) in 1969 have already

TABLE 12–2

The Saffir-Simpson Scale

Category	Pressure mb	Wind km/hr	Speed mph	Storm m	Surge ft	Damage
1	≥ 980	119–154	74–95	1–2	4–5	*Minimal.* No major damage to most building structures.
2	965–979	155–178	96–110	2–3	6–8	*Moderate.* Some roof, door, and window damage. Some trees blown down. Considerable damage to mobile homes.
3	945–964	179–210	111–130	3–4	9–12	*Extensive.* Some structural damage to small residences. Some large trees blown down. Some mobile homes destroyed.
4	920–944	211–250	131–155	4–6	13–18	*Extreme.* Some complete roof structure failures on small residences. Many shrubs, trees, and all signs are blown down. Complete destruction of mobile homes.
5	< 920	> 250	> 155	> 6	> 18	*Catastrophic.* Some complete building failures. All shrubs, trees, and signs blown down.

Category 5
Category 4
Category 3
Category 2
Category 1

▲ **FIGURE 12-23** Approximate level of storm surge height relative to coastal houses for different categories of hurricanes. Note the placement of houses on stilts, as is done on much beachfront property along the Gulf of Mexico coast.

been described. Since then, 16 Category 5 hurricanes have occurred in the Gulf of Mexico, Caribbean Sea, or western Atlantic, and only three of those struck land at full Category 5 intensity: Andrew in 1992, and Dean and Felix, both in 2007. But even though only a handful of Category 5 hurricanes made landfall at that maximum level, the majority of these extreme hurricanes have made landfall at some point in their lifetimes, often with catastrophic consequences. The great Galveston hurricane of 1900 that killed 6000 persons (see *Box 12–4, Focus on Severe Weather: The Galveston Hurricane of 1900*) is believed to have been a Category 4 hurricane. Table 12–3 lists all the Category 4 and 5 Atlantic–Gulf of Mexico–Caribbean hurricanes between 1900 and 2010, and *Box 12–5, Focus on Severe Weather: Recent Deadly Cyclones* describes some noteworthy cyclones elsewhere.

Recent (and Future?) Trends in Hurricane Activity

The extremely destructive seasons of 2004 and 2005 greatly heightened the public's awareness of the danger hurricanes pose. Florida made repeated headlines as four hurricanes hit it in 2004, and the Gulf Coast witnessed destruction of historic proportions in 2005, compliments of Hurricanes Katrina and Rita. These events stimulated public and media interest about whether the recent upsurge in Atlantic hurricane activity was related to natural periodic cycles, global warming, or both.

One thing we know for certain is that the **Atlantic Multidecadal Oscillation** (AMO), a 25- to 40-year oscillation in water temperatures (Figure 12–24), has been a major factor in the increase in Atlantic hurricane activity—and especially in strong hurricanes. Figure 12–25 plots the annual number of Atlantic-named systems (tropical storms and hurricanes), hurricanes, and Categories 3 through 5 hurricanes from 1851 through 2007. The period between the early 1970s and mid-1990s was one of relatively low Atlantic sea surface temperatures and hurricane activity. In fact, the years 1991 through 1994 had less Atlantic hurricane activity than any other 4-year period on record (despite the Category 5 Hurricane Andrew

in 1992). Then came an abrupt transition to a very active period beginning in 1995 that coincided with a shift in the AMO, and the 1995–1999 period proved to be the most active 5-year period on record (41 hurricanes)—at least until that record was surpassed in 2001–2005 (44 hurricanes).

Has global warming also influenced the increase in hurricane activity? Currently there is little, if any, evidence to indicate that the increase in Atlantic tropical storms and hurricanes is due to a longer-term trend in sea surface temperatures. In fact, the increase in Atlantic tropical storm and hurricane activity has not been observed in the other ocean basins around the globe—despite the fact that worldwide tropical ocean temperatures have increased by about 0.5 °C (1 °F) between 1970 and 2004.

The situation might be different with regard to the effect on *intense* hurricanes of long-term sea surface warming. Research published in 2005 showed a near doubling in the number of Category 4 and 5 hurricanes in the western North Pacific, western South Pacific, Indian, and Atlantic oceans since 1970 that coincides with increasing water temperatures. Interestingly, the North Atlantic experienced the smallest increase in major hurricanes among those basins. Another highly cited 2005 study pointed to a substantial increase in the overall energy released by Atlantic hurricanes in recent decades, reflected in both hurricane intensity and duration. That article noted, however, that only part of the increase in Atlantic hurricane activity could be ascribed to increasing sea surface temperatures (an assertion that was not widely reported by the popular press). And yet another 2005 article used the results of computer simulations to obtain the same conclusions based on empirical observations. Thus, there is better reason to suspect that global warming might influence the intensity of hurricanes than the number of hurricanes.

Based on the above considerations, it appears that an increase in hurricanes and intense hurricanes may be a fact of life for residents along the Gulf of Mexico and the Atlantic coast in the years to come—at least for a few decades. Whether we are witnessing a longer term trend is currently unknown. Factors beyond sea surface temperatures could affect the intensity of hurricanes in a world with higher air and sea surface temperatures, and some of these factors could make it harder for hurricanes to develop in a warmer world.

We are more vulnerable than ever to the destructive potential of hurricanes for the simple reason that there has been enormous population growth along the Atlantic and Gulf coasts in recent decades. This will continue to make evacuations of large populations more difficult in response to approaching hurricanes (as witnessed during the evacuation of southeast Texas as Hurricane Rita approached in 2005).

Without question, hurricanes are among the most exciting of natural phenomena, a fitting subject for concluding this section on weather disturbances. The remaining sections of this book will examine human activities and meteorology, and climate and climate change. In our next chapter we turn our attention to weather forecasting.

The Galveston Hurricane of 1900

Some natural disasters are so embedded in our folklore that virtually everybody knows about them. We have all heard about the San Francisco earthquake of 1906 and the Great Chicago Fire of 1871. Yet the single deadliest natural disaster in U.S. history, the Galveston (Texas) hurricane of 1900, seems to have been lost from the national memory. In just a few hours, rising sea waters and heavy surf drowned 6000 persons on Galveston Island—a narrow strip of land that peaks at less than 3 m (9 ft) above sea level (Figure 1).

The loss of life resulted not from lack of warning but rather from a failure to take the threat seriously. Two days earlier, a strong storm was reported moving westward into the Gulf of Mexico off Cuba, and ships returning from the Gulf of Mexico reported encountering the storm offshore the day before it made landfall. Furthermore, the local weather forecaster, Isaac Cline, observed the combination of winds and heavy surf along the local beach and deduced that the storm would move onshore. But evidence of the impending landfall seems to have been largely unheeded, in part because some meteorol-ogists erroneously believed it was virtually impossible for a storm in the Caribbean to track across the Gulf. Scientists (including Cline) were also erroneously convinced that the gently sloping seafloor offshore would protect Galveston from major flooding in the event of a hurricane.

There is some uncertainty as to when people started to take the hurricane seriously. According to Cline's account of the disaster, he rode through Galveston Island urging residents to evacuate, but recent research casts doubt about the degree to which he actually warned the populace. Regardless of how urgent Cline's warnings were, however, few people evacuated, and some residents even rode to the beach to watch the heavy waves crash against the shore.

When the hurricane arrived, the people of Galveston had no way to escape. Within hours the rising seas completely covered the island so that the only potential shelter was in taller, well-built structures. Even these failed to withstand the pounding of waves and debris. Cline later gave the following account of his ordeal:

> By 8 P.M. a number of houses had drifted up and lodged to the east and southeast of my residence, and these with the force of the waves acted as a battering ram against which it was impossible for any building to stand for any length of time, and at 8:30 P.M., my residence went down with about fifty persons who had sought it for safety, and all but eighteen were hurled into eternity. Among those lost was my wife, who never rose above the water after the wreck of the building.

Cline and his brother were luckier and grabbed onto floating debris that helped them stay afloat. After 3 hours, the floodwaters subsided and the Clines were on solid land, among the survivors.

The horror did not end with the passage of the hurricane. There were still 6000 bodies to deal with. Some were taken out to sea on barges, but many washed back to shore. Ultimately, most of the bodies were cremated where they were found.

With our current ability to track and forecast the movement of approaching hurricanes, there is no reason for a repeat of this type of loss of life in North America. When Hurricane Ike (Figure 2) struck the Galveston area in 2008 the loss of life was tragic, but at levels far below those of the 1900 hurricane. But hurricanes will always present a threat to Gulf and Atlantic coasts that must be respected.

▲ **FIGURE 1** The Galveston hurricane of 1900 was the deadliest natural disaster in U.S. history.

▲ **FIGURE 2** Damage to Galveston after Hurricane Ike in 2008.

TABLE 12–3

Category 4 and 5 Hurricanes 1990–2010

Category 4 Hurricanes					
Name	Year	Month	Maximum Sustained Winds (km/hr)	Maximum Sustained Winds (mi/hr)	Minimum Pressure (mb)
Claudette	1991	September	215	130	946
Felix	1995	August	220	140	929
Luis	1995	August	220	140	935
Opal	1995	September, October	240	150	919
Edouard	1996	August, September	230	145	933
Hortense	1996	September	220	140	935
Georges	1998	September	250	155	937
Bret	1999	August	230	145	944
Cindy	1999	August	220	140	942
Floyd	1999	September	250	155	921
Gert	1999	September	240	150	930
Lenny	1999	November	250	155	933
Isaac	2000	September, October	220	144	943
Keith	2000	September, October	220	140	941
Iris	2001	October	230	145	938
Michelle	2001	November	220	140	934
Lili	2002	October	230	145	940
Fabian	2003	September	230	145	939
Charley	2004	August	230	145	947
Frances	2004	August	230	145	937
Karl	2004	September	230	145	938
Dennis	2005	July	240	150	930
Gustav	2008	August	250	155	941
Ike	2008	September	230	145	935
Omar	2008	October	230	135	958
Paloma	2008	November	230	145	944
Bill	2009	August	230	135	943
Danielle	2010	August	215	135	942
Earl	2010	Aug-Sept	230	145	928
Igor	2010	September	250	155	920
Julia	2010	September	215	140	948

Category 5 Hurricanes					
Andrew	1992	August	280	175	922
Mitch	1998	October–November	285	180	905
Isabel	2003	September	270	165	915
Ivan	2004	September	270	165	910
Emily	2005	July	260	160	929
Katrina	2005	August	280	175	902
Rita	2005	September	285	180	895
Wilma	2005	October	295	185	902
Dean	2007	August	280	175	905
Felix	2007	September	280	175	929

12–5 FOCUS ON SEVERE WEATHER

Recent Deadly Cyclones

Hurricanes that make landfall over the United States can be deadly, but nothing on the scale of what has been witnessed in other parts of the world due to cyclones and typhoons. In 1970 Bangladesh (then part of Pakistan) was hit by a tropical cyclone that killed between 300,000 and 500,000 people. This disaster led to the construction of more than 2500 concrete shelters on pillars (Figure 1) to protect residents against future cyclone hits. These shelters have undoubtedly saved hundreds of thousands of lives, first in 1991 when a cyclone hit the country with heavy rain, Category 5 winds, and a 9 m (30 ft) storm surge. Despite the fact that this was the strongest cyclone to hit the country in more than a century, the death toll was about 70,000—a horrific number but far smaller than that of the 1970 cyclone. The shelters once again saved tens of thousands of lives in November 2007 when another Category 5 cyclone, Sidr, hit the country. Estimates of the number of fatalities have varied between 3000 and 10,000—another terrible number but far less than that which would have occurred without the shelters.

In May 2008 Tropical Cyclone Nargis (Figure 2) made international news when it hit Myanmar (formerly called Burma) at its peak strength, packing peak winds estimated at 213 km/hr (132 mph) and producing heavy rains and a 3.7 m (12 ft) storm surge. The disaster was intensified by the ruling military dictatorship that denied entry to international relief workers trying to bring food and medical supplies to the country. At least 77,000 people died from Nargis—perhaps as many as 100,000—which would make it the deadliest cyclone to hit Asia since the 1991 storm. Two to three million people were left homeless.

Much of the coastline of South Asia is perfectly situated for disasters such as these. Low-lying coastal areas, often near rivers that can overflow their banks, and a poorly developed infrastructure for shelter and evacuation put the lives of millions of people in great danger. It is a tragic fact that events such as the ones described here are not unlikely to be repeated at great human cost.

▲ **FIGURE 1** One of more than 2500 cyclone shelters set up in Bangladesh after the catastrophic cyclone of 1970.

▲ **FIGURE 2** Satellite image of Tropical Cyclone Nargis approaching Myanmar.

▶ **FIGURE 12-24** A time series of an index representing the Atlantic Multidecadal Oscillation from 1856 to 2009.

Monthly values for the AMO index, 1856–2009

▶ **FIGURE 12-25** The number of Atlantic tropical storms (red), Category 1–2 hurricanes (green), and Category 3–5 hurricanes (yellow) by year, 1851–2010.

Summary

Hurricanes (and their counterparts such as typhoons and tropical cyclones) are extremely powerful storms that originate in tropical regions and migrate into the middle latitudes. They bring enormous destruction and loss of life to many coastal regions of the world. The hurricane that hit Galveston Island, Texas, in 1900 was the greatest single natural disaster to hit North America, with a death toll of 6000. This figure pales in comparison to the hundreds of thousands of fatalities associated with individual tropical cyclones in southern Asia.

Most hurricanes begin their life cycles as uneventful tropical disturbances, small clusters of thunderstorms. When they intensify and organize into a rotating band of cloud cover and thunderstorm activity, they are called *tropical depressions*. Further intensification results in their being classified as tropical storms, or hurricanes if their sustained wind speeds exceed 120 km/hr. Because strong tropical storms can form only over oceans having high surface temperatures, tropical depressions most often become tropical storms and eventually hurricanes over the western portions of the ocean basins.

Hurricanes are smaller than midlatitude cyclones but much larger than tornadoes. They can last for a week or more and travel thousands of kilometers before dissipating. The heaviest thunderstorm activity occurs within bands of thick cloud cover that spiral toward the center of the system in a pinwheel pattern. The intensity of the storm increases toward its center until reaching the eye wall, the concentric zone of maximum activity that surrounds the eye. The eye of a hurricane is strikingly different from the rest of the hurricane because it is marked by generally clear skies, light winds, and higher air temperatures. Often it is hard to discern the true structure of a hurricane from above, because the anticyclonically rotating outflow in the upper troposphere creates a blanket of cirrostratus clouds overlying the thicker cumulus.

Hurricanes can produce damage in several ways. Copious amounts of rain can bring intense floods, and strong winds can bring down structures. The most serious threat posed by a hurricane is the storm surge, the elevated rise in sea level due to low atmospheric pressure and the piling up of water by strong winds. When the storm surge coincides with a high tide, the floodwaters (coupled with heavy surf) can penetrate considerable distances inland.

The National Hurricane Center of the National Weather Service uses a sophisticated network of satellites, research aircraft, and computer hardware and software to issue advisories on the likelihood of hurricane landfall. The erratic nature of hurricanes makes predicting them particularly difficult, but recent modernization at the National Weather Service has substantially increased forecast accuracy.

Hurricanes have become more frequent in the Atlantic Ocean since the mid-1990s, and several particularly devastating hurricanes left their mark on United States coastal areas in 2004 and 2005. This increased activity is partly due to a shift in the Atlantic Multidecadal Oscillation, a 25- to 40-year cycle on ocean temperatures, which occurred around 1995. While the potential impact of global warming on hurricanes is not entirely understood, it is likely that global warming will lead to more intense hurricanes rather than more frequent hurricanes.

Key Terms

hurricane *page 346*

typhoon *page 346*

cyclone *page 346*

trade wind inversion
 page 346

marine layer *page 346*

eye *page 349*

eye wall *page 349*

double eye walls *page 350*

eye wall replacement
 page 350

hot towers *page 350*

tropical disturbance
 page 352

easterly waves *page 352*

tropical depression
 page 353

tropical storm *page 353*

storm surge *page 362*

hurricane watch *page 368*

hurricane warning *page 368*

tropical storm watch
 page 368

tropical storm warning
 page 368

Saffir-Simpson scale *page 369*

Atlantic Multidecadal
 Oscillation *page 370*

Review Questions

1. Describe the geographic distribution of hurricanes, typhoons, and cyclones. What environmental conditions at these locations favor the development of such storms?

2. Which region has the greatest incidence of major tropical storms?

3. What is the trade wind inversion, and what impact does it have on the formation of hurricanes?

4. Describe the size, sea level air pressure, and wind speed of a typical hurricane.

5. When are hurricanes most likely to form?

6. Describe the cloud and precipitation patterns associated with hurricanes, including those associated with the eye and eye wall.

7. Describe the various ways in which hurricanes differ from midlatitude cyclones.

8. What are tropical disturbances, and how do easterly waves influence them?

9. Describe the characteristics that distinguish tropical disturbances, tropical depressions, tropical storms, and hurricanes from one another.

10. What ocean surface characteristics are required for the intensification of storms into hurricanes and the maintenance of hurricanes?

11. Is there a "typical" path that hurricanes take after forming? Explain.

12. What feature associated with hurricanes causes the greatest destruction to coastal regions? Is this also true of inland regions?

13. Why is the right-hand side of a hurricane (relative to its direction of movement) the most dangerous?

14. Where are tornadoes most likely to be embedded in a hurricane?

15. What are hurricane watches and warnings? Are they exact corollaries to tornado watches and warnings?

16. Why are forecasters concerned with issuing hurricane advisories for areas that do not eventually get hit?

17. What is the highest hurricane category on the Saffir-Simpson scale? How frequently do hurricanes of that magnitude occur?

Critical Thinking

1. Why don't hurricanes cross the equator?

2. If two hurricanes pass just to the west of Cuba over a 2-week period, what reasons might one have for expecting the second one to be weaker than the first?

3. How might previous drought conditions affect the intensity of a former hurricane as it passes over the southern United States?

4. El Niño conditions are believed to suppress hurricane development in the Atlantic. How might the phenomenon affect hurricane formation and movement in the Pacific?

5. It has been postulated that an increase in global temperatures could lead to an increase in the number and intensity of tropical storms and hurricanes. Global temperatures were particularly high during the 1990s and early 2000s, and there has been an increase in Atlantic hurricane activity since 1995. Does this association prove the connection between temperature and hurricane activity? Explain why or why not.

6. If global warming continues, thermal expansion of the oceans and the melting of glaciers will lead to a higher sea level. How would this affect the threat of storm surges relative to wind damage and flooding?

7. Experts believe that New York City is the third most dangerous city in the United States with regard to hurricanes, despite the fact that there has been no major hurricane-inflicted damage on the area (other than some wind damage and coastal erosion from Hurricane Gloria in 1985). What factors could be responsible for this vulnerability? After answering this question, refer to **www2.sunysuffolk.edu/mandias/38hurricane** for an informative discussion of this issue.

Problems and Exercises

1. Compare the area of a hurricane that measures 600 km in diameter to a midlatitude cyclone having a diameter of three times greater (1800 km).

2. During the tropical storm season, use the Web sites described below to note the positions of current systems and the probabilities of landfall at various coastal locations. Describe how successful the predictions proved to be.

3. Refer to the forecast for the upcoming tropical storm season at **tropical.atmos.colostate.edu**. What existing conditions have led the forecast team to make its prediction? Also, use this Web site to determine how successful last year's forecast was.

4. Refer to **www.ncdc.noaa.gov/oa/climate/severeweather/hurricanes.html** and read the Special Reports on the past year's hurricane activity. Were any tropical storms particularly noteworthy?

Quantitative Problems

You can gain a deeper understanding of hurricanes by working out some numerical problems. These are available from this book's Web site, **www.MyMeteorologyLab.com**.

Log on to the site and go to the Chapter 12 section for some thought-provoking problems.

Useful Web Sites

www.nhc.noaa.gov
Official Web page for the National Hurricane Center Tropical Prediction Center.

www.wunderground.com/tropical
Site opens with a map showing current tropical activity and provides numerous links for satellite images, advisories, outlooks, and discussions.

www.nrlmry.navy.mil/tc_pages/tc_home.html
Comprehensive information provided by the Monterey Naval Research Laboratory.

cimss.ssec.wisc.edu/tropic/tropic.html
Archived and real-time imagery and a large amount of text information from the University of Wisconsin–Madison Tropical Cyclone Research Team.

www.solar.ifa.hawaii.edu/Tropical/tropical.html
Current hurricane information and several data archives.

www.usatoday.com/weather/hurricane/hurricane-resources.html
Some basic information from *USA Today.*

www.ncdc.noaa.gov/oa/climate/severeweather/hurricanes.html
Climatological information and data on past hurricanes from the National Climate Data Center.

tropical.atmos.colostate.edu
Official Web site of the Tropical Meteorology Project at Colorado State University. Includes much information, such as the seasonal tropical storm forecast and a description of factors used to predict the upcoming season.

www2.sunysuffolk.edu/mandias/38hurricane
A comprehensive and interesting review of the hurricane of 1938 and what it tells us about the vulnerability of the New York area to future events.

www.ncdc.noaa.gov/oa/reports/tech-report-200501z.pdf
An excellent analysis of Hurricane Katrina.

www.cpc.noaa.gov/products/outlooks/hurricane.shtml
The latest seasonal hurricane forecast issued each spring by the NOAA Climate Prediction Center.

www.nasa.gov/mission_pages/hurricanes/main/index.html
Excellent source of storm images and recent data.

MyMeteorologyLab™

The *Understanding Weather and Climate,* Sixth Edition, MyMeteorologyLab™ web site contains numerous multimedia resources to aid in your study of **Tropical Storms and Hurricanes**.
Visit **www.MyMeteorologyLab.com** to:

- **Review** key chapter concepts.
- **Read** the Pearson eText, *In the News RSS feeds,* and other chapter-specific Web resources.
- **Visualize** challenging topics with Interactive Tutorials, Geoscience Animations, MapMaster interactive maps, and Weather in Motion movies and visualizations, and more.
- **Test Yourself** with self-study quizzes.

WEATHER IN MOTION
View movies and visualizations of meteorology patterns and processes.

Improving Hurricane Predictions
A Hurricane in the Middle Latitudes
Hot Towers and Hurricane Intensification
Hurricane Eye Wall
Interview with Chase Plane Pilot
The 2005 Hurricane Season
Hurricane Katrina 1
Hurricane Katrina 2
Hurricane Katrina 3

PART FIVE

Human Activities

13 Weather Forecasting and Analysis

TUTORIAL Forecasting

What resources are used by forecasters in predicting the weather a few days in advance?

14 Human Effects on the Atmosphere

What factors affect the distribution of acid precipitation?

Altitude of 300 mb level (in tens of meters)

Temperature at 300 mb (in °C)

Isotachs

Winds greater than 70 knots
Winds greater than 110 knots

300 MB ANALYSIS
HEIGHTS/TEMPERATURE
AT 7:00 A.M., E.S.T.
JUN. 10 1997

pH values

acidity

4.5–4.6
4.6–4.7
4.7–4.8
4.8–4.9
4.9–5.0
5.0–5.1
5.1–5.2
5.2–5.3
≥ 5.3

As important as weather is to our lives, people are not defenseless against it. One of the best tools people have in dealing with weather is a sophisticated system of forecasting weather days in advance. But people interact with the atmosphere in other ways as well. For centuries we have polluted the air in the course of industrial and even agricultural activities. We also alter the climate by building large cities that modify their own climate. This section looks at several ways people are connected to their atmospheric environment.

Times Square, New York City, during Tropical Storm Irene, August, 2011

13 Weather Forecasting and Analysis

After reading this chapter, you should be able to:

▶ Summarize the basic procedures of weather forecasting.

▶ Explain why weather forecasting is imperfect.

▶ Describe the main methods of weather forecasting and types of forecasts.

▶ Explain how meteorologists assess the quality of weather forecasts.

▶ List some major sources of data for weather forecasting.

▶ Describe forecast procedures, including numerical modeling, and explain how meteorologists develop short-, medium-, and long-range forecasts and seasonal outlooks.

▶ Explain how meteorologists use weather maps and satellite images in forecasting.

▶ Explain the use of thermodynamic diagrams in forecasting.

orecasters for the National Weather Service (NWS) spend a large portion of each workday fielding telephone inquiries about upcoming weather. Sometimes people want to know if their round of golf is likely to be rained out or if their outdoor plants might be vulnerable to overnight frost damage. But the most important aspects of a forecaster's job do not involve matters of simple convenience but questions that deal with life and death.

Such was the case for Mark Moede, a meteorologist with the NWS Forecast Office in San Diego. In late August 1998 much of southern California was threatened by brush fires triggered by unusually heavy thunderstorm activity. Firefighters battling the blazes received constant updates on weather conditions that could either suppress or enhance the spread of the fire. On September 2 the situation reached its most critical stage. Pete Curran of the Orange County Fire Department maintained close contact with Moede for constant updates on a line of approaching thunderstorms. At issue was whether the thunderstorms would continue to strengthen and whether they would pass through the areas where the firefighters were working. If the storms passed that way, then strong winds, deadly lightning, and blinding rains would place Curran's crews in jeopardy. With the aid of Doppler radar and information from automated weather stations, Moede made the right call. He advised Curran to evacuate his crews from the eventual path of the storm, where winds in excess of 95 km/hr (60 mph) created an uncontrollable firestorm. Afterward, firefighters reported that the fire line they had evacuated had been completely burned over. If there had been no call to evacuate, in all likelihood a number of firefighters would have been killed.

This chapter discusses the methods by which forecasters perform their job. We first look at some important issues regarding the general concept of weather forecasting and then discuss ways in which necessary data are obtained and processed. We then study the various types of weather maps and how they are used in weather analysis.

◀ Part of major wildfire in eastern Arizona, June 10, 2011.

Weather Forecasting—Both Art and Science

Weather forecasting is a highly sophisticated and data-intensive endeavor. Forecasters routinely employ state-of-the-art computer hardware and software systems that perform millions of calculations, based on a huge amount of input data and displayed at sophisticated workstations. The meteorologists who work with the information come armed with rigorous university training that includes intensive coursework in mathematics, physics, chemistry, and, of course, meteorology. But the scientists who work at the public and private forecasting agencies also rely heavily on their ability to subjectively analyze the current weather situation, thus making their task a blending of scientific principle and art.

Forecasters for the **National Weather Service (NWS)** in the United States and the **Meteorological Service of Canada (MSC)** work 8-hour shifts in offices that are open 24 hours a day, 7 days a week. All forecasters apply the same scientific underpinnings to their work and make use of the same types of resources, but there is no precisely defined routine by which a forecast is made. Every meteorologist has his or her own preferred set of procedures for analyzing the weather situation—and those procedures may vary considerably under differing conditions. For example, a forecaster concerned with the possibility of near-term tornado development might spend much of his or her time looking at radar images, while one concerned with the possibility of an advancing snowstorm might spend more time assessing the changing pattern shown by satellite images. Furthermore, meteorologists employ numerous rules of thumb that might help them assess the likelihood of a given storm producing rain instead of snow, and these rules of thumb often vary by geographic region.

Upon starting his or her shift, a forecaster will probably receive a verbal briefing about the weather of the last 24 hours from the meteorologist whose shift is about to end. The incoming meteorologist will also examine a large amount of textual and graphic information. Many professional meteorologists can still remember a time when most of the maps and information came in via a facsimile machine that produced black-and-white output. Each weather map or satellite image would take several minutes to print on a damp sheet of paper, and then it would be hung up on a wall atop the previously printed maps or images. If the meteorologist wanted to examine the changes that occurred in the weather over a certain period, he or she would have to manually flip through a sequence of maps. It would take a large section of wall to display the various types of maps and images to which the forecaster might want to refer. This situation has changed radically over the last several decades.

Figure 13–1 shows a typical National Weather Service office (a) and workstation (b), at which the forecaster will perform most of the job. Each workstation employs the **Advanced Weather Interactive Processing System (AWIPS)**, which allows forecasters to display maps of current weather conditions, output of computer forecast models displayed in map form, satellite and radar images, forecast advisories and discussions from other weather facilities, and a great deal more. Each AWIPS station has two graphical display monitors and a separate alphanumeric display. The graphical display monitor is typically set up to present one large panel and four smaller panels off to the left side. Forecasters can zoom in on or out of any image and even superimpose several different types of information simultaneously. For example, a forecaster might set the large panel to show a satellite image of North America, with superimposed arrows depicting the upper-level winds. At the same time, the four smaller panels on the left side of the screen might display the current temperature, humidity, surface winds, and air pressure distributions. With just a few clicks of a mouse, the forecaster can call up just about any information that might help decipher the current weather patterns and help produce a successful forecast. AWIPS can also display all maps and images as movie loops to show the recent movement of significant weather features.

(a)

(b)

▲ **FIGURE 13–1** A typical National Weather Service office (a) and workstation (b).

The forecaster goes through a sequence of procedures using AWIPS, depending on what the current conditions are and what type of weather may be developing. The first step is often to determine what type of cloud or precipitation conditions might be in the offing. This is of primary concern not only because of their intrinsic importance but also because such knowledge is necessary before any kind of temperature forecast can be made. (Obviously, a heavy overcast will reduce the amount of sunlight that can reach the surface, which will retard daytime warming.) If cloudy conditions currently exist, the question at hand is whether the clouds will move on and be replaced by clear skies. Forecasters rely heavily on visual inspection of satellite loops in these cases. Morning cloud cover can also "burn off" later in the day as the sun gets higher above the horizon. Meteorologists would want to look at the temperature and moisture profiles in these instances to determine just how thick the cloud layers are; if clouds are relatively thin, they are more likely to give way to clear skies. Clear days also include the possibility that clouds will move in from other regions or develop later on and create local rain showers. Vertical profiles showing the trend in temperatures and humidity from the surface to the upper stratosphere can be particularly useful in these situations.

Changes in the weather can occur even if there is little movement of major weather systems. Assume, for example, that a forecaster starting the morning shift at Charleston, South Carolina, determines that no major systems are moving toward or away from the southeastern United States. The meteorologist still has to look for a change in temperatures over the next few days. A good start in such instances would be to compare the current temperature to that of the same time the day before. If the morning temperature is 5 degrees warmer than it was 24 hours earlier, it is likely afternoon temperatures will undergo a similar increase. But additional information can contribute to an even more reliable forecast. Therefore, the meteorologist might compare other conditions, such as the distributions of pressure across the Southeast. A change in such a pattern could, for example, indicate a weakening of the wind blowing from the coast to the inland region, which might promote even further warming.

Whether skies are cloudy or clear, forecasters now rely heavily on the output of computer models. These highly complex programs apply the fundamental laws of atmospheric physics to a huge amount of data and produce maps showing the expected distribution of air pressure, temperature, and other variables at the surface and upper atmosphere. The meteorologist interprets the overall patterns depicted on the maps, infers the overall type of weather that can be expected at a particular location, and issues a forecast that is usable for the general public. For example, if the computer maps indicate a substantial decline in air pressure at the surface coupled with particular upper-level air flows, the forecaster may advise the public of increasing cloudiness and possible precipitation.

Forecast maps produced by the computer models often depict the predicted distribution of weather variables directly as an application of physical laws. Such a panel of maps appears in Figure 13–2a, showing the characteristics of the atmosphere at 200 mb, 500 mb, the surface, and 850 mb. Notice that each of the four panels maps different characteristics of the atmosphere in addition to the height of the given pressure level. Meteorologists can easily change the characteristics plotted or zoom in and out of the image as desired. Figure 13–2b shows 48-hour forecasts for the same pressure levels.

Other models rely on statistical methods to assist the forecaster. These products rely on output from the physical models. Thus, the physical models predict the state of the entire atmosphere over a wide area (such as North America), and the statistical models use that information to predict values such as maximum and minimum temperature at particular locations (Figure 13–3). Though the temperatures predicted by the computer can provide the forecaster with useful guidance, the professional meteorologist knows when to overrule such output by using his or her professional judgment.

Forecasters make more than a single forecast and various weather statements over the course of a shift, each intended to serve a particular person or constituency. Numerous types of forecasts are routinely issued at particular times during each shift, and a forecaster's activities assure these reports are disseminated on time. Scheduled reports include short-term forecasts, more detailed forecast discussions intended for people with a more advanced understanding of meteorology, local aviation advisories, hydrologic information (such as information on potential flooding), and numerous other statements.

Perhaps the most familiar type of weather statement to many people is the **zone forecast**, issued at designated times each day and extending out to a week into the future. The following forecast from Des Moines, Iowa, serves as a good example:

ZONE FORECASTS FOR IOWA

NATIONAL WEATHER SERVICE DES MOINES IA

356 AM CDT MON JUL 18 2011

POLK COUNTY-

INCLUDING THE CITIES OF . . . DES MOINES

356 AM CDT MON JUL 18 2011

. . . EXCESSIVE HEAT WARNING IN EFFECT UNTIL 9 PM CDT THURSDAY . . .

.TODAY . . . SUNNY. HOT AND HUMID. PATCHY FOG THROUGH MID MORNING. HIGH IN THE UPPER 90S. SOUTHWEST WIND 5 TO 10 MPH. HEAT INDEX READINGS 111 TO 116.

.TONIGHT . . . MOSTLY CLEAR. LOW IN THE MID 70S. SOUTHWEST WIND NEAR 10 MPH. HIGHEST HEAT INDEX READINGS 107 TO 112 THROUGH MIDNIGHT.

.TUESDAY . . . SUNNY. HOT AND HUMID. HIGH IN THE MID 90S. SOUTHWEST WIND NEAR 10 MPH. HEAT INDEX READINGS 108 TO 113.

.TUESDAY NIGHT . . . MOSTLY CLEAR. LOW IN THE UPPER 70S. SOUTH WIND 5 TO 10 MPH. HIGHEST HEAT INDEX READINGS 107 TO 112 THROUGH MIDNIGHT.

.WEDNESDAY . . . SUNNY. HIGH IN THE UPPER 90S. SOUTHWEST WIND 10 TO 15 MPH WITH GUSTS TO AROUND 25 MPH. HEAT INDEX READINGS 106 TO 111.

.WEDNESDAY NIGHT . . . MOSTLY CLEAR. LOW IN THE UPPER 70S.

(a)

▲ **FIGURE 13–2** Output from a computer model run showing the current patterns at different levels of the atmosphere (a). Weather maps for the same four levels of the atmosphere, but as predicted by computer models for 48 hours later in time (b), on the next page.

.THURSDAY . . . MOSTLY SUNNY WITH A 20 PERCENT CHANCE OF THUN-DERSTORMS. HIGH IN THE LOWER 90S.

.THURSDAY NIGHT AND FRIDAY . . . PARTLY CLOUDY. A 30 PERCENT CHANCE OF THUNDERSTORMS. LOW IN THE MID 70S. HIGH AROUND 90.

.FRIDAY NIGHT THROUGH SATURDAY NIGHT . . . PARTLY CLOUDY WITH A 20 PERCENT CHANCE OF THUNDERSTORMS. LOW IN THE MID 70S. HIGH IN THE LOWER 90S.

.SUNDAY . . . PARTLY SUNNY WITH A 40 PERCENT CHANCE OF THUNDER-STORMS. HIGH IN THE UPPER 80S.

The above forecast was issued at 3:56 A.M., in accordance with the routine schedule for that particular shift. But on that day conditions were such that the forecaster decided to issue a supplementary, unscheduled report about the potential for severe weather. The following hazardous weather outlook was issued slightly more than an hour and a half after the scheduled zone forecast:

HAZARDOUS WEATHER OUTLOOK

NATIONAL WEATHER SERVICE DES MOINES IA

536 AM CDT MON JUL 18 2011

IAZ004>007-016-017-026>028-038-039-191100-

EMMET-KOSSUTH-WINNEBAGO-WORTH-HANCOCK-CERRO GORDO-FRANK-LIN-BUTLER-

(b)

BREMER-GRUNDY-BLACK HAWK-

536 AM CDT MON JUL 18 2011

THIS HAZARDOUS WEATHER OUTLOOK IS FOR PORTIONS OF CENTRAL IOWA.

.DAY ONE . . . TODAY AND TONIGHT

ISOLATED THUNDERSTORMS ARE POSSIBLE TODAY AND TONIGHT. SEVERE STORMS ARE NOT EXPECTED.

WARNINGS FOR EXCESSIVE HEAT ARE IN EFFECT. IN ADDITION . . . RIVER

FLOOD WARNINGS REMAIN IN EFFECT FOR PORTIONS OF THE DES MOINES RIVER BASIN. PLEASE REFER TO THE NATIONAL WEATHER SERVICE WEBSITE AT WEATHER.GOV/DESMOINES FOR FURTHER DETAILS.

.DAYS TWO THROUGH SEVEN . . . TUESDAY THROUGH SUNDAY

AN EXCESSIVE HEAT WARNING IS IN EFFECT FOR THE AREA. PLEASE REFER TO THE NATIONAL WEATHER SERVICE WEBSITE AT WEATHER.GOV/DESMOINES FOR FURTHER DETAILS.

SCATTERED THUNDERSTORMS ARE POSSIBLE THURSDAY THROUGH SUNDAY.

SEVERE WEATHER CAN NOT BE RULED OUT AT THIS TIME.

Some forecast offices require the issuance of routine weather reports particularly important to the local environment. Thus, a forecaster in Honolulu routinely has to write a marine forecast such as the following:

(a)

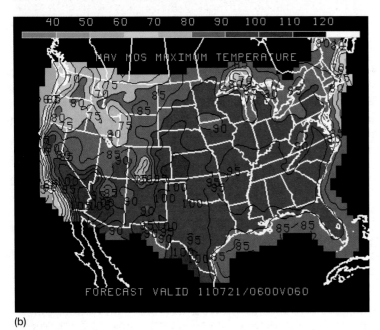

(b)

▲ **FIGURE 13-3** Example of map using statistical output to predict the distribution of minimum and maximum temperatures 2 days in advance.

Did You Know?

Meteorology offers a wide range of career possibilities to students who enjoy the physical sciences. The American Meteorological Society's Web site has several useful links for budding professionals. The Web page www.ametsoc.org/atmoscareers provides some background on the training required to be a professional meteorologist. Also, the site at www.ametsoc.org/amsucar_curricula/curriculaAlpha.cfm offers a listing of all the programs in the United States that offer meteorology or atmospheric science degrees.

OFFSHORE WATERS FORECAST FOR HAWAII

NATIONAL WEATHER SERVICE HONOLULU HI

1200 PM HST MON JUL 18 2011

HAWAIIAN OFFSHORE WATERS BEYOND 40 NAUTICAL MILES OUT TO 240

NAUTICAL MILES INCLUDING THE PORTION OF THE PAPAHANAU-MOKUAKEA

MARINE NATIONAL MONUMENT EAST OF FRENCH FRIGATE SHOALS

SEAS GIVEN AS SIGNIFICANT WAVE HEIGHT...WHICH IS THE AVERAGE HEIGHT

OF THE HIGHEST 1/3 OF THE WAVES. INDIVIDUAL WAVES MAY BE MORE THAN

TWICE THE SIGNIFICANT WAVE HEIGHT.

From these examples, you can see that not all meteorologists have the same set of procedures to follow on a given day. Variations from one forecast office to another affect a meteorologist's schedule, and differing weather conditions may alter the usual routine.

Checkpoint

1. What is a zone forecast?
2. What are the roles of physical models and statistical models in weather forecasting? Explain.

Why Is Weather Forecasting Imperfect?

We've all had careful plans upset by a bad weather forecast and are understandably quick to find fault when actual conditions depart from the forecast. Implicit in such criticism is the premise that forecasts ought to be accurate and that there is no excuse for a miss. So why are forecasts often so far from correct? After all, with powerful computers, satellites, weather radar, and global communication networks, it seems as if making a good forecast ought to be easy. But however much the public might think so, this is definitely not the case—in fact, accurate weather forecasting is extraordinarily difficult.

To see why, imagine you want to forecast tomorrow's temperatures and think about just a few of the factors you must consider. First, remember that the temperature structure of the atmosphere depends in part on absorption and emission of radiation (shortwave and longwave), which itself depends on the vertical and horizontal distribution of atmospheric gases, clouds, and so on. So, to compute the temperature at a point, you need to begin with detailed information about the composition of the atmosphere in three dimensions.

Of course, with water constantly shifting between the liquid, solid, and vapor phases, atmospheric composition is hardly constant, so you will need to forecast those changes over time. Remember also that as water changes phase, latent heat is added or removed from the atmosphere; thus, you will have to keep track of that as well as radiation transfer. But the phase changes are influenced by small-scale updrafts and downdrafts, so you will need to somehow forecast vertical motions as part of your overall effort. Furthermore, horizontal motion can't be neglected—you will need to allow for warm or cold air advection (see Chapter 10).

With regard to wind, near the surface you face the problems of flow around complex terrain and somehow quantifying frictional effects between the atmosphere and Earth's surface. Above the friction layer things are less complicated, but not by much. The basic problem is one of continual interaction: The motions of the atmosphere change the motions that subsequently develop. In other words, after a short time the winds change the winds. So, even though you're only interested in temperature, you can't pretend the winds are unchanging; you are forced into the business of forecasting atmospheric motion. Unfortunately, this is very difficult, because the atmosphere is dynamically unstable. Small disturbances often grow into large features that eventually dominate the field of motion. Thus, whereas you might only care about motion on large scales for the purposes of heat advection, you must consider small-scale motions to know how the large-scale pattern will evolve.

Obviously, weather forecasting involves a set of interlocking problems, each difficult to solve in isolation, let alone in combination. In light of all these difficulties, it's remarkable that forecasts show any accuracy at all. Rather than wondering why they fail, we're more likely to wonder how they are able to succeed!

Weather forecasting by the U.S. government began in the 1870s, when Congress established a National Weather Service under the authority of the Army Signal Corps. In 1890 the organization was renamed the National Weather Bureau and transferred to the Department of Agriculture. There it remained until it was shifted to the Department of Commerce in 1942. The **National Oceanic and Atmospheric Administration (NOAA)**[1] was established in 1970 to include a number of environmental agencies, including the National Weather Service (NWS), which reverted to its original name. The **Meteorological Service of Canada (MSC)**, located in Downsview, Ontario, assumes all forecasting duties for that country and provides local and regional information to its 14 **regional weather centres**.

Checkpoint

> 1. What factors complicate the prediction of temperature at a given point?
> 2. Why is the prediction of water vapor in the atmosphere so complicated?

---

[1]Employees of NOAA sometimes claim that the letters NOAA stand for National Organization for the Advancement of Acronyms.

Forecasting Methods

There is no single "correct" way to forecast the weather. Depending on the length of forecast, the type of information desired, and how much is known about the current state of the atmosphere, any one of a number of methods might be used. One can even attempt a forecast in the absence of any data about current weather, provided that long-term information is available. For instance, a forecast of hot, muggy conditions with a chance of afternoon thundershowers in Orlando, Florida, in mid-August has a reasonably good chance of proving correct. Such prognoses based on long-term averages are known as **climatological forecasts**. Obviously, the reliability of a climatological forecast depends on year-to-year variability in weather conditions for the forecast day. Thus, a forecast based on climatology might have a reasonable chance of being accurate in Orlando during the summer, but only the truly daring would try this for Chicago in April, when almost any kind of weather can occur.

Another type of forecast, called a **persistence forecast**, relies completely on current conditions with no reference to climatology. Actually, a special case of persistence forecasting is used by all of us in everyday life. When we see clear skies and leave the umbrella behind, we are betting that the prevailing conditions will continue and are making a short-term forecast on that basis. This simple procedure might work for a little while, but it will eventually fail to catch changes in weather. A more sophisticated version might assume that a decrease in pressure over the past few hours will continue through time and serve to predict the approach of a low-pressure system and its associated increase in cloud cover. Here, too, one extrapolates present behavior forward in time, again with certain knowledge that the forecast will fail when that behavior ceases. Of course, it is often precisely those breaks from past behavior we want to forecast; a method that can't provide such information is not terribly useful.

Until the 1950s, weather analysis and forecasting depended entirely on the experience of meteorologists and their interpretation of current conditions and the recent past. A meteorologist would base a forecast largely on the comparison of the current situation to similar conditions encountered previously (but not necessarily in the immediate past). This approach led to development and use of numerous "rules of thumb," which attempted to capture repeatable patterns and relations among various weather elements. For example, winter precipitation over the eastern United States and Canada tends to be snow north of the 5640 m contour for the 500 mb level and rain south of that curve. In this so-called **analog approach**, one tries to recognize similarities between current conditions and similar well-studied patterns from before. There are many variations of the analog approach, some subjective (depending on the forecaster's expertise), others objective (depending on statistical relations). All assume that what happened sometime in the past holds a clue about the future.

In the last few decades, **numerical weather forecasting** has come to occupy a dominant position. The term is somewhat misleading, because the methods just described all produce numerical forecasts. What is different is that this method is based on computer programs that attempt to mimic the actual behavior of the atmosphere. Numerical weather models explicitly compute the evolution of wind, pressure, temperature, and other elements over time. By examining the output for a given point in time, one obtains a depiction of the three-dimensional state of the atmosphere for that moment. (This is in contrast to forecasting the surface values for just a few weather elements, as might be done with other methods.) The numerical models typically used in weather forecasting are very large and can only be run on supercomputers. As reflects their importance in modern forecasting, these models are described in more detail in the section called "Forecast Procedures and Products" and in still more detail in the appendix to this chapter. For now, we simply want to draw a contrast between other methods and this physically based ("numerical") approach.

Types of Forecasts

The product, or result, of a forecast method can take a variety of forms, which we are calling the *type* of forecast. We are all familiar with quantitative forecasts, in which the "amount" of the forecast variable is specified. For example, a forecast that says "An inch of rain is expected" is a **quantitative forecast**. Similarly, forecasts of the expected high or low temperature are quantitative, because a value for the forecast variable is provided. In contrast, **qualitative forecasts** provide only a categorical value for the predicted variable. Examples of this are "rain/no rain," "hurricane/no hurricane," "above/below normal," or "cloudy/partly cloudy/ mostly clear." In these examples, the predicted variable is assigned to a particular class, or category; hence it is a qualitative forecast.

In the preceding examples, the forecasts were provided without qualification. **Probability forecasts** are an alternative, in which the chance of some event is stated. For example, a categorical hurricane forecast might be stated as a probability, rather than as a certainty. Probability forecasts can take a variety of forms, the most common of which is undoubtedly the probability-of-precipitation forecast (PoP forecast). When the broadcaster says "The rain chance today is 70 percent," or "There is a 60 percent chance of afternoon showers," he or she is reporting a PoP forecast. Note that these forecasts don't specify an amount of rain. Rather, a PoP forecast means that a randomly chosen point in the forecast area has the stated probability of receiving measurable precipitation. For example, a 75 percent PoP forecast means that the odds of precipitation are 3:1, or equivalently, you have only one chance in four (25 percent) of staying dry throughout the forecast period. Figure 13–4 shows a map of the probability of precipitation over the United States for a particular 6-hour time period.

▲ **FIGURE 13–4** An example of a probability of precipitation (PoP) map.

Checkpoint

1. What is a climatological forecast?
2. What are some advantages of numerical weather forecasting, as compared with other approaches to forecasting?

Assessing Forecasts

Regardless of which method is used, or what form it takes, we obviously need some way of deciding how good a forecast is likely to be. Measures are needed, for example, to compare one forecast method to another or to decide how much to consider a forecast when making plans. Most importantly, assessment measures are needed by those responsible for developing and administering forecasting programs. As attempts are continually made to improve data-gathering and forecast procedures (at an ever-increasing cost), methods are needed for judging the value of changes, justifying future expenditures, and determining the return on investment. Although as consumers we don't hear much about forecast assessment, it is a routine and integral part of professional forecasting.

Over the years, many evaluation measures and practices have been developed, each with particular advantages and disadvantages. To sort them out, we must first think about the purpose of the assessment. Do we want information about **forecast quality** or **forecast value**? The former refers to the agreement between forecasts and observations, whereas the latter refers to the usefulness of a forecast. These sound similar but are quite different. Because there is no simple relation between quality and value, different measures are needed for each. For example, a high-quality, 100 percent accurate forecast of rain might have zero value for scheduling crop irrigation, if knowing the total amount of rainfall is also essential. A second issue is the type of forecast: quantitative or qualitative, probability or unqualified, and so forth. Clearly, the appropriate assessment measure will vary with the type of forecast variable. Finally, do we want an absolute measure of performance or a relative, comparative measure?

Forecast *value* necessarily depends on applying a forecast to a particular problem or decision. Most measures of value are based on loss/payoff tables, which attempt to capture the risks and rewards associated with various forecasts and responses to those forecasts. For example, knowing the cost of a ruined concrete job, the money earned when things go right, and the probability of a correct forecast, you could assign monetary value to the forecast. The details involve probability concepts that are beyond an introductory text; therefore, we won't discuss forecast value any further, except to note that a single forecast can have widely varying value, depending on how it is used.

With regard to the *quality* of a forecast, an obvious question to ask concerns **forecast accuracy**. That is, on average, how close is the forecast value to the true value? There are many ways to answer this simple question, each leading to a different accuracy measure. At the broadest level, we might want information about **forecast bias**, which concerns systematic over- or underprediction. A biased forecast method is one whose average forecast is above or below the true average. An unbiased method, by contrast, shows no tendency for over- or underprediction. Of course, that is not to say the method is perfect—it simply means the average overprediction is as large as the average underprediction, yielding an average error of zero. For example, if you wanted to predict the number of spots showing after the roll of a die, you could forecast 3.5 dots on every roll. Over many rolls (of a fair die), the average would be 3.5, agreeing with your predictions—the predictions are therefore unbiased. Of course, it is impossible for this method to yield even a single correct prediction. It is clear that although bias is a useful measure, we also need accuracy measures that don't allow positive errors to cancel negative errors. The simplest is the *mean absolute error* (MAE), which ignores the sign (positive or negative) of the errors. That is, over- and underpredictions are treated the same, and we simply find the average error without regard to sign.

For laypeople using a forecast, accuracy is probably the main quality issue. But professionals who develop forecasting methods are more likely to report **forecast skill**. Skill can be measured in various ways, but the concept is defined as the *improvement* a method provides over what can be obtained using climatology, persistence, or some other "no-skill" standard. If the method is no better than relying on past climate, the skill score would be zero—a climatology forecast requires no special knowledge of atmospheric behavior and thus does not contain any skill. For example, no measurable rainfall has been recorded for July in Jerusalem for the last 100 years. Any "no-skill" method that forecasts "no rain" for July is certainly going to be accurate most of the time. Likewise, if you predict that a hurricane will be present somewhere in the Atlantic

Ocean next September 10, you have a 90 percent probability of success (using past behavior as a guide). But these forecasts have no skill, because they do not improve on "chance" (climatology in this case). Only if predictions were correct more than 90 percent of the time would we say that the forecast method possesses any skill. In the case of air temperature, we might compare the MAE of the forecast method to the MAE obtained using the climatological mean temperature.

Data Acquisition and Dissemination

The starting point for almost all weather forecasting is information about the current state of the atmosphere. To know the future, we begin with information about the present. Thus, the first process in operational weather forecasting is acquiring the necessary data. For reasons that will become clear, this requires an international effort, even when making forecasts for "small" areas, such as individual countries.

The **World Meteorological Organization (WMO)**, under the auspices of the United Nations, oversees the collection of weather data across the globe from its 179 member nations. The WMO collects data from about 10,000 land observation stations, 7000 ship stations, 300 moored and drifting buoys with automatic weather sensors, and several weather satellites. It also obtains upper-air data from about 1000 weather balloon sites twice a day and on a continuous basis from instruments aboard wide-bodied commercial aircraft. The data from all countries are sent to the three **World Meteorological Centers** at Washington, D.C.; Moscow, Russia; and Melbourne, Australia, which in turn disseminate the data to all member countries.

The member nations of the WMO maintain their own meteorological agencies that obtain and process the data and issue regional and national forecasts. In the United States, the **National Centers for Environmental Prediction (NCEP)** of the Weather Service performs these tasks, while in Canada they are handled by the **Canadian Meteorological Centre** of the **Atmospheric Environment Service (AES)**.

Not surprisingly, the United States has a relatively dense network of surface observation stations. Of the approximately 1000 locations where surface conditions are recorded, about 120 are National Weather Service offices; the rest are Federal Aviation Administration (FAA) airport sites. The Canadian AES operates about 270 surface stations. Together, these sites record temperature, humidity, pressure, cloud conditions (including type and height, and the percentage of sky obscured by cloud), wind direction and speed, visibility, the presence of significant weather, such as fog or rain, and accumulated precipitation readings at ground level. As part of an ongoing modernization program (see *Box 13–1, Special Interest: Modernization of the National Weather Service*), the FAA and NWS have installed a network of more than 800 automated sensors, called **Automated Surface Observing System**, or **ASOS**, for measuring and recording these variables (Figure 13–5, page 391). The AES operates more than 100 data

13–1 SPECIAL INTEREST

Modernization of the National Weather Service

For several decades the National Weather Service has dedicated considerable resources to improvements in its data acquisition, analysis, and dissemination, with considerable success. One of the first steps in the modernization process was improving computer hardware and refining numerical models. Advances in computer hardware have been staggering. Now the agency uses supercomputers that can process almost 70 trillion calculations each second—34 times faster than was possible a decade earlier (Figure 1).

GOES-1 and GOES-2 (short for Geostationary Orbiting Earth Satellite), the first in the current series of geostationary orbiting satellites, have been replaced by the currently operating GOES-11 and GOES-12, which provide excellent coverage over the western and eastern portions of North America, respectively (GOES-13 is in orbit as a backup if GOES-11 or GOES-12 needs replacement). The newest generation of GOES satellites offers superior resolution and obtains better temperature and moisture information for various levels of the atmosphere.

Forecasters have been using Doppler radar since the 1980s. A new version has begun implementation in the fall of 2011. Called polarimetric radar (or dual-polarization radars), these units transmit pulses of energy with both horizontal (i.e., nearly parallel to the surface) and vertical pulses that provide additional information on cloud characteristics, such as droplet size, the relative mixture of rain and snow, and precipitation rate.

Wind profilers obtain measurements of horizontal winds for up to 72 different levels, between 0.5 and 10 km (0.3 to 6 mi) above ground level. They work by simultaneously emitting three radar beams. One beam travels directly upward from the surface. The other two beams are directed 16 degrees away from the vertical, one oriented toward the north and the other toward the east. Changes in the wavelength of the backscattered radar beams occur in response to the movement of various backscattering agents (such as dust, bugs, and air molecules) and indicate the speed and direction of the wind. NOAA currently operates 35 profilers throughout the central United States. If the system proves valuable to

▲ **FIGURE 1** The National Weather Service runs some of the most powerful computers in the world.

forecasters, the network will expand to about 120 profilers.

Modernization has involved not only applying new technology but also streamlining operations. Formerly there were 52 Weather Service Forecast Offices (WSFOs) and 204 smaller Weather Service Offices (WSOs). Today, the daily operations take place at 119 **Weather Forecast Offices (WFOs)**, providing a significant savings to the public.

---

observation platforms across Canada similar to ASOS, called **Automated Weather Observation Systems (AWOS)**.

In addition to the surface observations made at these sites, the NWS launches hydrogen-filled balloons[2] carrying weather instrument packages called **radiosondes** (Figure 13–6). Twice a day—at 0000 and 1200 UTC (Universal Coordinated Time[3])—about 750 radiosondes are launched worldwide, about 80 within the United States and Canada. Radiosondes continually observe and transmit to ground recording stations the pressure, air temperature, and wet bulb temperature (from which dew point temperatures are determined). Some radiosondes are tracked by radar as they ascend through the atmosphere, which enables a determination of the wind speed and direction of the middle and upper atmosphere. Radiosondes tracked by radar are called **rawinsondes**.

Most radiosondes rise high into the stratosphere to about the 4 mb level—or about 30 km (19 mi)—whereupon

the balloon bursts and the package parachutes back to the surface. Usually a complete radiosonde ascent will take about 1 hour and 50 minutes, by which time the instrument package may have been carried many kilometers downwind from the launch site. Interestingly, radiosondes are often found by nonmeteorologists who happen to stumble across them. Many people who find them follow the instructions on the packages requesting that the package be dropped in a mailbox for return to the Weather Service, and the recovered radiosondes are then refurbished for subsequent reuse. Most of them are never recovered, however, landing in oceans or remote areas on land.

Upper-level information is obtained from other sources as well. Many commercial aircraft are equipped with weather sensors that automatically monitor the atmosphere throughout their flights, and weather satellites supplement the upper-level database by determining temperatures and humidities at several depths throughout the atmosphere. Together, the aircraft and satellites provide data from locations far away from any radiosonde sites and play a particularly important role in gathering information over the oceans. In addition, surface data are collected by sensors on buoys and relayed to land via satellites.

---

[2]Rather than using helium in the balloons, the Weather Service uses hydrogen, which it extracts locally, directly from the surrounding air.

[3]These times correspond to 1900 (7:00 P.M.) and 0700 (7 A.M.) Eastern Standard Time.

◄ **FIGURE 13-5** A typical ASOS unit.

Visibility sensor

Wind tower (tilting)

North

Precipitation
identification
sensor

Ceilometer

Freezing rain
sensor

Temp/dew point
sensor

Conduit

Rain
sensor

Gravel walkway

Forecast Procedures and Products

As mentioned earlier, numerical models are the preeminent tool of modern weather forecasting. Weather agencies around the world develop their own models and typically maintain a suite of models rather than a single program. For example, the National Weather Service currently uses three primary models. The models are updated continually,

and from time to time new models are introduced and older ones retired.

Phases in Numerical Modeling

Although there are large differences among models, the general procedure is the same for all numerical models. The three phases include analysis, prediction, and postprocessing.

◄ **FIGURE 13-6** A radiosonde launch outside the Topeka, Kansas, Weather Service Office.

Analysis Phase First is the **analysis phase**, in which observations are used to supply values corresponding to the starting ("current") state of the atmosphere for all the variables carried in the model. Remember, the models are three-dimensional, meaning that values are needed throughout the depth of atmosphere, not just at the surface. Moreover, some of the models cover the entire planet, and therefore require starting values everywhere, over land and ocean.

Unfortunately, the network of weather stations and radiosonde launches is highly irregular and does not come close to providing even coverage. Part of what analysis accomplishes is converting those irregular observations into "uniform" initial values. Though only a preparatory step, this is a difficult task. There are millions of data values from a variety of sources (satellite, ships, and so on), representing various moments in time. None of the measurements is completely free of error, and many are subject to large error. It is necessary to remove as much error as possible while at the same time producing fields[4] that are self-consistent. For example, when wind velocities are assigned, the resulting wind field must satisfy the conservation of mass.

Observed values also need to be consistent with the particular model being used. All the models are approximations to the real atmosphere, and it is important that the initial field not contain features that cannot be represented by the model. Otherwise, early in the forecast period the model will adjust to the mismatch between the initial conditions and what it perceives to be plausible. This adjustment will be superimposed on whatever changes arise from processes within the scope of the model and will possibly contaminate the forecast. The various models therefore rely on different analysis methods. For all of them, analysis is a complicated procedure involving the integration of the latest available observations, past values (such as snow cover), and output from other models.

Prediction Phase Fundamentally, the job of a numerical model is to solve the basic equations describing atmospheric behavior: the equation of motion, continuity equation, energy equation, and so on. Collectively, these are known as the *governing equations*. Beginning with values delivered by the analysis phase, the model uses the governing equations to obtain new values a few minutes into the future. The process is then repeated, using the output from the first step as input for the next set of calculations. This procedure is performed over and over as many times as necessary to reach the end of the forecast period (24 hours, 48 hours, or whatever). This is called the **prediction phase** of the model run. Again, we emphasize that huge computational resources are needed for this. The governing equations cannot be solved directly but must be broken down into simple operations that computers can perform, such as multiplication and addition. This results in many billions of calculations for each time step, despite the fact that there are just a handful of fundamental atmospheric variables (temperature, pressure, wind velocity vector, density, and moisture).

[4]A *field* refers to all the values of a given variable, such as wind velocity, within a defined area.

Postprocessing Phase The conditions forecast by the model at regular intervals (for example, every 12 hours) are represented on a grid for mapping and other display purposes. For example, in this **postprocessing phase**, a series of maps might be produced for each of the 12-, 24-, 36-, and 48-hour periods, depicting the forecast distributions of

1. Sea level pressure and 1000 to 500 mb thicknesses
2. 850 mb heights and temperature
3. 700 mb heights and vertical velocities (such as the speed at which air rises or sinks)
4. 500 mb heights and absolute vorticity values
5. Precipitation amounts

These products are used in various ways, some of which are described later in this chapter. Speaking generally, forecasters study maps for each period and interpret the conditions that would probably be associated with such patterns. They compare the maps with one another and with corresponding maps from other models for the same time periods. Of course, the model forecasts differ from one another—the forecaster uses model output as guidance, weighting the results differentially according to what is known about each model's strengths and weaknesses and supplementing model guidance with other analyses and observations.

Often the actual (final) forecast will not match any of the models exactly. For example, Figures 13–7a, b, and c show precipitation forecasts from three models for June 2, 1992. Although all three forecasts call for heavy rain in the Texas–Oklahoma area, the location and amount vary considerably from model to model. (Will central Oklahoma get 5 inches, or half an inch?) In Figure 13–7d, the forecaster produced a "manual" forecast (based on his or her subjective analysis) significantly different from any of the models, with the area of maximum rain displaced south and east of the models. Sadly, nature was perverse that day, confounding both machine and human—the heaviest rain was found in extreme southeastern Texas (Figure 13–7e). Despite this forecast's lack of success, you can see that the forecaster coupled model output with other information in producing official forecasts. Numerical models are certainly superior to purely subjective methods, but even better is a combination of model output and other information (including subjective judgment).

It is interesting that experience and "rules of thumb" have not been rendered obsolete by computer forecasts, but instead are applied to those forecasts in hopes of improving on them. It is also interesting that as the numerical models evolve and change for the better, old rules must be constantly reevaluated in light of new model behavior. In other words, model improvements allow us to update conventional wisdom.

In addition to gridded fields of model variables, forecasts for a number of secondary variables are produced. Examples include maximum and minimum temperature, dew point, wind conditions, and the probability of precipitation. These forecasts are constructed using statistical relationships between model output and observed surface conditions from the past. The output products are called *model output statis-*

◄ FIGURE 13-7 Twenty-four-hour precipitation forecasts from three numerical models (a–c), the final (manual) forecast (d), and observed rainfall (e) for June 2, 1992.

—2— Lines represent precipitation in inches

tics (MOS) and are designed to capture the effect of topography and other factors that influence local weather conditions. Numerical models have only a limited ability to represent processes occurring near the surface, and they provide a rather generalized picture of the atmosphere. Thus, a statistical approach has considerable appeal. MOS is most effective for the places from which statistical relations were derived and is somewhat less effective at intermediate locations having a different topographic setting.

How good are today's forecasts? There is no single answer to this: It depends very heavily on the variable in question, the forecast lead time, the model used, and the place and season. For example, there is no doubt that temperature, wind, and pressure distributions are forecast far better than precipitation. An example of precipitation skill is shown in Figure 13–8, which gives skill scores from several models for various amounts of precipitation for spring and summer of 1995.[5] Skill scores decline significantly with increasing precipitation amount, indicating the difficulty of forecasting heavy precipitation, which tends to be highly localized. Also evident are significant differences between models, although the skill rankings of the models are the same for all precipitation amounts. For example, model *A* always has the highest skill, whereas model *D* shows lowest skill at all precipitation levels. (This is not generally

the case—using other measures of skill, the models would change position relative to one another.)

There can also be significant seasonal variations in skill, as indicated in Figure 13–9, which shows 1-inch forecasts for a decade (1983–1993). Clearly, winter precipitation is predicted with much more skill than summer rain. This is partly due to seasonal changes in the precipitation processes. As we go from large, synoptic-scale systems in winter to smaller-scale

▲ FIGURE 13-8 Twenty-four-hour forecast skill (equitable threat score) for various NCEP models and precipitation amount. The details of the equitable threat score are not relevant to this discussion, but the score relates to the number of grid points correctly forecast correctly relative to what would likely occur by chance alone. Forecasts are for warm season only (March through August) of 1995.

---

[5]In this figure, skill is measured by the equitable threat score, described in the appendix to this chapter. Here we simply note that this score, like that shown in Figure 13–6, measures the area of correct forecasts. A method that perfectly forecasts rain and rainfree areas would obtain a skill score of unity, representing 100 percent skill.

Did You Know?

The National Weather Service has implemented a *National Digital Forecast Database* (NDFD). Unlike conventional forecasts that focus only on large, well-known sites (such as Indianapolis, Indiana) and give only general information about surrounding areas, the system sets up a grid and extrapolates weather forecasts for all individual points on the grid—even if they are not part of a large metropolitan area (for example, Beanblossom, Indiana, with a population of 2740). You can use this system by going to **www.weather.gov/ forecasts/graphical/sectors** and scrolling on a sequence of maps. (You may have to experiment a bit at first, but the results are well worth it.) More details on this emerging technology can be found at **www.weather.gov/ndfd**.

convection in the summer, it becomes more difficult to forecast the exact location of precipitation events. Second, winter precipitation tends to be much less intense, with storms lasting many hours. Finally in the winter it is more likely that there will be just a few large centers of precipitation associated with fronts or midlatitude cyclones. These might receive more attention from forecasters than numerous "pop-up" thunderstorms and other small disturbances that dot the summer landscape.

Advances in theory, computer technology, observing networks (especially satellites), and data integration have all contributed to steadily increasing forecast skill since the 1960s. Consider Figure 13–10(a), which shows the trend in skill scores for NCEP forecasts of 2.5cm (1 in.) rainfall events. Skill for 1-day forecasts has nearly doubled since 1965 and has

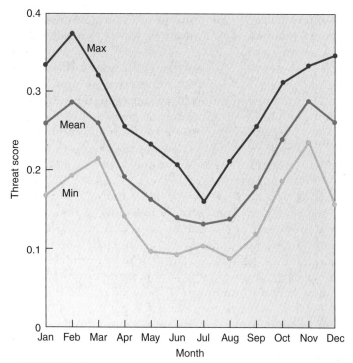

▲ **FIGURE 13–9** Twenty-four-hour, 1 in. (2.5 cm) NCEP precipitation forecast skill averaged over 1983–1993. Skill is for the manual forecast, not any particular model.

almost quadrupled over the same period for 2-day forecasts. Although the record is much shorter, skill for 3-day forecasts has been increasing even faster. Despite technological advances, there remains a large role for human judgment in forecasting. In fact, the forecasts underlying Figure 13–10(a) have a large human component as forecasters blend guidance from various models with their own experience when producing a forecast. Figure 13–10(b) exposes the role of humans by comparing actual forecasts with those that would be obtained from model guidance alone. Exactly how much skill a human adds to a forecast depends on the model used for comparison, but it is interesting to see that the gap has remained fairly constant over the last 15 years or so, with judgment contributing about 30% to forecast skill. Clearly, local knowledge (and experience regarding model failings) continues to be useful and adds significant value to purely objective forecasts.

Checkpoint

1. What are the three phases of numerical modeling?
2. Why do you think skill scores decrease for forecasts further into the future? Explain.

Medium-Range Forecasts

Going beyond "short-term" forecasts (72 hours or less), so-called **medium-range forecasts (MRFs)** are receiving considerable attention. For example, the **European Center for Medium-range Weather Forecasting (ECMWF)** has a model that generates forecasts for up to 7 days. In the United States, the Global Spectral Model (see this chapter's appendix) is used at NCEP to prepare 15-day forecasts. Although the models are tailored somewhat to MRF, the procedure is fundamentally the same as for short-term forecasting. That is, to the extent that data, theory, and computer resources will permit, the approach is numerical, based on known physical laws rather than statistical relations.

Rather than making just a single forecast, **ensemble forecasting** is widely employed, in which a number of different runs are performed for the same forecast period. The reasons have to do with something we mentioned earlier, that small disturbances can grow into large disturbances. If two model runs are made with slightly different initial values, the results might be very different after a week or so. Recognized in the early 1960s by E. N. Lorenz, this behavior is now known to be typical of many natural and human systems and is usually called *chaotic* behavior.

Chaos is a condition that occurs in physical systems that makes it impossible to precisely predict how a system (such as the state of the atmosphere) will appear some time in the future. Chaos arises because small errors in the input value of a variable (or even round-off error) can become magnified in time. This situation is referred to as sensitive dependence on initial conditions. It is a serious problem for weather forecasting because the initial conditions are never known exactly. For example, if an upper-level trough appears

(a)

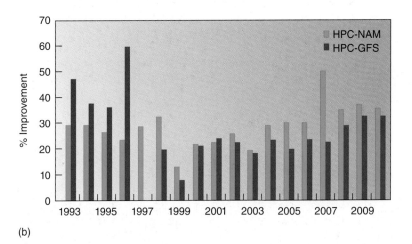

(b)

◄ **FIGURE 13-10** (a) Annual skill scores for 2.5 cm (1 in.) precipitation forecasts issued by the Hydrometeorological Prediction Center of NCEP. (b) Percent difference in skill between HPC forecasts in (a) and two numerical weather prediction models. HPC forecasts result from subjective forecaster judgments applied to a combination of model output and other data.

in the 15-day forecast, there is no way to know if it is "real" or if it arose because of errors in the initial data. Ensemble forecasting uses multiple runs starting with slightly different initial values. The different initial fields are created by introducing small changes (perturbations) in the best-guess field. There are several methods for assigning perturbations, but all attempt to mimic errors that might reasonably appear in the data. From this ensemble of initial conditions, an ensemble of forecasts is produced, each different.

Figure 13–11 shows a 10-day ensemble (sometimes referred to as a *spaghetti diagram*) for the Northern Hemisphere from NCEP. The map is a kind of 500 mb map, except that only one contour is shown: the 5700 m contour. Plotted are the 17 NCEP ensemble members, plus the unperturbed (control) run. Notice that all ensemble members call for a trough in the eastern Pacific Ocean. The implication is that whatever errors might exist in the data, they don't affect the forecast there. Thus, we might have some confidence that a real trough will develop. But over central Asia, western Europe, and the north Atlantic, ensemble spread is much larger: Ensemble members show much less consistency with one another, suggesting we should be less confident about the forecast for these regions. (Of course, we would not base the final assessment on a single contour, and we would look at other variables in addition to 500 mb height.)

Figure 13–11 illustrates the primary use of ensembles, which is to provide information about forecast uncertainty. With that, one has an estimate of reliability to go along with the forecast and won't pay much attention to forecasts deemed unreliable. Ensembles also can be used in other ways, including to generate the forecast itself. In particular, the mean of all ensembles can be treated as a forecast, even though it doesn't arise from a model run. Moreover, we might expect this forecast to be reasonably good, on the grounds that averaging will smooth away features found in just one or two members.

Yet another use of ensembles is to diagnose failings of a model. Suppose all the ensemble forecasts agree in some area, yet all depart from what is ultimately observed. The departure between the forecast and observation cannot be explained by data error because the ensemble suggests the forecast is insensitive to data error. This leaves model error as a likely culprit, meaning that one or more processes is poorly treated in the model and needs improvement. The ensemble doesn't identify the problem, of course, but exposes the situation in which the problem influences the forecast. By studying the details of that situation, it might be possible to learn which aspect of the model needs work.

At the present time, there is little skill evident in MRF output beyond a week or so, especially for precipitation. However, the MRF often does produce accurate (and valuable) forecasts

▶ **FIGURE 13–11** Ten-day ensemble forecast from the NCEP MRF model. The 5700 m contour is plotted for all 17 ensemble members. Each of the contours fall into somewhat different positions because of the effect that minor changes in the initial conditions of each model run exert on the model output.

500 mb heights, 5700 m contours
—— Ensemble runs
—— MRF run

and can provide useful guidance about general tendencies. In addition, it has been shown that ensemble spread correlates with forecast error for as long as 10 days, suggesting that estimates of forecast reliability also have value at medium range.

Long-Range Forecasts and Seasonal Outlooks

Forecasts at still longer lead times are called **long-range forecasts**. In the United States, the **Climate Prediction Center (CPC)** of NCEP is charged with preparing forecasts for periods ranging from a week to the limits of technical feasibility. The methods used include climatology, statistics, numerical models, and subjective judgment. For example, because of its role in the global climate system, sea surface temperature (SST) in the tropical Pacific is routinely forecast for up to a year in advance. The SST forecasts are based on a combination of three models:

1. An analog statistical model
2. A "canonical" statistical model based on correlations of temperature and precipitation patterns with prior sea surface and atmospheric conditions measured over space and time
3. A numerical model

In the numerical model, ocean and atmosphere are coupled so that the ocean responds to changes in the atmosphere, and vice versa. Output from the three models is combined statistically to yield the final forecast.

Another CPC product is the **seasonal outlook**, a kind of forecast for an entire season, often misunderstood. In contrast to long-range forecasts that predict conditions *for particular days*, seasonal outlooks predict *average conditions* for an entire season. Figure 13–12 shows the format, in which seasons are classed as "above normal," "near normal," or "below normal." "Above normal" includes the upper third of the distribution; it occurs about 33 percent of the time, when a variable is at the 66th percentile or higher. "Near normal" and "below normal" are defined similarly, as the middle and bottom thirds, respectively. Thus, without any special knowledge we could forecast "above normal" with 33 percent accuracy and likewise for "near normal" and "below normal." Suppose we believe "near normal" conditions are quite likely, say 45 percent likely. In that case we would issue a "near normal" forecast. But if we believe the chances are only 34 percent, we would probably refuse to make a forecast on the grounds that above or below normal conditions are just about as likely. Looking at the maps, you see that for most of the area, the probabilities are close to climatology

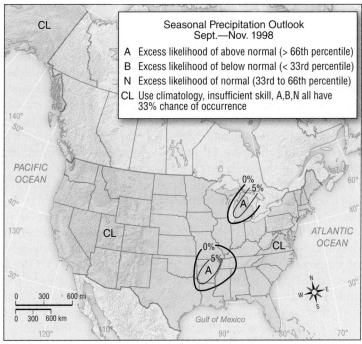

▲ **FIGURE 13-12** The seasonal outlook for September–November 1998, prepared 1 year earlier, September 15, 1997.

(labeled "CL"). There are only a few regions for which there is reason to think anomalous conditions will occur.

How are these forecasts obtained? Again, forecasters use a combination of methods, both numerical and statistical. For example, SST forecasts from the coupled ocean–atmosphere model are used as boundary conditions for repeated runs of an atmospheric general circulation model. Another set of runs is made using SST based on persistence rather than physical principles. The result, once again, is an ensemble of atmospheric

forecasts. The statistical techniques are mainly based on persistence in anomaly patterns. That is, departures from average are analyzed for stability and/or consistent patterns of evolution. Although the details are beyond an introductory text, the basic idea is not: The past is the key to the future.

Keep several things in mind when interpreting seasonal forecasts. First, they are forecasts for the entire period (season), not a particular day. One certainly does not expect above-average conditions throughout the forecast period, no matter what the "above average" probability might be. Also, at the present time, these forecasts do not exhibit much skill, and what skill exists varies by season, location, and the forecast variable (temperature vs. precipitation). Finally, even if the skill is relatively high (say 20 percent), the associated probability is not likely to be large. In fact, it is unusual to have probabilities much above 50 percent, even in areas of highest confidence.

We end this section by pointing out that although recent improvements in computer power and model sophistication are impressive, we are immeasurably far from being able to forecast the weather with 100 percent accuracy. In the first place, it is impossible to develop a model that perfectly simulates all natural processes at all spatial and time scales. This is partly a matter of insufficient knowledge—for some processes the underlying theory is not complete, so any computer program based on that theory will be imperfect (turbulence is a good example). But even with perfect understanding, it would not be possible to build a perfect model—the computational requirements are too great.

Earth has a circumference of about 40,000 (4×10^4) km, so this is the largest size a complete model needs to represent. What is the smallest feature necessary? For the sake of discussion, let's say that in a perfect model, cloud drops 4 microns (4 μm) in diameter must appear, but nothing smaller. That gives a range of features covering 13 orders of magnitude in size. Today's models span scales over roughly 3 orders of magnitude. To reach 4 orders of magnitude, about 5 km resolution, will require a computer a thousand times more powerful. Even that will leave us 9 orders of magnitude short of the goal, or a factor of 1 billion. Given this immense computational gulf, together with the lack of complete theory, it's hard to imagine that model error will ever disappear.

Did You Know?

In 1961, Edward Lorenz of MIT discovered that his primitive prediction model was infuriatingly sensitive to initial conditions. In other words, just like the real atmosphere, future states (predictions) were easily contaminated by seemingly insignificant differences in starting values. This became evident when he reran a primitive forecast model with input data slightly different from those of a previous run. To his surprise, the model output was very different from that which derived from the initial input. What he discovered was the concept of *sensitive dependence on initial conditions*, which tells us that over time, minor perturbations in model input can compound to produce very different outputs. This factor inevitably defeats any attempt to numerically forecast weather beyond a certain time period—currently believed to be about 14 days.

But model error is only part of the problem; chaos is the other major difficulty. Earlier we briefly mentioned dynamic instability, in which disturbances grow in size over time. With that there is necessarily a propagation of energy upward from smaller scales to larger scales. The growth of small updrafts into cumulus towers is another example of upward energy propagation, and indeed, such events are routine in the atmosphere.

What chaos means for numerical weather forecasting is that one must have detailed and accurate initial values to make a reliable forecast. Even a small error might have dire consequences: A large feature fails to materialize in the forecast, or perhaps a fictitious feature unlike anything in the real atmosphere appears. This extreme sensitivity to initial conditions is characteristic of chaotic systems in general and the atmosphere in particular. Unless we are prepared to begin with perfect data, there is no hope for perfect prediction. Note that the problem of chaos is independent of any defects in the model, such as processes ignored or improperly treated. Even if we could build a perfect model, chaos rules out the possibility of a perfect forecast.

Checkpoint

1. What is ensemble forecasting and how is it used?
2. Which type of forecasting does the production of a seasonal outlook most resemble: climatological forecasting, persistence forecasting, or analog forecasting? Explain.
3. Apart from the limitations of data collection and computing power, what is one reason that weather forecasting will never be 100 percent accurate? Explain.

Weather Maps and Images

Although computers play a critical role in weather analysis, ultimately the meteorologist applies his or her knowledge to produce the forecast that is issued to the general public. Probably no tool is as valuable to a forecaster as a weather map. Newspapers and television news segments often just show surface maps, but weather forecasting requires analyzing conditions in the middle and upper atmosphere as well. Not only do clouds exist well above the surface, but the middle and upper atmosphere are closely intertwined with the air near the surface. As a result, accurate weather analysis requires using a series of maps representing different layers of the atmosphere. In this section we review resources commonly used in general weather forecasting. See *Box 13–2, Focus on Aviation: Special Forecasts and Observations for Pilots* to see some resources of particular interest to pilots.

TUTORIAL

FORECASTING

Use the tutorial to review weather map basics, then follow the instructions to make your own forecasts and compare them to what actually happened.

Surface Maps

Surface maps of prevailing conditions (such as that shown in Figure 13–13) present a general depiction of sea level pressure distribution and the location of frontal boundaries. The pressure is shown by isobars drawn every 4 mb, with zones of locally highest and lowest pressure labeled *H* and *L,* respectively.

Large-Scale Features Even a nonprofessional can make a number of inferences from briefly inspecting a surface weather map. General wind speeds and directions obey the rules discussed in Chapter 4. That is, wind speed varies according to the spacing of the isobars, and Northern Hemisphere winds rotate clockwise out of high-pressure systems and counterclockwise into lows. High-pressure systems favor downward vertical motions that promote clear skies, whereas low pressure promotes updrafts that lead to adiabatic cooling and formation of cloud cover. Surface maps become even more valuable when viewed in sequence. Because new maps are compiled every 3 hours, one can easily track the movement of individual weather systems as they move cross-country. By assuming that the systems will continue to behave as they have in the past few hours, we can infer their movement and intensification or dissipation over the next few hours.

Station Models More detailed knowledge of the conditions at particular locations can be obtained from **station models**. Well over a dozen weather elements, including temperature, dew point, and pressure, are represented on each station model. A complete station model contains some information beyond the needs of most readers, so Figure 13–14 on page XX describes only the most important symbols. A more complete description of the surface station model, along with the basics of the 500 mb station model, is presented in Appendix C.

The surface station model indicates cloud coverage by the amount of shading inside the central circle. A completely open circle indicates cloud-free conditions, a fully darkened one represents complete overcast, and intermediate amounts of shading correspond to varying fractions of cloud cover. A line extending from the circle with tick marks or flags at the end is a wind "arrow," or "barb," showing the wind speed and direction. The free end of the arrow corresponds to the direction that the wind is blowing *from.* Thus, for instance, an arrow at the top of the circle indicates winds from the north, whereas one on the right side of the circle means that the wind is from the east. The wind speed is represented by the number of full or half tick-marks, and/or flags, as represented in Figure 13–14. An arrow having one complete tick-mark and one half tick-mark, for example, has winds between 15 and 20 mph (23 to 32 km/hr).

Temperature and dew point values (in degrees Fahrenheit) are listed to the top left and bottom left of the circle, respectively. Symbols representing common weather conditions (such as rain or fog) are located between the two.

The sea level pressure (in millibars) is given in a shorthand manner to the top right of the central circle. To convert

13-2 FOCUS ON AVIATION

Special Forecasts and Observations for Pilots

The National Weather Service offers comprehensive weather information and forecasts for aviators that is readily accessible from the Aviation Weather Center (AWC), Aviation Digital Data Service (ADDS). These data are readily accessible at www.aviationweather. gov/adds (Figure 1). Some of the more commonly used resources are described below. In some cases information is presented using abbreviations and codes. For a complete description of the codes and conventions used in these reports, you can download at no cost a copy of *Aviation Weather Services, Advisory Circular 00-45G Change* 1 (the URL is a long one, so it's easiest to access if you do a Web search for the document and click on the link).

Aviation Routine Weather Reports (METARs)

These reports are issued by airports on an hourly basis and provide basic weather observations. METARs present observations in a string of "groups," as illustrated in Figure 2. Each airport is represented by a 4-digit identifier, followed by the time of observation and a modifier report describing the data source. This coding system used is somewhat archaic but usually easy to decipher; at other times it is less obvious. But this problem is now remedied by the option to click on a "translation" of the data that presents the information in real words. The Web site also has an interactive Java tool that allows the user to scroll over a map and have the data immediately displayed.

▲ **FIGURE 1** The home page for the Aviation Weather Center's Aviation Digital Data Service.

Aviation Selected Special Weather Reports (SPECIs)

Though METARs are issued every hour, weather conditions can sometimes change dramatically between reports to a level that might be important for aviators. When this happens, the AWC issues Aviation Selected Special Weather Reports. These are formatted the same way that METARS are and are differentiated from them by the use of SPECI as the type of report preceding the station identifier.

Pilot Weather Reports (PIREPS)

Weather data from airports help pilots prepare for takeoffs and landings, but aviators also need information on flight conditions along their flight route. This type of information is provided to the Aviation Weather Center by pilots themselves and then disseminated to the rest of the aviation community. PIREPs report on icing, turbulence, wind, or other conditions at designated flight levels. As with METARs and SPECIs, the Web site offers decoded versions of the reports and a Java tool for easy access.

▲ **FIGURE 2** An example of a traditional METAR, with a brief description of the data in each group.

Significant Meteorological Information (SIGMETs)

SIGMETs alert pilots to existing or expected weather conditions that could impact aircraft safety. The threats could involve severe icing, heavy turbulence, sandstorms or dust storms, or airborne volcanic ash. These are issued by regional Meteorological Watch Offices (MWOs) and remain valid for a periods of up to 4 hours (they can be reissued an unlimited number of times). Most SIGMETs are classified as nonconvective; convective SIGMETs are issued when thunderstorms may lead to extreme turbulence, icing, or low-level wind shear.

Short-Range Surface Prognostic Charts (PROG) and Low-Level, Mid-Level, and High-Level Significant Weather (SIGWX) Charts

You have seen numerous examples of surface and upper-air maps in this and previous chapters of this book, all of which are useful to pilots. In addition to these, the AWC compiles specialized maps for the aviation community—both for current conditions and forecasts up to 48 hours. These maps are tailored for the needs of pilots and highlight weather conditions of special importance to aviators.

---

the three-digit number to the true pressure, you first assume a decimal point before the last digit and place a number 9 or 10 at the beginning of the number. Thus, 997 represents 999.7 mb, while 104 corresponds to 1010.4 mb. How do you know whether to add a leading 9 or a 10? A simple answer, which nearly always works, is to add the number that yields a value closest to 1000 mb (in other words, add 9 if the coded value is more than 500 or add 10 if it is less than 500).

The change in pressure (pressure tendency) over the last 3 hours is shown just to the right of the circle. Again, it is necessary to assume a decimal point before the last digit so that

▲ **FIGURE 13-13** A typical surface weather map, in this case for June 10, 1997.

−10 indicates a pressure drop of 1.0 mb. A symbol to the right of the number indicates in a qualitative sense how the pressure has been changing (for example, initially rising but then falling).

Upper-Level Maps

Twice a day, at 0000 and 1200 UTC, the National Centers for Environmental Prediction disseminate maps of the observed 850, 700, 500, 300, and 200 mb levels. Forecast maps are also produced for those levels for a variety of lead times. Whether the maps represent current or predicted conditions, each provides its own unique combination of advantages for the weather analyst.

850 mb Maps The **850 mb level** resides at an average height of 1.5 km (1 mi) above sea level. Because friction is often considered to be negligible at heights of about 1.5 km above the surface, gradient or geostrophic flow can exist at this level over terrain with elevations near sea level. On the other hand, much of the Rocky Mountain region of the western United States and Canada has elevations above this height, so in those areas the 850 mb map actually represents near-surface conditions. At high elevations, friction retards the wind and the air flows somewhat across the height contours. Figure 13–15 on page 402 shows a typical 850 mb map, in this case for June 10, 1997. (Figures 13–13, 15, 18, 19, and 21 all represent the same observation day and time.)

Heights of the 850 mb level are plotted with solid lines, analogous to the isobars found on surface maps.

Contours are spaced at 30 m intervals and labeled in units of decameters (10 m). Thus, a value of 150 represents a height of 1500 m. Though not specifically drawn, frontal boundaries are distinguishable at the 850 mb level, where the height contours are packed closely together. Air temperatures (in °C) at the 850 mb level (and all higher levels) are plotted with dashed lines.

Figure 13–16 on page 402 shows a hypothetical pattern of 850 mb heights and temperatures over an assumed nonmountainous region. With this assumption, the 850 mb level is far enough above the surface for gradient flow, and the wind blows parallel to the height contours. As the air approaches the trough axis, it moves from a region below 10 °C to a region above 15 °C. Conversely, as the air flows away from the axis, it moves toward a colder region. Thus, the region upwind of the trough axis is marked by cold air advection, while warm air advection occurs downwind of the axis. These patterns are important because warm air advection at the 850 mb level indicates upward motion, which favors the formation of clouds and precipitation. Thus, in this example uplift is probably downwind of the trough, making the region favorable for cyclone development or intensification. Notice the zone of warm air advection in Figure 13–17a (page 403) from Georgia to southern Ohio. This region experienced significant rainfall, as shown in the radar image in Figure 13–17b.

The distribution of temperatures at the 850 mb level provides forecasters with some useful rules of thumb. During the morning, for example, the 850 mb temperature often

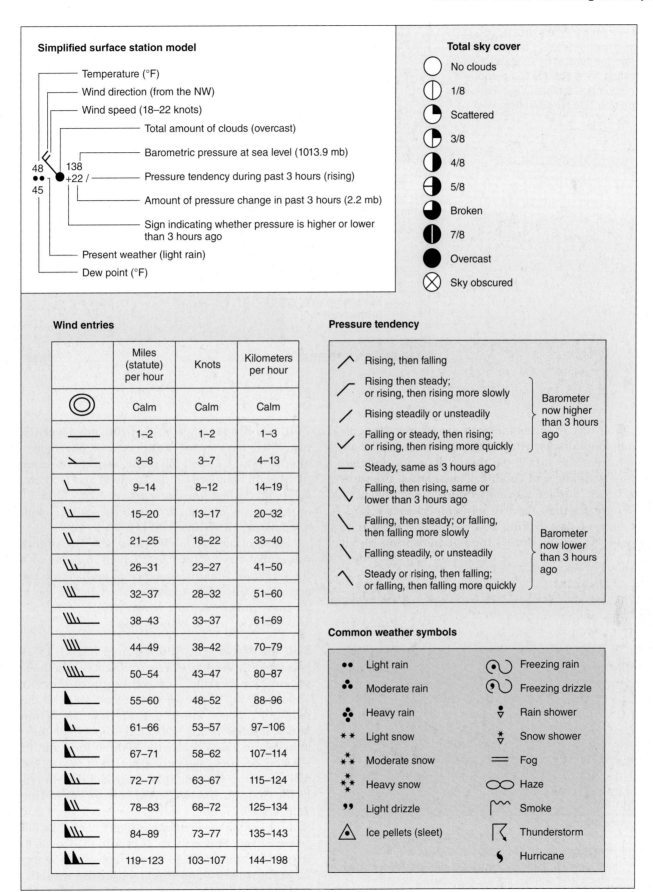

▲ **FIGURE 13-14** The arrangement of a surface station model and some important symbols.

▶ **FIGURE 13–15** An 850 mb map for June 10, 1997. The solid lines represent the altitude of the 850 mb level in tens of meters above sea level. The dashed lines plot the temperature at that level. These maps are useful for identifying areas of warm and cold air advection.

850 MB ANALYSIS
HEIGHTS/TEMPERATURE
AT 7:00 A.M., E.S.T.
JUN. 10 1997

Altitude of 850 mb level
(in tens of meters)

Temperature at 850 mb
(in °C)

provides a good way to guess the daily maximum temperature over nonmountainous areas. At the 850 mb level, the air is far enough from the surface so that it does not undergo daily cycles of warming and cooling. Thus, during the summer, the maximum surface air temperature is usually about 15 °C (27 °F) greater than the 850 mb temperature, regardless of the time of day. During the winter, the difference will be about 9 °C (16 °F); during the fall and spring, about 12 °C (22 °F).

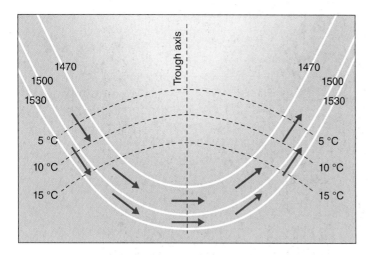

▲ **FIGURE 13–16** Map view showing warm and cold air advection at the 850 mb level. Outside of mountainous regions, air at the 850 mb level flows parallel to height contours. If these lines cross temperature contours (the dashed lines), warm air advection occurs where the air flows toward colder regions, and cold air advection happens where the air flows toward warmer areas. Warm air advection at the 850 mb level favors cloud development.

700 mb Maps Maps of the **700 mb level** (Figure 13–18, page 404) have many of the same applications that the 850 mb maps do. Like 850 mb maps, maps of observed conditions plot height contours (in decameters) and isotherms (°C), with solid and dashed lines, respectively. The 700 mb level occurs in the vicinity of 3 km (2 mi) above sea level and has a mean temperature of about –5 °C (23 °F).

Maps of the 700 mb level are best for observing the short waves that were shown in Chapter 10 to be so important in the formation and maintenance of midlatitude cyclones. They are also particularly valuable in predicting the movement of air mass thunderstorms, which usually move with about the same velocity as the 700 mb winds.

500 mb Maps The **500 mb map** (Figure 13–19, page 404) is commonly used to represent conditions in the middle atmosphere. The globally averaged height of the 500 mb level is about 5.6 km (18,000 ft) above sea level, and the mean temperature is about –20 °C (–4 °F). Because the mean pressure at sea level is nearly 1000 mb, about half the mass of the atmosphere exists below the 500 mb level and half above.

For decades meteorologists have made special use of 500 mb maps. For example, there is a certain type of pattern at the 500 mb level called an *omega high*. This feature, so named for its resemblance to the Greek letter Ω, often signals that the upper-level pattern is likely to change only slowly for several days. This pattern is clearly evident in Figure 13–20 (page 405), with a single omega high covering the central part of North America and troughs over the western and eastern regions. The height contours of the 500 mb maps are spaced at 60 m intervals instead of the 30 m intervals used for the 850 and 700 mb maps.

(a)

(b)

▲ **FIGURE 13-17** The 850 mb map for 4 P.M., EST, March 4, 2008 in (a) shows warm air advection in a band from Georgia to southern Ohio. The precipitation occurring along that band is shown in the radar image (b).

▶ **FIGURE 13–18** A 700 mb map for June 10, 1997; 700 mb maps are often used to observe short waves and to predict the movement of air mass thunderstorms.

300 and 200 mb Maps Lying near the tropopause, the **300 mb** (approximately 9 km above sea level and having mean temperatures of about –45 °C, or –50 °F) and **200 mb levels** (12 km; –55 °C, or –65 °F) have the strongest jet streams. During the colder months, the 300 mb map works best for identifying the jet stream; during the summer, the 200 mb map is best. (In this section, all references to the 300 mb level pertain equally to the 200 mb level.) Rossby waves show up best on the 300 mb maps, which makes the charts useful for determining the rate at which the waves are likely

to migrate downwind (or in rare instances, upwind). (See Figure 13–21.)

In addition to height contours (at 120 m intervals) and isotherms, 300 mb maps also plot **isotachs**, which are lines of equal wind speed. These are drawn at intervals of 20 knots, beginning with the 10-knot isotach. Areas where the wind speed is between 70 and 110 knots are indicated by shading. Superimposed areas having winds between 110 and 150 knots are unshaded, and zones having winds above 150 knots are again shaded. This makes it easy to note the regions of

▶ **FIGURE 13–19** A 500 mb map for June 10, 1997. The 500 mb level marks the midpoint of the atmosphere, with about half its mass below this level and the other half above that level.

◄ **FIGURE 13-20** An omega high. This pattern often remains fixed for days at a time, suggesting little change in weather condition for several days.

Satellite Images

increasing and decreasing jet streams. This is important because air flowing into or out of local areas of high wind speeds (called *jets,* or *jet streaks*) generates local regions of upper-level convergence and divergence, as shown in Figure 13–22.

Anybody who watches the weather segment of any news broadcast is certainly familiar with radar and satellite imagery. **Visible images** (Figure 13–23a) view the atmosphere the way an astronaut in space would, simply by registering the intensity of reflected shortwave radiation. Obviously, these images are available only during the daytime.

Infrared images (Figure 13–23b) are based on measurements of longwave radiation *emitted* (not reflected) from below. If dense clouds are present, the source of the radiation will be the cloud top; otherwise the surface and

Checkpoint

1. Draw a station model, showing where the temperature, dew point, wind speed and direction, cloud conditions, and sea level pressure are indicated.
2. How are maps of the 850 mb level used? The 700 mb level? The 500 mb level?

◄ **FIGURE 13-21** A 300 mb map for June 10, 1997. The purple dashed lines are isotachs contoured at 20-knot intervals. Shaded areas have winds in excess of 70 knots. Open areas within the shaded regions have winds greater than 110 knots. These maps are best for identifying Rossby waves in the atmosphere.

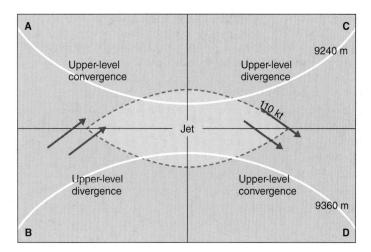

▲ **FIGURE 13–22** Map view of a jet streak. Air flowing into and out of jet streaks (areas of locally fastest winds) creates patterns of upper-level convergence and divergence. Air entering the jet streak sets up convergence along its left flank (quadrant A), and divergence on its right (B). Exiting the jet streak, there is divergence to the left (quadrant C) and convergence to the right (D).

lower atmosphere supply most of the upwelling radiation. Cumuliform clouds result from condensation associated with the adiabatic cooling of rising air. A deep cloud formed by air parcels rising great distances will therefore have a lower cloud-top temperature than will a mid-level cloud formed by just a kilometer or two of ascent. It follows that a satellite sensor will receive less radiation from a tall cloud than a low cloud and less from a low cloud than from a cloud-free region. Satellite images show the colder (higher) cloud top as whiter than warmer (low) clouds. Clear areas are very dark on these images.

Water vapor images (Figure 13–23c) provide a unique and often beautiful perspective on the atmosphere. Water vapor is a very effective radiator at wavelengths near 6.7 μm. Relative humidities above about 50 percent result in a high output of radiation in this part of the spectrum, and the sensors translate high values of this radiation into bright regions on the imagery. Water vapor images are particularly useful for tracking the flow of moisture across wide regions and helping identify the location of frontal boundaries.

Radar Images

Radar images generated by conventional systems have been extremely valuable for routine weather analysis for decades. Radar observes the internal cloud conditions by measuring the amount of radiation backscattered by precipitation (both liquid and solid). A transmitter sends out brief pulses of electromagnetic energy with wavelengths on the order of several centimeters. A receiver records the intensity of the echoed pulses (indicative of the number and size of droplets and crystals) and the time elapsed between pulse transmission and return (which indicates the distance to the scattering agent).

(a)

(b)

(c)

▲ **FIGURE 13–23** A visible (a), infrared (b), and water vapor (c) image obtained from the GOES-8 satellite. Visible images are based on sunlight reflected off cloud tops, and can therefore only be obtained for regions having daytime conditions. Infrared images are based on the intensity and wavelengths of radiation actually emitted by clouds. Water vapor images show whiter areas where either substantial cloud cover or high water vapor contents exist.

The radar continually emits these pulses as the transmitter/receiver rotates 360°, giving a two-dimensional representation of the cloud conditions surrounding the unit. After each rotation, the transmitter angle is increased slightly, and the radar scans a higher slice of the atmosphere. This procedure is repeated until a large volume of the surrounding atmosphere has been scanned, and the meteorologist can observe the distribution of the clouds, the heights of their tops and bases, and the relative intensity of precipitation. The information is then displayed in color-coded map form. A sequence of these images can be put together into a loop that shows the movement and changes in weather activity.

Each radar unit covers distances up to about 400 km (250 mi). The Weather Service assembles the returns from all the radar sites in the national network and compiles the information onto maps of the 48 contiguous United States and southern Canada (Figure 13–24).

While it is relatively easy to determine the big picture of local activity from a radar image, a meteorologist has to beware of potential errors in its use. For instance, the curvature of Earth's surface causes a horizontally emitted beam of radiation to assume greater heights above the surface as it moves far from the transmitter. Also, radar waves are refracted (bent) somewhat as they travel through the atmosphere. The extent to which the refraction occurs depends on the stability. Thus, the meteorologist not accounting for this effect can obtain false readings of cloud-top height.

Checkpoint

1. Which type of satellite imagery would be most useful in detecting differences in the altitude of cloud tops?
2. How is a radar composite map produced and how could it be useful in forecasting?

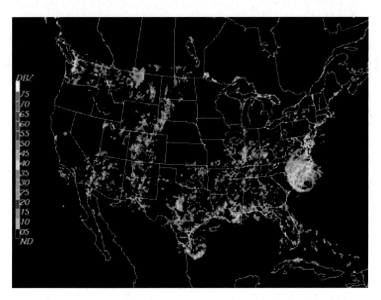

▲ **FIGURE 13-24** A radar composite map. These maps assemble data obtained from regional radar units across the country and combine the information onto a single map.

Thermodynamic Diagrams

The maps and images previously described provide two-dimensional views of atmospheric conditions, but they fail to provide detailed vertical information. Vertical profiles of temperature and dew point data observed by radiosondes are plotted on **thermodynamic diagrams** (also called *pseudo-adiabatic charts*). The simplest thermodynamic diagram is the Stuve chart (Figure 13–25), on which the air temperature is scaled along the horizontal axis and the pressure on a nearly logarithmic vertical axis. The straight, solid lines that slant upward to the left are called *dry adiabats.* These show the rate of temperature change for an unsaturated parcel of air that is lifted or lowered (in other words, they plot the dry adiabatic lapse rate). The dashed, slightly curved lines are called *wet adiabats,* showing temperature changes experienced by a rising saturated parcel.

Soundings are plotted on the charts by marking the temperature and dew point data at numerous pressure levels and connecting the dots. Figure 13–26 shows an actual sounding for the Dallas–Ft. Worth airport on the morning of June 4, 1997. The heavy lines represent the temperature and dew point profiles (the temperature profile is the one on the right—which must be the case because the temperature is always equal to or greater than the dew point). Notice that the two profile lines nearly merge between about 800 and 660 mb, indicating that the layer is saturated and cloud covered. Throughout the rest of the atmosphere, the temperature is much higher than the dew point and the air is unsaturated.

The steepness of the temperature profiles relative to the dry and wet adiabats indicates the static stability for any portion of the atmosphere. Where the temperature decreases more rapidly with height than does the DALR (as in profile

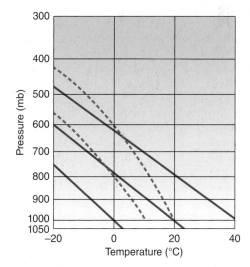

▲ **FIGURE 13-25** A Stuve thermodynamic diagram. The vertical lines across the horizontal axis represent temperature (°C); the horizontal lines, pressure. The solid and dashed lines sloping to the left with height are dry and saturated adiabats, respectively.

► **FIGURE 13-26** An example of a sounding on a Stuve diagram. The heavy line on the right plots the temperature; the one on the left shows the dew point profile. Wind barbs on the right of the diagram depict wind velocities at various levels.

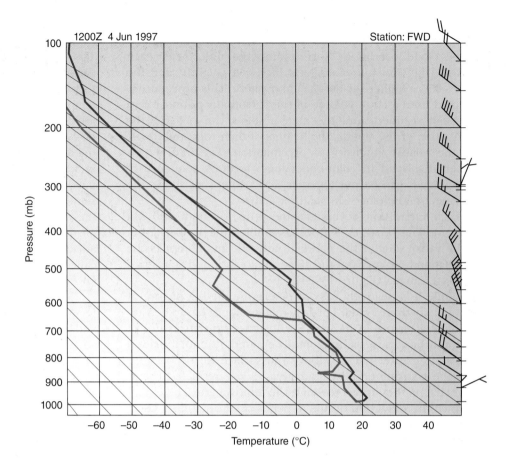

A in Figure 13–27), the air is absolutely unstable. Where the temperature drops less rapidly with height than adjacent wet adiabats (profile B), the air is absolutely stable. Temperature lapse rates intermediate between the two adiabatic lapse rates (profile C) indicate conditionally unstable conditions, as we discussed in Chapter 5.

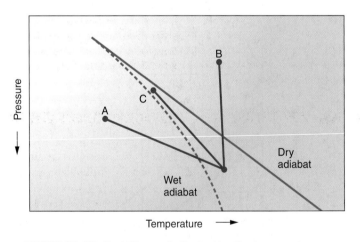

▲ **FIGURE 13-27** Stability, as indicated by the temperature profile on a Stuve diagram. Profile A has a more rapid drop in temperature with height than the dry adiabat, so the air is absolutely unstable. Profile B is less rapid drop in temperature than the wet adiabat, so the air is absolutely stable. Profiles with slopes intermediate between the dry and wet adiabats (such as C) indicate conditionally unstable air.

Displaying data on wind speed and direction at numerous heights (and thus the presence or absence of wind shear), thermodynamic diagrams provide a wealth of information to meteorologists, especially in predicting severe weather. Given the complexity of these diagrams, it is not surprising that a number of measures have been invented that combine the various elements related to severe weather into a single numerical value, or index. In the past these values were determined manually with the aid of thermodynamic diagrams; today they are calculated by computers. The lifted index and the K-index are two such summary measures.

Lifted Index

The **lifted index** was developed in the mid-1950s. It combines the average humidity in the lowest kilometer of the atmosphere, the predicted maximum temperature for the day, and the temperature at the 500 mb level into a single number. The magnitude and sign of the values together indicate the potential for thunderstorms; negative values indicate sufficient water vapor and instability to trigger thunderstorms. More specifically, lifted index values between –2 and –6 indicate a high potential for thunderstorms, whereas values lower than –6 suggest a threat of severe thunderstorms.

K-Index

The **K-index** is similar to the lifted index but works better for predicting air mass thunderstorms and heavy rain than

Television Weather Segments

Television weather broadcasters are often local or even national celebrities. In many cases people choose a particular news broadcast each morning or evening based on the person doing the weather segment. So who exactly are these people and how do they do their jobs? Some weather anchors have little formal training in the subject, but many others have professional backgrounds in meteorology. These backgrounds range from a few college courses to formal degrees in the subject.

Weather anchors ultimately base their forecasts on the same information used by National Weather Service meteorologists, but the level of external support they use varies. Some of them depend entirely on forecasts issued by the NWS; others rely on forecasts provided by private weather companies; and some with strong meteorological training rely solely on their own analyses.

Regardless of the level of meteorological training, weather broadcasters can use an impressive array of graphics. The weather anchors will often use specialized software that allows them to set up a sequence of text pages, maps, and satellite and radar loops. These images and loops appear to viewers to be behind the meteorologist, but in fact he or she is standing in front of a blank screen (Figure 1). So how does the presenter manage to point to spots on a weather image that are not really shown on

▲ **FIGURE 1** Television meteorologist Christina Russo doing a live, on-air weather segment.

the screen? This is done by looking at one of several strategically placed television monitors. Whichever direction the weather anchor is facing, he or she can see a monitor.

Since 1957 the American Meteorological Society has given the AMS Seal of Approval to television meteorologists meeting established criteria. Holders of the seal must have a certain level of scientific understanding, as demonstrated by the completion of at least 20 semester units of meteorology and the submission of three examples of their on-air work. Since its inception, more than 1400 broadcasters have been awarded the seal of approval.

In January 2005 the AMS initiated the Certified Broadcast Meteorologist Program (CBM). This program requires completing a college or university degree in meteorology or atmospheric science from an accredited institution, passing a written test, and submitting samples of televised work that are judged for their methodological soundness and the broadcaster's communication skills. In 2008 the AMS stopped issuing new seals of approval and began granting only the CBM certification. As of July 2011 there were 497 television and radio meteorologists holding the CBM certification.

---

for severe weather. The index uses values of temperature and dew point at the surface and the 850, 700, and 500 mb levels. Various rules of thumb for different geographic regions and times of year translate K-values to the probability of heavy rains and thunderstorms. In general, K-values less than 15 indicate no potential for thunderstorms; values above 40 suggest that they are highly likely.

The National Centers for Environmental Prediction compile maps of K-index and lifted index values across the United States and Canada each day, giving forecasters a quick reference for the possibility of thunderstorms. Though the indices are very useful as a first step in predicting thunderstorms, a meteorologist would never use them alone

in making a forecast. Rather, the indices help identify regions in which the forecaster should make more detailed analyses.

So far this chapter has given an overview of how government forecasters go about their tasks. For a look at how television meteorologists put their segments together, read *Box 13–3, Special Interest: Television Weather Segments.*

Checkpoint

1. What data are shown on a thermodynamic diagram?
2. In Figure 13–25, at what pressure levels would you expect to find clouds? Explain.

Summary

Weather analysis and forecasting depend on the cooperative efforts of nations belonging to the World Meteorological Organization. The weather services of these nations obtain surface data from land and sea-based stations, radiosondes, and satellites. For decades, weather forecasting involved subjective analysis of the available data. Although today's meteorologists still rely heavily on their own interpretation of current weather patterns, they also make wide use of output from numerical models. There is a wide variety of these models, and they undergo constant evolution, but all attempt to forecast the atmosphere by applying known physical laws. Short-term forecasts (up to 2 days) have improved considerably in recent years in response to improved data acquisition and processing, as well as model improvements. Medium-range forecasts offer some degree of skill out to about a week. On the other hand, long-term forecasts for 30- and 90-day periods have not displayed a high degree of accuracy or skill.

While information is available in many different forms, the weather map continues to be one of the most important tools for forecasting. The NWS updates current surface maps for the contiguous 48 states and southern Canada every 3 hours, and forecast maps are published at 12-hour increments. Maps of observed conditions plot the distribution of pressure with isobars and the location of fronts with heavy lines. Station models provide detailed information on temperature, dew point, pressure, wind, and other conditions at large number of stations.

Surface weather maps provide only part of the story and must be augmented by maps of atmospheric conditions at higher levels. In the United States, maps are routinely produced for the standard levels: the 850, 700, 500, 300, and 200 mb pressure levels. Each of these has its own set of advantages. For example, at the 850 mb level, analysis of air flow relative to the temperature distribution helps identify areas of low-level uplift. The 700 mb map has many of the same uses as the 850 mb map. In addition, it is particularly useful in predicting the movement of air mass thunderstorms. Of all the upper-atmosphere maps, 500 mb charts are probably the most widely used. Maps of the 300 and 200 mb levels are best for defining the number and position of Rossby waves and for analyzing jet stream patterns.

While weather maps provide two-dimensional views of the atmosphere at a number of levels, they do not provide much information on the vertical structure of the atmosphere at particular locations. This is where thermodynamic charts make a contribution. Profiles of temperature and dew point identify moist and dry layers within the atmosphere along with the location of clouds. The steepness of temperature profiles relative to dry and wet adiabats also indicate the stability of the atmosphere. Over the years, meteorologists have formulated several numerical indices that condense the information from these diagrams into single numerical values. Values above or below certain critical numbers indicate the need for further analysis of the possibility of heavy precipitation or severe weather.

The material in this chapter has relied heavily on the principles discussed in the previous 12 chapters and has applied these principles to an activity of great interest to society, forecasting. The next chapter looks at another human aspect of meteorology—the effects of human activity on weather.

Key Terms

National Weather Service (NWS) *page 382*

Meteorological Service of Canada (MSC) *page 382*

Advanced Weather Interactive Processing System (AWIPS) *page 382*

zone forecast *page 383*

National Oceanic and Atmospheric Administration (NOAA) *page 387*

Meteorological Service of Canada (MSC) *page 387*

regional weather centre *page 387*

climatological forecasts *page 387*

persistence forecasts *page 387*

analog approach *page 387*

numerical weather forecasting *page 388*

quantitative forecasts *page 388*

qualitative forecasts *page 388*

probability forecasts *page 388*

forecast quality *page 388*

forecast value *page 388*

forecast accuracy *page 389*

forecast bias *page 389*

forecast skill *page 389*

World Meteorological Organization (WMO) *page 389*

World Meteorological Centers *page 389*

National Centers for Environmental

Prediction (NCEP) *page 389*

Canadian Meteorological Centre *page 389*

Atmospheric Environment Service (AES) *page 389*

Automated Weather Observation Systems (AWOS) *page 390*

Automated Surface Observing System (ASOS) *page 389*

wind profilers *page 390*

Review Questions

1. Briefly describe some of the variables that complicate weather forecasting.

2. Describe the basic characteristics of climatological forecasts, persistence forecasts, the analog approach, and numerical forecasting.

3. What are the distinguishing characteristics of quantitative, qualitative, and probability forecasts?

4. Explain how weather data are obtained and disseminated to agencies across the globe.

5. What are radiosondes and rawinsondes? What other sources of upper-atmosphere information are available to forecasters?

6. Describe the analysis, prediction, and postprocessing phases in numerical forecasting.

7. What are model output statistics?

8. What are the primary characteristics of short-range, medium-range, and long-range forecasts? What types of information are needed to prepare the individual forecasts?

9. What is ensemble forecasting?

10. Describe the station model used for surface weather maps. How is the information presented on the station model? What measures must be used to convert the numerical data on the station model to real values?

11. Describe the characteristics of the 850 mb, 700 mb, 500 mb, 300 mb, and 200 mb maps that make each of them useful to forecasting.

12. What is the significance of cold air advection? Which weather map is most useful for locating it?

13. Which upper-level weather map would you use to locate short waves?

14. Why are omega highs significant? Which map is best for identifying them?

15. What is an isotach?

16. Which maps are most useful for locating the polar jet stream?

17. Describe the three types of satellite images discussed in this chapter. What characteristics make them useful?

18. Describe how radar works and how its information is presented.

19. Explain what a thermodynamic diagram does and how it is constructed.

20. Describe the lifted index and the K-index. How are they valuable to forecasters?

Critical Thinking

1. Why is it that the climatological, persistence, and analog approaches will never be entirely eliminated from the process of weather prediction?

2. If a forecast calls for a 70 percent chance of rain and no precipitation occurs, was the forecast actually wrong? What if this happens 2 days in a row? Ten days?

3. What would have to happen to the data acquisition network for the analysis phase of forecasting to be bypassed?

4. Are further improvements in weather forecasting more likely to occur for large-scale phenomena or for smaller-scale events? Explain your answer.

Problems and Exercises

1. Go to the Web site www.nws.noaa.gov/organization .php. (weatheroffice.ec.gc.ca/canada_e.html in Canada), and click on the National Weather Service office nearest to you. Make note of the 24-hour forecast and follow up the next day to see if the forecast was correct. Do this for extended forecasts as well. In general, do you find the forecasts to be accurate?

2. During times of unusual or inclement weather, visit the Web sites listed below to obtain weather-map, satellite, radar, and thermodynamic-diagram information. Is the weather you are experiencing consistent with what you would have expected, based on the information you obtained?

3. On a daily basis, make use of the available information from the Web sites listed below and make your own forecast (before reading the official forecast for your area). Then compare your forecast to that of the local weather office. Are your forecasts generally consistent with those of professional meteorologists?

4. Read the forecast discussion on the Web page of your local weather service office each day. These discussions explain the meteorologists' reasons for making their particular forecasts. Determine how this elaborates on the zone forecast for the same area.

Quantitative Problems

Weather forecasting is now strongly based on numerical techniques. This chapter is followed by an appendix that discusses some detailed aspects of numerical forecasting.

The book's Web site, www.MyMeteorologyLab.com, offers several quantitative problems to help you better understand some nuances of numerical techniques.

Useful Web Sites

ww2010.atmos.uiuc.edu/(Gh)/guides/mtr/fcst/home.rxml
An excellent primer on forecasting from the University of Illinois.

www.crh.noaa.gov/lmk/soo
Excellent information on forecasting tools available to meteorologists, courtesy of the Louisville, Kentucky, office of NWS. Includes pages describing AWIPS, Doppler radar, and numerous training modules.

meted.ucar.edu/dl_courses/nwp
Tutorials on using numerical weather prediction.

www.theweatherprediction.com/habyhints/index.html
An extensive array of forecasting tips and tidbits.

www.ssec.wisc.edu/data
Excellent introduction to the use of satellite imagery in forecasting. Use the panel on the left side of the screen for navigation.

www.wrh.noaa.gov/wrh/forecastoffice_tab.php
Provides links to local NWS forecast offices and various support centers.

weatheroffice.ec.gc.ca/canada_e.html
A great first stop for forecast information in Canada.

weather.uwyo.edu
Contains just about everything you need to make your own forecasts.

MyMeteorologyLab™

The *Understanding Weather and Climate,* Sixth Edition, MyMeteorologyLab™ web site contains numerous multimedia resources to aid in your study of **Weather Forecasting and Analysis**. Visit **www.MyMeteorologyLab.com** to:

- **Review** key chapter concepts.
- **Read** the Pearson eText, *In the News RSS feeds,* and other chapter-specific Web resources.
- **Visualize** challenging topics with Interactive Tutorials, Geoscience Animations, MapMaster interactive maps, and Weather in Motion movies and visualizations, and more.
- **Test Yourself** with self-study quizzes.

 TUTORIAL
FORECASTING

Use the interactive animations and quizzes in the tutorial to review this chapter's key concepts.

WEATHER IN MOTION

View movies and visualizations of meteorology patterns and processes.

Modeling the Atmosphere on Your Desktop
Forecasting Precipitation
Uncertainty in Numerical Models
Forecasting Upper-Level Winds
Forecasting Thunderstorms

APPENDIX

Numerical Forecast Models

Over the last four decades, weather forecasters have relied on several generations of computers and a variety of different models for guidance. As computers have increased in speed and capacity, the models have become increasingly complex, always straining the limits of computer power in an effort to achieve greater realism and accuracy. Even with today's computers, numerous compromises and approximations are necessary. In fact, it might be said that the spectrum of models arises not so much from differences in purpose or theory, but rather from differing approaches to the basic problems of abstraction and simplification. The result is tremendous variety in both the details and gross features of numerical models. In this appendix we first discuss the major features of numerical models, using the primary NCEP operational models as examples. We then describe some methods for assessing forecast quality.

Model Characteristics

Today there are three numerical forecast models (ignoring variants): the **Global Spectral Model**, the **Global Forecast System (GFS)**, and the **North American Mesoscale (NAM) Model** (formerly called the eta model).

Scale Considerations

Thinking first about gross features, perhaps the most fundamental difference among models concerns the model **domain**, the region of the globe to be represented. Of course, if one wants to forecast for the entire globe, no decision is necessary—the computational domain must be the entire globe. But what if the goal is a European or a North American forecast? Where should the edges be? The boundaries of the domain require special treatment, usually in the form of strong assumptions about mass and energy transfer. To minimize their effect, one wants as large a domain as possible, with boundaries far outside the forecast region. Just how far outside depends in part on the forecast lead time. As lead time increases, locations farther removed from the forecast region begin to affect the forecast. Here, then, is a classic trade-off. A larger domain is preferable but requires more computer resources, which means some other aspect of the model must be compromised. Of the three NCEP models mentioned, only the Global Spectral Model has a global domain; the others have domains centered on the United States and cover only part of the Northern Hemisphere.

Another issue is spatial resolution. The fundamental equations governing atmospheric behavior are continuous in space, meaning that they describe the evolution of the atmosphere everywhere. If the governing equations were solved directly, they would yield a solution for an infinite number of points (we could find forecast values at every location within the domain). But the equations are far too complex to solve directly by analytical means; no such solution exists. Instead, the equations are written and solved in approximate form, with the result that forecast values are available only at widely spaced locations.

The approximation is accomplished in a number of different ways, but the result is always the same: There is some minimum size below which explicit representation is impossible. Current models are limited to a few tens of kilometers and larger, so, for example, nothing as small as an individual thunderstorm cloud can possibly appear. If such "subscale" phenomena are to be considered, thcy can only be expressed in terms of features that *are* resolved, an error-prone process called *parameterization*. A cloud parameterization would typically use simple relations that compute ice and liquid water fractions on the basis of relative humidity and air temperature, without modeling the details of cloud growth. Obviously, one wants high resolution so that small-scale processes and phenomena can be modeled and appear in the forecast, but this can come only at the cost of more computation. (Roughly speaking, doubling the resolution leads to eight times the computation.) Increases in resolution can be subsidized by reducing the domain, but that creates its own problems, as previously described.

The resolution issue applies to vertical coordinates as well as horizontal coordinates. Modelers face difficult questions about how many vertical levels to include, as well as how they should be arranged (at what altitudes). As a point of comparison, the horizontal resolution of the Global Spectral Model is about 1° in latitude and longitude (about 60 mi). It has 28 levels in the vertical, ranging from the surface to the 2.7 mb level. These are far from equally spaced—there are eight levels below 800 mb, with increasing spacing at lower pressures (higher altitudes). Operational versions of the NAM model have been run with increasing resolution since its inception, ranging from 80 km initially to 29 km (17 mi) at the time of this writing. There are now 50 levels in the vertical. The GFS has 64 layers and two grids. The smaller inner grid, centered over the United States and Canada, has a resolution of about 70 km (42 mi). It lies completely within a larger, coarser outer grid that extends the domain to much of the Northern Hemisphere. The advantage of this scheme is that most of the computing is confined to the region of interest—computer resources are not wasted on detailed features far outside the forecast area. But at the same time, those areas are not ignored, as they would be with a smaller domain.

Horizontal Representation

Yet another major difference among models is the horizontal representation. Many models adopt a **grid representation**, in which the domain is subdivided into a lattice of grid points. (The grids at various levels need not coincide, nor need they be the same for all variables.) The equations are solved only at the grid points. As we've said, the finer the grid, the higher the model's resolution. Implicit in this is the idea that the grid captures horizontal variation in the atmosphere and that intermediate values can be inferred knowing values at nearby grid nodes.

The alternative is called a **spectral representation**. (Can you guess which representation is used in the Global Spectral Model?) Variables are represented as a series of "waves" in space, each having a characteristic wavelength. For example, there are waves in the model corresponding to Rossby waves, repeating just a few times around a parallel of latitude. Superimposed on these are other waves representing smaller-scale and larger-scale variations. To obtain the value for a particular variable at a point, its various wave functions are summed over all the wavelengths (the Global Spectral Model uses 126).

Spectral models have some advantages, especially for a global domain (the North and South Poles are easily handled in a spectral representation). There are other quite technical advantages as well, which we will not discuss. Instead we will make a few general points about spectral modes. First, horizontal resolution is determined by the smallest of the "harmonics" present (smallest wavelength), so there is no escaping the problem of how to represent subscale processes. Second, not all of the variables can be represented in spectral terms. Only quantities subject to advection, such as heat and moisture, are treated this way. Other variables, such as radiation, must be computed on a point-by-point basis. Thus, the Global Spectral Model uses a grid for these "physical" quantities, and there is constant transformation of information from spectral to grid representation. (Fortunately, there are fast computational methods for this.) Finally, the spectral representation applies only to the horizontal—spectral models are layered in the vertical.

Physical Processes

Numerical models also differ greatly in their "physics," which basically includes all processes other than those related to motion ("dynamics"). A model's physics package includes purely atmospheric processes (such as condensation), atmosphere–surface interactions (such as friction between the atmosphere and ground), and purely surface processes (such as soil moisture or depth of snow). Consider, for example, just some of the physical processes included in the Global Spectral Model:

- **Radiation:** Shortwave absorption and scattering in three wavelength bands (including the effects of ozone, water vapor, and carbon dioxide), with multiple reflections between clouds and ground. Effects of clouds, water vapor, CO_2, and O_3 on longwave absorption and emission are explicitly modeled, including overlap in absorption bands. Radiative properties of clouds depend on cloud thickness, temperature, and moisture content.

- **Convection, Clouds, and Precipitation:** Stratiform (nonconvective) clouds, as might be found with fronts and tropical disturbances, are determined from relative humidity using a statistical relationship. Convective clouds are either precipitating (deep convection) or nonprecipitating (shallow convection). Temperature and moisture profiles are used to determine which occurs. Downdrafts and evaporation of precipitation are simulated, as are entrainment and detrainment of updraft and downdrafts. Precipitation arises from both deep convection and from large-scale condensation (when the air at a point becomes saturated, regardless of convection). Precipitation is evaporated into unsaturated air below the cloud; only that which survives the descent is deposited on the surface.

- **Surface Properties and Processes:** Sea surface temperature and sea-ice distributions are fixed for the model run. Sea-ice temperature is computed from heat exchange between the atmosphere and ocean below the ice. Surface albedo depends on zenith angle, snow cover, and vegetation type. Snow cover is determined by accumulation from falling snow, snow melt, and sublimation. Land surface evaporation consists of evaporation from the ground and plant canopy, as well as transpiration by the canopy. Precipitation not intercepted by the canopy is partitioned into soil moisture recharge and runoff. (Melting snow also contributes to soil moisture.)

Notice that many of the preceding are subscale processes; thus, the physics package uses parameterization heavily. Different models not only include different processes but also employ different parameterizations for the same processes. Sometimes this is by necessity rather than preference, as parameterizations appropriate at one scale are not necessarily useful at another scale.

As we mentioned, there is a continual evolution of models (and no doubt many of the specifics in this appendix will have changed by the time you read this). For example, when it was introduced in 1973, the Global Spectral Model was the state of the art at NCEP. Today, the NAM model is generally considered to occupy that position and is itself evolving. Is this effort at improving models justified—has forecasting skill improved? It definitely has, as we will see later in this appendix.

The future of numerical prediction calls for ever more realistic models. For example, NCEP plans a 5 km/150-level NAM-like model in the near future. This is a daunting task: Such a model will require 1000 times the computing power and 100 times the memory of today's model. Even a 5 km model will not permit explicit calculation of convection and other extremely important processes. Estimates suggest that this will require about 10,000 times the power of today's supercomputers. Whether or not this is possible depends on both the public's willingness to pay and the development of computer technologies not yet on the drawing board.

Measures of Forecast Accuracy and Skill

For some variables, such as temperature or precipitation amount, accuracy measures, such as bias and mean absolute error (MAE), practically suggest themselves. Bias is most easily defined as the difference between the average forecast value and the average observed value. It reveals any tendency for the method to give forecasts values above or below the true value (for example, too hot on average, too much rain on average). A mathematically equivalent definition of bias is simply the average error. For each forecast, we find the departure from observed and compute the average of those errors.

Figure 1 illustrates how bias can be used as a means to assess the accuracy of models. The figure plots the bias for 1 in. (2.5 cm) precipitation events predicted by NCEP for the last few decades. Considering the Day 1 forecasts, the figure shows that in the early 1960s and again in the mid-1970s errors were large, as forecasters overpredicted rain areas (as indicated by values < 1. Otherwise, from 1980 onward, there is remarkably little improvement in bias over the entire period. Bias for Day 2 forecasts are slightly below unity from about 1980 onward, indicating a slight tendency for underprediction.

Does the absence of bias mean a forecast method is perfect? Certainly not—it is always possible for overprediction (positive errors) to nearly balance underprediction (negative errors). As a matter of fact, most statistical forecast methods are totally unbiased yet give far from perfect forecasts. To avoid this problem, one can use MAE, defined as the average

◄ **FIGURE 1** Changes in accuracy (bias) for 1 in. (2.5 cm) precipitation forecasts for Day 1 and Day 2. Day 1 forecasts are for the period 12 to 36 hours ahead, whereas Day 2 represents the period 36 to 60 hours ahead. A bias value of unity occurs when the size of the forecast and observed precipitation areas are equal.

of the absolute errors. Because absolute values are always positive, there is no cancellation of errors. MAE therefore provides information about how far an individual forecast is likely to be from true value, without regard to sign (positive or negative). If the MAE is 2 °C, we expect an individual forecast to be 2 °C away from the true value.

Note that in MAE we averaged the absolute values to remove the sign of each error. We could accomplish the same thing in another way: by squaring each error before taking the average. This is done during the calculation of the **root-mean-square error (RMSE)**, which has some statistical advantages over MAE and is therefore more common. Obviously, as we sum only positive values, there is no cancellation of error. The root-mean-square error is just the square root of average squared error. Like MAE, it has the same units as the forecast variable (°C, millimeters of snow, and so on) and provides an estimate of the error expected for a single forecast.

For a qualitative variable, the most common accuracy measure is the proportion of correct forecasts, known as the **hit rate**. In the case of a binary variable (just two classes), we would ask, of all the forecasts made, whether "yes" or "no," what percentage turned out correct? For example, suppose rain/no rain forecasts have been made for a 100 km² area, with the results shown in Table 1.

The same data are shown in map form in Figure 2a. Looking at either the table or the map, we see rain was correctly forecast for 10 percent of the region, and no rain was correctly forecast for 40 percent of the area, giving a hit rate of 0.5 (50 percent).

Somewhat similar is the **probability of detection (PoD)**, which is the proportion of occurrences that were correctly forecast. The difference is that nonevents are excluded in the PoD. Thus, for example, in Figure 2a, the PoD is 10/30 = 0.33 (one-third of the area receiving rain was correctly forecast). Because one can maximize the PoD by always forecasting "yes," the PoD is often accompanied by the **false alarm rate**, the proportion of "yes" forecasts that were wrong. In Figure 2a, 75 percent of the rain forecasts are false alarms (30/40).

Bias can also be computed for a binary variable; in this case, bias is simply the ratio of forecast to observed occurrences. Here, however, a perfect method has a bias value of 1. Values larger than unity imply a tendency to overpredict, whereas values less than 1 suggest underprediction. For our example (Figure 2a), the rain forecast area is too large by one-third: the bias is 1.33 = (40/30). As before, a perfect bias value does not indicate a perfect method. For example, if the rain forecast area were sliced down to 20 km², the size of forecast and observed areas would agree perfectly, even though the forecast would be wrong for a large part of the region.

TABLE 1

Rain Forecasting Accuracy

	Rain Observed	No Rain Observed	Total
Rain Forecast	10 km² (Correct)	30 km² (Error)	40 km²
No Rain Forecast	20 km² (Error)	40 km² (Correct)	60 km²
Total	30 km²	70 km²	100 km² Total Area

(a)

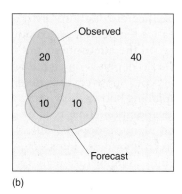

(b)

▲ **FIGURE 2** Hypothetical distributions of observed and forecasted precipitation. Numbers in single ovals refer to percentages of times that precipitation was observed but not forecast, or vice versa. Numbers contained in both ovals indicate the percentage of times precipitation occurred in an area forecasted to have precipitation. Numbers outside the ovals are for areas in which precipitation was not observed or forecasted.

Another widely used measure for binary variables is the **threat score**. This measure is similar to PoD in that it is concerned only with occurrences. However, it tries to penalize false alarms as well as missed occurrences (PoD responds only to the latter). Looking at Figure 2a, you can see the 10 km² area of correctly forecast occurrences is flanked by large areas of false alarm as well as missed occurrences. Given the size of those areas, the 33 percent PoD seems a rather inflated measure of success. The threat score accounts for this, and is given by

$$TS = \frac{\text{Correct}}{\text{Forecast} + \text{Observed} - \text{Correct}}$$

where the variables on the right-hand side are the area (or number of places) correct, area forecast, and area observed. The term *threat* is used because the denominator is the area "threatened" by an occurrence—either forecast, observed, or both. For Figure 2a, the threat score is 0.17 = 10/(40 + 30 − 10), half as large as PoD for the same map. Note that a perfect forecast will have a threat score of unity, because the forecast and observed areas will coincide perfectly.

Figure 3 illustrates the improvement in NCEP forecast model performance by plotting annual threat scores for 1 in. or more of precipitation. Unlike bias, threat scores over time have shown considerable improvement, reflecting better performance for the numerical models for both 1- and 2-day forecasts. Some of the variability is weather-related (wet years have higher scores), but significant upturns can be tied to new models coming online. Particularly striking is a narrowing of the gap between Day 1 and Day 2 forecasts, implying that Day 2 forecasts have improved more. In fact, by the early 1990s the 2-day forecasts were about as good as the 1-day forecasts of the early 1960s. In that sense, predictive ability doubled over the period. Looking at changes in the 2-day score alone, there is a threefold improvement.

It is a little more complicated to measure accuracy for probability forecasts. As we mentioned in the text, the most common probability forecast is a probability-of-precipitation, or PoP forecast. The PoP refers to the chance that a random place in the forecast area will receive 0.01 in. or more of precipitation in the forecast period. Equivalently, considering just a single location, a 60 percent chance of rain means that out of 10 forecast days, 6 will be rainy. A forecast might say, for instance, "the rain chance is 70 percent." How can we assess the accuracy of this statement? The forecast itself means that there is a 70 percent probability of rain for a randomly chosen location in the forecast area. Obviously, there is no way to assign accuracy on the basis of just a single precipitation measurement. (Failing to observe rain on a single day does not mean the forecast was wrong.) But if we were to count the number of times rain is recorded on many of these "70 percent" days, we could see if the actual frequency is above or below the forecast frequency, and thereby assess the method's accuracy. To fully verify the forecast method, we would compare all the forecast classes, not just the 70 percent class. In effect, we compare the forecast probability distribution with the observed probability distribution.

As we mentioned, forecast skill refers to the improvement provided by a forecast over and above some reference accuracy. The reference, or "no-skill," system is often taken to be persistence, or alternatively, climatology. But these aren't the only choices for the reference system; we could compare one numerical model with another, or with purely random values. For example, the threat score is often adjusted by the number of hits expected by random assignment of forecasts. In Figure 2a, rain was observed over 30 percent of the area. If we were

to randomly forecast rain for various places, we would expect considerable success simply because so much of the area is wet. (If you forecast a single point by throwing a dart at the diagram, you have a 30 percent chance of being correct—assuming you can hit the diagram.) As the rain forecast area grows with no change in observed rain, the threat score increases. To account for this, we first find the expected area of correct forecasts, assuming purely random forecasting. This is given by

$$EC = \text{Forecast} \left(\frac{\text{Observed}}{\text{Total}} \right)$$

The term in parentheses is the proportion of the total area where rain fell—multiplying this by the forecast area gives the area expected to be correctly forecast. With this, the so-called **equitable threat score** is

$$ETS = \frac{\text{Correct} - EC}{\text{Forecast} + \text{Observed} - \text{Correct} - EC}$$

For Figure 2a, $EC = 40(30/100)$, or 12 km^2. That is, knowing nothing about the atmosphere, we would expect rain to be correctly forecast for 12 km^2, given that we are forecasting rain to occur over 40 km^2. For this example, ETS is $(10 - 12)/(40 + 30 - 10 - 12)$, or -0.04, a negative number! Does this make sense? Well, as our forecast method generated fewer successes than would occur by chance, it certainly seems reasonable to treat this as "negative" skill.

Figure 2b shows a similar situation, but with a smaller rain forecast area. The PoD is exactly the same as before (0.33), and the threat score is only slightly larger (0.25 vs. 0.17). However, where in Figure 2a we saw negative skill compared to chance, here the equitable threat score is 0.12. Although there are no more successful rain forecasts than before, this situation has fewer false rain forecasts, and therefore shows positive skill.

We see that there is no single measure of forecast quality. Rather, various measures provide different kinds of information, so each has its own uses. The other side of this is that whether or not a forecast model shows improvement over time, or whether it is better than some other competing model, depends in part on the way we measure accuracy and skill. Careful assessment always requires that we compare a number of measures.

14 Human Effects on the Atmosphere

After reading this chapter, you should be able to:

▶ Describe the different types of atmospheric pollutants.

▶ Explain the role of atmospheric conditions in determining pollution concentrations.

▶ Describe the causes and effects of photochemical smog.

▶ Explain the factors that produce urban heat islands and describe their effects on local weather.

With 20 million people and 4.2 million cars, it is little wonder that Mexico City is one of the smoggiest places on Earth. Surrounded by mountains that confine the polluted air and subject to frequent temperature inversions that inhibit vertical dispersion of pollutants, Mexico City has all the right ingredients for a serious smog problem—and some consider this the smoggiest city in the world. Much of the problem relates to a large number of motor vehicles with less than state-of-the-art emission control devices, persistent inversions that trap emissions, and high levels of solar radiation that allow chemical reactions that convert emissions to a particular type of pollution, called *photochemical smog* (discussed later in this chapter). But the situation is more complex due to factors unrelated to human activities within the city. For example, in the spring of 1998 a rash of forest fires in southern Mexico brought huge amounts of smoke into the city. To make matters worse, Popocatepetl volcano, 50 km (30 mi) southeast of the city, spewed tons of smoke and ash into the air. Wind transported the pollution south to Honduras and north all the way to Florida and Texas, where people were advised to stay indoors to mitigate health hazards.

The situation can be perilous indeed. A 2007 study revealed that smog irreversibly reduces the growth of lungs in children in Mexico City. But Mexico City also provides a lesson in how air pollution can be drastically reversed if a serious commitment is made to doing so. The government has mandated that lead be removed from gasoline; old, high-emitting vehicles have been taken off the road; and public transportation has been improved. The result is a 50 percent or more reduction or more in most categories of pollutants—some by 70 percent or more between 1990 and 2010! In 1999 smog levels exceeded acceptable standards on all but 10 days of the year. In 2009 about half the days had air quality officially deemed as unhealthful.

◄ Mexico City has made great progress in the fight against air pollution but still suffers from many episodes of unhealthful air.

Air pollution is not the only way people affect the atmosphere. The effects of human activities are not restricted to air quality issues. We change the atmosphere in more subtle ways as well. For example, the construction of cities influences the way energy and water are exchanged near the surface. Every time a subdivision is laid out, natural soil and vegetation are replaced by concrete or asphalt. This greatly reduces the amount of water that can evaporate into the air and thereby increases the sensible heat flux (Chapter 3) to the atmosphere. We also build structures with vertical walls that receive sunlight at a more direct angle than the original absorbing surface. These processes work to increase the temperature of urban areas relative to their rural counterparts, creating the heat islands that we describe later in the chapter.

Atmospheric Pollutants

Nowhere is the air entirely pristine. Small suspended solids and liquids (called *particulates*) enter the atmosphere from natural and human sources. Likewise, many gases that are considered pollutants also arise naturally from processes such as lightning-induced forest fires and volcanic eruptions. Nonetheless, natural dilution and removal of these gases and particulates makes them relatively unimportant to the air quality experienced by most people. More important are the effects of human activity, especially in and around urban and industrial centers where anthropogenic emissions are concentrated into much smaller areas. In this chapter all references to **air pollution** will concern the introduction of undesirable gases and particulates by humans. The varying sources of particulates and other pollutants in the United States and their relative concentrations are shown in Figure 14–1. (We present some background on major pollution episodes of the past in *Box 14–1, Focus on the Environment: Severe Pollution Episodes.*)

In the most general sense, pollutants can be divided into two categories. Some, called **primary pollutants**, are emitted directly into the atmosphere. Others, called **secondary pollutants**, do not go directly into the atmosphere but result from one or more chemical transformations. Thus, a chemical emitted into the atmosphere may be innocuous in its original state but becomes a noxious gas or particulate after combining with other emissions or naturally occurring compounds. Several primary and secondary pollutants figure most prominently in the degradation of air quality.

Particulates

Particulates (also called *aerosols*) are solid and liquid materials in the air that are of natural or anthropogenic (human-made) origin. Though always small, particulates come in a wide range of sizes ranging from about 0.1 to 100 μm. Some of the particulates are primary pollutants put directly into the atmosphere, while others are secondary pollutants formed by the transformation of preexisting gases or from the growth of smaller particulates into larger ones by coagulation.

Sources of Particulates Particulates introduced directly into the air can originate from natural fires, volcanic eruptions, the ejection of salt crystals by breaking ocean waves—and as any sufferer of hay fever can tell you, by the entrainment of pollen by wind. Human activities, especially those involving combustion, produce primary and secondary particulates.

Some secondary particulates form by the coagulation of gases. This process is most rapid when the humidity is high, which creates an interesting situation. Recall from Chapter 5 that water droplets in nature always form on condensation nuclei—with large, hygroscopic aerosols being particularly effective at attracting water and lowering the relative humidity needed for droplet formation. Thus, the introduction of particulates, especially large ones, promotes the formation of fog or cloud droplets. At the same time, high humidities favor the conversion of certain gases into secondary particulates, which in turn promote the condensation of water vapor into liquid droplets. As a result, humid areas with a high concentration of industrial activities can become foggy at relative humidities considerably below 100 percent. These processes worked together to make the London type of smog ubiquitous in eastern, industrial North American cities in previous years.

Removal of Particulates Though particulates are always present in the air, no individual particulate stays in the atmosphere forever. As we have seen in Chapter 7, terminal velocities increase with the size of falling objects. Thus, particulates, which are always small, can remain suspended in the atmosphere for considerable lengths of time. Larger ones remain in the air for perhaps just a few hours, while smaller particulates can exist for weeks.

Several different processes remove particulates from the air. *Gravitational settling,* the process wherein they fall from the air (even if very slowly), effectively removes larger particulates. The smaller ones are less susceptible to this process because even very small eddies can keep them in suspension. Precipitation, on the other hand, removes both large and small particulates in two ways. First, the particulates that served as condensation nuclei in clouds are removed when the droplets that they are part of fall as rain or snow. Other particulates are removed by *scavenging,* the process in which falling droplets and crystals collide with particulates in their path. Upon collision, the precipitation incorporates the particulate and carries it to the surface. The scavenging of particulates largely explains why the air is so much cleaner and visibility is enhanced after a rain shower.

Effects of Particulates Particulates reduce visibility by increasing scattering of visible radiation, but their effect on visibility is of less importance than their impacts on health. Perhaps this is not surprising, given that we are bathed in these tiny objects every minute of the day. By 1987 it had become clear that a certain class of particulates—those smaller than 10 μm in diameter (called PM_{10})—most readily enter the lungs and bring about the most serious tissue damage. Although the lungs have cilia that can remove these small (less than one-tenth the diameter of a human hair) particulates, the particulate removal occurs very slowly—even on the order of several months.

A large body of research analyzing the effects of particulates has shown that a more specific class of particulates—those smaller than 2.5 μm (called **PM$_{2.5}$**)—also present serious health problems. For this reason, the Environmental Protection Agency (EPA) in July 1997 revised its regulations regarding particulates so that in the future they will be based on these so-called *fine particles*. But the recent focus on PM$_{2.5}$ should not be taken to imply that larger particulates are not dangerous. For example, one study has shown a strong correlation between hospital admissions in the Los Angeles basin and the levels of large particulates in the air. The increase in hospital admissions is nearly equally divided between patients with acute respiratory illness and those with cardiovascular disease.

Checkpoint

1. Look at Figure 14–1. Why are transportation and electrical utilities major sources of air pollutants?
2. What are the major sources of particulates?
3. How are the processes of gravitational settling and scavenging similar? How are they different?

Carbon Oxides

Carbon oxides (also called *oxides of carbon*) include **carbon monoxide** (CO) and **carbon dioxide** (CO$_2$). The latter has been discussed in Chapter 1 as one of the important variable gases that make up the atmosphere, and in Chapter 16 we expand on its possible role in climate change. Though

14–1 FOCUS ON SEVERE WEATHER

Severe Pollution Episodes

Though many of us live in places where poor air quality is a disturbing fact of life, much progress in solving the problem has been made in the developed world in recent decades, with the result that the most disastrous types of smog events are a thing of the past. Consider, for example, what is probably the most famous air pollution episode in history—the one that hit London, England, between December 5 and 9, 1952. In this 5-day period, a combination of stagnant, damp air and the burning of low-quality coal produced a lethal mixture of smoke and fog. An estimated 3500 to 4000 people—mostly children, elderly, and the already infirm—died as a direct result of the episode.

The most famous air pollution disaster in North America occurred in Donora, Pennsylvania, 50 km (30 mi) from Pittsburgh. Between October 26 and 31, 1948, sulfur, carbon monoxide, and heavy metal dusts emitted from the American Steel & Wire's Zinc Works mixed with a dense radiation fog to create what has been called the "Hiroshima of air pollution." Four days of intense smog took on even greater proportions by Saturday, October 30. Fans at a high school football game were unable to see the events happening on the field. Others left the game early as word came that family members at home had died or

(a)

(b)

▲ **FIGURE 1** Like some other former industrial centers, Pittsburgh's air quality has undergone a huge improvement due to the closure of foundries and factories. These photos show Pittsburgh in 1906 (a), and the same scene in 1986 (b).

were hospitalized from respiratory problems brought on by the smog. Those who tried to evacuate the town were unable to leave because near-zero visibility completely stalled traffic. By Sunday morning firefighters were bringing oxygen to people who were unable to breathe, but relief was only temporary as the departing firefighters felt their way over to the next victim requiring assistance. On Sunday morning, officials finally closed down the Zinc Works, and later that day the smog was finally washed away by rain—but only after 20 people had died and 7000 people had been hospitalized.

Except for its magnitude, this infamous event was not unique. Many industrial cities have endured severe air pollution as a result of local industrial manufacturing, smelting, petroleum refining, or other activities. However, it is widely believed that the Donora event was the principal catalyst in the enactment of antipollution legislation in the United States. Since 1948, economic changes, along with greater attention to environmental issues, have greatly improved air quality in many cities (Figure 1).

important for its role in the transfer of energy within the atmosphere, carbon dioxide is officially considered a pollutant by the United States Environmental Protection Agency because of its role in climate change. High levels of CO_2 have no short-term deleterious effect on people or the environment. The same cannot be said for carbon monoxide, however.

Carbon monoxide is a colorless, odorless gas. In the natural environment it is released as a primary pollutant by volcanic eruptions, forest fires, bacterial action, and other processes. Though natural processes emit far more CO into the environment than do human activities, soil microorganisms consume it effectively, and background values are very low. In cities, however, inputs can greatly exceed the rate of removal, and unsafe concentrations can accrue. In the United States, the most important source of CO is the automobile (see Figure 14–1), which releases the gas as a by-product of incomplete combustion. In well-maintained vehicles, carbon

monoxide emissions are low, but poorly operating engines can cause CO concentrations to accumulate to unsafe levels. This is particularly true in confined areas, such as garages and tunnels. In the home, an improperly vented or malfunctioning furnace can release lethal doses of CO very quickly. Carbon monoxide is also released in home fires, where it probably is responsible for a high percentage of fire-related fatalities. Cigarette smoke also releases carbon monoxide as a by-product sufficient to greatly increase its concentration in the bloodstream.

Carbon monoxide is extremely toxic. Even low levels can cause a person to immediately experience slowed reflexes, drowsiness, and a reduction or loss of consciousness. Exposure for 3 hours at 400 parts per million (ppm) is life threatening, and at 1600 ppm death comes within an hour. Over the long haul, it can contribute to heart disease. Table 14–1 highlights some effects of varying CO levels.

TABLE 14–1

Threshold Levels of Carbon Monoxide

Carbon Monoxide Concentration (ppm)	Comment
50	Maximum allowable OSHA dose for 8-hour exposure
200	Headache, fatigue, dizziness, nausea in 2-3 hours
400	Headache in 1-2 hours, life threatening after 3 hours
800	Dizziness, nausea, and convulsions within 45 minutes; death in 2-3 hours
1600	Headache, dizziness, nausea in 20 minutes; death in 1 hour
3200	Headache, dizziness, nausea in 5-10 minutes; death in 25-30 minutes
6400	Headache, dizziness, nausea in 1-2 minutes; death in 10-15 minutes
12,800	Death within 1-3 minutes

Unlike other pollutants that act mainly on the pulmonary system, carbon monoxide's toxicity arises from its effects on the bloodstream. Hemoglobin (the agent that gives red blood cells their characteristic color) absorbs oxygen in the lungs and circulates it throughout the body. Under ideal conditions the hemoglobin releases oxygen to cells and then returns to the lungs, whereupon the process is repeated again and again. Carbon monoxide in the bloodstream completely disrupts this process. As it turns out, hemoglobin has a two-hundredfold greater affinity for carbon monoxide than O_2. In other words, with carbon monoxide and oxygen both available in the lungs, the blood will far more readily absorb the carbon monoxide. Thus, the inhalation of carbon monoxide reduces the cardiovascular system's ability to circulate oxygen to the rest of the body.

Sulfur Compounds

Sulfur compounds in the atmosphere can occur in gaseous or aerosol form. The majority—roughly two-thirds—of all the sulfur compounds emitted into the atmosphere originate from natural processes. Steam vents, such as those at Yellowstone National Park in Wyoming or Lassen National Park in California, provide interesting examples of the emission of sulfur compounds. The most important of these is the bacterial release of hydrogen sulfide (H_2S), a particularly noxious gas that smells like rotten eggs. Volcanic eruptions and sea spray also play an important role in releasing sulfur compounds. Fortunately, sulfur gases are readily dispersed in the atmosphere, so background concentrations are extremely low (on the order of one-half a part per billion) and their environmental and health impacts are minimal.

Of the anthropogenic sulfur compounds released to the atmosphere, the most important are **sulfur dioxide** (SO_2) and **sulfur trioxide** (SO_3). These oxides of sulfur fall under the collective designation of SO_x. Sulfur dioxide is a primary pollutant released mainly by the burning of sulfur-containing fossil fuels, particularly coal and oil used for heating and electric

power. Other industrial activities, such as petroleum refining and ore smelting, also contribute SO_2 (see Figure 14–1). Unlike natural processes, human activities tend to be concentrated over relatively small areas, allowing SO_x to attain high values over urban and industrial areas.

Sulfur dioxide is a colorless but highly corrosive gas that irritates human respiratory systems. High concentrations are associated with a number of lung problems, and even low concentrations can cause asthmatic subjects to experience severe bronchial constriction during exercise. Though it is widely blamed for causing respiratory problems, scientists are not sure what role high SO_2 concentrations directly play in their onset. It may be that the occurrence of respiratory illness during high SO_2 episodes is not due directly to the presence of the gas, but rather to the particulates that often accompany high sulfur dioxide concentrations.

Sulfur trioxide can be put directly into the air as a primary pollutant, but it more commonly builds up as a secondary pollutant following reactions involving SO_2. Sulfur trioxide is not by itself a major component of air pollution. However, it readily combines with water droplets to form *sulfuric acid,* H_2SO_4. If this process occurs near the surface, it forms **acid fog**; if it occurs in clouds, subsequent precipitation of the acid compound produces **acid rain**.[1] Not surprisingly, acid fog and rain are both capable of causing extreme environmental harm and through time can wear down human structures. Acid fog can be particularly dangerous to people because it is so easily inhaled. Buildings and monuments made of limestone are especially vulnerable to weathering from acid rain and fog (Figure 14–2).

Acid precipitation reaching the surface eventually feeds into the hydrologic system. Though some water falls directly onto lakes and rivers, the majority gets into them indirectly as soil or groundwater. Despite the indirect input into the

---

[1]The term *acid deposition* refers to the acidification of the surface environment by either contact with fog or the accumulation of acidic precipitation.

▲ **FIGURE 14–2** Acid deposition can gradually wear down the surfaces of monuments and buildings.

▶ **FIGURE 14-3** Acid precipitation is a widespread problem across eastern North America. The different colors indicate the average pH values for precipitation. Low pH values represent greater acidity. For reference, the pH for normal rainwater is 5.6. Tomatoes have a pH near 4.2. Data obtained from Central Analytical Laboratory, 2009.

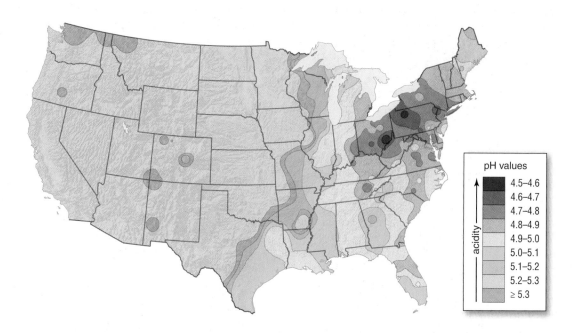

pH values

acidity		pH values
		4.5–4.6
		4.6–4.7
		4.7–4.8
		4.8–4.9
		4.9–5.0
		5.0–5.1
		5.1–5.2
		5.2–5.3
		≥ 5.3

surface waters, however, the water retains its acidity as it flows beneath the surface and eventually enters lakes and rivers. Through time, the surface water system becomes so acidic that it is inhospitable to life. At its worst, acidification can render lakes and rivers completely devoid of fish and aquatic birds. Unfortunately, this problem is neither hypothetical nor abstract. In the eastern United States, nearly 1200 lakes and 4700 streams have become acidified—some to the point where they no longer support any fish. In the Canadian province of Ontario, 1200 lakes are now essentially lifeless. Staggering as these figures may be, they pale in comparison to the 6500 lakes similarly impacted in Norway and Sweden.

As shown in Figure 14–3, acid precipitation is a much greater problem in the eastern United States and Canada than in the West, primarily because of the greater use of coal and heating oil. A huge proportion of the sulfur dioxide contributing to the acid rain originates from a relatively small number of sources. It is estimated that the 50 largest sulfur emitters in the region (all of which are power-generating plants) account for half of the acid deposition.

Interestingly, one of the measures undertaken to improve air quality near sulfur-emitting industries and power utilities may have exacerbated the acid deposition problem farther downwind. To help in the dispersion of sulfur oxides from industrial plants, many industries and utilities have built large smokestacks to release the pollutants well above ground level (Figure 14–4). The idea behind the stacks is that, by releasing the smoke far above the surface, sulfur compounds will be transported considerable distances downwind before settling near the ground. While the stacks have successfully reduced sulfur concentrations near their source, they have had the unintended consequence of allowing the sulfur compounds to be transported hundreds of kilometers downwind, where they react to form acid deposition. Thus, the acid problem over much of the eastern United States and Canada is due to

transported, rather than locally generated, pollutants. This has led to many years of litigation between states in the Midwest and Northeast and between the United States and Canada.

While most acid deposition in eastern North America is associated with sulfur compounds, this is not always the case in other areas. Some acid deposition, especially in the western United States and Canada, is related to compounds made of nitrogen and oxygen.

Nitrogen Oxides (NO$_x$)

Nitrogen oxides (also called *oxides of nitrogen*) are compounds consisting of nitrogen and oxygen atoms. The two most important of these from an air pollution viewpoint are

▲ **FIGURE 14-4** Smokestacks on manufacturing and power plants are designed to keep emissions away from the ground near the source. Unfortunately, pollutants are carried downwind for hundreds of kilometers, where they can exacerbate acid deposition.

nitric oxide (NO) and **nitrogen dioxide** (NO_2). Together the two gases are commonly referred to as NO_x. Nitric oxide is a nontoxic, colorless, and odorless gas that forms naturally by biological processes in soil and water. While millions of tons of the material are introduced into the atmosphere each year, it is highly reactive and tends to break down very quickly. Nitric oxide also forms as a by-product of high-temperature combustion associated with automobile engines, industrial manufacturing, and electric power generation. Its primary importance from an air quality perspective is that it oxidizes to form nitrogen dioxide, a major component of smog in many places.

Nitrogen dioxide is a toxic gas that gives polluted air its familiar yellow to reddish brown color (Figure 14–5), as well as a pungent odor. It is an important component in air pollution in that it is relatively toxic and corrosive and undergoes transformations that contribute to acid deposition and other secondary pollutants. As with nitric oxide, nitrogen dioxide breaks down very readily and, as a result, NO_2 concentrations in urban areas tend to rise and fall in accordance with vehicular traffic patterns. Furthermore, the rapid decay of nitrogen dioxide prevents large concentrations from occurring in rural areas surrounding source areas.

Like sulfur compounds, nitrogen oxides can cause serious pulmonary health problems. Clinical studies have shown that NO_2 easily passes through bronchial passages and irritates tissue deep within the lungs. Laboratory tests have shown animals to experience severe lung damage and reduced immunity to infection when exposed to high levels of NO_2.

Volatile Organic Compounds (Hydrocarbons)

Volatile organic compounds (VOC), also called **hydrocarbons**, are materials made entirely of carbon and hydrogen atoms. These compounds, including methane, butane, propane, and octane, occur in both gaseous and particulate forms. Globally, the vast majority of hydrocarbons arrive in the atmosphere via natural processes, including plant and animal emissions and decomposition. In the United States, industrial activities account for the greatest proportion of anthropogenic hydrocarbons, with automobiles also contributing a major share. The emissions associated with automobiles arise primarily from incomplete fuel combustion and the evaporation of gasoline (often while filling gas tanks).

Even in cities that have high VOC concentrations, there is little evidence that these chemicals have any direct adverse health impacts. Nonetheless, they are extremely important because in the presence of sunlight they recombine with nitrogen oxides and oxygen to produce photochemical smog.

Photochemical Smog

If you have ever visited Los Angeles in the summer, you have probably heard the term **photochemical smog** and know what it feels like. Burning eyes, sore lungs, and a subtle but unpleasant odor accompany an atmosphere with poor visibility. Photochemical smog consists of secondary pollutants that include ozone (O_3), NO_2, peroxyacyl nitrate (PAN), formaldehyde, and other gases that occur in very minute quantities. As the name implies, this type of smog forms when sunlight triggers numerous reactions and transformations of gases and aerosols. Unlike the **London-type smog** found in many places where smoke combines with damp air (the word *smog*, in fact, originally derived from the terms *smoke* and *fog*), this **Los Angeles–type smog** usually involves dry air.

Ozone has been designated by the Environmental Protection Agency as the most important agent of photochemical smog.

◄ **FIGURE 14–5** Nitrogen dioxide gives polluted air a yellowish to reddish brown color, as in this photo of Hong Kong.

14–2 FOCUS ON THE ENVIRONMENT

The Counteroffensive on Air Pollution

Regulations designed to improve air quality have made a substantial impact on the lives of people in the United States and Canada. Although federal regulations regarding air pollution in the United States did not come into being until the 1950s, certain cities and states have had laws on the books regulating smoke emissions since the late nineteenth century. In some cases, such as in Pittsburgh, fairly stringent controls were in effect by the 1940s.

The first major U.S. initiative to clean up the nation's air was the 1963 Clean Air Act, which, among other things, expanded the role of the federal government in interstate air pollution control and authorized increased research and technical development initiatives. Major extensions of government involvement were subsequently added to the Clean Air Act by the adoption of major amendments in 1970, 1977, and 1990.

The original Clean Air Act and its subsequent amendments established air pollution standards and created government agencies to ensure that those standards were being met. Maximum concentrations were established for PM_{10}, sulfur dioxide, carbon monoxide, nitrogen oxides, ozone, and lead (formerly an ingredient in gasoline but now outlawed by federal regulations). Individual states were required to establish agencies to ensure enforcement of the standards,

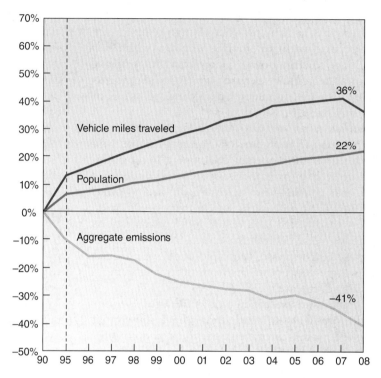

▲ **FIGURE 1** Trends in aggregate pollutant levels, motor vehicle miles travelled annually, and population in the United States relative to 1990.

and regions where air quality fell short of the standards were designated nonattainment areas and required to take appropriate actions. The act and its amendments also required automobile manufacturers to install emission reduction devices, such as the catalytic converter, that have lowered individual vehicle emissions by about 95 percent since the 1960s. The law also ordered the phaseout of open burning of refuse, the installation of filters on industrial smokestacks, and other emission-reducing measures.

It can cause serious physical and environmental harm at surprisingly low concentrations, so low that the EPA established a concentration of only 0.12 ppm averaged over a 1-hour period as the maximum allowable without exceeding federal standards.

Exposure to ozone causes inflammation of air passages that can reduce lung capacity by as much as 20 percent. The EPA estimates that perhaps 20 percent of all respiratory-related hospital visits in the northeastern United States during the summer result from exposure to ozone. Although acute symptoms usually subside fairly quickly after ozone concentrations decline, research has shown that long-term exposure to ozone can cause permanent damage to lung tissue and impairment of the body's ability to resist bronchitis, pneumonia, and other diseases. Of course, ozone causes even greater problems for people with asthma and other preexisting

pulmonary problems. For those people, ozone constricts lung passages to the point where breathing becomes nearly impossible, and the gas may contribute considerably to the number of asthma-related fatalities that occur in the United States annually. Currently, an average of 5000 Americans die each year during acute asthma attacks.

Not only are people directly harmed by ozone, but high levels also result in serious environmental degradation. Damage to agricultural crops by photochemical smog (mainly ozone) was identified in southern California grape fields in the late 1950s. Since that time, plant damage from oxidants (and to a lesser extent, PAN) has been widespread across North America, with conifers being particularly vulnerable. (*Box 14–2, Focus on the Environment: The Counteroffensive on Air Pollution*, summarizes recent efforts to combat air pollution.)

Overall, the act and its amendments have resulted in substantial reductions in pollutant levels in urban areas, despite a large increase in motor vehicle miles driven each year. Figure 1 shows the reduction in aggregate pollution levels relative to 1990 values that occurred despite increases in population and vehicle miles traveled. Though progress has been dramatic, some counties still fail to meet minimum standards for at least one of the major pollutants (Figure 2).

Recent research has shown that the standards adopted for various pollutants were not always based on the appropriate targets. For example, scientists now know that 1-hour exposures to ozone do not have as great an effect on human health as do longer-term exposures at lower levels. Thus, the EPA set a new standard in which compliance is based on 8-hour concentrations above 0.08 ppm, as opposed to the previous 1-hour standard of 0.12 ppm. The EPA believes the enactment of the new standards will annually prevent thousands of premature deaths and hundreds of thousands of lung disorders in children.

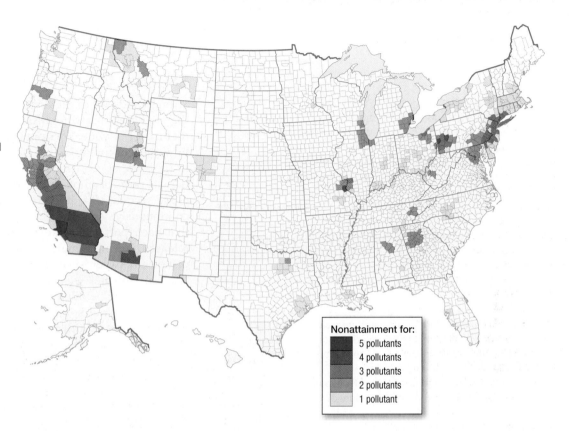

Nonattainment for:
- 5 pollutants
- 4 pollutants
- 3 pollutants
- 2 pollutants
- 1 pollutant

▲ **FIGURE 2** Map of counties that have failed to meet air quality standards as of April 2010.

Air Quality Index

The Environmental Protection Agency (EPA) has created a uniform index that is useful for air pollution monitoring across the United States called the **Air Quality Index** (AQI). The AQI is calculated each day for locations across the United States by applying a particular formula for ozone, particulates, carbon monoxide, sulfur dioxide, and nitrogen dioxide, and expressing each pollutant on a scale that ranges from 0 to 500. The official AQI for any place at a particular time and place is the highest of the five individual pollutant values. The EPA has also established a color scheme to represent different levels of AQI ranges for better communication to the public. Table 14–2, taken directly from the EPA, illustrates the color scheme used and summarizes the significance of AQI values within designated ranges. Current maps of AQI values and next-day forecast maps can be obtained at **www.airnow.gov**.

Did You Know?

Severe air pollution is not a phenomenon restricted to large, urban centers. Indeed, even national parks can have major smog problems. At the Great Smoky Mountains National Park in North Carolina and Tennessee, pollutants—mostly from regional industry—have caused the average summer visibility to be reduced from a maximum summertime distance of 187 km (117 mi) to only 32 km (20 mi). Acid precipitation is also a problem, with rainfall acidity being as much as ten times that of natural precipitation. The park has exceeded Environmental Protection Agency standards for ozone an average of 30 days per year from 1991 to 2003.

The situation is even worse for Sequoia/Kings Canyon National Park in California. The smoggiest of all U.S. national parks, this otherwise magnificent natural resource routinely has more days in excess of EPA ozone standards (370 days in the period from 1999 to 2003) than do most large U.S. cities. Visitors and staff there are frequently advised to limit their physical activities because of unhealthful air.

TABLE 14-2

Air Quality Index

Air Quality Index Levels of Health Concern	Numerical Value	Meaning
Good	0 to 50	Air quality is considered satisfactory, and air pollution poses little or no risk
Moderate	51 to 100	Air quality is acceptable; however, for some pollutants there may be a moderate health concern for a very small number of people who are unusually sensitive to air pollution.
Unhealthy for Sensitive Groups	101 to 150	Members of sensitive groups may experience health effects. The general public is not likely to be affected.
Unhealthy	151 to 200	Everyone may begin to experience health effects; members of sensitive groups may experience more serious health effects.
Very Unhealthy	201 to 300	Health alert: everyone may experience more serious health effects
Hazardous	301 to 500	Health warnings of emergency conditions. The entire population is more likely to be affected.

Checkpoint

1. Pick one of the following atmospheric pollutants and explain its causes and effects: carbon oxides, sulfur compounds, nitrogen compounds, volatile organic compounds.
2. Is acid precipitation a primary or secondary pollutant? Explain.
3. What are two important reasons why ozone levels in photochemical smog should be reduced? Explain.

Atmospheric Conditions and Air Pollution

As we have seen, a large portion of the chemicals that we consider pollutants occur naturally in the environment. These emissions do not create high concentrations, however, because their release into the atmosphere is over such a large area that they are immediately diluted. Urban emissions, on the other hand, are concentrated over much smaller areas and can thereby lead to significant pollution episodes.

Atmospheric conditions play a major role in determining pollution concentrations in several ways. Atmospheric stability and wind conditions control the vertical and horizontal dispersion of pollutants, and cloud conditions can influence the rate of photochemical reactions taking place. Furthermore, unusually cold or warm conditions encourage the increased use of heaters or air conditioners, which can increase emissions.

Effect of Winds on Horizontal Transport

Strong winds aid in the dispersal of pollutants two ways. First, they rapidly transport emissions from their source and spread them over a wide horizontal extent. Figure 14-6 illustrates how the concentration of pollution is inversely proportional to the wind speed (to make this easier to visualize, the figure depicts puffs of smoke being released from a stationary source every second, though in reality pollutants would be released continuously). In (a), the wind blows at 5 m/sec (11 mph) so that each puff of smoke travels 5 m before the next one is released. In (b), the wind flows twice as fast as in (a), and the distance between successive puffs of smoke likewise doubles. Thus, the greater wind speed in (b) causes the same amount of pollution to be diluted within twice as large a volume of air.

Greater wind speeds also lower the pollution concentration indirectly. Recall from Chapter 3 that air does not flow uniformly in a given direction. Instead, it contains small, swirling motions, called *eddies,* that mix the air vertically. This forced convection increases with wind speed and, as a result, strong winds favor greater vertical dispersion.

Short-term variations in wind direction also affect dispersion. If the wind direction varies only slightly through time, pollution will be concentrated within a relatively narrow area downwind of the source. If wind directions are highly variable, the pollutants will spread out over a wider area. More people will be subjected to the pollutants, but the concentration will be lower than it would under a more constant wind regime.

Effect of Atmospheric Stability

Just as the stability of the air (Chapter 6) influences lifting and cloud formation, it also affects the vertical movement of pollutants. Recall that when the air temperature decreases slowly with height (or if it increases with height), the air is said to be stable. Stable air resists vertical displacement and leads to

▶ **FIGURE 14-6** The effect of wind speed on pollutant dispersal. In (a), the 5 m/sec (11 mph) wind moves individual puffs of smoke downwind slowly, so each successive puff is only 5 m behind the previous one. In (b), the 10 m/sec (22 mph) wind causes twice as great a distance between puffs, and thus only one-half the concentration of smoke.

Wind speed = 5 m/s

X = 1 puff/s

(a)

Wind speed = 10 m/s

X = 1 puff/s

(b)

higher pollutant concentrations near the ground. Unstable air, on the other hand, enhances vertical mixing, and any material introduced near the surface is easily displaced upward. This reduces pollution concentrations near the surface.

Inversions, the situation in which air temperature increases with height, make the air extremely stable and impose the greatest restraint on vertical mixing. Radiation inversions (described in Chapter 6) originate at the surface in response to cooling of the lower atmosphere (Figure 14–7a). These inversions usually dissipate in the late morning after the Sun has warmed the surface and lower atmosphere (Figure 14–7b). As a result, they tend to have the greatest impact on pollution concentrations in the early morning. These inversions are most important in areas subject to a London-type smog.

Subsidence inversions are often important where photochemical smog is the major problem (Figure 14–8). The base of a subsidence inversion marks the maximum height to which the air below can be easily mixed. An inversion with a base at

▲ **FIGURE 14-8** Subsidence inversions occur some distance above ground level (unlike radiation inversions whose bases are always right at the surface). The strong stability of the inversion layer prevents air from mixing above its base and thereby confines pollutants to the layer of air below the inversion.

(a)

(b)

▲ **FIGURE 14-7** Radiation inversions develop at night when cooling lowers the air temperature more rapidly near the ground than aloft. The stability associated with the inversion inhibits vertical mixing of air pollutants, keeping them close to the ground (a). Daytime heating warms the air closest to the ground most rapidly and eventually breaks up the inversion (b).

500 m (0.3 mi) above the surface will result in a *mixing depth* of 500 m, doubling the concentration of pollutants that would accompany a mixing depth of 1000 m (0.6 mi). Figure 14–9 shows how distinct that base of an inversion layer can be.

Just as the semipermanent Hawaiian high-pressure system accounts for the mostly dry summers of southern California, subsidence from the same system also plays an important role in the region's poor air quality. As air rotates clockwise out of the eastern margin of the high, air in the middle troposphere descends and creates an inversion. During the summer, the base of the inversion level over Los Angeles typically occurs at about 700 m (2300 ft) above sea level, but the base of the inversion can also occur at lower levels and lead to particularly bad smog events. (*Box 14–3, Focus on the Environment: Smog in Southern California*, presents further information on atmospheric and other controls on air pollution in the region.)

▲ **FIGURE 14-9** The base of an inversion is clearly evident above the smog layer in Los Angeles.

14-3 FOCUS ON THE ENVIRONMENT

Smog in Southern California

Los Angeles has long had a reputation for extremely bad air quality—and for good reason. Of all the cities in the United States, Los Angeles is the only one classified by the Environmental Protection Agency as an "extreme area" of noncompliance of ozone standards. A number of factors work together to make the air quality bad enough to earn this dubious distinction. As shown in Figure 1, Los Angeles occupies part of a basin bounded by mountains to the north and east that block the free movement of pollutants by the sea breeze, while the presence of a subsidence inversion during the warmer months restricts vertical dispersion. Add to that the typically cloud-free conditions during the mid-day period that trigger photochemical reactions. And finally there is the city's well-known love affair with the automobile, which contributes much of the estimated 2 million kilograms (2200 tons) of hydrocarbons and 1 million kilograms (1200 tons) of NO_x released each day into the four-county South Coast Air Basin.

Fortunately, a number of new regulations have improved the situation. For example, beginning in 1984 all automobiles were required to undergo biannual smog checks. More recently, regulations have been enacted requiring gas pump nozzles to have rubber sleeves to capture fumes that would otherwise escape into the air as people fill

▲ **FIGURE 1** The topography of the Los Angeles basin.

their tanks. Also, a cleaner-burning type of gasoline that releases fewer hydrocarbons has been phased in at all area gas stations. To illustrate how much progress has been made, consider the fact that during the 5-year period from 1976 to 1980, there was an average of 112 stage-1 smog alerts (ozone $\geq$ 0.20 ppm) each year in the South Coast Air Basin. From 1999 through 2010 there was only a single such smog event!

This was due to the reformulation of gasoline to reduce emissions, requirements that vehicles undergo smog checks, and laws limiting pollution from nonvehicular sources.

During the summer, daily concentrations of photochemical smog vary on a regular basis in the course of a day. Prior to the morning rush hour, residual primary and secondary pollutants from the previous day leave a background level of air pollution.

Checkpoint

1. What are two main factors that can disperse or concentrate air pollutants?
2. What atmospheric conditions and other factors help to create Los Angeles's air pollution problems? Explain.

Urban Heat Islands

Not all human impacts on the atmosphere are as dramatic as the pollution of the atmosphere. The well-known **urban heat island** is an excellent case in point. For centuries it has been known that urbanized areas often have higher temperatures than do adjacent countrysides. These differences can be quite dramatic, with temperatures in major metropolitan areas sometimes exceeding those of their hinterlands by as much

as 12 °C (22 °F). Although the exact nature of the urban heat island varies from one city to another, in general the urban–rural temperature differences are greatest during the late evening and night and during the winter months.

Urban heat islands occur because of modifications to the energy balance (Chapter 3) that result when natural surfaces are paved and built upon and when human activities release heat into the local environment. Though it is not possible to generalize the relative importance of these processes for every city, all of them probably play some role in causing the phenomenon to occur.

Several variables influence the magnitude of the heat island effect. Some are related to the local setting, others to the activities undertaken within it. But the most important are the size of the city itself and the density of its population, with large, densely populated cities having the largest heat island effect. Figure 14–10 illustrates the relationship

As traffic increases during the morning, emissions increase substantially. Early morning winds are usually weak, which leads to little movement of pollutants. At the same time, the low Sun angle and common presence of early morning fog and low clouds inhibit photochemical activity. The situation normally changes by late morning. A sea breeze usually develops along the coast and moves pollutants inland, while clearing skies and increasing Sun angles increase photochemical conversions.

As the sea breeze develops, a boundary called a **sea breeze front** separates the relatively clean marine air from the more polluted, drier air ahead. As the sea breeze front moves inland, it pushes the emissions eastward or northeastward. This often creates a strong gradient in ozone concentrations near the sea breeze front, with relatively clean air behind it and increasing concentrations to the east or northeast (Figure 2). By late afternoon, the cities in the eastern portion of the basin get the full onslaught of the advected pollutants, while local commuters add their own contribution to the photochemical smog. As a result, pollution levels can become extraordinarily high in the areas downwind of Los Angeles.

The area to the south of Los Angeles—including Orange and San Diego counties—also has a significant air pollution problem. Usually the air pollution in San Diego, some 150 km (100 mi) to the south, is

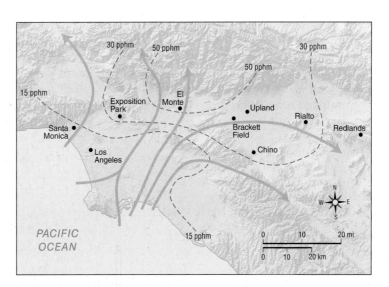

▲ **FIGURE 2** Typical pattern of smog on a summer afternoon in the Los Angeles basin. Highest concentrations (shown by dashed lines) occur in the northeast, ahead of the sea breeze front. Solid lines show the wind direction.

of local origin. During severe episodes, however, most of the pollution originates over Los Angeles and Orange counties and gets carried into San Diego by the wind. These episodes often occur as Santa Ana winds die out. During the Santa Ana, the easterly winds force the basin's pollutants offshore. As the Santa Ana begins to weaken, a thermally induced low-pressure system over the eastern desert stretches into the San Diego area. This creates a northwesterly flow that transports pollution originating over Los Angeles and Orange counties. Not surprisingly, the trend toward decreasing pollution in Los Angeles has also occurred in San Diego.

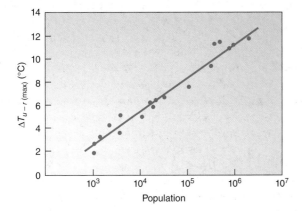

▲ **FIGURE 14–10** Urban heat islands vary with city population. The vertical axis plots temperature differences between urban centers and surrounding areas in °C against city population. Note that the horizontal axis (population) is on a logarithmic scale.

between the population of North American cities and the maximum urban and rural temperature differences (note that the horizontal axis is plotted on a logarithmic scale).

The intensity of an urban heat island varies spatially across a city, with highest temperatures normally found within the city core. Figure 14–11 illustrates this effect by plotting temperatures over Vancouver, British Columbia, on a July evening. The downtown area is located on the southeastern part of the peninsula that juts into Burrard Inlet. Immediately to the northwest of downtown (on the northwestern part of the peninsula) lies Stanley Park, a wooded area with few buildings. As expected, temperatures are greatest over the downtown region and decrease substantially over less-populated areas. Temperatures on the peninsula decrease dramatically between downtown and the middle of the park, with a difference of about 9 °C (16 °F) occurring over a distance of only about 1.5 km (1 mi). Wind speeds also play an important role.

During windy conditions, cooler air from the surrounding countryside displaces the warmer urban air and thus reduces the magnitude of the urban heat island.

Radiation Effects

Urban particulates can also affect the intensity of the urban heat island through their effect on the radiation balance. Increased particulates associated with urban activity can absorb and scatter incoming solar radiation and also increase the amount of absorption and reradiation of longwave energy in the atmosphere. Although it is hard to generalize for all cities, it is believed that the particulates tend to decrease the amount of incoming solar radiation at the urban surface, but the increase in net longwave radiation probably offsets the reduction in absorbed solar energy. Thus, the direct effect of particulates on urban temperatures is probably negligible.

Particulates also affect the radiation balance indirectly. Recall that water droplets in the atmosphere form onto condensation nuclei. The increase in particulates due to human activity can increase cloud cover, as in the case of London, England, which has been shown to receive 270 fewer hours

▶ **FIGURE 14-11** (a) Temperatures in Vancouver, British Columbia, at 9 P.M. on July 4, 1972. Note the sizable temperature gradient between downtown and Stanley Park. (b) Photo showing the Vancouver downtown area with Stanley Park in the background.

(a)

(b)

of bright sunlight annually than the surrounding area. Particulates have long been known also to increase precipitation downwind of urban regions. Interestingly, studies have also shown that precipitation can decrease downwind of major industrial centers, perhaps because cloud water is spread over many condensation nuclei, which lessens the chance of growth to precipitation size.

More important than the effect of increasing particulate concentrations is the impact that buildings have on the radiation balance. Consider the impacts the construction of buildings with vertical walls might have on the receipt of solar radiation. When the sun is low in the sky—near sunrise and sunset, and during much of the day at high latitudes in the winter—direct sunlight that would otherwise reach the horizontal surface hits the vertical walls of buildings. This causes the angle of incidence to become closer to perpendicular and increases surface heating, which leads to a higher temperature.

The presence of buildings also affects the rate of heating by changing the surface albedo. Darker buildings, of course, absorb more sunlight than lighter ones, and, in general, urban surfaces (asphalt streets, roofing materials) have lower albedos than the natural surfaces they replace. The presence of buildings also affects the amount of absorption by causing multiple reflections to occur, as shown in Figure 14–12. As sunlight penetrates the urban landscape and hits the side of a building, some of the energy is absorbed and some is scattered back as diffuse radiation. Some of the scattered radiation strikes an adjacent building, where once again a portion is absorbed. This process goes on repeatedly, with each successive reflection at least partially absorbed upon contact with another wall. This increases the total absorption so the albedo of the urban area is actually lower than the albedo of the individual surfaces.

The presence of tall buildings also affects the transfer of longwave radiation in a way that favors higher nighttime temperatures. Essentially, the process is very similar to the multiple reflection of solar radiation just discussed. Longwave radiation emitted from an open, rural surface travels upward without being impeded by buildings. Urban areas, in contrast, reduce the amount of longwave radiation that freely escapes to space because walls absorb a portion of the outgoing radiation. The resultant reduction in longwave radiation loss slows the rate of nocturnal cooling and promotes higher daily minimum temperatures.

Changes in Heat Storage

As explained in Chapter 3, radiation is not the only mechanism that transfers energy from one place to another; conduction and convection are also important heat transfer mechanisms. In the middle of a sunny day, absorption of solar radiation warms the surface of the ground. Conduction within a very thin layer of the atmosphere and convection transfer much of this energy to the air. At the same time, a gradient develops in which soil temperature decreases with depth, and heat is conducted downward.

During the late afternoon, the surface begins to cool when energy losses by radiation and convection exceed the absorption of shortwave and longwave radiation. The soil temperature profile eventually reverses, with temperature increasing with depth. Thus, during the evening hours, heat that has been stored within the soil is transferred to the surface.

The same processes just described occur in the walls and roofs of urban buildings. As the surface is warmed during the day, a temperature gradient develops that conducts heat toward the building interior. When cooling occurs in late afternoon, the stored heat is released to the surface. What is different from the rural setting is that materials used in building construction have a much greater capacity to store heat than most natural surfaces. As a result, more stored heat is available for transfer to the lower atmosphere during the evening and nighttime than for natural surfaces, and nocturnal temperatures are increased.

The release of heat from buildings just described is supplemented by the anthropogenic heat produced for comfort

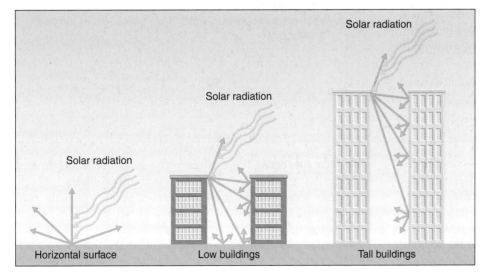

◄ **FIGURE 14–12** Effect of buildings on solar radiation receipt. As incoming radiation contacts a building, some is scattered off in all directions and some is absorbed. The scattered radiation may in turn hit an adjacent building, where further absorption can take place. This lowers the urban albedo.

Solar radiation

Solar radiation

Solar radiation

Horizontal surface   Low buildings   Tall buildings (e.g., space heating) or as a by-product of other activities (e.g., waste heat from a hot car engine). Anthropogenic heating is greatest during the winter, which partially explains why heat islands are most pronounced during the low sun season. Anthropogenic heat can be a surprisingly large component of the urban energy balance. In Vancouver, British Columbia (49° N), for example, the amount of anthropogenic heat released in the winter has been estimated to be nearly four times that received as net radiation. That estimate, based on 1970 data, probably understates the importance of anthropogenic heat currently released because the city has grown significantly since then.

Sensible and Latent Heat Transfer

In Chapter 3 it was shown that most of the global surface has a surplus of net all-wave radiation on an annual basis. That surplus is transferred to the atmosphere as sensible and latent heat. When moisture is available near the surface, the transfer of energy as latent heat can exceed sensible heat transfer, indicating that most of the surplus is consumed by evaporation. On the other hand, if the surface is completely dry, surplus energy raises the surface temperature far above the air temperature, and sensible heat dominates. All other things being equal, the higher the ratio of sensible to latent heat, the greater the temperature.

Urbanization affects the routing of precipitation in a way that favors increased sensible heat transfer. Unlike natural surfaces that allow rainfall or snow melt to permeate the soil and be retained below ground, city streets and sidewalks are almost impervious to water. So when precipitation occurs, most of the water runs off the surface and ultimately flows out through the flood-control system. The reduction in available water increases the input of sensible heat to the atmosphere at the expense of latent heat and helps increase the temperature of the urban environment.

Urban Heat Islands and the Detection of Climate Change By now everyone has heard much discussion about the possibility of impending climatic change resulting from the anthropogenic emission of greenhouse gases. Many atmospheric scientists believe this change has already begun to take place. To support this notion, they point to an overall

Did You Know?

It has long been known that cities create their own urban heat islands. But cities can also affect precipitation patterns—in conflicting ways. The urban areas around the San Francisco Bay Area, Los Angeles, and San Diego increase the amount of small particulates in the air that are transported downwind. The increase in particulates causes clouds over the foothills of the mountains to the east to have their liquid or ice content distributed among a larger number of droplets or crystals. The smaller size of the cloud constituents makes it harder for precipitation to develop and may reduce mountain snowfalls by as much as 20 percent in some places.

But a separate NASA study shows that in cities more prone to thunderstorm activity, the urban heat island effect—combined with other urban factors such as surface convergence and increased aerosol concentrations—can *enhance* precipitation downwind by as much as 51 percent. Summertime precipitation increases have been observed downwind of numerous cities, ranging from Atlanta, Georgia, in the Southeast, to Dallas, Texas, in the South, and as far north as Chicago, Illinois.

warming of 0.3 to 0.6 °C (0.5 to 1 °F) occurring since the late nineteenth century for weather stations having records going back more than a century. However, we cannot use these recorded temperature changes at face value because many of the data come from urban weather stations, and most cities with long-running weather stations have undergone considerable growth over the last century. Thus, we must contend with the problem of enhanced urban heat islands influencing the data, which means that records from large urban areas are not representative of the surrounding region. Atmospheric scientists are well aware of this source of bias in temperature records—which happens to be relatively small—and routinely account for its effect, either by discarding contaminated data or by adjusting values downward for affected stations.

Checkpoint

1. Why does the intensity of an urban heat island vary spatially across a city?
2. What are some ways in which a city could reduce the heat island effect?

Summary

As the human population has grown in size and become more industrialized, societies have increased their impact on the atmospheric environment. The most dramatic effects result from the release of numerous gases and particulates. Although many of the emissions that we consider pollutants result from natural processes, rapid dispersion of these materials prevents them from causing negative impacts. In industrial and urban areas, on the other hand, these emissions are concentrated into smaller areas and often lead to serious problems.

Some atmospheric pollutants are released directly into the atmosphere (primary pollutants), while others result from transformations of other gases (secondary pollutants). Particulates are solid and liquid aerosols that can be produced as either primary or secondary pollutants. Recent research has shown that the health effects of the smallest particulates are most critical because they are easily lodged in lung tissue. Pollutant gases include carbon oxides, sulfur compounds, nitrogen oxides, volatile organic compounds (also called *hydrocarbons*), and photochemically formed gases (the most notable of which is ozone). Each gas presents its own set of health problems, ranging from reduced immunity to permanent lung damage to cardiovascular problems.

Until the middle part of the twentieth century, efforts to control these emissions were instituted on a local scale. Beginning in 1955, the U.S. government began to enact laws designed to improve air quality across the nation and reduce the number of major health problems that result from air pollution. The original Clean Air Act and its amendments have had a dramatic impact on air quality. This legislation has required the formation of local agencies to monitor pollution levels and ensure compliance with federal standards. It has also required the automobile industry, power utilities, ore smelting plants, and manufacturing industries to reduce the amount of their emissions. While reductions in many pollutants have been dramatic, numerous cities in the United States still have not met clean air goals.

The amount of air pollution does not depend entirely on the activities of people; atmospheric conditions also affect the dispersal of pollutants. If winds are strong and constantly shift direction, pollutants are distributed over a larger area and concentrations decrease. Statically unstable air also favors the dilution of gases and particulates by enhancing vertical mixing. On the other hand, stable conditions and, in particular, the presence of an inversion, can greatly restrict vertical motions and concentrate pollutants near the ground.

Human impacts on the atmosphere are not restricted to pollution. The urban heat island is a well-known phenomenon in which changes in the surface (such as the replacement of vegetated surfaces with concrete and asphalt), the existence of buildings with vertical walls, and the release of heat as a by-product of human activity combine to increase temperatures. These increases are most notable during the evening and nighttime hours and in the winter season.

We have now looked at the natural processes that make up daily weather, and the ways humans analyze, predict, and alter the resultant patterns. The next two chapters of this book concern the longer-term state of the atmosphere—the climate. Chapter 15 looks at general patterns across the globe, and Chapter 16 examines past and possible future changes in climate.

Key Terms

air pollution *page 420*

primary pollutants
 page 420

secondary pollutants
 page 420

particulates *page 420*

PM$_{10}$ *page 420*

PM$_{2.5}$ *page 421*

carbon oxides *page 421*

carbon monoxide
 page 421

carbon dioxide
 page 421

sulfur dioxide *page 423*

sulfur trioxide *page 423*

acid fog *page 423*

acid rain *page 423*

nitrogen oxides *page 424*

nitric oxide *page 425*

nitrogen dioxide *page 425*

volatile organic compounds
 (hydrocarbons) *page 425*

photochemical smog
 page 425

London-type smog
 page 425

Los Angeles–type smog
 page 425

Air Quality Index
 page 427

urban heat island *page 430*

sea breeze front *page 431*

Review Questions

1. Explain the distinction between primary and secondary pollutants.

2. What are particulates and how are they introduced into the atmosphere?

3. What are the two processes most responsible for removing particulates from the atmosphere?

4. What are PM$_{10}$ and PM$_{2.5}$? Does one pose a greater health risk than the other?

5. List the most important gases that contribute to air pollution.

6. What are the primary sources of carbon monoxide in the atmosphere? If these sources are nonanthropogenic, why is it that CO is considered a pollutant?

7. In what way does carbon monoxide harm the human body?

8. What are the primary sources of sulfur dioxide and sulfur trioxide in the atmosphere?

9. Would a person be more likely to notice the presence of high CO or high SO$_2$ contents in the ambient air?

10. Which primary pollutant is most likely to promote the formation of acid fog or acid rain?

11. Why is it that nitric oxide is much less common in the atmosphere than nitrogen dioxide?

12. Describe the general composition of volatile organic compounds.

13. How do London-type and Los Angeles–type smog differ from each other?

14. Which pollutant gases cause a noticeable coloration of the atmosphere? Which have a characteristic odor?

15. Describe the various atmospheric controls that affect the concentration of air pollutants.

16. How does the construction of buildings in cities alter the radiation exchange near the surface and contribute to the urban heat island effect?

17. Describe the way in which heat storage in cities differs from that of surrounding rural areas.

18. What effect does urbanization have on the exchange of sensible and latent heat?

19. Does the urban heat island manifest itself equally during the day and night? Are there seasonal differences in the magnitude of the heat island effect?

Critical Thinking

1. There has been much improvement in air quality for North America as a whole. Further improvements will only arise from measures that may be costly, both directly through the application of technology and by reductions in certain economic activities. Do you personally believe that further improvements can be brought about at a cost that people are willing to bear?

2. Visit the Web page at **earthobservatory.nasa.gov/ GlobalMaps**, and form your own conclusions about whether air pollution is now primarily a local or a global process.

3. It is believed that global warming over the last few decades might have been even greater, were it not for the effect of certain types of air pollution. Explain how this could come about.

Problems and Exercises

1. Examine the map in *Box 14–2, Focus on the Environment: The Counteroffensive on Air Pollution.* How is the air pollution situation in your area? Is the information on the map consistent with your perception of the local air quality? What factors do you think lead to the type of air quality that your area has?

2. Visit the Web page at **www.epa.gov/air/partners .html** for a listing of EPA, state, and local agencies that provide data on air quality. Refer to one of the agencies for a daily report on the air quality in your area. How do the changes in air quality fluctuate with different weather patterns where you live?

3. If you live in or near an urban area, make note of the daily maximum and minimum temperatures in your region. Do you detect a significant urban heat island? How does the magnitude of this heat island compare for maximum and minimum temperatures? Are there seasonal differences?

Quantitative Problems

The companion Web site to this textbook, **www .MyMeteorologyLab.com**, has some interesting problems related to the concentration of pollutants and precipitation pH values. We suggest you solve those problems to further your understanding of air pollution.

Useful Web Sites

www.epa.gov/oar/airpollutants.html
Environmental Protection Agency Web page. Includes a wide range of information on various types of air pollution.

www.cdc.gov/nceh/airpollution
National Center for Environmental Health, an agency of the Centers for Disease Control. Includes numerous links to reports on the effect of air pollutants on human health.

www.ec.gc.ca/Air/default.asp?lang=En&n=14F71451-1
Information on air pollution in Canada.

www.ars.usda.gov/Main/docs.htm?docid=12462
Describes the effects of ozone on plants.

www.epa.govair/caa
Presents comprehensive information on the Clean Air Act.

www.nrdc.org/air/pollution/default.asp
Web page of the Natural Resources Defense Council, a nonprofit public interest group that advocates environmental causes.

www.nasa.gov/topics/earth/features/health-sapping.html
Reports on how NASA has been monitoring global air pollution. Includes some very interesting movies.

MyMeteorologyLab™

The *Understanding Weather and Climate,* Sixth Edition, MyMeteorologyLab™ web site contains numerous multimedia resources to aid in your study of **Human Effects on the Atmosphere**. Visit **www.MyMeteorologyLab.com** to:

- **Review** key chapter concepts.
- **Read** the Pearson eText, *In the News RSS feeds,* and other chapter-specific Web resources.
- **Visualize** challenging topics with Interactive Tutorials, Geoscience Animations, MapMaster interactive maps, and Weather in Motion movies and visualizations, and more.
- **Test Yourself** with self-study quizzes.

WEATHER IN MOTION

View movies and visualizations of meteorology patterns and processes.

The Smog Bloggers
Global Carbon Monoxide Concentrations
Tropospheric Ozone
Particulates over the Pacific in 2003
Urban Heat Islands

Climatic conditions are extremely varied across Earth, with large areas of tropical rainforest, extremely dry deserts, and cold polar and high-elevation regions. But variations through time have been (and will continue to be) equally dramatic. We now turn to an examination of climatic distributions across the globe and the reasons behind their existence, followed by an examination of climatic changes and how atmospheric scientists infer past conditions and predict future changes.

Lowell Glacier in Kluane National Park, Yukon, Canada

15 Earth's Climates

After reading this chapter, you should be able to:

▶ Define climate.

▶ Explain the climatic normal and its limitations as the sole indicator of climate.

▶ Identify the main climate groups in the Koeppen system.

▶ Describe the characteristics of tropical climates.

▶ Describe the characteristics of dry climates.

▶ Describe the characteristics of midlatitude climates.

▶ Describe the characteristics of polar and highland climates.

Although we don't always know in advance what type of weather will occur at any particular place on a particular day, we do have some idea of what type of weather is considered normal for a given location. The scene in the opening photo probably fits with everyone's mental image of northern Russia's frigid winters. The relative warmth of Miami, Florida, in March makes it an attractive destination for countless Midwestern students on spring break, whereas a vacationer planning to be there in August knows to expect heat and stifling humidity (along with correspondingly low hotel room rates). Along the same lines, the searing dryness of summer in Death Valley, California, is entrenched in nearly everyone's mind and could not be more different. However loosely formed, these are conceptualizations of climate, our focus for this chapter. There are two key aspects to such thinking. First, the impression about a place's climate is based on many years. We would not revise our impression of Siberia based on a single unusually warm January. Second, the weather events of a particular day are of no real interest. In describing the Siberian winter it makes no difference whether a storm comes on a Tuesday or Wednesday; we know winter storms will come, and that when they do it will mean snow, and we know the extreme cold of the place means it will not readily melt. Thus climate deals with what is typical of a location, as opposed to the passage of daily weather occurrences. In this chapter we examine the issues involved in classifying climate, and we describe the broad climatic regions across the globe.

◄ Russia's subarctic climate stretches from this village near Finland to the Pacific coast thousands of kilometers to the east.

Climate and Controlling Factors

We define **climate** formally as the statistical properties of the atmosphere. By far the most common climatic statistic is an average calculated over a number of consecutive years. Notice that in using an average value we deliberately ignore differences from one year to the next. Using temperature as an example, one year in the averaging period might have been abnormally warm and another very cold, but the average does not reflect those differences nor does it represent the most extreme value observed. Rather, the average reflects the middle of the distribution of temperatures and is a valid definition of "typical" for the period. Two obvious questions arise with regard to climatic averages. First, how long a period should be used? Too short a record will fail to remove enough year-to-year variation, and too long a record runs the risk of missing any changes in climate that occurred over the period. Standard practice regards a 30-year record as the appropriate trade-off. A second question is "What time period?" That is, what 30-year period do we select? It seems natural to use the most recent 30 years in order to capture the present climate, but that would call for yearly updates in climate statistics. Instead the convention is to use the most recent 30-year periods ending on a decade, such as the period 1981–2010. Taken together these two conditions define the **climatic normal,** which has been in use world-wide for about 75 years. When a TV reporter says "normal for the date is 73 °F," he or she is almost certainly referring to the climatic normal.

It is worth realizing that the climatic normal is merely a convention and may not be suitable for some uses. For example, electric energy companies trying to estimate power demands find the normal to be a relatively poor predictor of the future. Nevertheless, the climatic normal is preeminent and reinforces the idea of climate as "average weather." But climate is more than average values. It also includes variation from one year to the next. For example, two places might have similar long-term average rainfall. But if one place typically has both very high and very low values, while the other tends to receive nearly the same total year in and year out, we would surely say they have different climates. After all, in one location the average provides a good estimate regarding the amount of rain that will fall, whereas in the other it is a very poor indicator. This year-to-year variability is another statistical property and is thus a measure of climate.

In addition to variability, we might be interested in the degree to which above-average or below-average values tend to come in runs or be clustered in successive years. For example, do periods of drought tend to be followed by wet periods? Or are extended episodes of extreme heat and cold a common occurrence? These are questions about another statistical property, namely, the correlation between values in successive years. Again, this is part of climate. The point we want to make is that climate consists of many statistical properties, not just average values. For our purposes, a consideration of average values will be sufficient, but a complete description of climate would include much more.

Notice that the definition of climate is not limited to just one or two variables. Temperature and precipitation are surely important elements, but climate also includes wind, cloudiness, and net radiation, to name just a few others. Thus the climate of a place involves a wide array of variables. It should be obvious that because no two places will have identical statistical properties across the full range of atmospheric variables, in a technical sense every location has a unique climate. We might speak of Chicago's climate, but its Miracle Mile along Lake Michigan has a different climate than the area surrounding O'Hare Airport just to the west. Likewise, no place on Earth has a climate exactly like that of the Miracle Mile. But for many purposes small differences in climate can be safely ignored. We know rain-fed corn can be grown across wide swaths of the Midwest United States with little risk of complete crop failure, and we know the same attempt would be well beyond foolhardy for the entire African Sahara. In this book we will therefore work with climates on a large spatial scale, encompassing areas that are typically hundreds to thousands of kilometers wide.

On large spatial scales the factors determining climates can be readily identified. The change of solar radiation input with latitude comes to mind immediately. Thus to a first approximation, symmetric climatic bands centered on the equator is a reasonable guess and has some basis in reality. However, other controls exert their influence to create a considerably more complicated pattern. As discussed in Chapters 3 and 4, the uneven distribution of land and ocean is an obvious factor. Similarly, the role of elevation is self-evident, with high-altitude locations colder than would be expected solely on the basis of latitude. The importance of prevailing atmospheric and ocean circulations can hardly be overstated, and for many places they exert a dominant influence on the resulting climate. Notice that depending on location, controlling factors can interact with one another. Thus prevailing winds blowing across a cold ocean surface toward land yield a very different climate than similar winds moving over a warm tropical sea (see Figure 15–1 for an example). Climatic

▲ **FIGURE 15–1** Elevation interacts with prevailing winds and cold sea-surface temperatures to create fogs that nourish desert vegetation in the bone-dry Atacama Desert of South America.

gradients on the windward side of mountains are very different from those on the leeward side, but the effect varies by latitude. The variety of combinations and interactions is huge, so much so that a general discussion of climatic controls is not very useful. Therefore, throughout our discussion of individual climates we will make reference to underlying controlling factors. We will rely on the reader to be familiar with the relevant background material as presented in previous chapters.

Although the delineation of distinct climates may seem like a very straightforward endeavor, establishing the criteria by which climates are separated requires considerable subjectivity. Consider what you would do if called on to devise a scheme by which Earth's surface would be covered by distinct climatic zones, each having properties that set them apart from the others. The job would require you to set boundaries that separate one climate zone from another. Yet in nature such clear boundaries are rare. Thus, there is a considerable difference in temperature and precipitation along the east coast of North America from Florida to the Maritime Provinces of Canada, and you certainly would not want to put St. John's, Newfoundland, in the same climatic zone as Tallahassee, Florida. But exactly where would you draw the lines that separate the various climates? And how would you decide how many climates? Too many would make the system too complex; too few would fail to capture the patterns you would want to identify.

Climatologists over the years have made numerous attempts to establish useful climatic classification schemes. Some are based on the obvious properties of temperature and precipitation. Others use the frequency with which air mass types occupy various regions, differences in energy budget components, or seasonal characteristics of the water balance at the surface. Each has its own advantages, depending on the purpose of the classification. Agriculturalists, for example, would probably be most interested in using a classification that yields information on water availability relative to plant needs, reflecting gains and losses of water in the soil column (precipitation, evapotranspiration,[1] runoff, and losses to groundwater). For more on this topic, see *Box 15–1, Special Interest: Different Climate Classification Schemes for Different Purposes.*

Checkpoint

1. What is climate?
2. What considerations enter the decision regarding the length of period to use in a climatic average?
3. What are some of the factors that determine the climate of a location?
4. What are some of the problems in dividing the Earth into climatic zones?

---

[1]Evapotranspiration is the combination of water directly evaporated at the ground surface and that absorbed by plants and evaporated through their leaves.

The Koeppen System

For many people, a climatic classification scheme based on temperature and precipitation is useful because it yields information on the two meteorological variables of greatest general interest. The most widely used systems based on these variables have followed on the work of Vladimir Koeppen,[2] a German citizen of Russian ancestry. The **Koeppen system** was developed over the period from 1918 to 1936 in a process of almost continual revision and refinement. Koeppen looked at the world distribution of natural vegetation types, located the boundaries that separated them, and determined what combinations of monthly mean temperature and precipitation were associated with those boundaries. Thus, although its climatic types are determined by temperature and precipitation, Koeppen's system is inherently tied to natural vegetation. Contrary to what one might assume, it does not begin with the idea of "natural" temperature/precipitation regimes. The boundaries are associated with plant associations, which may or may not coincide with what seem to be obvious or striking gradients in temperature and precipitation.

This chapter will use a version of the Koeppen system as modified by Trewartha. Various versions of the Koeppen system apply different names for each of the climates and may have slightly varying criteria for distinguishing them. Thus, our maps and descriptions are necessarily somewhat different from some other portrayals of the Koeppen scheme. Moreover, this scheme is totally descriptive and does not attempt to explain the causes of the various climates. However, as seen in Figure 15–2 on page 446, in most cases climates can be related to the various controlling factors mentioned above.

Koeppen used a multitiered classification system that delineated primary climates by capital letters ranging from A through E. These five broad categories tend to arrange themselves across Earth's surface in response to the latitude, degree of continentality, and location relative to major topographic features. In addition to these climates, the version we are using includes an additional one for mountain environments, designated by H. The main climate groups (designated by first letter only) can be briefly described as follows:

- **A—Tropical.** Climates in which the average temperature for all months is greater than 18 °C (64 °F). Almost entirely confined to the region between the equator and the tropics of Cancer and Capricorn.
- **B—Dry.** Potential evaporation exceeds precipitation.
- **C—Mild Midlatitude.** The coldest month of the year has an average temperature higher than –3 °C (27 °F) but below 18 °C (64 °F). Summers can be hot.
- **D—Severe Midlatitude.** Winters have at least occasional snow cover, with the coldest month having a mean temperature below –3 °C (27 °F). Summers are typically mild.
- **E—Polar.** All months have mean temperatures below 10 °C (50 °F).

---

[2]Also spelled Köppen.

15–1 SPECIAL INTEREST

Different Climate Classification Schemes for Different Purposes

The Koeppen system is undoubtedly the most widely used for distinguishing and mapping the climates of the world, in large part because it is based on easily obtainable data and attempts to have boundaries that coincide with observable vegetation borders. But there really is nothing magical about it—it is simply one of many schemes that have been developed over the years. Like all the others, it is a human construct, created with certain goals in mind. It follows, therefore, that the climates that emerge from Koeppen's classification rules are also human constructs, rather than objective entities waiting to be discovered by Koeppen and his followers. In other words, although the mechanics of climatic classification involve the development and use of objective rules, in a larger sense climate classification is a decidedly subjective process. What constitutes a "good" classification depends very much on judgment and purpose. By no means, then, should one take Koeppen's scheme as "correct" or even "best." In fact, even its admirers readily admit to several shortcomings.

One of the most important shortcomings of the system is that it is based on vegetation boundaries that have been associated with monthly values of mean temperature and precipitation. This is problematic because these two variables alone do not directly determine the geographic limits of natural vegetation. A superior system would be based on the factors that play a more direct role in determining the geographic limits of vegetation, the most prominent of which are precipitation and potential evapotranspiration.

In tandem, the opposing effects of evapotranspiration and precipitation determine the *water balance*. Wherever evapotranspiration exceeds precipitation, the amount of moisture in the soil is reduced, as its loss is not offset by soil moisture inputs. When precipitation exceeds evapotranspiration, the soil moisture is replenished until it reaches the maximum amount of water that can be retained by the soil against the force of gravity (field capacity). Typical plots of monthly water budgets reflecting these inputs and outputs are shown in Figure 1.

Thornthwaite's classification system, based on the principle of the water balance, evolved through decades of work that culminated in its final form in 1955. The system uses four criteria for delimiting climate regions. The first criterion is a *moisture index* that compares the amount of average precipitation each month to the potential evapotranspiration. The latter value derives from a formula using mean temperatures and the monthly values of average period of daylight (a function of latitude) for each station to establish a monthly moisture index. These monthly values are then summed to produce an annual moisture index, the value of which distinguishes arid, semiarid, subhumid, humid, and perhumid (very humid) climates. These divisions are based on arbitrary percentage changes in the moisture index (20 percent changes for the humid climates, 33 percent changes for the dry climates), not from associations with plants or other nonclimatic phenomena.

The second criterion is the *thermal efficiency* of a location, or the total amount of potential evapotranspiration. The remaining two criteria are based on the seasonality of precipitation and potential evapotranspiration. When combined, the four criteria create a more physically based climate classification scheme than that of Koeppen.

So why hasn't the Thornthwaite system supplanted Koeppen's as the most popular? This is partly explained by its greater complexity. Compared to Koeppen's system, the water-balance computations needed by Thornthwaite's method are laborious, and the resulting regions are therefore quite removed from the underlying climatic data. Thus, the pattern of climates that emerges is harder to interpret in terms

Did You Know?

Climate classifications have been derived based on some unconventional but nonetheless significant climatological elements. For example, scientists have calculated and mapped the average apparent temperatures (an index of temperature discomfort due to a combination of high temperature and humidity, which was discussed in Chapter 5) for the hottest 12 days of the year for weather stations across the United States. The maps indicate that typically the most severe heat occurs in the region extending from south Texas northeastward to about St. Louis, Missouri. There the heat index on the hottest days of the year exceed those found in much of southern Arizona or the Florida panhandle.

Climatic zones A, C, D, and E are based on temperature characteristics. The A climates (tropical) tend to straddle the equatorial regions; C, D, and E climates usually occur sequentially further from the tropics and toward the polar regions. The sole primary climate type that considers precipitation is the B climate, designating deserts and semideserts.

The A through E climates are subdivided into smaller zones represented by a second letter, which are further subdivided. We will use subdivisions only through the third letter; thus, each individual climate is represented by a three-letter combination describing its temperature and precipitation characteristics. The descriptions for each of the three-letter climates are presented along with their distributions in Figure 15–3 on page 446. While this system has been taught to countless numbers of students over the years, there is little insight to be gained by memorizing the exact temperature and precipitation values that define the climate boundaries. Thus, the critical values are omitted from the table.

of large-scale processes. Also, although the basic concept of potential evapotranspiration is widely accepted, the method developed by Thornthwaite does not follow from physical principles, but instead relies on data collected mainly in the eastern United States. The data were used to construct empirical (observation-based) equations for potential evapotranspiration. Unfortunately, Thornthwaite did not publish details regarding how the equations were developed, and there are questions as to how well they work in other areas. Thus, if Thornthwaite's method were applied to the entire globe, it is not clear how meaningful the resulting regions would be.

Of course, the Koeppen and Thornthwaite systems are not the only systems that have been devised. Most of the others are designed for more specific applications than either of these two. Some, for example, have identified regions of human comfort, whereas others consider the dominance of different types of air masses. The familiar map of plant-hardiness zones found on seed packages is another example. More sophisticated classifications have used energy budget considerations. Regardless of the premises on which they are based, each has its own set of advantages and disadvantages.

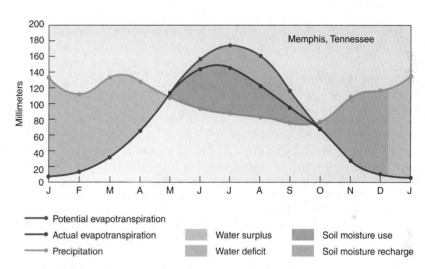

▲ **FIGURE 1** Water balance diagrams for Irkutsk, Russia, and Memphis, Tennessee.

For the A climates, the second letters *f, m,* and *w* indicate if and when a dry season occurs. Af climates have no dry season at all. Am climates are the monsoonal climates in which a short dry season is normally experienced while the rest of the year is rainy. Aw climates have a distinct dry season, usually coinciding with the seasonal presence of the subtropical high pressure of the Hadley circulation. The dry climates are divided into two classes: the true deserts (BW) and the semideserts (BS). The second letter of the C and D climates signifies the timing of the dry season. The letter *f* indicates no dry season at all (as with the A climates), while *s* and *w* represent dry summers and winters, respectively. The second letter of the E climates (capitalized) distinguishes polar tundra (ET) regions from areas covered by glaciers (EF). The third letter of each climate represents the temperature regime. As you can see, the meaning of the third symbol varies among the major groups.

Tropical Climates

The name of this climatic group could not be more straightforward or accurate. Tropical climates exist almost entirely between the tropics of Cancer and Capricorn. The tropical group consists of three climates, each of which is warm year-round, with only minor—and in some cases, minimal—variation in temperature throughout the year. The three climates are distinguished by their different degrees of precipitation seasonality. The **tropical wet** climate has significant rainfall every month of the year, the **tropical wet and dry** climate has a pronounced dry season, and the **monsoonal** climate undergoes relative dryness for 1 to 3 months but receives sufficient moisture that vegetation need not be adapted to seasonal drought. All three tropical climates are dominated by the seasonal movement of the Hadley cells, described in Chapter 8.

▶ **FIGURE 15–2** A generalized map of Koeppen climates for the continents and important controlling factors. Note the importance of prevailing pressure and winds, ocean currents, and topography.

Legend (Figure 15-2):

- Tropical
- Dry
- Moist, mild winter
- Moist, severe winter
- Polar
- Mountains

→ Warm ocean current
→ Cold ocean current
→ Wind direction
---- Intertropical Convergence Zone (ITCZ)

A TROPICAL HUMID CLIMATES

	Af	**Tropical wet** No dry season
	Am	**Tropical monsoonal climate** Short dry season
	Aw	**Tropical wet and dry climate** Winter dry season

B DRY CLIMATES

	BWh	**Subtropical desert** Low-latitude dry
	BSh	**Subtropical steppe** Low-latitude semi-dry
	BWk	**Midlatitude desert** Midlatitude dry
	BSk	**Midlatitude steppe** Midlatitude semi-dry

C MILD CLIMATES

	Csa **Csb**	**Mediterranean** Summer dry
	Csb	**Midlatitude** Dry, warm summer
	Cfa	**Humid subtropical, no dry season** Hot summer, no dry season
	Cwa **Cwb**	**Humid subtropical, winter-dry** Hot summer (Cwa) or warm summer (Cwb), winter dry season
	Cfb **Cfc**	**Marine west coast** Mild throughout year, no dry season, warm summer (Cfb) or cool summer (Cfc)

D SEVERE MIDLATITUDE CLIMATES

	Dfa **Dfb**	**Humid continental, hot summer** Severe winter, no dry season, hot summer (Dwa) or warm summer (Dfb)
	Dwa **Dwb**	**Humid continental, warm summer** Severe winter, winter dry season, hot summer (Dwa) or warm summer (Dwb)
	Dfc **Dfd**	**Subarctic, no dry season** Cool summer (Dfc) or cool summer and extremely severe winter (Dfd)
	Dwc **Dwd**	**Subarctic, winter-dry** Cool summer (Dwc) or cool summer and extremely severe winter (Dwd)

E POLAR CLIMATES

	ET	**Tundra** No true summer
	EF	**Polar ice cap** Perennial ice

H HIGHLAND

	H	**Highland** Cold climates due to elevation

▲ **FIGURE 15–3** World map of Koeppen climates. Notice how some climates are much more extensive over ocean than land, whereas others are almost exclusively continental.

Tropical Wet (Af)

The three largest tropical wet climates are found in the Amazon Basin of South America, over western equatorial Africa, and on the islands of the East Indies (see Figure 15–3). As the map shows, the majority of these locations exist within about 10° on either side of the equator, though some are found as far poleward as 20°. Tropical wet climates have no dry period because their position near the equator puts them under the constant influence of the intertropical convergence zone. For this reason, precipitation is almost always convectional, with strong solar heating of the surface triggering brief but heavy thundershowers in the mid- to late afternoon. In the more poleward areas in which Af climates occur, rain often results from orographic uplift of the predominant trade winds. The tropical wet climate of the Atlantic coast of Central America serves as an excellent example of this phenomenon.

Figure 15–4 presents two **climographs** (a climograph depicts monthly mean temperatures and precipitation, with line and bar graphs plotted simultaneously) for Singapore and for Belém, Brazil, typical tropical wet climate stations. Notice that the rainfall is distributed nearly uniformly throughout the year for most Af locations (although Belém has greater seasonality than Singapore), and that all months average at least 22 cm (5 in.) of precipitation. Even more striking is the uniformity of average monthly temperatures, which vary in these examples by only about 2 °C (4 °F).

While temperatures are often high throughout the year, these climates are not among the hottest on Earth. The ever-present moisture availability at the surface allows a large portion of the incoming solar radiation to be expended on evaporation rather than increasing the surface temperature. Furthermore, the convection of humid air promotes the formation of cumulus clouds that scatter much of the incoming solar radiation back to space. Thus, maximum temperatures never come close to those found in the subtropical deserts. On the other hand, the high humidity retards nighttime cooling, so diurnal temperature ranges are low compared to drier

climates. Unlike most climates, the diurnal range in tropical wet climates often exceeds the annual range. Minimum and maximum temperatures normally range from about the low 20s Celsius (low 70s Fahrenheit) in the morning to the low 30s Celsius (high 80s Fahrenheit) in the afternoon.

In addition to the unsurpassed consistency in temperature and precipitation, areas having tropical wet climates tend to have the same type of natural vegetation—the tropical rainforests—which house a very dense canopy of tree cover and a tremendous species diversity in both the plant and animal kingdoms (Figure 15–5).

Monsoonal (Am)

The monsoonal climate can be thought of as a transition between the tropical wet to the tropical wet and dry climates. Monsoonal climates usually occur along tropical, coastal areas subjected to predominant onshore winds that supply warm, moist air to the region throughout most of the year. Such areas are found along northeastern South America; southwest India, near the eastern Bay of Bengal; and in the Philippines.[3] These areas do not extend nearly as far inland as the tropical wet climates, because their existence depends largely on the effect of speed convergence that occurs as offshore air reaches the coast. Rainfall in these climates is also enhanced by orographic uplift. Thus, localized convergence from surface heating is much less a factor in causing precipitation than it is in the tropical wet climates. During the low sun season, some precipitation may occasionally result from the passage of midlatitude cyclones migrating unusually far equatorward. Near the end of summer and into early fall, tropical cyclones and hurricanes can also bring heavy deluges.

As shown in Figure 15–6, precipitation does not occur nearly as steadily throughout the year in a monsoonal climate

---

[3]Note that the designation *monsoonal climate* is not synonymous with locations subject to the reversal of the winds discussed in Chapter 8.

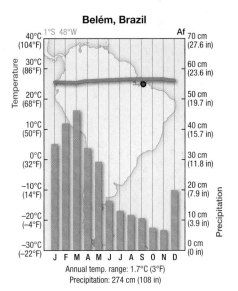

Belém, Brazil

1°S 48°W — Af

Annual temp. range: 1.7°C (3°F)
Precipitation: 274 cm (108 in)

Singapore

1°N 104°E — Af

Annual temp. range: 1.7°C (3°F)
Precipitation: 241 cm (94 in)

◀ **FIGURE 15-4** Climographs for Belém, Brazil, and Singapore, representative of tropical wet climates. The bars plot the monthly mean precipitation (scaled on the right vertical axis). The lines represent mean monthly temperature (scaled on the left).

▲ **FIGURE 15–5** This rainforest in Panama is a response to the year-round warmth and rainfall of tropical moist climates.

as it does in a tropical wet climate. Some months can experience exceedingly heavy rainfall while others are nearly dry. In many cases, the wet months in monsoonal climates yield far more rain than does the wettest month for tropical wet climates. In fact, annual precipitation totals in some monsoonal regions are among the highest in the world, with monthly precipitation values during the peak rainfall periods easily exceeding 80 cm (33 in.). Seasonal totals can even surpass 10 m (400 in.)!

Despite the presence of a brief dry season, monsoonal climates usually support dense forests. In these environments, the soil retains sufficient moisture to maintain the lush vegetation even in the absence of heavy rain for part of the year. In other words, the annual total precipitation is large enough that plants do not experience pronounced moisture stress during the dry season; therefore, vegetation generally

does not require adaptation to drought. Thus, the Am climate is said to have a "noncompensated" dry season, unlike the tropical wet and dry climate. While not as luxuriant or as abundant in species diversity as the tropical wet environments, these locales contain much more living matter than those of the drier, tropical wet and dry climates.

Tropical Wet and Dry (Aw)

Tropical wet and dry climates often occur along the poleward margins of the tropics and border dry climates on one side and tropical wet climates on the other. They are most extensive in South and Central America and southern Africa. Because they are farther from the equator, they undergo much greater seasonality in precipitation than do the tropical wet and the monsoonal climates, and they have somewhat greater seasonal variation in temperature as well.

As with the other two climates of the humid tropics, tropical wet and dry climates owe their existence largely to the Hadley cell. During the high sun season, the Intertropical Convergence Zone favors the formation of afternoon thundershowers. As the position of the overhead sun shifts to the opposite hemisphere, however, the subtropical high arrives to bring descending air and the resultant lack of precipitation. These periods of dryness are more pronounced and longer lasting than those of the monsoonal climate because their distance farther from the equator puts them closer to the mean position of the subtropical high.

Localized convection by solar heating within the ITCZ is not the only process that brings precipitation to tropical wet and dry climates. Tropical depressions can bring widescale precipitation. Along coastal areas, occasional tropical storms and hurricanes can increase average accumulations. Figure 15–7 illustrates the seasonality of temperature and precipitation for typical tropical wet and dry climates. In Acapulco, for example, each of the months between May and October receives an average of at least 12 cm (5 in.) of rain. September is by far the wettest month, with an average of about 36 cm (15 in.) of rain.

▶ **FIGURE 15–6** Climographs for Akyab, Myanmar, and Monrovia, Liberia, representative of monsoonal climates.

Akyab, Myanmar

Annual temp. range: 7.8°C (14°F)
Precipitation: 521 cm (203 in)

Monrovia, Liberia

Annual temp. range: 2.2°C (4°F)
Precipitation: 303 cm (118 in)

◄ **FIGURE 15–7** Climographs for Bamako, Mali, and Acapulco, Mexico, representative of tropical wet and dry climates.

This is largely due to the occasional passage of tropical storms and hurricanes that can dump huge amounts of precipitation. Although most years go by without none of these storms, their occasional occurrence increases the monthly average. During the fall, the monthly precipitation decreases, until the dry months of February–April. On an annual basis, these climates receive less precipitation than do either the tropical wet or the monsoonal climates.

Unlike the other two tropical humid climates, the tropical wet and dry undergo considerable year-to-year variability. Thus, drought episodes can reduce even further the amount of precipitation received in the dry season—often with fatal consequences, as we have witnessed in the Sahel of Africa (see Chapter 8). Likewise, unusually wet rainy seasons can lead to severe flooding and erosion.

Within the year, monthly mean temperatures exhibit more variability than is found in the other tropical climates, but the variability is nonetheless low compared to most others. The annual temperature range is usually between about 3 and 10 °C (5 and 18 °F). Diurnal variations are likewise greater in these parts of the tropical humid environment than in the wetter regions. This is especially true during the dry season, when the absence of clouds facilitates greater daytime heating and nighttime cooling, and daily lows and highs might range between 15 and 30 °C (59 and 86 °F). During the rainy season, the combination of high humidity and cloud cover reduces the diurnal ranges to values similar to those of the tropical wet regions (about 10 °C, or 18 °F).

Tropical wet and dry climates are associated with a natural vegetation type unique among the tropical regions—the **savanna**. This vegetation consists mainly of grasses interspersed with widely separated trees or clumps of trees (Figure 15–8). While it is tempting to attribute the lack of forest to the presence of the dry season, many ecologists doubt the causality of this relationship. Instead they believe that the vegetation complex results from numerous factors, including recurrent fire, waterlogged soils, and the development of hard layers within the soil.

Checkpoint

1. What data are shown in a climograph?
2. In what ways are tropical moist climates and tropical monsoons similar? How do they differ?
3. What is a savanna, and with what climate is it associated?

Dry Climates

It may surprise many people to learn that the definition of a dry climate is not based on precipitation alone. In other words, there is no set value of annual precipitation (for example, 10 cm) that makes a region considered to be arid. Instead, aridity also depends on the potential evapotranspiration, as well as the timing of the precipitation relative to the period of peak potential evapotranspiration. For our purposes, it is sufficient to state that climates are dry when annual potential evaporation exceeds the annual precipitation. These climates occupy about 30 percent of Earth's land surface, which is considerably more than any other climate group.

The dry regions of the world can be divided two ways: by the degree of aridity (that is, true deserts versus the less arid semideserts) and by their latitudinal position (hot versus cooler dry areas). **Semideserts** are transitional zones that separate the true deserts from adjacent climates. They are also called

▲ **FIGURE 15–8** Forests of tropical moist climates give way to savannas in the tropical wet and dry climate.

steppe climates, with reference to the associated vegetation type consisting of short grasses. True deserts are so dry that only a sparse vegetation consisting entirely of xerophytic species (that is, adapted to drought conditions) can take hold.

Deserts and semideserts can be classified as either *subtropical* or *midlatitude*. Subtropical dry regions extend across wide expanses of land between the latitudes of about 10° to 30° in either hemisphere and result from large-scale sinking air motions during most of the year. Notable examples include the desert of southern California–Arizona–Baja California, the Sahara Desert of North Africa and the Arabian Peninsula, and most of the Australian interior.

Midlatitude deserts and semideserts usually occur to the east of major topographic barriers that create a strong rain shadow or over interior continental regions well removed from moisture sources. They are found over large portions of the western United States and interior Asia.

The two-tiered system of categorization yields four types of dry climates: **subtropical desert**, **subtropical steppe**, **midlatitude desert**, and **midlatitude steppe**.

Did You Know?

One of the driest desert climates of the world is separated from a tropical wet climate by a mere 250 km (150 mi). The coastal area of northern Peru occupies part of the Atacama Desert and receives on average less than 2.5 cm (1 in.) of precipitation annually. The desert is abruptly truncated by the rugged Andes Mountains, which separates it from the large Af climate to the east, where precipitation exceeds 285 cm (114 in.) per year.

Subtropical Deserts (BWh)

As shown in Figure 15–3, the most extensive areas of desert exist in the subtropical regions, particularly within the western portions of the continents. The most important factor in the formation of the subtropical deserts is the subsidence associated with the Hadley circulation. As we saw in Chapter 8, the subtropical highs of the Hadley circulation do not appear as continuous bands at the surface, but rather as semipermanent cells. Although the Hawaiian and Bermuda–Azores high-pressure systems at the surface contract and weaken during the winter, subsidence nonetheless dominates within the middle troposphere. This causes stable conditions to exist in the middle troposphere, which restricts uplift and inhibits precipitation, creating the major subtropical deserts exemplified by the scene in Figure 15–9.

Some deserts occur in subtropical regions as narrow strips along the west coast of continents, adjacent to cold ocean currents. The Atacama Desert along the west coast of Chile has the lowest average precipitation on Earth and provides an excellent example of this phenomenon (Figure 15–1). As air flows out of the subtropical high pressure system in the eastern part of the South Pacific, it flows over the cold Peru current, and the lower portion of the atmosphere is cooled. This cooling can bring the air temperature down to the dew point so the air becomes damp and foggy. It also lowers the environmental lapse rate and causes the air to be extremely stable. Though the air advected off the coast is damp, stable conditions suppress uplift so several years can go by without any precipitation at all in some areas. Similar west-coast deserts are found in Baja California, Namibia, and Morocco.

Over many subtropical deserts, the precipitation that does occur often comes in the form of localized showers from summertime convectional activity. This is not the case for all subtropical deserts, however, as illustrated in Figure 15–10. Yuma, Arizona, and Cairo, Egypt, are both located at about the same latitude, in the midst of subtropical deserts. While both receive scant precipitation over the course of the year, Cairo's precipitation occurs mainly in the winter while Yuma's is about equally divided between the late summer and winter months. As is the case over much of southern Arizona, August usually marks the peak of what is locally known as the *Arizona*

▶ **FIGURE 15–9** Climographs for Yuma, Arizona, and Cairo, Egypt, representative of subtropical deserts.

▲ **FIGURE 15–10** A hot subtropical desert scene from the Sahara desert of North Africa.

Did You Know?

In some parts of South America and Baja California, villagers in remote desert areas use a simple but ingenious system for capturing drinking water. Nylon mesh screens, each about one square meter in area, are extended between supporting posts so that passing advection fogs condense water onto their surfaces. As the fog accumulates on the mesh screens, the surplus drips down into collecting systems and provides enough drinking water to sustain small communities.

monsoon. Although it has little similarity to the wet season of the Asian monsoon, the Arizona version does undergo a shift in the airflow that brings damp air into the area in late summer. This influx of moisture is subject to strong surface heating that can lift the air sufficiently to trigger isolated thunderstorms. While these thunderstorms can be intense and cause flash flooding, they are neither frequent nor strong enough to make the area anything other than the true desert that it is. The winter precipitation at Yuma results from the passage of midlatitude cyclones. While these systems can bring precipitation across a wide swath of the country, often in the form of heavy rain showers or snowstorms, over the desert the moisture supply is usually too low to allow much precipitation to fall. This is because Pacific moisture is blocked by the western mountains, and Gulf of Mexico air does not usually flow this far westward.

Daytime summer temperatures in subtropical deserts can be extremely high. In fact, the hottest locations in the world are all found in these regions. During the summer, the combination of low humidities, high sun, and clear skies allows high inputs of solar radiation to be absorbed at the surface. Furthermore, the lack of soil moisture (except after recent rain showers) causes the ground temperatures to become extremely high as little heat is expended in the evaporation of water. Thus, it is not uncommon for daytime temperatures to reach as high as 45 °C (113 °F) or higher. After the Sun sets, the clear skies and low humidities that led to rapid heating also allow the air to cool considerably. As a result, diurnal temperature ranges can be very large.

The same applies to the annual temperature ranges. At Baghdad, Iraq, for example, the mean monthly temperature in August is 35 °C (95 °F), which is 25 °C (45 °F) higher than the January mean of 10 °C (50 °F). This type of temperature range is greater than that found in any other climate of the tropics or subtropics. As we will see, however, midlatitude deserts commonly have even greater temperature ranges.

There is a widely heard axiom in regional climatology that areas of low annual precipitation also have the greatest amount of interannual variability. This certainly applies to subtropical deserts, where many years can go by with only minimal rainfall, only to be followed by a season having numerous rain showers, delivering several years' worth of precipitation. Thus, the annual average is a poor indicator of how much rain is likely to fall in a given year. Figure 15–11 plots annual rainfall at Borrego Springs, California, using a 45-year period of record as an example.

Subtropical Steppe (BSh)

All the conditions that distinguish a subtropical desert also apply to subtropical steppe climates—though to a lesser degree. Like their more extreme desert counterparts, subtropical steppes are marked by aridity, high year-to-year variations in precipitation, extreme summer temperatures, and large annual and daily temperature ranges. It is therefore not surprising that subtropical steppe climates commonly border the subtropical deserts. And though they are transitional between deserts and nonarid regions in terms of their climatological characteristics, they are not necessarily narrow buffer zones. Much of the southwestern United States and northern Mexico, for example, is occupied by subtropical steppe.

Cloncurry, Australia, and Monterrey, Mexico (Figure 15–12), illustrate the greater precipitation totals, somewhat cooler conditions, and lower annual temperature ranges associated with these climates. These two examples also reveal a distinct seasonality in the precipitation regime typical of subtropical steppes located on the equatorward side of deserts. In these regions, precipitation occurs more often during the summer months than during the winter, as a result of localized convection and tropical disturbances. In contrast, steppe regions on the poleward side of subtropical deserts experience most precipitation in the winter in response to the passage of midlatitude cyclones.

Midlatitude Deserts (BWk)

Midlatitude deserts result from extreme continentality. Such regions occur deep within continental interiors or downwind of orographic barriers that cut off the supply of moisture from the ocean. The greatest expanse of midlatitude desert occurs in Asia—which is not surprising considering the immense size of that continent. The two major areas of midlatitude desert in Asia are found just east of the Caspian Sea and north of the Himalayas. Both are far to the east of the Atlantic Ocean, so much of the moisture associated with eastward-moving cyclones

▶ **FIGURE 15–11** Annual precipitation at Borrego Springs, California, 1951–1995. Dry climates such as this one normally have a wide year-to-year variability in precipitation.

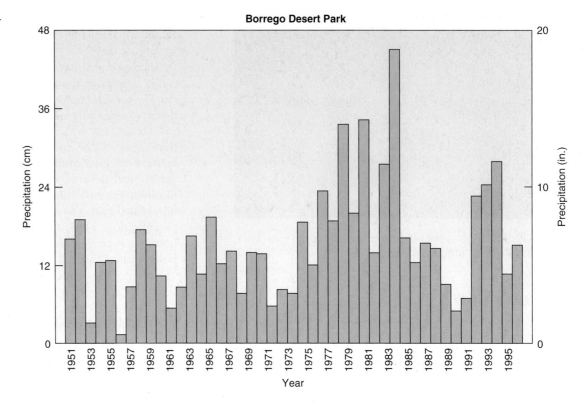

is depleted before reaching either area. Mountains to the south block the northward flow of moisture out of the Indian Ocean.

The second greatest expanse of midlatitude desert occurs in the western United States. The midlatitude desert extends southward and merges with the subtropical desert of the Southwest. Although the midlatitude desert lies poleward of the subtropical desert, it should not be assumed that it is neither as dry nor as hot in the summer as the desert to the south. In fact, Death Valley, California, one of the hottest and driest places in the world, lies within the midlatitude desert.

Midlatitude desert in the Southern Hemisphere is confined to a narrow strip in South America, east of the Andes. A quick look at Figure 15–3 reveals that the midlatitudes of the Southern Hemisphere are almost completely covered by ocean. Thus, it is rare to see any type of land climate in this region, let alone a midlatitude desert.

Figure 15–13 shows two typical climographs for midlatitude deserts. Midlatitude deserts have a greater range of temperatures, both on a daily and annual basis, than do their subtropical counterparts. While both types of deserts become

▶ **FIGURE 15–12** Climographs for Cloncurry, Australia, and Monterrey, Mexico, representative of subtropical steppe climates.

◄ **FIGURE 15–13** Climographs for Lovelock, Nevada, and Kuqa, China, representative of midlatitude desert climates.

extremely hot during summer days, midlatitude deserts have more rapid nighttime and winter cooling. With more precipitation and lower potential evapotranspiration, they are generally more humid than subtropical deserts. Thus, although vegetation is adapted to dry conditions, ground cover is likely to be continuous, without large patches of bare ground common in hotter (subtropical) desert regions.

Midlatitude Steppe (BSk)

The midlatitude steppe accounts for most of the arid regions of western North America. It flanks the northern portion of the midlatitude desert and merges with the subtropical steppe to the south. A large swath of steppe also extends from the Great Plains, east of the Rocky Mountains, all the way from northeast Mexico into western Canada. The midlatitude steppe region of Asia is in many places a fairly narrow strip of land surrounding the true deserts.

Midlatitude steppes have the same temperature characteristics as the midlatitude deserts. The primary difference

between the two is the greater amount of precipitation in the steppes, which commonly totals about 50 cm (20 in.) annually (Figure 15–14).

Checkpoint

1. Explain why low rainfall is an inadequate indicator of dry climates.
2. What accounts for the band of subtropical deserts found between north and south latitudes 10° and 30°?
3. What accounts for midlatitude deserts and steppes?

Mild Midlatitude Climates

The mild midlatitude climates are located in parts of the latitude range between 30° and 60° in either hemisphere. They occur as long, narrow strips of land along the west coasts of North and South America and southern Australia. In North and

◄ **FIGURE 15–14** Climographs for Denver, Colorado, and Semey, Kazakhstan, representative of midlatitude steppe climates.

South America, the eastern border of the climate is delimited by mountains. Another west coast mild midlatitude climate—in fact, the largest—surrounds the Mediterranean Sea.

Mild midlatitude climates also cover large areas of the eastern portions of the continents, especially in North and South America and Asia. While these areas extend farther inland than do their counterparts along the west coasts, these mild midlatitude climates have narrower latitudinal extents. While west coast midlatitude climates can be found as far poleward as just a few degrees shy of the Arctic Circle, most on the eastern sides of continents do not extend poleward of 40° latitude.

The individual climates within this general group do not share similar precipitation patterns. In fact, the precipitation regime can vary greatly from one region to another. To get an idea of the extent of this variability, consider the fact that along the west coast of the United States, the mild midlatitude includes San Diego, California, which receives about 25 cm (10 in.) of annual precipitation, and the Olympic rainforest in Washington State, where the precipitation locally exceeds 375 cm (150 in.) per year.

The term *mild* refers to the winter temperatures and not necessarily those of the summer. In North America, for example, this climate group is found over inland areas of California where summer temperatures routinely exceed 38 °C (100 °F), as well as in the Gulf states of Florida, Alabama, Mississippi, and Louisiana—places hardly noted for their mild summers. Thus, in effect, *mild* refers to little or no snow cover.

This climate group is subdivided into three climates, two of which exist along the west coasts of continents and the other on the eastern sides. **Mediterranean** climates can be found along the west coasts between about 25° and 40° latitude. Within about the same range of latitude on the eastern side of continents are the **humid subtropical** climates. The **marine west coast** climates lie adjacent to and poleward of the Mediterranean climates.

Mediterranean (Csa, Csb)

Mediterranean climates are the only extensive climates with a distinct summer dry season and a concentration of precipitation in the winter. Figure 15–15 clearly depicts this pattern for two typical sites, Athens, Greece, and Los Angeles, California. The summer aridity in mediterranean climates is attributable to the presence of semipermanent subtropical high-pressure systems offshore. For North America, the Hawaiian high-pressure system to the west "blocks" the eastward migration of summertime midlatitude cyclones and deflects them to the north. This deflection of the storms, coupled with subsidence along the eastern portion of the subtropical high, deprives coastal southern California of the uplift mechanisms necessary for precipitation. During the winter months, the Hawaiian high weakens, shrinks in size, and migrates toward the equator. This opens the way for the arrival of Pacific cyclones.

Annual precipitation increases with latitude and with elevation along windward slopes in mediterranean climates. The latitudinal gradient results from the more frequent passage of midlatitude cyclones at the higher latitudes. At the same time, the mountain slopes induce orographic uplift of the predominantly westerly airflow.

While the climographs in Figure 15–15 reveal the mean precipitation patterns for the two sites, they do not reveal the extreme variability inherent with winter rainfall. Indeed, some winters can be quite dry, while others bring an onslaught of sequential storms that produce major flooding and hillside erosion. This climatic feature has been particularly prominent since the 1970s in southern California, which has experienced several record drought years along with extremely wet winters.

Winter temperatures in mediterranean climates are usually mild, especially right along the coast. Inland temperatures occasionally drop below freezing and sometimes threaten fruit-growing areas with widespread crop damage. On the other hand, precipitation results mainly from the passage of midlatitude cyclones, which transport relatively warm,

▶ **FIGURE 15–15** Climographs for Los Angeles, California, and Athens, Greece, representative of mediterranean climates.

Los Angeles, California

40°C (104°F) · 34°N 118°W · Csb · 70 cm (27.6 in)
30°C (86°F) · 60 cm (23.6 in)
20°C (68°F) · 50 cm (19.7 in)
10°C (50°F) · 40 cm (15.7 in)
0°C (32°F) · 30 cm (11.8 in)
−10°C (14°F) · 20 cm (7.9 in)
−20°C (−4°F) · 10 cm (3.9 in)
−30°C (−22°F) · 0 cm (0 in)
J F M A M J J A S O N D
Annual temp. range: 9.4°C (17°F)
Precipitation: 38 cm (15 in)

Athens, Greece

40°C (104°F) · 38°N 24°E · Csa · 70 cm (27.6 in)
30°C (86°F) · 60 cm (23.6 in)
20°C (68°F) · 50 cm (19.7 in)
10°C (50°F) · 40 cm (15.7 in)
0°C (32°F) · 30 cm (11.8 in)
−10°C (14°F) · 20 cm (7.9 in)
−20°C (−4°F) · 10 cm (3.9 in)
−30°C (−22°F) · 0 cm (0 in)
J F M A M J J A S O N D
Annual temp. range: 17.8°C (32°F)
Precipitation: 38 cm (15 in)

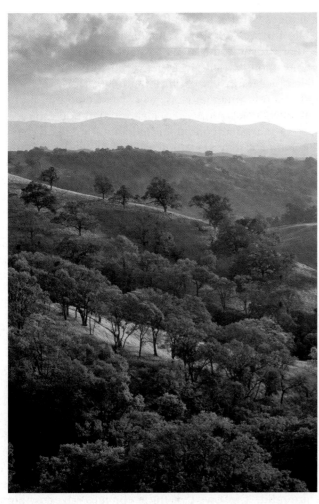

▲ **FIGURE 15-16** An oak woodland typical of the Mediterranean climate of coastal California.

maritime polar air. Thus, precipitation in the lower elevations falls almost exclusively as rain and not snow. Figure 15–16 shows vegetation common in the California version of the Mediterranean climate.

Did You Know?

Mediterranean climates are far from uniform. Summer temperatures range from mild to hot, with daily highs typically decreasing toward the coast and with increasing latitudes. Central California provides a good example of how variable the summer temperatures can be in mediterranean climates. San Francisco is noted for particularly cool summers, while about 120 km (70 mi) to the northeast, Sacramento experiences hot, dry conditions.

Humid Subtropical (Cfa, Cwa)

Humid subtropical climates occur within the lower middle latitudes of eastern North America, South America, and Asia. Typical temperature and precipitation patterns are shown in Figure 15–17.

Though they exist in the middle latitudes, these climates have a distinct tropical feel during their long summers. Located to the west of large semipermanent anticyclones and warm ocean currents, the prevailing winds circulate hot, humid air into these climatic zones. Summer daytime temperatures are usually in the lower 30s Celsius (high 80s to low 90s Fahrenheit), and dew points in the mid-20s Celsius (mid-70s Fahrenheit) help retard nighttime cooling. As a result, hot, muggy conditions remain throughout the day and night, especially along the equatorward boundaries of the climates. Fortunately, afternoon convectional thundershowers are common in these areas and bring temporary relief from the extreme heat.

Winter temperatures are typically lower than those of mediterranean climates farther to the west because of their greater continentality, and subfreezing temperatures are not uncommon. The occurrence of frost and snow decreases toward the lower latitudes, but even south Florida is not completely immune.

Humid subtropical areas receive abundant precipitation, ranging from about 75 to 250 cm (30 to 100 in.) per year. Over most areas the maximum precipitation is concentrated in the

◄ **FIGURE 15-17** Climographs for Dallas, Texas, and Guangzhou, China, representative of humid subtropical climates.

summer, but this generalization does not always hold. Over most of the southeastern United States, for example, summer is the wettest season. But the area extending from east Texas into Tennessee and Kentucky has a winter precipitation maximum. Regardless of which season receives most precipitation, summer is always the season of moisture deficit, because of greater potential evapotranspiration.

Precipitation during the summer is largely convectional in nature and tends to be scattered and brief. Winter precipitation, on the other hand, is usually associated with midlatitude cyclones. Over the coastal areas, tropical storms and hurricanes are capable of bringing extreme rainfall from time to time during the late summer and early fall.

Marine West Coast (Cfb, Cfc)

Marine west coast climates (Figure 15–18) normally occur poleward of mediterranean climates and, as the name would suggest, along the west coasts of climates. But inspection of Figure 15–3 shows that the latter is not always the case, as areas having the climate are found covering New Zealand and along southeastern Australia and southeast Africa. While not truly located on west coasts, these areas are located east of fairly narrow strips of land that do not greatly modify the maritime nature of the predominantly westerly flow. This exposure to air passing over cold ocean currents and their location in the path of eastward migrating midlatitude cyclones gives these climates their characteristic features.

Both summers and winters are typically mild. Extremely high summer temperatures are certainly not unknown in these climates, but they are the exception rather than the norm. From northern California through Alaska, for example, low and midlevel cloud decks frequently keep daytime temperatures down and often bring drizzle or light rain to coastal regions. Along the lower elevations, the percentage of days experiencing precipitation can be high, but the total amount of rainfall is usually light. On the other hand, the

Coast Ranges parallel the Pacific coastline and create a major barrier to the westerly flow. This creates a strong orographic effect that causes some areas to have extremely heavy precipitation amounts that rival or exceed those of the tropics. The Olympic Peninsula of Washington State is an excellent example of this phenomenon, where extremely heavy rainfalls are sufficient to support a lush, midlatitude rainforest.

Just as maritime conditions moderate temperatures during the summer, they also allow for mild winter conditions at surprisingly high latitudes. Even as far north as Sitka, Alaska (57° N), the coldest month of the year has a mean temperature above the freezing point of water. At many lower-latitude locations, low-elevation snowfall is a rarity; when it does occur, it usually melts away within a short time. In Europe, where the marine west coast climate extends farther inland than it does in North America, snow is more frequent and remains on the ground for a longer period of time.

Did You Know?

Seattle, Washington, has a reputation for experiencing a lot of rain each year. But it actually receives less annual precipitation on average—93 cm (37 in.)—than does New York City, with 124 cm (50 in.). The difference is that Seattle has an average of 150 rainy days each year, but these are usually of low intensity. New York averages only 120 days with measurable precipitation annually, but that city usually experiences more intense precipitation events.

Given the fact that both summer and winter temperatures are mild, it follows that these climates have low annual temperature ranges. This contrasts sharply with the situation for climates in the eastern portion of continents at the same range of latitudes. These climates have moderately high summer temperatures, but when combined with the extreme cold of winter, they yield extremely high temperature ranges of the severe midlatitude group of climates.

▶ **FIGURE 15–18** Climographs for Greenwich, England, and Sitka, Alaska, representative of marine west coast climates.

Chicago, Illinois

Shenyang, China

Annual temp. range: 26.1°C (47°F)
Precipitation: 85 cm (33 in)

Annual temp. range: 38.3°C (69°F)
Precipitation: 64 cm (25 in)

◄ **FIGURE 15-19** Climographs for Chicago, Illinois, and Shenyang, China, representative of humid continental climates.

Checkpoint

1. In the context of the Koeppen classification system, what does it mean to say that a midlatitude region has a mild climate?

2. How are mediterranean and marine west coast climates similar? How are they different?

Severe Midlatitude Climates

The severe midlatitude climate group includes two climates, humid continental and subarctic, both of which are marked by very cold winters. These climates require large continental areas within the high-middle latitudes—between about 40° and 70°—in order to avoid the moderating effects of an ocean. Thus, they are restricted to Europe, Asia, and North America and are not found at all in the Southern Hemisphere. As expected from climates resulting from strong continentality, they exhibit large annual temperature ranges.

Both of the severe midlatitude climates receive precipitation throughout the year and have no true dry season. In many locations, there is greater precipitation in the summer than winter. Precipitation during the summer can result from local convection or by the passage of midlatitude cyclones; winter precipitation (mostly as snow) results almost entirely from cyclonic activity.

Humid Continental (Dfa, Dfb, Dwa, Dwb)

A huge segment of the population of the United States, Canada, eastern Europe, and Asia live in **humid continental** climates (Figure 15–19), including the inhabitants of New York, Chicago, Toronto, Montreal, Moscow, Warsaw, and Stockholm. This is the more temperate of the two severe midlatitude climates, normally found between about 40° N and 55° N in the eastern parts of continents. Summers are warm and often hot. New York City, for example, has an average

high temperature in August of 29 °C (84 °F). But winter is another matter altogether, with a mean February maximum temperature of 4 °C (40 °F). It should be noted that New York City is located near the southern boundary of the climate and on the coast, so temperatures are milder than those of most other locations with a humid continental climate.

Mean annual precipitation in these climates usually ranges between 50 and 100 cm (20 to 40 in.). Over the United States and southern Canada, there is a conspicuous decrease in precipitation with increasing latitude and distance from the Atlantic shoreline, reflecting a reduced moisture content in the atmosphere.

Subarctic (Dfc, Dfd, Dwc, Dwd)

Subarctic climates occupy the northernmost extent of the severe midlatitude regions, with more than half of the area of Alaska and Canada having this type of climate. In North America, the coniferous forest that dominates the region is referred to as the **boreal forest** (Figure 15–20); in Asia, it

▲ **FIGURE 15-20** Vast tracts of the Canadian and Siberian subarctic climates are covered by evergreen coniferous forests.

goes by the name **taiga**. Summer temperatures are somewhat lower than those of the adjacent humid continental regions, but the major difference occurs in winter, when mean monthly temperatures can be extremely low. At many locations, the monthly mean temperature can remain below the freezing level for up to seven months. Thus, winter is long and separated from the brief summer only by a short-lived autumn and spring. Figure 15–21 highlights the very large annual temperature ranges that result from the mild summers and severe winters.

Typically, precipitation is greater in the summer than winter, mostly because of the more poleward displacement of midlatitude cyclone tracks in summer. Nonetheless, annual precipitation is usually low, ranging from about 12 to 50 cm (5 to 20 in.).

Polar Climates

Polar climates exist in the highest latitudes—typically areas poleward of about 70°. Such regions occur in the Northern Hemisphere, across northern Canada, Alaska, Asia, and Greenland. In the Southern Hemisphere they are almost entirely confined to the continent of Antarctica. As we mentioned earlier, the climate boundaries of the Koeppen system were set to coincide with boundaries of differing natural vegetation types. In this case, polar climates begin at the high latitude boundaries of the subarctic climates, where vast expanses of coniferous forest give way to a generally treeless landscape. The polar climate group consists of two distinct types. The most equatorward and milder of the two is the **tundra,** which has at least one month with an average temperature above 0 °C (32 °F). At the most poleward regions of the globe lie the true ice cap climates.

Did You Know?

The polar front is a vitally important factor in much midlatitude weather and climate. Yet it can be found year-round in only one continent: North America. Continental polar air can be blocked from making contact with maritime polar air over Eurasia by the Himalayan Mountains. Australia and South America often do not have continental polar air because of their small amount of land mass at high latitudes.

Tundra (ET)

Tundra climates are named for the associated vegetation type that consists primarily of low-growing mosses, lichens, and flowering plants, with few woody shrubs and trees. The boundary dividing the tundra from the subarctic climate occurs where the mean monthly temperature of the warmest month does not exceed 10 °C (50 °F). In the Northern Hemisphere, this occurs in the general vicinity of 60° N. This climate is almost entirely missing from the Southern Hemisphere, where there is only minimal land coverage at this latitude.

Tundra climates have severe winters in which the Sun rises only briefly each day and never gets very high above the horizon. Under these conditions, radiational cooling at the surface leads to low temperatures and strong stability—both factors that inhibit precipitation. Thus, as shown in Figure 15–22, low winter precipitation is the rule. Surprisingly, perhaps, the winters in the tundra regions are often less severe than those of the adjacent and lower-latitude subarctic climates. This is because the tundra regions of North America, Greenland, and Asia are located nearer to major water bodies and thus have a lesser degree of continentality. (Even when ice-covered, an ocean has a somewhat moderating effect on winter temperatures.)

▶ **FIGURE 15–21** Climographs for Verkhoyansk, Russia, and Dawson, Yukon, Canada, representative of subarctic climates.

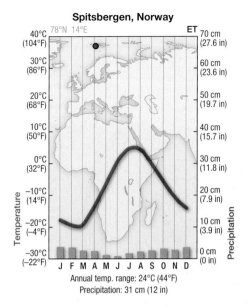

Spitsbergen, Norway

Annual temp. range: 24°C (44°F)
Precipitation: 31 cm (12 in)

Barrow, Alaska

Annual temp. range: 34°C (61°F)
Precipitation: 13 cm (5 in)

◄ **FIGURE 15–22** Climographs for Spitsbergen, Norway, and Barrow, Alaska, representative of tundra climates.

During the summer, tundra regions have very long periods of daylight, but again the Sun never gets very high above the horizon. As a result, temperatures are normally mild in the midafternoon, not very much greater than those of the predawn period. Thus, despite the fact that annual temperature ranges are fairly high, daily temperature ranges are low.

One very conspicuous feature of tundra regions is the existence of **permafrost**, a perennially frozen layer below the surface. During the winter, conditions are so cold that the entire soil is completely frozen, in some places to a depth of several hundred meters. When summer arrives, there is enough warming right at the surface to melt the uppermost soil, and a relatively shallow layer of thawed soil overlies the completely impermeable frozen layer several tens of centimeters below. As a result, any rain or snowmelt that occurs at the surface is unable to permeate deeply into the soil, and the upper part of the surface becomes saturated. This combination of water-logged soil near the surface and a solid ice layer below precludes the establishment of any deeply rooted plants, and only low-growing vegetation can take hold (Figure 15–23a).

Ice Cap (EF)

As the name rightfully suggests, **polar ice cap** areas exist where ice is present throughout the entire year (Figure 15–23b). On land they are confined to the Greenland interior and most of Antarctica, and over ocean they are found in the interior of the Arctic sea. They exist where the mean temperature of the warmest month does not go above the melting temperature for ice, 0 °C (32 °F). On the continents ice accumulations over these regions can exceed several kilometers, so in addition to being found at high latitudes they also occupy high elevations—a situation perfect for maintaining extraordinarily low winter temperatures (Figure 15–24).

The cold air overlying polar ice caps (also called *continental glaciers*) becomes extremely dense and frequently flows down the margins of the ice under the force of gravity. When funneled through narrow canyons, these katabatic winds (discussed in Chapter 8) can become extremely strong.

Most areas of ice cap receive little precipitation because of the intense cold. On the other hand, all precipitation falls as snow and has the potential to accumulate onto existing ice. Except along the margins, there is no melt and only a

(a)

(b)

◄ **FIGURE 15–23** (a) Long days in the tundra provide a short growing season suitable for flowers, grasses, and low-growing plants, but woody vegetation is nearly absent. (b) In most of the ice cap climate snow persists throughout the year, building large continental ice sheets.

▶ **FIGURE 15-24** Climographs for Little America, Antarctica, and Eismitte, Greenland, representative of ice cap climates.

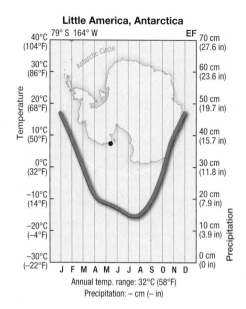

Little America, Antarctica

79° S 164° W

Annual temp. range: 32°C (58°F)
Precipitation: – cm (– in)

Eismitte, Greenland

70° N 40° W

Annual temp. range: 36°C (65°F)
Precipitation: 10 cm (4 in)

small amount of sublimation to remove the ice. Yet some ice does get removed. Beneath the surface, the individual ice crystals merge into an almost solid block of ice. When subjected to the constant pressure of the overlying mass, the ice gradually deforms and expands outward toward the margins of the ice sheet. At the continental margins, the ice often breaks off into large icebergs that float in the Arctic and Antarctic waters.

Highland Climates (H)

Highland climates are unique among those in the Koeppen system because their distribution is not governed by geographic location but rather by topography. These climates are found in large mountain or plateau areas. As we saw in Chapter 1, the temperature tends to decrease with height in the troposphere. Thus, in high mountains there can be large changes in mean temperature over short distances, simply as a function of elevation. The temperature situation becomes more complicated when one considers that changes in slope angle and aspect over short distances influence the intensity of solar radiation received.

Precipitation type and intensity also vary spatially across highlands. Mountain slopes can enhance precipitation on their windward sides and simultaneously create a rain shadow downwind. Elevation difference also affects the ratio of precipitation falling as snow versus rain, with higher elevations favoring a greater amount of snow. Mount Kilimanjaro in Tanzania provides a striking example of this phenomenon. Located very near the equator at 3° S, Kilimanjaro possesses an extraordinarily wide range of climates from its base near sea level to its peak at 5895 m (19,340 ft). Near the surface, the mountain has a tropical

wet and dry climate similar to the surrounding area. But the climate changes with height and the mountain eventually becomes covered with perennial ice over its higher reaches. This is an example of *vertical zonation,* the layering of climatic types with elevation in mountainous environments. Thus, although we use a single designation H, it must be understood that this category contains an extremely rich collection of climates. This is true when considering zones within a single H climate and also when comparing one H climate with another. Thus the vertical zonation of the tropical Andes is distinctly different from that in the Canadian Rockies.

Did You Know?

Mount Washington, New Hampshire, set three world wind speed records on a single day—April 12, 1934. The peak gust of 373 km/hr (231 mph) and the 5-minute average speed of 303 km/hr (188 mph) were new records, while the mean wind speed for the entire day—206 km/hr (128 mph)—tied the record set the day before at the same location. To put these wind speeds in perspective, recall that the minimum wind speed for a Category 5 hurricane is 250 km/hr (155 mph).

Checkpoint

1. What are the characteristics of the subarctic climate such as Verkhoyansk, Russia (see Figure 15-21)?
2. Why is there relatively little precipitation in polar climates?
3. Why are highlands treated as a separate climate zone?

Summary

People have long recognized the desirability of a classification system by which distinct climates could be delineated and mapped. This pursuit is neither as straightforward nor as easy as one might expect, because climate consists of a number of different elements. Thus, a certain degree of ingenuity is necessary to create a scheme that includes the major climatic variables and still retains enough simplicity to make the climates comprehensible. The Koeppen system has become the most widely used of all of these techniques.

Koeppen devised a system in which temperature and precipitation characteristics along vegetation boundaries were used as criteria for distinguishing one climate from another. The scheme is a hierarchical, multitiered system that uses combinations of capital and lowercase letters.

The first level of categorization in the Koeppen system uses a capital letter from A through E, along with a special category, H, for highland climates. All but one of the A through E climates are based on temperature characteristics. The one exception, the B climates, represent areas of aridity and are called *dry climates*. The temperature-based climates—A, C,

D, and E—are referred to as tropical, mild midlatitude, severe midlatitude, and polar climates, respectively.

The second level of categorization depends in most cases on the seasonality of precipitation, although for the dry climates the second level reflects the level of aridity. The third level of classification typically refers to temperatures during particularly critical months. Having established the various climates, one can refer to a map such as that shown in Figure 15–3 to visualize their geographic distribution and understand the processes that give rise to the various types.

At the outset of this chapter we distinguished short-term changes in the atmosphere (weather) from long-term conditions (climate). This might lead us to believe that climates are permanent characteristics that never change. This is not true, however, as the climate is forever undergoing slow changes through time. In fact, looking over the history of the planet, we see that there have been many time periods in which the climates were far different from what we know today. The next chapter of this book addresses those temporal changes in climate.

Key Terms

climate *page 442*

climatic normal *page 442*

Koeppen system *page 443*

Thornthwaite's classification system *page 444*

tropical wet *page 445*

tropical wet and dry *page 445*

monsoonal *page 445*

climograph *page 447*

savanna *page 449*

semidesert *page 449*

steppe *page 450*

subtropical desert *page 450*

subtropical steppe *page 450*

midlatitude desert *page 450*

midlatitude steppe *page 450*

mediterranean *page 454*

humid subtropical *page 454*

marine west coast *page 454*

humid continental *page 457*

subarctic *page 457*

boreal forest *page 457*

taiga *page 458*

tundra *page 458*

permafrost *page 459*

polar ice cap *page 459*

highland *page 460*

Review Questions

1. Describe what is meant by *climate*.

2. Describe the general criteria by which the Koeppen system delineates climates.

3. The first order grouping of climates in the Koeppen system is based mainly on temperature. Which climate type departs from that rule?

4. Describe the geographical distribution of tropical climates. What features distinguish this particular group?

5. Briefly describe the fundamental differences between Af, Am, and Aw climates.

6. Of the three types of tropical climates, which occupies the smallest portion of Earth's land surface?

7. Despite their low latitudes, tropical climates are not among the hottest on Earth. Why not?

8. Where are the various dry climates located, and what geographical characteristics cause them to occur where they do?

9. Describe the four types of dry climates and explain how they differ from each other.

10. What factor other than annual precipitation is involved in a climate being defined as dry?

11. Describe the various types of mild midlatitude climates and their distribution. Why is it that two of them locate mostly along the west coast of continents, while the other tends to be on the eastern side?

12. Are the mild midlatitude climates really mild? Explain.

13. What are the two types of severe midlatitude climates, and how do they differ?

14. Why are severe midlatitude climates missing from the Southern Hemisphere?

15. Describe the three types of polar climates and their distributions.

Critical Thinking

1. Although the Koeppen system is intended to create distinct climate classes, its boundaries are based on the boundaries between vegetation types. Is this really a problem? Can you think of any alternative methods to delineate climates?

2. Do you anticipate that global warming, if it continues as expected, will substantially alter the location of the boundaries of the various Koeppen climates?

3. Western Kansas is located near the junction of B, C, and D climates. Do you suppose that a person driving around this region would notice substantial climatic differences as she crossed from one climate zone to another? What does this tell us about the applicability of large-scale classification schemes to smaller-scale analysis?

Problems and Exercises

1. Check Figure 15–3 and determine the type of climate you live in. Then log on to **ggweather.com/normals/ index.htm** and find the climate information for the location nearest to where you live. Create a climograph for that location and compare it to the example given in the book for the type of climate you live in. How closely do they match?

2. Go to the Web page for the National Weather Service office nearest to where you live. Compare the monthly temperature and precipitation values observed there over the last year to the average for that location. Were they markedly different? How much variance do you expect to encounter between monthly observed values and climatological averages?

Useful Web Sites

www.fao.org/WAICENT/FAOINFO/SUSTDEV/EIdirect/ climate/EIsp0002.htm
Offers a large number of climate maps compiled by the United Nations Food and Agriculture Organization.

ggweather.com/normals/index.htm
A great source of information for monthly average temperature and precipitation data.

www.ncdc.noaa.gov/oa/climate/research/monitoring.html
Online access to many reports from the National Climatic Data Center.

www.ncdc.noaa.gov/oa/climate/severeweather/extremes .html
Provides data on extreme weather events.

www.cpc.ncep.noaa.gov
Home page of the NOAA Climate Prediction Center.

MyMeteorologyLab™

The *Understanding Weather and Climate*, Sixth Edition, MyMeteorologyLab™ Web site contains numerous multimedia resources to aid in your study of **Earth's Climates.**
Visit **www.MyMeteorologyLab.com** to:

- **Review** key chapter concepts.
- **Read** the Pearson eText, *In the News* RSS feeds, and other chapter-specific Web resources.
- **Visualize** challenging topics with Interactive Tutorials, Geoscience Animations, MapMaster interactive maps, and Weather in Motion movies and visualizations, and more.
- **Test Yourself** with self-study quizzes.

 TUTORIAL

GLOBAL CLIMATE: CONTROLS AND PATTERNS

Use the tutorial to explore Earth's climate regions and the factors responsible for them.

WEATHER IN MOTION

View movies and visualizations of meteorology patterns and processes.

Operation IceBridge in Greenland
Lightning Seasonality
Studying Fires Using Multiple Satellite Sensors
Climate, Crops, and Bees
Diurnal Variability in Global Precipitation

16 Climate Changes: Past and Future

After reading this chapter, you should be able to:

▶ Define climate change and explain how it can be described in terms of changes to boundary conditions in Earth's climate system.

▶ Explain methods used for determining past climates.

▶ Explain the time scales over which climate changes can occur.

▶ Describe major climate changes that have occurred over geologic time.

▶ Identify millennial-scale oscillations and annular modes.

▶ Explain factors involved in climate change, including changes in solar output; Earth's orbit; land configuration and surface; atmospheric turbidity; and radiation-absorbing gases.

▶ Explain feedback mechanisms and Earth-system interactions involved in climate change.

▶ Describe the role of general circulation models in identifying causes of climate change and projecting future changes in climate.

The opening photo pair offers just one indication of recent climate change. Glaciers in many other locales have experienced similar retreat, and many other types of climatic indicators also signal change over the last century or so. These include direct measurements of temperature, as discussed in Chapter 3, as well as a host of indirect measures typified by the opening photos. For example, satellite measurements show that Arctic sea ice has decreased in recent years to the point that the fabled Northwest Passage connecting the Atlantic and Pacific oceans has opened repeatedly since 2007. Indeed, in the last 5 years no fewer than a dozen craft ranging from inflatable boats to cruise ships have made the crossing in a single season. The fact that no openings could be inferred on the basis of satellites prior to 2007 was widely reported in the press and fueled the stories of global warming. A little more analysis of the Northwest Passage story illustrates some of the issues regarding climate change. First, the satellite measurements go back only to 1978. Is it possible the Arctic was open in prior years? Extrapolating a short record is obviously fraught with difficulty. Second, there were repeated crossings between 1978 and 2007, which shows the satellite record is not perfect, and illustrates the general problem of drawing conclusions from indirect measures of climate. Third, even accepting the recent record as correct, how can we be sure the imputed warming is climate change and not merely a string of somewhat warmer years, soon to be followed by more typical cold conditions? Finally, assuming we are confident a climate change has occurred, how can we identify the underlying cause?

This chapter will explore all of these issues. In particular, we will review past climates and the methods used to infer them, the factors that might be responsible for climate change, and the use of general circulation models to study potential human effects.

◀ Photos of Argentina's Upsala Glacier taken in 1928 (top) and 2004 (bottom) reveal a dramatic loss of ice consistent with global warming. Once the largest glacier in South America, at present its length decreases by about 200 meters per year.

Defining Climate Change

We saw in Chapter 15 that climate is defined as the statistical properties of atmospheric variables, including temperature, precipitation, and wind. Thus, climate change can be defined as a change in any statistical property of the atmosphere, such as a change in mean temperature. We also saw in Chapter 15 that climate is more than just the average, or mean, value. Year-to-year variations, seasonal variations, and the tendency for above- or below-normal years to occur in sequence are also a major component of climate. Thus, changes in climate may occur even though the mean values of precipitation, temperature, and wind remain the same over time. For example, even with no change in annual average precipitation, changes in the timing of drought and heavy rainfall years could have profound consequences for people and would certainly be considered climatic change.

Consider how changes in climate might occur, thinking first in very general terms. In some ways a planet's climate is like a system that responds to the configuration of external factors, often called **boundary conditions**. For example, the importance of boundary conditions is clear when we compare Earth and Titan, a moon of Saturn somewhat larger than the planet Mercury. Like Earth's, the Titan atmosphere is mostly nitrogen, but as Table 16–1 shows, its boundary conditions and resulting climate could hardly be more different.

Whereas Earth's surface temperature averages about 15 °C (59 °F), on Titan the global mean is –180 °C (–292 °F)! Earth has a hydrological cycle that replenishes atmospheric water every 10 days or so. That is, Earth clouds shed a few centimeters of water every couple of weeks. On Titan methane plays the role of water, forming clouds, rain, snow, lakes, and ice. However, the hydrological cycle is many thousands of time slower on Titan, with centuries of drought interspersed by relatively brief flood-producing downpours. As you know, the Hadley cells on

(a)   (b)

▲ **FIGURE 16–1** A desert scene on Titan. The dunes are similar in shape to wind-blown deposits found in Arabia, but the "sands" comprising them are more like coffee grounds. It is thought that they develop from hydrocarbons precipitating from the sky and joining to form larger grains.

Earth create moist equatorial climates bounded by deserts in the subtropics. But on Titan equatorial latitudes are covered by vast deserts containing dunes 100 m (300 ft) or more in height, with lakes and flowing liquids found in middle and polar latitudes (Figure 16–1).

These differences are largely explained in terms of different boundary conditions. With solar radiation 100 times weaker than on Earth, Titan's low temperature is no surprise. Likewise, the weak solar constant gives an evaporation rate only 1/100th that of Earth. Because methane is chemically very different from water, Titan's atmosphere is able to hold much more liquid—about 10 m (30 ft)—despite its lower temperature. The low evaporation rate and huge store of liquid combine to give the very slow "hydrologic" cycle. On Earth the large overturning Hadley cells carry heat from areas of intense solar heating poleward, but the Coriolis force confines these vigorous circulations to the low latitudes. With rising air common over the equator and sinking air in the subtropics, Earth has a moist equator and dry subtropical zones. Something analogous to Hadley circulation develops on Titan, but Titan's slow rotation rate makes it very different. The resulting weaker Coriolis force, coupled with the long year, gives rise to one huge cell extending from the summer hemisphere midlatitudes to the winter pole. Rising air does not persist over Titan's equator, and thus the equatorial zone is a desert (albeit one with hydrocarbons at the surface rather than the silica sands found on Earth).

TABLE 16–1

Approximate Values for Climatically Important Variables on Earth and Titan

	Earth	**Titan**
Boundary Conditions		
Main gas	Nitrogen	Nitrogen
Surface pressure	1013 mb	1455 mb
Cloud-forming gas	Water vapor	Methane
Rotation rate	1 per day	1 per 15 days
Length of year	365 days	10,759 days
Solar constant	1400 W/m^2	15 W/m^2
Selected Climate Variables		
Global surface temperature	15 °C	−180 °C
Precipitation/evaporation rate	1 m/year	0.01 m/year
Liquid inventory of atmosphere	.025 m	10 m
Precipitation cycle	9 days	1000 years

Notice the importance of external factors in Titan's climate: weaker solar radiation, slow rotation, a different gas (methane) involved in the precipitation cycle. If any of these were to change, we would expect the system to adjust accordingly. Obviously, this is true for Earth as well. Thus, for example, if the Sun's output were to increase, we would expect an increase in global average temperature. Some years might be cooler than before, but on average we would expect a warmer climate, which is, of course, a change in a statistical property. External conditions might vary too quickly for any sort of climatic equilibrium to be obtained, but we can nevertheless think of the external conditions as driving climatic change, or as acting as **forcing agents**.

It must be noted that the response to external factors can be heavily modified by internal **feedback processes**. Think, for example, about the factors governing winter climates in northern Canada. It should be obvious that as solar radiation inputs decline following the summer solstice, temperature declines as well. Once the surface becomes snow-covered, a smaller fraction of solar radiation is absorbed, contributing to further cold. The high albedo off snow amplifies the effect of decreasing incoming solar radiation (see Figure 16–2). This is an example of a *positive* feedback, one that amplifies a given change. A *negative* feedback does just the opposite, namely, it damps out or at least partially offsets a change. In our Canada example the increasing temperature gradient across the middle latitudes gives rise to a negative feedback. A strong gradient favors more storminess (chiefly midlatitude cyclones), which acts to move heat northward. This makes Canadian winters warmer than they would be otherwise and clearly works contrary to decreasing sunlight. As will be seen later, a host of amplifying and damping feedbacks are involved in climate change.

Recognizing there are both direct and indirect (feedback) effects, climate change can be defined as the response of the Earth–atmosphere system to changes in boundary conditions.

Before adopting this very appealing (and widely used) view of climate change, we need to ask two questions. First, we wonder if a given set of boundary conditions uniquely determines Earth's climate. Putting the same question differently, we could ask, "Is more than one climate possible for a single set of boundary values?" Interestingly, both theory and observations suggest the answer might be "yes." However counterintuitive it might seem, this definitely occurs in systems far less complicated than Earth's climate. The technical term for this is **intransitivity**. If Earth's climate is intransitive, that obviously complicates the problem of knowing if boundary conditions have forced an observed change and also makes it difficult to ascribe changes to particular causes.

A second question is, how can we detect climatic change? Imagine a perfect instrument, one that could measure the state of the atmosphere at any moment in time throughout Earth's history. Measurements from such a device could be used to compute statistical properties for various periods, which in turn would provide evidence for climatic change. We might, for example, average temperature values over successive 100-year periods to get an idea of century-to-century change. We wonder, "Is it possible for statistics computed like this to change over time, even with no change in boundary conditions?" The answer, again, is "yes." Using elementary statistical theory, it is easy to show that statistical measures computed like this are subject to variations independent of those arising from changes in boundary conditions. Thus, for example, even though the true mean is unchanging, the individual 100-year means rise and fall from one century to the next. (Other statistics behave similarly.) In physical terms, the problem arises because short, unpredictable variations associated with weather events spawn variations on longer time scales.[1] Short-term departures from average become large-term departures and contaminate whatever changes

---

[1]We discussed how this happens in Chapter 13 under the topic of chaos.

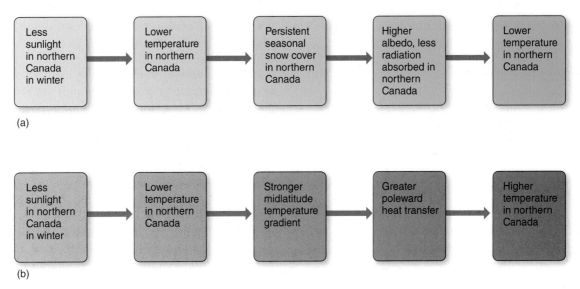

(a)

(b)

▲ **FIGURE 16-2** Positive (a) and negative (b) feedbacks affecting winter temperatures in Canada.

16–1 PHYSICAL PRINCIPLES

Blaming Climate Change

When unusual weather events occur—record heat, a Category 5 hurricane, prolonged summer drought or flooding—the public naturally asks if climate change is the reason. There is no good answer to this kind of question. For example, although it might be tempting to fault global warming for a heat wave or blame ENSO for a powerful storm, there is almost always no way to connect climate change to a particular event.

We know that the intrinsic variability of the atmosphere ensures that both routine and unusual events will occur in the absence of any long-term change. For example, global warming might increase the mean temperature at a location and also the occurrence of week-long runs of high temperature, say, greater than 100 °F. If so, we can be sure that more frequent heat waves are part of global warming. But knowing that heat waves are more common does not allow one to say that a particular

heat wave would not have occurred anyway. It is a little like switching from a fair coin to one that is more likely to turn up tails. If you observe a run of 5 tails, it's not certain that the coin's bias is at fault, because there is 1 chance in 32 of seeing a run like that even with a fair coin. This same reasoning applies to climate change, and, indeed, the fundamental goal is to understand analogous changes in statistical properties.

---

are caused by external conditions. So, if we observe the 100-year mean temperature changing, it is not clear that we are seeing climatic change—it might be an artifact of working with finite-length samples. Obviously, there is no way around this problem; using longer averaging periods reduces the sampling error but increases the likelihood of blurring truly distinct climates (those arising from changing external conditions). *Box 16–1, Physical Principles: Blaming Climate Change* discusses a related problem, that of relating individual events to climate change.

In light of these difficulties, we will need to broaden our view of climate change to include all change, whether driven by boundary values or not. In other words, in this book we do not distinguish between change generated by internal processes and change generated by external processes. Regardless of the "instrument" we use (thermometer, ice accumulation in glaciers, tree-ring width), if the statistical values change over time, we will call that climatic change. Admittedly, this is not ideal (change now depends in part on the sample length), but it does have great practical appeal and avoids the complications just described.

Checkpoint

1. What are the boundary conditions of a system?
2. How do positive and negative feedbacks affect changes in climate induced by outside agents? Explain.

Methods for Determining Past Climates

Though huge gaps exist in our knowledge of past climates, scientists have learned much about the climatic history of Earth. It is really quite remarkable that we are able to describe past climates as well as we can. Consider the fact

that although Earth formed about 4.5 billion years ago, people are believed to have inhabited the planet for only about 2 million years, or about 0.05 percent of the planet's existence.

You can also view the relative time spans another way: Imagine the planet formed at midnight on January 1 of some vast cosmic year, and it is now midnight 1 year later. In this scenario people would have first appeared on the scene about $3^{1}/_{2}$ hours ago, at 8:27 P.M., December 31. The first thermometer would have appeared about 2 seconds ago. Thus, throughout the vast majority of Earth's history, nobody was around to observe and record the climate. The lack of firsthand information is further worsened by the fact that, throughout most of human history, no direct records of climate were preserved.

We continue to gain insights into past climates based on information left in the geological and biological records through what are called **proxy** indicators. Proxy measures have been obtained in different places, using different techniques, and brought together by scientists in various disciplines. Taken all together, proxy data gives us some idea of what **paleoclimates**, the climates of the past, must have been like. Not surprisingly, the further back we go in time, the less detailed is the information provided by the various indicators.

We now briefly describe some methods for studying past climates, beginning with techniques useful for uncovering change on the longest time scales.

Oceanic Deposits

Scientists have been drilling into the ocean floor since the 1970s. They extract deep cores of material that has been deposited over very long time periods, with more recent material constantly burying older material previously laid down (see Figure 16–3). Included in the deposited material are the bones and shells of plankton and other animal life, made largely of calcium carbonate ($CaCO_3$). The information contained in the oxygen in the calcium carbonate is most important for determining past climates.

▲ **FIGURE 16-3** Ocean sediment cores contain the remains of long-dead animals that reveal climate conditions prevailing during their lifetime. In addition, rocks deposited by melting icebergs give clues about continental ice sheets.

Most oxygen atoms have an atomic weight of 16 (^{16}O), but a small percentage of oxygen atoms contain two additional neutrons per atom, making their atomic weight 18 (^{18}O). Both *isotopes* exist in ocean water, and both are incorporated into the shells and bones of marine animals. If the ratio of ^{18}O to ^{16}O in the water is relatively high, it will also be high in the sea life living in that water. Because ^{16}O is lighter than ^{18}O, water containing ^{16}O evaporates more readily than does water containing ^{18}O. Thus, if glaciers are expanding, the oceans (and the oxygen-containing calcium carbonate in shells and bones) will have relatively high $^{18}O/^{16}O$ ratios, as more of the ^{16}O water is removed from the ocean and deposited as snow onto the growing ice sheets. When the organisms die, they sink to the ocean bottom, where their calcium carbonate is deposited. The ocean bottom thereby maintains a record of climate through the varying ratios of $^{18}O/^{16}O$ in its layers. Scientists extract cores of the ocean-bottom material, note the isotope ratios, and infer past changes in global ice volume. Information obtained from sea cores was instrumental in overturning a long-held, but mistaken, idea that there were four glaciation episodes in the last 2 million years.

Other ocean indicators of past climates come in the form of material removed from land surfaces by glaciers. Icebergs shed by continental glaciers carry rocks, pebbles, and other debris equatorward, and this material is deposited on the ocean floor when the iceberg melts. Called **ice-rafted debris**, this material was a key in discovering a number of very rapid changes in northern hemisphere ice sheets.

Ice Cores

Scientists have also determined $^{18}O/^{16}O$ ratios for deep **ice cores** obtained from the Greenland and Antarctic ice sheets and from alpine glaciers at lower latitudes. When snow is deposited onto existing glaciers, the oxygen in the H_2O may be either ^{18}O or ^{16}O. Snow that falls under relatively warm conditions contains a higher ratio of the heavier isotope. On the ice caps, scientists from Europe and the United States have been drilling through the ice and extracting cores nearly 3 km (1.8 mi) deep to infer past temperature patterns.

In addition to the temperature data obtained from isotope ratios, ice cores provide information on the past chemistry of the atmosphere and on the incidence of past volcanic eruptions. As new snow falls onto a glacier, bubbles of the ambient air become permanently trapped in the ice. The concentration of carbon dioxide and other trace gases in these bubbles yields a long-term record of their varying levels in the air. Among the more interesting results from chemical analyses at the cores is the strong correlation between past temperatures and concentrations of carbon dioxide and methane (see Figure 16–4). Past periods of high temperature coincide with high concentrations, whereas glacial periods coincide with reduced concentrations of these greenhouse gases. Generally the changes in gas concentration lag behind the temperature changes by 800 to 1000 years, which implies that changes in composition are not the root cause of the temperature changes. Instead, this indicates a positive feedback process in which changes in the greenhouse gases amplify the climate change. Thus, for example, warming triggers release of the gases to the atmosphere, and the elevated greenhouse concentration produces more warming.

Did You Know?

Dripping water in caves leaves behind mineral deposits such as stalagmites growing upward from the floor. Known collectively as speleothems, these rocks can be dated and analyzed for changes in oxygen isotope ratio and other indicators of climate. They have recently become an important source of information about past climates on continents, where (obviously) ocean cores are unavailable.

Ice sheets also provide valuable information about major volcanic eruptions. When such eruptions occur, some of the dust ejected into the atmosphere settles on the tops of glaciers. Researchers can determine when heavy volcanic activity occurred by noting the depth of the dust layers within the cores. The chemical composition of aerosols deposited in glacial ice also provides information on past events, with high acidities implying increased volcanic activity. This information may prove to be valuable in determining the importance of volcanic activity as a causal factor for climatic change.

Remnant Landforms

All of Earth's landforms are the end result of processes that build up and wear down features at the surface. The largest features, such as mountains and valleys, are produced by *tectonic* forces, those that produce a deformation of Earth's crusts. Once formed, all such features become subject to

▶ **FIGURE 16–4** Temperature, carbon dioxide, and methane variations for the past 600,000 years based on ice cores. Shaded periods are times of elevated greenhouse gases and warmth.

erosional processes that remove material at the surface and transport it to other locations. When the forces transporting the material are no longer capable of moving the material, *deposition* occurs. There are several mechanisms for eroding and depositing material, including the movement of water, the slow-moving ice sheets expanding across the surface, wave action along coastlines, wind, and floating icebergs carrying land debris. Each of these mechanisms leaves certain telltale characteristics, and trained field scientists can use the evidence to infer climatic conditions at the time of erosion or deposition.

Features Associated with Ice and Water Obviously, glaciers can exist only under cold climatic conditions. Fortunately for the earth scientists, erosional and depositional features caused by the movement of glaciers often leave very distinct signatures on the landforms that last long after the glaciers have fully retreated. Some of the features associated with glaciation are related to erosion. For example, as alpine glaciers expand down preexisting valleys, they widen out the lower reaches of the valleys. This can transform typical V-shaped valleys (looking up or down the valleys), associated with those cut by running water, to U-shaped valleys. Glaciers also often leave scratch marks on solid rock walls and valley floors, or polish exposed rocks to a very smooth finish.

Unlike running water, flowing ice sheets are capable of moving very large-sized sediment as effectively as small particles of sand, silt, and clay. When the ice sheet melts away, its deposits (called *till*) can contain a very wide assortment of sediment sizes, from microscopically small clays to very large boulders. Thus, poorly sorted deposits often exist along the past margin of an ice sheet. Where glaciers terminate in an

ocean, land materials are rafted away and eventually deposited on the seafloor as icebergs melt.

Streams and rivers also provide useful evidence for reconstructing past climates. Depositional features can be particularly useful in this regard, because the size of material that can be transported by running water depends on the speed of flow. Rapidly moving streams can carry large rocks. If the flow of water in a stream begins to slow down, the maximum size of sediment it can transport decreases, and the largest material it contains will be deposited. This allows us to infer that layers made up of very large sediment must have been deposited by large streamflows. Thus, by examining the layering, or *stratigraphy,* of stream banks or road cuts, one can get some idea of the sequencing of high and low precipitation episodes.

Waves along coastlines also leave distinct features that can be left behind after sea level rises or lowers. Thus, by examining the elevation of terraces or submerged coastal platforms, we can infer how much glacial ice has accumulated or melted.

Coral Reefs **Coral reefs** are hard ridges extending from the ocean floor to just below the water surface, along the shallow margins of warm, tropical oceans. They form by the growth of small marine organisms (coral) having hard shells, composed largely of calcium. Colonies of coral live together atop the hard material left behind by past coral. When these coral die, their shells remain at the top of the reef and provide the foundation upon which subsequent coral grow.

Coral reefs have been useful in several ways for the inference of past conditions. Because they exist along shallow waters, relic coral reefs can provide information on the location of past sea levels. Moreover, because the chemical

Weber Lake, northeastern Minnesota

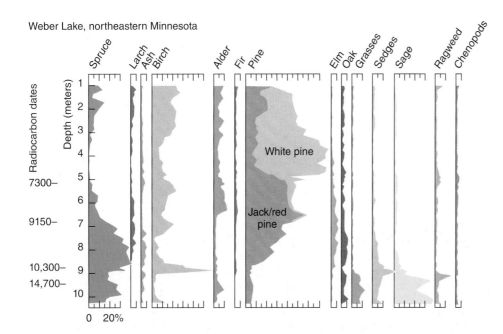

◀ **FIGURE 16–5** Pollen diagrams provide information on past vegetation at a site, which is useful for determining past climates.

composition of growing coral is affected by the water temperature, analysis of the changing chemistry of coral reefs with depth provides useful information on past conditions. They have been particularly useful in providing information on past El Niño events.

Past Vegetation

Climate is by no means the only important factor affecting the distribution of vegetation, but it does exert a strong influence on the distribution of vegetation communities. When a vegetation community occupies a region, some of its pollen and spores can be deposited and preserved indefinitely in lake beds or bogs. This preserved pollen can be extracted and identified by *palynologists*. Organic material deposited with the pollen is often subjected to a technique called **radiocarbon dating**. Radiocarbon dating provides a good estimate of the age of material younger than about 50,000 years and, along with the pollen data, allows a determination of the distribution of vegetation species that existed at various times during the past. This information can then be displayed in pollen diagrams (Figure 16–5) that depict the sequence of past vegetation assemblages. Because many types of vegetation stands are identified with particular climate types, these diagrams provide useful information for deciphering the climatic history of an area.

Much information about past climates extending back for several thousand years can also be obtained from **tree rings**. Each year many trees increase the width of their trunks by the growth of concentric rings, each distinct from the previous rings (Figure 16–6). The width of each ring depends on how favorable temperature and/or moisture conditions were during a given year for the particular tree species. Under climatic stress conditions resulting from a lack of moisture or excessive warmth, the growth of these rings will be retarded. When conditions are favorable for growth, the rings will be relatively thick.

Some tree species are sensitive to temperature variations, whereas other species are affected primarily by changing

moisture conditions. In either case, the extraction of cores from the trunks of very old trees (the oldest ones date back more than a millennium) yields a continuous record of annual tree growth, which correlates with precipitation and/or temperature. The correlation depends on the species and on the environmental setting. At high elevations, for example, low temperatures might slow annual growth, while for the same species at lower elevation, warm temperatures might create slower growth through increased moisture stress. Although

▲ **FIGURE 16–6** Tree rings.

obviously a complicating factor, differential climatic response can be used to advantage by permitting researchers to isolate various types of climatic change from one another. Indeed, such records have been obtained from old stands of trees around the world and provide climatologists with a wealth of information about past conditions. In recent years isotopic analysis of tree cores has added to what can be learned from ring widths alone, and thereby greatly expanded the usefulness of this general method.

Checkpoint

1. How is it possible to infer past climate conditions for times before climate records were kept?
2. Explain how the following can provide evidence for past climates: ocean-floor rocks, ice cores, relict landform, tree rings.

Temporal and Spatial Scales of Climate Change

Meteorologists and climatologists are frequently asked a very straightforward question: Is the climate warming? Although the question is simply phrased, there is no simple response, in part because the answer depends on the time scale involved. Look again at Figure 16–4 and note what was happening 200,000 years ago. A short warming period began a few thousand earlier, thus one could say the climate was warming 200,000 years ago. However, that warm interval appears in the middle of a longer trend toward colder temperatures that didn't end for another 60,000 years. Thus on a longer scale the planet was cooling 200,000 years ago. So, in this example our visual inspection reveals temperature changes occurring on two time scales at once. A more careful statistical analysis would find changes on other time scales, both longer and shorter than what we can easily see.

This is typical of the climatic record. There is no hard-and-fast relationship between the size of a climate change and its time scale; however, as a general rule, oscillations that take place over longer time periods typically have greater magnitudes than those occurring on shorter time scales.

The warming question is also confounded by the fact that climate change varies spatially across the globe. Thus, for example, a temperature trend in the Northern Hemisphere might be absent, or even of opposite direction, in the Southern Hemisphere. Again, there is no fixed relation, but tropical latitudes typically change less than high latitudes. Significant longitudinal variations can also occur, so that large differences emerge *within* a latitude band. This might happen, for example, if the mean position of Rossby waves shifts east or west. In contrast to changes over time, there is a widespread tendency for changes on small spatial scales to be larger than those measured across big areas. Thus, we see that answers to questions about future climate change must be referenced to both a spatial and a temporal scale. Similarly, in looking at past climates, very different pictures emerge, depending on the area considered and time scale adopted.

Past Climates

Earth scientists have devised a widely used scheme to divide the planet's natural history into distinct time frames. The geologic column shown Figure 16–7 uses a hierarchical system dividing time into *eras, periods,* and *epochs.* These time segments are not based on climatic characteristics but rather on geologic and fossil evidence indicative of past environmental conditions and events. Thus, while climatic episodes are sometimes associated with particular eras, periods, or epochs, the time segments should not be considered to have uniform climatic conditions. In addition, significant climate events sometimes cross boundaries in the column, so the divisions do not mark starting or ending times. Finally, the geologic column is not absolute—the names, number, and duration of the time intervals are revised periodically to reflect improved knowledge. Figure 16–7 shows the most recent geologic scale, which was adopted in 2009. For our purposes, the terminology is best used as a kind of calendar, to identify points in time, rather than to identify particular climates or events. With this in mind, we can discuss some significant episodes of Earth's climatic history.

Did You Know?

When sand grains are thrown into the air, solar radiation resets an internal radiometric clock that can be used to date the time of exposure. Relict dunes provide evidence of dry episodes, and their age tells scientists when those episodes occurred. The dating technique, optically stimulated luminescence, was developed in the mid-1990s. It can date sediments with ages ranging from 300 to 100,000 years ago.

Warm Intervals and Ice Ages

We tend to think of our own time as "normal," but this is really not the case at all. Looking at a broad span of Earth's history, we would have to describe the present climate as highly unusual, because most of the time our planet has been considerably warmer than it is today. Unlike today, when the Arctic Sea is mostly frozen all year and huge ice sheets cover the bulk of Antarctica and Greenland, for most of its life Earth has been largely free of permanent (year-round) ice. A more accurate depiction is one of a warm planet punctuated by multiple relatively brief **ice ages**, as seen in Figure 16–8. Though varying, the warm times persist for hundreds of millions of years to billions of years, whereas the ice ages last on the order of tens of millions of years to perhaps a hundred million years. All of human existence, including the historical period, has been spent in the most recent of these great ice ages. Thus, if someone asks if an ice age is coming, the answer is, "No, it's already here."

Erathem Era	System Period	Series Epoch	Duration in Millions of Years	Millions of Years Ago
CENOZOIC	Neogene	Holocene	0.0117	0.0117
		Pleistocene	2.6	2.588
		Pliocene	2.7	5.332
		Miocene	17.7	23.03
	Paleogene	Oligocene	10.9	33.9± 0.1
		Eocene	21.9	55.8± 0.2
		Paleocene	9.7	65.5± 0.3
MESOZOIC	Cretaceous		80	
				145.5± 4.0
	Jurassic		54.1	
				199.6± 0.6
	Triassic		51.4	
				251.0± 0.4
PALEOZOIC	Permian		48	
				299.0± 0.8
	Carboniferous	Penn-sylvanian	19.1	318.1± 1.3
		Missis-sippian	41.1	359.2± 2.5
	Devonian		56.8	
				416.0± 2.8
	Silurian		27.7	
				443.7± 1.5
	Ordovician		44.6	
				488.3± 1.7
	Cambrian		53.7	
				542.0± 1.0
PRECAMBRIAN				

▲ **FIGURE 16-7** The geologic column.

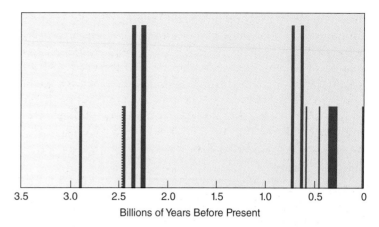

▲ **FIGURE 16-8** Warm intervals and major ice ages throughout Earth history. Heights of bars indicate the spatial extent of glaciations.

The earliest known ice age dates to about 2.8 billion years ago, followed by three others beginning at 2.4 billion years ago. Then we see a billion-year gap of warm conditions, ultimately broken by multiple ice ages punctuating the last billion years of Earth history. Our own, the latest, covers the last 15 million or so years.

Although precise estimates are impossible, it is clear the majority of Earth's history has been marked by conditions warmer than those of today. Especially prominent in this regard was the warmth of the mid-Cretaceous period, extending from about 120 to 90 Million Years Ago (MYA). In those times, dinosaurs roamed beyond the Arctic Circle. Coral reefs, which thrive only in warm waters, grew up to 15° latitude poleward of their present locations, as did a number of continental plant communities. With very little water held in land ice, sea level was perhaps 150 to 200 m (490 to 650 ft) higher, flooding about 20 percent of continental areas. On a seasonal basis climates were probably more equitable, especially at middle and higher latitudes. Global average temperature is thought to have been anywhere from 5 °C to 15 °C (9 °F to 27 °F) warmer than at present, and the equator-to-pole temperature gradient perhaps 15 °C (27 °F) smaller.

Another prominent warm period existed from about 45 to 55 MYA, just before the start of the most recent ice age (Figure 16–9a). Along the way toward the broad maximum, Earth experienced a natural warming event unrivaled by any other. Over the course of about 20,000 years, global temperatures warmed by more than 5 °C (9 °F). This is a huge warming rate, about a thousand times faster than that of the Cretaceous. Massive releases of carbon dioxide and methane from multiple sources led to the so-called **Paleocene-Eocene Thermal Maximum** (PETM) and set in motion environmental changes affecting planetary life nearly everywhere. All ice ages stand in stark contrast to "warm ages" such as the mid-Cretaceous and Eocene. At one extreme, nearly the entire planet might have been ice covered during the ice age of about 700 MYA, in a condition aptly called "snowball Earth." Evidence for a global snowball comes from a variety of diverse indicators of biological

▲ **FIGURE 16-9** Climate indicators for the last 70 million years. Curves (a), (b), and (c) indicate mainly global changes in ice volume. Curve (d) reflects climatic conditions over the North Atlantic but is broadly consistent with global changes, as can be seen by comparison with later portions of (c).

productivity, which are thought to show a decline to almost zero with growing ice volume, followed by a slow rebound during warmer climates that followed. Short of nearly global ice cover, it is hard to see why productivity would be so low. There is other evidence pointing toward snowball Earth: Snow lines were near sea level in the tropics during this period, and we find glacial deposits are capped by particular rock deposits suggestive of rapid warming

following intense cold. For many, the power of the snowball hypothesis is its ability to account for all of these along with still other indicators of a global deep-freeze. Other interpretations are possible, however, and the snowball hypothesis remains controversial insofar as the amount of ice is concerned.

Regardless of how the controversy is resolved, it is clear all ice ages have abundant year-round ice. For example, there is no dispute that during the most recent ice age, ice sheets advanced to within 40° of the equator, covering the ground to a depth of several kilometers. With so much water locked in land ice, sea level is much lower during ice ages. Temperature differences between ice ages and warm ages vary greatly by latitude, with high latitudes showing greater changes than the tropics. For example, during an ice age, polar sea surface temperatures (SSTs) might be 10 °C (18 °F) colder but tropical sea surface temperatures only 1 °C to 5 °C (2 °F to 9 °F) colder. Beyond these generalities, little can be said with certainty about the six earliest ice ages. On the other hand, we have substantial information regarding conditions during the most recent ice age.

The Current Ice Age

One of Earth's recent epochs, the **Pleistocene**, is often referred to as the *Ice Age*. However, as we have already seen, Earth has undergone at least several ice ages, and we are in one today.

Moreover, the term *Pleistocene Ice Age* is misleading, because the roots of the last glaciation go back considerably farther than the start of the Pleistocene. In fact, the most recent ice age had its origins some 55 MYA, when the global climate began to cool following the Eocene warm episode (Figure 16–9a). The first major ice accumulation began in Antarctica about 38 MYA. But Antarctic cooling was interrupted by a thawing episode, so even as recently as 20 MYA the climate was warm enough that areas of forest could be found on Antarctica. Reglaciation eventually began and by about 14 MYA east Antarctica was glaciated. Nevertheless, tundra vegetation was growing elsewhere until about 12.8 MYA; it wasn't until 10 MYA that the Antarctic sheet had reached its present size. Glaciation in the Northern Hemisphere lagged behind, as widespread glaciations in southern Greenland did not occur until 7 MYA. Still more glaciations occurred 2.4 MYA, but inland areas of Greenland remained unglaciated until later, perhaps 1 MYA. With glaciers finally covering Greenland, the last ice age was firmly established by any measure.

Did You Know?

Large-scale human impact on climate might have begun 8000 years ago. Following the depths of the last glacial period, the concentration of carbon dioxide and methane rose significantly, contrary to the downward trend that would have occurred if they followed cycles observed to hold for the previous 250,000 years. The expansion of irrigated (paddy) rice cultivation and later widespread land clearing are believed to have increased these greenhouse gases enough to cause global warming of about 0.8 °C (1.4 °F), long before the Industrial Age.

Within our ice age, climate has been anything but uniform, with numerous oscillations clearly evident (Figure 16–9b). Beginning about 2.5 MYA, these oscillations began increasing in amplitude, and about 800,000 years ago (Figure 16–9c) the amplitude increased dramatically, becoming about twice

as large as in the preceding million years. These oscillations in temperature and ice cover are called **glacial/interglacial cycles**. As can be seen in the diagram, they are quite irregular. For most cycles ice volume increases slowly and then terminates rapidly in a warming event. In addition, neither ice growth nor decay is uniform; "quivering change" is a better descriptor, with short-term oscillations superimposed on the longer cycles. The last three-quarters of a million years have been dominated by cycles lasting about 100,000 years, with shorter-term quivering present on a subdued scale. A relatively small part of each cycle, typically only 20 percent, is spent in the warm interglacial phase. Taking the last 2 million years as a whole, there have been about 30 cycles altogether, with associated global temperature changes of perhaps 5 °C (9 °F).

Ice volume changes have been largest in the Northern Hemisphere, with the size of ice sheets growing and shrinking by a factor of three or so with each glacial cycle. Although the Antarctic sheet has changed far less, Antarctic air temperature changes are believed to match those of the north polar latitudes, with glacial period temperatures about 10 °C (18 °F) lower than an interglacial. Evidence from mountain snow lines suggests that other Southern Hemisphere locations change temperature as much as their northern counterparts between glacial and interglacial times, and there is other evidence implying that the timing of major warming and cooling events has been generally in step over the last 150,000 years. Although the data are insufficient to construct detailed chronologies, we must conclude that the Southern Hemisphere, as a whole, has participated in glacial/ interglacial cycles just as the rest of the planet has.

As is clear from Figure 16–9, the planet is now in a warm interglacial, rivaled only a few times in the last 2 million years. Interestingly, one of those times was the last interglacial, which reached its peak about 125,000 years ago and might have set the record for Pleistocene warmth. Global sea level was about 6 m (20 ft) higher than now, and there is evidence that midlatitude continental areas were 1 °C to 3 °C (2 °F to 5 °F) warmer. In contrast, SSTs were not too different from what they are now. Between these two warm periods sits the most recent glaciation, an event that reached its maximum about 20,000 years ago.

Checkpoint

1. What does the term "snowball Earth" refer to?
2. What is one major trend in Earth's climate since the Eocene epoch? Explain.
3. Should Earth be considered a "warm" planet or a "cold" planet? Explain.

The Last Glacial Maximum

Following the last interglacial, ice volume increased, but not uniformly. There were two main pulses of glaciation, one about 115,000 years ago and another about 75,000 years ago. It seems that most ice was added to polar caps during the first pulse and to ice caps in North America and Eurasia during the later pulse. In the depths of the last glaciation, around 20,000 years ago, many aspects of the Earth–atmosphere system were different. Most dramatically, of course, land ice covered much more area, as seen in Figure 16–10.

In North America, it is certain that ice reached about as far south as present-day St. Louis, but only to the latitude of New York and Seattle on the East and West Coasts. Paradoxically, there is considerable doubt about the polar boundary of the northern sheet, with some arguing that much of the Arctic rim was unglaciated. Regardless, tremendous quantities of water were transferred from ocean to land, building sheets 3500 to 4000 m thick. Given enough time, this would be enough to depress the continental crust by more than 800 m. When the ice melted, the land surface

gradually expanded upward toward its original level. (Even now, continents have yet to fully rebound from glacial depression.) The Laurentide ice sheet in North America was in some ways equivalent to a huge mountain range, running from the Rocky Mountains to the Atlantic. Sea level was about 120 m (394 ft) lower than it is now, so that a land bridge existed between Siberia and Alaska. (Given enough time, movement of water from the ocean would cause oceanic floors to rise by about 35 m—115 ft.) There were also significant changes in sea ice, especially in the Antarctic Ocean, where winter sea ice covered about twice the area it now does. Of course, the sea ice changes had little effect on sea level, because water did not move between the land and ocean reservoirs.

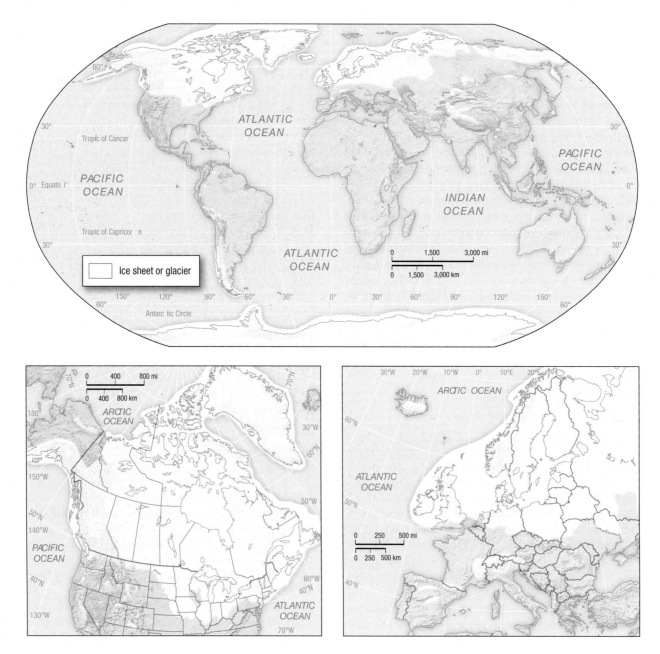

▲ **FIGURE 16–10** Map showing the maximum extent of ice and coastlines at the time of the last glacial maximum.

On land, temperature changes varied greatly by proximity to the ice sheets and to the ocean. For example, in western North America, maritime air masses kept temperatures to within 4 °C to 5 °C (7 °F to 9 °F) of modern values. This contrasts sharply with the area that is now Tennessee and South Carolina, where temperatures were 15 °C to 20 °C (27 °F to 36 °F) colder than they are now. These two examples might represent the extremes for midlatitude changes—a number of other midlatitude locations were in the range of 5 °C to 8 °C (9 °F to 14 °F) cooler. Temperature changes in the tropics were smaller, perhaps 4 °C to 5 °C (7 °F to 9 °F). Snowline was about 1000 m (3300 ft) lower, which translates into a temperature decrease of 5 °C to 6 °C (9 °F to 11 °F) for elevations above 2000 m (6600 ft).

Most places were not only colder but they seem also to have been drier. This is especially true for the high latitudes, where precipitation amounts were about 50 percent below today's. Some desert areas of both South America and Africa were larger, and lake levels were lower in tropical Africa and Central America. A cold desert covered much of the section of western Europe that was not under ice. Although increased dryness was the norm, a number of wetter areas could be found on all continents, as reflected in lake level changes and other indicators. Precipitation changes were undoubtedly determined to a large degree by circulation changes. Where prevailing winds shifted such that they blew from high latitudes (as for eastern North America), conditions were drier. What little evidence exists suggests that wind speeds were greater, as might be expected with a stronger equator-to-pole temperature gradient.

It must be emphasized that this glacial period was hardly uniform. In fact, abrupt climate changes were very common throughout this period, with polar temperatures changing by 8 °C to 16 °C (14 °F to 29 °F) over the course of just decades to centuries (Figure 16–9d). Because changes such as these are not confined to the glacial period, they are covered in more detail in a later section.

The Holocene

Like other glacial-to-interglacial transitions, warming following the last glacial maximum was fast, at least compared to times of cooling. Warming began about 15,000 years ago (prior to the start of the **Holocene**), only to be interrupted about 2000 years later when colder conditions returned. The cold event was most extreme (almost glacial) in the North Atlantic area, but it can be found in records as far away as Antarctica. Called the **Younger Dryas**, it lasted for about 1200 years. Following the Younger Dryas, starting about 11,800 years ago, came another period of abrupt warming, with temperatures in Greenland increasing by about 1 °C per decade, bringing climate into the interglacial we enjoy today. Spanning the end of the glacial to the mid-Holocene was the so-called **African Humid Period**, running from about 15,000 to 5000 years before present (BP). Northern Africa was nearly completely vegetated, with rainfall far above what is seen today. Savanna landscapes were abundant, including

lakes: Antelope, giraffe, elephant, hippopotamus, crocodile (and human) remains have been found in places that now have virtually no measurable precipitation. The transitions in and out of this wet period were very abrupt, requiring a few hundred years or less.

Did You Know?

Current levels of carbon dioxide and methane are higher than at any time in the last 650,000 years. A nearly 3 km (2 mi) long ice core from the interior of Antarctica shows that today's concentrations of CO_2 and CH_4 are respectively 27 percent and 130 percent higher than what occurred naturally over that time span. Other evidence suggests that current levels of CO_2 have not been matched in the last 10 million or more years. Thus, our time has seen the shattering of a truly long-standing record for this important greenhouse gas.

Two other significant events mark the Holocene. First, the early warmth was interrupted by yet another abrupt and short cooling 8200 years ago. Temperatures fell rapidly throughout the Northern Hemisphere Atlantic region and remained low for 300 to 400 years, with SSTs in the subtropical Atlantic suppressed by 7 °C to 8 °C (13 °C to 14 °F). Second, there was a long, severe drought in midcontinental North America from perhaps 4300 to 4100 years BP. Water tables fell, sand dunes became active, wildfires increased in frequency and/or intensity, and there were widespread changes in forest vegetation. Droughts lasting decades are common for this part of the world in the Holocene, but the 4200 BP drought stands unmatched for duration and depth of dryness. Many other Northern Hemisphere locations experienced persistent drought at this time, including the Middle East, northern Africa, the Mediterranean, and southern Europe. Despite limited data, it appears that much of the low to midlatitudes of the Northern Hemisphere experienced severe drought at the same time, suggesting that this was a hemispheric event.

Global mean temperature during the middle Holocene was about average for the epoch (meaning a little cooler than present). Northern Hemisphere summers were warmer, and tropical locations somewhat cooler. Since then global temperatures have been modulated by a number of events of various durations and expressions, depending on the time and place. For example, there is evidence that the period A.D. 900–1200 was warm in the North Atlantic. Called the **Medieval Climate Anomaly**, it coincides with the Viking settlement of Greenland. Mountain glaciers in Europe advanced before and after, but not during, this time, and there is evidence for glacial retreat elsewhere as well (including the Canadian Rockies). But there is also much physical and historical information indicating that this was not a global event of any significance.

Less ambiguous is another celebrated event, the so-called **Little Ice Age**. Spanning perhaps 1400 to 1850, this was a cold period for western Europe. During these years, alpine glaciers advanced as temperatures fell by about 0.5 °C to 1 °C

▶ **FIGURE 16-11** Temperature reconstructions for the last 1000 years, expressed as departures from the 1961–1990 average. The heavy pink line is believed to be the most reliable.

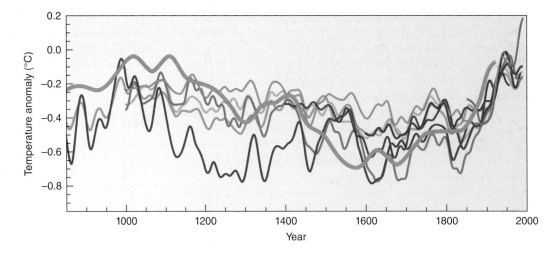

(0.9 °F to 1.8 °F). Historical records indicate this seemingly small decrease in mean temperature had a considerable effect on living conditions throughout Europe. Shortened growing seasons led to reductions in agricultural productivity, especially in northern Europe. In contrast to the Medieval Climate Anomaly, the Little Ice Age is expressed in mountain records from around the world. Although in no sense is it a true "ice age," it does represent the largest temperature change during historical times and is considered a global event.

Within the Holocene, considerable attention has been devoted to understanding the last 1000 years, in large part because it sets the context for presumed global warming occurring in the last hundred or so years. There have been a number of climate reconstructions for this, as seen in Figure 16–11. The latest of those shown (published in 2005) used what are believed to be more reliable indicators and more advanced data analysis methods. Compared to the others, it suggests ample warmth for the Medieval period followed by a more intense Little Ice Age than most other curves imply.

The Last Century

During the twentieth century, a growing network of meteorological stations was established around the world. Although there are many problems with the resulting measurements, such as the movement of meteorological instruments at a given site, the data give firsthand accounts of temperature and precipitation patterns. Augmenting the land-based measurements are observations from ships, and much more recently, satellite estimates of temperature, precipitation, wind, and other parameters. As discussed in Chapter 3, these measurements give an unambiguous picture of rapid warming over the last few decades. Figure 16–12 shows the geographic distribution of the rate of warming for the last 25 years or so. (Recall that this was the period of strongest warming.) The map depicts the change in annual temperature in degrees C per decade, assuming a linear change over the entire period. Figure 16–12 (and other diagrams in this chapter) was published in a 2007 report by a group of climate experts known as the Intergovernmental Panel on Climate Change. The IPCC

panel is described in *Box 16–2, Focus on the Environment: Intergovernmental Panel on Climate Change.*

Notice that many ocean locations show no warming at all or even slight cooling, whereas the interiors of Asia and northwestern North America show warming much above global average values. Clearly, there is considerable spatial variation in warming on an annual basis. Similar conclusions hold for individual seasons. That is, different places have warmed at different rates in different seasons (again, with some places cooling in some seasons). All this is to say that the term "global warming" must be used carefully, as a reference to change in the global average value.

Did You Know?

Snowpacks in western North America have declined over the last few decades and are projected to fall more with global warming. Growth rates for some tree species are sensitive to the snowpack that accumulates in the preceding winter. Using rings from trees collected at various elevations, scientists have used this relationship to estimate snowpacks in Rocky Mountain headwater areas of the Columbia and Missouri Rivers. Their analysis shows recent twentieth-century droughts have been rivaled only twice in the last 800 years.

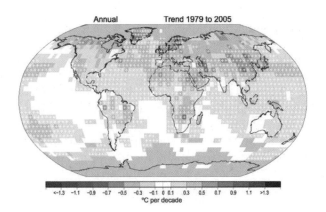

▲ **FIGURE 16-12** Rate of temperature change (°C per decade) for 1979–2005 as published in the 2007 IPCC report. Areas with insufficient data are shown in grey.

16–2 FOCUS ON THE ENVIRONMENT

Intergovernmental Panel on Climate Change

The IPCC was established in 1988 under the auspices of the World Meteorological Organization (WMO) and the United Nations Environment Programme (UNEP). Its mandates included the assessment of the current knowledge about climate changes (including an analysis of what is not known), discussion of how these changes impact the global environment and socioeconomic activity, and the development of policy recommendations.

The panel is composed of three working groups: Working Group I (WGI) analyzes the state of scientific knowledge of climate change, Working Group II (WGII) reports on its human and environmental impacts, and Working Group III (WGIII) offers mitigation strategies. None of the working groups is charged with undertaking new research; their responsibility is to present an overview of the state of knowledge in each of their respective areas. This is particularly significant because the IPCC reports do not reflect the views of just a select group of people but rather a consensus of the scientific community at large about what it knows and does not know about climate change.

The assessment report released in 2007 was the fourth since the inception of the panel. Like the earlier reports, it incorporated the most recent findings into its analysis and made use of improvements in computer modeling, enhanced satellite data acquisition, and further observational evidence. The report was a massive collective effort; it was authored by more than 150 expert climate scientists from 30 countries and was subject to review by another 600 experts. Extensive opportunity was provided for comment by governments, organizations, and individuals. Indeed, more than 30,000 written comments were submitted, and editors were required to ensure an adequate response to all substantive submissions. Draft documents were given two rounds of review. Each assessment report also described the level of uncertainty in its analyses, with explicit definitions for qualitative characterizations. (The "very likely" role of humans in the observed warming mentioned above is one such example.)

In summary, the IPCC reports are not position papers prepared by a group of scientists intent on espousing a particular viewpoint. Rather, they are the scientific community's response to a request for informed analysis of the current state of the physical science surrounding climate change. The IPPC reports represent the most comprehensive and authoritative summary available on a topic of intense interest and great importance for both humans and the natural realm. In 2007 the group was collectively awarded the Nobel Peace Prize, along with former United States Vice President Al Gore, who has campaigned to spread information on the seriousness of global warming.

Checkpoint

1. How have climates of the Holocene generally differed from those in the preceding epoch?
2. How do you think the climate of the African Humid Period may have affected human populations in northern Africa?
3. How does global climate in the last century compare with average conditions over the last 1000 years?

TUTORIAL
GLOBAL WARMING

Use the tutorial to observe the mechanism of greenhouse warming and important feedback processes. Go to section 9.2 to see the spatial pattern of temperature change for any season of the year.

Narrowly construed, the term "warming" refers to temperature. However, over the last century many other variables have participated in changes consistent with a warmer planet. A few of those are mentioned in the (IPCC) report are the following:

- The number of days with frost has decreased over many parts of the midlatitude regions.

- There has been a decrease in the number of extreme cold events across much of the world, and extreme warm events have become more frequent.

- Snow cover has decreased in most areas (especially in the Northern Hemisphere), and those decreases have mostly been driven by increasing temperature. In places where snow cover increased, increasing precipitation (rather than cooling) was the cause.

- The breakup date for river and lake ice occurred earlier by an average of 6.5 days per century, and the freeze-up date occurred later by an average of 5.8 days per century.

- From 1901 to 2002 the maximum extent of seasonally frozen ground declined by about 7 percent in the Northern Hemisphere.

Global warming is undoubtedly not the only climate trend in the last hundred years, but it is probably the least ambiguous. For example, precipitation—the other major climate variable—exhibited nothing like the progressive change seen for temperature (Figure 16–13). Although there are episodes of apparently wetter and drier than average global conditions, there is no strong trend for the record as a whole. Spatial patterns of precipitation trends are even less clear (Figure 16–14). Looking at the map, notice how much of the world lacks sufficient data to calculate a trend (grey areas).

▲ **FIGURE 16–13** Global average precipitation as published by the 2007 IPCC report. Values are plotted in mm as departures from the long-term mean. The curves correspond to various data processing methods and meteorological organizations.

Notice also that of those places where data are sufficient, the calculated trend value is statistically significant in only a few locations (places marked with a black cross). The difficulty is that precipitation exhibits extreme variability from year to year and from place to place. If they are to be detected statistically, long-term changes must be very large to stand out against this background of variation. For most of the world that simply did not happen over the period of record.

Variations on Characteristic Time Scales

The preceding sections mostly took a sequential view, tracking climate through time. Another way of looking at climate is to focus on individual time scales of change, without regard to the state of the system (warm vs. cold, etc.). Suppose, for

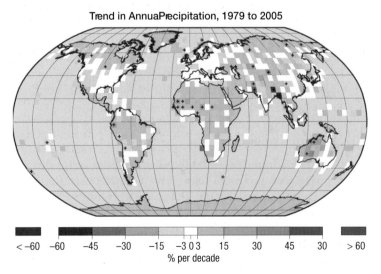

▲ **FIGURE 16–14** Rate of precipitation change (% per decade) for 1979–2005 as published in the 2007 IPCC report. Areas with insufficient data are shown in grey.

example, we believed that sunspot number is somehow connected to global temperature. Knowing that sunspots rise and fall roughly every 11 years, we would expect temperature to rise and fall on this same time scale. Global temperatures do not show such an oscillation, but there are variations on several other distinctive time scales.

Millennial-Scale Oscillations   One set of persistent oscillations appear for every time period examined so far, going back 500,000 years. The oscillations are called *millennial-scale* because they appear at intervals of thousands of years. During the most recent glacial period, the oscillations have been quite similar to one another; each begins with a rapid increase in temperature, taking from just a few decades up to a century to develop. Air temperatures over Greenland increase by 8 °C to 16 °C (14 °F to 29 °F), and North Atlantic SSTs increase by about 3 °C (5 °F). Temperatures remain high for 1000 to 2000 years, after which they return quickly to former low values. Thus, rather than rounded cycles, these oscillations are more like square waves, with climate jumping between two states. The cycles appear in clusters, as shown schematically in Figure 16–15. Each flip-flop is a *Dansgaard-Oeschger* (D-O) cycle, thought to arise from circulation changes in the North Atlantic ocean. There is a progressive cooling from one D-O cycle to the next, sometimes ending in an especially cold period called a *Heinrich event*. Together the multiple D-O cycles form a single *Bond cycle* of about 5000 years' duration. These cycles, which are named after the researchers who discovered them, are revealed by different indicators and have somewhat different causes.

Oscillations of similar length occurred during other glacials, as well as during interglacials and transitions between glacials and interglacials. Oddly, they are largest during interglacial-to-glacial transitions, with SST changes in the range of 4 °C to 4.5 °C (7 °F to 8 °F). During the interglacials, the oscillations are just as persistent but much smaller, with North Atlantic SST warming and cooling by about 0.5 °C to 1 °C (1 °F to 2 °F). They appear throughout the Holocene and are expressed in a variety of climate variables. For example, during the Holocene the frequency of very large storms in the northeastern United States shows pronounced millennial-scale cycles. The Little Ice Age cooling is widely considered to be the most recent example.

Millennial-scale oscillations are significant because they suggest that the Earth–atmosphere system has a tendency to flip back and forth between warm and cold states, independent of whatever the long-term climate is doing. Thus, rather than a regular progressive change that might follow changes in some boundary condition, it is as if the system suddenly reorganizes itself and moves quickly to another mode. No one knows why this happens, although a widely held view is that it arises from processes internal to the system, most likely involving feedbacks between atmosphere and ocean. The alternative would call for something external to turn on and off at more or less regular intervals, with effects that are amplified during transitional and glacial times. Regardless of

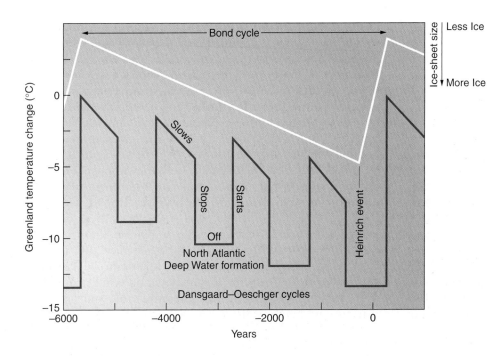

◄ **FIGURE 16–15** Idealized millennial-scale climate cycles. Four Dansgaard-Oeschger cycles of increasing intensity terminate in a Heinrich event. The entire sequence forms a single Bond cycle. The time axis covers an arbitrary period of about 7000 years, with the Heinrich event appearing as the last of five oscillations. Changes in deep water formation rates depicted on the graph occur in all cycles.

the cause, the size and speed of change is a matter of some concern. If the planet really does have this bipolar behavior, a switch from the present warm state to the other could have profound effects on many social systems.

Annular Modes Other much shorter oscillations are connected to hemispheric patterns of climate variability known as **annular modes**. The word "annular" means a ring or band of latitudes experiencing similar changes from normal, and although similar-sounding, it has nothing to do with "annual." There is one annular mode in the Northern Hemisphere and another in the Southern Hemisphere. By definition, annular modes are meant to capture variations in atmospheric winds or pressure that are not part of the seasonal cycle. In other words, the annular modes represent anomalous flow: flow that is faster or slower than normal for a particular time of year, or 500 mb heights that are higher or lower than normal than the seasonal mean. Given a pattern of winds or pressure and the corresponding seasonal average, a scientist can calculate a number that indicates the overall state of the wind or pressure field; that is, the atmosphere's circulation "mode." In terms of wind, the annular index tracks north–south shifts in prevailing winds. In terms of pressure, the annular index reveals north–south movements of atmospheric mass. This might seem reminiscent of the Arctic Oscillation described in Chapter 8, and indeed, the Arctic Oscillation is one expression of the Northern Annular Mode (NAM). It and the Southern Annular Mode (SAM) are significant because they account for about 25 percent of the variability in the flow outside the tropics. They are persistent, which means that you could make a forecast for the future knowing the state of the NAM or SAM. To be more specific, on time scales from weeks to years, more variability in extratropical flow is associated with the NAM and SAM than with anything else yet discovered.

Both the NAM and SAM have shown a trend toward higher index values in recent decades, meaning below-normal polar sea level pressure and stronger westerlies. Although defined in terms of pressure or wind, annular modes are connected to other variables. For example, the NAM and SAM have been related to sea ice and sea surface and land temperatures in their respective hemispheres. For much of the western United States, the latest increase in NAM index values has been linked to a reduction in storminess, warmer spring temperatures, and earlier seasonal drying. These changes are superimposed on those occurring on other times scales, such as recent global warming. In part because the NAM and SAM modulate changes occurring for other reasons, much research is directed at understanding the variation in these modes and their connection to climate more broadly.

Before concluding this section, we must stress again that climatic changes are not restricted to mean values; there can also be changes in the frequency of rare events. Thus, for example, computer models suggest that many areas may experience an increase in the incidence of heavy precipitation and drought in association with an increase in temperature. A recent study has shown that just such an increase has occurred over North America during the last century. The percentage of annual precipitation accounted for by heavy rainfall events has risen substantially over most of the United States in conjunction with the observed increase in temperature. Extreme events now yield about 12 percent of annual precipitation, whereas in 1910 they accounted for about 9 percent.

Checkpoint

1. What are Dansgaard-Oescher cycles?
2. Why might these oscillations occur?
3. What are NAM and SAM and what do they help to explain?

Factors Involved in Climatic Change

As we have seen, Earth's climate has undergone significant changes of varying magnitudes and time scales over the course of its existence. The big question, of course, is *why?* Several possible causes are easy to identify. These include variations in the intensity of radiation emitted by the Sun, changes in Earth's orbit, land surface changes, and differences in the gaseous and aerosol composition of the atmosphere. Each operates on a different time scale, as we will now see.

While examining the list of factors affecting climate change, it is important to understand that many of them do not operate independently. This means, first of all, that they operate simultaneously. Thus, for example, while one agent might be leading to warming, another might counteract or enhance that warming. Second, it means that agents of change might interact with one another, so that the effects are not merely additive. For example, the effect of tropospheric aerosols produced by humans might differ depending on whether or not they are overlaid by a plume of stratospheric volcanic aerosols. The issues of simultaneous and interacting factors will become especially evident in the discussion on changes in Earth's orbital characteristics.

Variations in Solar Output

As we implied earlier, Earth's climate is quite sensitive to the Sun's output. Considering that the amount of energy emitted by the Sun is not truly constant, this mechanism of climate change has considerable theoretical appeal. For example, some changes in the solar output, on the order of 0.1 to 0.2 percent, appear to be related to the occurrence of sunspots. As mentioned in Chapter 2, sunspots are relatively cold regions of the photosphere, about the size of Earth's diameter. The abundance of sunspots rises and falls on several time scales, including the very striking 10.7-year cycle. (As is customary, we will refer to this as the 11-year cycle.)

Satellite measurements show that when measured over the course of a few weeks, solar radiation decreases as sunspots increase. This is consistent with sunspots being cold—when more of the Sun is covered by cold regions, less radiation is emitted. But at longer time scales, such as those of a complete 11-year cycle, increasing sunspots are correlated with more radiation. Clearly, at these longer time scales, there must be solar changes that compensate for the increased area of sunspots. (The most likely explanation is an increase in surrounding bright areas.) The link between sunspot activity and solar output has led to considerable speculation that such phenomena could account for some of the climatic changes that have occurred on Earth. For example, droughts in the Great Plains of the United States have shown some tendency to recur at an interval that roughly corresponds to a double sunspot cycle, and earlier research has noted similar periodicities in Nile River flows. In addition, air temperatures over eastern North America rise and fall by about 0.2 °C (0.4 °F) in apparent synchronicity with the 11-year cycle. Some supporting evidence for the connection between climate and solar activity is also given by the fact that the **Maunder Minimum**, the period of minimal sunspot activity between about 1645 to 1715, coincided with one of the coldest periods of the Little Ice Age (see *Box 2–2, Physical Principles: The Sun,* on page XX). However, there have been other episodes in which variations in sunspot activity did not coincide with changes in climate, and alternative explanations have been offered for the apparent 22-year climate changes.

The equivocal evidence for a Sun–climate connection became stronger in the late 1980s, when several scientists observed that the relationship between tropospheric conditions and sunspot activity was much stronger when the direction of stratospheric winds over the tropics was taken into account. These winds tend to reverse their direction in approximately 2-year cycles in a pattern known as the **quasi-biennial oscillation (QBO)**. When the QBO is in its west-to-east mode, for example, there appears to be a relationship between the number of sunspots and winter conditions over northern Canada. Surface pressure rises and falls with sunspot number, and the mean storm track shifts north and south. When the QBO is in its east-to-west phase, however, no such connection is evident. Although these associations are intriguing and statistically very strong, no causal mechanisms have been proven to explain these relationships.

Somewhat stronger evidence for an Earth–Sun connection comes from consideration of millennial-scale variations in solar output and climate. Data for the last 12,000 years show a definite 1500-year cycle in solar radiation that matches well with debris deposited on the floor of the North Atlantic by icebergs. This has led some researchers to conclude that solar forcing underlies at least the Holocene portion of the North Atlantic's millennial oscillation.

On an entirely different time scale, it is believed that the rate of solar output from the Sun has increased by about one-third since the formation of the solar system. This brings up an interesting paradox because, as we saw earlier in this chapter, Earth is believed to have been warmer than at present throughout most of its early history. This counterintuitive association between a less radiant Sun and a warmer Earth is called the "early faint Sun paradox." Two very different explanations of the paradox are currently in vogue. One calls for a massive CO_2 greenhouse effect, with an early atmosphere having CO_2 partial pressures up to ten times the current total surface pressure. Organic molecules are much more difficult to generate in such an atmosphere; thus, this idea makes the appearance of life hard to fathom. In addition, there is geologic evidence that says such levels were never reached. Thus, an alternative view is popular as well, one calling for elevated ammonia levels creating an early greenhouse. Disputed for many years on the grounds that ammonia would be broken up by ultraviolet radiation, this was considered an implausible explanation. However, in 1997 it was shown that high-altitude shielding by other gases may have allowed ammonia to accumulate at lower levels. Regardless of which (if either) hypothesis is correct, it is interesting that the Sun–climate connection runs from the extreme of wondering whether there is any effect at all (sunspots) to wondering why the effect was formerly so weak (early faint Sun).

Changes in Earth's Orbit

In Chapter 2 we saw that the seasons occur primarily because of the tilt of Earth's axis relative to the Sun. If we imagine a plane on which Earth makes its revolution around the Sun, we can see that the axis of rotation is oriented 23.5° from the perpendicular to the plane (that is, it has an obliquity of 23.5°). The orientation of the axis is constant through the course of the year, so no matter where Earth is relative to the Sun, its axis points toward the North Star, Polaris. For the 6 months following the March equinox, the Northern Hemisphere is inclined toward the Sun, and during the rest of the year the Southern Hemisphere has a greater exposure to the Sun. This is the primary factor in causing the seasons. Obviously, if the axis of rotation were greater than 23.5°, this effect would be stronger and lead to a greater seasonality. Likewise, the effect would disappear if the axis were exactly perpendicular to the plane of the orbit.

We also saw that Earth's orbit is elliptical, rather than circular, so that on about January 4 Earth is about 3 percent closer to the Sun than on July 4 (see Figure 2–11). This change in Earth–Sun distance causes the planet as a whole to receive about 7 percent more solar radiation at the top of the atmosphere in early January (perihelion) than during early July (aphelion). It should be readily apparent that a greater eccentricity would result in greater differences in incoming radiation available at the top of the atmosphere during the course of a year.

As far as the Northern Hemisphere is concerned, the Earth–Sun distance is largest during the summer and smallest during the winter, causing winters to be somewhat warmer and summers to be cooler than they otherwise would be. Thus, not only is the amount of eccentricity important with regard to seasonality, but so is the timing of the minimum and maximum Earth–Sun distances with respect to the equinoxes and solstices (as will soon be explained).

In sum, three astronomical factors influence the timing and intensity of the seasons: eccentricity in the orbit, the tilt of Earth's axis off the perpendicular to the plane of the orbit, and the timing of aphelion and perihelion relative to the timing of the equinoxes. As it happens, these three factors all change slowly over time on a variety of time scales.

Eccentricity The **eccentricity** of Earth's orbit changes cyclically on several time scales, with a cycle of about 100,000 years being especially prominent. Though the Earth–Sun distance at aphelion is currently about 3 percent greater than at perihelion, the relative distance has varied between about 1 and 11 percent over the last 600,000 years. Over about the last 15,000 years, there has been a steady decrease in eccentricity, which will continue for about another 35,000 years.

Obliquity The tilt of the Earth's axis, **obliquity**, also varies cyclically, but with a dominant period of about 41,000 years, during which it varies between 22.1° and 24.5° off the perpendicular. While the range of the axis tilt may seem small, it is capable of producing substantial differences in summer and winter insolation. In particular, high-latitude regions can undergo changes in available solar radiation at the top of the atmosphere of about 15 percent due to variations in obliquity. The most recent peak in obliquity occurred roughly 10,000 years ago. Thus, we are about midway in the half cycle from maximum to minimum obliquity.

 TUTORIAL
ORBITAL VARIATIONS

Use the tutorial to explore past and future changes in Earth's orbit and understand how those changes affect radiation reaching Earth.

Precession Though the summer solstice for the Northern Hemisphere currently occurs near the time of aphelion, this changes through time because the axis wobbles on a 27,000-year cycle. In other words, the axis of rotation gyrates so that in about 13,500 years it will point to a different star, Vega, instead of Polaris (Figure 16–16). This change in orientation of the Earth axis, called **precession**, directly alters the timing and intensity of the seasons. Combined with changes in the orientation of the elliptical orbit, the result is a 23,000-year cycle in radiation. If orientation of the axis toward Vega were to exist today along with the current timing of aphelion and perihelion, the winter solstice for the Northern Hemisphere would nearly coincide with aphelion. The resultant increase in seasonality would cause warmer summers and cooler winters in the Northern Hemisphere. At the same time, the Southern Hemisphere would experience less seasonality because its summer solstice would occur near aphelion.

It is important to note that the importance of precession in influencing radiation receipts depends on the magnitude of the eccentricity of the orbit. Low values of eccentricity (nearly circular orbits) mitigate the importance of precession; greater eccentricity amplifies it. Given the current trend toward decreasing eccentricities, we can expect that this effect will be relatively small over the next 50,000 or so years.

These three cycles are collectively called the **Milankovitch cycles**, in honor of the early twentieth-century astronomer who expounded on their potential influence on Earth's climate. Most scientists believe the Milankovitch cycles have played an

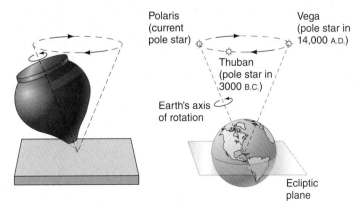

▲ **FIGURE 16–16** Precession. Over long periods of time Earth precesses on its axis like a spinning top, and thus the north pole points toward different stars.

important role in the expansion and retreat of glaciers during the Pleistocene because of the way they work together to influence seasonality. Large glaciers, such as those currently occupying most of Greenland and Antarctica, are most likely to expand when seasonality is low. With less seasonality, warmer winter temperatures foster a greater amount of snowfall over much of the ice sheets due to a greater availability of water vapor. Cooler summers also promote glaciation because the rate of melt along the margins of the ice sheets is slowed.

Considerable observational evidence for the role of Milankovitch cycles is seen in the climate record of the Pleistocene. Using radiation in the northern middle latitudes as a surrogate for orbital forcing, there is good agreement between the timing of ice advances and retreats. In addition, when the variability in the climate record is broken down according to various time scales, the three main Milankovitch periods emerge as containing most of the climate variation. There is also some understanding of the pathways by which orbital forcing influences climate. For example, computer modeling suggests that increased summer radiation during the early Holocene led to enhanced monsoon circulations, especially in North Africa. These would have brought increased moisture to the continent, as is observed in the climate record. For these and other reasons, changing orbital parameters are widely accepted as driving glacial/interglacial cycles.

To the extent that episodes of glacial advance and retreat are affected by Milankovitch cycles, the scenario is for us to return to another episode of glacial advance sometime during the next few millennia. But the Milankovitch theory is not without its problems, chief of which is one that concerns the eccentricity cycle. The 100,000-year cycle has the smallest effect on radiation reaching the planet but emerges as the strongest in the climate record. If this is the true cause, processes yet to be fully explained must amplify the relatively small radiation cycle. A related question concerns the observed amplitude increase in the 100,000-year climate cycle.

If eccentricity cycles are an explanation, why did they suddenly become more important about 800,000 years ago? Finally, there is the question of precession, which is out of phase between the Northern and Southern Hemispheres. Precession effects that would cause cooler conditions in one hemisphere should produce the opposite response in the other hemisphere. Yet, as we have seen, climate changes have been similar in both hemispheres, not opposed to one another.

Changes in Land Configuration and Surface Characteristics

Many climatologists believe that climatic changes occurring over the longest time spans were at least partly in response to changes in the size and location of Earth's continents (see Figure 16–17). Support for an older view, that continental drift was the primary forcing variable, has eroded in the face of quantitative estimates of change obtained from computer simulations. The breakup of the early supercontinent Pangaea and the slow movement of the resultant continents undoubtedly caused major climatic changes, even if not as large as observed in the geologic record. Such would have to be the case, of course, because all the factors that affect temperature and other climate variables (such as latitude and continentality) were themselves greatly affected by the movement of the continents. Though dramatic, the climatic changes resulting from continental displacement would be extremely slow. Like changes in position, episodes of continental mountain building have almost certainly produced significant climate change. For example, computer modeling suggests that the presence of large mountain regions (the Rockies, Himalayas, and Andes) would amplify Rossby waves during the winter season and promote enhanced monsoon circulations in the summer, both of which are consistent with observational data.

At shorter time scales, modification of Earth's surface, especially by human activity, can greatly influence the disposition

▶ **FIGURE 16–17** Distribution of land and ocean 150 MYA.

North America
South America
Eurasia
Africa
Antarctica
Australia

of solar radiation. One such activity is deforestation, in which large tracts of land are cleared of trees. The loss of vegetation reduces evapotranspiration from the surface. This in turn leads to higher temperatures near the surface, as the amount of energy channeled into the latent heat of evaporation is reduced and also decreases precipitation. In addition, decomposition of cleared vegetation directly increases atmospheric CO_2, an important greenhouse gas. Compounding this is the loss of vegetative surfaces formerly providing photosynthesis, a process that removes CO_2 from the atmosphere. Even if reforestation occurs, the conversion of mature to immature trees still reduces the rate of CO_2 removal from the atmosphere.

Alteration of arid and semiarid land surfaces by the overgrazing of cattle may also lead to changes in the regional climate. Soil compaction associated with grazing can increase the amount of runoff, thus making less water available for evaporation into the atmosphere. Also, one hypothesis suggests that the removal of vegetation leads to an increase in the surface albedo. As the albedo increases, according to this line of reasoning, the reduction in the energy absorbed by the ground causes cooling of the surface and a reduction in the environmental lapse rate. Because a lowered lapse rate would make the air more stable and less susceptible to convectional precipitation, overgrazing could enhance the vulnerability of these regions to drought. On the other hand, the reduction in vegetation due to overgrazing could have the opposite effect, in which a reduction in evaporation leads to a general warming of the surface and an increase in the environmental lapse rate. Of course, the instability associated with the increased lapse rate will have little effect if there is not enough water available to yield precipitation. Yet another effect arises from the change in surface roughness, which alters the momentum transfer between atmosphere and ground.

Checkpoint

1. Define the following terms: eccentricity, obliquity, precession.
2. How do cyclical variations in Earth's orbit, including Milankovitch cycles, affect climate? Explain.

Changes in Atmospheric Turbidity

Atmospheric **turbidity** refers to the amount of suspended solid and liquid material (aerosols) contained in the air. Some aerosols are released into the atmosphere through natural processes, such as large-scale volcanic eruptions; others are released by human activity, such as smokestack emissions. Aerosols can enter the atmosphere directly as in the examples just named. They can also enter the atmosphere indirectly as a result of chemical processes in which certain gases—usually sulfates, but also nitrates and hydrocarbons—react in the presence of sunlight to form solid and liquid aerosols. Some aerosols are no more than tiny clumps of molecules, whereas others are millimeters across. This is a huge range in size, roughly comparable to the difference between tennis balls and planets. Regardless of their source, composition, or size,

however, prolonged variations in aerosol contents can have important ramifications for climate.

Existing both in the stratosphere and the troposphere, aerosols directly affect the transmission and absorption of both solar and infrared radiation. By absorbing incoming sunlight they can cause a heating of the atmosphere around the aerosols, but they can also increase the amount of backscattering and thereby reduce the amount of radiation reaching the surface. The relative effects of aerosol absorption vs. backscatter are hard to assess and depend on numerous factors, including the albedo of the aerosols and the underlying surface. Note, however, that both absorption and backscatter reduce solar radiation reaching the surface. For example, during the period from about 1970 to 1990, the amount of solar radiation reaching Earth's surface *decreased* by several percent—a phenomenon referred to as **global dimming**. This dimming is attributed to heavy aerosol releases during that era. More recently, as increasingly stringent pollution controls took hold, dimming was replaced by modest "brightening."

To a lesser extent, aerosols can increase the absorption of outgoing longwave radiation that would otherwise escape to space and thereby increase the longwave radiation radiated from the atmosphere to the surface. In this way, aerosols can have the direct effect of increasing nighttime temperatures.

Aerosols can also affect climate indirectly, through their ability to serve as cloud condensation nuclei. A cloud with a greater number of condensation nuclei might contain as much liquid water as one with few nuclei, but it will have more droplets of mostly smaller sizes. Clouds containing a large number of small droplets are less likely to yield precipitation and, therefore, are likely to persist for longer periods. In addition, the total droplet surface area of such clouds is greater, and this greater surface area increases the reflectivity of the cloud (assuming the same amount of liquid water). Thus, tropospheric aerosols might give rise to more extensive, longer-lived, brighter clouds. Aerosols can also *reduce* cloud amount through their warming effect in the troposphere. As we saw in Chapter 7, the presence of warmer air aloft inhibits the vertical motions that lead to cloud development. Returning to the case of global dimming, the current scientific understanding is that indirect effects of aerosols were more important than direct effects in the observed dimming.

Observational support for the indirect effect of aerosols comes from satellite images of ship trails, which are tracks of low-level clouds produced by oceangoing vessels (Figure 16–18). In the relatively clean ocean atmosphere, smokestack exhaust aerosols enhance condensation, creating long cloud trails that stand out against the surrounding clear sky. With their high albedo, these clouds reduce the amount of solar radiation reaching the surface. Additional support for indirect aerosol effects comes from a recent study of pollution tracks on land, which appear on satellite images as plumes of brighter clouds downwind of cities. By analyzing data from a number of sensors, it was discovered that aerosols substantially reduce precipitation in downwind areas by suppressing both ice formation and droplet coalescence. (Recall that these are important precipitation growth mechanisms.)

▲ **FIGURE 16–18** Ship tracks appear as white streaks embedded in a low-level cloud deck of speckled light gray clouds. The image shows an area just offshore of the northwestern United States (colored green), with small amounts of cloud-free ocean (blue).

Moving beyond these generalities, let us now look in more detail at the climatological effects of tropospheric and stratospheric aerosols as agents of climatic change.

Tropospheric Aerosols Natural sources of tropospheric aerosols include the spraying of salt particles by ocean waves and bubbles, soot and gases from fires, the erosion of soil by wind, the dispersal of spores and pollen, the emission of sulfides by marine plankton, and other biospheric processes. Of course, volcanic eruptions can also discharge a huge amount of material into the troposphere, but it settles and precipitates out soon after the eruption and thus exerts no long-term effects on climate (unless many volcanoes erupt sequentially). Humans produce particulate matter in a number of ways, such as combustion of fossil fuels, burning wood and other biofuels, burning forests for conversion to agriculture, and plowing fields that become sources of dust. For example, we have all seen soot—a mixture of dark carbon-bearing particles—billowing out the exhaust of diesel trucks. Many tropospheric aerosols originate from combustion gases that are subsequently converted to particles by processes within the atmosphere. For example, coal burning releases sulfur dioxide gas (SO_2) that is converted to sulfuric acid (H_2SO_4), which ultimately condenses to form sulfate aerosols. Some aerosols of natural origin also begin as gases.

Although we have no direct long-term records of tropospheric aerosols, human activities have unquestionably increased their concentrations. This is especially true over industrialized land areas. Unlike some gases, however, which have very long residence times (Chapter 1), individual aerosols do not have long life spans in the troposphere. Aerosols produced over urban or industrial centers settle out before they can be widely dispersed across the globe, thereby leading to concentrations that vary widely both temporally and spatially. As a result, their climatic effects tend to be isolated to areas near the pollution sources.

We have seen that aerosols both absorb and scatter incoming solar radiation. The combination of aerosol backscatter and absorption results in the brown haze we recognize as a evidence of a polluted atmosphere and gives rise to the term *Atmospheric Brown Cloud* (*ABC*), which simply refers to a local accumulation of aerosols. Thus one immediate impact of aerosols is to reduce solar radiation at the surface, which promotes surface cooling. Another direct impact of ABCs is to warm the lower atmosphere through increased absorption of solar radiation, and for larger particles, increased absorption of longwave radiation. Another direct affect of tropospheric aerosols is to modify absorption of solar radiation by the surface. Where black carbon from aerosols is deposited on snow or ice, the surface albedo falls, which leads to warming. Indeed, recent studies have shown that for some Arctic locations, the warming effect of aerosols is several times greater than the warming effect of increasing CO_2.

Until recently it was not known whether the direct effects of tropospheric aerosol increases would lead to an overall warming or cooling near the surface. Numerical models now tell us that increased aerosol contents have the net effect of reducing surface temperatures globally. In fact, the radiative cooling due to anthropogenic aerosols appears to have been half as large as the radiative warming that would theoretically accompany observed increases in greenhouse gases over the industrial period. Thus, the warming effect of additional greenhouse gases has been masked somewhat by the cooling effects of aerosols. But we cannot expect the aerosol effect to offset the greenhouse gas effect indefinitely. If there were to be a stabilization in the release of anthropogenic emissions, atmospheric CO_2 contents could take up to 200 years to reach a new equilibrium level, whereas aerosol levels would respond almost immediately. Thus, the greenhouse effect would continue to increase while the aerosol effect would stabilize and cease to offset the opposing effect. Despite the possibly beneficial effects of aerosols on climate change, it should be remembered that aerosols exert their own negative effects on humans and the environment (Chapter 14).

Stratospheric Aerosols Unlike tropospheric aerosols, those in the stratosphere result primarily from natural processes; human activities are not as important. The stratosphere maintains a background level of aerosols introduced by the upward diffusion of sulfur gases from the troposphere, which then undergo gas-to-particle conversion. Though background levels remain fairly constant through time, significant increases can occur in the months following volcanic eruptions.

Did You Know?

Huge fires triggered by even a small nuclear exchange could lead to climate changes larger than any experienced in recent human history. Research published in 2007 and 2008 shows that smoke (lifted in part by heating through enhanced absorption of solar radiation) would reach the upper stratosphere and darken the skies for years. Cooling would be immediate and would lower planetary temperatures below those of the Little Ice Age.

Stratospheric aerosols can remain in the stratosphere longer than their tropospheric counterparts for two reasons: First, they tend to be smaller and therefore have lower terminal velocities. More importantly, the most effective agent of aerosol removal, precipitation, is not important in the stratosphere.

As previously stated, aerosols affect the energy balance by absorbing and backscattering solar radiation and by absorbing and radiating longwave radiation. The reduction in solar radiation reaching the surface would favor a lowering of tropospheric air temperatures, while the absorption of outgoing longwave radiation and the increase in downward emission would promote surface warming. So which actually dominates in the case of stratospheric aerosols? It turns out that the relative importance of shortwave and longwave radiation changes depending on the size of the aerosols. If the stratospheric aerosols are small, the reduction in solar radiation reaching the surface exceeds the gain in longwave radiation. In fact, the aerosols *are* small and promote lower temperatures for up to several years after volcanic eruptions.

Several recent volcanic eruptions have provided scientists with useful observational data on the effects of stratospheric aerosols. Mount St. Helens in Washington underwent a major eruption on May 18, 1980. Despite the massive ejection of solid material (much of one side of the mountain was blown away), it caused little if any major impact on hemispheric weather because it released relatively small amounts of sulfuric gases, which can ultimately transform to aerosols. Though the daytime sunlight was blocked out downwind of the eruption for a few days, no effects were noted much beyond that time frame.

The April 4, 1982, eruption of El Chichón in Mexico was far more violent than that of Mount St. Helens; and, more significantly, it released a particularly large amount of sulfuric gases. The result was an increase in atmospheric albedo, which decreased global temperatures by about 0.2 °C (0.4 °F) for several months. The Mt. Pinatubo (Philippines) eruption of June 12, 1991, is believed to have released about twice as much sulfuric gas as did El Chichón, along with the predictable effect of an even greater reduction in global temperatures. The eruption increased the atmospheric albedo both directly (by enhanced aerosol backscatter) and indirectly (by an increase in the reflectivity of clouds). The cooling effect was greatest by about August 1992, when the global mean tropospheric temperature decreased by about 0.73 °C (1.6 °F) from that of June 1991 (despite the fact that August is climatologically warmer than June in the Northern Hemisphere). Figure 16–19 shows the change in aerosol content over the globe prior to (a) and during periods (b), (c), and (d) following the eruption. Aerosol contents increased 100-fold during the peak concentration.

We must emphasize that the role of anthropogenic (and natural) aerosols in climate is subject to great uncertainty. This comes partly from the fact that the sources are highly variable over space and time and are not known with great precision. Also, aerosols are not at all well mixed. They remain in the troposphere from just a few days to a month before being rained out or deposited. Thus, their concentration varies greatly by proximity to a source, prevailing winds, and rainfall patterns. In addition, even if the distribution of aerosols were known perfectly, there would remain large uncertainty about their impact because their role in important atmospheric processes is extremely complicated and only partly understood. As we have seen, aerosols have both cooling and warming effects, and these vary by location, season, and altitude. Thus, while it is believed that globally ABCs have had a cooling effect during industrial times, in some places warming effects have dominated (particularly in parts of Asia). Understanding the climatic response to aerosol loading is the subject of intense research, as it is so closely related to the issue of greenhouse warming. There is, however, one very important difference. The short lifetime of aerosols means that changes in release rates have an almost immediate impact. This stands in strong contrast to greenhouse gas emissions, whose residence times are on the order of decades to centuries. In the case of greenhouse gas warming, our actions of the last few decades have committed the planet to future climate changes that can only slowly be controlled by reductions in emission.

Changes in Radiation-Absorbing Gases

The Mechanism of Greenhouse Warming The issue of potential climatic warming due to increases in carbon dioxide and other greenhouse gases has been the subject of intense scientific scrutiny and political controversy. On both human and geologic time scales, there is little doubt that large changes in CO_2 have occurred. For example, ice cores show that CO_2 rises and falls in concert with glacial/interglacial cycles. (The CO_2 changes lag behind the ice volume changes, and thus they are believed to amplify the glacial cycles, but they are not thought to be the fundamental cause of those cycles.) On human time scales the burning of fossil fuels and the clearing of forested areas have led to a steady increase in the carbon dioxide content of the atmosphere (see Figure 1–5).

What is the physical mechanism behind greenhouse warming? The simple answer is that because greenhouse gases absorb longwave radiation, less surface emission escapes to space. This is true, but a more complete explanation is somewhat more involved. Recall that temperature in the troposphere decreases with altitude, and that as temperature decreases, emitted radiation also decreases. With this in mind, suppose your eyes were sensitive only to longwave

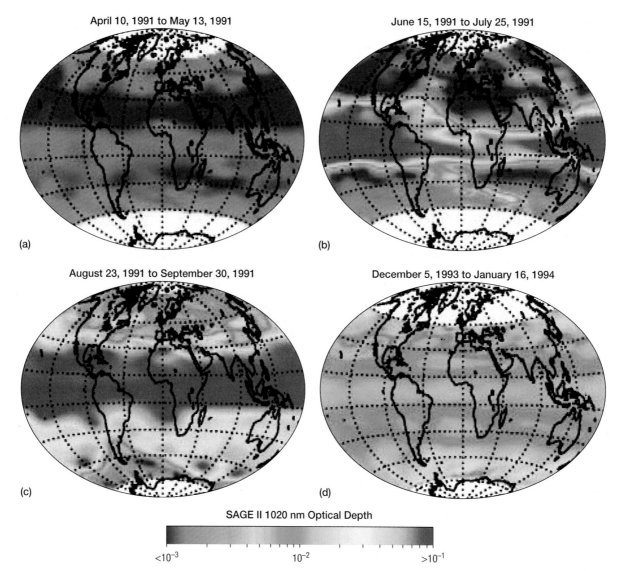

April 10, 1991 to May 13, 1991

(a)

June 15, 1991 to July 25, 1991

(b)

August 23, 1991 to September 30, 1991

(c)

December 5, 1993 to January 16, 1994

(d)

SAGE II 1020 nm Optical Depth

$<10^{-3}$　　　　10^{-2}　　　　$>10^{-1}$

▲ **FIGURE 16-19** The average aerosol content over the globe prior to and after the 1991 Mt. Pinatubo eruption. Redder areas indicate greater aerosol contents.

radiation and that you looked down from space on a planet like ours. Because the atmosphere absorbs so strongly at longwave wavelengths, you would see little of the surface. Instead, because most of the upwelling radiation originates high in the atmosphere, you would see a colder radiation source, as depicted in Figure 16–20a. In other words, emission to space is less than would be expected from knowing the warm surface temperatures enjoyed by us surface dwellers. The presence of a greenhouse gas raises the effective radiating altitude, and because temperature decreases with altitude, there is less emission to space than would otherwise occur. This is therefore consistent with the simple explanation of greenhouse warming. Of course, assuming thermal equilibrium, the amount of emitted radiation does balance absorbed solar radiation, as we saw in Chapter 3. Now suppose greenhouse gases increase. The immediate effect is to darken the atmosphere even more at longwave wavelengths, which raises the effective radiating altitude further, which in turn reduces

longwave losses to space (Figure 16–20b). Absorbed solar radiation now exceeds emission, which means the planet warms and emits more longwave radiation. Increases in both temperature and emission of longwave continue to increase until a new radiation balance is achieved (Figure 16–20c). Obviously, that new equilibrium will be a warmer climate.

This basic mechanism is modified by feedback mechanisms described in a later section. Before discussing those, we provide an overview of recent changes in greenhouse gas concentration.

Did You Know?

A significant greenhouse gas not regulated by the Kyoto Protocol is released in manufacturing flat-panel displays. Because nitrogen trifluoride (NF_3) is far more absorbing of longwave radiation than CO_2, the release of just 4000 tons each year is equivalent to 67 million tons of CO_2 (about 4 percent of the CO_2 release).

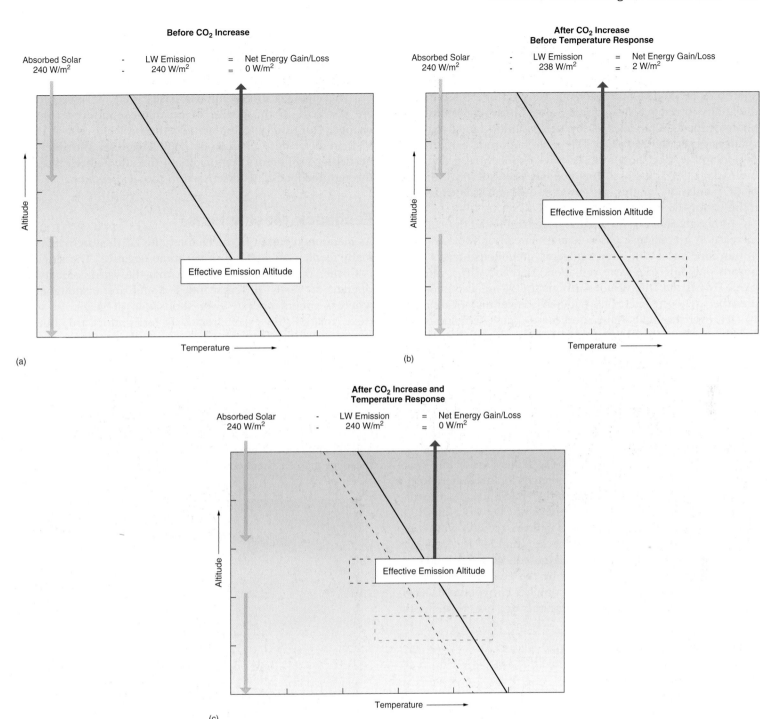

Before CO₂ Increase

| Absorbed Solar | - | LW Emission | = | Net Energy Gain/Loss |
| 240 W/m² | - | 240 W/m² | = | 0 W/m² |

Effective Emission Altitude

(a)

After CO₂ Increase
Before Temperature Response

| Absorbed Solar | - | LW Emission | = | Net Energy Gain/Loss |
| 240 W/m² | - | 238 W/m² | = | 2 W/m² |

Effective Emission Altitude

(b)

After CO₂ Increase and
Temperature Response

| Absorbed Solar | - | LW Emission | = | Net Energy Gain/Loss |
| 240 W/m² | - | 240 W/m² | = | 0 W/m² |

Effective Emission Altitude

(c)

▲ **FIGURE 16-20** Simplified depiction of how increasing greenhouse gases lead to warming. In (a), before a greenhouse gas increase, the planet has a radiation balance and the effective altitude of emission is low. In (b), an increase in some greenhouse gas raises the effective emission altitude. With emission from a higher, colder layer, the planet emits less radiation. In (c), the temperature warms sufficiently to restore a radiation balance.

TUTORIAL
GLOBAL WARMING

Use the simulations in section 2 to understand the mechanism of greenhouse warming.

Recent Changes in Greenhouse Gases

Since the middle of the nineteenth century, there has been an exponential increase in the input of carbon dioxide to the atmosphere by industrial activities (chiefly fossil fuel combustion and cement manufacture). Not surprisingly, most emissions have come from the developed nations, with 90 percent originating in the Northern Hemisphere. The growth rate has varied with business cycles but has averaged a few percent per year. Over the period 1999 to 2005, the growth rate was 3 percent per year. These industrial releases have been augmented by smaller but still significant releases resulting from deforestation. Deforestation adds CO_2 to the atmosphere through direct burning of the logged trees, decomposition of biomass, and other processes. Moreover, the removal of trees reduces the ability to remove subsequent inputs of carbon dioxide by photosynthesis. CO_2 releases by deforestation (and land-use changes generally) were relatively more important early in the period and do not show the explosive exponential growth of industrial CO_2. Thus fossil fuel use has been the main driver of CO_2 emission for the last few decades, and it is expected to be so for the foreseeable future.

Although much of the media discussion about the current increase in greenhouse gases centers on carbon dioxide, CO_2 is only one of several anthropogenic greenhouse gases that absorb outgoing longwave radiation. Methane (CH_4), nitrous oxide (N_2O), and chlorofluorocarbons (CFCs) are also effective absorbers whose contents have increased over historical times. In fact, over the industrial period these gases are responsible for about 35 percent of the greenhouse-induced increase in net radiation.

Of course, greenhouse gas emission is not the whole story; in some cases removal processes are extremely important in governing the amount of a particular gas in the atmosphere. For example, much of the anthropogenic CO_2 released to the atmosphere has been taken up by the oceans and lost from the atmosphere in other ways. Indeed, the best current estimate is that only about 60 percent of the CO_2 emission for any year remains in the atmosphere. Thus changes in atmospheric concentration do not necessarily mirror the emission patterns. Still, as seen in Figure 16–21, measurements of CO_2, CH_4, and N_2O show the unmistakable imprint of human activity, with concentration for all gases far above background levels. Although difficult to see in the graph, methane concentrations slowed beginning about 1990, leveled off by the end of the century, but have been rising since 2006. A number of hypotheses have been advanced, but the reasons for this are not understood,

and it is not clear what the future holds for methane. CFCs (not shown on the graph) represent a bright spot in this picture; their abundance is actually decreasing as a result of regulations intended to protect the ozone layer.

The changes in greenhouse gases seen in Figure 16–21 are just part of the puzzle of greenhouse-induced climatic change. Our ability to understand their role in past changes and predict future changes is complicated by the interconnections of different components of the atmospheric systems through feedbacks, as discussed in the next section.

Feedback Mechanisms

As mentioned earlier in the chapter, the Earth system involves many feedbacks, some positive, some negative. The situation is further confounded by the fact that the various feedbacks operate at vastly different rates: Some are instantaneous, whereas others operate over thousands of years. In some cases relationships seem relatively straightforward, but for others it is difficult to determine if a single connection between two variables is positive or negative. With these points in mind, we will examine some of the important feedback mechanisms related to climatic change.

Ice-Albedo Feedback   Much of Earth's surface is occupied by large continental ice sheets and floating sea ice (Figures 16–22 and 16–23). If the atmosphere were to cool along the margins of these ice masses, there could be an expansion of the ice-covered

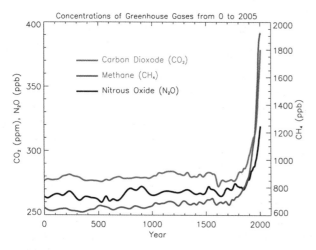

▲ **FIGURE 16–21** Time series of important greenhouse gases from A.D. 0 to the present as published by the 2007 IPCC report.

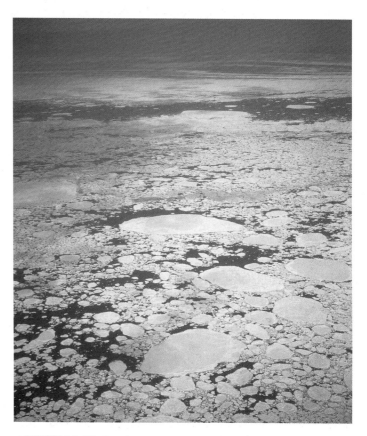

▲ **FIGURE 16–22** Sea ice.

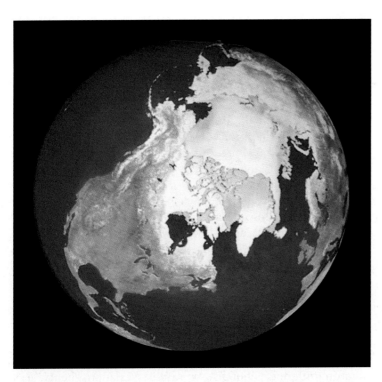

▲ **FIGURE 16-23** The distribution of Northern Hemisphere sea ice, as obtained by satellite-based radar imagery.

area. Similarly, any warming would likely lead to melting and a retreat of the ice margin. Because ice has a higher albedo than most other natural surfaces, an expansion of the ice would lead to a reduction in the amount of insolation absorbed by the surface, which in turn would lead to further cooling (a positive feedback). Likewise, a retreat of the ice associated with a warming climate would increase the amount of surface no longer covered by the reflective ice and, therefore, lead to enhanced warming. Thus, there is a positive feedback mechanism that favors continued warming or cooling once such a temperature trend is initiated. It is largely for this reason that middle and high latitudes show larger changes than low latitudes. Where there is little prospect for ice (the tropics), this feedback does not operate. Of course, ice advances and retreats do not continue unchecked, because other feedback mechanisms prevent the ice from totally covering the globe or from completely disappearing.

 TUTORIAL
GLOBAL WARMING

Go to section 4.1 if you have a hard time understanding the difference between positive and negative feedbacks. Run the climate model in section 6.1 to see ice-albedo feedback in action.

Water-Vapor and Cloud Feedbacks There are two important feedbacks involving water vapor in the atmosphere. First, observations and calculations show that relative humidity tends to be surprisingly constant when temperature changes by even large amounts. Thus, for example, the cold winter of a middle latitude might have a relative humidity not too

different from that of the summer. This means, of course, that the summer has a higher specific humidity—there is more moisture in the summer atmosphere. The same thing plays out when temperature increases from one year to the next. Again, the warmer atmosphere contains more water vapor. As already mentioned, water vapor is a very strong greenhouse gas, so this contributes to further warming. This direct water vapor effect is a positive feedback, because the initial change (warming) is amplified by the process.

The second water-related feedback concerns changes in the lapse rate of the troposphere. If the planet warms and its atmosphere contains more water, the lapse rate at low latitudes will decrease. Air aloft will remain colder than air below, but the difference will not be as great as in the prior cooler climate. The reason for this lapse rate change is related to the value of the saturated lapse rate. Recall that the saturated rate is variable and is smaller for warmer air. That is, a warm saturated parcel cools more slowly during ascent than a cold parcel. Therefore, in a warmer climate, rising saturated parcels are expected to cool more slowly. In moist tropical climates the lapse rate of the troposphere is largely governed by the saturated rate, thus the lower latitudes will show the largest decrease in lapse rate if the planet warms. If the lapse rate is smaller, then everything else being equal, the temperature at the effective radiating altitude is higher. A higher temperature means more emission, thus the surface does not need to warm as much in order to achieve a radiation balance. We see that this is a negative feedback because it counteracts the initial change.

 TUTORIAL
GLOBAL WARMING

Use the tutorials in sections 7.1 and 7.2 to observe the water vapor and lapse rate feedbacks.

Although there is near unanimous agreement on the direction in which the two water vapor feedbacks act and agreement that the lapse rate feedback is less important, there are considerable differences between various calculations regarding their magnitude. However, because they are related physically to each other, a calculation indicating a large positive direct water vapor feedback will necessarily suggest a large negative lapse rate feedback. This greatly narrows the differences between estimates of the overall feedback effect. According to the calculations referenced in the IPCC report, the water vapor feedback (direct plus lapse rate) is positive, and it amplifies expected global warming by about 50 percent.

A final water-related feedback that needs mentioning is clouds. We have seen that clouds are very active from a radiative standpoint. Thick clouds strongly reflect solar radiation, which obviously promotes cooling. On the other hand, a cloud of even modest depth absorbs all upwelling longwave radiation and itself radiates as a nearly perfect blackbody down to the surface. Whether cloud feedback is positive or negative depends on which of these effects is larger, and whether the amount of cloud increases or decreases. Although most calculations suggest a positive cloud feedback, the range is large—from 10 percent to 50 percent, according to the IPCC report.

Checkpoint

1. Describe the ice-albedo feedback. As a positive feedback, does it necessarily lead to global warming? Explain.
2. How do changes in water vapor and clouds affect atmospheric temperatures? Use the concepts of positive and negative feedback in your answer.

Atmosphere–Ocean Interactions One obvious relationship between the oceans and the atmosphere relates to the effect warming temperatures have on sea levels. Increases in global temperatures can raise the mean sea level, primarily by the expansion of warmer waters. Increases in atmospheric temperature can also contribute to rising sea levels by causing glaciers to melt and release water back to the oceans. A recent study estimates that increasing temperatures associated with a projected doubling of carbon dioxide within the next half century could cause the average sea level to increase by 19 cm (7 in.), but the rise in sea level would not be uniform throughout the oceans, and the rise could be as much as 35 cm (14 in.) off the coast of Europe.

On longer time scales, 500 to 1000 years, melting of the West Antarctic ice sheet becomes a possibility, which would raise sea levels by 5 m (16 ft) or more. Unlike any other ice sheet, the West Antarctic sheet is a concern because its base is below sea level. It is therefore susceptible to melting by ocean waters eroding its periphery. Until lately this idea was not widely accepted, in large part because there was no evidence that such a thing had ever happened before. However, recent information gleaned from sediments beneath the ice sheet strongly suggested that part or all of the sheet collapsed sometime in the last 2 million years, most likely in the interglacial period 400,000 years ago. Sea level changes, whether large or small, feed back on climate patterns through a number of mechanisms, including effects on surface water currents and the movement of deep water below the surface, the proportion of land and ocean, and marine and terrestrial productivity, to name a few.

A very important feedback, only recently recognized, concerns global-scale ocean circulations. As we discussed in Chapter 8, there is a large flow of water northward in Atlantic surface waters that delivers tremendous heat to the overlying atmosphere. After cooling and sinking in the North Atlantic, water flows southward in a much wider stream at great depth to the rim of Antarctica, where it joins deep water forming at the edge of the ice cap. Other branches of deep water flow northward into the Indian and Pacific Oceans, eventually rising and returning to the southern ocean at the surface. This huge, beltlike circulation system is partly driven by temperature differences but also by salinity variations; hence, it is called **thermohaline circulation** (see Figure 16–24).

Much attention has focused on the Atlantic portion of this system, known as the Atlantic Meridional Overturning Circulation (MOC). There is now convincing evidence to suggest that the MOC is not steady but instead makes rapid jumps from one mode of operation to another. Rates and locations of sinking water change abruptly, which in turn affect other aspects of the pattern. The new configuration lasts for some time before changing abruptly again. Numerous reorganizations of such ocean circulation have been found in a marine

▶ **FIGURE 16–24** The ocean's thermohaline circulation.

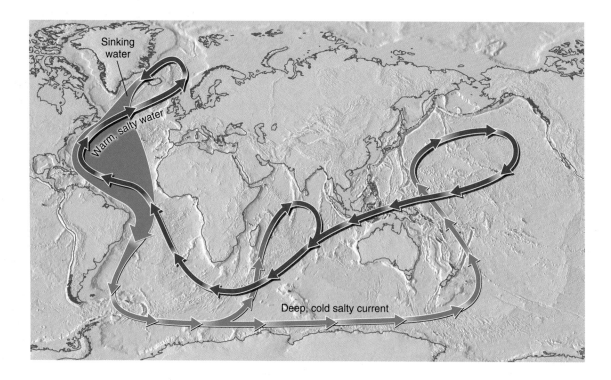

record running from about 60,000 years ago until about 10,000 years ago. Because they appear at nearly the same times as large swings in Greenland air temperature, they are associated with the millennial-scale climate oscillations described earlier. In this way, changes in ocean circulation are implicated in abrupt climate jumps, at least during the last glacial period. For example, Heinrich events are believed to be triggered by pulses of freshwater released during melting episodes of the Laurentide ice sheet. A layer of less dense freshwater at the surface shuts off downwelling, which in turn interrupts the flow of warm surface waters northward across the Atlantic. By implication, the same processes are responsible for millennial-scale changes in other periods, though that has yet to be demonstrated.

Did You Know?

Reservoirs around the world have partially counteracted sea level rise from melting ice caps. By storing roughly twice the water volume of Lake Michigan, the roughly 30,000 artificial lakes dotting Earth have prevented about 30 mm of sea level rise in the last 50 years.

Feedback from atmosphere to ocean arises as changing climate alters patterns of runoff, precipitation, and evaporation, which in turn influence ocean temperature and salinity. In addition, changing winds and other atmospheric processes affect the distribution of sea ice, which has a huge impact on the ocean heat balance. For example, modeling suggests that removal of sea ice can account for the magnitude of the D-O warmings in Greenland. If these changes are large enough, ocean circulation shifts to a new mode, which in turn influences global climate. If true, this feedback mechanism provokes a concern that relatively modest global warming might cause another rapid reorganization perhaps 100 years from now, when world population will demand about three times today's food supply. If the resulting climate change turns out to be as large as the one 8200 years ago (which occurred when the climate was even warmer than now), the results could be disastrous.

It should be noted here that the huge mass and high specific heat of the oceans, along with their vertical and horizontal mixing, create a strong thermal inertia. This means that warming of the ocean in response to increasing atmospheric temperatures would be very slow. Thus, oceanic temperature changes would be retarded in response to global warming, and there could be a time lag of several decades before any increase in atmospheric temperature would be observed in the oceans. Once the warming is realized, it would modify atmospheric pressure and wind distributions.

Oceans are implicated as affecting climate in other ways as well. For example, there is a form of water-methane ice stored in ocean sediments at cold temperatures and high pressure. If seafloor temperature rises above a threshold, say, because of changes in deep-water circulation, these methane hydrates decompose into water and gaseous methane. A sudden release of methane to the atmosphere by this process

matches spikes in atmospheric methane observed with rapid temperature change of D-O cycles. Similarly, but with vastly larger amounts involved, it has been suggested that a huge release of oceanic methane led to a strong warming episode that occurred 55 million years ago. Paradoxically, very recent analysis suggests that so much carbon was added to the atmosphere (about 80 percent of all fossil fuel reserves) that methane hydrates could not be the sole source. In other words, the most recent evidence confirms the role of carbon in that warming but requires a supply larger than seemingly possible from methane hydrates.

Another oceanic process is thought to operate in the Antarctic Ocean. During glacial times the deep ocean was much more stable than now, with a salty, dense layer found in bottom waters around Antarctica. Excessively briny conditions are attributed to salt left behind as water ice froze at the margin of the ice sheet. Regardless of how it formed, a dense layer like that would suppress vertical mixing, and it is therefore certain that the glacial ocean was very stable. Furthermore, it likely would have remained especially so at times when the thermohaline circulations was shut down. But gradually, over thousands of years, geothermal heat from the Earth below would have warmed the salty bottom layer. It is thought that a temperature change of 2 °C (3.6 °F) would have been enough to destabilize the ocean, leading to a pulse of warm water and salt delivered to the upper ocean and surface. The salt pulse would restart the thermohaline circulation, which would lead to warming in the Greenland and North Atlantic area. After the pulse, the ocean stabilizes, which leads to another cycle of slow deep-water warming and eventual discharge. The calculated timing for this—roughly every 10,000 years—fits with Bond cycles, as do deep ocean warmings known to have occurred before a number of Heinrich events. This hypothesis is attractive for its ability to explain the sudden warmings so characteristic of glacial times and for its explanation of why the warm periods following Heinrich events are longer and warmer than others.

Atmosphere–Biota Interactions Changes in climate are linked with land–vegetation patterns. The influence of climate on vegetation is easy to comprehend—banana trees do not grow in Greenland and there is no tundra vegetation in the Amazon Basin. Less obvious, perhaps, is the fact that vegetation likewise influences the climate in a number of ways. By transpiring moisture to the air, the presence of plants affects the moisture content of the atmosphere and the likelihood of precipitation. Plant assemblages also affect the surface albedo and the transfer of heat at the surface. It is easy to see why this is important in arid and semiarid locations, where the abundance and type of vegetation are very sensitive to precipitation. Less obviously, vegetation–albedo feedback effects are also important at high latitudes, because forest cover greatly reduces the high albedo that would otherwise occur during winter in a snow climate. For example, it is believed that poleward migration of forests following the last glacial maximum had a significant amplifying effect on Holocene warming.

One of the most important feedback mechanisms involved in the possible warming of the atmosphere involves the influence of CO_2 concentration on photosynthetic rates. It has been known for some time that many plant species undergo accelerated growth and enhanced photosynthesis in an environment rich in carbon dioxide, and some species also do better under warmer conditions. This could create a negative feedback (with potentially positive results as far as people are concerned) in which increasing CO_2 contents and temperatures allow plants to suppress further increases in the greenhouse gas.

Scientists believe that this "fertilization" process may already be appearing, especially in the latitudes between about 45° and 70° N. Satellite observations indicate that the region's supply of green vegetation is increasing and that its growing season is beginning about a week earlier and ending a week later than it did prior to the 1980s. These effects would be consistent with those expected from increased warmth and CO_2 levels observed in the area. At the same time, the seasonal oscillations in Northern Hemisphere carbon dioxide contents described in Chapter 1 have undergone changes in the intensity and timing of their cycles. The spring decrease in carbon dioxide associated with the leafing of deciduous plants is also occurring about a week earlier, and the difference between springtime maxima and late-summer CO_2 minima is increasing. Changes in the seasonal cycle of CO_2 are most conspicuous near the Arctic, where the magnitude of the oscillations has increased by 40 percent over the past few decades.

On the other hand, a lack of nutrients can become a limiting factor in plants' ability to respond to an enriched CO_2 atmosphere. This has recently been demonstrated in the Alaskan Arctic, where tundra vegetation was artificially subjected to twice the carbon dioxide level of the normal atmosphere. At first, the vegetation responded to the fertilization with increased growth rates. But by the third year, elevated CO_2 ceased to have any effect.

In a separate experiment, it appeared that high CO_2 contents might even *harm* tropical rainforest plants by causing a loss of nutrients in the soil. And another recent study has shown that trees in tropical rain forests have been growing, maturing, and dying more rapidly than in the past. This leads to a more open environment that favors replacement of trees with vines requiring greater amounts of sunlight. Ironically, the vines are less effective at photosynthesis than the trees they replace, and they do not even consume enough CO_2 to offset that released by the decaying trees.

Increases in atmospheric CO_2 could lead to other unwelcome results. It is possible, for example, that certain weeds would benefit more from higher carbon dioxide concentrations than would agricultural plants. Insects and other pests might also enjoy a warmer environment associated with increased CO_2, becoming more troublesome to agriculturalists than they are today.

A warming climate is also likely to have a large impact on fire regimes. Higher temperatures wick additional moisture out of both live and dead woody vegetation during dry seasons, making them more flammable. Coupled with a shortened winter, naturally occurring fire would therefore become more common. Calculations show that for parts of the western United States and Canada, a 1 °C (1.8 °F) increase in global temperature would increase the median area burned by more than a factor of six (Figure 16–25). (For more information on climate change impacts on vegetation, see *Box 16–3, Focus on the Environment: Plant Migrations and Global Change*.)

Checkpoint

1. What is the thermohaline circulation?
2. Climatologists have hypothesized that, in the past, pulses of fresh water from melting glaciers in what is now Canada may have disrupted the flow of warm surface waters across the North Atlantic Ocean toward western Europe. How might this have affected that region's climate?

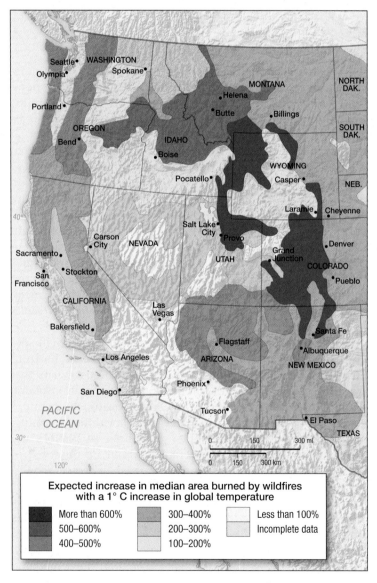

▲ **FIGURE 16–25** Expected percent increase in median area burned by wildfires with a 1 °C increase in global temperature.

16–3 FOCUS ON THE ENVIRONMENT

Plant Migrations and Global Change

As global temperatures have warmed and cooled, plant communities have responded by migrating poleward during periods of glacial retreat and equatorward as glaciers advanced. For example, large tracts of species such as spruce that once inhabited much of what is now the northern and northeastern United States have shifted northward into Canadian regions formerly covered by glacial ice.

The rate at which plant communities can migrate in response to changing climates is important to their survival. If plant communities are too slow to adapt to changing conditions, the vegetation type faces possible extinction. Because of the possibility of future rapid increases in global temperatures, scientists are concerned about the

potential for major changes in Earth's plant communities. But recently obtained evidence suggests that plants migrate more rapidly in response to climate changes than previously believed. Thus, the threat of extinctions may be less menacing than once feared.

On the other hand, it is clear that some species in a plant community might be better able to expand their boundaries than others. Such changes in the rate of migration for individual species would lead to a change in the plant community, in which the overall composition of the plants in the new environment is different from what it previously was. Moreover, the expansion of suburbs and agriculture into previously undisturbed areas has caused some plant communities to become fragmented. In other words, instead of having extensive areas of a particular plant community, those vegetation associations now occur in isolated

patches. The fragmentation of these communities hinders their ability to migrate in response to climate change and could make some future migrations impossible.

An interesting example of rapid vegetation change due to increasing temperatures has been observed in the grasslands of northeastern Colorado. A trend toward increasing nighttime temperatures has caused the average date of the last killing frost to occur earlier in the spring. This has put the normally dominant grass species—the blue grama, which historically has accounted for 90 percent of the ground cover—at a competitive disadvantage compared to various weeds. The weeds that are taking over the grasslands landscape are far more susceptible to drought. This is highly significant to ranchers who rely on the grass cover to supply most of their livestock's food needs.

General Circulation Models

Identifying the Causes of Climate Change

Perhaps the most significant conclusion of the Fourth Assessment Report of the IPCC can be summarized in the following two paragraphs from the Synthesis Report (available at **www.ipcc.ch/ipccreports/ar4-syr.htm**):

> MOST OF THE OBSERVED INCREASE IN GLOBALLY-AVERAGED TEMPERATURES SINCE THE MID-20TH CENTURY IS VERY LIKELY DUE TO THE OBSERVED INCREASE IN ANTHROPOGENIC GHG [GREENHOUSE GAS] CONCENTRATIONS. THIS IS AN ADVANCE SINCE THE TAR'S [THIRD ASSESSMENT REPORT'S] CONCLUSION THAT "MOST OF THE OBSERVED WARMING OVER THE LAST 50 YEARS IS LIKELY TO HAVE BEEN DUE TO THE INCREASE IN GHG CONCENTRATIONS."

> THE OBSERVED WIDESPREAD WARMING OF THE ATMOSPHERE AND OCEAN . . . SUPPORT THE CONCLUSION THAT IT IS EXTREMELY UNLIKELY THAT GLOBAL CLIMATE CHANGE OF THE PAST 50 YEARS CAN BE EXPLAINED WITHOUT EXTERNAL FORCING, AND VERY LIKELY THAT IT IS NOT DUE TO KNOWN NATURAL CAUSES ALONE. DURING THIS PERIOD, THE SUM OF SOLAR AND VOLCANIC FORCINGS WOULD LIKELY HAVE PRODUCED COOLING, NOT WARMING. . . . IT IS LIKELY THAT INCREASES IN GHG CONCENTRATIONS ALONE WOULD HAVE CAUSED MORE WARMING THAN OBSERVED BECAUSE VOLCANIC AND ANTHROPOGENIC AEROSOLS HAVE OFFSET SOME WARMING THAT WOULD OTHERWISE HAVE TAKEN PLACE.

The above statement indicates that the IPCC paid close attention to the influence of volcanic eruptions, solar radiation variability, and other factors that could influence global climates. Moreover, the panel concluded that these other factors not

only do not account for the observed warming of the last half century, but that those factors operating by themselves would probably have combined to cool the climate rather than warm it. How can anyone know what might have happened to climate had there been no release of greenhouse gases? How can the role of one variable be disentangled from another?

In both attributing observed changes to particular causes and in making predictions about possible futures, climate science relies heavily on so-called **Atmosphere–Ocean General Circulation Models** (GCMs). GCMs are mathematical representations of the Earth atmosphere–ocean–land system that run on supercomputers. They use well-established physical theory (such as conservation of energy, mass, and momentum and radiative transfer laws) to calculate the three-dimensional motion of the atmosphere and ocean for the entire globe. GCMs are somewhat similar to the models used in numerical weather prediction, but they differ greatly in data needs, internal structure, and purpose. In particular, GCMs estimate the response of Earth climate to a set of given external conditions known as "boundary" conditions. Examples of such boundary conditions include Earth topography, tilt of Earth's axis, solar output, volcanic aerosol loading, and greenhouse gas emissions. The programs compute the state of the atmosphere, oceans, and land surface conditions at discrete points in time (for example, every 30 minutes), using the current state as a starting point for the next time step.

Because of the enormous computing resources required, GCMs have rather limited spatial detail. The atmospheric component of a GCM, for instance, might provide values only every degree or two in latitude and longitude. The limited spatial detail means that small-scale processes (e.g., storm development) can be represented only approximately as so-called *parameterizations*. Similarly, the need to minimize computation means that many other processes are highly parameterized or even ignored. For example, although GCMs compute soil moisture inputs and outputs, the role of vegetation is highly simplified, and water movement within the soil column is often not modeled at all. Note that GCMs do not compute climate directly, they compute the state of the atmosphere at various points in time. The climate of a GCM is found by running the model for many years of simulated time and averaging over the end of the simulation period. Multiple years are needed because just as with the real atmosphere, there is considerable variability from year to year even when boundary conditions are unchanging. The early part of a simulation is discarded so that arbitrary starting values do not taint the climate statistics.

Numerous GCMs are in use at research institutions and universities throughout the world. Indeed, the IPCC report relied on results from 23 different models. Although all are based on the same fundamental laws of physics, they employ different parameterizations, different computational methods, and have different temporal and spatial resolution. For example, there is wide variation in how sea ice is modeled, in assumptions about cloud formation, in how vertical overturning in the ocean is treated, in freshwater inputs from land surfaces to the ocean, in how the terrestrial biosphere is represented, and in how the atmosphere and oceans are coupled (to name just a few of the many differences!).

In light of all these differences, and considering the obvious deficiencies of GCMs, what reason is there for confidence in their output? First, GCMs have advanced to the point that they do a good job of reproducing many aspects of today's large-scale climate. They successfully simulate observed seasonal patterns of air temperature, major rainfall and desert regions, ocean currents, the seasonal migration of storm tracks and tropical monsoon circulations, and the extent of sea ice. In addition, GCMs exhibit various forms of year-to-year variability that are typical of the observed Earth system (including ENSO events). A third source of confidence comes from GCM success at simulating past climates formed under different boundary conditions. For example, models have simulated the response of tropical monsoon circulations to the greater Northern Hemisphere seasonality in solar radiation that prevailed 6000 to 10,000 years ago. Fourth, GCMs have had success at reproducing the observed global temperature curve seen in Figure 3–32. Although the greenhouse gas and other boundary forcings were relatively modest during this period, their effects are captured by GCMs. Finally, some confidence in GCM output comes from realizing that they rest, as much as possible, on physical theory. Their development has been a story of increasing progression of realism in processes represented, and that has resulted in a steady progression in the quality of their output. It is important to realize that GCM projections are not mere extrapolations of past behavior, nor do they depend on purely statistical associations, which may or may not reflect causal relations.

GCMs provide what many see as the strongest evidence for the role of humans in the warming of the last century. As seen in Figure 16–26a, the models successfully reproduce the observed warming. The simulated curve is the average from 58 simulations using 14 different models, all driven by a combination of natural and human forcings. Agreement between the curves implies that observed changes have resulted from both natural and anthropogenic factors. Natural factors included in the models are changes in volcanic activity and solar output. Human agents included greenhouse gases, aerosols, and some runs also considered changes in land use. There is considerable uncertainty in some of these forcings—aerosols are a good example—so the inputs should be considered as merely the best available estimates for potential drivers of recent climate. The models differed significantly in what boundary condition changes were included (e.g., some included tropospheric ozone while others did not) and in the details of their representation (e.g., there were differences in the treatment of aerosol effect on radiation). Figure 16–26b

(a)

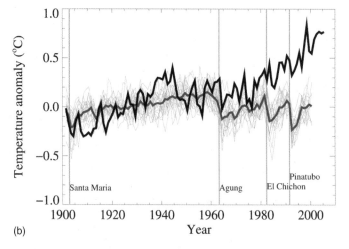

(b)

▲ **FIGURE 16-26** Global mean temperature observed (black curve) and modeled using a suite of GCMs as published by the 2007 IPCC report. In (a), natural and anthropogenic forcings were used, whereas in (b) only natural forcings were included. Light yellow and blue curves show individual models. Heavy lines are averages from group of models.

shows a similar curve from 19 simulations using five models but with only natural forcings included. We see that warming in the last few decades is absent, and in fact the models show modest cooling in that period. In other words, without the anthropogenic inputs the planet would have cooled, not warmed. Of the two human inputs, greenhouse gases warmed the planet, whereas aerosols were an overall cooling agent.

Projecting Climate Change

General circulation models have been used in a very large number of scientific studies to examine how the atmosphere might respond to increasing greenhouse gas concentrations in the future. But one of the inherent problems in doing such studies is that we do not know what the rate of input of greenhouse gas emissions will be in the future. Although most of the world's countries have agreed to reduce their emissions (see *Box 1–2, Focus on the Environment: Photosynthesis, Respiration, and Carbon Dioxide*), the rate of future CO_2 and other greenhouse gas emissions may fall anywhere within a fairly wide range of values. So the IPCC examined climate forecasts put out by multiple GCMs, each considering a variety of greenhouse emission scenarios over the next century.

Predicted Temperature Trends through the Twenty-First Century
Figure 16–27 plots the global temperature departures from the 1980–1999 averages projected by different GCMs. In each of the plots the solid colored lines represent the output of a particular GCM, and the black dotted line is the average of the individual models. The panels, from top to bottom, illustrate modeled output from examples of high-, medium-, and low-emission scenarios, respectively (the exact scenarios involve more assumptions than just CO_2 emissions, and their complete description falls beyond the scope of this discussion). Not surprisingly, the model outputs differ from each other, and the differences become greater further into the future. But they all lead to the same conclusion that the warming experienced over the last half century or so will continue into the future. They also show that the human-induced warming rate would dwarf what occurred during the PETM. During that warming event temperatures rose by about 0.025 °C/century. With values ranging from 1 to 4 °C/century, the anthropogenic greenhouse warming rates are a factor of 4–150 times larger. Again, the rapidity of possible near-term change is as worrisome as the magnitude.

Table 16–2 shows the average amount of warming predicted by the models for three 20-year time periods in the twenty-first century for each of the three emission scenarios. For 2011–2030 there is little difference in the warming predicted by the models under the three scenarios, ranging from between 0.64 °C (1.15 °F) and 0.69 °C (1.24 °F). The minimal difference is due to the fact that the expected warming is the result of the emissions that have already been put into the atmosphere rather than those that will be added between now and 2030. But by the time we get to the 2080–2099 period, the three emission scenarios yield very different amounts of warming. The average model prediction under the high-emission situ-

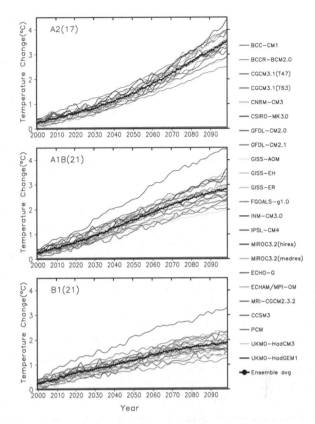

▲ **FIGURE 16–27** IPCC global mean temperature projections for various emission scenarios. The dark curve at bottom represents the warming rate seen during the PETM (compare with Figure 16-7a).

ation yields 3.13 °C (5.63 °F) of warming, significantly higher than the 1.79 °C (3.22 °F) under the low-emission situation. In other words, the actions we take in the near future in terms of controlling emissions will not greatly affect the amount of warming experienced in the early part of the twenty-first century, but it will have very significant impacts on conditions later in the century (and beyond).

Geographic Variations in Warming The amount of warming experienced over the twentieth century has not been uniform across the globe, nor will it continue to be in the twenty-first century. Figure 16–28 illustrates the global pattern of warming relative to the 1980–1999 period for the three emission scenarios for three upcoming 20-year periods. By the end of the century temperatures over land

TABLE 16–2

Global Mean Temperature Changes for a Range of Emission Scenarios Published in the 2007 IPCC Report

	Global Mean Warming (°C)		
	2011–2030	**2046–2065**	**2080–2099**
A2	0.64	1.65	3.13
A1B	0.69	1.75	2.65
B1	0.66	1.29	1.79

Each value represents the mean change relative to the 1980-1999 average as computed from a suite of models.

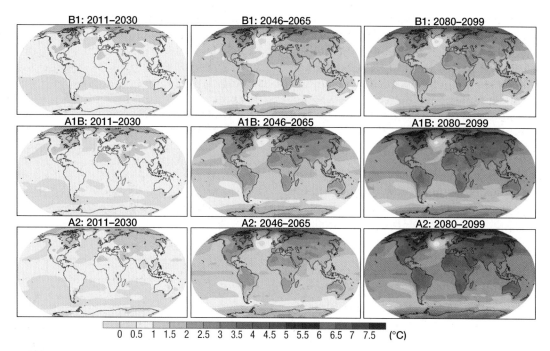

B1: 2011–2030   B1: 2046–2065   B1: 2080–2099

A1B: 2011–2030   A1B: 2046–2065   A1B: 2080–2099

A2: 2011–2030   A2: 2046–2065   A2: 2080–2099

0 0.5 1 1.5 2 2.5 3 3.5 4 4.5 5 5.5 6 6.5 7 7.5   (°C)

▲ **FIGURE 16–28** IPCC air temperature projections for selected time periods and emission scenarios.

masses and at the high latitudes of the Northern Hemisphere show pronounced warming. Other areas of intense warming are projected to occur in interior North America and South America, the western Sahara, and western Australia. Lesser amounts of warming are expected over the North Atlantic south of Greenland and over much of the Southern Hemisphere oceans outside of the tropics.

The details of these projections are undoubtedly wrong, both because the greenhouse inputs are unknown and because GCMs are far from perfect. In other words, the IPCC projections are not the final word; much research is under way to better understand the impact of human activity, and much new information will undoubtedly be obtained that will change the picture presented here. Although it is tempting to hope that scientific uncertainty means the negative consequences might never occur, it could be that new discoveries

reveal a future that is graver than we now think. Indeed, research published since the IPCC report shows the Arctic is warming faster than expected. Whether or not this continues remains to be seen, but it points out that the lack of perfect knowledge is no reason to feel secure about the threat of global warming.

Checkpoint

1. What physical principles and boundary conditions are used in building general circulation models of the atmosphere?
2. Look at the two graphs in Figure 16-26. How do the differences between the two graphs help in deciding whether the global warming is attributable to natural or human causes?

Summary

The climate of Earth has undergone numerous changes that have occurred over differing time scales, and there is every reason to expect that the climate will continue to change long into the future. Most of what we know about past climates has been inferred through indirect evidence of past conditions. Such evidence includes remnant landforms, botanical information, the presence of old soils, oceanic deposits, and cores obtained from the Greenland and Antarctic ice sheets. Climatic changes have occurred with differing magnitudes over varying time frames. In general, the longer period oscillations have had more extreme changes than those over shorter time intervals.

At the longest time scales, Earth has been mostly warm, with little permanent ice. But at least seven times the planet has experienced an ice age, the most recent of which continues to the present. Within an ice age, glaciers grow and shrink in what are called glacial/interglacial cycles. Over the last 5 million years, these cycles have varied considerably, both in amplitude and length. Present times correspond to an interglacial, which followed the last glacial maximum of 20,000 years ago. Within this period (the Holocene), the most notable changes were a large abrupt cooling 8200 years ago and another much smaller one from 1400 to 1850 (the Little Ice Age). Over the historical period, temperature records suggest global warming by a few tenths of a degree Celsius. Throughout both glacial and interglacial times, there have been rapid changes in climate as Earth jumps back and forth between two more or less stable states (millennial-scale oscillations).

We know of several processes that can cause the climate to change. For example, changes in solar radiation receipts over hundreds of thousands of years are caused by variations in Earth's orbit around the Sun—the Milankovitch cycles. Radiation receipts also vary in response to changes in radiation emitted by the Sun. Changes in the land surface of the planet may have been responsible for shifts in climate over both very long and relatively short time scales. The absorption of solar and thermal radiation by aerosols and radiation-absorbing gases may be important determinants in recent and future climatic conditions. All of these factors that influence climatic conditions do not act separately, but as part of complex feedback mechanisms connecting many parts of the Earth system. The effects of these feedbacks can be examined by the use of General Circulation Models (GCMs). The use of GCMs has advanced to the point that they can now simulate major features of present climate as well as global temperature change over the last 100 years. Taken with other theory and observations, they leave little doubt that most of the warming over the industrial period has been the result of increasing anthropogenic greenhouse gases. GCMs are also key in projecting future climates resulting from various assumptions about greenhouse emissions.

Key Terms

boundary conditions *page 466*

forcing agents *page 467*

feedback processes *page 467*

intransitivity *page 467*

proxy *page 468*

paleoclimates *page 468*

ice-rafted debris *page 469*

ice cores *page 469*

coral reefs *page 470*

radiocarbon dating *page 471*

tree rings *page 471*

ice ages *page 472*

Paleocene-Eocene Thermal Maximum *page 473*

Pleistocene *page 475*

glacial/interglacial cycles *page 475*

Holocene *page 477*

Younger Dryas *page 477*

African Humid Period *page 477*

Medieval Climate Anomaly *page 477*

Little Ice Age *page 477*

millennial-scale oscillations *page 480*

annular modes *page 481*

Maunder Minimum *page 482*

quasi-biennial oscillation (QBO) *page 482*

eccentricity *page 483*

obliquity *page 483*

precession *page 483*

Milankovitch cycles *page 483*

turbidity *page 485*

global dimming *page 485*

thermohaline circulation *page 492*

Atmosphere–Ocean General Circulation Models *page 495*

Review Questions

1. Define *climatic change* and discuss the practical difficulties of identifying climatic change from observations.

2. Explain how cores taken from ocean deposits and ice sheets can be used to infer past climate conditions.

3. Describe two types of remnant landforms that can provide information on past climates in a region.

4. How are tree cores used as indicators of past climates?

5. Describe how pollen samples obtained from old soils provide information on past climates.

6. Explain how it is possible that the global climate can be both cooling and warming at the same time.

7. How does the present climate compare to past climates over the course of geologic history? Is it correct to say the ice age is over?

8. Describe the frequency at which glacial/interglacial cycles have occurred during the Pleistocene.

9. What time frames constitute the Pleistocene and Holocene epochs?

10. What magnitude of mean temperature differences coincided with the various glacial/interglacial episodes of the Pleistocene?

11. Which regions of North America experienced the greatest temperature differences from current values during the last glacial episode of the Pleistocene?

12. List the factors that can lead to climatic change. At what time scales do each of these occur?

13. Describe the factors that can lead to variations in the amount of solar radiation available at the top of Earth's atmosphere.

14. What evidence is there that variations in sunspot activity do or do not lead to climate changes on Earth?

15. How do changes in eccentricity and obliquity and precession interact to influence Earth's climate? What time scales apply to each?

16. What types of occurrences on Earth can affect atmospheric turbidity? How do turbidity differences affect global temperatures?

17. Describe positive and negative feedbacks and provide examples of each.

18. What is a general circulation model?

19. What reasons are there to think projections from GCMs can be trusted?

Critical Thinking

1. The opening of this chapter discussed the concept of intransitivity. What would be the implications if Earth's climate were indeed found to be intransitive? Do you think that the question of whether the climate is intransitive can ever be answered?

2. There is little doubt most of the warming of recent decades is due to the activities of humans, but some are not convinced that the effect of this warming will be problematic for society. How would you argue for and against this viewpoint?

3. In addition to changes in average temperature and precipitation, it is believed that the frequency of unusual events (such as droughts and floods) might be more common as a result of global warming. What regions of North America are more vulnerable to economic losses from these events than from increased temperatures?

4. As recently as 20,000 years ago a continental glacier extended as far south as the central United States. Describe what impacts the ice might have had on the average position and magnitude of the polar jet stream.

Problems and Exercises

1. View the Web site at **www.pewclimate.org** and check to see if it includes any new press releases. If so, what are the major findings contained in them? Are these findings actually new or do they expand on known information?

2. Open the Weather in Motion module, "Retreat of Continental Ice Sheets," on this book's Web site. Move the cursor on the map so that you can observe the retreat of the Northern Hemisphere glaciers over the last

21,000 years. Did the ice retreat uniformly across North America? Where in the Northern Hemisphere did the ice first begin to retreat?

3. Check your favorite newspaper or news magazine to see if there are any articles describing new findings about climate change or political issues related to the topic. How would you rank the importance of climate change relative to other major political issues?

Quantitative Problems

The atmosphere is an extremely complex system with numerous feedbacks. Atmospheric scientists use sophisticated models to better understand the interrelationships between its different components. The Quantitative

Problems section for this chapter on the book's Web site at **www.MyMeteorologyLab.com** allows you to use a very simplified model to see how some atmospheric variables may respond to changes in other variables.

Useful Web Sites

www.ipcc.ch
Contains a vast amount of information from the latest and previous reports of the Intergovernmental Panel on Climate Change. An extremely valuable resource for anybody interested in any aspect of climate change.

www.pewclimate.org
Web page for the Pew Center for Global Climate Change, an advocacy group to promote action on issues related to the topic.

www.globalchange.gov
Comprehensive site of the U.S. Global Change Research Pro-gram. Offers numerous links, including summary reports of major research initiatives.

www.gcrio.org/ipcc/qa/index.htm
A useful primer on climate change from the U.S Global Change Research Information Office.

www.ncdc.noaa.gov/paleo/pollen/viewer/webviewer.html
Animations of North American vegetation change over the last 20,000 years. Based on pollen analysis, many plant taxa are represented.

www.exploratorium.edu/climate/index.html
An interesting and well-designed site with pages dedicated to various aspects of global climate change.

www.unep.org/climatechange
Published by the United Nations Environment Programme, this page links to 30 climate information pages discussing various aspects of climate change and its impacts.

climatechange.gc.ca/default.asp?lang=En&n=E18C8F2D-1
The Climate Change Web site produced by the Canadian government.

climate-l.iisd.org
Web-based newsletter providing information on climate change as it affects Canada.

MyMeteorologyLab™

The *Understanding Weather and Climate*, Sixth Edition, MyMeteorologyLab™ Web site contains numerous multimedia resources to aid in your study of **Climate Changes: Past and Future.**
Visit **www.MyMeteorologyLab.com** to:

- **Review** key chapter concepts.
- **Read** the Pearson eText, *In the News* RSS feeds, and other chapter-specific Web resources.
- **Visualize** challenging topics with Interactive Tutorials, Geoscience Animations, MapMaster interactive maps, and Weather in Motion movies and visualizations, and more.
- **Test Yourself** with self-study quizzes.

 TUTORIAL
GLOBAL WARMING

ORBITAL VARIATIONS

Use the animations and quizzes in these tutorials to review the chapter's key concepts.

WEATHER IN MOTION
Climate Change through Native Alaskan Eyes
Sea Level Rise
The Thermohaline Circulation
18,000 Years of Pine Pollen
Retreat of Continental Ice Sheets

PART SEVEN

Special Topics and Appendices

17 Atmospheric Optics

What role do vertical temperature gradients play in forming a mirage?

The previous sections of this book have covered the issues of the mass and energy of the atmosphere, water in its various states, the distribution and movement of air, disturbances, and human activities. We now conclude the discussion of atmospheric phenomena with a chapter on atmospheric optics and several appendices that provide additional detail for some of the topics covered earlier.

Rainbow, Torres del Paine National Park, Chile

17 Atmospheric Optics

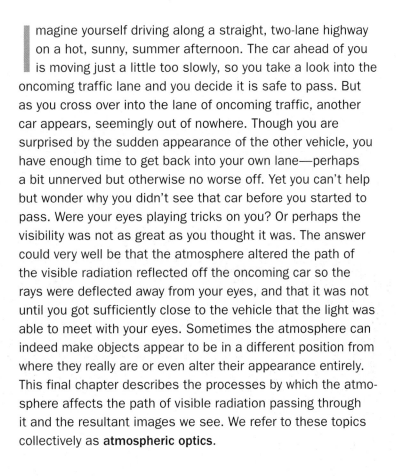

LEARNING OUTCOMES

After reading this chapter, you should be able to:

▶ Explain the processes that produce optical effects that can be seen in clear air.

▶ Explain the processes that produce optical phenomena involving clouds and precipitation.

Imagine yourself driving along a straight, two-lane highway on a hot, sunny, summer afternoon. The car ahead of you is moving just a little too slowly, so you take a look into the oncoming traffic lane and you decide it is safe to pass. But as you cross over into the lane of oncoming traffic, another car appears, seemingly out of nowhere. Though you are surprised by the sudden appearance of the other vehicle, you have enough time to get back into your own lane—perhaps a bit unnerved but otherwise no worse off. Yet you can't help but wonder why you didn't see that car before you started to pass. Were your eyes playing tricks on you? Or perhaps the visibility was not as great as you thought it was. The answer could very well be that the atmosphere altered the path of the visible radiation reflected off the oncoming car so the rays were deflected away from your eyes, and that it was not until you got sufficiently close to the vehicle that the light was able to meet with your eyes. Sometimes the atmosphere can indeed make objects appear to be in a different position from where they really are or even alter their appearance entirely. This final chapter describes the processes by which the atmosphere affects the path of visible radiation passing through it and the resultant images we see. We refer to these topics collectively as **atmospheric optics**.

◄ A mirage gives the false impression of a wet road.

Clear Air Effects

In Chapter 3, we saw that the atmosphere scatters and absorbs incoming solar radiation. Both processes have an important effect on the energy obtained by the surface and the atmosphere. In addition to scattering and absorbing, the atmosphere also *refracts* solar radiation, where **refraction** is defined as the bending of rays as they pass through the atmosphere.

Refraction occurs whenever radiation travels through a medium whose density varies or whenever it passes from one medium to another having a different density. Refraction in air occurs because radiation speed varies with density—the denser the atmosphere, the slower the radiation. To visualize how differential speeds cause the radiation to bend, imagine two people in a canoe, with one of them paddling on the right side of the canoe and the other on the left. If the person on the right paddles more vigorously, the canoe steers to the left. The same thing happens to a wave of electromagnetic energy passing through the atmosphere. Under most circumstances the density of the atmosphere decreases with height above the surface. This causes the radiation to refract slightly, forming an arc with the concave side oriented downward (Figure 17–1). The rate at which density changes with height varies considerably from place to place and from day to day because of differences in the temperature profile above the surface (the relationship between temperature and density was discussed in Chapter 4). Thus, the amount and direction of refraction vary with atmospheric conditions. Let's look now at several notable results of refraction.

Refraction and the Setting or Rising Sun

Refraction of incoming solar radiation is greatest when the Sun is low over the horizon, because the low solar angle causes the rays to pass through a greater amount of atmosphere (as described in Chapter 3). At sunset, refraction is sufficient to cause direct rays to be visible even after the Sun has dropped below the horizon (this is also true just prior to sunrise). In Figure 17–2, the Sun is positioned below the horizon. Without an atmosphere this would bring nightfall, but refraction causes the Sun to appear to be just above the horizon. When the Sun is positioned slightly farther below the horizon, its direct rays cannot be seen at the surface, but diffuse radiation can illuminate the sky to create **twilight** conditions. A further lowering of the Sun puts it far enough below the horizon to bring total nighttime. The period of twilight varies; it is longest during the high Sun season and increases with latitude.

In addition to slightly shifting the apparent position of the Sun near the horizon, refraction can also affect its apparent shape and color. You may have noticed that near dawn or sunset the Sun seems to have horizontal bands of different colors, with redder coloration near the bottom. This occurs because longer wavelength colors (such as reds and oranges) undergo less refraction than do shorter wavelength colors (such as blue and green), as exemplified by light passing through a prism (Figure 17–3). The shorter wavelengths, undergoing a greater amount of refraction, concentrate near the top of the apparent sun, while the longer wavelengths locate near the bottom. Under some atmospheric conditions the Sun appears momentarily to be capped by a bright green spot, known as the **green flash** (Figure 17–4).

Mirages

We are all familiar with movie and cartoon images of a parched, emaciated man crawling through a hot desert who perceives an oasis on the horizon, only to be disappointed by the reality of it being a mere **mirage**. Such mirages are caused not by excessive optimism but rather by the refraction of visible light when the temperature decreases rapidly with increasing height. And contrary to what some people think, the false oasis is not the only type of mirage; that term applies to any apparent upward or downward displacement of an object due to refraction.

To understand how mirages form, let's first consider a midday situation in which the air temperature near the surface decreases rapidly with increasing height, as shown in Figure 17–5a. Near the surface, vertical changes in air density with height are controlled primarily by vertical temperature gradients. With a very high temperature near the surface,

▶ **FIGURE 17-1** Refraction due to air density differences causes light to be refracted. In this case, the light from the top of the tall building is bent downward, so its path is concave downward. The light reaches the viewer's eye at an angle slightly greater than what it would be without refraction, making the top of the building appear higher up than it really is.

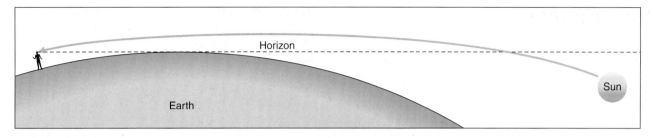

▲ **FIGURE 17-2** When the Sun is slightly below the horizon, it can still be seen at Earth's surface because of refraction. The various wavelengths are refracted differentially so that the bottom of the Sun appears redder than the top.

density is low. The steep temperature gradient in this example results in increasing density with height, so the path of rays moving through the atmosphere are curved upward, as indicated by the yellow lines on Figure 17–5b.

In Figure 17–5b the viewer at the left (let's call her Lauren) perceives distant objects to be slightly lower than they actually are. This is because the light rays reflected off distant objects approach Lauren's eyes at an angle slightly below horizontal. Thus, a person standing at position A (we will call him William) appears slightly shorter than he really is. Despite the minor distortion in the perceived height of William, Lauren has no trouble seeing him in his entirety.

Now look at how one type of mirage develops as William moves over to position B. The lower portion of his body appears to have disappeared because the light reflected off his legs is bent all the way to the ground, where it is

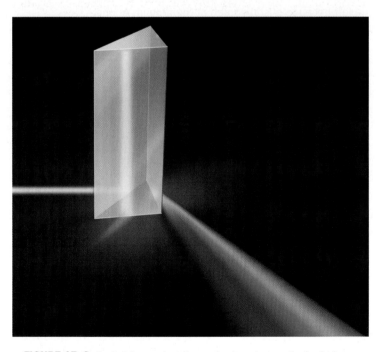

▲ **FIGURE 17-3** As light passing through air enters a typical glass prism, the greater density of glass causes refraction. The same thing occurs as light traveling through the prism exits it. But shorter wavelengths of light undergo more refraction than longer wavelengths, resulting in a separation of colors.

absorbed before it can reach Lauren. As William walks farther away, more and more of his body disappears from the bottom upward until he completely vanishes from sight at position C.

A different type of mirage appears from still more intense heating near the surface, such as that over an asphalt road on a hot afternoon. The heated air in the shallow layer just above the surface has an extremely steep temperature profile, while the air immediately above the shallow layer is somewhat cooler and has a less steep vertical temperature profile (Figure 17–6a). The steeper temperature gradient of the lower layer causes it to refract air more strongly than does the air above it. When this happens, a two-image **inferior mirage** can be seen, in which the viewer perceives not only a true image of an object but also an inverted image directly below. This can be seen in Figure 17–6b, where light is reflected off the top of the tree in all directions. Some of the light is directed toward the viewer's eyes after undergoing only a small amount of refraction. This produces the regular image of the tree. But some of the reflected light is also directed toward the ground, where the steep temperature gradient causes very strong refraction. This light is refracted upward and reaches the viewer's eye from below, making the top of the tree appear below the ground and upside-down as if the viewer were looking at the tree's reflection on the calm surface of a pond. Light reflected off all parts of the tree undergoes this type of refraction to produce a mirror image in the "pond."

So what actually happens when one sees a mirage that resembles a puddle of water? In that case a viewer perceives an inferior mirage of the sky, which looks very similar to a water surface. This chapter's opening photo of the car on the hot desert road shows an inverted image of the car apparently immersed in a puddle of water.

Note that the amount of refraction need not be the same for light leaving the top and bottom of the object. If light reflected by the bottom is refracted more strongly, the image will appear vertically stretched, or taller than reality. But if refraction increases vertically, the object will appear compressed. Obviously, the question of whether an object is stretched or compressed is independent of displacement, which is always downward for an inferior mirage.

A **superior mirage** forms when images are displaced upward. Light rays are bent concave downward as a result of

▶ **FIGURE 17-4** A green flash.

(a)

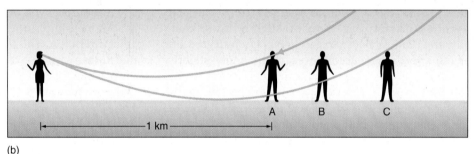

(b)

▲ **FIGURE 17-5** A steady, steep drop in temperature with height (a) can cause a refraction pattern in which rays are bent concave upward (b). Though there is some distortion of the perceived image, a person standing at position A can be viewed in his or her entirety by the person on the left. As the person on the right moves to position B, the visible light reflected off the lower legs does not reach the viewer on the left because it is absorbed at the ground. At position C, the person's image disappears entirely to the viewer on the left. Note the extreme vertical exaggeration of the diagram.

▶ **FIGURE 17-6** If there is a very steep decrease in temperature immediately near the surface along with a lesser temperature gradient just above (a), a two-image inferior mirage can occur (b). Diffuse, visible radiation off the top of the tree goes in many directions. Some of the light passes directly to the viewer with minimal refraction (the upper arc from the treetop to the viewer). Some of the light approaches the surface, where intense refraction (the lower arc) gives a second image of the tree. This creates an inverted image beneath the normal image.

(a)

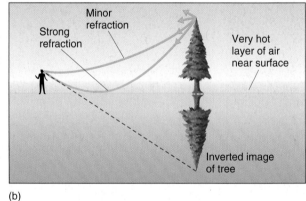

(b)

decreasing density with increasing height. This is the normal situation described earlier that caused the Sun to appear higher above the horizon than it really is. But for a mirage to be noticeable, the normal density gradient must be enhanced by a temperature profile in which warm (less dense) air lies

above cold air. This may happen, for example, over cool water surfaces with warmer air just above. As with inferior mirages, there may be stretching or compression of the image, depending on how refraction varies with altitude. Here, however, there is compression when refraction decreases with altitude

◄ **FIGURE 17-7** A superior mirage.

and stretching when refraction increases vertically. In extreme situations, small surface features can be lifted and stretched, giving the appearance of floating cities or mountains. In Figure 17–7 small whitecaps are stretched to the point where they appear as small waterfalls.

Checkpoint

> 1. What are inferior mirages and superior mirages?
> 2. Can mirages only be seen in the desert? Explain why or why not.

Cloud and Precipitation Optics

Refraction is not limited to clear air, nor is it the only process that can create interesting optical effects. In the remainder of this chapter, we will see how refraction and two other processes produce some very familiar (and not so familiar) optical effects.

Rainbows

One of the most striking features to appear in the atmosphere is the familiar **rainbow** (Figure 17–8). Rainbows are sweeping arcs of light that exhibit changes in color from the inner part

◄ **FIGURE 17-8** Rainbows. Note that the brighter primary rainbow is surrounded by a dimmer secondary rainbow.

▶ **FIGURE 17–9** Sunlight from behind the viewer undergoes reflection and refraction (a) to produce a primary rainbow. The amount of total refraction is different for each wavelength, causing the familiar color separation of a rainbow. A viewer at ground level observes two concentric arcs creating a primary and a secondary rainbow (b).

(a)

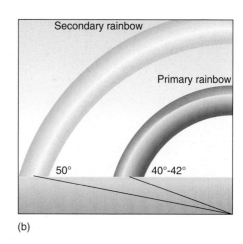

(b)

of the ring to the outer part. Rainbows only appear when rain is falling some distance away and with a clear sky above and behind the viewer that allows sunlight to reach the surface unobstructed. You might have observed that rainbows are always located in exactly the opposite direction of the Sun. In other words, if the Sun is to your back in the southwest, your shadow will point toward the center of the rainbow to the northeast.

The brightest and most common rainbows are **primary rainbows**. These rainbows are always the same size, so at that horizon the angular distance from one end to the other extends about 85 degrees of angle (for visualization purposes, think of this as an angle that extends almost all the way from due north to due east). In a primary rainbow, the shortest wavelengths of visible light (violet and blue) appear at the innermost portion of the ring, and the longer wavelengths (orange and red) frame the outermost portion. A primary rainbow is often surrounded by a less distinct **secondary rainbow** that covers about 100 degrees of arc at the horizon and has the reverse color scheme of the primary rainbow (that is, the reds appear on the inner portion of the ring and the blues on the outer portion). Of course, if the precipitation shaft is not large enough or is too far in the distance, only a partial rainbow will appear.

The big question, of course, is how do these form? The answer lies in the way in which sunlight is refracted (bent) and reflected as it enters and penetrates a raindrop. When light passes through a medium of varying density, it is subject to refraction. The same phenomenon occurs as light penetrates a boundary separating substances of dissimilar density, such as air surrounding a raindrop. Let's first look at how this creates a primary rainbow. As sunlight enters a raindrop, it undergoes some refraction, with longer wavelengths being refracted less than shorter wavelengths. The refracted light penetrates the raindrop, with the majority exiting at the opposite side from which it entered. However, a small portion of the light hitting the back of the raindrop is reflected back from the interior of the surface, penetrates the droplet once again, and is refracted a second time as it exits the front side of the droplet at a position somewhat lower than where it entered the droplet. This process is shown for two hypothetical raindrops in Figure 17–9a. Because each wavelength is refracted

differentially, only one particular wavelength of light exiting a raindrop is directed toward a particular viewer at a particular location. Thus the upper raindrop directs red light toward the viewer, while the lower raindrop directs violet light to that person. Because the lower raindrop appears at a lower angle above the horizon, its violet light forms the lower (inner) side of the ring, and the red light from the upper raindrop appears at the upper (outer) portion. The red light is refracted 42.3° and the violet light is refracted 40.6°. As a result, the ring is only 1.7° wide.

Secondary rainbows are formed in much the same manner as are primary rainbows, except that two reflections occur at the back of the raindrop, as shown in Figure 17–10. This results in a reverse color scheme than that of the primary rainbow, with longer wavelengths of light situated on the inner portion of the band and with the shorter wavelengths on the outer portion. Sunlight exits the same side of the raindrop as it enters but is directed downward at a 50° angle, thus putting the top of the rainbow at 50° above the horizon (Figure 17–9b).

Checkpoint

1. Describe the spatial relationship among sun, raindrops, and observer for a primary rainbow.
2. What accounts for the fact that a primary rainbow spans about 85 degrees of arc?
3. Why is a secondary rainbow dimmer or less distinct than a primary rainbow?

Halos, Sundogs, and Sun Pillars

Cirrostratus clouds produce circular bands of light that surround the Sun or Moon, called **halos** (shown in Figure 6–19), with radii of 22° or 46° (46° halos are less common and not as bright as 22° halos). Unlike rainbows, whose appearance requires that the Sun be directly behind the viewer, halos occur when ice crystals are between the viewer and the Sun or Moon. Figure 17–11a illustrates the refraction within ice crystals that produces a 22° halo. Sunlight (or moonlight)

▲ **FIGURE 17–10** A secondary rainbow requires two reflections within raindrops.

passes through the sides of column-shaped and platelike ice crystals, in which each of the six edges form a 60° angle. The ice crystal acts as a prism that refracts the sunlight 22°. Crystals that are 22° away from the sight path of the Sun or Moon and have the necessary orientation will refract light toward the viewer. Because ice crystals are so numerous and randomly aligned within the cloud, a sufficient number will direct the light toward the observer to make the halo bright enough to be visible from the ground. Figure 17–11b shows how column-shaped ice crystals refract light at 46° when the crystal is oriented lengthwise toward the incoming light. The 46° halo can only result from column-shaped crystals, not from platelike crystals.

Platelike ice crystals larger than about 30 µm across tend to align themselves horizontally. If the Sun is slightly above the horizon and behind these crystals, bright spots appear 22° to the right and left of the Sun (Figure 17–12). These **sundogs** (or *parhelia*) often appear as whitish spots in the sky, but sometimes they exhibit color differentiation,

with redder colors located on the side of the sundog nearest the sun and the blues and violets located on the outer side. Platelike crystals between a low Sun and an observer can also *reflect* (as opposed to refract) sunlight off their tops and bottoms to produce **sun pillars** (Figure 17–13). The many ice crystals are aligned almost, but not exactly, horizontally, with each reflecting a portion of the incoming light differently to produce the apparent columns stretching upward and downward from the Sun.

Coronas and Glories

Coronas and glories are optical phenomena resulting from the bending of light as it passes around water droplets (**diffraction**). The **corona** (Figure 17–14) is a circular illumination of the sky immediately surrounding the Moon—or in rarer instances, the Sun. Clouds having uniform droplet sizes cause highly circular coronas that concentrate shorter wavelength (bluish) colors on their innermost portions and longer (redder) wavelengths on their outer margins. When the cloud contains a wide assortment of droplet sizes, the illumination appears white and irregularly shaped. The size of the corona is also related to droplet size, with larger droplets producing smaller coronas.

If you are ever in an aircraft flying above a cloud deck, look for the plane's shadow on the clouds. You may see a series of rings called a **glory** (Figure 17–15a). Glories occur when sunlight entering the edge of a water droplet (Figure 17–15b) is first refracted, then reflected off the inside of the back of the droplet, and refracted again as it exits the droplet. In this regard, the process that produces a glory is very similar to that which creates a primary rainbow. However, a glory requires that the combination of these processes redirect the incoming light a full 180°, so the viewer sees the returned light from the top of the cloud with the Sun behind her. To accomplish the additional bending, diffraction must occur along the edge of the droplet as the light makes its return back toward the Sun.

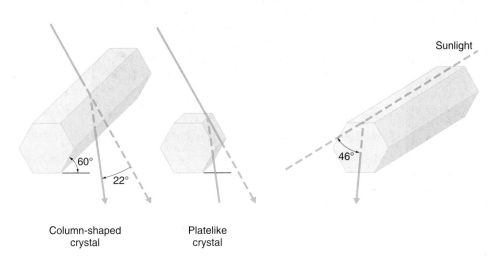

◀ **FIGURE 17–11** Column-shaped and platelike crystals refract light to produce a 22° halo (a). Refraction where ice crystals have 90° angles produces a 46° halo (b).

Sunlight

60°
22°

46°

Column-shaped crystal

Platelike crystal

(a)

(b)

(a)   (b)

▲ **FIGURE 17–12** Nearly horizontally oriented ice crystals refract light from a setting or rising Sun to produce sundogs (a), so named because of the way they accompany the Sun (b).

Checkpoint

1. What factors account for the differences among haloes, sun dogs, and sun pillars?
2. What process produces coronas and glories? Explain.

▲ **FIGURE 17–13** A sun pillar. These columns of bright light can occur when looking in the direction of the Sun near the horizon through clouds composed primarily of platelike crystals arranged horizontally.

▲ **FIGURE 17–14** A corona. Clouds consisting of fairly uniform size water droplets can cause circular zones of illumination around the moon or Sun.

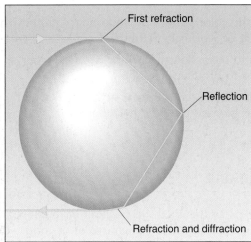

▲ **FIGURE 17-15** Glories can appear as rings of light surrounding an aircraft's shadow on the top of a cloud (a). Glories require diffraction along the edge of a cloud droplet as the sunlight exits the droplet (b). The bending from refraction returns the sunlight almost 180° from the direction at which it entered the droplet.

Summary

As rays of light travel through the atmosphere, they undergo a certain amount of bending, or refraction, as a result of air density differences away from the surface. Refraction causes some interesting sunrise and sunset effects. First, the bending of solar radiation allows the Sun to be "visible" even though it is truly below the horizon. Also, the fact that different wavelengths of sunlight are refracted differentially causes a setting or rising Sun to display a sequence of horizontal bands with different colors. At its most extreme, refraction can cause only the green light from the Sun that is below the horizon to reach a viewer, resulting in a green flash.

Cloud droplets and ice crystals, along with precipitation, can produce their own unique optics. Rainbows form by the combination of refraction and reflection in raindrops that cause multicolored bands to be seen when the viewer stands between the Sun and the drops. Refraction within ice crystals causes halos and sundogs, while reflection on ice-crystal surfaces can produce sun pillars that seem to emanate from the setting or rising Sun. Diffraction by water droplets creates a corona around the Sun or Moon, while a combination of refraction, reflection, and diffraction produces glories.

Key Terms

atmospheric optics
 page 505
refraction *page 506*
twilight *page 506*
green flash *page 506*

mirage *page 506*
inferior mirage
 page 507
superior mirage *page 507*
rainbow *page 509*

primary rainbow *page 510*
secondary rainbow
 page 510
halo *page 510*
sundog *page 511*

sun pillar *page 511*
diffraction *page 511*
corona *page 511*
glory *page 511*

Review Questions

1. What is refraction and why is it related to variations in atmospheric density?

2. Describe the way refraction alters the apparent position of the setting or rising Sun.

3. Do longer or shorter wavelengths of light undergo greater refraction when passing through the atmosphere? How does the differential refraction cause an apparent banding of the Sun near the horizon?

4. Which type of vertical temperature gradients promotes the appearance of superior and inferior mirages?

5. How do some mirages create the appearance of standing water on hot days?

6. Explain why the Sun must be behind you when you see a rainbow.

7. Describe the difference in the way primary and secondary rainbows form.

8. How does the color pattern of a secondary rainbow differ from that of a primary rainbow?

9. In addition to refraction, what process must occur within raindrops to produce a rainbow?

10. Why is it that some halos appear at 22° angles and others at 46° around the Sun or Moon?

11. How are sundogs formed? Describe the color patterns associated with them.

12. Describe the formation of sun pillars. Does refraction play a role in their formation?

13. Explain how coronas are formed around the Sun or Moon. What factor or factors determine their size?

14. What are glories, and how are they formed? Are they the result of refraction alone or is another process also involved?

Critical Thinking

1. Consider the way the apparent position of the Sun sweeps across the sky over the course of the day (see Chapter 2). How will the period of twilight vary between summer and winter where you live? Will twilight conditions generally last longer in the tropics or in the high latitudes? After answering this question, go to Problems and Exercises question 1, and check to see if your answer was correct.

2. Can falling ice crystals produce rainbows? Explain why or why not.

3. Can altostratus clouds produce halos? Explain why or why not.

4. Which of the optical phenomena described in this chapter are most likely to occur where you live? Are they equally likely to appear at all times of the year?

5. In Chapter 3 we discussed Rayleigh, Mie, and nonselective scattering. What similarities and dissimilarities exist between those scattering processes and the optical effects caused by refraction, reflection, and diffraction that were discussed in this chapter?

6. Explain why superior mirages do not occur over land on hot, sunny days.

Problems and Exercises

1. Refer to the Web site **aa.usno.navy.mil/faq/docs/ RST_defs.php#top**, and look up the definitions of civil, nautical, and astronomical twilight. Then use the available tables to determine the length of day where you live for March 21, June 21, September 21, and December 21. Does the length of day show significant differences using each of the three definitions of twilight? How do these differences vary through the year?

2. On hot, sunny days, look for the presence of mirages. Are they equally apparent in all directions? If not, why do you think that might be the case? Also, check to see how long they remain visible. Do they still persist at sunset?

Useful Web Sites

aa.usno.navy.mil/faq/docs/RST_defs.php#top
Provides tables of sunrise and sunset for any location, incorporating the effects of twilight. Also defines three types of twilight.

www.weather-photography.com/gallery.php?cat=optics
Offers photos depicting many different types of optical phenomena.

ww2010.atmos.uiuc.edu/(Gh)/guides/mtr/opt/home.rxml
An interesting site from the University of Illinois with much information on various aspects of atmospheric optics.

www.atoptics.co.uk
Extensive information on the effects of ice crystals, water droplets, and other topics.

finland.fi/public/default.aspx?contentid=160069&content lan=2&culture=en-US
A Finnish site with explanations and photos of different types of mirages.

www.polarimage.fi
Contains numerous photographic examples of the phenomena discussed in this chapter, as well as other interesting images.

MyMeteorologyLab™

The *Understanding Weather and Climate,* Sixth Edition, MyMeteorologyLab™ web site contains numerous multimedia resources to aid in your study of **Atmospheric Optics**.
Visit **www.MyMeteorologyLab.com** to:

- **Review** key chapter concepts.
- **Read** the Pearson eText, *In the News RSS feeds,* and other chapter-specific Web resources.
- **Visualize** challenging topics with Interactive Tutorials, Geoscience Animations, MapMaster interactive maps, and Weather in Motion movies and visualizations, and more.
- **Test Yourself** with self-study quizzes.

WEATHER IN MOTION
Weather Satellites in Orbit

APPENDIX A
Units of Measurement and Conversions

SI UNITS

I. Basic

Length	meter (m)
Mass	kilogram (kg)
Time	second (s)
Electrical Current	ampere (A)
Temperature	kelvin (K)

II. Derived

Force	newton ($N = kg\ m/s^2$)
Pressure	pascal ($Pa = N/m^2$)
Energy	joule ($J = N\ m$)
Power	watt ($W = J/s$)
Electrical Potential Difference	volt ($V = J/C$)
Electrical Charge	coulomb (C)

III. Some Useful Conversions

Length

1 centimeter = 0.39 inches

1 meter = 3.281 feet
    = 39.37 inches

1 kilometer = 0.62 miles

1 inch = 2.54 centimeters

1 foot = 30.48 centimeters
    = 0.305 meters

1 mile = 1.61 kilometers

Mass/Weight

1 gram = 0.035 ounces

1 kilogram = 2.2 pounds

1 ounce = 28.35 grams

1 pound = 0.454 kilograms

Speed

1 meter/second = 2.24 miles/hour
       = 3.60 km/hour

1 mile/hour = 0.45 meters/second
      = 1.61 km/hour

Temperature

Celsius Temperature = (°F − 32)/1.8
        = K − 273.15

Fahrenheit Temperature = 1.8 °C + 32

Kelvin Temperature = °C + 273.15

Energy

1 joule = 0.239 calories

1 calorie = 4.186 joules

APPENDIX B

The Standard Atmosphere

Altitude (km)	Temperature (°C)	Pressure (mb)	p/p_0*	Density (kg/m³)	ρ/ρ_0*
30.00	−46.60	11.97	0.01	0.02	0.02
25.00	−51.60	25.49	0.03	0.04	0.03
20.00	−56.50	55.29	0.05	0.09	0.07
19.00	−56.50	64.67	0.06	0.10	0.08
18.00	−56.50	75.65	0.07	0.12	0.09
17.00	−56.50	88.49	0.09	0.14	0.12
16.00	−56.60	103.52	0.10	0.17	0.14
15.00	−56.50	121.11	0.12	0.20	0.16
14.00	−56.50	141.70	0.14	0.23	0.19
13.00	−56.50	165.79	0.16	0.27	0.22
12.00	−56.50	193.99	0.19	0.31	0.25
11.00	−56.40	226.99	0.22	0.37	0.30
10.00	−49.90	264.99	0.26	0.41	0.34
9.50	−46.70	285.84	0.28	0.44	0.36
9.00	−43.40	308.00	0.30	0.47	0.38
8.50	−40.20	331.54	0.33	0.50	0.40
8.00	−36.90	356.51	0.35	0.53	0.43
7.50	−33.70	382.99	0.38	0.56	0.45
7.00	−30.50	411.05	0.41	0.59	0.48
6.50	−27.20	440.75	0.43	0.62	0.50
6.00	−23.90	472.17	0.47	0.66	0.54
5.50	−20.70	505.39	0.50	0.70	0.57
5.00	−17.50	540.48	0.53	0.74	0.60
4.50	−14.20	577.52	0.57	0.78	0.63
4.00	−11.00	616.60	0.61	0.82	0.67
3.50	−7.70	657.80	0.65	0.86	0.70
3.00	−4.50	701.21	0.69	0.91	0.74
2.50	−1.20	746.91	0.74	0.96	0.78
2.00	2.00	795.01	0.78	1.01	0.82
1.50	5.30	845.59	0.83	1.06	0.86
1.00	8.50	898.76	0.89	1.11	0.91
0.50	11.80	954.61	0.94	1.17	0.95
0.00	15.00	1013.25	1.00	1.23	1.00

*p/p_0 of air pressure to sea level value; ρ/ρ_0 = ratio of air density to sea level value

Weather Map Symbols

Explanation of Codes

$$\begin{array}{c} \text{ff} \nearrow \\ T T_{dd} \ \ C_H \\ \ \ \ \ C_M PPP \\ VV_{ww} \ Ⓝ \ \pm pp a \\ T_d T_d \ C_L N_h W R_t \\ h \ \ RR \end{array}$$

Symbol station model

N	Total cloud cover
dd	Wind direction
ff	Wind speed
VV	Visibility in miles
ww	Present weather
W	Past weather
PPP	Barometric pressure reduced to sea level (add an initial 9 or 10 and place a decimal point to the left of last number)
TT	Current air temperature in °F
N_h	Fraction of sky covered by low or middle clouds
C_L	Low clouds or clouds with vertical development

$$\begin{array}{l} 34 \diagdown \text{⌣} \ 247 \\ ^3\!/_4 \text{✷✷} \bullet +28 \diagup \\ 30\text{---}6 \ \bullet 4 \\ \ \ \ \ 2 \ \bullet 45 \end{array}$$

Sample report

h	Height in feet of the base of the lowest clouds
C_M	Middle clouds
C_H	High clouds
$T_d T_d$	Dewpoint temperature in °F
a	Pressure tendency
pp	Pressure change in mb in preceding 3 hr (+28 = +2.8)
RR	Amount of precipitation in last 6 hr
R_t	Time precipitation began or ended (0 = none; 1 = <1 hr ago; 2 = 1–2 hr ago; 3 = 2–3 hr ago; 4 = 3–4 hr ago; 5 = 4–5 hr ago; 6 = 5–6 hr ago; 7 = 6–12 hr ago; 8 = >12 hr ago; 9 = unknown)

Air Pressure Tendency

∧	Rising, then falling; same as or higher than 3 hr ago	\	Falling, then rising; same as or lower than 3 hr ago
/ (flat top)	Rising, then steady; or rising, then rising more slowly	\ (flat bottom)	Falling, then steady; or falling, then falling more slowly
/	Rising steadily, or unsteadily	\	Falling steadily, or unsteadily
✓	Falling or steady, then rising; or rising, then rising more rapidly	∧	Steady or rising, then falling; or falling, then falling more rapidly
—	Steady; same as 3 hr ago		

Fronts

Fronts are shown on surface weather maps by the symbols below. (Arrows—not shown on maps—indicate direction of motion of front.)

↓ ▼▼▼	Cold front (surface)
↑ ●●●	Warm front (surface)
↑ ●▲●▲	Occluded front (surface)
● ● ▼	Stationary front (surface)
↑ ⌒⌒⌒	Warm front (aloft)
↓ ▽▽▽	Cold front (aloft)

Cloud Abbreviations

St	stratus
Fra	fractus
Sc	stratocumulus
Ns	nimbostratus
As	altostratus
Ac	altocumulus
Ci	cirrus
Cs	cirrostratus
Cc	cirrocumulus
Cu	cumulus
Cb	cumulonimbus

Height of Base of Lowest Cloud

Code	Feet	Meters
0	0–149	0–49
1	150–299	50–99
2	300–599	100–199
3	600–999	200–299
4	1000–1999	300–599
5	2000–3499	600–999
6	3500–4999	1000–1499
7	5000–6499	1500–1999
8	6500–7999	2000–2499
9	8000 or above or no clouds	2500 or above or no clouds

Cloud Cover

○	No clouds
⦶	1/8
◑	Scattered
◓	3/8
◐	4/8
⊖	5/8
◕	Broken
◑	7/8
●	Overcast
⊗	Sky obscured

Cloud Types

⌒ Cu of fair weather, little vertical development and seemingly flattened

⌂ Cu of considerable development, generally towering, with or without other Cu or Sc, bases all at same level

⍟ Cb with tops lacking clear-cut outlines, but distinctly not cirriform or anvil-shaped; with or without Cu, Sc, or St

⊶ Sc formed by spreading out of Cu; Cu often present also

⏝ Sc not formed by spreading out of Cu

— St or StFra, but no StFra of bad weather

- - - StFra and/or CuFra of bad weather (scud)

⏝ Cu and Sc (not formed by spreading out of Cu) with bases at different levels

⏛ Cb having a clearly fibrous (cirriform) top, often anvil-shaped, with or without Cu, Sc, St, or scud

⟋ Thin As (most of cloud layer semitransparent)

⫽ Thick As, greater part sufficiently dense to hide sun (or moon), or Ns

ω Thin Ac, mostly semitransparent; cloud elements not changing much and at a single level

⟋ Thin Ac in patches; cloud elements continually changing and/or occurring at more than one level

⟋ Thin Ac in bands or in a layer gradually spreading over sky and usually thickening as a whole

⋈ Ac formed by the spreading out of Cu or Cb

⊗ Double-layered Ac, or a thick layer of Ac, not increasing; or Ac with As and/or Ns

Ⲙ Ac in the form of Cu-shaped tufts or Ac with turrets

⟋ Ac of a chaotic sky, usually at different levels; patches of dense Ci usually present

⟋ Filaments of Ci, or "mares' tails," scattered and not increasing

⟑ Dense Ci in patches or twisted sheaves, usually not increasing, sometimes like remains of Cb; or towers or tufts

⟍ Dense Ci, often anvil-shaped, derived from or associated with Cb

⟋ Ci, often hook-shaped, gradually spreading over the sky and usually thickening as a whole

⟋ Ci and Cs, often in converging bands, or Cs alone; generally overspreading and growing denser; the continuous layer not reaching 45° altitude

⟋ Ci and Cs, often in converging bands, or Cs alone; generally overspreading and growing denser; the continuous layer exceeding 45° altitude

⟋ Veil of Cs covering the entire sky

⟋ Cs not increasing and not covering entire sky

⟋ Cc alone or Cc with some Ci or Cs, but the Cc being the main cirriform cloud

Wind Speed

	Miles per hour	Kilometers per hour
◎	Calm	Calm
(symbol)	1–2	1–3
(symbol)	3–8	4–13
(symbol)	9–14	14–19
(symbol)	15–20	20–32
(symbol)	21–25	33–40

Wind Speed

	Miles per hour	Kilometers per hour
(symbol)	26–31	41–50
(symbol)	32–37	51–60
(symbol)	38–43	61–69
(symbol)	44–49	70–79
(symbol)	50–54	80–87
(symbol)	55–60	88–96

Wind Speed

	Miles per hour	Kilometers per hour
(symbol)	61–66	97–106
(symbol)	67–71	107–114
(symbol)	72–77	115–124
(symbol)	78–83	125–134
(symbol)	84–89	135–143
(symbol)	119–123	192–198

Weather Conditions

Cloud development NOT observed or NOT observable during past hour

Clouds generally dissolving or becoming less developed during past hour

State of sky on the whole unchanged during past hour

Clouds generally forming or developing during past hour

Visibility reduced by smoke

Light fog (mist)

Patches of shallow fog at station, NOT deeper than 6 feet on land

More or less continuous shallow fog at station, NOT deeper than 6 feet on land

Lightning visible, no thunder heard

Precipitation within sight, but NOT reaching the ground

Drizzle (NOT freezing) or snow grains (NOT falling as showers) during past hour, but NOT at time of observation

Rain (NOT freezing and NOT falling as showers) during past hour, but NOT at time of observation

Snow (NOT falling as showers) during past hour, but NOT at time of observation

Rain and snow or ice pellets (NOT falling as showers) during past hour, but NOT at time of observation

Freezing drizzle or freezing rain (NOT falling as showers) during past hour, but NOT at time of observation

Slight or moderate dust storm or sandstorm, has decreased during past hour

Slight or moderate dust storm or sandstorm, no appreciable change during past hour

Slight or moderate dust storm or sandstorm has begun or increased during past hour

Severe dust storm or sandstorm, has decreased during past hour

Severe dust storm or sandstorm, no appreciable change during past hour

Fog or ice fog at distance at time of observation, but NOT at station during past hour

Fog or ice fog in patches

Fog or ice fog, sky discernible, has become thinner during past hour

Fog or ice fog, sky NOT discernible, has become thinner during past hour

Fog or ice fog, sky discernible, no appreciable change during past hour

Intermittent drizzle (NOT freezing), slight at time of observation

Continuous drizzle (NOT freezing), slight at time of observation

Intermittent drizzle (NOT freezing), moderate at time of observation

Continuous drizzle (NOT freezing), moderate at time of observation

Intermittent drizzle (NOT freezing), heavy at time of observation

Intermittent rain (NOT freezing), slight at time of observation

Continuous rain (NOT freezing), slight at time of observation

Intermittent rain (NOT freezing), moderate at time of observation

Continuous rain (NOT freezing), moderate at time of observation

Intermittent rain (NOT freezing), heavy at time of observation

Intermittent fall of snowflakes, slight at time of observation

Continuous fall of snowflakes, slight at time of observation

Intermittent fall of snowflakes, moderate at time of observation

Continuous fall of snowflakes, moderate at time of observation

Intermittent fall of snowflakes, heavy at time of observation

Slight rain shower(s)

Moderate or heavy rain shower(s)

Violent rain shower(s)

Slight shower(s) of rain and snow mixed

Moderate or heavy shower(s) of rain and snow mixed

Moderate or heavy shower(s) of hail, with or without rain, or rain and snow mixed, not associated with thunder

Slight rain at time of observation; thunderstorm during past hour, but NOT at time of observation

Moderate or heavy rain at time of observation; thunderstorm during past hour, but NOT at time of observation

Slight snow, or rain and snow mixed, or hail at time of observation; thunderstorm during past hour, but NOT at time of observation

Moderate or heavy snow, or rain and snow mixed, or hail at time of observation; thunderstorm during past hour, but NOT at time of observation

Weather Conditions

∞	Haze	S	Widespread dust in suspension in the air, NOT raised by wind, at time of observation	↑	Dust or sand raised by wind at time of observation	⚇	Well-developed dust whirl(s) within past hour	(S→)	Dust storm or sandstorm within sight of or at station during past hour
)•(	Precipitation within sight, reaching the ground but distant from station	(•)	Precipitation within sight, reaching the ground, near to but NOT at station	⍑	Thunderstorm, but no precipitation at the station	∨	Squall(s) within sight during past hour or at time of observation	)(	Funnel cloud(s) within sight of station at time of observation
• ∇]	Showers of rain during past hour, but NOT at time of observation	✻ ∇]	Showers of snow, or of rain and snow, during past hour, but NOT at time of observation	△ ∇]	Showers of hail, or of hail and rain, during past hour, but NOT at time of observation	≡	Fog during past hour, but NOT at time of observation	⍑]	Thunderstorm (with or without precipitation) during past hour, but NOT at time of observation
⌇S	Severe dust storm or sandstorm has begun or increased during past hour	⊣+	Slight or moderate drifting snow, generally low (less than 6 ft)	⊢+	Heavy drifting snow, generally low	↟	Slight or moderate blowing snow, generally high (more than 6 ft)	↟+	Heavy blowing snow, generally high
≡	Fog or ice fog, sky NOT discernible, no appreciable change during past hour	⊨	Fog or ice fog, sky discernible, has begun or become thicker during past hour	⊫	Fog or ice fog, sky NOT discernible, has begun or become thicker during past hour	⩛	Fog depositing rime, sky discernible	⩛	Fog depositing rime, sky NOT discernible
,,,	Continuous drizzle (NOT freezing), heavy at time of observation	(∿)	Slight freezing drizzle	(∿)	Moderate or heavy freezing drizzle	;	Drizzle and rain, slight	;	Drizzle and rain, moderate or heavy
••••	Continuous rain (NOT freezing), heavy at time of observation	(∿•)	Slight freezing rain	(∿••)	Moderate or heavy freezing rain	•✻	Rain or drizzle and snow, slight	•✻	Rain or drizzle and snow, moderate or heavy
✻✻✻	Continuous fall of snowflakes, heavy at time of observation	←→	Ice prisms (with or without fog)	—△—	Snow grains (with or without fog)	—✳—	Isolated starlike snow crystals (with or without fog)	△.	Ice pellets or snow pellets
✻∇	Slight snow shower(s)	✻∇	Moderate or heavy snow shower(s)	△∇	Slight shower(s) of snow pellets, or ice pellets with or without rain, or rain and snow mixed	△∇	Moderate or heavy shower(s) of snow pellets, or ice pellets, or ice pellets with or without rain or rain and snow mixed	▲∇	Slight shower(s) of hail, with or without rain or rain and snow mixed, not associated with thunder
•/✻ ⍑	Slight or moderate thunderstorm without hail, but with rain, and/or snow at time of observation	△ ⍑	Slight or moderate thunderstorm, with hail at time of observation	•/✻ ⍑	Heavy thunderstorm, without hail, but with rain and/or snow at time of observation	S ⍑	Thunderstorm combined with dust storm or sandstorm at time of observation	△ ⍑	Heavy thunderstorm with hail at time of observation

Weather Extremes

This appendix presents various weather extremes and the geographic locales of the extremes on world maps. The weather extremes are presented on the following pages:

U.S.A.'s Lowest Annual Mean Temperature
9°F (-12.8°C)
Barrow, Alaska
(1941 to 1970)

U.S.A.'s Coolest Summer Average
36°F (2.2°C)
Barrow, Alaska
(1941 to 1970)

North America's Lowest Mean Temperature for a Month
(exluding Greenland)
-54°F (-47.8°C)
Eureka, Nunavut
(February 1979)

Canada's Lowest Annual Mean Temperature
-3°F (-19.4°C)
Eureka, Nunavut
(1947 to 1980)

U.S.A.'s Lowest Temperature
-79.8°F (-62.1°C)
Prospect Creek, Alaska
(23 January 1971)

U.S.A.'s Coldest Winter Average
-16°F (-26.7°C)
Barter Island, Alaska
(1941 to 1970)

Greenland's Lowest Temperature
-87°F (-66.1°C)
Northice
(9 January 1954)

North America's Lowest Temperature
(excluding Greenland)
-81.4°F (-63°C)
Snag, Yukon
(3 February 1947)

Canada's Highest Temperature
113°F (45°C)
Midale and Yellow Grass, Saskatchewan
(5 July 1937)

Difference Between Lowest and Highest Recorded Temperatures
177°F (98.3°C)
Mayo, Yukon
(period-of-record unknown)

U.S.A.'s Lowest Temperature
(excluding Alaska)
-69.7°F (-56.5°C)
Rogers Pass, Montana
(20 January 1954)

U.S.A.'s Largest 24-Hour Temperature Fall
100°F (56°C)
Browning, Montana
(23 and 24 January 1916)

U.S.A.'s Largest 2-Minute Temperature Rise
49°F (27.2°C)
Spearfish, South Dakota
(22 January 1943)

U.S.A.'s Warmest Winter Average
73°F (22.8°C)
Honolulu, Hawaii
(1941 to 1970)

Western Hemisphere's Highest Temperature
134°F (56.7°C)
Greenland Ranch, California
(10 July 1913)

Western Hemisphere's Hottest Summer Average
98°F (37°C)
Death Valley, California
(1941 to 1970)

NORTH AMERICA

SOUTH AMERICA

South America's Highest Temperature
120°F (48.9°C)
Rivadavia, Argentina
(11 December 1905)

South America's Lowest Temperature
-27°F (-32.8°C)
Sarmiento, Argentina
(1 June 1907)

South Pole's Highest Temperature
7.5°F (-13.6°C)
(27 December 1978)

60°

30°

0° Equator

30°

60°

150° 120° 90° 60°

TEMPERATURE EXTREMES

Difference Between Lowest and Highest Recorded Temperatures
183.2°F (101.8°C)
Verkhoyansk, Russia

Europe's Lowest Temperature
-67°F (-55°C)
Ust'-Shchuger, Russia
(31 December 1978)

Europe's Highest Temperature
122°F (50°C)
Seville, Spain
(4 August 1881)

Mean Annual Temperature Range
146°F (81.1°C)
Eastern Sayan Region, Russia

Northern Hemisphere's Lowest Temperature
-90°F (-67.8°C)
Verkhoyansk and Oymyakon, Russia
(5 and 7 February 1892
and 6 February 1933)

Australia's Highest Temperature
128°F (53.3°C)
Cloncurry, Queensland
(16 January 1889)

Asia's Highest Temperature
129°F (53.9°C)
Tirat Tsvi, Israel
(22 June 1942)

World's Highest Temperature
136°F (57.8°C)
El Azizia, Libya
(13 September 1922)

World's Highest Annual Mean Temperature
94°F (34.4°C)
Dalol, Ethiopia
(October 1960 through
November 1966)

Africa's Lowest Temperature
-11°F (-23.9°C)
Ifrane, Morocco
(11 February 1935)

Australia's Longest Hot Spell
≥100°F (37.8°C) on 162 consecutive days
Marble Bar, Western Australia
(30 October 1923 to 7 April 1924)

Australia's Lowest Temperature
-9.4°F (-23.0°C)
Charlotte Pass,
New South Wales
(29 June 1994)

Mean Monthly Temperature
-99.8°F (-73.2°C)
Plateau Station, Antarctica
(July 1968)

Lowest Annual Mean Temperature
-71°F (-57.2°C)
Sovietskaya, Antarctica
(1957 and 1958)

World's Lowest Temperature
-129°F (-89.4°C)
Vostok, Antarctica
(21 July 1983)

Antarctica's Highest Temperature
59°F (15°C)
Vanda Station, Antarctica
(5 January 1974)

EUROPE

ASIA

AFRICA

AUSTRALIA

ANTARCTICA

Equator

60°

30°

0°

30°

0°   30°   60°   120°   150°   60°

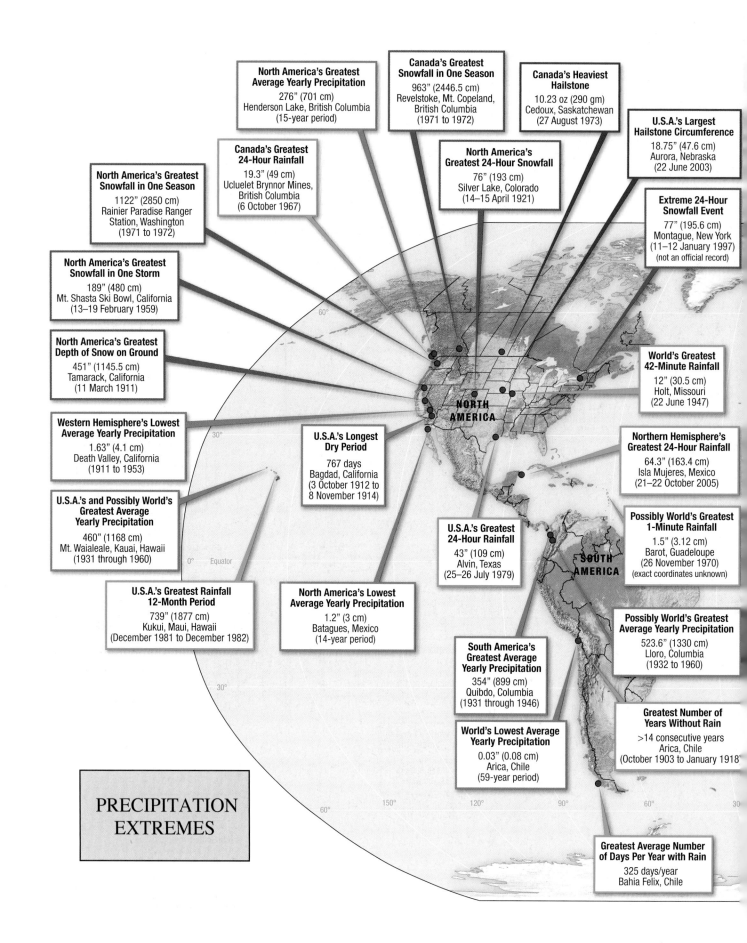

North America's Greatest Average Yearly Precipitation

276" (701 cm)
Henderson Lake, British Columbia
(15-year period)

Canada's Greatest Snowfall in One Season

963" (2446.5 cm)
Revelstoke, Mt. Copeland,
British Columbia
(1971 to 1972)

Canada's Heaviest Hailstone

10.23 oz (290 gm)
Cedoux, Saskatchewan
(27 August 1973)

U.S.A.'s Largest Hailstone Circumference

18.75" (47.6 cm)
Aurora, Nebraska
(22 June 2003)

Canada's Greatest 24-Hour Rainfall

19.3" (49 cm)
Ucluelet Brynnor Mines,
British Columbia
(6 October 1967)

North America's Greatest 24-Hour Snowfall

76" (193 cm)
Silver Lake, Colorado
(14–15 April 1921)

Extreme 24-Hour Snowfall Event

77" (195.6 cm)
Montague, New York
(11–12 January 1997)
(not an official record)

North America's Greatest Snowfall in One Season

1122" (2850 cm)
Rainier Paradise Ranger
Station, Washington
(1971 to 1972)

North America's Greatest Snowfall in One Storm

189" (480 cm)
Mt. Shasta Ski Bowl, California
(13–19 February 1959)

North America's Greatest Depth of Snow on Ground

451" (1145.5 cm)
Tamarack, California
(11 March 1911)

World's Greatest 42-Minute Rainfall

12" (30.5 cm)
Holt, Missouri
(22 June 1947)

Western Hemisphere's Lowest Average Yearly Precipitation

1.63" (4.1 cm)
Death Valley, California
(1911 to 1953)

U.S.A.'s Longest Dry Period

767 days
Bagdad, California
(3 October 1912 to
8 November 1914)

Northern Hemisphere's Greatest 24-Hour Rainfall

64.3" (163.4 cm)
Isla Mujeres, Mexico
(21–22 October 2005)

U.S.A.'s and Possibly World's Greatest Average Yearly Precipitation

460" (1168 cm)
Mt. Waialeale, Kauai, Hawaii
(1931 through 1960)

U.S.A.'s Greatest 24-Hour Rainfall

43" (109 cm)
Alvin, Texas
(25–26 July 1979)

Possibly World's Greatest 1-Minute Rainfall

1.5" (3.12 cm)
Barot, Guadeloupe
(26 November 1970)
(exact coordinates unknown)

U.S.A.'s Greatest Rainfall 12-Month Period

739" (1877 cm)
Kukui, Maui, Hawaii
(December 1981 to December 1982)

North America's Lowest Average Yearly Precipitation

1.2" (3 cm)
Batagues, Mexico
(14-year period)

South America's Greatest Average Yearly Precipitation

354" (899 cm)
Quibdo, Columbia
(1931 through 1946)

Possibly World's Greatest Average Yearly Precipitation

523.6" (1330 cm)
Lloro, Columbia
(1932 to 1960)

World's Lowest Average Yearly Precipitation

0.03" (0.08 cm)
Arica, Chile
(59-year period)

Greatest Number of Years Without Rain

>14 consecutive years
Arica, Chile
(October 1903 to January 1918)

Greatest Average Number of Days Per Year with Rain

325 days/year
Bahia Felix, Chile

PRECIPITATION EXTREMES

Europe's Greatest Average Yearly Precipitation

183" (465 cm)
Crkvica, Bosnia-Herzegovina
(22-year period)

World's Greatest 20-Minute Rainfall

8.10" (20.6 cm)
Curtea de Arges, Romania
(7 July 1889)

Europe's Lowest Average Yearly Precipitation

6.4" (16.3 cm)
Astrakhan, Russia
(25-year period)

Record Snowfall

67.7" (172 cm)
Bessans, France
(5–6 April 1969)

Relative Variability of Annual Precipitation

94%
Themed, Israel
(1921 to 1947)

Relative Variability of Annual Precipitation

108%
Lhasa, Tibet, China
(1935 to 1939)

Asia's and Possibly World's Greatest Average Yearly Precipitation

467.4" (1187.3 cm)
Mawsynram, India
(1941 to 1979)

World's Greatest 60-Minute Rainfall

15.78" (40.1 cm)
Shangdi, China
(3 July 1975)

World's Greatest 1-Month Rainfall

366" (930 cm)
Cherrapunji, India
(July 1861)

World's Greatest 12-Month Rainfall

1042" (2647 cm)
Cherrapunji, India
(August 1860 to July 1861)

World's Heaviest Hailstone

2.25 lbs (1.02 kg)
Gopalganj District, Bangladesh
(14 April 1986)

Asia's Lowest Average Yearly Precipitation

1.8" (4.6 cm)
Aden, Yemen
(50-year period)

Africa's Greatest Average Variability of Annual Precipitation

75" (191 cm)
Debundscha, Cameroon

Africa's Lowest Average Yearly Precipitation

<0.1" (<0.25 cm)
Wadi Halfa, Sudan
(39-year period)

Australia's Greatest Average Yearly Precipitation

340.16" (864 cm)
Bellenden Ker, Queensland
(9-year period)

Africa's Greatest Average Yearly Precipitation

405" (1029 cm)
Debundscha, Cameroon
(32-year period)

World's Greatest 12-Hour Rainfall

46" (117 cm)
Grand Ilet, Reunion
(26 January 1980)

World's Greatest 24-Hour Rainfall

72" (182.5 cm)
Foc-Foc, Reunion
(7–8 January 1966)

World's Greatest 5-Day Rainfall

196.06" (498 cm)
Cratere Commerson, Reunion
(24–28 February 2007)

Australia's Greatest 24-Hour Rainfall

44.9" (114 cm)
Bellenden Ker, Queensland
(4 January 1979)

Australia's Lowest Average Yearly Precipitation

4.05" (10 cm)
Troudaninna, South Australia
(42-year period)

EUROPE

ASIA

AFRICA

AUSTRALIA

ANTARCTICA

60°

30°

Equator 0°

30°

0°   30°   60°   120°   150°   60°

Canada's Maximum Hourly Wind Speed

125 mph (201 kph)
Cape Hopes Advance, Quebec
(18 November 1931)

Canada's Highest Average Annual Wind Speed

22 mph (35 kph)
Cape Warwick,
Resolution Island, Nuvavut
(1962 to 1973)

Greenland's Peak Wind Gust

207 mph (333 kph)
Thule
(8 March, 1972)

Canada's Highest Frequency of Days with Thunderstorms

34 per year average
Windsor, Ontario

North America's Highest Sea-Level Air Pressure

31.85" (1067.7 mb)
Northway, Alaska
(31 January 1989)

U.S.A.'s Foggiest Place (West Coast)

average 2,552 hours/year
Cape Disappointment,
Washington
(10-year period or more)

U.S.A.'s Greatest Average Number of Days with Hailstorms

9.4 days per season
Cheyenne, Wyoming
(40-year period)

North America's Highest Frequency of Days with Thunderstorms

100 per year average
Tampa International
Airport, Florida

North America's Lowest Sea-Level Air Pressure

26.35" (892.3 mb)
Matecumbe Key, Florida
(2 September 1935)

U.S.A.'s Foggiest Place (East Coast)

average 1,580 hours/year
Moose Peak Lighthouse,
Mistake Island, Maine
(10-year period or more)

North America's Highest Surface Wind Peak Gust

231 mph (372 kph)
Mt. Washington, New Hampshire
(12 April 1934)

World's Highest Surface Wind 5-Minute Speed

188 mph (303 kph)
Mt. Washington, New Hampshire
(12 April 1934)

Mean Wind Speed for a Month

70 mph (113 kph)
Mt. Washington, New Hampshire
(February 1939)

Mean Wind Speed for 24 Hours

128 mph (206 kph)
Mt. Washington, New Hampshire
(11 and 12 April 1934)

Highest Average Annual Wind Speed

35 mph (56 kph)
Mt. Washington, New Hampshire
(1934 to 1983)

MISCELLANEOUS EXTREMES

World's Highest Dew Point
95°F (35°C)
Dhahran, Saudi Arabia
(8 July 2003)

Sea-Surface Temperature
96°F (35.6°C)
Persian Gulf
(5 August 1924)

World's Highest Sea-Level Air Pressure
32.01" (1083.8 mb)
Agata, Russia
(31 December 1968)

World's Lowest Sea-Level Air Pressure
25.69" (870 mb)
(estimated by dropsonde in eye of Typhoon Tip)
area of 17N 138E
(12 October 1979)

Average Afternoon Dew Point in June
84°F (28.9°C)
Assab, Eritrea
(no dates given)

Average Number of Thunderstorm Days Per Year
242 days/year
Kampala, Uganda
(10-year period)

Average Number of Thunderstorm Days Per Year
322 days/year
Bogor, Indonesia
(1916 to 1920)

World's Highest Surface Peak Wind Gust
253 mph (407 kph)
Barrow Island, Australia
(10 April 1996)

Average Daily Insolation in December
955 ly (463 W/m)
South Pole
(1958 to 1966)

Mean Wind Speed for 24 Hours
108 mph (174 kph)
Port Martin, Antarctica
(21 and 22 March 1951)

Mean Wind Speed for a Month
65 mph (107 kph)
Port Martin, Antarctica
(March 1951)

EUROPE

ASIA

AFRICA

AUSTRALIA

ANTARCTICA

60°

30°

Equator   0°

30°

60°

0°   30°   60°   90°   120°   150°

Glossary

A

Absolute Humidity Mass of water vapor per unit volume of air, usually expressed in grams per cubic meter (g/m^3).

Absolute Vorticity The sum of vorticity relative to the surface and vorticity arising from Earth's rotation.

Absorption A process in which radiation is captured by a molecule. Unlike reflection, absorption represents an energy transfer to the absorbing molecule.

Acceleration A change in velocity: a change in speed or direction, or both.

Acceleration of Gravity Acting alone, gravity would accelerate all objects by the same rate, about 9.8 m/sec/sec. *See* gravity.

Adiabatic Term for processes in which no heat is added or removed. For example, a rising air parcel cools adiabatically as it expands.

Advanced Weather Interactive Processing System (AWIPS) A system for the display and manipulation of weather information at Weather Service Offices.

Advection Horizontal transport of some atmospheric property (heat, moisture, etc.).

Aerosol Brown Cloud Tropospheric haze caused by particulates absorbing and back-scattering solar radiation.

Aerosols Small, suspended particles in the atmosphere.

Aggregation The process in which ice crystals join together to form snowflakes. If these snowflakes melt as they fall, they arrive as rain.

Air Mass A large body of air having little horizontal variation in temperature and moisture.

Air Mass Thunderstorms Relatively small, short-lived thunderstorms that do not produce very strong winds, large hail, or tornadoes.

Albedo The fraction of solar radiation arriving at a surface that is reflected.

Aleutian Low A semipermanent cell found in the North Pacific in winter.

Altocumulus A midlevel layered cloud with some rolls or patches of vertical development.

Altostratus A midlevel layered cloud.

Aneroid Barometer A device used to measure air pressure. The barometer's elastic chamber expands and contracts in response to the surrounding pressure.

Annular Mode A measure of hemispheric wind and pressure patterns. Annular modes exist in both hemispheres and encapsulate more extratropical variability in circulation than any other known phenomenon.

Antarctic Circle The line of latitude 65.5° S, marking the northern limit of Southern Hemisphere locations that can receive 24 hours of daylight or darkness.

Aphelion Earth's position when it is farthest from the Sun (~July 3).

Arctic Circle The line of latitude 65.5° N, marking the southern limit of Northern Hemisphere locations that can receive 24 hours of daylight or darkness.

Arctic Oscillation A see-saw in pressure between the arctic and northern midlatitudes. Similar to the North Atlantic Oscillation, but confined to the middle and higher latitudes.

Atlantic Multidecadal Oscillation A periodic reversal of North Atlantic sea surface temperature anomalies on the order of several decades. Associated with the frequency of Atlantic hurricanes.

Atmosphere The gases, droplets, and particles surrounding Earth's surface.

Atmospheric Window The range of wavelengths (about 8 to 12 μm) that are not readily absorbed by the gases of the atmosphere.

Aurora Borealis or Aurora Australis An illumination of the sky found in the high northern (borealis) or southern (australis) latitudes, which is produced as charged particles arriving from the Sun react with the upper atmosphere.

AWIPS Acronym for Advanced Weather Interactive Processing System, the computer system used by NWS forecasters for the display of weather maps, satellite and radar imagery, and other types of data.

B

Banner Clouds A cloud formed near the top of a topographic barrier by orographic uplift.

Barometer An instrument for measuring air pressure.

Beam Spreading The process whereby a beam of radiation is distributed over a larger horizontal area as the angle of incidence departs from vertical. Reduces the intensity of radiation absorption by the surface.

Bergeron Process The primary mechanism for precipitation formation outside the tropics; this process involves the coexistence of ice crystals and supercooled water droplets.

Bermuda-Azores High A semipermanent cell found in the Atlantic in summer.

Blackbody An object or substance that is perfectly efficient at absorbing and radiating radiation. Blackbodies do not exist in nature, but represent an ideal.

C

Calorie Amount of heat required to raise the temperature of 1 gram of water 1 °C (about 4.2 J).

Carbon Dioxide An important variable gas in the atmosphere, made up of one atom of carbon bound to two atoms of oxygen. An important greenhouse gas.

Celsius Scale The temperature scale that designates 0° as the freezing point and 100° as the sea level boiling point of water.

Charge Separation The separation of positive and negative ions into different parts of a cloud. A necessary precursor for lightning.

Chromosphere The layer of the Sun immediately surrounding the photosphere.

Cirrocumulus A high cloud composed of ice that is generally layered but with some rolls or pockets of vertical development.

Cirrostratus A high, layered cloud consisting of ice crystals.

Cirrus A high cloud made up entirely of ice crystals.

Climate The statistical properties of the atmosphere, including measures of average conditions, variability, etc.

Climatology   The study of long-term atmospheric conditions.

Cloud   An area of the atmosphere containing sufficient concentration of water droplets and/or ice crystals to be visible.

Cloud-to-Cloud Lightning   Lightning that flows from one part of a cloud to another or from one cloud to another. Distinct from cloud-to-ground lightning.

Cloud Seeding   An attempt to stimulate precipitation by introducing certain materials into existing clouds.

Cold Cloud   A cloud with a temperature below 0 °C from top to bottom.

Cold Front   Transition zone between cold and warm air masses; forms when a cold air mass advances on a warm air mass.

Collector Drop   A relatively large falling raindrop that collides and coalesces with smaller, slower-moving droplets beneath it.

Condensation   Change from vapor to liquid phase. Condensation releases the energy required for evaporation. *See* latent heat of vaporization.

Condensation Nuclei   Small, airborne particles that enhance condensation. Without condensation nuclei, condensation would occur only at very high relative humidity (at about 200 percent or more), while condensation nuclei allow condensation to occur at or slightly below 100 percent relative humidity.

Conduction   Heat transfer from molecule to molecule, without significant movement of the molecules.

Confluence   A type of horizontal convergence that occurs when streamlines come closer together in the downstream direction.

Convection   Heat transfer by fluid flow (movement of a gas or liquid).

Convective Outlooks   Information on the probability of different types of severe storms issued by the Storm Prediction Center.

Convergence   Horizontal motions of air resulting in a net inflow of air (with more air imported than exported). Causes rising or sinking motions.

Conveyor Belt Model   The modern description of air flow through midlatitude cyclones.

Cool Cloud   A cloud in which the lower reaches have temperatures above 0 °C and in which the temperatures in the upper portions are below 0 °C.

Cooling Degree-Day   An index of the amount of seasonal air conditioning required for a location. Cooling degree-days are calculated by subtracting a base temperature (usually 65 °F) from the daily mean temperature and summing the differences. Days with mean temperature below the base are ignored in the calculation.

Core   The interior of the Sun, where nuclear fusion produces energy that is ultimately radiated to Earth.

Coriolis Force   An imaginary deflective force arising from Earth's rotation that is necessary to account for motions measured relative to the surface.

Corona   A circular illumination of the sky immediately surrounding the Moon, or in some instances the Sun, caused by diffraction.

Cumulonimbus   A cumulus cloud with very deep vertical development extending into the lower stratosphere, distinguished by an anvil at its top consisting of ice crystals.

Cumulus   Any cloud having substantial vertical development.

Cutoff Low   An upper-level area of low pressure that takes on a circular flow distinct from the general flow around it.

Cyclogenesis   The beginning of cyclone formation.

Cyclone   A region of low pressure relative to the surrounding area.

D

Dalton's Law   This law of physical science states that the pressure of a combination of gases is equal to the sum of the partial pressures of each of the gases.

Dart Leader   A zone of ionized air that serves as a conduit for a lightning stroke subsequent to the initial stepped leader in a lightning flash.

Density   The mass of a substance per unit volume, expressed as kilograms per cubic meter (kg/m^3) in the International System of Units (SI).

Deposition   Change from the vapor phase to the solid phase (frost is an example). Deposition releases the energy of vaporization and fusion. *See* latent heat.

Derechos   Powerful, large-scale winds that flow in a straight line.

Dew Point   The temperature to which the air must be cooled to become saturated.

Dew Point Temperature   Also called *dew point*. The temperature at which saturation will occur, given sufficient cooling.

Diabatic   Processes that involve the addition or removal of heat. For example, air in contact with a cold surface loses heat diabatically by conduction.

Diffluence   A type of horizontal divergence that occurs when streamlines spread apart in the downstream direction.

Diffuse Solar Radiation   Sunlight that is scattered downward to the surface. *See* scattering.

Direct Solar Radiation   Sunlight that passes through the atmosphere without absorption or scattering.

Divergence   Horizontal motions of air resulting in a net outflow of air (with more air exported than imported). Causes rising or sinking motions.

Doppler Radar   A type of radar that can measure horizontal motions as well as the internal characteristics of clouds.

Double Eye Wall   A feature on a hurricane in which there are two eye walls—an inner one and an outer one. The appearance of a double eye wall is often a precursor to hurricane intensification.

Dry Adiabatic Lapse Rate (DALR)   Temperature decrease experienced by a rising unsaturated parcel (about 1 °C/100 m). Sinking parcels warm at the same rate. The DALR is a constant.

Dry Line   A boundary between humid air and denser dry air. A favored location for thunderstorm development.

Dynamic Low   A low-pressure system created by divergence in the middle or upper troposphere.

E

East Greenland Drift   An ocean current that flows southward in the North Atlantic.

Ecliptic Plane   The imaginary surface swept by Earth as it orbits the Sun.

El Niño   A recurrent event in the tropical eastern Pacific in which sea surface temperatures are significantly above normal. The inverse event (cold sea surface temperatures) is called a *La Niña*.

Electromagnetic Radiation   Energy emitted by virtue of an object's temperature. Radiation is unique in that it does not require a transfer medium and can travel through a vacuum. The energy transfer is accomplished by oscillations in an electric field and a magnetic field.

Emissivity   The property of a substance or object that expresses, as a fraction or percentage, how efficient it is at emitting radiation.

Entrainment   The incorporation of surrounding, unsaturated air into a cloud.

Environmental Lapse Rate (ELR)   The rate of vertical temperature decrease in the air column. The value is highly variable, depending on local conditions. For the troposphere, the global average is about 0.65 °C/100 m.

ENSO   An acronym for the El Niño Southern Oscillation phenomenon. Involves the interaction of Tropical Pacific Sea Surface Temperatures and atmospheric pressures.

Equation of State   The equation relating air pressure to temperature and density.

Equinoxes   The two days of the year on which Earth's axis is not tilted toward or away from the Sun. On the equinoxes every latitude receives 12 hours of sunlight, and the Sun is overhead at the equator. The equinoxes occur March 21–22 and September 22–23.

Escape Velocity   The rate of movement required of an air molecule to escape Earth's gravity.

European Center for Medium-Range Weather Forecasting   Weather forecasting agency for the European Union.

Evaporation   The change in phase of liquid water to water vapor.

Evapotranspiration   The combined processes of evaporation and transpiration; the delivery of water to the atmosphere by vegetation and by direct evaporation from wet surfaces.

Extraterrestrial Radiation   The solar radiation incident at the top of Earth's atmosphere.

Eye   The center of a hurricane, marked by generally clear skies and light winds.

Eye Wall   Very cloudy portion of a hurricane immediately adjacent to the eye; usually the region of highest wind speed and most intense precipitation.

F

Fahrenheit Scale   A temperature scale that assigns values of 32° to the freezing point of water and 212° to the sea level boiling point of water.

Fetch   Distance traveled by wind over a uniform surface, such as a water body.

First Law of Thermodynamics   Most generally, the law that states that energy is a conserved property. In a meteorological context it states that heat added to a gas results in some combination of expansion of the gas and an increase in its internal energy.

Flares   Intensely hot eruptions on the solar surface.

Foehn Wind   A synoptic-scale wind that flows downslope and warms by compression.

Fog   Air that is adjacent to the surface and contains suspended water droplets, usually formed by diabatic cooling.

Force   The product of mass and acceleration, as expressed in Newton's law ($F = ma$).

Freezing Rain   A form of precipitation in which rain droplets freeze as they fall below an inversion and pass into air having a temperature below 0 °C.

Friction   Force that acts to slow wind but does not change its direction. Friction develops between the atmosphere and surface and between layers of air moving at different velocities.

Front   A transition zone between two dissimilar air masses (that is, air masses with differing temperature, moisture, or density).

Frost   A coating of ice crystals on a surface when the air adjacent to the surface becomes saturated at temperatures below 0 °C.

Frost Point   The temperature at which saturation occurs, provided that temperature is less than 0 °C.

Frozen Dew   A coating of ice on a surface that occurs when a layer of dew freezes as temperatures drop below 0 °C.

Fujita Scale   The scale for categorizing tornado intensity.

Funnel Cloud   A column of rapidly rotating air similar to a tornado, except that the column has not extended to the ground.

G

Gamma Rays   Electromagnetic radiation at wavelengths far shorter than those of visible light (from about 0.0000001 μm to 0.000001 μm). Gamma rays, which make up only a tiny proportion of the Sun's energy, are absorbed hundreds of kilometers above the surface.

General Circulation   A term that refers to planetary-scale winds and pressure, features that appear in the time-averaged state.

Geopotential Height   Loosely defined, the altitude at which atmospheric pressure takes on a particular value, as in "500 mb height." Because pressure reflects the mass of overlying atmosphere, geopotential heights reflect the potential energy of the atmosphere above that height.

Geostrophic Flow   An idealized condition in which the upper-level air flows at constant speed and direction, parallel to straight isobars. There is no acceleration in geostrophic flow, and frictional forces are negligible.

Global Dimming   A measured decrease in solar radiation reaching the surface, which lasted from about 1970 until 1990. It was probably not global in extent, and it did not persist beyond the early 1990s.

Glory   A series of rings formed around the shadow of an aircraft on the top of a cloud, formed by refraction, reflection, and diffraction.

Gradient   A change in some quantity (temperature, moisture, pressure) over space. For example, the temperature gradient is the rate of change of temperature per unit distance and can be expressed in degrees Celsius per kilometer (°C/km).

Gradient Wind   Wind flowing parallel to curved isobars. Frictional forces are negligible. With gradient flow there is a constant adjustment between the pressure gradient force and Coriolis force, causing the wind to change speed and direction as it flows along the isobars.

Graupel   Ice crystals that have grown by riming to produce a spongy, somewhat translucent particle.

Gravity   The force that attracts objects to Earth's surface. Although the acceleration of gravity is constant, the force of gravity varies from object to object. The force of gravity per unit volume of air is directly proportional to density.

Graybodies   Bodies or substances that are not 100 percent efficient at absorbing or radiating energy. In reality, all bodies are graybodies.

Green Flash   The brief appearance of green light near the top of the Sun sometimes observed at sunrise or sunset.

Greenwich Mean Time   Also called *universal time* (*UT*). An international reference for time-keeping used for weather observations, satellite imaging, etc. It corresponds to local time at 0° longitude, a meridian passing through Greenwich, England.

Growing Degree-Day   An index for estimating when crops will have undergone enough growth to send them to market. Growing degree-days are calculated by subtracting a base temperature for a particular crop from the daily mean temperature and summing the differences.

H

Hadley Cell   A somewhat idealized, large-scale wind and pressure pattern found in tropical latitudes of both hemispheres. Air rises above the equator, flows poleward to about 25° latitude, subsides, and flows back to the equator at low levels.

Hail Precipitation in the form of ice crystals, almost always associated with thunderstorms. Hail falls rapidly to the surface and thus does not melt during its descent.

Halo A circular band of light surrounding the Sun or Moon, caused by ice crystal refraction.

Hawaiian High A semipermanent cell found in the Pacific, most notably during the summer months.

Heat The kinetic energy of atoms or molecules that make up a substance.

Heat Index A measure of apparent temperature used for warm conditions, incorporating temperature and humidity.

Heating Degree-Days An index of the amount of seasonal heating required for a location. Heating degree-days are calculated by subtracting the daily mean temperature from a base temperature (usually 65 °F) and summing the differences. Days with mean temperature above the base are ignored in the calculation.

Heterogeneous Nucleation The condensation of liquid droplets or the deposition of ice crystals onto condensation or ice nuclei.

Heterosphere The high atmosphere (above 80 km or so), where gases are not well mixed but rather are stratified according to molecular weight. Vertical motions are too weak to overcome gravitational settling, so heavier gases are found beneath lighter gases.

High Another term for *anticyclone*.

Homogeneous Nucleation The condensation of water droplets or deposition of ice crystals without condensation or ice nuclei.

Homosphere The lowest 80 km (50 mi) of the atmosphere in which the relative abundance of the permanent gases is constant.

Horse Latitudes Areas associated with the oceanic subtropical highs, generally characterized by clear skies and light winds.

Hot Tower An area of intense convection within a hurricane. Often associated with hurricane intensification.

Humidity An expression of the amount of water vapor in the air.

Hurricane An intense tropical cyclone (warm-core low), with sustained winds of at least 120 km/hr.

Hydrologic Cycle The perennial movement of water in its three phases between the atmosphere, Earth's surface, and groundwater.

Hydrostatic Equilibrium When the vertical pressure gradient force is balanced by the force of gravity. Because the forces balance, there is no acceleration upward or downward.

Hygrometer An instrument that measures humidity.

Hygroscopic Nuclei Airborne particles having an affinity for water, serving as condensation nuclei.

Hygrothermograph An instrument that records humidity and temperature.

I

Ice Nuclei Particles onto which ice crystals can form when the air becomes saturated. In the absence of freezing nuclei, water droplets freeze only at very low temperatures (near −40 °C). Ice nuclei allow ice to form at relatively "high" temperatures (around −10 °C).

Ice Storm A weather event that results in freezing rain.

Icelandic Low A semipermanent cell found in the North Atlantic in winter.

Ideal Gas Law Also known as the *equation of state*. Important law describing the relationship between pressure, temperature, and density.

Infrared Radiation Electromagnetic radiation at wavelengths longer than visible radiation, from about 0.7 μm to 1000 μm.

Insolation Incident, or incoming, solar radiation.

Interglacial A warmer segment of time between episodes of glaciation.

Intergovernmental Panel on Climate Change (IPCC) The agency of the World Meteorological Organization charged with assessing and disseminating current knowledge about climate change.

Inversion *See* temperature inversion.

Ionosphere Region in the upper atmosphere from about 80 to 500 km (50 to 300 mi) where charged particles (ions) are relatively abundant.

Ions Electrically charged atom or group of atoms.

Isobar A line on a weather map connecting points of equal pressure. Moving along an isobar, there is no change in pressure. The pressure gradient force acts perpendicular to isobars.

Isotherm A line on a weather map connecting points of equal temperature. Moving along an isotherm, there is no change in temperature. Temperature gradients are perpendicular to isotherms.

J

Joule Basic unit of energy in the International System of Units (SI). A joule (J) is the energy needed to accelerate 1 kg at a rate of 1 m/sec/sec across a distance of 1 m. A joule is equivalent to about 0.25 calories. See *calorie*.

K

Katabatic Winds Airflow down a slope under the influence of gravity.

Kelvin Scale An absolute temperature scale, where a value of 0 K implies an absence of thermal energy. The Kelvin scale assigns 100 units between the melting and sea level boiling points of water.

Kilopascal A unit of pressure equal to 1000 pascals or 0.1 millibars.

Kinetic Energy Energy of motion.

L

La Niña The opposite pattern to an El Niño, in which below-normal sea surface temperatures exist in the tropical eastern Pacific.

Land Breeze A wind that blows from the land toward the water along the coastal zone during the night and early morning. The result of differential cooling between land and sea.

Latent Heat (1) Energy present in water vapor, used in converting water from liquid to gas. Latent heat is released upon condensation. (2) Energy associated with the change of phase of a substance. *See* latent heat of fusion *and* latent heat of vaporization.

Latent Heat Flux Heat transfer that occurs whenever water vapor moves from one place to another. Energy used to evaporate water travels with the water vapor and is released upon condensation.

Latent Heat of Fusion Also called *latent heat of melting*. Energy released when a substance freezes, consumed when a substance melts. For water, the latent heat of fusion is about 334,000 J/kg.

Latent Heat of Vaporization Also called *latent heat of condensation*. Energy consumed when a substance evaporates, released when a substance condenses. For water, the latent heat of vaporization is about 2,500,000 J/kg.

Leader A column of ionized air that approaches the surface and precedes cloud–ground lightning.

Lenticular Cloud A lens-shaped cloud that usually forms downwind of topographic barriers.

Level of Free Convection The level to which conditionally unstable air must be lifted so that it can continue to rise due to its own buoyancy.

Lifting Condensation Level Altitude to which an air parcel would need to be lifted for condensation to occur.

Limb-Darkening The phenomenon in which the edge of the Sun appears darker than its center.

Little Ice Age A period in Earth's history from about 1400 to 1850 characterized by low temperatures.

Long-Range Forecast A weather prediction extending beyond 7 days.

Long Waves *See* Rossby waves.

Longwave Radiation Another term for *infrared radiation*.

Low Another term for *cyclone*.

M

Mammatus A feature on parts of some cumulonimbus clouds consisting of round, downward-extending protrusions.

Mature Stage The stage in an air mass thunderstorm marked by heavy storm activity, with strong updrafts, lightning, and heavy precipitation.

Maunder Minimum A period of Earth's history between about 1645 and 1715 characterized by minimal sunspot activity.

Mean Free Path The average distance traveled by molecules before colliding with adjacent molecules; increases with altitude.

Mechanical Turbulence Also called *forced convection*. Mixing of the air caused by horizontal movements (wind).

Medium-Range Forecast Weather forecasts for predictions 3 to 7 days in advance.

Mercury Barometer The standard instrument for the measurement of atmospheric pressure.

Meridional Wind Wind flowing north–south parallel to a line of longitude. Actual winds are seldom completely meridional, but usually have both a meridional and *zonal* component.

Mesocyclone A rotating region within a cumulonimbus cloud where tornadoes often form.

Mesoscale A scale of meteorological phenomena typically having horizontal extents of several tens of kilometers.

Mesoscale Convective Complex (MCC) A type of mesoscale convective system having an oval or nearly circular shape.

Mesoscale Convective System (MCS) A general clustering of thunderstorms on the order of a few hundred kilometers across.

Mesosphere Region of the atmosphere from about 50 km to 80 km (30 to 50 mi), characterized by decreasing temperature with increasing altitude.

Meteorological Service of Canada The official meteorological agency for Canada.

Meteorology The science that studies the atmosphere.

Microburst A small but severe downburst, whose wind shear is capable of causing air crashes.

Microscale The smallest scale of meteorological phenomena, such as that which might surround a leaf.

Microwave Radiation Electromagnetic radiation with wavelengths between about 1000 and 1,000,000 μm. Weather radars use microwave radiation for imaging.

Midlatitude Cyclone A low-pressure system characterized by the presence of frontal boundaries.

Mie Scattering Scattering of visible radiation caused by particulates.

Milankovitch Cycles Variations in Earth's orbital characteristics having periodicities of tens of thousands of years and longer.

Millibar A unit of atmospheric pressure, abbreviated as mb. Sea level pressure is about 1013 mb.

Mirage The apparent displacement of an object's true position due to refraction.

Mixed Layer That part of the lower atmosphere in which vertical motion (convection) is strong enough for even dispersal of pollutants.

Mixing Ratio A measure of atmospheric moisture: the mass of water vapor per unit mass of dry air, usually expressed in grams per kilogram (g/kg).

Moist Adiabatic Lapse Rate Another term for the *Saturated Adiabatic Lapse Rate (SALR)*.

Monsoon A regional circulation pattern in which there is a seasonal reversal of wind and pressure, generally characterized by onshore flow during the summer and offshore flow during the winter.

Monsoon Depressions Areas of low pressure superimposed in the southeasterly airflow out of the Bay of Bengal.

Mountain Breeze A breeze that flows down a hill at night.

N

Nacreous Clouds Multicolored, pearlescent clouds found in the stratosphere. Also called *mother-of-pearl clouds*, these consist of ice crystals or supercooled water.

National Weather Service The official meteorological agency for the United States.

NCDC Stands for National Climate Data Center.

NCEP The initials of the National Centers for Environmental Prediction.

Nested Grid Model A particular numerical weather prediction model.

Neutral Stability A condition in which a lifted parcel of air does not return to its original position nor continue to rise. Neutral stability occurs when the environmental lapse rate is equal to the appropriate adiabatic rate, so that temperatures inside the parcel match those of the surroundings.

Newton's Second Law An expression of the conservation of momentum that is stated as: net force equals mass times acceleration ($F = ma$).

NEXRAD Stands for Next Generation Weather Radar, a network of Doppler radar units established by the U.S. National Weather Service.

NHC Stands for National Hurricane Center.

Nimbostratus A low, layered cloud that yields light precipitation.

NOAA Stands for National Oceanic and Atmospheric Administration.

Noctilucent Cloud A type of cloud that exists in the mesosphere, visible just after sunset (or before sunrise), when the surface and the lower atmosphere are in Earth's shadow.

Nonselective Scattering Scattering of radiation in which all wavelengths are scattered about equally. This type of scattering causes clouds to appear white.

North Atlantic Oscillation A see-saw in pressure between the *Icelandic Low* and the *Bermuda-Azores High*. Similar to the *Arctic Oscillation*, but confined to middle and sub-tropical latitudes.

Northeaster or Nor'easter A winter weather condition of the Atlantic Coast of the United States and Canada associated with the passage of midlatitude cyclones. The strong northeasterly winds are usually coupled with blizzard conditions.

NSSFC Stands for National Severe Storms Forecast Center.

Nuclear Fusion The thermonuclear process in which extreme heat and pressure cause atoms to combine, forming a different (heavier) element. A small part of the original mass is converted to tremendous quantities of energy and released to the environment.

Numerical Weather Prediction Weather prediction based on equations representing physical processes (as opposed to statistical relations).

NWS The initials of the National Weather Service.

O

Obliquity The degree of tilt of Earth's axis relative to the ecliptic plane, currently about 23.5°.

Occluded Front A front found in the late stages of a midlatitude cyclone.

Ocean Current The horizontal movement of surface waters caused by prevailing winds.

Omega High A pressure pattern depicted on upper-level weather maps by a pattern resembling the Greek letter Ω.

Orographic Lifting Rising motions caused by airflow over a mountain range or other topographic barrier.

Outgassing The emission of gases that accompanies volcanic eruptions.

Overrunning Warm air sliding over a dense cold air mass; the characteristic flow associated with a warm front.

Oxides of Carbon A general class of air pollutants consisting of oxygen and carbon.

Ozone Molecules consisting of three oxygen atoms, most abundant in the middle and upper stratosphere.

Ozone Hole Ozone depletions found at high latitudes (especially over Antarctica) in the spring of each year.

Ozone Layer The portion of the stratosphere where ozone is relatively abundant, reaching a few parts per million.

P

Pacific Decadal Oscillation An alternating pattern of sea surface temperature in the Pacific that reverses itself over periods of several decades.

Particulates *See* aerosols.

Pascal The standard unit of pressure in most scientific applications, equal to $1 N/m^2$.

Perihelion Earth's closest approach to the Sun (~January 3).

Permanent Gases Those gases whose relative abundance is constant within the homosphere.

Persistence Forecast A weather forecast made by assuming some existing trend continues into the future.

Photochemical Smog Secondary air pollutants formed by chemical reactions in the presence of sunlight.

Photodissociation Splitting of molecules into atoms or sub-molecules by radiation. For example, in the thermosphere, ultraviolet radiation dissociates molecular oxygen (O_2) into atomic oxygen (O).

Photosphere That part of the Sun that emits most of the energy reaching Earth. It is the "visible" part of the Sun, a layer representing about 0.05 percent of the solar radius.

Photosynthesis The growth process of green plants, whereby water and carbon dioxide are converted to carbohydrate, releasing oxygen.

PM$_{2.5}$ Designation given to particulates smaller than 2.5 μm in diameter. Major attention has recently been given to this class of particulates as possibly the most damaging to human health.

PM$_{10}$ Designation given to particulates with diameters smaller than 10 μm, which are believed to have major health consequences for humans.

Polar Easterlies Low-level winds originating in the polar highs, a feature of the general circulation of the atmosphere, often very weak or absent.

Polar Front Transition zone between cold polar air and warmer air of the midlatitudes.

Polar Front Theory The theory postulated in the early part of the twentieth century describing the formation, development, and dissipation of midlatitude cyclones. Many of the features of the theory are still considered valid.

Polar Highs Low-level anticyclones of the Arctic and Antarctic. A feature of the general circulation of the atmosphere, often absent or weakly developed.

Polar Jet Stream A jet stream found in the upper troposphere above the polar front, a result of the strong temperature contrast across the front.

Polaris The North Star.

Potential Energy Energy possessed by virtue of an object's position above some reference level. Potential energy is available for conversion to kinetic energy.

Potential Instability The condition in which a layer of air can become statically unstable if lifted sufficiently.

Power The rate at which work is done or energy expended. The standard unit is the watt, equal to 1 J/sec.

Precession The wobble of Earth's axis that has a periodicity of about 27,000 years. Combined with changes in the orientation of Earth's orbit, it yields radiation cycles of about 23,000 years.

Precipitable Water Vapor A measure of the total water vapor content of the atmosphere. The depth of water that would result if all the water in the column were to condense. Global average precipitable water vapor is about 2.5 cm (1 in.).

Precipitation Liquid water or ice that falls to Earth's surface. Rain is considered precipitation, but dew is not.

Precipitation Fog A type of fog that develops when falling raindrops evaporate enough water vapor into the air to saturate it.

Pressure Force exerted per unit area. In most sciences the standard unit of measurement is the pascal (Pa), equal to $1 N/m^2$. In daily meteorological applications, however, the *millibar* (mb) is frequently used in the United States and the kilopascal in Canada.

Pressure Gradient Force A force that arises from spatial variation in pressure. Acting alone, the pressure gradient force would cause air to blow from an area of high pressure toward an area of low pressure. The vertical pressure gradient force is always present but is nearly balanced by gravity most of the time. Much weaker horizontal pressure gradients are the ultimate cause of wind.

Primary Pollutants Substances that pollute the atmosphere upon release. See *secondary pollutants*.

Psychrometer An instrument for measuring atmospheric moisture.

Pyranometer An instrument for measuring solar radiation.

R

Radar A device that uses microwave radiation for imaging the atmosphere.

Radiation Another term for *electromagnetic radiation*.

Radiation Fog   A low-level cloud formed diabatically when the atmosphere loses heat by radiation upward.

Radiosonde   An instrument package carried by balloon, used to measure vertical profiles of temperature, moisture, and pressure. Measurements are radioed to the ground from the instrument cluster.

Rain   Precipitation arriving at the surface in the form of liquid drops, usually between 0.5 and 5 mm (0.02 and 0.2 in.). Outside of the tropics, rain usually begins in the ice stage and melts before reaching the surface. Rain that freezes on contact with the surface, forming a layer of ice, is called *freezing rain*.

Rain Shadow   An area on the lee (downwind) side of a mountain barrier having relatively low precipitation.

Rainbow   A wide, sweeping band of light caused by refraction of sunlight by raindrops.

Rawinsonde   A radiosonde tracked by radar to provide wind information.

Rayleigh Scattering   The scattering of radiation by agents substantially smaller than the radiation's wavelength. In the case of the atmosphere, this applies to the scattering of visible radiation by air molecules.

Reflection   A process in which radiation arriving at a surface bounces back, without being absorbed or transmitted. Reflection does not heat the reflector, because there is no net energy transfer to the surface.

Refraction   The bending of light within a medium or as it passes from one medium to another. Refraction results from density differences within/between the transfer media.

Relative Humidity   The measure of the amount of water vapor in the air as a fraction of saturation, often expressed as a percentage. Because the saturation point is temperature-dependent, relative humidity depends on both the moisture content and the temperature of the air.

Return Stroke   Synonymous with *lightning stroke*.

Ridge   An elongated axis of high pressure.

Riming   The growth of a falling ice particle as it collides with nearby water droplets that freeze onto the particle.

Roll Cloud   A rotating cloud often found within the leading edge of a gust front.

Rossby Waves   Also known as *long waves*. Waves in the midlatitude westerlies having wavelengths on the order of thousands of kilometers. Often a series of Rossby waves circle the planet, forming a pattern of ridges and troughs.

S

Saffir-Simpson Scale   A scheme for classifying the intensity of hurricanes.

Santa Ana Wind   A local name for a foehn wind in California.

Saturated Adiabatic Lapse Rate (SALR)   Rate of temperature decrease for a rising saturated parcel of air. The value ranges from about 4 to 10 °C/km, depending mainly on temperature.

Saturation   The maximum amount of water that can exist in the atmosphere as a vapor. More precisely, saturation occurs when a flat surface of pure water is in equilibrium with the overlying atmosphere. The evaporation rate equals the condensation rate, so the vapor content of the air is unchanging. The saturation point increases with increasing temperature.

Saturation Mixing Ratio   The mixing ratio of the atmosphere when it is saturated.

Saturation Specific Humidity   The specific humidity of the atmosphere when it is saturated.

Saturation Vapor Pressure   The vapor pressure of the atmosphere when it is saturated.

Scalar   A quantity or property that possesses magnitude but has no direction. Examples are temperature, pressure, and density.

Scattering   The dispersion or redirection of radiation by gases, dust, water drops, ice, and other particulates. Scattering does not heat the atmosphere because there is no energy transfer to the scattering agent.

Sea Breeze   A flow of air from the water toward land along a coastal region.

Sea Level Pressure   The pressure that would presumably exist at a point if it were at sea level. This involves a conversion of observed surface air pressure.

Secondary Pollutants   Pollutants that form in the atmosphere from reactions involving anthropogenic or natural substances.

Semidesert   A type of dry climate that annually receives enough precipitation to distinguish it from a true desert.

Semipermanent Cell   Large area of high or low pressure present year after year, usually with size and location changing seasonally.

Sensible Heat   The energy contained in air that can be sensed via its temperature.

Severe Thunderstorm   A thunderstorm that produces either very strong winds, large hail, or tornadoes.

Severe Thunderstorm Warning   An advisory issued by a local office of the National Weather Service indicating that severe thunderstorms are occurring or imminent.

Severe Weather Warning   Advisory issued when severe weather is observed. In the case of a hurricane warning, the potential for landfall exists within a 24-hour period.

Severe Weather Watch   Advisory issued when atmospheric conditions are favorable for severe weather. In the case of a hurricane watch, the potential for landfall exists outside a 24-hour period.

Shelf Cloud   A portion of a severe thunderstorm cloud that protrudes ahead of the main portion of the cloud and above a gust front.

Short Wave   A small wave in the midlatitude westerlies. Often superimposed on Rossby waves, they move more quickly, and thus travel through the large-scale pattern.

Shortwave Radiation   Electromagnetic energy having wavelengths shorter than about 4 µm.

Siberian High   A semipermanent cell found in North Asia during winter.

Sleet   Precipitation in the form of ice pellets, resulting when raindrops freeze before reaching the surface.

Snow   Frozen, crystalline precipitation that forms and remains in the ice stage throughout its descent.

Solar Altitude   The angle between the horizon and the Sun. When the Sun is overhead, the solar altitude is 90°. *See* zenith angle.

Solar Constant   The amount of radiation reaching the top of the atmosphere when Earth is at its average distance from the Sun. This is not a pure constant, but rises and falls with changes in solar emission. Its value is about 1376 W/m².

Solar Declination   The latitude of overhead Sun; the place where one would go to find the Sun directly overhead at noon.

Solar Wind   A continuous stream of particles (mostly protons and electrons) emitted by the Sun, traveling about one-third to one-half the speed of light.

Solstices   The two times each year that mark the northern and southern limits of the latitude of overhead Sun. On the June solstice (approximately June 21) Northern Hemisphere latitudes have their longest day of the year. On the December solstice (approximately

December 22), Southern Hemisphere latitudes have their longest day of the year.

Source Region A large area of land or ocean of more or less uniform characteristics, above which an air mass can form.

Southern Oscillation The reversal of surface pressure patterns over the tropical Pacific associated with El Niño events.

Specific Heat The amount of energy required to raise the temperature of a given mass of a substance by a given amount.

Specific Humidity A measure of atmospheric moisture. The mass of water vapor per unit mass of air, usually expressed in grams per kilogram (g/kg).

Speed A scalar property representing the rate of motion.

Speed Convergence The compaction of air due to decreasing wind speed in the downwind direction.

Speed Divergence The spreading of air due to increasing wind speed in the downwind direction.

Squall Line A linear band of thunderstorms, often found several hundred kilometers ahead of a cold front.

Stable Air Air that, when displaced vertically, returns to its initial position. Stable air resists uplift.

Standard Atmosphere The mean structure of the atmosphere with regard to temperature and pressure.

Static Stability The condition of the atmosphere that inhibits or favors vertical displacement of air parcels due to the effects of buoyancy.

Station Model A plotting on weather maps for individual locations depicting current temperature, dew point, pressure, and other meteorological information.

Stationary Front A transition zone between dissimilar air masses (a front) showing little or no tendency to move.

Steam Fog Fog that forms when cold air moves over a warmer water surface.

Stefan-Boltzmann Law The law for blackbody emission stating that the total energy emitted over all wavelengths is proportional to the fourth power of absolute temperature.

Steppe Another term for *semidesert*.

Stepped Leader A narrow zone of ionized air that serves as a conduit for an initial lightning stroke.

Storm Surge A potentially damaging influx of coastal waters brought about by high winds and low pressures associated with hurricanes.

Stratocumulus A low, layered cloud having superimposed rows or cells of vertical development evidenced by areas of differing whiteness.

Stratopause Upper limit of the stratosphere; the transition between the stratosphere and mesosphere.

Stratosphere A layer of the atmosphere between about 16 and 50 km (10 and 30 mi), characterized by generally increasing temperature with increasing altitude.

Stratus A cloud with a layered structure.

Streamlines Lines that depict the direction of airflow. Air parcels are envisioned to flow along streamlines. *See* confluence *and* diffluence.

Structure The layering of the atmosphere combined with the reduction in density with altitude.

Stuve Diagram A particular type of thermodynamic diagram used for plotting temperature and moisture profiles.

Sublimation Change from a solid into a vapor without passing through the liquid phase. Also used to describe the reverse (vapor to solid).

Subpolar Low A belt of low pressure in the three-cell model, between the polar easterlies and midlatitude westerlies.

Subtropical High A semipermanent cell that occupies large areas of the midlatitude oceans, especially in the warm season.

Subtropical Jet A jet stream common in the upper troposphere on the poleward side of the Hadley cells, produced by the conservation of angular momentum.

Suction Vortex A zone of intense rotation within a large tornado that often causes the most devastation.

Sulfur Oxides A general class of pollutants consisting of sulfur and oxygen.

Sun Pillars Bands of light stretching vertically from the Sun caused by reflection off almost horizontally aligned ice crystals.

Sundogs Paired bright spots found 22° to the right or left of the Sun caused by ice crystal refraction.

Sunspots Magnetic storms of the Sun, appearing as dark (Earth-sized) spots on the photosphere.

Supercell Thunderstorm A very large thunderstorm formed from an extremely powerful updraft.

Supercooled Water Water existing in the liquid phase with a temperature less than 0 °C.

Supersaturation A relative humidity greater than 100 percent, when the atmosphere is more than saturated with water vapor. Requires a very clean atmosphere, where condensation nuclei are lacking.

Synoptic Scale The scale of meteorological phenomena having areas on the order of hundreds or thousands of square kilometers.

T

Teleconnection Relationship between weather or climate patterns at two widely separated locations.

Temperature An index of the average kinetic energy of the molecules comprising a substance.

Temperature Gradient Temperature change per unit distance. A strong temperature gradient implies that temperature changes rapidly over a short distance.

Temperature Inversion Condition in which temperature increases with increasing altitude.

Terminal Velocity The final speed obtained by an object falling through the atmosphere, when friction with the surrounding air balances the force of gravity.

Thermal Low Low-pressure cell produced by heating of the surface.

Thermistor An object whose electrical resistance changes with temperature, thus allowing temperature to be determined by measuring changes in electrical current.

Thermodynamic Diagram A diagram showing the relationship between pressure, temperature, density, and water vapor content, such that characteristics of air parcels can be determined as they ascend and descend.

Thermoelectric Effect A theory of lightning formation in which separation of charge is produced by positive ions migrating from warmer particles to colder ice crystals.

Thermohaline Circulation A movement of surface waters in the oceans due to variations in temperature and salt content.

Thermometer Instrument used to measure temperature.

Thermosphere Outermost reaches of the atmosphere, beginning at about 80 km (50 mi), characterized by increasing temperature with increasing altitude and by extremely low density.

Thornthwaite's Classification System The most widely used system for classifying general climate zones.

Threat Score A measure of precipitation forecast skill that considers the area correctly forecast relative to that under threat of precipitation.

Three-Cell Model A generalized description of global-scale circulation that calls for three large cells in each hemisphere. The cells rotate on a vertical plane with axes parallel to latitude lines, thereby moving heat and moisture in a north–south direction.

Thunder Sound produced when lightning discharges heat into the surrounding air, causing pressure waves to emanate outward.

Tibetan Low A semipermanent cell found in southern Asia in summer.

Tipping-Bucket Gage A type of automated rain gage.

Tornado A rotating column of air with extreme horizontal winds.

Tornado Outbreak A severe storm that has spawned at least six tornadoes.

Tornado Warning An advisory issued by a local office of the National Weather Service indicating that a tornado is occurring or imminent.

Trade Wind Inversion A temperature inversion (layer of air having increasing temperature with altitude) commonly found at subtropical and tropical latitudes.

Trade Winds Prevailing lower troposphere winds of the tropics, associated with Hadley circulation. Strongest in the respective winter season, the trades blow from the northeast in the Northern Hemisphere and from the southeast in the Southern Hemisphere.

Transpiration Transfer of water to the atmosphere by vegetation, mostly by water evaporating and escaping plant tissues through leaf pores.

Tropic of Cancer A line of latitude at 23.5° N, the northern limit of solar declination.

Tropic of Capricorn A line of latitude at 23.5° S, the southern limit of solar declination.

Tropical Depression A closed zone of low pressure with wind speeds less than 60 km/hr (35 mph).

Tropical Disturbance A disorganized group of thunderstorms with weak pressure gradients and little or no rotation.

Tropical Storm A storm that originates in tropical regions and has wind speeds between 60 and 120 km/hr (35 and 70 mph).

Tropopause Upper limit of the tropopause; the transition between the troposphere and stratosphere.

Troposphere The lowest temperature layer of the atmosphere, from the surface to about 16 km, characterized by generally decreasing temperatures with increasing altitude.

Trough An elongated axis of low pressure.

Turbidity Loosely speaking, the "dustiness" of the atmosphere, including the effect of all particulates that reduce visibility.

U

Ultraviolet Radiation Electromagnetic radiation at wavelengths too short to be visible, from about 0.001 to 0.4 μm.

Unstable Air Air that experiences a buoyant force following a vertical displacement, causing it to rise. Uplift is promoted by instability.

Upwelling Movement of ocean or lake water from lower levels toward the surface.

Urban Heat Island Increased local temperatures that result from urbanization.

V

Valley Breeze A low-level movement of air in an upslope direction, developing during the daylight hours as result of solar heating.

Vapor Pressure A measure of atmospheric moisture, the partial pressure exerted by water vapor.

Variable Gases Gases present in amounts that vary greatly in abundance, either vertically, horizontally, or seasonally. Water vapor is the most important variable gas.

Vault An apparently empty area on a radar display, where moist air enters a supercell thunderstorm. Water droplets are abundant but are too small to provide a strong radar echo.

Vector A quantity or property possessing both magnitude and direction. Examples are wind velocity, the Coriolis force, and the pressure gradient force.

Veering Wind Wind that changes direction in a clockwise sense.

Velocity A vector property that includes speed and motion of an object or substance.

Vertical Pressure Gradient Force The upward-directed force that arises because pressure always decreases with increasing altitude.

Visible Radiation Electromagnetic radiation between about 0.4 and 0.7 μm, which is detectable by the human eye.

Volatile Organic Compounds Carbon-hydrogen molecules, both anthropogenic and naturally produced, which can be gaseous or particulate. A precursor to photochemical smog.

Vorticity The turning of an object (such as an air parcel), usually with respect to the vertical direction. This is important to meteorology because of its association with areas of divergence and convergence.

W

Walker Circulation An east–west circulation pattern of the tropics, characterized by several cells of rising and sinking air connected by horizontal motions along more or less parallel lines of latitude.

Wall Cloud Thick cloud beneath a rotating thunderstorm, a place where severe weather often develops.

Warm Advection Heat carried by airflow across isotherms from warm to cold.

Warm Core High High-pressure cells with higher temperature than surrounding air. Also called *warm core anticyclones*, they originate from atmospheric motion, not from differences in heating.

Warm Core Low Low-pressure cells that are warmer than surrounding air, produced by heating. Also called *warm core cyclones*.

Warm Front Transition zone between two air masses of different temperature, produced when warm air advances on and overruns cold air.

Water Vapor Water in its gaseous phase, not to be mistaken for small water droplets. Colorless and odorless, it seldom amounts to more than a few percent of the total atmospheric mass.

Waterspout A rather weak whirlwind (narrow rotating column of air) forming over a water surface. Rising and condensing air makes the waterspout visible.

Watt The SI unit of power, abbreviated as W, with dimensions of energy per unit time; 1 W = 1 J/sec.

Wave Cyclone Low-pressure storm common in midlatitudes, typically formed in a favored position along a Rossby wave.

Wavelength Distance between successive peaks of a wave, or successive troughs, or between any two corresponding points along a wavetrain.

Weather Day-to-day conditions of the atmosphere.

Weather Forecast Office A United States Weather Service facility that issues local and area forecasts.

Weighing-Bucket Gage A type of automated gage for the measurement of precipitation.

West Greenland Drift A branch of a current flowing southward in the North Atlantic.

Westerlies Winds belts found in the middle latitudes of both hemispheres that have a strong west-to-east component.

Wet Bulb Depression The difference between air temperature and wet bulb temperature. Large values (large differences between dry and wet bulb temperatures) are indicative of low humidity.

Wet Bulb Temperature The minimum temperature achieved as water evaporates from a wick surrounding a thermometer's bulb.

Wien's Law Law for blackbody emission that states that the wavelength of maximum emission is inversely proportional to absolute temperature.

Wind The horizontal movement of air.

Wind Chill Temperature Index An index of apparent temperature used for cold conditions that incorporates air temperature and wind speed.

Wind Profiler A type of Doppler radar unit that provides information about vertical changes in wind speed and direction.

Wind Shear Large change in wind direction or speed over a short distance.

Wind Vane An instrument for measuring or indicating wind direction, often a pivoting arrow that continually points into the wind.

WMO Stands for the World Meteorological Organization, the agency of the United Nations charged with collecting and disseminating meteorological data.

X

X-Ray Radiation Electromagnetic radiation at wavelengths between about 0.00001 μm and about 0.01 μm.

Z

Zenith Angle The angle between the Sun and the vertical direction. When the Sun is overhead, the zenith angle is zero. *See* solar altitude.

Zonal Wind (Flow) Wind flowing east–west, parallel to a line of latitude. Actual winds are seldom completely zonal but usually have both a meridional and zonal component.

Zone Forecast A weather forecast issued at regular intervals for a particular region.

Credits

Chapter 1

Photographs
Part 1 Opener, © Westend61 GmbH/Alamy
Chapter 1 Opener, CBE/ZOB/WENN.com/Newscom
1–B1–1, REUTERS/Jessica Rinaldi
1–2, NASA
1–B2–1, AP Photo/Mark Schiefelbein
1–4a, Space Science and Engineering Center/SSEC
1–4b, Space Science and Engineering Center/SSEC
1–6, George Holton/Photo Researchers, Inc.
1–8, Koos van der Lende Africapictures.net/Newscom
1–10, Andrew Lambert Photography/Photo Researchers, Inc.
1–13, © All Canada Photos/SuperStock
1–14, © NaturePL/SuperStock
1–15a, NOAA
1–15b, NOAA

Illustrations
1–5, From National Oceanographic and Atmospheric Agency and ucsd.edu Agency
1–7, From National Oceanographic and Atmospheric Agency

Chapter 2

Photographs
Chapter 2 Opener, UPI/Brian Kersey/Newscom
2–4a, NASA
2–4b, National Optical Astronomy Observatories
2–9, NOAA

Illustrations
2–1, Modified from Nebel and Wright, *Environmental Science*, Sixth Edition, ©1998, Prentice Hall, Inc., Upper Saddle River, NJ
2–7, Modified from Chaisson and McMillan, *Astronomy Today,* Second Edition, ©1997, Prentice Hall, Inc., Upper Saddle River, NJ
2–B1–1 From Chaisson and McMillan, *Astronomy Today,* Second Edition, © 1997, Prentice Hall, Inc., Upper Saddle River, NJ
2–B2–1 From Chaisson and McMillan, *Astronomy Today,* Second Edition, © 1997, Prentice Hall, Inc., Upper Saddle River, NJ
2–B2–2, Data from Solar–Terrestrial Physics Division, National Geophysical Data Center, Boulder, Colorado, and John A. Eddy, "The Maunder Minimum," *Science*, Vol. 192, pp. 1189–1202

Chapter 3

Photographs
Chapter 3 Opener, NOAA/Satellite and Information Service
3–1, DesignPics Inc./Photolibrary
3–4, NASA/Johnson Space Center
3–6, Lawrence Wee/Shutterstock
3–12b, Stephen Mcsweeny/Shutterstock
3–21, Edward Aguado
3–26, Qualimetrics, Inc.
3–27, Edward Aguado

Illustrations
3–7a, 3–7b National Renewable Energy Laboratory
3–10b, J. N. Howard, *Proc. I.R.E.*, v. 47, 1959, and R. M. Goody and G. D. Robinson, *Quart. J. Roy. Meteor. Soc.*, v. 77, 1951, as modified by R. G. Fleagle and J. A. Businger, *An Introduction to Atmospheric Physics*, Second Edition, © 1980, Academic Press, New York, NY
3–17, 3–19, Modified from Christopherson, *Geosystems, An Introduction to Physical Geography*, Third Edition, ©1997, Prentice Hall, Inc., Upper Saddle River, NJ
3–29b, From *The Atlas of Canada*, Natural Resources Canada

Chapter 4

Photographs
Chapter 4 Opener, Mike Siegel/MCT/Newscom
4–2, momesso/Fotolia
4–4b, Charles D. Winters/Photo Researchers, Inc.
4–5a, Edward Aguado
4–5c, Edward Aguado
4–25, Edward Aguado

Illustrations
4–5b, Modified from Moran and Morgan, *Meteorology, The Atmosphere and the Science of Weather*, Fifth Edition, © 1997, Prentice Hall, Upper Saddle River, NJ

Chapter 5

Photographs
Part 2 Opener, © Robert McGouey/agefotostock
Chapter 5 Opener, Purdue Aviation Education Multimedia Library
5–3, iStockphoto/Thinkstock
5–9a, NOVALYNX Corporation, Grass Valley, CA
5–9b, NOVALYNX Corporation, Grass Valley, CA
5–12, oksana.perkins/Shutterstock
5–15, REUTERS/JP Moczulski
5–19a, Yaroslav/Shutterstock
5–19b, © Anette Linnea Rasmussen/Istockphoto.com
5–19c, © age fotostock/SuperStock
5–20, Michael Hare/Shutterstock
5–21b, NASA
5–22, Kavram/Shutterstock
5–26, REUTERS/ERIC THAYER

Illustrations
5–10, National Climatic Data Center

Chapter 6

Photographs
Chapter 6 Opener, © Ryan McGinnis/Alamy
6–1, Howard Bluestein/Photo Researchers, Inc.
6–2a, Sasha Buzko/Shutterstock
6–2b, Jose Antonio Santiso Fernández
6–3a, Edward Aguado
6–3b, Edward Aguado
6–3c, Edward Aguado
6–B3–1a, © Darlene Christensen/Alamy
6–B3–1b, AP Photo/The Palm Beach Post, Bill Ingram
6–16, Joyce Photographics/Photo Researchers, Inc.
6–17, © Jerome Wilson/Alamy
6–18, Maurice Nimmo/Frank Lane Picture Agency
6–19, John R. Foster/Photo Researchers, Inc.
6–20, © AfriPics.com/Alamy
6–21, Robert Gill; Papilio/Photolibrary
6–22, Pekka Parviainen/Photo Researchers, Inc.
6–23, Lincoln Nutting/Photo Researchers, Inc.
6–24, © Warren Faidley/Weatherstock
6–25, James Steinberg/Photo Researchers, Inc.
6–26, Cebolla4Cebolla4/istockphoto
6–27, Day Williams/Photo Researchers, Inc.
6–28, Howard B. Bluestein/Photo Researchers, Inc.
6–29, Kent & Donna Dannen/Photo Researchers, Inc.
6–30, Richard J. Green/Photo Researchers, Inc.
6–31, © LorenRyePhoto/Alamy
6–32, z03/ZUMA Press/Newscom
6–33, Pekka Parviainen/Photo Researchers, Inc.
6–34, University Corporation for Atmospheric Research

Chapter 7

Photographs
Chapter 7 Opener, AP Photo/Jeff Roberson
7–2, David Greedy/Getty Images
7–7, Doug Millar/Photo Researchers, Inc.
7–11a, Clyde H. Smith/Photolibrary
7–11b, Scott Camazine/Photo Researchers, Inc.
7–11c, Ted Kinsman/Photo Researchers, Inc.
7–13, REUTERS/Joe Traver
7–19a, NOAA
7–19b, University Corporation for Atmospheric Research/ Photo Researchers, Inc.
7–23, Larry W. Smith/EPA/Newscom
7–24a, Edward Aguado
7–24b, Qualimetrics, Inc.

Illustrations
7–1, National Oceanographic and Atmospheric Agency
7–3, Adapted from McDonald, *The Physics of Cloud Formation, Advances in Geophysics*, v. 5, Academic Press, New York, NY, as modified by Rogers, *A Short Course in Cloud Physics*, © 1979, Pergamon Press, New York, NY
7–25, From David R. Legates

Chapter 8

Photographs
Part 3 Opener, George Steinmetz/National Geographic Stock
Chapter 8 Opener, AP Photo/Tony Gutierrez
8–2, AP Photo/John Bazemore
8–5, NASA
8–19, O. Brown, R. Evans, and M. Carle, University of Miami Rosenstiel School of Marine and Atmospheric Science, Miami, Florida
8–26, Edward Aguado
8–27, AP Photo/Chris Carlson
8–28, Image courtesy Jacques Descloitres, MODIS Rapid Response Team at NASA GSFC
8–31, NASA/Johnson Space Center

Illustrations
8–1, NOAA/NESDIS/NCDC
8–6, Modified from Christopherson, *Geosystems, An Introduction to Physical Geography*, Third Edition, © 1997, Prentice Hall, Inc., Upper Saddle River, NJ
8–8, Adapted from Miller et al., *Elements of Meteorology*, © 1983, Merrill Publishing Co., Columbus, OH
8–17, Modified from Trujillo and Thurman, *Essentials of Oceanography*, Sixth Edition, © 1999, Prentice Hall, Inc., Upper Saddle River, NJ
8–33, Modified from Peixoto and Oort, Physics of Climate, New York: American Physics Inst., © 1992, p. 419
8–35, 8–36, 8–37, 8–40, 8–42, Climate Diagnostics Center/ NOAA
8–39, Climate Prediction Center/NOAA
8–41, Courtesy of NASA/JPL/California Institute of Technology

Chapter 9

Photographs
Chapter 9 Opener, REUTERS/Scott Audette

Illustrations
9–1, Modified from Moran and Morgan, *Meteorology, The Atmosphere and the Science of Weather*, Fifth Edition, © 1997, Prentice Hall, Inc., Upper Saddle River, NJ
9–8, Modified from Christopherson, *Geosystems, An Introduction to Physical Geography*, Third Edition, © 1997, Prentice Hall, Inc., Upper Saddle River, NJ

Chapter 10

Photographs
Part 4 Opener, © Ted Foxx/Alamy
Chapter 10 Opener, AP Photo/Milwaukee Journal–Sentinel, Kristyna Wentz–Graff

Illustrations
10–7, Adapted from Lutgens and Tarbuck, *The Atmosphere, An Introduction to Meteorology*, Ninth Edition, 2004, Prentice Hall, Upper Saddle River, New Jersey

10–8, Adapted from Petterson, *Weather Analysis and Forecasting*, Vol. 1, Second Edition, © 1956, McGraw–Hill, New York, NY

Chapter 11

Photographs

Chapter 11 Opener, REUTERS/Mark Blinch
11–1, NOAA Central Library Photo Collection
11–3, Rick Rhay/Istockphoto.com
11–5, © World History Archive/Alamy
11–6, U.S. Air Force photo by Master Sgt. Timothy Lancaster
11–7, Daniel L. Osborne, University of Alaska/Detlev Van Ravenswaay/Photo Researchers, Inc.
11–8, Figure reprinted by permission from Nature:Figure 2c, vol. 416, pp. 152–154, 2002; V.P. Pasko, M.A. Stanley, J.D. Mathews, U.S. Inan and T.G. Wood. Nature Publishing Group/McMillian Publishers Ltd.
11–B3–1, © Dennis Hallinan/Alamy
11–10, Larry Miller/Photo Researchers, Inc.
11–16, Howard Bluestein/Photo Researchers, Inc.
11–18, Howard Bluestein/Photo Researchers, Inc.
11–B4–1a, b, National Center for Atmospheric Research
11–B4–3, Herb Stein/Center for Severe Weather
11–B6–1, © Craig Ruttle/Alamy
11–B6–2, LARRY W SMITH/EPA /Landov
11–26a, Charles Doswell III/Getty Images
11–26b, © Ryan McGinnis/Alamy
11–28a, Philippe Bourseiller/The Image Bank/Getty Images
11–28b, Howard Bluestein/Photo Researchers, Inc.
11–36b, Howard Bluestein/Photo Researchers, Inc.
11–41, National Oceanic and Atmospheric Administration

Illustrations

11–2, 11–14, 11–19a, Adapted from Wallace and Hobbs, *Atmospheric Science, An Introductory Survey*, © 1977, Academic Press, New York, NY
11–9, Adapted from Gedzelman, *The Science and Wonders of the Atmosphere*, © 1980, Wiley, Inc., New York, NY
11–12, Image from Roger Edwards/NOAA Storm Prediction Center
11–B4–1, National Weather Service/NOAA
11–19, Adapted from Djuric, *Weather Analysis*, © 1994, Prentice Hall, Inc., Upper Saddle River, NJ
11–24, NOAA
11–25, "The North American Lightning Detection Network (NALDN)—First Results: 1998–2000" by R. Orville in *Monthly Weather Review 130*(8): 2098–2109, © 2002 American Meteorological Society.
11–30, 11–31, Adapted from Brandes, in Church et al., eds., *The Tornado: Its Structure, Dynamics, Prediction, and Hazards*, © 1993, American Geophysical Union, Washington, DC
11–32, Modified from McKnight, *Physical Geography, A Landscape Appreciation*, © 1996, Prentice Hall, Inc., Upper Saddle River, NJ
11–33, Modified from Christopherson, *Geosystems: An Introduction to Physical Geography*, Sixth Edition, 2006, Prentice Hall, Upper Saddle River, NJ
11–34, Source: NOAA
11–36a, Adapted from Fujita, *Journal of Atmospheric Sciences*, © 1981, American Meteorological Society, Boston, MA, as modified by Battan, *Fundamentals of Meteorology*, © 1984, Prentice Hall, Inc., Upper Saddle River, NJ
11–40, Storm Prediction Center/NOAA

Chapter 12

Photographs

Chapter 12 Opener, Zuppa, Chris/St. Petersburg Times/PSG/Newscom
12–1, NOAA via CNP/Newscom
12–18, Marvin Nauman/FEMA
12–19, USGS
B12–3–4, Rick Wilking/Reuters
12–21a, AP Photo/Wilfredo Lee
12–21b, AFP/Newscom
12–23, Bob Pearson/Newscom
B12–4–1, © CORBIS
B12–4–2, Eric Kayne/ Rapport Pres/Newscom
B12–5–1, © Neil Cooper/Alamy
B12–5–2, NASA

Illustrations

12–4, From *Severe and Hazardous Weather* by Rauber et al., copyright © 2002 Kendall/Hunt Publishing Company.
12–5, Adapted from Miller, *Science*, v. 157, 1967, American Association for the Advancement of Science, as modified by Eagleman, *Meteorology: The Atmosphere in Action*, Second Edition, © 1985, Wadsworth, Inc., Belmont, CA
12–6, Anthes et al., "Three–Dimensional Particle Trajectories in a Model Hurricane," *Weatherwise*, v. 24, 1971
12–7, NASA Earth Observatory
12–9, NASA Tropical Rainfall Measuring Mission (TRMM)
12–11, 12–12, National Oceanic and Atmospheric Administration
12–13, Modified from Moran and Morgan, *Meteorology: The Atmosphere and the Science of Weather*, Fifth Edition, © 1997, Prentice Hall, Inc., Upper Saddle River, NJ
12–14, 12–B2–2, 12–B3–1, National Hurricane Center/NOAA
12–16, Adapted from Novlan and Gray, *Monthly Weather Review*, © 1974, American Meteorological Society, Boston, MA, as modified by unpublished National Weather Service Training Center pamphlet
12–B3–2, 12–B3–3, National Weather Service/NOAA

Chapter 13

Photographs

Part 5 Opener, Timothy A. Clary/AFP/Getty Images/Newscom
Chapter 13 Opener, AP Photo/Marcio Jose Sanchez
13–1a, Edward Aguado
13–1b, Eduardo Aguado
13–B1–1, NOAA
13–6, Edward Aguado
13–B3–1, Christina Russo

Illustrations

13–5, Adapted from NOAA and DeFelice, *Introduction to Meteorological Instrumentation and Measurement*, ©1998, Prentice Hall, Inc., Upper Saddle River, NJ
13–7, Adapted from Olson et al., *Weather and Forecasting*, v. 10, 1995
13–8, 13–9, Adapted from Mesinger, *Bulletin of the American Meteorological Society*, v. 77, 1996
13–11, 13–12, Source: NOAA
13–24, Nowrad™ radar mosaic. Used by permission of Weather Services International Corporation.

Chapter 14

Photographs

Chapter 14 Opener, DanitaDelimont.com "Danita Delimont Photography"/Newscom
14–B1–1a, Carnegie Library of Pittsburgh
14–B1–1b, Arthur G. Smith
14–2, Adam Hart–Davis/Science Photo Library/Photo Researchers, Inc.
14–4, Martin Bond/Science Photo Library/Photo Researchers, Inc.
14–5, Tony Craddock/Science Photo Library/Photo Researchers, Inc.
14–26, © Corbis Flirt/Alamy
14–B3–1, The MapFactory/Photo Researchers, Inc.
14–11, Edward Aguado

Illustrations

14–1, From Nebel and Wright, *Environmental Science*, Sixth Edition, © 1998, Prentice Hall, Inc., Upper Saddle River, NJ
14–10, 14–11a, Adapted from Oke, *Boundary Layer Climates*, © 1978, Methuen & Co., Ltd., New York, NY
14–12, Adapted from Hidore and Oliver, *Climatology, An Atmospheric Science*, © 1993, Macmillan Publishing Co., New York

Chapter 15

Photographs

Part 6 Opener, Howie Garber/DanitaDelimont.com "Danita Delimont Photography"/Newscom
Chapter 15 Opener, Evgenii Zadiraka/age fotostock
15–1, Joel Sartore/National Geographic Stock

15–5, Will & Deni McIntyre/Photo Researchers, Inc.
15–8, Eric LeFranc/Photolibray
15–10, © McPhoto/agefotostock
15–16, © Gary Crabbe/age fotostock
15–20, Radius Images/Photolibray
15–22a, Thomas Sbampato/Photolibrary
15–22b, © Colouria/Alamy

Illustrations

15–3, 15–4, 15–6, 15–7, 15–9, 15–12, 15–13, 15–14, 15–15, 15–17, 15–18, 15–19, 15–21, 15–22, 15–24, Modified from McKnight, *Physical Geography, A Landscape Appreciation*, Fifth Edition, © 1996, Prentice Hall, Inc., Upper Saddle River, NJ
15–11, NOAA
15–B1–1, Adapted from Critchfield, *General Climatology*, Third Edition, © 1974, Prentice Hall, Inc., Upper Saddle River, NJ

Chapter 16

Photographs

Chapter 16 Opener, Daniel Beltra/ZUMA Press/Newscom
16–1a, ESA/NASA/Univ. of Arizona /UPI/Newscom
16–1b, NASA/JPL
16–3, Chris Walker/Chicago Tribune/MCT/Newscom
16–6, Matthijs Wetterauw/Shutterstock
16–22, © Johner Images/Alamy
16–23, NASA/Goddard Space Flight Center Scientific Visualization Studio, The Blue Marble data is courtesy of Reto Stockli (NASA/GSFC)

Illustrations

General Circle Models, Based on Climate Change 2007: The Physical Science Basis. Working Group I Contribution to the Fourth Assessment Report of the Intergovernmental Panel on Climate Change, Summary for Policymakers, p. 10. Cambridge University Press. Used by permission of the IPCC.
16–7, Modified from Hamblin, *Earth's Dynamic Systems*, Eighth Edition, © 1998, Prentice Hall, Inc., Upper Saddle River, NJ
16–9b, c, d, From "Past Environmental Changes: Characteristic Features of Quaternary Climate Variations" by Patrick J. Bartlein from *Past and Future Rapid Environmental Changes*, edited by B. Huntley et al., copyright © 1997.
With permission of Springer Science+Business Media and P. J. Bartlein.
16–10, Modified from McKnight, *Physical Geology, A Landscape Appreciation*, Fifth Edition, © 1996, Prentice Hall, Inc., Upper Saddle River, NJ
16–11, From "Millennium's Hottest Decade Retains Its Title, for Now" by Richard A. Kerr in SCIENCE 307(5711): 828–829, February 11, 2005. Copyright © 2005 AAAS
16–12, 16–13, 16–14, Based on Climate Change 2007: Report of the Intergovernmental Panel on Climate Change, IPCC.
16–15, Modified from R. B. Alley, Nature, 392 (1998), 335–337
16–16, Modified from Chaisson and McMillan, *Astronomy Today*, Second Edition, © 1997, Prentice Hall, Inc., Upper Saddle River, NJ
16–19, NASA
16–21, 16–26, 16–27, 16–28, Based on Climate Change 2007: Report of the Intergovernmental Panel on Climate Change, IPCC.

Chapter 17

Photographs

Part 7 Opener, Jon Cornforth/DanitaDelimont.com "Danita Delimont Photography"/Newscom
Chapter 17 Opener, Jeremy Woodhouse/Getty Images
17–4, Pekka Parviainen/Science Photo Library/Photo Researchers, Inc.
17–6, Pekka Parviainen/Science Photo Library
17–7, Jim Steinberg/Photo Researchers, Inc.
17–12, © Danita Delimont/Alamy
17–13, © Clint Farlinger/Alamy
17–14, Pekka Parviainen/Science Photo Library/Photo Research
17–15, Franz Kerschbaum/NASA

Index

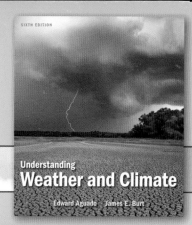

SIXTH EDITION

Understanding
Weather and Climate

Edward Aguado · James E. Burt

JIM LEE/NOAA

SHUTTERSTOCK

Cumulus Congestus These clouds have considerably more vertical development than cumulus humilis. They may produce heavy precipitation, but not the severe weather associated with some cumulonimbus clouds.

Cumulonimbus These clouds result from very strong updrafts that may push the cloud tops up to several kilometers into the stratosphere. Their characteristic feature is the anvil, a zone of ice crystals extending outward from the main portion of the cloud.

PUBLIC DOMAIN

NOAA

Cumulonimbus with Mammatus These are dramatic features associated with some cumulonimbus, resulting from strong downdrafts and turbulence along the bases or margins of the clouds.

Cumulonimbus with Wall Cloud A feature associated with some cumulonimbus clouds. When wall clouds are present, heavy rain, hail, and sometimes tornadoes can be expected.

© 2013 by Pearson Education, Inc.

Cloud Guide

SHUTTERSTOCK

Cirrus These clouds are made exclusively of ice crystals. They are not as horizontally extensive as cirrostratus clouds.

SHUTTERSTOCK

Cirrocumulus These high clouds can produce striking skies. Composed of ice crystals, they often contain linear bands, numerous patches of greater vertical development, or both.

JIM LEE/NOAA

Cirrostratus These are thin layered clouds composed of ice crystals. They are relatively indistinct and give the sky a whitish appearance.

DENNIS TASA

Contrails A contrail is a long, narrow cloud that is formed as exhaust from a jet aircraft condenses in cold air at high altitude. Upper level winds may gradually cause contrails to spread out.

Middle Clouds: cloud bases 2-6km (6,500–20,000 ft)

Understanding
Weather and Climate

Edward Aguado · James E. Burt

SHUTTERSTOCK

Altocumulus These midlevel clouds are horizontally layered but exhibit varying thicknesses across their bases. Thicker areas can be arranged as parallel linear bands or as a series of individual puffs.

SHUTTERSTOCK

Altocumulus (Lenticular) These clouds are marked by their lens-shaped appearance. They usually form downwind of mountain barriers as horizontal airflow is disrupted into a sequence of waves.

JIM LEE/NOAA

Altostratus These are midlevel, layered clouds that produce gray skies and obscure the Sun or Moon enough to make them appear as poorly defined bright spots. In this example, the setting sun brightens the clouds near the horizon but the gray appearance remains elsewhere.

JIM LEE/NOAA

Altostratus (Multilayer) These are midlevel layered clouds that are dense enough to completely hide the Sun or Moon.

© 2013 by Pearson Education, Inc.

Cloud Guide

JIM LEE/NOAA

Cumulus These clouds often have flat bottoms, rounded tops, and a "cellular" structure made up of individual clouds. (The word "cumulus" comes from the Latin word for "heap.") Cumulus clouds tend to grow vertically.

JIM LEE/NOAA

Cumulus Humilis Often called fair-weather cumulus, these small white individual masses lack conspicuous vertical development and rarely produce precipitation.

JIM LEE/NOAA

Nimbostratus These low clouds are thick gray layers that contain sufficient water to yield light-to-moderate precipitation.

JIM LEE/NOAA

Stratocumulus These are low, layered clouds that have regions of some vertical development. Differences in thickness create varying degrees of darkness when seen from below.

January Precipitation and Ocean Currents

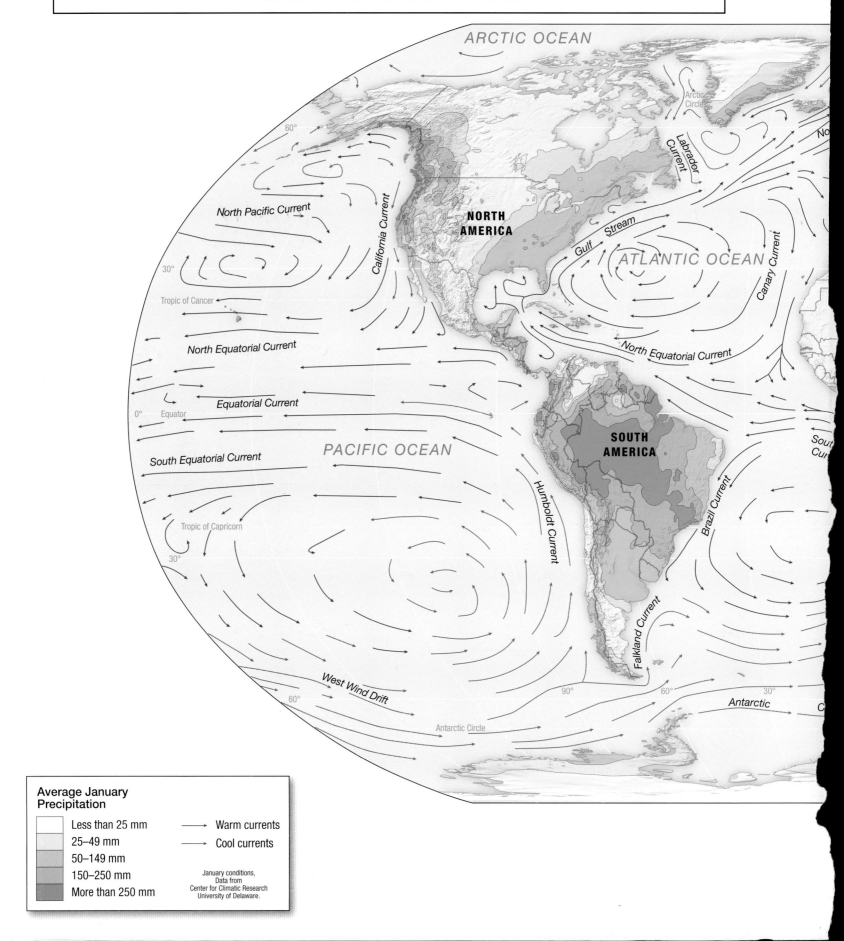

ARCTIC OCEAN

60°

Arctic Circle

Labrador Current

North Pacific Current

California Current

NORTH AMERICA

Gulf Stream

ATLANTIC OCEAN

Canary Current

30°

Tropic of Cancer

North Equatorial Current

North Equatorial Current

Equatorial Current

0° Equator

PACIFIC OCEAN

SOUTH AMERICA

South Equatorial Current

Humboldt Current

Brazil Current

Tropic of Capricorn

30°

Falkland Current

West Wind Drift

90°

60°

30°

Antarctic

60°

Antarctic Circle

Average January Precipitation

Less than 25 mm	→ Warm currents
25–49 mm	→ Cool currents
50–149 mm	
150–250 mm	January conditions, Data from Center for Climatic Research University of Delaware.
More than 250 mm	

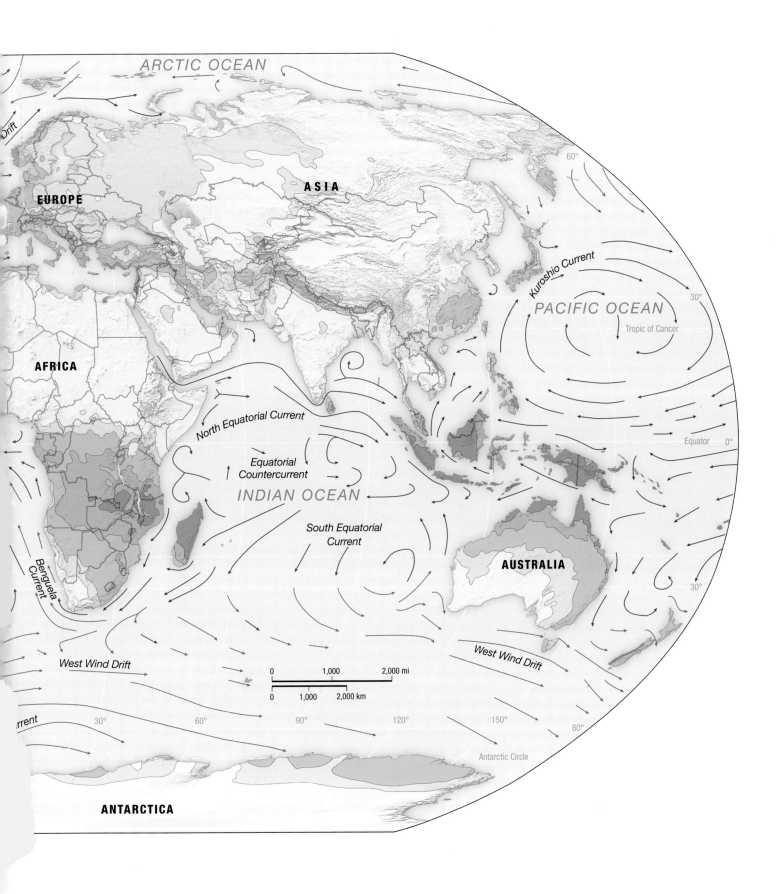

ARCTIC OCEAN

EUROPE

ASIA

Kuroshio Current

PACIFIC OCEAN

60°

30°

Tropic of Cancer

AFRICA

North Equatorial Current

Equatorial
Countercurrent

INDIAN OCEAN

Equator   0°

South Equatorial
Current

Benguela
Current

AUSTRALIA

30°

West Wind Drift

West Wind Drift

0   1,000   2,000 mi

0   1,000   2,000 km

rrent   30°   60°   90°   120°   150°   60°

Antarctic Circle

ANTARCTICA